AF830886

A Treatise on the Garden of Jiangnan

Hongxun Yang

A Treatise on the Garden of Jiangnan

A study on the Art of Chinese Classical Garden

Springer

Hongxun Yang
Architectural History
Chinese Society of Science & Technology
Architectural History
Beijing, China

Translated by
Aibin Yan
East China University of Science and Tec
Shanghai, China

Biyu Wu
East China University of Science and Tec
Shanghai, China

Jianguo Wang
University of International Business
and Economics
Beijing, China

ISBN 978-981-16-6923-1 ISBN 978-981-16-6924-8 (eBook)
https://doi.org/10.1007/978-981-16-6924-8

Translated by Wu Biyu, Yan Aibin, Wang Jianguo
Revised by Sun Hui, Karl Stefan.
Sponsored by Chinese Fund for the Humanities and Social Sciences.

© Springer Nature Singapore Pte Ltd. 2022
This work is subject to copyright. All rights are reserved by the Publisher, whether the whole or part of the material is concerned, specifically the rights of reprinting, reuse of illustrations, recitation, broadcasting, reproduction on microfilms or in any other physical way, and transmission or information storage and retrieval, electronic adaptation, computer software, or by similar or dissimilar methodology now known or hereafter developed.
The use of general descriptive names, registered names, trademarks, service marks, etc. in this publication does not imply, even in the absence of a specific statement, that such names are exempt from the relevant protective laws and regulations and therefore free for general use.
The publisher, the authors and the editors are safe to assume that the advice and information in this book are believed to be true and accurate at the date of publication. Neither the publisher nor the authors or the editors give a warranty, expressed or implied, with respect to the material contained herein or for any errors or omissions that may have been made. The publisher remains neutral with regard to jurisdictional claims in published maps and institutional affiliations.

This Springer imprint is published by the registered company Springer Nature Singapore Pte Ltd.
The registered company address is: 152 Beach Road, #21-01/04 Gateway East, Singapore 189721, Singapore

Preface

The Classical Garden of China—An Art of Time and Space that Portrays Nature

Jiangnan, "the south of the River," refers to the water network region along the lower reaches of the Changjiang River,[1] primarily Jiangsu Province and the northern part of Zhejiang Province. The Jiangnan garden relates to the classical Chinese gardens built in this region during the sixteenth and seventeenth centuries of the late Ming and early Qing dynasties.

The garden of Jiangnan represents the elite of the classical Chinese gardens and serves as a prime exemplar for its northern counterpart, the Ming and Qing imperial gardens. Differing from the traditional Western concept of "garden," which is intended for the growth of herbs, fruits, flowers, or vegetables, or "park," which is mainly a tree-planted space, the Jiangnan garden is a comprehensive creation of natural landscapes that includes shaping land surface (for topographical features such as mountains and water bodies), planting vegetation, placing ornamental animals, and disposing architectural structures. A Jiangnan garden is an architectural space where artificial and natural elements are combined. While unable to get a precise English translation, *zao yuan*, "garden making" in Chinese, is in fact closer to "landscape architecture" in concept.

Constitution of the Scenic Imagery

As *image* is the basic unit in art, *scenic image* is the basic unit in the art of garden. A private Jiangnan garden, whether it is a *zhai yuan* (residential garden) that is built

[1] Formerly known as Yangtze River

adjacent to and relatively independent of a residential compound or a *ting yuan* (court garden) that is arranged in the courtyard within a residential compound, is an integral part of everyday living space and an extension and supplement to the residence. It is a pleasure-seeking space intended for relaxation and recreation that is reminiscent of a natural environment. The scenic imagery of a Jiangnan garden comprises two antithetical and yet unified aspects: scenic constituents and scenic guidance.

Scenic constituents are the material foundation of the garden's scenic structure. A garden's functional nature and facility setup as well as its landscape nature—whether it is a mountainscape garden, a waterscape garden, or a combination of the two—are primarily determined by the presence of the garden's scenic constituents. Simply put, scenic constituents are the building materials of the garden, or the material means of garden building.

Scenic constituents include natural and artificial constituents. Natural constituents, namely, topography (such as mountain and water features), vegetation[2], and animal life[3], are the dominant factors that determine the natural features of the scenic imagery. They are the means to manifestations of the natural ecologic environment and constitute the basic content of the garden's viewing capacity, without which architecture alone—no matter how ingeniously planned and executed—cannot bring a landscape garden to fruition.

Artificial constituents, on the other hand, refer to buildings and all other architectural treatments, including paths, grounds, courtyards, walls, fences, trellises, bridges, stepping stones, staircases, bank revetments, and so on. Artificial constituents provide the garden with functional value that affords the utilitarian efficiency as in traffic facilitation, sun and rain shelters, daily routine such as dining, resting, and entertaining, as well as general garden recreational activities like hill climbing, vale exploring, boating, and fishing, all of which rely directly on the placement of the artificial constituents. If a garden is not even furnished with the most elemental architectural feature, such as a footpath, it is a garden bereft of utility and not meant for any human activity. The implementation of artificial constituents, however, should follow the naturalistic principle of garden making so as to bring the artificial constituents in harmony with the natural constituents of the garden.

Scenic guidance is an organizing factor that represents the structural relationship of the garden. From the perspective of a visitor to the garden, scenic guidance is an arranger of and a guide to the sequential and non-sequential scenic images. It is a pathfinder that steers the movement of the visitor, that determines how he sees and appreciates the scenery, and that ultimately shapes his experience of the garden.

---

[2]Flowers, plants, and trees

[3]Birds, beasts, fish, and insects

Some of the popular scenic effects, such as *qu jing tong you,*[4] *yin ren ru sheng,*[5] *feng hui lu zhuan,*[6] and *kai men jian shan,*[7] are but a few examples effected by means of scenic guidance. This has also demonstrated that scenic constituents can only be organized into coherent scenic imagery by means of scenic guidance; otherwise, they would simply be a casual and meaningless material buildup that fails to convey the functionality and artistic ideology of the garden. In other words, it is through scenic guidance that scenic constituents are configured into a functional and purposeful artistic space.

Scenic constituents and scenic guidance are inter-premised and interdependent, and the roles of the two are constantly interchanged. For instance, a bridge or a pavilion, as a scenic constituent, is an object of viewing when admired from distance but becomes scenic guidance when serving as a medium of sightseeing—a passageway or a lookout and resting point. This is to say that a scenic image can be both an aesthetic viewing object and simultaneously a spatial entity that can be entered and experienced personally. This is where the landscape garden lies superior to landscape painting or potted landscape.

Special mention should be made of that while scenic constituents carry the garden's aesthetic value and scenic guidance the functional value, scenic constituents are also the basis for garden's functionality and scenic guidance is the organizing agent of garden's aesthetic experience. Scenic constituents and scenic guidance are mutually inclusive and inter-penetrable; they are inseparable organisms. Scenic constituents are finite, whereas scenic guidance is infinite. It is the unity of the two antitheses that brings forth the rich and varied scenic imagery, whose inspired and imaginative wonders never cease to captivate and intrigue its admirers.

An Art of Time and Space

Chinese landscape painting—the ink wash painting known as *shanshui hua*[8]—is a representation of the three-dimensional natural scenery on a two-dimensional picture scroll by means of painting artistry and techniques. In other words, it is manipulation of space on a flat surface. The garden of Jiangnan, described as the "natural-type" or "landscape-type" garden, portrays a space in the space, re-creating natural landscapes in a limited garden space with natural materials of stone, sand, water, soil, and vegetation, and animal life.

Creation of the Jiangnan garden is a process of artistic condensation and even necessary formulization as practiced in Beijing opera or *xie yi* painting. Opposite to

[4] A winding path leading to a secluded spot

[5] Enticing one into a wonderful place

[6] Where the peak stands brings the next turn of the mountain path

[7] A mountain springing to view as the door opens

[8] Mountain-and-water painting

realism, both Beijing opera and *xie yi* painting are art forms that express the artistry in an abbreviated and abstract style, aspiring to capture the spirit and essence of the object rather than its physical likeness. In the art of the Jiangnan garden, the abstractionism and symbolism are equally evident: a mound of a few meters high is used to evoke a full-scale mountain and a stretch of pond water an expansive lake, and a single piece of upright *taihu* stone (a waterworn rock indigenous to the Taihu Lake famous for its irregular and fantastical shapes) is suggestive of a mountain peak, all of which is what is described as *xiao zhong jian da*, evoking the large in the small.

Jiangnan gardens are mostly built adjacent to a private residence. It is in effect a continuation of the living space, which gives rise to the practice of *yuan ju*, a lifestyle of leisure and refined pursuits sought after by intellectual elites that features idealized daily routine, including reading, painting, poetizing, playing musical instruments, tasting tea, savoring wine, playing chess, appreciating opera performances, banqueting, and entertaining. The practical nature of garden living hence requires the garden to be functional with more architectural structures of a greater variety, all designed to be in tune with the grace of the natural landscape of the garden. As such, making of the Jiangnan garden has always striven to create a utilitarian as well as aesthetic living space where natural and artificial worlds are closely incorporated.

The Jiangnan garden is not only a space of reality but also a space of artistic re-creation. It is a space of art. Jiangnan gardens, in their artistic virtues, are pictures and poems expressed in the vocabulary of garden making, and the experience with a Jiangnan garden is oft likened to stepping through the picture frame and taking "a stroll in the painting." Though usually inspired by or modeled after nature, Jiangnan gardens are idealized nature and a product of human subjectivity. Nature is grander, richer, and more dynamic while the garden—the re-creation of nature—is more condensed, more epitomized, more poetical, and more intriguing.

The difference between the art of garden making and the art of painting lies not only in that the garden portrays a space in the space, but also in that the experience with a garden is not confined to static viewing as with a painting. On the contrary, the aesthetic enjoyment of a garden, to a large extent, transpires in the physical movement of the viewer, which is described as *you lan*, or perambulatory viewing.

Sequential and non-sequential proceeding of garden viewing is an art of temporal planning. Rigorously speaking, a single scenic image has an infinite number of scenic planes that are to be progressively revealed to the viewer as he moves and his perspective shifts. The aspect, duration, and sequence of the revelation of each scenic plane represent the intrinsic fabric of the garden art. In music, notes, chords, tempo, timbre, etc. are interwoven into tunes and rhythms that express a certain motif or interest and are played over the course of time for the listener to enjoy. Similarly in a garden, with the passage of time, various scenic planes gradually unfold in a sequence of views, ambience, and perspectives along with the movement of the viewer, thus completing the artistic impression of the garden.

Manifestations of the scenic imagery are also dictated by nature-obliged factors: change of seasons, passing of time, or variations in climate such as rain, snow, sunshine, or clouds, all of which, to varying degrees, cast a different light on the artistic effect of the garden's scenic imagery. This, too, is an issue of temporal planning that is to be taken into consideration in garden creation. As with a piece of sculpture that produces the most aesthetic effect when placed in specific lighting, some of the garden scenes render the best artistic appeal, or *yi jing*, only at a certain time of the day, during a certain season of the year, or under a certain condition of the weather. Making of the classical Jiangnan garden has precisely mastered the temporal use of these natural assets. Therefore, creation of the Jiangnan garden is an art of both spatial and temporal planning. It is an art of time and space with functional value.

Yi Jing: The Ultimate Criterion of Garden Creation

In the Chinese classical art theory, *yi jing* is explained as "*xiang wai zhi xiang, jing wai zhi jing*," which literally says "the image beyond the image, and the scenery beyond the scenery," purporting the non-material and intangible quality of a scenic image or garden space that is capable of evoking an aesthetic atmosphere or emotional appeal. It is the supreme principle and ultimate criterion of Chinese garden making.

As shown in this illustration of the Free Roaring Pavilion (Shuxiao Ting) in the garden that was preserved miraculously (Liu Yuan) of Suzhou, perched on a tree-planted earthen hill, the light of the low setting sun comes through the tall-growing trunks and casts long, narrow shadows that stretch far on the hilltop, creating a pensive mood tinged with quiet solitude and somber aesthetics, redolent of ancient poetic thoughts such as expressed in this quatrain by Du Mu, a noted poet of the Tang dynasty:

Up on the chill mountain where a stone path slants,
  A bower looms far where the white clouds arise.
  Stopping the carriage I sit to admire the maples in the dusk,
  Redder are the frosted leaves than the February blossoms.

Completion of the construction of a garden or garden scene does not necessarily mean the completion of its artistic creation. Only when the garden or the scenery is given the poetic and picturesque quality and artistic profundity to express the inspiration of nature and life does it come into possession of the quintessential artistry and arrive at the highest precinct of the garden art—*yi jing*.What kind of garden scenery is capable of effecting yi jing?

Creation of the garden's natural landscapes, including those of a pastoral interest as manifested in the mountain village or canal-filled country scenery, should follow the laws and principles of the natural ecologic world, namely, the natural structural relationship among the mountains, water bodies, vegetation, animal life, etc. and approximate the garden landscape features to their natural prototypes.

In regard to "approximation" in artistic creation, there is a significant similarity in the creative principle of the painting and that of garden making. In the Chinese classical painting theory, two antithetical concepts are used in defining the artistry of a landscape painting: xing si (formal likeness) and shen si (essential likeness). The former relates to the painting's physical resemblance to its natural subject, which can be described as the "realistic truth" of nature, while the latter refers to the "artistic truth" that aims to capture the spirit or essence of nature rather than its outward similitude.

Shen si is premised on xing si, and shen si is nonetheless regarded as the higher aesthetic criterion in judging the artistry of a natural landscape painting. Similarly in garden making, shen si is the desired quality in natural landscape re-creation, especially given the dilemma of the usually restricted garden space in Jiangnan's urban residential districts.

Renowned poet Bai Juyi of the Tang dynasty once remarked: "There is no fixed rule in painting, and the rules lie in similarity. There is no set examples in learning, and the examples are truth itself." Master painter Qi Baishi of the Qing dynasty commented that the artistry of a painting lay "between likeness and non-likeness" of the painted subject, which resonated with the observation by great German poet J.W. Goethe that beauty existed "between truth and non-truth," both of which serve as an excellent footnote to Bai Juyi's argument. The "likeness and non-likeness" or "truth and non-truth" of an artistic image is precisely the knack for the artistic creation of the garden.

A landscape garden work embodies the logical thinking of the garden maker that is formed on his aesthetic judgment and outlooks on nature and gardens; it is also a product of his imaginative thinking fostered by his fascination with and love of beauty and life. As such, the accomplishment and profundity of a garden—whether it is a living epitome of nature and whether it is relishing of rich and deep yi jing—rests largely with the rich and broad life and creative experience as well as the artistic erudition of the garden maker. In other words, the makings of a garden maker have an immediate bearing on his logical and imaginative thinking, which exerts a direct impact on the outcome of the garden creation.

From the perspective of garden appreciation, the empathetic experience of yi jing is predicated upon the visitor's knowledge of nature and life, cultural attainment, aesthetic capability, and understanding of the garden art language. The depth and extent of one's experience with yi jing is in direct proportion to those of his life experience, cultural sophistication, and artistic aptitude. Therefore, to a garden visitor, to fully appreciate the garden's *yi jing* hinges upon the enhancement of his cultural and artistic accomplishment.

Jiangnan gardens are often created by imitating landscape paintings or idyllic writings as well as famous scenic landscapes from nature. Regardless of the form of its inspiration, creation of the garden scenes always follows a certain thematic subject, in consequence of which the classical Chinese gardens are more or less "subject gardens." This notion is further reinforced by the practice that each scenic cluster or scenic space cell is customarily given a name or title by means of placards with calligraphic literary inscriptions to infuse the poetic sentiment into the scenic imagery. This is also an effective contrivance to tap the *yi jing* potential of the scenery and bring the enjoyment of the garden up on a spiritual plane.

Contents

Chapter 1
Introduction

1.1 Demand for a Garden

Gardens, a man-made natural environment for relaxation and enjoyment that is born of man's desire for nature, are an object of aesthetic pleasure. Traditionally, gardens are a luxury for the privileged stratum of a society. To build a garden, one will first need land; even the "Half-acre Garden"[1] will need a land of half acre to build upon. Even Hanging Gardens of Babylon, another case in point, needed a tract of land upon which the garden was strutted. Land means wealth, as do the finances, manpower and materiel necessitated for building the garden on the land. Therefore, gardens were historically owned and enjoyed by the affluent. Admittedly, it is out of the question for the destitute who did not have "a place to stick an awl" to even think of owning a garden; even families comfortably off that could afford building a garden were scant in number. Such was the situation in China in its former days, so was it in other nations around the world. As for the modern times marked by increasingly advanced civilization, however, enjoying a garden has become a necessity in life for the broad masses of the people. As a form of public welfare, many countries have set up public gardens, or parks, of various types. In more developed countries, gardens of private households have also become increasingly common.

Countries of different parts of the world have their own aesthetic demands as regards natural beauty. Nations with culture developed to a certain degree also have

[1]Half-acre Garden is a classical garden built in the Qing Dynasty. The home garden of Jia Hanfu, Shangshu of the Ministry of War in the early Qing Dynasty, is located in Huangmi Hutong, Dongcheng district, Beijing, and only remains today. According to records, the park is surrounded by rocks and mountains, diverts water as marshes, and the terraced chambers are quiet and spacious; the structure is tortuous, the furnishings are quaint, and the magnificent and bookish, exuding the spirit, temperament and charm of traditional culture.

© Springer Nature Singapore Pte Ltd. 2022

H. Yang, *A Treatise on the Garden of Jiangnan*,
https://doi.org/10.1007/978-981-16-6924-8_1

their own creation of gardens. From the perspective of the human race as a whole, man's desire for nature is born of man's attachment to nature. Our fulfillment of the means of livelihood—food, clothing, shelter, and transportation—relies on nature. For man, nature is the subject of labor and thinking. Man cannot for a second dispense with nature, whose *modus operandi* itself depends on the metabolic principle. Nature is both the opposite of man and the unity of his that embodies man himself. Man is but a component part of the external natural world.

Advances in civilization have drifted man away from Mother Nature, where he originally came from. When the hub of man's social activities have evolved to be metropolises that are congested with artificial structures, man will inescapably develop nostalgia for nature. This nostalgic feeling was first expressed in the form of hunting by chief slave owners in the primitive society, a way of returning to mountains and woods that was not necessarily related to production. One may well say that it was man's nascent enlightenment on seeking pleasure in gardens. The primeval ages in remote antiquity were a dark world of horror and inclemency for man that was fraught with vipers and beasts, thunder and lightning, and wild fires in an unpredictable jungle. Only after eons of arduous efforts and generations of struggle and sacrifice has nature become tamed, charming, and genial to the human race. To be able to enjoy the beauty of nature is the reward nature returns to man for his hard work. Man has overcome nature, and, regaled with her beauties in manifolds—lush mountains, clear waters, sweet flowers, singing birds, fragrant grass, shady woods, and so forth, who wouldn't fall in love with nature and want to indulge in her magnificence? Restricted by social conditions, however, for the past generations only the wealthy and the powerful could afford to pursue such refined interest. That people admire nature has led them to developing an affinity for her. They climb the mountains, wade the waters, walk the country, fish in the river, boat in the lake, and hunt in the forest, basking in nature and abandoning themselves to her wonderment of many facets. These non-production related leisure activities, be they for physical exercise or spiritual enhancement, have in nature served as a vehicle for man to "recollect his childhood" and relive the "childhood" dreams from his ancient past. For man, who have civilized and long severed from a life once at home with mountains and forests, this is already an enjoyment.

Garden is an epitome of nature that man has triumphed over. Naturalistic landscapes processed and represented through the art of garden making are idealized garden scenery that teems with life interest of mankind. It potentially brings a sense of gratification to people who have labored to tame nature and, more directly, affords aesthetic pleasure that delights both the eyes and the mind of the beholder, as do painting and poetry. The period of a society adverse to reason and common sense is only ephemeral in the long history of civilization; it will ultimately become the past and be replaced with an ideal and scientific society that nurtures the well-being of the people. Naturalistic landscapes and the art of garden making eventually belong to their creators and mother earth, where they will play a broader role. Today, the classical gardens in the Jiangnan region have become historical vestiges. The living examples remaining today are cultural heritage for artistic appreciation and a testimony of China's magnificent ancient civilization, which, in this day and age, do not fit entirely the relaxing purpose in contemporary life as does a public park nowadays. Nevertheless, the hidden principles and techniques applied to constructing the landscapes of natural beauty still provide invaluable experience and expertise that today's garden builders can draw on. They are the foundation on which we study these garden examples from the past. The reason we now study the gardens that used to entertain the privileged minority is not only for summarizing our cultural heritage, but more importantly, it is for creating a garden-friendly living space that the whole society can enjoy at present and in the future.

1.2 Basic Concept of Chinese Classical Gardens: Integration of Human Environment and Natural Environment

Chinese gardening has a long history. Archaeologists have uncovered remains of the pool in the backyard of the Royal Palace of Shang Dynasty in Yanshi City, Henan Province, which proves that Chinese gardening with a natural living space for relaxation and enjoyment appeared at least 3000 years ago. As an integrated art form of the landscape, plants, animals and buildings, Chinese gardening matured about 2000 years ago in the Zhou Dynasty according to the patterns and documents recorded on the bronze casting at that time. Chinese classical garden, different from the west in taste and interest, has its own philosophical basis.

1.2.1 Philosophical Foundation of Chinese Classical Garden

Two systems of human civilization, Eastern civilization and Western civilization, the former originated from the Yellow River and the Yangtze River basins, the latter from the Euphrates and Tigris river basins. The "Eastern" and "Western" I'm talking about here are the concepts that represent two cultures, which constitute the perfection of human wisdom complementing each other, as shown in the Taiji diagram, (Fig. 1.1). To illustrate with religion, which is actually the reflection of social culture. According to Clifford Geertz, religion is the expression of symbols first, and what symbols carry is "meaning", that is, general thinking including cognition, emotion and morality, which are the core contents of culture. For example, philosophically speaking, although we believe that God dominates the world, we still regard man as the subjective factor in the objective world. The Bible can represent the fundamental idea of the western world that has not yet been exposed to Oriental culture. God created the world (Fig. 1.2) and created human beings in his own way, which means that man is the image of God; God gives the world to man to manage and enjoy. In

Fig. 1.1 Taiji diagram

Fig. 1.2 God created the world

fact, man is the master of the world. So what we emphasize is the power of human beings. The relationship between human beings and nature is opposite. Human beings should conquer nature. Our art emphasizes the praise of human beings,

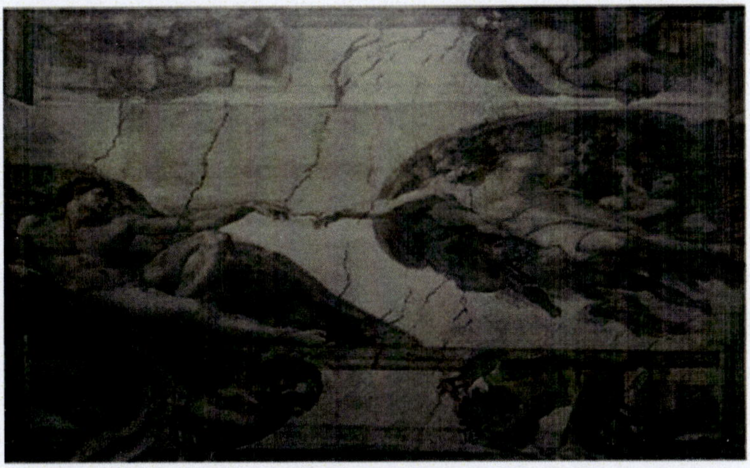

Fig. 1.3 Physical beauty

such as traditional painting, sculpture etc. describing the beauty of human body (Fig. 1.3), while Zhouyi, which is called "the true Oriental philosophy" by Hegel, represents the Oriental philosophy. The basic idea is that everything in the universe comes from nature: Wuji[2] generates Taiji[3]; Taiji generates two complementary forces[4]; Two complementary forces generate four aggregates[5]; Four aggregates generate eight trigrams[6]; Eight trigrams determine myriads of phenomena.[7] Therefore, Oriental philosophy believes that man is an integral part of nature and emphasizes the unity of objective nature and subjective human beings. Based on the idea of nature worship, Chinese literature and art usually focus on praising the nature (Fig. 1.4).

Classical garden arts in the Eastern and Western worlds differ a lot before massive communication. Classical Chinese gardens are an artistic representation of natural beauty for manifesting and praising nature (Fig. 1.5), which can be shown in the management of water surface while the Western classical gardens with geometric designs admire artful beauty. The forms of water in Classical Chinese gardens are scenes describing natural landscape, such as rivers, lakes, pools, waterfalls, streams and gullies (Fig. 1.6) while pools with geometric patterns and fountains violating the laws of nature, such as "water flows upward" in western gardens (Fig. 1.7). As for plants, Chinese gardens are configured with trees, flowers and plants that simulate

[2]Mathematically expressed as 0.

[3]Mathematically expressed as ∞, proved to be at the same point with 0.

[4]Yin and Yang, positive or negative in Mathematics.

[5]Four quadrants of the Taiyang, Shaoyin, Shaoyang and Taiyin.

[6]Eight basic elements of the arrangement and combination of the universe.

[7]Everything between heaven and earth, including human beings.

Fig. 1.4 Chinese Landscape Paintings

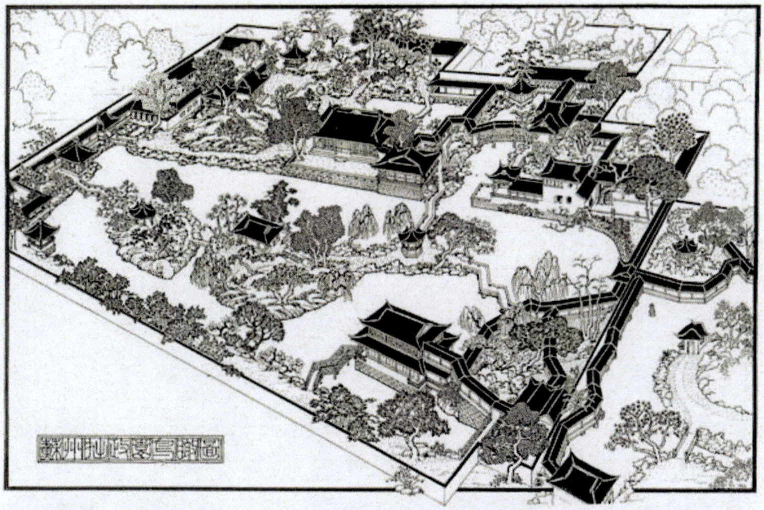

Fig. 1.5 The Classical Chinese Garden describing natural landscape

natural scenery (Fig. 1.8), on the contrary, in western gardens, designs such as "embroidered flowers" and "green carvings" are employed to intensify artful beauty (Fig. 1.9).

1.2.2 Integration of Man and Nature Embodied in Scenes of Chinese Classical Gardens: Integration of Man-Made and Natural Environment

Due to pursuit of living in nature, Chinese make gardens. Therefore, many buildings for living and pleasure can be found in classical Chinese gardens. In a Chinese garden recreating nature, the temporal and spatial organization topographical sculpturing (simulating the geomorphic landscapes such as mountains, water and fields), vegetal and fauna disposition, architectural arrangement for garden tours basically lie in Architecture. The making of Chinese garden strives to integrate architecture with natural scenery, which contributes to the concept of integrating artificial environment with natural one.

The extension of this concept is the overall environment design of living space combined with nature, whose crux lies in borrowing of scenes among its creation principles.

Borrowing of scenes, which intensifies the depth of scenes, refers to borrowed scenes, that is, borrowing the scenes that do not fall within the scope of garden planning and design—involving the beautiful scenery outside the garden into the garden landscape.

Fig. 1.6 Management of water surface in Classical Chinese Garden—a description of natural stream scene

In the fifteenth century, in his treatise on garden making, *The Craft of Gardens* (Yuan Ye), noted garden maker Ji Cheng of the Ming dynasty first proposed the concept of "borrowing of scenes", advocating that garden making should be "skillful by borrowing". At the beginning of the eighteenth century, Qing scholar Li Yu, who was erudite in garden making, also proposed "key to scenes is borrowing" in his treatise *Yi Jiayan* (*My Own Points*). The theory and technique of "borrowing of scenes" is an important contribution to classical garden making. More than 1000 years ago, Chinese garden making were learned by Japan, as well as the theory

Fig. 1.7 (**a**) Western classical gardens with interesting geometric designs; (**b**) Western classical gardens decorated with sculptured fountains

Fig. 1.8 Chinese classical garden imitating the natural plant configuration–the phlox peaches leaning out of the water in Zhuozheng Yuan, Suzhou

of borrowing of scenes, which is equal to *shakkey* in Japanese, and some western scholars learned borrowing of scenes from Japan.

Borrowing of scenes serves as not only a technique to intensify the depth of the scene, but also an important principle. By borrowing of scenes, the garden design, which is usually confined to the selected space, is based on the environment where it is located and its time and condition, and makes full use of favorable factors to improve the artistic effect of the garden. The concept of integrity contained in this means is an extremely valuable and fundamental idea in the design of classical gardens. Therefore, Ji Cheng concluded that "the most important thing in the garden making is borrowing of scenes".

Dating back to the Qin Empire in the third century BC, as the essence of gardens, the design concept of integrating man-made environment and natural environment has matured. One of its typical representative work is the New Dynasty Palace in Xianyang (the capital of the Qin Empire), which, also known as E pang Palace, blends into natural mountains and rivers and "covers over 300 li[8] of land"; The central axis of its main building, the "E pang" front hall, directly faces the interspace between the two peaks of the southern mountains, which is the so-called "mark the peaks of the southern mountains as the watchtowers (the sign of entrance)". "Jieshi Gate", the "gateway" of the Qin Empire in the Eastern Sea, also uses a pair of natural stone pillars (Jieshi) in the sea as watchtowers, facing which a walkway leading directly to the shore is paved with rubbles. Furthermore, "Jieshi Palace" is built on the coast opposite the natural "Jieshi Gate" as the main building, and two

[8] About 93 miles.

a

b

Fig. 1.9 (**a**) The artificial beauty of plant pruning in Western classical garden, so called "green carving"—trim plants into geometric shapes; (**b**) The artificial beauty of plant pruning, so called "green carving"—trim plants into adult puppets or animal shapes

symmetrical watchtowers are also set on two headlands 1000–2000 meters away from the main building as the auxiliary buildings. These are all ways of integrating man-made environment into natural environment and taking advantage of nature's imposing momentum. During the reign of Emperor Wu in the Western Han Dynasty, a vast "Taiyechi" (Calm Lake) symbolizing the residence of immortals is created in the grand imperial garden "Shanglin Yuan", and stone statues of "Niulang" and "Zhinv" were set up on both sides of a natural river to symbolize the Milky Way in the sky and the corresponding two constellations on both sides, which is also a representative work of the unified planning and design of the man-made environment and the natural environment in early times.

From January 16 to 19, 2004, an international academic seminar on the theme of "the 'borrowing of scenes'" was held in Nara, Japan, where I delivered a keynote speech, comprehensively expounding this outstanding theory in Chinese garden. The convening of Nara international conference shows that the design concept of "the 'borrowing of scenes'" has attracted the attention of international academic community.

Speaking of the relationship between man-made and natural environment, "Fengshui" or "geomancy omen" has been held in esteem from time immemorial in China, which has long been spread to the Korean peninsula, Japan, Vietnam and neighboring countries and regions in southeast Asia. Therefore the relationship is a special content worth studying in East Asian architecture system with Chinese architecture as the main body. Fengshui is called "Jiaxiang" in Japan,[9] which means "the choice of houses". In later ages, many secular superstitions and delusions have been blended into the theory of Fengshui, and our research should smash it and absorb its scientific core, that is to discard the dregs and take the essence. The essence of the concept can be summed up as follows:

1. the organic connection between man-made environment and natural environment (for example, when choosing a residential area, its relationship with sun and natural landscape should be taken into consideration, and it thus should be located to the south where the sun is located, backed by mountains and facing water, so that residents can get good sunshine and avoid the north wind in severe winter).
2. Man-made environment should not only meet physical needs, but also meet mental needs.
3. The living environment should be suitable for the healthy and sustainable development of people-the construction should not only benefit the present people, but also make future generations prosperous.
4. All the components in the universe are "numbers"(modern numerology has proved that everything can be reduced to numbers), and there is a "number" relationship between human settlement form and nature as well as with human beings.

The living environment should meet the needs of living, working and relaxing, as well as the pleasure and aesthetic enjoyment. This has become the core of modern "architectural psychology" and "environmental psychology". A living environment that can meet the needs of life but cause psychological discomfort will still have a great negative impact on people's health and behavior. Therefore, we conclude,

[9] Its Japanese pronunciation is kassou.

buildings' environment is not simple, and it has a kind of comprehensive field effect of material and spirit. The essence of *Feng shui* (geomancy omen) is to explore the unified functional field of matter and spirit, which should also be the fundamental problem of environmental art. People attach importance to the practical and sanitary functions of buildings' environment, but the spiritual function cannot be ignored. Modern living environment requires environmental art to provide comfortable and warm feelings, and further to have the aesthetic feeling and pleasure compatible with nature.

1.2.3 Chinese Concept of Landscape Design Contributing to the Harmony Between Human Residence and Natural Environment: The Ecologicalization of Living Environment

The theory of garden design goes far beyond the significance of garden making. In fact, it reveals the relationship between man and nature. Different garden construction theories reflect different views of universe and nature, thus determining the construction ideas of the whole human living environment. In western traditional geometric gardens which show artificial beauty, it is against the nature to spray water upwards and modify the natural form of plants to trim the tree crown into the shape of geometry or other animals and dolls. If these can be regarded as interesting ornamental objects, it will be a problem to carry out major environmental engineering under the same guiding principle. Taking the water conservancy project as an example, Dujiangyan, Sichuan, a water conservancy project built by the Chinese in the Qin Dynasty more than 2000 years ago, due to its design that respects and conforms to nature and reasonably guides the river diversion, the economic benefits of irrigating 10,000 mu of farmland have been achieved, benefiting the people so far. However, The Turkmen Grant Canal, built by the former Soviet Union under the guidance of the ideas of "transforming nature" and "conquering nature", forced the river to flow backwards, and the Aswan Dam, an ecological destruction project built by the former Soviet Union with the same guiding ideology in Egypt, both have caused serious consequences of disasters such as earthquakes and droughts. Following the same mindset, modern China has carried out the huge Three Gorges Reservoir Project, which has broken the original ecological balance. This cannot but be worrying!

Since the Europeans invented the steam engine and led to the industrial revolution 250 years ago, the west has taken the lead in the world's trend and the idea of "conquering nature" has gained popularity all over the world. After the global industrialization and urbanization in the twentieth century, human beings, relying on their intelligence and growing social productivity, seized natural resources and destroyed natural ecology without restraint/in an unbridled way. By the end of the twentieth century, human beings had been punished by nature, resulting in serious air pollution, water pollution, noise, vibration, land sinking, photochemical smog, greenhouse effect, ozone hole and ecological imbalance. As a result, epidemic diseases are prevalent and some biological species are extinct. The "environmental crisis" that some prophetic scientists have proposed is by no means alarmist. At present, countries all over the world have generally realized that development can no longer be achieved at the expense of environment, and "sustainable development" has become the basic national policy of governments. Visionary people in the international academic community have begun to think seriously about China's thought of "Heaven and Man Are United as One" and the thought of respecting nature and using nature to benefit mankind on the basis of conforming to nature. China's ancient wisdom,[10] including the thoughts of "Heaven and Man Are United as One" and the balance of yin and yang, has advanced guiding significance. Therefore, at the end of the twentieth century, some scientists predicted that the twenty-first century would be the century of the East, the century of Asia and even the century of China.

The development of European ocean navigation since the eighteenth century, especially the emergence of modern mechanical ships powered by steam engines since the nineteenth century, has broken the barriers between nations and between regions, greatly promoted the exchanges of eastern and western cultures, and contributed to an unprecedented leap in human wisdom. In the last 100 years of twentieth century, the economy has begun to merge the wisdom of east and west, and became a common culture of all mankind. In physics and psychology, Einstein's relativity theory and Freud's sixth sense are remarkable results of the fusion of eastern and western wisdom. In terms of garden, the cultural exchanges between the east and the west have contributed to the emergence of Anglo-Chinese gardens and landscape gardens in Europe, which contributed to the emergence of "Garden City" theory in Europe in the early twentieth century.

Modern science has made people generally realize that human beings are not only the opposite of nature, which takes objective world as the object of thought and labour, but the product of nature, and a part of nature. Since the beginning of humanity, they have been exchanging materials with earth through breath, diet and excretion. Human beings cannot live without the ecological chain of earth for a moment. However, the earth is not owned by human beings. It is the common home for all lives on earth. For the survival and development of human beings, we must also respect other lives on earth. Therefore, the construction of human living environment should take into account that animals and plants are in their proper places, and that human beings coexist with other lives on earth. This is the true essence of "all beings are equal" in Buddhism. The great development of material production in the last century not only caused the "environmental crisis", but caused the loss of human spirit—a vicious expansion of selfishness and greed. Purifying the living environment and human soul is a great and arduous task in the twenty-first

[10] Which has long been the common spiritual wealth of the East Asian cultural circle of Chinese characters in history.

century. The high-tech and ecologicalization of living environment which combines science and art—strengthening the unified function of man-made environment, will be conducive to the transformation of objective environment of human existence and subjective world. In this regard, gardens will play an important role in shaping the affection of people and nature. Therefore, the philosophy of respecting nature in eastern culture and using nature for the benefit of mankind on the basis of conforming to nature, especially experience and theory embodied in the garden system of east Asia, as well as the achievements of landscape gardens developed in the west over the past hundred years have become indispensable guiding ideology.

In this way, it can be said that achievements in gardening have a promising future. It is by no means limited to the creation of recreational environment, it's more important role lies in the management of the whole living environment. The sublimation of landscape gardening will become a systematic science of environment design guided by construction concept which achieves harmony between man-made and natural environment. This systematic science is vital to the future of mankind, because as "direct production and reproduction of life", "people's eating, dressing, housing, and the production of necessary tools for production" are "decisive factors of history". Proper settlement of human housing problem can play a decisive role in promoting the development of human history.

1.3 International Influence of the Classical Garden of China

Garden making in China boasts a long history and profound accomplishments. In the world history of garden making, the art of the Chinese garden stands out with a unique style and has exerted a certain influence on garden making in both Eastern and Western worlds, which earned China the reputation of "the mother of world gardens". In the following sections, we will focus on Japan and some major countries in Europe to take a brief look at the Chinese influence on garden making in these countries.

1.3.1 Influence on Japanese Gardens

In the course of over three thousand years, Chinese garden construction formed a unique system of its own and spread to its neighboring countries early in history like Korea, Vietnam, and other continental states. The island nation Japan from across the sea also drew on ancient Chinese art and techniques of garden construction during its nascent formation.

Japanese scholars believe that, before the Asuka period (A.D. 593–709),[11] very little information was available on garden construction and that, based on a domicile image carved on an ancient bronze mirror recovered at an archaeological excavation site, only a few trees were planted nearby the house. It was after the influx of continental culture—with Chinese culture as the main current— during the Asuka and Nara (593–793) periods did Japanese garden construction start to make considerable headway. The remains of the *liubeiqu* stones—stones on which water passages were carved for floating wine cups— that belonged to this period were excavated in Japan in recent years. These gullied stones have revealed that some of the Chinese garden thematic contents already spread to Japan by this time, such as the *qushui liushang* theme—"floating wine cups down a winding stream", which was originally an intellectuals' pastime of drinking wine while composing poetry fancied by those who sought after *linquan yinju*, the "secluded living amid forests and streams" life-style first pursued by Tsin (265–420) Chinese.

It is believed that Taoism thought and mythology reached Japan from China in the late seventh and early eighth centuries (late Asuka period) while, in fact, it should be considerably earlier. Apparent influence of Taoist ideology exerted on Japanese garden construction was observed around the early ninth century and was widely reflected in the garden works done during the Heian period (794–1185). The so-termed *chiquan tingyuan,*[12] common to Japanese imperial gardens and residences of the nobility at the time, resembled the garden prototype called *shanchi yuan*[13] of Tang China (618–906), where hills were arranged in the pool. Such a composition is also designated as *shenxian dao*[14] or *yichi sanshan,*[15] referring to the of Penglai, Fangzhang, and Yingzhou mountains, the legendary abode of immortals. An example of such creation was the Toba Rikyu, an imperial summer palace built in Japan in 1086. This idiom of garden composition was much in favor during the Momoyama and Edo periods (1574–1867). *Sanbaoyuan Tingyuan*[16] at the Daigo-ji Temple, an acclaimed masterpiece of the Momoyama period built in 1598 in the fashion of *yichi sanshan*, also belonged to this category. Up to the Edo period (1603–1867), the theme of *pengdao shenshan*[17] was broadly adapted by imperial, ducal, monastery, as well as private gardens. Although later in the period, the number of the islands arranged was no longer confined to the original three—with now anywhere from one

[11]Asuka period, in Japanese history and art, the era from 552 to 645 CE, which began with the introduction of Buddhism from Korea and culminated in the adoption of a Chinese pattern of government. (https://www.britannica.com/event/Asuka-period)

[12]Pool-and-spring garden.

[13]Hill-and-pool court.

[14]Island of the immortals.

[15]One pool and three mountains.

[16]Garden of the Three-treasure Court.

[17]Fairy islands and immortal mountains.

to numerous appearing in the layout, its symbolic significance remained. Further evolvement of the "fairy islands and immortal mountains" led to the formation of the typically Japanese "turtle island" and "crane island", a trend whose influence on Japanese garden construction continued until modern times.

Another ideological trend in the history of Chinese garden construction was Buddhist ideals, which reached Japan at the turn of the seventh and eighth centuries. As a matter of fact, the depth and profundity of the Buddhist impact on garden making in Japan seemed to have surpassed those in China. The Japanese techniques of "*xumi shan*"[18] and "*jiushan bahaishi*"[19] were a manifestation of such impact. The incarnation of the Buddhist influence on Japanese garden construction began approximately in the middle of the Heian period.[20] The Motsu-ji Temple Garden, a "Jodo Garden" as termed by Japanese scholars, was considered a typical work built under such influence.[21]

With the eastbound migration of Chinese culture, the art and techniques of garden construction were also directly and comprehensively brought to Japan. Apart from the above-mentioned "fairy islands and immortal mountains" and the "Jodo Land", there were furthermore direct imitations of Chinese ancient creations. During the Heian period, for instance, the design and construction of the Heian capital and the palatial gardens were modeled after Chang'an (today Xi'an), the capital of Tang China, and among those palatial gardens, the forbidden Shinsen-en garden was built under the inspiration of the Ling You animal park of King Wen of the Zhou Dynasty (c. 1066-c. 221 B.C.).

Ideological trends in Chinese garden art historically had a bearing on the making of Japanese gardens, so did the changes in social thinking of China. Admittedly, the Buddhist thinking in the art of Chinese gardens infected Japanese creations during the Heian period, but more profound impact on its garden making as regards garden's *yi jing*—the non-material and intangible quality of a scenic image or garden space that is capable of evoking an aesthetic atmosphere or emotional appeal—was found during the Kamakura period (1186–1333) when Chinese Chan Buddhism and the neo-Confucianism of the Song Dynasty (960–1279) penetrated Japan. Adopted by the Japanese contemporaneous ruling class—the Buka, these philosophies were venerated and encouraged for political reasons and consequently prevailed rapidly and widely. During the Muromachi period (1333–1573), Chan Buddhism[22] and neo-Confucianism were deeply rooted among the populace and became the guiding factors in social conduct. This period was also the golden age of Japanese garden construction.

---

[18] Maitreya Mountain.

[19] Nine mountains and eight sea rocks.

[20] From the tenth to the mid eleventh centuries.

[21] In Japanese garden making, *Jodo*, literally Pure Land, refers to the Western Paradise of Buddha Amida.

[22] Known as Seng in Japan though using the same character.

In the Kamakura and Muromachi periods, Zen Buddhist monks were particularly fond of reciting Chinese poems with naturalistic outlook and mystical recondite Zen undertone as found in this couplet by Su Dongpo, a Song neo-Confucian poet-scholar:

Babbles of the stream are a tongue broad and long.
And hues of the mountains the body pure and clean.

This also demonstrates how profound the influence of the Song and Ming (1368–1644) neo-Confucianism had been in Japan. It was a time when Zen monks held a leading position in the Japanese academic circles, as a result of which the Zen Buddhism and the Song-Ming neo-Confucianism became the dominant ideology in Japanese art and literature. As to the art of gardens, such ideology was evidently mirrored in not only the creation of *yi jing* but also the specific technical treatment of the garden. Examples are everywhere to be seen: in exploiting the acoustic properties of pine trees (in wind), bamboo groves (in rain), and waterfalls (in motion) to effect the sense of remoteness and seclusion as felt in the depth of the mountains; in disposing rock peaks to symbolize Sakyamuni, Guanyin (a Bodhisattva), and Luohans (Arhats); in carving Buddhist figures on cliff faces; in the popularly used "Buddhist triad stones" (*sanzun yizu*) arrangement, and so on and so forth.

With the increasingly frequent visits between Chinese and Japanese diplomats, monks, and merchants, cultural interflow of the two countries escalated. As a result, the garden art and techniques as well as the life-style of secluded living amid woods and water were directly and promptly introduced to Japan. For instance, Japanese Zen monk Eisai[23] of the Kamakura period revisited Southern Song China (1127–1279) for the second time and studied there for four years. When he returned to Japan, he brought back with him tealeaves and tea-sipping custom, which was a signature of the "forests and streams" living. One may well say that monk Eisai sowed the seed of the custom of *chado*—the tea ceremony, which rose to popularity during the Muromachi period (around the mid Ming Dynasty) and which in turn brought about immense transformation in Japanese garden construction. The emergence of *Chatei* was one such consequence, among other garden structures.

Again, during the Muromachi period, the consummate garden builder, Master Muso Kokushi,[24] started to attend to the Tenryu-ji Temple Garden. He engaged the Tenryu-ji Temple vessel in trading with China in order to raise funds for the garden.

[23] Japanese tea ancestor: Zen Master Ei Sai. It was Ei Sai who first introduced it to the Japanese. The history of Japanese tea kicked off with the spread of Buddhism by Ei Sai.

[24] Mengchuang Guoshi (1275–1351) was a famous Zen monk in the Ming Dynasty in Japan. He founded Tenryuji Temple located at the foot of the Arashiyama Mountains in Kyoto, and the world-famous Sangoku Temple of Japan was named the founder of Mengchuang Guoshi. Mengchuang Guoshi has deep attainments in Buddhism and has a high position in the history of Zen Buddhism in China and Japan. Master Mengchuang has longed for nature throughout his life, loved deep mountains and valleys, and cultivated many Zen disciples. In addition to his achievements in Buddhism, Meng Chuang Guo Shi is also a well-known poet monk. He once wrote a poem "Nothing Affects a Bed Wide" rich in philosophy of life, which has been circulated in Buddhism for a long time. Poetry has a deep philosophy.

While doing so, he brought back from China well-known produce, handicraft, paintings, and many more. In another case, the prominent Japanese painter Setsu Shu's[25] visit and studying in Ming China between 1467 and 1469 facilitated the Japan-bound movement of Ming Chinese culture. These activities, especially the influx of the Song and Ming landscape paintings, cast a great impact on the Japanese painting in the Muromachi period. Tensho Shubun is considered to be the founder of the Chinese style of suiboku ink painting in Japan. He was influenced by Chinese landscape painters Ma Yuan and Xia Gui of the Southern Song Dynasty (1127–1279), to give one example. Their paintings broke fresh ground by adopting the Zen naturalistic perspective and established the guiding style for Japanese ink-wash landscape paintings, which played a direct role in Japanese garden construction of this period. The Japanese garden builders took much inspiration from the landscape paintings, and many garden landscapes were copied straight from the Song and Ming brushwork. In creating garden scenes, the flair of Song and Ming paintings was also evident in both the tranquil and leisure *yi jing* of the environment as well as the use of multistoried buildings in the garden. (In fact, since late Kamakura period, the multistoried structures were already coming in favor under the influence of Song paintings.)

The Japanese academia have observed that, although there are various causative factors of the symbolism and abstractionism in the art of Japanese gardens—manifested through the abstract *xieyi* technique,[26] the fundamental and most compelling factor came from the Chan Buddhism and the Song neo-Confucianism imported from China. To their influence was attributed the further development in Japanese garden building and hence the emergence of the extreme *xieyi* garden forms, such as *shi ting* (stone court) and *ku shanshui* (dry landscape), also known as the Tang landscape that uses white sand as a metaphor for water. Since the Tang and the Song eras, the pastoral ideals aspired after by the Chinese intellects and artists and especially the naturalistic outlook of the Song and Ming neo-Confucianism once articulated in the art of Chinese gardens, now as a creative ideological force, had directly or otherwise produced a far reaching influence on the advancement of Japanese garden construction (Fig. 1.10).

Another telling influence of China on Japanese gardens in the later period could be found in the experience of Zhu Shunshui, a one-time official of the Ming court who was exiled to Japan in 1665 at the death of the Ming Dynasty. During the more than ten years of his remaining life in Japan before his death there, Zhu was accorded the respected status of master by the Japanese rulers. Giving lectures aside, Zhu was mostly engaged in building gardens, which contributed conspicuously to the garden

[25] Sesshu (1420–1506) Japanese painter. The name Deng Yang. Born in Bichu Akahama (now Soja City, Okayama Prefecture). He once entered Xiangguo Temple as a monk, and may have painted with the landscape painter Zhou Wenxue of the temple. His works widely absorbed the painting styles of the Song, Yuan and Tang dynasties in China. He was later recognized as a world cultural celebrity by a decision passed by the Vienna World Peace Conference.

[26] Technique which aims at capturing the spirit of the object rather than its physical likeness as seen in the freehand brushwork that "condenses thirty thousand miles into a few feet or inches".

Fig. 1.10 (**a**) Japanese garden designed according to the garden making idea of Chinese natural garden-lake-shaped water surface, arranged with banks laid with natural stones, shoal and promontory, stone footbridge, stone lamps, etc. (Katsura Imperial Villa, Kyoto); (**b**) Scenery of stone footbridge in Tenryu-ji Temple Garden, Kyoto, Japan; (**c**) Peninsula composed of huge rocks protruding from the water's edgein Tenryu-ji Temple Garden, Kyoto, Japan

construction in Japan. This was a time when the Japanese sovereigns highly venerated the Song neo-Confucianism and set great store by its practice of "restoring ancient ways", which consequently set the tone and style for the art of garden building in Japan. The famous ducal garden, the Koraku-en—"Garden of Deferred Enjoyment" (Hou Le Yuan)—in Kyoto, was so named by Zhu (1668–1669) to evoke the meaning that "A man of virtue takes his enjoyment after the others", from *Meng Zi: Liang Hui Wang.*[27] In the process of revising and refining this famed garden, Zhu followed the artistic principles of Chinese garden building throughout. He designed and constructed, among other scenes for the garden, the Full-Moon Bridge (Yuanyue Qiao), a single-arch stone bridge in the pastoral vein of the Jiangnan region in China (Fig. 1.11).

Making its advent in Japanese garden construction, this brand-new garden structure was the first to employ the Chinese technique of constructing arch bridges in Japan. Later, many garden scenes were built in the wake of the Full-Moon Bridge, as seen in the execution of the Taikobashi Bridge built in the Shukukei-en Garden in Hiroshima in 1781–1788, to give just one such example (Fig. 1.12).

[27] Mencius: King Hui of the state of Liang.

Fig. 1.11 The Yuanyue Qiao in Koraku-en Garden, Kyoto (Li Yuhong photographed)

There were other architectural works with Chinese motives added to the Koraku-en Garden. The Hall of Attaining Benevolence, dedicated to Bo Yi and Shu Qi, was built under the ideological guidance of "revering Confucianism and restoring ancient ways" (Fig. 1.13).

The Little Mount Lushan, created in quest of natural landscapes, was a copy from the scenery of Mount Lushan in China. The West Lake Dyke was fashioned after the Su Dyke and the Bai Dyke of the West Lake in China's Hangzhou (Fig. 1.14).

Such a design concept of aspiring to natural scenery also gained widespread currency and led to the formation of the characteristic style of the Daimyo garden, a large-scale garden favored by the influential stratum of the society. Among the works during this period that emulated nature—mountainscapes in particular, many of them were modeled after Mount Lushan (Fig. 1.15). Under such elicitation, many garden builders began to take inspiration from the famous scenic spots within their own native land Japan itself.

In the use of plant life for gardens, Japan was long influenced by the design concept where a variegated display of collected rare and well-known species was arranged for appreciation. Such a plant concept was adopted in Chinese garden construction since an early stage. Many varieties in Japanese gardens were introduced from legendary scenic spots in China and cultivated and domesticated to become garden materials in Japan. The first such instance on record showed that, during the Nara period, the founding master of the Toshodai-ji Temple and eminent monk Jianzhen brought back pine seeds from Gu Mountains near Hangzhou and nurtured them into trees as ornamental plants in his monastery garden. Up to the Edo period, the practice of introducing plant species from famous scenic areas for use in

Fig. 1.12 Chinese arch stone bridge–the Taikobashi Bridge built in the Shukukei-en Garden in Hiroshima

garden appreciation already became the rage. Varieties that were derived from scenic regions in China included the Sichuan willow and the West Lake plum raised in the Rikuenkan Garden.

Throughout the history, China exerted an extraordinarily profound influence on Japanese garden construction. Major changes in ideological trends of art and literature in Chinese history directly affected the course of Chinese garden construction, which often cast their impact on its Japanese counterpart. Even Chinese folk legendry was often assimilated into the themes of Japanese gardens. For example, as early as the Heian period, the Dragon Gate Waterfall was built based on the Chinese folklore "The Silver Carp Leaps over the Dragon Gate", which was later

Fig. 1.13 The Hall of Attaining Benevolence in Koraku-en Garden, Kyoto (Li Yuhong photographed)

Fig. 1.14 The West Lake Dyke in Koraku-en Garden, Kyoto (Li Yuhong photographed)

Fig. 1.15 Scene of the Little Mount Lushan in Koraku-en Garden, Kyoto (Li Yuhong photographed)

stylized into one of the waterfall prototypes. In the Muromachi period, the clustered rockery arrangement of "Tigress Ferrying Cubs" in the Ryoan-ji Temple Garden took its theme from a Chinese folktale about a migrating wild tigress who ferries her cubs across the river.[28] The legend of Chinese ancient Eight Battle Formations, namely the Fish Scales, Crane Wings, Long Snake, Supine Moon, Sharp Arrow, Square Circle, Cross Yoke, and Flying Geese, reached Japan during the Tang Dynasty and was since adopted and elaborated into various rockery compositions that represented the formations invented by Sun Zi, Wu Daozi, Zhuge Liang, and other ancient Chinese military strategists. Examples of such works include the recently constructed Eight Battle Formation Plan credited to Zhuge Liang, a group rockery arrangement that consists of the General along with Heaven, Earth, Wind, Cloud, Dragon, Tiger, Bird, and Snake; it was built for the Kishiwada Garden in 1930s.

Nevertheless, throughout the development of its garden construction, Japan has maintained an independent and intact style of its own while assimilating elements from other countries, China in particular. On the basis of its traditions, Japanese garden construction in the modern times has made tremendous progress in both the art of management and techniques of engineering; and their high standards have been recognized worldwide. At such, Chinese garden construction, which has lagged behind since the modern age, shall find plenty to draw on the Japanese experience.

[28] See *Houhan shu: Liu Kun zhuan*, or Book of the Later Han Dynasty: biography of Liu Kun.

1.3.2 *Influence on European Gardens*

Chinese garden construction also cast a conspicuous impact on the development of European garden construction. But before we proceed to the discussion, circumstances that pertain to this subject will have to be touched upon first. Since the first road penetrated the Western Regions off central China in A.D. first century, especially after the Silk Road rose to glory during the Tang Dynasty, Chinese merchandise such as silks and porcelains streamed into the Western Regions and reached as far as the Roman Empire. For a relatively long period of time in the history thereafter, contacts between merchants and travelers of China and Europe continued intermittently. Favored though the Chinese goods by the Europeans may be, the social conditions and cultural achievements of China were very little known in Europe for a considerably long time. During the thirteenth century, the military expansion of the Yuan Dynasty (1271–1368) brought forth monumental changes to the Sino-European relations. *The Travels of Marco Polo*, published in 1298, brought China to the Europeans and aroused a widespread interest in China and the desire to know her more. At this time, although the overland travel on the Eurasian continent was still fraught with difficulties, direct sea routes that were initiated during the fourteenth century opened up a new prospect to the cultural interflow between the two sides. However, more direct contact between Europe and China did not begin until the rise of European capitalism and oversea expansion. Since 1515, when the first Portugese merchant vessel dropped anchor at a Chinese port and started trading, a great amount of Chinese porcelains, lacquerware, silks, tea, and handicrafts were imported into Europe, which brought with it Chinese culture at the same time. Holland followed suit in developing seafaring and ocean trading with the Far East; their merchants and missionaries traveled in China and in turn spread Chinese culture back in their own homeland. Up to the seventeenth century, talk about China, combined with the mysterious and exotic merchandise from China, diffused widely in Europe, which created a European crave for things Chinese. During this time, merchants from Holland, France, and other European countries began to manufacture for sale a great deal of imitation Chinese porcelains, silks, handicrafts, and the like, which further magnified the Chinese influence. Prior to this point, Turkey had embodied the Orient to European minds. All decorative art and handicraft items manufactured to assume the "Oriental" traits were made in accordance with the "Turkish style"—*turquerie*, which was by now replaced by *chinoiserie*—the Chinese style. The Chinese influence first showed in France, a country considered the fashion setter in Europe. Initially, it was seen in the décor of banquets and the costumes at masquerades, later it spread gradually to decorative arts and handicrafts. The trend of such a fad in France owed its inception to the French rulers; Louis XIV was deeply infatuated with *chinoiserie*. He collected in his palace a great number of Chinese silks, brocades, porcelains, gold vessel, lacquerware, garments, furniture, and other handicrafts, as well as their imitations. The craze for *Chinoiserie* was further deepened when French merchant ship *Amphitrite* made two trade voyages to China in 1698 and 1703.

When studying the influence of Chinese artistic style, one should first bear in mind that there is a fundamental difference in the nature and consequence of such influence between Europe and Japan, Korea, Vietnam, or other Asian countries. ·The Chinese influence took a divergent path in Europe despite the breadth and intensity of the *chinoiserie* craze, which spanned the seventeenth through the early nineteenth centuries—almost two hundred years, which alone may well demonstrate the profundity of such influence. However, it was a time when Europeans had scant knowledge of China's history and social conditions and only surface understanding of her art. European designers could only base their imagination and appreciation of things "Chinese" on hearsay and often fallacious accounts bout China or on the mediocre handiwork and inferior imitations that were brought to them by the merchants. Thus, the depth and accuracy of their discernment of Chinese art—or rather the lack of them—can well be imagined. One may say that the rage of *chinoiserie* in Europe during that period is a result of novelty hunting. Europeans did not particularly have a handle on Chinese art, nor would they necessarily attempt a serious study in order to command such understanding; alternatively, they simply interpreted it from their own impressions. What Europeans perceived in those days as the "Chinese idiom" was interrelated with the Rococo style that pervaded all aspects of art across Europe in the early eighteenth century. From the standpoint of artistic creation, the interest that Europeans showed toward the Chinese idiom stemmed from their fascination for its freedom of approach, seen as a strange and novel lack of constraint that was disparate from Europe's own traditional geometric-style gardens. Therefore, the imported Chinese objets d'art, to a certain degree, gave inspiration and substance to the Rococo style and contributed to its development. The eighteenth-century Rococo artists often referred to the figures and scenes depicted on the Chinese objects—porcelains, lacquerware, textile fabrics, embroidery, wallpapers, New Year illustrations, garments, furniture, etc.—for their creative inspiration and diverged into pursuing the peculiar and indulged in a superficial play of techniques. At such, we recognize that the so-termed *chinoiserie* rage since the seventeenth century was nothing truly Chinese but a result of an intrinsic evolution of European art under the Chinese influence. Some European scholars segmented the development of *chinoiserie* into three indistinct and mutually overlapping phases: the "unrestrained phase" at the turn of the sixteenth and seventeenth centuries when *turquerie* was being replaced; it then evolved into the "baroque phase", which further developed into the final "imitative phase" in artistic creation. Such a pattern, to a large extent, also reflected the development of the art of European garden construction.

The sixteenth century first saw the rise of the art of garden construction in Italy. The culture of the Italian Renaissance expanded to France, England, and elsewhere in Europe, so did its art of garden construction. In France, the monarchal autocracy was being intensified during the second half of the seventeenth century, as a result of which the palatial construction became more developed. The French making of new palatial gardens first reformed the garden making techniques originally introduced from Italy and later established a form of garden landscape of their own that emphasized the rationalistic, geometric beauty, labeled as the "classical" art of garden. Its principal founder, Andre le Notre (1613–1700), was a renowned garden builder in Europe. France was the trendsetter in European culture; its garden art of classicism was rapidly becoming the prototype for formal palatial gardens at the royal courts of England, Germany, and Russia. The popularity of this new garden form eventually found it way to the backyards of the manor houses of the wealthy and the influential and continued to proliferate in Europe and became the mainstream garden construction of this period. At the same time, the influence of the art of Chinese architecture began to be felt in Europe, exemplified by Porcelain Trianon, a Chinese pavilion, built during 1670–1671 at Versailles outside Paris. In form, the structure was an imaginary concept of neither fish nor fowl, but it sufficed to demonstrate the French fancy for Chinese palatial gardens at the time of Louis XIV. It is true that the Chinese effect in architectural art and art of garden construction first debuted in France, but a profound transformation in the art of garden construction truly began in England.

In the middle of the eighteenth century, the upsurge of the bourgeois ideological trend of the French Revolution was in the making, and the Chinese *mores* and political thinking provoked French interest, which deepened the pursuit of Chinese culture in France. On the other hand, England had accomplished its own bourgeois revolution by this time and was in the height of the industrial revolution, which primed the nation for cultural changes. The country took the lead in developing new art of garden design and formulated the English Landscape style in the early eighteenth century. Under the shock wave of the Romantic Movement in art and literature in the second half of the eighteenth century, the naturalistic Landscape style was further evolved into the untamed Picturesque style, which were mostly found in the gardens of the manors and mansions of the nouveaux riche at the time. These gardens, characterized by natural landscaping, owed their inception mainly to the influence of the art of Chinese garden making in terms of the external causative factor. Chinese garden art perhaps made its formal entrance in the English garden community with the publication of Sir William Temple's *Upon the Gardens of Epicurus* in 1685. In this writing, Temple provided comparative observations on the European "regular" formal garden and the Chinese "irregular" natural-style garden, which "for aught I know, have more beauty than any of the others". The author was apparently taken with the beauty of this garden "without order", in contrast to the symmetry, uniformity, and precision of the European garden. He accredited such accomplishments to that Chinese was "a people, whose way of thinking seems to lie as wide of ours in Europe, as their country does" and that the natural-style garden must have owed its beauty "to some extraordinary dispositions of nature in the seat, or some great race of fancy or judgment in the contrivance". Temple's commentaries were believed to consequently give the impetus to the formation of the English Landscape garden design. Subsequently in 1772, Sir William Chambers gave an emphatic introduction on the art of Chinese garden construction in *A Dissertation on Oriental Gardening*. He vigorously advocated that the English Landscape-style garden ingest the Chinese taste from China's creations. While admiring the artistic accomplishments in Chinese garden making, Chambers deplored the devastation and bareness of the English Landscape design, describing it as artless, raw, and unrefined. It was a time when Lancelot "Capability" Brown—the pioneer English Landscape garden designer—was in his heyday, and Chambers' book dealt a sharp criticism to the pastoral landscapes done by the Brownians. Chambers believed that it was ascribed to the erudition and artistic cultivation of the Chinese garden makers that gardens in China, also taking after nature, could achieve such artistic profundity. By this time, garden making in Europe was not entirely professionalized; the Brownians, being vegetable gardener and horticulturists, only concurrently took up landscaping projects. Therefore, Chambers maintained that English landscape designers aspire to greater erudition and strive to learn from the Chinese accomplishments in the course of developing the English Landscape garden. Chambers was a supercargo to the Swedish East India Company and frequented China in his earlier years. During his sojourn there, he wrote many sketches on Chinese architecture, furniture, apparel, etc. and published in 1757 *Design of Chinese Buildings, Furniture, Dresses, Machines and Utensils*, which stirred a considerable impact in Europe. His *Dissertation on Oriental Gardening*, which came off the press fifteen years later, provided valuable guidance of a more direct nature for the creation of the English Landscape garden. Chambers' opinions were controversial in Europe at this time, but the Chinese influence by no means diminished thereof; on the contrary, it escalated and contributed to the advancement of the English Landscape garden. As Chambers remarked, the English Landscape design, when in its nascent stage, was rather dilettantish—for example, the creation of the idyllic pastoral scenery was little more than a mechanistic replication of the Scottish grazing ground. Once taking further in the art of Chinese garden making, the English Landscape garden became richer in substance and more epitomized in style.

As the English Landscape garden, as well as the Picturesque garden, had largely assimilated Chinese elements, they were described by the French as "Chinese garden" or "Anglo-Chinese garden" (Fig. 1.16). The Anglo-Chinese garden grew to be all the mode throughout Europe, and Chinese garden making turned into such a sensation that Christian Cajus Lorenz Hirschfeld, a German professor of aesthetics, griped in his 1779 book, *Theorie der Gartenkunst* (Theory of Garden Making):

> People who build gardens nowadays do not following their own inspiration, nor do they respect the more refined taste from the former days. All they care for is whether it is the Chinese or Anglo-Chinese style.

Nevertheless, Anglo-Chinese gardens proliferated to almost every country in Europe. Chambers had Kew Garden in Surrey, London, retrofitted between late 1758–1759, and experimented in small scale architectural ornament, adding on a good many scenes "in the Chinese taste", the most famous being the Pagoda at Kew Garden, which was a surviving exemplar of its kind (Fig. 1.17).

As to garden making in France, although signs of Chinese influence were already apparent as early as late seventeenth century, Chinese-style gardens did not come into vogue until the eighteenth century when spurred on by the growth of the Anglo-Chinese garden, which, once coming ashore in France, grew to be the prevalent influence in French garden making of the time. The novel trend went to extremes,

Fig. 1.16 "Anglo-Chinese garden" pavilion at Hidcote Manor Garden, Gloucester, England

however, as Exoticism (termed "Sensibility" by Western scholars) was infused into the pastoral scenery. What was called Exoticism in French gardens was mostly meant the architectural structures fashioned after the Chinese style. Nevertheless, building Anglo-Chinese gardens under the generic heading of the "Chinese" style was often mixed in with a streak of Japanese, Indian, Turkish, and even ancient Egyptian or Greco-Roman traits. As pursuing Exoticism was *au courant* in garden building, these other-than-Chinese traits were also brought into Anglo-Chinese gardens. When the Anglo-Chinese style entered its late stage, garden scenes created in the name of the Chinese fashion were already architecturally unrecognizable as "Chinese" as a result of being elaborated or even mutated by the original-minded European architects and garden designers. Pavilions, pagodas, bridges, walkways,

Fig. 1.17 The Pagoda at Kew of London in England

and other such architectural structures were transmuted into some sort of outlandish monstrosities that were a far cry from the Chinese style or that of other Oriental countries and that bore no resemblance to European own traditions as well (Fig. 1.18).

Such a trend fared worst in France. According to a count by Eleanor von Erdberg, there were 25 well-known gardens in France during this period that featured architectural structures done in what-was-called "Chinese taste".[29] Although few actual examples are preserved today, one may still learn about them from written accounts such as *Jardins Anglo-Chinois*, by Georges-Louis le Rouge, published in 1774; *Description des Nouveaux Jardinsde la France er de ses Anciens Chteaux*, by Alexander La Borde, 1809; *Maisons de Compagne*, by Johann Carl Krafft, 1876; and *La Chinesen France au XVIII siecle*, by Henri Cordier, 1910. Under the influence of *chinoiserie*, Landscape-style garden making in France during the eighteenth century diverged into seeking after the baroque and the bizarre in the name of the Anglo-Chinese style, unlike its English and German counterparts, which developed in a more wholesome manner. In view of this situation, German scholar in history of garden making H. Jager had the following remarks:

> The French Landscape garden design lacked the noble English taste born of the love of nature and was short of the profound German observations of nature; instead, it merely imitated the Chinese oddities and deviants.

Despite Jager's dubious apprehension of Chinese garden art, his statement certainly attested that, under the charge of Chinese influence during the eighteenth century, the development of the Landscape-style garden in various European countries were treading along different paths.

---

[29] See Erdberg's *Chinese Influence on European Garden Structures.*

Fig. 1.18 Anglo-Chinese garden scenery of the frescoe painted in 1800 at the French Art Capital Museum—the forest stream, rock and a so-called "Chinese style" pavilion bridge

By and large, the chinoiserie garden style or Anglo-Chinese style gardens that emerged in Europe in the modern age were merely simulations of certain ostensible features of Chinese gardens culled from curious impressions. Judged by their garden works, only the materials for naturalistic landscaping were employed; and except for the topography, water bodies, vegetation, and walkways that were executed rather spontaneously and a few Chinese-style structures—halls, pavilions, bridges, pagodas, and boats—that were appended to the scenes as garden embellishments,[30] these Chinese-style gardens in actuality failed to assimilate the laws and principles of classical Chinese garden construction in all their entirety and profundity.

Although growing to enjoy furious popularity in European garden making, the Anglo-Chinese style garden did not perpetuate due to its superficial faddish nature. Gardens built under such trend, as going out of vogue, were mostly altered or demolished, very few remaining intact today. On the other hand though, the naturalistic Landscape garden drew on the experience of Chinese garden making and constantly improved based on Europeans' own understanding of and interest in nature, was able to transcend the superficial norm and enjoyed a healthy transformation. Works of these gardens attained ingenuity in both thematic conception and the scenic *yi jing* that could rival Chinese classical gardens. The English rock garden was an example of such accomplishment. Since the sixteenth century, England had been introducing alpine plant variety that was domesticated and cultivated to be garden ornamental plants. They were simply potted plants in the early days and later used to bedeck parterres in the topiary tradition, which became known as the "wall garden". At the close of the seventeenth century, Chinese (and in some instances Japanese) art of artificial hills reached Europe, where, initially, the hills constructed were merely massive meaningless conglomerates of earth and stone—even with some stiff and awkward caves and cascades—that rendered no *yi jing* or any atmospheric appeal. But in the nineteenth century, the English drew lessons from Chinese gardens and apprehended the true essence of creating natural beauty. Since then the English landscape garden went on the right track and began its exploration of integrating ornamental alpine plants with the art of piled-rock mountains, whence the appearance of the rock garden, which matured into a distinguished garden type in 1940s after long-term creative practice, summing up, and improvement. Back in 1910, the alpine plants in the botanical garden at Kew of London were still arranged in the fashion of a topiary retaining wall. After thirty years of study and transformation that further enhanced the creative standards of hilling techniques and scenic artistry, the garden makers successfully achieved the scenic unity between the alpine plant disposition and natural-form artificial hills, thus completed the world-renowned masterpiece, Rock Garden at Kew, known for its typical scenery of bright mountain blossoms in full bloom along with the water cascading down the mountain slopes.

The artistic accomplishment of classical Chinese garden construction was historically used for reference and assimilated by other nations. It still maintains infinite attraction even to this day (Fig. 1.19).

However, due to a sequence of historical reasons since the modern times, both external and internal, this splendid heritage has regrettably not been properly studied, inherited, and carried forward. Today at a time when China is being revitalized, one should further delve into this legacy to benefit the construction of

[30] Some emulated architecture even went to the extent as to pattern after different regional styles, such as the "Beijing-style" and "Guangzhou-style" halls built in Holland.

Fig. 1.19 Chinese waterscape garden depicted by the British (from Viscountess Wolselve, Gardens, Their Form and Design, New York, Longmans, Green and Co., 1919)

a modernized society and the achievement of both material and spiritual refinement. In the meantime, it is the author's hope that the art of classical Chinese gardens be brought to the international academic arena on the level of a profound discipline and to facilitate a broader cultural exchange worldwide in the new historical era.

1.4 Jiangnan Gardens: The Elite of the Classical Chinese Garden

During the Ming and Qing (1368–1912) Dynasties, the word *Jiangnan*—the south of the River—referred primarily to the region along the lower reaches of the Changjiang[31] River within the borders of Jiangsu Province. Typically, the region is further defined as the urban and rural areas south of the River, except for Yangzhou and a few other cities and towns that are located on the north bank of the River. In the Ming Dynasty, "Jiangnan Province" was once set up and directly affiliated to the central imperial government. In *Nanxun shengdian,*[32] compiled by

[31] Also known as Yangtze.

[32] A grand record of the imperial itinerant inspection to the South.

the imperial court during the Qianlong reign of the Qing Dynasty, the Jiangnan area was mentioned in juxtaposition with Zhejiang Province, indicating a same regional and administrative hierarchy between the two. Since the modern times, "Jiangnan" has been given a broader definition—based on the general character of the anthropography across the region and its immediate environs—to include Jiangsu Province, the south of Anhui Province, and the Taihu Lake valley, a water network district in northern Zhejiang Province (Fig. 1.20).

The term "Jiangnan garden" mainly refers to those known existing private gardens located in the historic cities and towns in southern Jiangsu and northern Zhejiang, where more of these gardens survive. These residential gardens are usually moderate in size, which in most cases ranges from one to ten *mu.*[33] Even the largest house garden covers not much larger than one and half acres. The development of this style of garden can be traced back to around A.D. third century with a history of at least 1500 or 1600 years. The original gardens preserved until the modern times were usually the creations of Qing, some of which might have been reconstructed on the remnant foundations of Ming, or even further back to the Five Dynasties (907–960), Song, and Yuan Dynasties. Those that are in existence today have all been repeatedly refurbished or rebuilt by later generations, and still some others are simply fragmentary sites or ruins. Nevertheless, these remaining material gardens, as the main objects of our research and study, will suffice us to discover and appreciate the artistry and techniques that had created classical Jiangnan gardens.

The fertile region of Jiangnan is historically celebrated as *yu mizhi xiang,*[34] a land of abundance. It is also a land of flourishing handicraft industry and high achievement in art and culture. Optimal natural conditions invigorate growth of vegetation, and a rich variety of plant life is made available for garden use. The *taihu* stone and mountain rocks produced nearby provide ideal materials for making stone mountains. In addition, Jiangnan is a land of a water network, blessed with rivers and canals that ramify over the region, affording convenient source of water supply. Furthermore, the long accomplished artistry in architecture and the deep-rooted tradition in handicraft art as reflected in furniture, decorative art, and the like, all offer many advantages and material prerequisites for the development of garden art. Therefore, it comes without a surprise that some of the highly recognized gardens were created in the Jiangnan region at a very early stage in history. Examples of such gardens are plentiful, which include the palatial Wutong Yuan and Lingyanshan Guan Wagong at the Mount Lingyan scenic region in Suzhou, both belonging to the king of the Wu State during the Eastern Zhou Dynasty (770–256 BC); Huijing Yuan in Jiaxing of Zhejiang Province, made also during the Eastern Zhou period; Fanglin Yuan, Luoxing Yuan, and Guilin Yuan, in Jianye (today Nanjing), the capital city of the Wu State during the Three Kingdoms (A.D. 220–280); Hualin Yuan and other imperial gardens in the same city, which changed the name to Jinling at this time during the Eastern Jin Dynasty (317–420); and of the same period Gupijiang Yuan, a

[33] One *mu* is approximately 0.16 acre.

[34] A land of fish and rice.

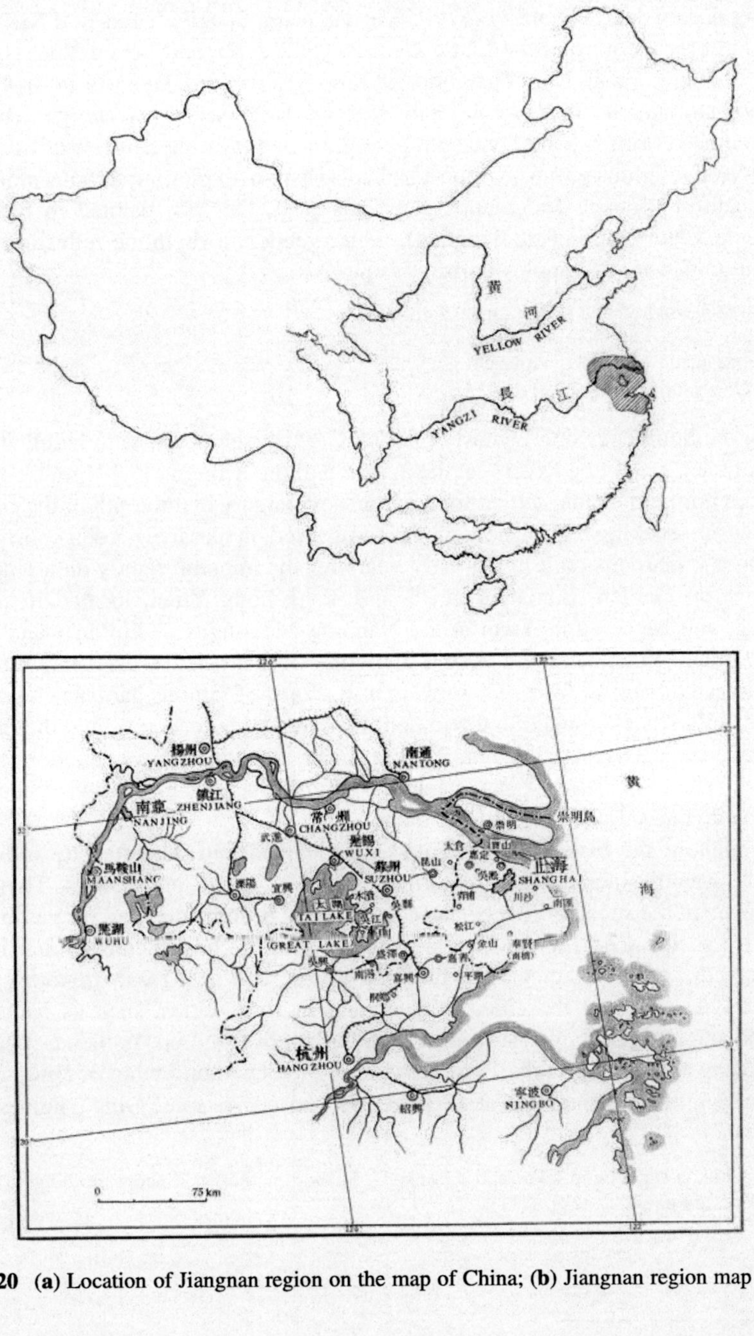

Fig. 1.20 (**a**) Location of Jiangnan region on the map of China; (**b**) Jiangnan region map

historied private garden in Suzhou. During the sumptuous Six Dynasties (222–589 A.D.), garden making became all the more rage and the gardens that survive from that period include Xuan Pu built at the foot of Mount Zhongshan in

Nanjing during the Qi Dynasty (479–502); Tu Yuan, Fanglin Yuan, and Xiangdong Yuan built for the king of the Liang Dynasty (502–557); and Leyou Yuan, Qinglin Yuan, Shanglin Yuan, Nan Yuan, etc., during the Liu Song Dynasty (420–479). In the faddish pursuit of the chic and elegant, small-scale private gardens also mushroomed during the Six Dynasties period. In the meantime, the rise of Buddhism empowered religious groups, which also had a share in exploiting famous mountains and beautiful spots in Jiangnan. As a Tang poet, Du Mu, painted in his poem "*Jiangnan chun*" (spring of Jiangnan), which renders a rhythmic reflection of the popular garden-motif temples during this period:

> With four hundred and eighty monasteries.
> Interspersed the Southern Dynasty,
> Shrouded in the misty rain.
> Were myriad towers and terraces.

During the Southern Song Dynasty (1127–1279), the emperor, who had fled to the south of the Changjiang River[35] at the invasion of the Xiongnu (the Hun) troops that occupied northern China, settled for a partial sovereignty in the south of the country. The pleasure-seeking regime nevertheless embarked on massive garden construction in Jiangnan. Gardens built in the early years for the imperial family alone included the Imperial Garden, Baxian Yuan,[36] and Yangzhong Yuan, located in today's Nanjing, and later, Yujin Yuan at the Nanlong Mountains in Lin'an[37] and Jujing Yuan, Jifang Yuan, and the Xizhu Imperial Garden inside the Lin'an palatial complex, so on and so forth. There were also scores of famous gardens constructed nearby Lin'an that belonged to the imperial kin, feudal bureaucrats, and the rich and powerful, such as Huanbi Yuan, Yinxiu Yuan, Zesheng Yuan, Yundong Yuan, Shuiyue Yuan, Zhenzhu Yuan, Meipo Yuan, Xiuye Yuan, Zongyi Yuan, Nan Yuan, Gan Yuan, Lu Yuan, and Pei Yuan.

Throughout the history, Jiangnan was frequented and inhabited by influential officials, wealthy merchants, dignitaries, as well as noted intellectuals. They were proprietors of the social affluence and culture; their hedonic demand for gardens had created a social need for the development of garden construction, which in turn encouraged the growth of garden building artists, and as a result private gardens were thriving all over the cities in Jiangnan. In some cities, such as Yangzhou, Suzhou, and Hangzhou, private gardens multiplied by hundreds in their heyday. Up to the Ming and Qing periods, especially during the Qianlong reign of Qing, garden construction in Yangzhou was at the peak of its prime. As some Qing commentators observed:

> Hangzhou is superior in lakes and mountains, Suzhou markets and shops, and Yangzhou gardens and pavilions (Fig. 1.21).

---

[35] Yangtze River.

[36] Garden of eight immortals.

[37] Today Hangzhou, the capital of the Southern Song.

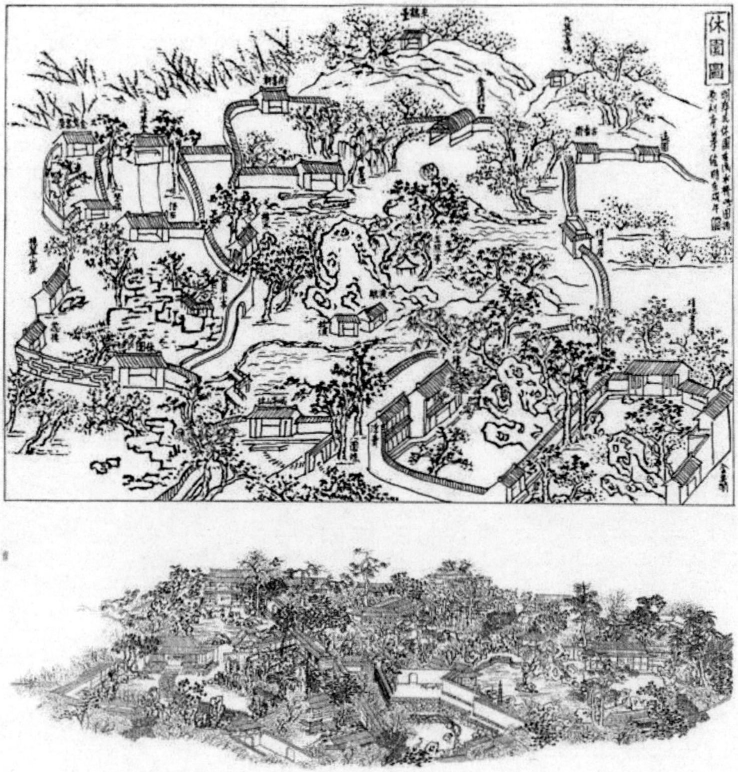

Fig. 1.21 (**a**) Picture of Xiu Yuan—a famous private garden of Ming Dynasty; (**b**) Picture of Li Yuan—a famous private garden of Qing Dynasty

The outstanding creative practice in and theoretical study of garden construction by many notable garden makers—namely Yu Cheng of the Southern Song Dynasty; Ni Zan (a.k.a. Yun Lin) of the Yuan Dynasty; Zhang Nanyang (Xiao Xizi or Wo Shisheng), Wen Zhenheng, Zhou Bingzhong (Shi Chen or Dan Quan), Ji Cheng (Wu Fou) of the Ming Dynasty; and Li Yu (Li Weng), Shi Tao (Shi Daoji), Zhang Lian (Nan Yuan), Zhang Ran, Ge Yuliang, and Qiu Haoshi of the Qing Dynasty—contributed tremendously to the development of Chinese garden construction as a whole, and their great amount of practice continually raised the bar for Jiangnan garden construction. Today, as we study the heritage of their masterpieces, we are also paying a tribute to our gifted garden artist generations before.

Chapter 2
On the Art of the Landscape Design

2.1 Introduction: A Form of Time Space Art of Practical Value

Chinese landscape painting—the ink-wash painting known as *shanshui hua*[1]—is a representation of the three-dimensional natural scenery on a two-dimensional picture scroll by means of painting artistry and techniques. In other words, it is manipulation of space on a flat surface. Described by the West as the "natural-type" or "landscape-type" garden, the Chinese garden is not only a space of reality but also a space of artistic re-creation. It is a space of art. The art of Chinese garden making is sometimes paralleled with the art of theatrical performance. Following the theatrical principles, actors portray the dramatic characters in the stage setting to project a true-to-life experience for the audience that is both artistically affecting and realistically convincing. Likewise, following the garden making principles, garden makers re-create natural landscapes by means of stone, sand, water, soil, and plant and animal life in a limited garden space to provide a true-to-nature experience to the viewer that is both artistically appealing and realistically enjoyable. Suffice it to say that the theatrical art represents life and the garden art re-creates nature.

In the Jiangnan region, naturalistic landscapes, even if created for a small garden tucked in a dense urban area, are often imbued with natural interests. Frequently, a piece of lake stone and a few clumps of flowers and trees can embellish a "natural state" where the garden stands regardless how small the garden space may be. Therefore, the Jiangnan garden has long enjoyed the name of *chengshi shanlin.*[2] To capsulize a natural space in a garden space on a restricted urban plot, a Jiangnan

[1]Mountain-and-water painting.

[2]Mountains and forests in the city.

© Springer Nature Singapore Pte Ltd. 2022

H. Yang, *A Treatise on the Garden of Jiangnan*,
https://doi.org/10.1007/978-981-16-6924-8_2

garden is said to be a landscape painting that "condenses rivers and mountains of a thousand miles into a silk scroll of a foot", which is also an epitome and typification of nature. As such, "mountains and forests in the city" are inevitably *zhichi shanlin.*[3] If a drama is "miniature life", a garden is "miniature nature". Creation of the Jiangnan garden is a process of artistic condensation and even necessary formulization as practiced in Beijing opera or *xie yi* painting. Opposite to realism, both Beijing opera and *xie yi* painting are art forms that express their stagecraft or brushwork in an abbreviated and abstract style, aspiring to capture the spirit and essence of the subject rather than its physical likeness. In the Beijing opera tradition, faces are painted in abstract patterns and colors that correspond to the personalities of the different characters in the play, which nonetheless projects a strong artistic appeal to the audience, and singing and dancing are performed symbolically with minimal properties such as using a horsewhip to represent riding a horse or an oar rowing a boat. Similarly, as in the Jiangnan garden, a span of *yunqiang*[4]—a low wall built by following the contour of the rolling hill—can be suggestive of a mountain villa while a stretch of clear water with a few mountain rocks can evoke an impression of remote mountains and deep rapids. Jiangnan gardens are often created by imitating landscape paintings or idyllic writings as well as famous scenic landscapes from nature. Regardless of the form of its inspiration, creation of the garden scenes—whether it is a mountain stream running through a remote ravine, a quiet lake nestling to a tranquil hillside, a narrow path winding through a mountain village, or a pond of lotus swaying by a waterside pavilion—always follows a certain thematic subject, which is why the garden's scenic imagery is said to be idealized nature and a product of human subjectivity. Nature is grander, richer, and more dynamic while the garden—the re-creation of nature—is more condensed, more epitomized, more poetical, and more intriguing. Garden scenes are affective and emotive. A fine Jiangnan garden, in its artistic qualities, is a picture and poem expressed in the vocabulary of garden making, which is of the oft-described "*shi qing hua yi*".[5] When you stroll in a Jiangnan garden where the buildings and the natural landscape set off one another and where the hills, lake, trees, and flowers are interlaced with pavilions, terraces, houses, and bridges, the scenery takes on a different view along with each step you take and each turn you make, and the rhythmic and melodic quality of the garden is reminiscent of the adage "Architecture is solidified music". The Jiangnan garden is not only solidified music but also solidified poetry. To create a poetically charged garden space with a natural interest is a scrupulous and elaborate work that involves topographical sculpturing, vegetative disposition, architectural arrangement, and fauna embellishment with creatures like birds, beasts, fish, insects, etc. Compared with the classic school of garden construction in Europe, which emphasizes rationalistic, geometric structures as

[3]Literally "foot-long mountains and forests".

[4]Cloud hill-climbing wall.

[5]Poetic sentiment and picturesque overtone.

reflected in embroidered parterres and green sculptures, the art of the Jiangnan garden has achieved an idealistic profundity unmatched by its European counterpart. No wonder Sir William Chambers, the renowned British architect and garden builder, deplored back in the mid eighteenth century that "The garden composition done by the Chinese is outstanding. The interest expressed by the garden is what the English have long pursued but been unable to attain".[6]

Garden making is like landscape painting in a three-dimensional space with stone, sand, water, soil, vegetation, and animal life. In other words, the Jiangnan garden is a three-dimensional picture that one can enter. Constructed to a human scale, the Jiangnan garden landscape allows the viewer to step through the "picture frame" and take a stroll in the painting. The spatiality of the garden affects its functionality; even the restricted garden space typical in the Jiangnan region does not efface this functional value of the garden. At such, the art of the Jiangnan garden is a functional art, and the functionality that assures the daily life and recreation is the very basis of the existence of the Jiangnan garden. Nevertheless, not all three-dimensional gardens are functional. The potted landscape, for instance, including the grand-style potted landscape of the Lingnan garden in the Guangdong and Guangxi regions, is also spatial, but these scaled-down potted gardens carry no functional value and are for viewing only. Guo Xi, a classical painting theorist of the Song dynasty, once observed:

> [Some] landscape paintings [are so vividly painted] that one could step in and go for a stroll, [some project such a deep spatial perspective] that one could look far into its distance, [some are so picturesque] that one could sport [amid its mountains and streams], and [still some others present such an affable prospect] that one would want to live [in the picture]. Any brushwork that achieves such perfection is a work of marvel.[7]

This being said, with a painted landscape, however, it is merely in the viewer's imagination that one strolls in the depicted scenery, gazes out at the distant mountains, sports in the wooded valley, or dwell in the tranquil nature. With the Jiangnan garden, these imaginary activities quickened by the "work of marvel" can be realized in the actual spatial circumstance by means of the spatiality of the garden.

As regards the Jiangnan garden, pundits often relish the virtue of *xiao zhong jian da*[8] and regard it as a design principle. Nevertheless, *xiao zhong jian da*, which is synonymous with *yi shao sheng duo,*[9] is merely an illusive effect of the garden making technique. The basic principle, though, is to establish a unity between two opposites: the limited garden space and the unlimited natural landscape; in other words, to create an infinite space in a finite space. Similarly, people frequently use *qu jing tong you,*[10] another noted feature of the Jiangnan garden, as an artistic principle of garden making, while in fact, it is merely a phenomenon, not a concept. It helps

---

[6] (*Designs of Chinese Buildings, Furnitures, Dresses, Machines and Utensils*).

[7] Linquan gaozhi ji, or Essays on refined interests amid forests and streams.

[8] Evoking the large in the small.

[9] Achieve more with less.

[10] A winding path leads to a seclusion.

demonstrate, however, the artistic principle at work, as reflected in the unity of the two antitheses—*you* (the seclusion), which is a natural attribute, and *jing* (the path), which is an architectural contrivance. In other words, the two opposites—the unaffected charm of *you* (natural beauty) and the architectural artifice of *jing* (artful beauty)—are unified in the single entity of the garden scene. Beauty lies in her geniality, which is more so in the art of garden. People see beauty in objects that are functional and accessible, and precisely because of this, "*qu jing tong you*", or a planned architectural or landscaped space for that matter, is more beautiful than merely a desolate mountain or rank forest that one is unable to set foot in. The beauty of *qu jing tong you* lies in its functional value shown by way of the winding path that guides the visitor to the "secluded" beauty through unexpected turns and twists and varying natural delights. Suzhou gardens are the quintessence of the Jiangnan garden, where the art of *xiao zhong jian da*[11] and *qu jing tong you*[12] are at its highest form. Some garden scholars conclude though that the antitheses in each of the two features[13] are a contradiction in design, which is an inextricable flaw in Suzhou garden making. What this conclusion is unwitting of, however, is that the "contradiction" is precisely the peculiarity that makes the "Suzhou garden" and the most essential to its existence. Should the "contradiction" were to be "eliminated" according to the opposite advice, there would have been no Suzhou gardens.

The difference between the art of garden making and the art of painting lies not only in that the garden portrays a space in the space, but also in that the experience with a garden is not confined to static viewing as with a painting. On the contrary, the aesthetic enjoyment of a garden, to a large extent, transpires in the physical movement of the viewer, which is described as "*you lan*", or perambulatory viewing. Sequential and non-sequential proceeding of garden viewing is an art of temporal planning. Rigorously speaking, a single scenic image has an infinite number of scenic planes that are to be progressively revealed to the viewer as he moves and his perspective shifts. The aspect, duration, and sequence of the revelation of each scenic plane represent the intrinsic fabric of the garden art. In music, notes, chords, *tempo*, *timbre*, etc. are interwoven into tunes and rhythms that express a certain motif or interest and are played over the course of time for the listener to enjoy. Similarly in a garden, with the passage of time, various scenic planes gradually unfold in a sequence of views, ambience, and perspectives along with the movement of the viewer, thus to complete the artistic impression of the garden. Manifestations of the scenic imagery are also dictated by nature-obliged factors: change of seasons, passing of time, or variations in climate such as rain, snow, sunshine, clouds, etc., all of which, to varying degrees, cast a different light on the artistic effect of the garden's scenic imagery. This, too, is an issue of temporal planning that is to be taken into consideration in garden creation. As with a piece of sculpture that produces the most aesthetic effect when placed in specific lighting, some of the garden

[11] Evoking the large in the small.

[12] A winding path leads to a seclusion.

[13] "Large" and "small" in *xiao zhong jian da* and "path" and "seclusion" in *qu jing tong you*.

scenes render the best artistic appeal only at a certain time of the day, during a certain season of the year, or under a certain condition of the weather. Making of the classical Jiangnan garden has precisely mastered the temporal use of these natural assets. Therefore, creation of the Jiangnan garden is an art of both spatial and temporal planning. It is a functional art of time and space.

Creation of the Jiangnan garden with pleasuring and living functions is also a comprehensive engineering endeavor, which involve earth-, stone-, and hydraulic works as for raising mountains and treating waters; architectural engineering as for building pavilions, terraces, towers, belvederes, halls, palaces, corridors, and bridges; horticultural work as for planting vegetation; and so on and so forth. These engineering techniques were implemented for structuring the garden, but their technological advances offered much enlightenment and innovative medium for the fulfillment of garden's artistic demands. The discussion in this chapter, however, will concentrate on the artistic treatment of the Jiangnan garden environment, as its engineering techniques are mostly obsolete and that it is not the intent of this study to write history, and therefore will not delve into engineering technical issues unless required on the part of the artistic relevance to the garden structure.

2.2 Functional Aspect of the Garden of Jiangnan: The Manner of Garden Living the Concept of Landscape Design a Working Philosophy

Most of the Jiangnan gardens are built adjacent to a residential compound. It is in effect a continuous and integral part of everyday living space, which gives rise to the practice of *yuan ju* (garden living), a lifestyle of leisure and refined pursuits sought after by intellectual elites that features idealized routine of daily living, including reading, painting, poetizing, playing musical instruments, tasting tea, savoring wine, playing chess, appreciating opera performances, banqueting, and entertaining. The practical nature of garden living hence requires the garden to be functional with more architectural structures of a greater variety, all designed to be in tune with the grace of the natural landscape of the garden. At such, making of the Jiangnan garden has always striven to create a utilitarian as well as aesthetic living space where natural and artificial worlds are closely incorporated.

The functional content of the garden that satisfies the lifestyle of garden living can be divided into two aspects: material and spiritual. The material aspect denotes the utilitarian content of the garden, in other words, the material prerequisites that guarantee the fulfillment of the garden living proceedings, such as pedestrian traffic amenities, sun and rain shelters, sitting and resting areas, recreation and sport grounds, and so forth. These conveniences, though necessary to the make of a garden, are nonetheless not the essence of the garden's attribute as they can avail themselves of the spatial environment—natural or man-made—other than the garden space. However, once established within the garden space and integrated with the ideas and interest of the garden, they become an inseparable component of the garden. The spiritual aspect, on the other hand, relates to the thematic content of the garden. This is the basic content of the garden, the theme that unifies, and is embodied in, the garden's scenic imagery and that incarnates the garden with the poetic and pictorial *shiqing huayi* quality. Also called the "aesthetic content", the thematic content affords the visitor the pleasuring experience by means of aesthetic viewing. Together, the material and spiritual aspects of the garden's functional content constitute the make of a garden and complete the capability of garden living.

Although the success or failure of creating a garden that effectively represents both aesthetic and utilitarian values is largely defined by the merit of the garden's construction, the planning and design of the garden are nevertheless the decisive factor. In his treatise on garden making, *The Craft of Gardens* (Yuan Ye), noted garden maker Ji Cheng of the Ming dynasty quotes an old saw from the architectural trade *Sanfen jiang, qifen zhuren*, which says that "three tenths (of the work is accomplished) by the workmen and seven tenths by the principal".[14] In other words, the role-play of the design and the construction in an architectural undertaking is in the ratio of seven to three. However, when it comes to garden making, Ji Cheng asserts that, based on his creative experience over the years, the design (comprising the patron's suggested ideas) and the workmanship are in a more polarized ratio of nine to one, which implies that much of the design is called for extemporaneously on the worksite during the construction. The style is the man. A garden, to a large extent, reflects the standard and taste of its designer. In the case where the garden designer is commissioned by the property owner, the creative standard and artistic taste of the garden will be more or less restrained by those of the patron's, especially the garden's creative ideology, which has to be complied with the patron's view of the garden. Therefore in actuality, the garden is always a mirror that reflects the aesthetic taste, cultural accomplishment, and garden living style of the garden owner. Moreover, following the completion of a garden, the use of the garden is a process of constant, continuous making of the garden. An old garden has often lived through multiple changes of hands, repairs, or even extensive reconstruction. The vegetational disposition of a garden is not the only area that can be altered easily; the hills, waters, and architectural structures that bear greater impact on a garden's framework often undergo makeovers as well. As to concomitant additions and removals of minute scenic details or other decorative appurtenances made to the garden that occur in keeping with the taste of the current owner and his accomplishment in garden art, example are ubiquitous.

Garden is undoubtedly a crystallization of spiritual culture. Jiangnan garden proprietors were mainly made up of high officialdom, incumbent or retired, and the wealthy and powerful magnates of the society; smaller gardens saw owners of mostly literati with enough wealth. Culturally sophisticated court officials who wanted to eschew the sink-or-swim feudal officialdom often aspired after secluded living in natural surroundings of mountains and forests to exhibit their detached

[14] Referring to the architect and often, the opinionated patron.

refinement. At such, urban residential gardens had duly become a way out for them to satisfy such needs and fulfill their desire of both material and spiritual pleasures. Many influential gentry and wealthy merchants, who usually had intimate ties with government's high official ranks, also built gardens. Confucian scholar Su Dongpo of the Song dynasty expressed his living philosophy in this symbolic epigram, relished by the affluent "meat-consumption" class in the Jiangnan region who also attempted cultural refinement:

> I'd rather dine without meat stew.
> Than dwell without a grove of bamboo;
> Meals without meat only make me pine.
> While an abode without bamboo turns me philistine.

Obviously, what was truly desired was to both "dine with meat"—indulgence in creature comforts—and "admire bamboos"—pleasure in artistic life embodied in garden living; in other words, to have the best of both worlds, a luxury that found no financial obstacle in this social stratum. Wang Shizhen, a noted scholar of the Jiangnan region in the Ming dynasty and the owner of the Yanshan Garden, shared the following observations with his contemporary Chen Meigong:

> Mountain living is much too solitary while urban living much too bustling; only garden living fares perfect in between.

His remarks alluded to a lifestyle pursuing both urban sophistication and bucolic delights, and the concept of *chengshi shanlin*[15] had just solved this dilemma for him. While pleasuring in their private gardens, a rare number of the owners also made the gardens accessible to those who admire garden living and extended their hospitality to unexpected visitors who called out of admiration, which was considered estimably open-minded and magnanimous at the time. In most cases though, the gardens were walled up and kept out the prying glimpse of the passersby, and only on special occasions did the owner necessarily open the garden to receive the kith and kin as a means of social intercourse. Still there were others who posed the garden as a stage for swanking of wealth and mingling with the culturally refined, and their possession of a garden purported more of a bait for fame and compliments than an indication of true garden appreciation. The Jiangnan private garden in the form of *chengshi shanlin*, or *zhichi shanlin,*[16] is itself a characteristic reflection of the social stratum that the garden owners belonged to. To create an illusion of infinite natural landscape in a constricted urban space, *chengshi shanlin*[17] is an epitomized art in its highest form. The scaled-down landscape of mountains and water spares the hardship of the trudge for the visitor and rewards him with the leisure enjoyment of the natural beauty just a stone's throw away. It thus fulfills the garden living ideal that literati and officialdom have aspired to assume: sauntering amid mountains and water, or simply *woyou* or *shenyou*—wandering by contemplation, as a pleasurable pastime.

---

[15]Mountains and forests in the city.

[16]Literally "foot-long mountains and forests".

[17]Mountains and forests in the city.

From a contemporary perspective on gardens, the design of a classical Jiangnan garden scene apparently emphasizes its viewing value called for by our visual sense rather than its dynamic quality meant for exercise-oriented “pro-active relaxation”.

In terms of its functional variety, the classical Jiangnan garden can be roughly divided into the following types: *ting yuan*, or court garden, arranged in the courtyard within the compound of a residence, temple, government bureau, ancestral shrine, school, guildhall, and so forth; *zhai yuan (residential garden)*, or residential garden, constructed abutting on, but relatively independent of, a residential compound; and *bieshu yuan*, or villa garden, which often integrates the everyday living content into the garden scenes whose scenic structure more or less resembles that of *zhai yuan (residential garden)*‘s. There is still another type, the relatively free-standing garden that is auxiliary to some of the temples, government bureaus, ancestral shrines, schools, guildhalls, etc., and its artistic structure is analogous to that of *zhai yuan (residential garden)*‘s. The architectural structure of the latter usually occupies a smaller proportion of the garden only because of less demand of day-to-day living on this type of garden. The configuration of an upper-class urban residence in the ancient Jiangnan region is similar to the basic setup of its counterpart elsewhere in the country. Observing the patriarchal-feudal order, its primary living quarters take on the layout of the strictly symmetrical *siheyuan*, or quadrangle compound. In an extensive residential compound with one or multiple sub-compounds that flank the central axial line and where the study and the private drawing room are usually located, the *ting yuan* (court garden) is usually arranged in the courtyard of the sub-compounds. In the case of smaller residences, the *ting yuan* (court garden) is commonly set up in the central courtyard. Occasionally, a small number of the *ting yuan* (court garden) are laid out symmetrically in keeping with the landscape configuration of a large house or a temple, guildhall, school, and suchlike. Relations to the central axial line of the architectural complex are taken into consideration in planting trees or displaying flowers, plants, mountain stones, potted landscapes, and so on. Yielding to the foursquare architectural environment, pools are customarily geometrically shaped, typically in rectangle, and positioned on the central axis. Comparatively, planning of the *ting yuan* (court garden) for the central courtyard of a smaller dwelling or the lesser courtyards[18] of a large residence enjoys more liberty. *Zhai yuan* (residential garden), as well as *bieshu yuan,*[19] on the other hand, is relatively separate in structure and therefore able to enjoy its own independent landscape integrity. As such, *zhai yuan* (residential garden) is the typical garden category of the classical Jiangnan garden, which will be the exemplar of our breakdown in the discussion to come (Figs. 2.1, 2.2, 2.3, 2.4, 2.5, and 2.6).

The use of a *zhai yuan* (residential garden) is primarily for family’s impromptu recreations or festival entertainment. For female members of the family who have rare chance of leaving the residence and are confined to the rear dwelling quarters all the year round, the garden becomes a “paradise” they come for diversions or a

[18] those flanking the central axis and/or located to the rear.

[19] villa garden.

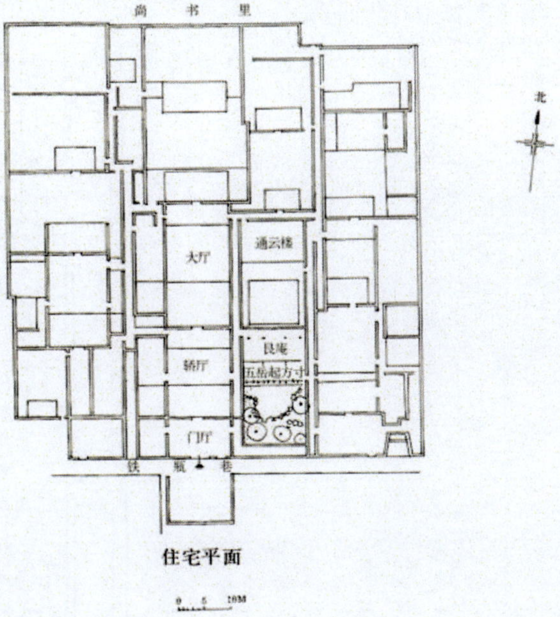

Fig. 2.1 The location of Jiangnan residential garden in the complex of private residence (from *Suzhou Classical Gardens*)

rendezvous. As to the patriarch of the household, the garden is a sanctuary where he can retire at intervals from the contentions for fame and gain in the outside world. Besides the customary strolling and touring the garden, the typical garden living style of a Confucian scholar or official is manifested in reading, painting, playing chess, plucking a *qin* zither, reciting poems, engaging a leisure conversation, drinking wine, sipping tea, as well as sitting in meditation or *yiwo shenyou*—wandering by contemplation while resting or napping. Varied by the owners' social stations and propensities, garden dwelling takes on other fashions for pleasure as well, such as watching traditional operas, gambling, and all such. At the same time, the garden is also a venue where social gatherings take place.

For the purpose of illustrating the garden living style that the scholar officials of the late feudal period typically assumed or aspired after, the following passages are quoted from *Michuan huajing,*[20] written by Chen Fuyao, a scholar from Hangzhou during Emperor Kangxi's reign of the Qing dynasty. The excerpt cited below, entitled "huajian rike"[21]—a dainty expression for otherwise-described "everyday garden living", portrays an idyllic life in the four seasons supported by the ideal garden living style:

[20]The esoteric flower mirror.

[21]Daily event amid flowers.

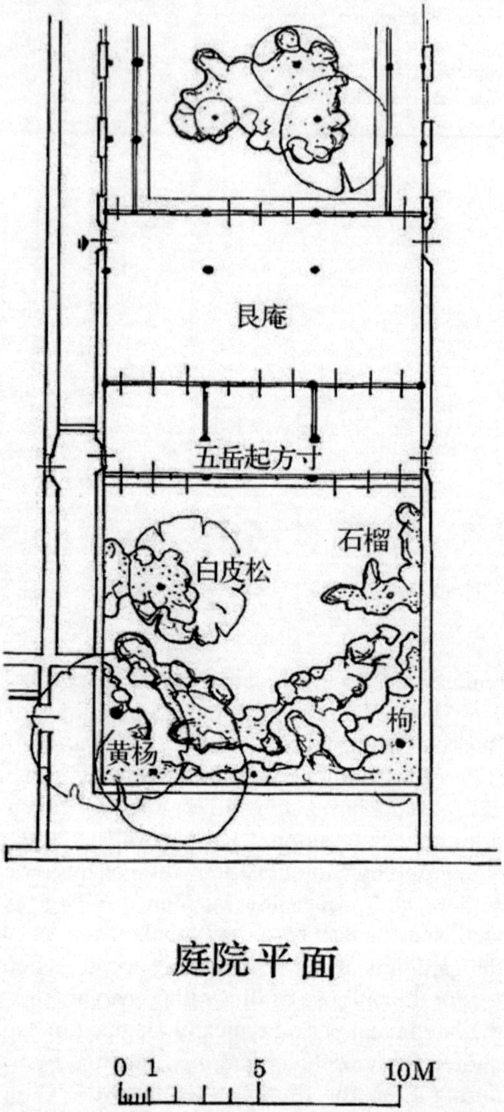

Fig. 2.2 No. 12, Tieping Street, Suzhou

Spring

Rise in the morning, refresh with sips of plum blossom soup, and ask the houseboy to sprinkle and sweep the retreat room and flower path; review the gardening calendar and tend the mosses on the steps. In the forenoon, rinse hands with rose fragranced water, cense the apparel with the fine *rui* herb scent, and read garden poems. At noon, pick bamboo shoots and pteridophytes, serve northern sesame, and draw water from the well to (brew and) sample new tea. After midday, amble on the

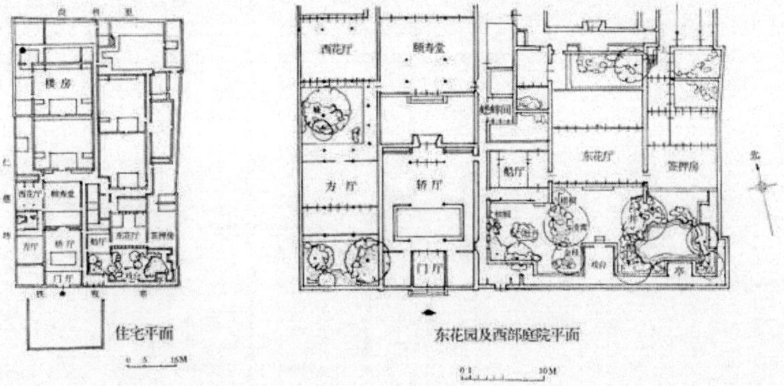

Fig. 2.3 No. 22, Tieping Street, Suzhou

horse, flick a soft-tip whip, take along a jug of wine and a pair of mandarins, and off to listen to the orioles. In late afternoon, sit in the willow breezes, lay open the colorful writing paper, and poetize and intone rhythmically to my heart's content. At dusk, meander the path, instruct the gardener in minding flowers, feed cranes, and raise fish.

Summer

Rise in the morning, clad in water caltrop and lotus leaves, imbibe the dewy air by the tree branches in blossom to moisten lungs; teach the parrot to speak poetry. In the forenoon, skim through a few pages of *Lao-Zhuang*'s work, or unfold the master calligraphy script to copy. At noon, remove the headdress, lounge on the screened bed and ramble discursively with a bosom friend; anon feeling weary, grab a dainty pillow and drift soundly into the dreamland. After midday, hollow a coconut to make a beaker, float (to refresh) melons in the clear spring, sink (to chill) plums in the cool stream, pestle lotus flowers, and drink aromatic jasper wine. In late afternoon, bathe with warm orchid water, and row a skiff to fish in a winding stream by the ancient rattan. At dusk, stand on an earthen hill with a bamboo-shoot sheath hat on my head and a cattail leaf fan in my hand, and watch the gardener holding a pottery urn to water flowers.

Autumn

Rise in the morning, lower the curtains; pick up a twig (from the yard), scoop some flower extract, and grind (and mix) the vermilion ink for adding punctuation and commentary (to the reading). In the forenoon, play a *qin* zither, tease the cranes, and relish some bronze and jade sacrificial vessels. At noon, wash the inkstone with a (seeded) lotus seedpod, clean the tea set, and polish the Chinese parasol and bamboo. After midday, don a hermit gown and a Baijieli cap, admire the maple leaves that are turning, think up a line or two, and inscribe them on a fallen leaf. In late afternoon, take crabs, chelae, perches, and Chinese herring; drinking from a big conch, taste a new brew; and get intoxicated while listening to the cricket chirring in the open fields and the woodman and buffalo boy singing from afar. At dusk, burn incense toward

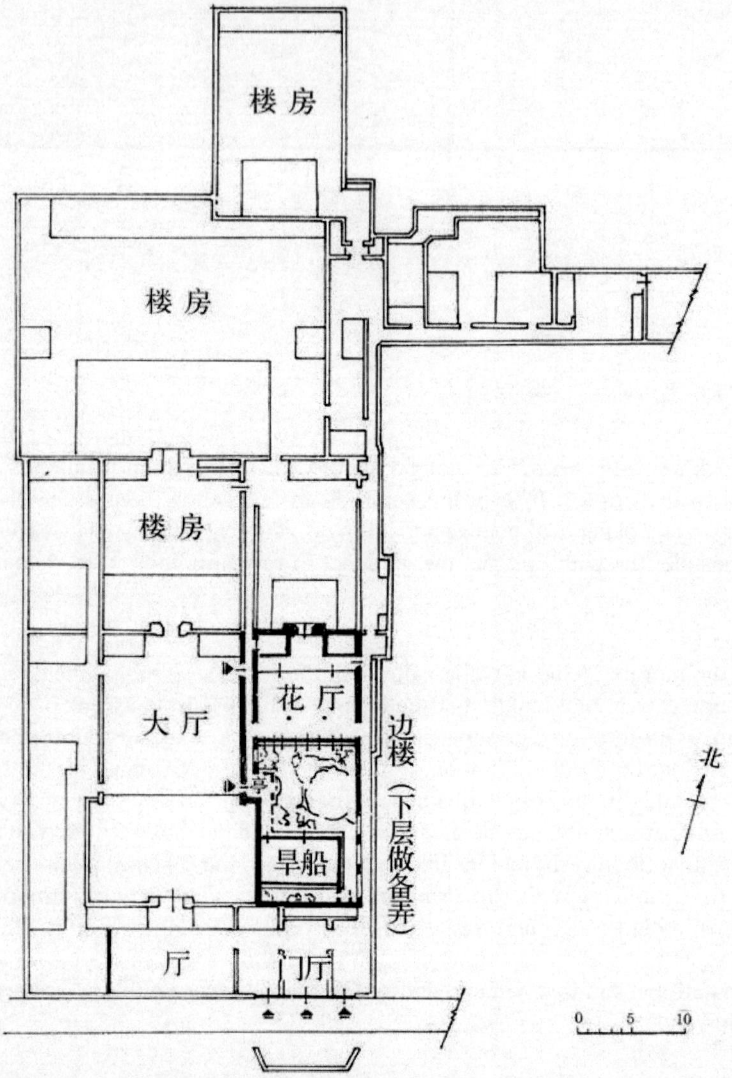

Fig. 2.4 A house in Fanmen Bridge Alley, Suzhou

the rising moon, earth up chrysanthemums, look up at the swan geese (flying south), and play a few tunes on the *qin* zither.

Winter

Rise in the morning, drink mellow fermented rice wine (to warm up); bask in the sun while having hands washed and hair combed. In the forenoon, lay felt mattresses, burn charcoal (in a brazier), and get together with eminent intellectuals to make poems and link verses. At noon, take out the bamboo chest to organize old

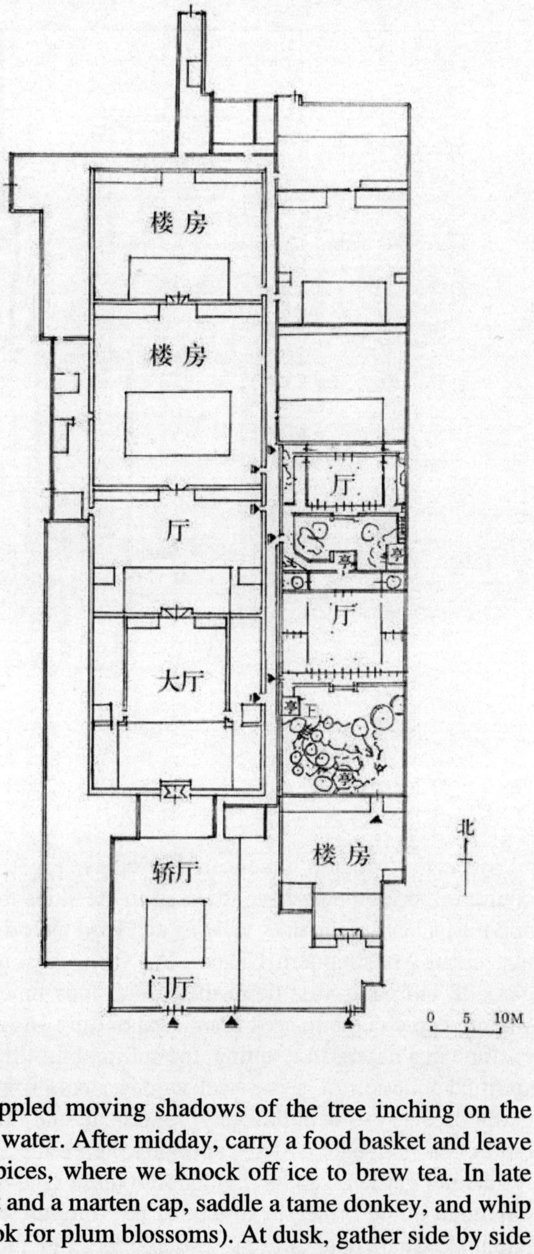

Fig. 2.5 A house on Jingde Road, Suzhou

manuscripts, contemplate the dappled moving shadows of the tree inching on the doorstep, and bathe feet in warm water. After midday, carry a food basket and leave for ancient pines amid the precipices, where we knock off ice to brew tea. In late afternoon, put on a lambskin coat and a marten cap, saddle a tame donkey, and whip it on to inquire after plums (to look for plum blossoms). At dusk, gather side by side around the stove, roast giant taros in fresh cinders, and talk Zen epigrams and witticism. Later, trim the snuff, read tales of chivalrous swordsmen and biographies of immortals, and deplore that the art of swordplay has been lost.

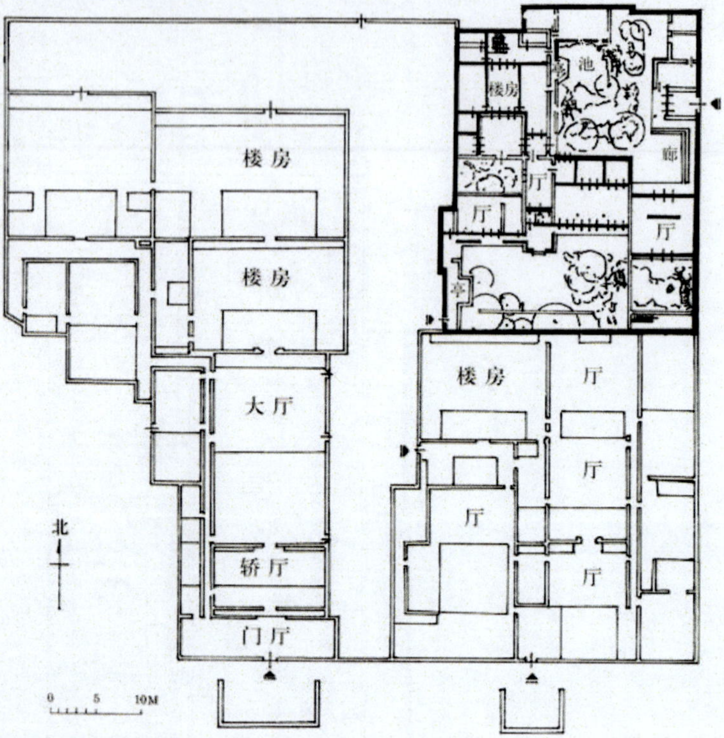

Fig. 2.6 (Tingfeng Yuan) No. 4, Jing Tai Shi Chang, Suzhou

Since the modern times, in big cities, especially those of the industrialized countries, skyscrapers have towered to the skies that shut out much of the sunlight and forms man-made dark valleys, and loud factories gush out smoke and pollutants that create "postmodern Chaos". As such, reasons behind the demand for urban gardens and parks first lie in their salubrious functions. As to the classical garden where people come to seek emotional nurture and sustenance or simply pleasurable pastime in a nature-like setting, the spiritual fulfillment is its primary function. Such spiritual function of a classical garden stems from the ideological content of the garden scenery—its philosophy, sentiment, and taste. Based on the scenic configuration, the Jiangnan garden, especially the exemplary *zhai yuan* (residential garden) residential garden, can be divided into these types: *shanjing yuan,*[22] gardens where mountain is the principle structure, as exemplified by Villa of beautiful surrounds (Huanxiu Shanzhuang) in Suzhou and the Yan Yuan garden in Changshu; *shuijing yuan,*[23] gardens where waters forms the principal view of the landscape, as seen in

[22]Mountainscape garden.

[23]Waterscape garden.

the Master-of-Nets Garden (Wangshi Yuan) in Suzhou and the Yi Yuan in Nanxun; and *shanshui yuan,*[24] gardens where mountain and waters have comparable presence, as shown in the Humble Administrator's Garden (Zhuozheng Yuan) in Suzhou and the Garden for Ease of Mind (Jichang Yuan) in Wuxi. There is yet another type, *hua yuan,*[25] consisting mainly of ornamental vegetation, which nonetheless will be omitted in this discussion as *hua yuan* does not fall into the naturalistic landscape garden category and that there is no surviving example today of this garden type.

The artistic image of a garden is defined by its creative ideology, which is in turn delineated by the concept of the garden. During the Qin (221–207 BC) and Han (206 BC–220 AD) dynasties, the rulers were deeply engrossed in the ideology of immortals and the art of longevity, and the "garden" they had constructed was patently a utopian abode of celestial immortals created by means of garden art. The "Donghai Shenshan"[26] setting, took form in Han gardens and evolved in Tang gardens into the typical composition of *yichi sanshan*, or "one pool and three mountains", namely Taiyechi (Calm Lake), and the three legendary immortal mountains entitled Penglai, Fangzhang, and Yingzhou. Such creative ideology and garden concept, though, had laid the cornerstone of the development of Chinese naturalistic landscape gardens for the next millennium to come.

Since the Wei (220–265 AD) and Jin dynasties, the reverence of nature and the ideology of recluse living amid mountains and forests that obsessed the officialdom and literati became the main current in the ideological trend of garden making. Under the enlightenment gained from the naturalistic and idyllic painting and literature that were burgeoning at the time, garden making saw the reform and enrichment in the former mountain-and-water setting that insinuated immortal habitat and the surfacing of realistic garden creations that described natural landscapes with more vitality.

Up to the Song and Ming dynasties, natural landscape gardens, which were founded on the naturalistic perspective of Confucianism and Chan-Buddhism, further matured in both theory and technique. Inheriting such a tradition, Jiangnan garden landscaping mostly patterns itself on an idealized nature as is refined in poetry and painting, with creating peaceful retreats from reality as the ultimate utopia. Toward the late imperial era, especially since the mid Qing dynasty, indulgence in creature comforts and hedonism—a reflection of the waning empire mentality—replaced the pursuit of lake-and-mountain tranquility. Rendered in garden landscaping, architectural structures used for daily living and pleasure seeking were considerably scaled up while the vegetation that once was the garden's primary compositional element was relegated to serving as a foil. Judging by some of the living examples, garden making of this period gave great prominence to the interest of architectural construction rather than the artlessness of rusticity and tranquility as seen in gardens built prior to Ming, where hills, waters, and woods were dominant and only speckled with a few thatched huts or pavilions, or bamboo fences or

[24]Mountain-and-water garden.

[25]Flower garden.

[26]The East Sea and Immortal Mountains.

trellises. In some cases of Qing gardens, such naturalistic elements were relegated to only embellishing the spaces between buildings, as witnessed in the eastern section of the Garden that was preserved miraculously (Liu Yuan) in Suzhou. Stressing the utilitarian aspect of the garden consequently enhanced the role of the architecture; organizing scenic spaces now largely relied on architectural execution. Large addition of building structures practically effaced the naturalistic vein of the garden's scenic imagery. The scale-down of hills, water bodies, plants, and other natural constituents impelled the change of the garden's artistic fabric and disposed the garden's naturalistic landscaping for symbolism that often seeks the presence of nature in a piece of rock, a stream of water, or a clump of bamboo. Works of such a symbolic trend in Jiangnan garden making can be found in Suzhou's Garden of Pleasure (Yi Yuan) where a brace of solitary rock peaks, named "Listening to the music" (Ting Qin), stand reticently outside the windows of Poxian Music Studio (Poxian Qinguan), and a solitary rock peak entitled "Stone Worship" (Bai Shi), named after a man with a passion for stones, is located at Veranda of Stone Worship (Baishi Xuan), also known as Thatched Hut of the Cold Season (Suihan Caolu). In Shanghai's Garden to delight one's parents (Yu Yuan), a house reminiscent of a boat image "anchored" in a small, enclosed courtyard, named "Boat Too" (Yi Fang), manifests a symbolic expression of a stationary land boat.

In the eighteenth century, the English Landscape style took form and gained great popularity in European garden making. The monumental-scale English Landscape garden that depicts Scottish pastureland, for instance, exhibits a garden style that aspires to great fidelity to nature, stressing the importance of *xing si*, the formal likeness to nature, or the "realistic truth". Contrary to this trend, Chinese garden making emphasizes *shen si*, the essential likeness or the "artistic truth", aiming to capture the spirit or essence of nature rather than its similitude and magnitude. Nevertheless, the art of the classical Chinese garden also experienced the realistic ideological trend, which to certain extent resembles the English Landscape garden style. While there are gardens, like the Humble Administrator's Garden (Zhuozheng Yuan) in Suzhou, that are noted for their condensed and epitomized landscape of hills and lakes, the Jiangnan garden repertoire is not without the near-realistic, life-sized mountain-and-water landscape creations,[27] as exemplified by the Garden of Art (Yi Pu) in Suzhou before it was tampered with. The former garden example was a work of *xie yi* where the scenic scale is ostensibly reduced and the scenic images greatly stylized. The latter is of *xie shi*, or realism, the antithesis of the former, which attempts to recreate the magnitude of nature. Zhang Nanyuan, a famous Jiangnan garden maker in the early Qing dynasty and a realistic garden making advocate, opposed full-bodied landscaping on a small plot and explained rather assertively as recorded in *Zhang Nanyuan zhuan*[28]:

> Lofty mountains that soar to the sky and deep precipices that obscure the sun are the works of the divine creator and not attainable by human power. Nature's landscape readily stretches

[27] Although created only in segments to be customized to the restricted garden space.

[28] *Biography of Zhang Nanyuan.*

> over hundreds of miles, and we yet attempt to emulate it with a ground of merely a few yards and a ditch a few feet, isn't this any different from lumping clays to fool a child?

Instead, Zhang laid out his perspectives on landscaping as follows:

> (To construct) an earthen hillock or hill slope, first, tamp up the earth, which takes only several days, then intersperse a few natural stones, encompass the area with a low garden wall, and conceal the background with a grove of dense bamboo, all of which impart the impression of ranges of lofty mountain ranges that stretches from beyond the wall.... Thus, it is like being next to the foot of a big mountain that reveals only part of its ravine or rivulet, which I come to own (as a garden) simply by enclosing a few mountain rocks. (*Zhang Nanyuan zhuan*)

In other words, under Zhang's realistic guiding ideology, when making artificial mountains, build only a segment of the mountains—a slope or a foot—to its natural scale rather than build the entire mountains but have to scale them down to fit a cramped garden space. Within the garden perimeter, suffice it to see simply a flank of the hill or a foot of the mountains evocative of ranges of mountains and forest concealed beyond the garden wall. On the same principle, when creating a water landscape, although the natural scale is applied, only a cove of the lake or an arm of the valley stream is configured as a reminiscence of the full-body water, provided other concomitant garden and architectural treatments are properly executed. Such a segmental landscaping approach seems to satisfy both *xing si*, the realistic resemblance, and *shen si*, the artistic truth, of nature. The Garden of Art (Yi Pu) in Suzhou may well be an exemplar of Zhang's realistic ideal, except for the rockeries skirting along the lake, which are inopportune superfluous addition from later periods. Garden scenes in this manner are often reminiscent of the landscape painting by Xia Gui and Ma Yuan of the Southern Song dynasty, which were ridiculed at the time as *can shan sheng shui*, literally "broken mountains and remnant waters", which traces only the ends and edges of the landscape.

Regardless of an approach of *xie shi*[29] or *xie yi*,[30] the highest criterion in the making of the Jiangnan garden is always its inspiring quality where *yi jing* stems, or the artistic appeal stemmed from the intangible atmospheric property. The creative ideology of Romanticism as witnessed in the modern-time European natural-form gardens (which often aspired to emulating ancient chateaux, mythical palaces, mausoleums, and ancient ruins or were laden with historical anecdotes, legends and mythology, or exoticism) also exists in the history of the Jiangnan garden. In the extant examples, scenes based on mythical and religious subject matters are ubiquitous. Among the mythical themes, piled-stone formations that predicate "the Eight Immortals crossing the sea" and the stories from the novel *The Journey to the West* are most common. Nevertheless, these predications might well be the interpretations of the viewers of later generations and not necessarily be the intention of the original scene creators. For instance, the Summer Hill[31] of the Garden of Bamboo (Ge Yuan)

---

[29] The realistic description.

[30] The artistic rendition.

[31] One of the four Hills named after the four seasons.

in Yangzhou is said to be modeled after Shuilian Dong[32] from *The Journey to the West*, and still some others interpret it as "seventy-two caves with a monster residing in each of them", an allusion also to the time-honored mythology, all of which have reflected a popular-culture based garden connoisseurship. Creations under religious themes include Lion Grove Garden (Shizi Lin) in Suzhou, taken after the metaphorical "Shi Lin" mountains in the Buddhist Scripture; the Buddhist stone *tachuang* or *tasha*[33] in the pond of the Subsidiary Garden (Bu Yuan) and the Garden that was preserved miraculously (Liu Yuan), both in Suzhou; and the garden of Xiling Seal-engravers' Society (Xiling Yinshe) (a modern creation) in Hangzhou, where Buddhist statues and epigraphy are carved on the perpendicular rock face and where the stone pagoda, slabstone-walled pavilion, and stone steles intersperse the premises.

As with art in other forms, the art of garden is defined by its ideological content, which not only determines the theme of a garden but also the configuration of the scenery.

2.3 Scenic Imagery: The Basic Unit of the Garden Art and the Form of Garden Living

As an image is the basic unit of art in general, a scenic image is the basic unit of the garden art. The ideological and functional contents that constitute the entirety of a garden are the basis on which a garden exists. The scenic imagery makes up the intrinsic structure of a garden, and it is the way in which a garden is present. In other words, the ideological and functional contents of a garden are incarnated through scenic images.

Scenic imagery is a concept of space, and its spatiality is manifested through the correlation among the structural constituents of the scenic imagery and through the intrinsic omnidirectional extensions of these structural constituents. The scenic space in the Jiangnan garden, be it a *zhai yuan* (residential garden) or a *ting yuan* (court garden), while serving as a recreational ground, is a continuous and integral part of everyday living space; it is a supplement to and an extension of the living space. Scenic imagery is, at the same time, a concept of time. The temporal properties are evident in the permutations of the appearance of the scenic constituents due to different times of the day (morning, evening, etc.), seasonal changes of the year (spring, autumn, etc.), and climatic variations of nature (sunshine, rain, snow, etc.). They are also palpable in the sequence and continuity of the scenic guidance. Hence, the making of the Jiangnan garden involves not only arrangement of scenic spaces but also orchestration of scenic time, such as the length and direction that each scenic plane, as well the entire scenic sequence, is intended to be revealed to the viewer and the transformations of the scenic imagery that are

[32] Water-curtain cave.

[33] A combination of a stela and a miniature pagoda that is engraved with Buddhist text or images.

consequent on the passage of time and the rhythm of Mother Nature. In this, the making of the Jiangnan garden reflects an artistry that reifies both space and time.

Scenic imagery contains two basic aspects: scenic constituents and scenic guidance. The unity of the two constitutes the entity of the scenic imagery. The creation of a Jiangnan garden is largely a process of unifying the constituents and the guidance under certain thematic subject and functional requirements of the garden.

Scenic constituents include natural and artificial constituents, and they are the material foundation of the garden's scenic structure. Specifically, natural constituents are topography, vegetation, and animal life, and artificial constituents are buildings and all other architectural treatments. A garden's functional nature and facility setup as well as its landscape nature—whether it is a mountainscape garden, a waterscape garden, or a combination of the two—are primarily determined by the presence of the garden's scenic constituents, which is why they are described as "the foundation of the scenic structure". A garden is formed when scenic constituents are organized by means of scenic guidance into coherent scenic imagery that carries a certain function and embodies a certain ideological and affective content. Simply put, scenic constituents are the building materials of the garden, or the material means of garden building.

Scenic guidance is in concomitant existence with scenic constituents. It is a rather dynamic factor in the constitution of the scenic imagery. Scenic guidance is an organizing factor that represents the structural relationship of the garden; it serves as a montage editor of the garden's various scenic planes. From the perspective of a visitor to the garden, scenic guidance is an arranger of and a guide to the sequential and non-sequential scenic images. It is a pathfinder that steers the movement of the visitor, that determines how his sees and appreciates the scenery, and that ultimately shapes his experience of the garden. Some of the popular scenic effects, such as *qu jing tong you,*[34] *yin ren ru sheng,*[35] *feng hui lu zhuan,*[36] and *kai men jian shan,*[37] are but a few examples effected by means of scenic guidance. This has also demonstrated that scenic constituents can only be organized into coherent scenic imagery by means of scenic guidance, otherwise they would simply be a casual and purposeless material buildup that fails to convey the functionality and artistic ideology of the garden. In other words, it is through scenic guidance that scenic constituents are configured into a functional and meaningful artistic space.

Scenic constituents and guidance are interdependent and inter-premised, which concurs in the old trade adage: "yinjing shelu; yinlu dejing", or "build a path by following the scenes and capture scenes by following the path". Like everything else in the universe, the correlation of the constituents and the guidance—the two unified antitheses that constitute the garden's scenic imagery—is relative, not absolute, and the functions of the two are constantly transposed. For instance, a bridge or a

[34] A winding path leading to a secluded spot.

[35] Enticing one into a wonderful place.

[36] Where the peak stands brings the next turn of the mountain path.

[37] A mountain springing to view as the door opens.

pavilion, as a scenic constituent, is an object of viewing when admired from distance but becomes scenic guidance when serving as a medium of sightseeing—a passageway or a lookout and resting point. Take touring the Humble Administrator's Garden (Zhuozheng Yuan) in Suzhou as an example: if the viewer stands at Hall of Distant Fragrance (Yuanxiang Tang) and looks north toward the lake-and-mountain scenery Snow-Like Fragrant Prunus Mume Pavilion (Xuexiang Yunwei), Hall of Distant Fragrance (Yuanxiang Tang) at this point acts as the guidance, serving as an aid to sightseeing and a base point for viewing, while on the other side, the lake-and-mountain scenery Snow-Like Fragrant Prunus Mume Pavilion (Xuexiang Yunwei), being an object of viewing, plays the role of a scenic constituent. However, if the viewer stands on the same side of Snow-Like Fragrant Prunus Mume Pavilion (Xuexiang Yunwei) and looks south toward Hall of Distant Fragrance (Yuanxiang Tang), the roles will be reversed: Snow-Like Fragrant Prunus Mume Pavilion (Xuexiang Yunwei) becomes the guidance and Hall of Distant Fragrance (Yuanxiang Tang), being a viewing object, a scenic constituent. This is to say that a scenic image can be both an aesthetic viewing object and simultaneously a spatial entity that can be entered and experienced personally. This is where the landscape garden lies superior to landscape painting or potted landscape. Special mention should be made of that, while scenic constituents carry the garden's aesthetic value and scenic guidance the functional value, scenic constituents are also the basis for garden's functionality and scenic guidance is the organizing agent of garden's aesthetic experience. Scenic constituents and scenic guidance are mutually inclusive and interpenetrating; they are an inseparable organism. Whereas scenic constituents are finite, scenic guidance is infinite. It is the unity of the two antitheses that brings forth the rich and varied scenic imagery whose inspiring and imaginative wonders never cease to captivate and intrigue its admirers.

The finest exemplars of the Jiangnan garden never fail to show the ingenuity in incorporating scenic constituents and scenic guidance. The ensuing chapter will further break down and discuss in greater detail the constituents and the guidance, the two aspects of the scenic imagery.

Chapter 3
On the Design of Garden

3.1 Formation of Scenic Imagery

3.1.1 The Elements of Scenic Imagery: Structural Foundation of the Scenery

As mentioned in the last chapter, scenic constituents are the material basis of the garden's scenic structure, and they consist of two antitheses—natural and artificial constituents. Natural constituents—the lay of the terrain, plant life, and animal life—are the dominant factor that determines the natural features of a scenic image. Mountains, water, flowers, trees, birds, beasts, insects, and fish are the essential means to emulating a natural environment, and they form the basic content of the garden's viewing capacity. Therefore, natural constituents are the basic ingredients of the scenic constituents and hence the basic material that constitutes a garden. Without the natural constituents, architecture alone—no matter how ingeniously planned and executed—cannot bring to fruition a landscape garden. A perfect garden is always outfitted with the natural constituents—topography, fauna, and flora—that help create an animated garden landscape intended to reflect the natural ecological environment. Vegetation is the core of the natural constituents, and gardens that feature flowers and plants only are omnipresent throughout the world. *Hua yuan*, or flower gardens, and ornamental botanical gardens belong to this garden type. In the Jiangnan region, classical Chinese gardens such as Mei Zhuang in Yangzhou of the Qing period[1] is an example of the *hua yuan* that accentuates the display of flowers and plants.[2] Vegetation and water, due to their salubrious properties, become the most congenial natural ingredients favored by people. However, many gardens are free of water resources or even feature no vegetation. Sand and stone in their raw state are also natural ingredients; therefore, sand and stone alone can be garden-

[1] During the reign of Emperor Qianlong.

[2] See Zheng Banqiao's *Mei Zhuang ji— Notes on the Mei Zhuang garden*, by Zheng Banqiao.

© Springer Nature Singapore Pte Ltd. 2022
H. Yang, *A Treatise on the Garden of Jiangnan*,
https://doi.org/10.1007/978-981-16-6924-8_3

making materials as well. The abstract and symbolic dry landscape garden is a singular example in the art of Japanese garden making. The solely sand-and-stone composed dry landscape garden (or the sand garden, as otherwise termed), however, bears no manifestation of waste desert, bare mountains, or dry ravines, but rather the varied scenes are significative for dynamic themes—mountain peaks shrouded in mist, rocks withstanding the rapids, or the "Eastern Sea and immortal mountains"—that in effect are latent with profound suggestiveness. This is a refined garden type that is impregnate with philosophical content. As such, the view that a garden is not a garden without greenery is therefore confronted with challenge. What about potted landscape? With its miniature plants, rockery, and cascade, a potted landscape is still a potted landscape, and not a garden. As far as the world garden making experience is concerned, one can only imagine that a garden made of no natural constituents does not exist.

The making of the Jiangnan garden adopts all three natural constituents. While premised on natural beauty, shaping of land surface, planting of vegetation, and placement of fauna often resort to constructional artifice. Piling of natural stones, treatment of water styles, and shaping and disposition of plants, all are but pregnant with artful beauty. Artificial constituents—buildings and all other architectural treatments—include paths, grounds, courtyards, walls, fences, trellises, bridges, stepping stones, staircases, bank revetments, and so on. They provide the garden with functional value that affords the utilitarian efficiency as in traffic facilitation, sun and rain shelters, daily life routine such as dining, resting, and entertaining, as well as general garden recreational activities like hill climbing, vale exploring, boating, and fishing, all of which rely directly on the placement of the artificial constituents. The natural constituents, though a dominant factor in defining the garden's natural features, can only play its role by means of the artificial constituents. If a garden is not even furnished with the most elemental architectural feature, such as a footpath, it is a garden bereft of utility and not meant for any human activity. The implementation of artificial constituents, however, should follow the naturalistic principle of garden making so as to bring the artificial constituents in harmony with the natural constituents of the garden.

Various constituents of the garden make the material base of the garden's scenic structure by means of their basic features. A basic feature of topography, for instance, is not rock, sand, soil, or water, but rather peak, valley, hill, slope, plain, lake, pond, spring, stream, waterfall, or other forms of land configuration. A basic feature of building, to give another example, is not stone, lumber, earth, or brick, but rather pavilion, tower, belvedere, terrace, corridor, bridge, wall, hedge, trellis, stepstone path, walkway, or other forms of landscape architecture. It is worth noting, though, that, in making of the Jiangnan garden, various scenic constituents are apt to fall in readily reproducible formulae, as to be further illustrated in the following sections.

Various scenic constituents are constructed based on the ideology and functional content of the garden. In making a garden, such construction presupposes the composition of scenic imagery and lays the foundation for the garden structure. At such, to recognize the structural principle of various constituents in the correct perspective bears a great significance to a fundamental understanding of the art of the Jiangnan garden.

3.1.1.1 Shaping of Ground Surface

Topography is the essential element of the scenic constituents, and shaping ground surface is the fundamental means to creating the topographic features of the garden's scenic imagery. The distribution of hills, water, and level ground establishes the topographical basis for the scenic imagery. That is to say, if an entire tract is sculpted into an undulate terrain, then the garden's basic appearance will be established as a mountain-related theme and rules out the possibility of creating a flat environment (though customarily a certain level ground is reserved). If, on the other hand, the entire land is primed to be covered with water and leave out the undulation treatment, then the garden's basic features will be characterized with a water-accentuated landscape (though customarily a certain area is set aside for use of hill-and-forest scenery). Although the lay of the land prescribes the basic contour of scenic surroundings, it cannot satisfy specific scenic effects, such as the look of mountains, the style of waters, or the mood of a scenic setting, which requires comprehensive planning and diverse devices to achieve, like disposition of plants, placement of creatures, and arrangement of buildings. In shaping a desired topography, natural materials—sand, stone, water, and soil—are used to emulate landscapes as they exist in nature. To create a nature-aspiring landscape on a garden site constricted by limited urban space, the garden artist will have to draw on the ingenuity of a painter, a sculptor, and an architect all combined in order to accomplish a feat that condenses and epitomizes nature into a pocket-edition work of art. As such, the crafted topography is logically an artistic integration of both natural and artificial beauty.

In the process of shaping ground surface in the naturalistic style, mountains and water bodies are the two basic contents. From the perspective of the plastic arts, the solid, dignified mountains and the pliable, compliant waters aptly form a contrast (which is redolent of "The benevolent delights to the mountains and the wise the waters", an ancient epigram from the Warring States period, 475–221 BC); therefore, in garden making, the two elements are often used in tandem to set off each other. Even though in a scenic setting where mountains are the principal structure, waters are often brought into play to serve as a relief; and vice versa, in which case small-dimension piled-rocks are also used in addition to the mountains. The bond between mountains and waters in the Jiangnan garden can be best analogized by the classical painting theory, which gives an animated exposition by Guo Xi in his *Linquan gaozhi ji*.[3]

[3]Essays on refined interest amid forests and streams.

> Waters are arteries to mountains, ... hence the mountains are alive with the waters; ...mountains are the countenance to waters, ... hence the waters are charming with the mountains.

Garden making is like landscape painting, whose creative ideology was further explained in *Linquan gaozhi ji*:

> (In painting a landscape), extend the range of mountains in keeping with the flow of the waters; run the stream of waters in accord to the form of the mountains.

The relationship between mountains and waters is co-dependent and reciprocal, and the scenic characteristics of each of the two aspects are relieved against the presence of the other.

Shaping ground surface incurs substantial earth-and stonework especially in the construction of the Jiangnan garden where typically mountains and waters make up the main content of the garden. Hence the economic issues arise and tend to restrain the liberty of artistic application. In contrast to the making of imperial gardens, that of a private garden is confined to the individual financial and material resources of the patron and the artistic maneuver is contingent upon the construction budget; therefore, ground shaping in Jiangnan garden making is premised on sparing earthwork. Nevertheless, the economic concerns do not necessarily inhibit as creation of the topography has great flexibility to adapt to the scale of a budget and the creative approach often hinges upon the existing topographical conditions of the garden site. This is what Ji Cheng of the Ming dynasty explicates in *Yuan Ye: xiang di*:[4]

> Siting a garden should not be restrained by the direction; regardless of the highs and lows of the landform, ... configure the scenes by conforming to the natural contours of the land.

A garden maker always strives to capitalize on a given terrain where heights or waters are already present to then define his creative approach to topographic sculpturing. The heights are raised higher into a mountain and the low-lying spots dug deeper into a pond, this way a considerable amount of earthwork will be retrenched. This is a highly cost-effective way for artistic creation of the mountain-and-water scenic imagery. During the construction, every key factor of the existing site conditions is taken into consideration. Take Lou Yuan (Tower Garden) on Mayike Lane in Suzhou as an example: before the garden was built in late Qing dynasty, the site was piled high with heaps of rubbles from collapsed buildings (possibly remains from the warfare of the Taiping Rebellion, 1851–1864), and thus the vertical compositional technique was adopted to create an elevated tableland where the garden is situated, hence the name Tower Garden. Another Suzhou garden, Couple's Retreat Garden (Ou Yuan), had a flat terrain and low groundwater level when its East Garden was being built. In order to economize the earthwork, a ditch was dug to expediently shape the low-lying land into the lay of a gorge. To construct the mountain-and-water configuration on a level ground, same principle is followed. Garden makers consciously seek the balance of earthwork between

[4] The Craft of Gardens: survey of the land.

digging and piling in order to preclude unnecessary out-transport of excessive soil from the site or import of needed soil from outer regions. While shaping ground surface has direct bearing on the future garden features as a whole, the existing topography of the garden site often dictates the planning of the overall garden's scenic imagery, therefore the garden makers set great store by the survey of the land before choosing the site. The conception of the garden's *yi jing*, although usually devised by the garden maker and perhaps combined with the patron's intention, ultimately pivots upon the existing factors on site. This is precisely what Ji Cheng describes in *The Craft of Gardens* (Yuan Ye) in discussion of making a garden:

> The ingenuity lies in conformity (to the natural conditions of the site) and borrowing (the existing features thereof), and the brilliantness rests with (a) rational and befitting (layout). This is not a deed contrivable by an artificer, nor is it a plan devisable by a garden designer.

Among the variety of Jiangnan gardens, *ting yuan* (court garden) and *zhai yuan* (residential garden) are mostly situated in the cities and towns amid the narrow streets and alleys. The garden site, typically a smooth open lot, is apt to lack the topography that features varied heights. For making a garden of mountain-and-water scenery, such a plane terrain is an unfavorable factor that demands considerable earthwork involving a great amount of digging and piling. On the other hand, however, as a blank sheet of paper is to a painter, the flat land leaves the garden maker with endless options of artistic conception. Raise a mountain here or insert a valley there at his own will, the garden designer—as endowed with the power of the Creator—is able to set free his creativity to no restriction (barring the construction budget) and, based on the theme of the scenic imagery and the requirements of the garden dwelling, scheme out the topographical composition that will transform the plane plot into an enchanting landscape of undulating hills and waters. As such, suffice it to say that, in the absence of topographical variations and sizable garden sites, the scarce urban land has precisely turned into a positive factor that makes private gardens in the Jiangnan region uniquely what they are and that has taken this garden genre up to an artistic precinct where the highly epitomized expression of nature reigns.

3.1.1.1.1 Building Earth Mound Stacking Stones

Mountains in the Jiangnan garden are commonly called *jia shan*, or artificial mountains. In this sense, should waters of the garden, as in the form of "sea", "lake", "river", "pond", or "waterfall"be also called *jia shui*, or artificial waters? Play-acting in the Beijing opera is also *jia*—artificial, or a mere pretence, as well as the minimal props and exaggerated facial makeup. The common expression of "*jia*" refers to "unreality", which implicates an artistic representation of the reality, and therefore the artifice in the abovementioned references is in fact an embodiment of the artistic truth.

Construction of artificial mountains is a big invention of the classical Chinese garden. Judging by the ancient expression recorded in *Shang Shu: Lü Ao*,[5] "Fail to build an earthen mountain for want of one final basket of soil" (which later became an idiom used to describe a task that falls short of success for lack of a final effort), construction of earthen mountains made its advent probably during the Spring and Autumn period (770–476 BC) at the latest. As to building stone mountains, significant development was witnessed during the Western Han dynasty (206–24 BC) according to *Sanfu huangtu*,[6] which has the recorded description of the private garden of Yuan Guanghan, a wealthy gentry in charge of guarding the Maoling mausoleum of Emperor Wu of the Han dynasty located in Shaanxi Province. Here at Maoling, the extant grave mound of the tomb of Huo Qubing, a famous general of the Han dynasty who was buried next to Emperor Wu's mausoleum, is actually an example of the earthen hill with implanted rocks built to simulate wild mountains in nature. The mountain construction technique of the Chinese garden started to spread to neighboring countries like Korea and Japan probably during the Northern and Southern dynasties (AD 386–589) and influenced European garden making since the eighteenth century. That the English "wall garden", which used to domesticate the alpine plants for display, transformed into the "rock garden", which integrated the ornamental alpine plants with the art of piled-rock mountains introduced from China, is an effective illustration of such influence. The technique of artificial mountain construction of the classical Chinese garden is the fruitage of the creative development of over two thousand years. Apart from the very few early examples (such as, in Suzhou, the earthen mountains in garden Canglang Ting—Garden of Surging Wave Pavilion—probably first built in the Five dynasties and later augmented in the Song and Yuan periods—and the earth—and stone mountains with Yuan and Ming foundations in gardens Lion Grove Garden (Shizi Lin) and the Humble Administrator's Garden (Zhuozheng Yuan), and in Shanghai, the stone mountains in the Garden to delight one's parents (Yu Yuan) that has basically preserved its original features built during Ming Emperor Wanli's reign), the majority of the extant gardens, however, are the creations of the Qing or later periods, which include most of the gardens that were repaired and altered or even rebuilt after the Taiping Rebellion. Mountain construction of the Jiangnan garden aims at creating an appeal of the mountain-forest *yi jing* atmosphere rather than pursuing the physical magnitude of the mountain. Such an inclination results from not only the spatial and financial restraints of an average private garden owner, but also the sophistication of artistic philosophy in garden making. During the Han dynasty, Emperor Wu drove massive serfs "to the palatial inner garden to raise earthen mountains that extended for ten *li* and consisted of nine hills" (see *Hanguan dianzhi*),[7] and the Maoling mogul Yuan Guanghan had "rocks stacked (in his private

[5] Ancient annals of imperial codes and records: Lü Ao.

[6] Records of the Imperial Architectural Constructions in the Han Capital and Its Environs.

[7] Compilatory Records of Han Institutions, Palatial Architecture and Gardens, and Imperial Anecdotage.

garden) to build mountains that topped ten *zhang* high and stretched several *li* long" (see *Sanfu huangtu*).[8] As the pioneers of the artificial mountain construction, these and other such early productions are unarguably remarkable; however, from the artistic perspective, they tend to replicate the enormity of the actual mountain, purporting a tendency of mechanical naturalism, and apparently lack the profundity couched in the creations of later periods. Mountain construction of the Jiangnan garden is a highly epitomized art that merits a scrupulous study. As far as the construction scale of the artificial mountains is concerned, it is apparently impossible to copy that of nature. In the historical perspective, such blind pursuit of the massiveness for the purpose of imitating the real mountains inevitably appears to be naïve at its nascent stage. Whether they are earthen mountains that "extended for ten *li* and consisted of nine hills" or stone mountains that "topped some dozen *zhang* and stretched for several *li*", they are nonetheless a scoop of earth and a basket of rocks against the grandeur of nature. Earthen mountain construction in the contemporary garden also has the tendency to lay undue emphasis on the height and size of the mountains, which often top hundreds of feet while in effect resemble nothing more than a big mound with trees planted on it, falling far short of the artistic effect projected by the classical production of the Jiangnan garden that, with simply 10- or 15-foot stature, relishes of a more realistic sense of the wooded mountains. Thus it can be seen that, without an artistic condensation, the expansiveness of the magnitude does not make the "mountains" but merely a big heap of earth or a huge pile of stones; and vice versa, a "basket of earth" is capable of imparting to people the impression of the mountains and forests by means of the epitomized manifestation of art. In the making of the Jiangnan garden, any ascent or descent in the terrain, no matter how slight, is suggestive of hills or mountains. As far as the art of the Jiangnan garden is concerned, what makes a mountain is not its height but its mountain-and-forest effect, which, concluding from the fine remaining examples of the Jiangnan garden, is attainable through the art of garden making.

Mountains are originally one of the primary contents of the natural landscape. In the domain of the nature-aspiring Jiangnan garden, the man made mountains with their rich and varied artistic allure are often adopted as the principal scenic image, and, in some instances, the sole mountainscape is the entire garden. Villa of beautiful surrounds (Huanxiu Shanzhuang) in Suzhou is an exemplary masterwork of *shanjing yuan*, or the mountainscape garden. The mountain-and-forest composition not only serves as a scenic image for viewing but also provides an important means to ascending a height to enjoy a distant or borrowed view or to acquire a bird's-eye view of the garden. At the same time, the mountain-and-forest configuration also bears practical functions and plays a significant part in organizing the garden space. For instance, the sequence of winding paths, peaks, and hollows allow people to clamber for sport or exercise while the cool, secluded rock overhangs or stone caves,[9] arguably a special form of garden architecture, afford a welcome shelter

[8] Records of the Imperial Architectural Constructions in the Han Capital and Its Environs.

[9] Cavities cut into piled-rock mountains.

from the sun or rain, where rustic stone furniture—a stone teapoy, a few stone seats, or even a stone *ta*[10] or two—can be arranged for visitors to take a rest, have a drink, or play a game of *weiqi*,[11] making them unique summer hideaways. In terms of the artistic structure of the garden, mountains compartmentalize the space and multiply the layers of the scenic imagery, hence they can be used to conceal garden walls and mask the depth of field, creating a sense of infiniteness of the scenic imagery. For the Jiangnan garden, which is usually limited in scope, mountains "spatialize" the touring route; in other words, they three-dimensionalize a level-ground road into an ascendant, circuitous mountain path. As a result, it not only enriches the sightseeing proceeding that intensifies the excursional interest—tramping up hill and down dale and venturing into the hidden, but also increases the sightseeing distance and duration, which fulfills the purpose of expanding the space.

In the language of the garden art, the mountain of the Jiangnan garden is not made of earth or stone, but rather peak, ridge, slope, mound, cliff, or valley. Based on its landscape feature, the mountain can be classified into two types: earthen mountain and stony mountain. All-earth and all-stone mountains, as two distinct types, did exist, but as of late Qing, all-earthen mountains became very rare in the surviving gardens as most of them were augmented with stones. For artistic purposes, areas of the earthen mountain are often laid with partially revealed rocks—termed *shi gu*,[12] alluding to the rocky frame of the mountain—to simulate the naturally denuded rocks. At the same, from the engineering-technical point of view, the placement of the denuded rocks also helps prevent unintended soil erosion of the earthen mountain. All-stone mountains, on the other hand, are more common than all-earthen mountains although, on occasion, soil is added in planned spots for planting. Like everything else in the world, there is no black-and-white distinction between an earth and a stone mountain, and between an earthen mountain that is implanted with a large amount of stones and a stone mountain piled with a large amount of soil, the definition is even more blurred. Apart from creating earth and stone mountains, shaping ground surface for a garden also involves disposition of dispersive stones and placement of solitary stone peaks—the upright single rocks, which will be dealt with in detail in the subsequent segments.

The stone material used in making stone mountains for the Jiangnan garden falls into two major categories: *shan shi*, or Mountain stone, and *hu shi*, or Lake stone, the water-eroded rocks. Occasionally, some will list the snowy *xuan shi*—also known as *xue shi*, or the "snow stone"—as the third category, which in fact can be grouped with the Lake stone. Mountain stone is multifaceted, multi-angulated rocks quarried from natural mountains. Most of those produced in the Jiangnan region take on a tawny tint and hence are commonly called the "yellow stone", which can be found in almost every mountain of the region, with the Yellow Mountain in Changzhou,

[10] A long, narrow, and low bed.

[11] An ancient Chinese board game.

[12] Rocky bones.

Mount Yaofengshan in Suzhou, and Mount Chuishan in Zhenjiang as the major producing areas. The Lake stone refers to the cavernous *karst* limestone, with the most reputed quarried from the Dongting Xishan island in the Taihu Lake.[13] As such, the limestone of various origins is jointly named *taihu shi*, or *hu shi*—Lake stone—for short. As recounted in *The Craft of Gardens* (Yuan Ye), Lake stones that are used in making Jiangnan gardens—besides the *taihu shi* excavated from the Taihu Lake floor—also include: *kunshan shi* produced from Mount Ma'anshan in Kunshan County, Jiangsu Province; *yixing shi* from the area of the Zhanggong Dong and Shanjuan Dong *karst* caves in Yixing, also in Jiangsu; *longtan shi* from the riparian strip nearby Nanjing between Qixingguan and the Shankou-Cangtou area; *qinglongshan shi* from Mount Qinglongshan in Nanjing; *lingbi shi*[14] from Mount Qingshan in Lingbi County, Anhui Province; *xianshan shi* from Mount Daxianshan in the southern suburbs of Zhenjiang, Jiangsu Province; *xuan shi*[15] from Ningguo County, Anhui Province; *hukou shi* from Hukou in Jiangzhou, Jiangxi Province,[16] *wukang shi* from Zhejiang Province[17] from the region between Yingde and Zhenyang Counties, Guangdong Province; *sanbing shi* from Chaohu, Anhui Province; and *jinchuan shi* (origin unverified), which is valuable for its patina of age. Stone, whether it is Mountain stone or Lake stone, always comes more precious and more costly than soil. Mainly for this reason, stone mountains were highly valued during the Qing period, and they convenience the garden owner to make a display of his wealth; merchants and landlords were particularly fond of giant, all-stone rockeries. Gardens like Wangshi Yuan (The Wang's Garden) in Yangzhou, where stone mountains were constructed, were claimed to have had ten thousand pieces of *taihu* Lake stone set (which is plausible as Yangzhou does not produce stone and such material is mostly shipped in small portions) and designated the garden as "Wanshi Yuan" (Ten-thousand Stone Garden) to flaunt the abundance of the stone. To minimize the usage of stone while attaining the sought-after effect equivalent to a stone mountain, garden makers contrived an expediential arrangement named *shi bao tu*,[18] literally "stone-covered earth", which is otherwise an earth-filled stone mountains. Due to the technical limitations of exploitation, transport, lifting, and piling of the material, the rocks used for all-stone mountains as well as earth-filled

[13]Tang scholar Bai Juyi once remarked: "Stone comes in different grades: the taihu stone ranks the finest and the luofu and tianzhu stones the second".—See Changqing ji— a collection of essays by Bai Juyi.

[14]It is compact and used for a bonsai setting.

[15]Of which the delicate, diminutive maya xuan stone is commonly made objects d'art for display on desks or stand (small pieces of which are often used in potted landscaping for its exquisite formation).

[16]Again, the pocket editions of this stone make valuable, aesthetic display for their dainty characteristic, as relished by Tang scholar Su Dongpo in one of his poems: " . . . the little delicacy I've acquired with a hundred taels of gold. . . .".

[17]Three grades of which—*jinluo*, *guimian*, and *diexue*—are recorded in *Yu Yuan ji*, Notes on the *Yu Yuan*, by Pan Yunduan; *ying shi* (only for potted landscaping as they come in miniature sizes).

[18]Stone-covered earth.

stone mountains are usually moderate in size, which poses a challenge for constructing stone mountains that have to appear an integral whole. Such an undertaking requires the garden maker not only to take into account the mountain's configuration but also painstakingly and resourcefully fit together piece by piece the stones that come in all shapes, proportions, and veins. Admittedly, composition of stone mountains makes an uphill task as they are more difficult to be incarnated, and a garden artificer, if without an artistic vision, would not be equal to the job of crafting stone mountains. Even among the survived works of the Jiangnan garden, those of rambling structure that resembles little more than jumbles of rocks are not uncommon. Relatively speaking then, using soil alternating with stone makes it easier to effect good results of mountains. At such, garden making masters like Ji Cheng of the Ming dynasty and Zhang Lian and Li Yu of the Qing dynasty were mostly in favor of mountain making that took soil as the dominant ingredient while interspersed with stones. Making of earthen mountains draws on home resources: dig the ground that yields the earth for raising mountains and concurrently renders the pitted area a lakelet or a fishpond, saving both labor and material. Moreover, earthen mountains allow convenient disposition of vegetation, which helps reach better mountain-and-forest effects. Stone mountains, on the other hand, are technically more demanding to make but relish of more manifestations if executed properly; they are best at expressing precipices, ravines, grottos, and such. Examples of notable works of various mountain types are plentiful. Of the stone mountains that are made of lake stones, Villa of beautiful surrounds (Huanxiu Shanzhuang) in Suzhou, created by Ge Yuliang during Qing Emperor Qianlong's reign, comes first on the list. Of those that are built with all yellow stone, the Autumn Hill of the Garden of Bamboo (Ge Yuan) in Yangzhou takes the lead. Among the *karst* cave landscapes, the most successful creation goes to the grotto of Xiao Lin Wu[19]—popularly called *shui jiashan*[20]—in the Qiayin Yuan garden in Suzhou; it was designed by Zhou Bingzhong of Ming, who modeled it after the natural grotto named Lin Wu (Cabin in the Woods) on the island of Dongting Xishan in the Taihu Lake. The best representatives of large-scale earth-filled stone mountains are the Garden to delight one's parents (Yu Yuan) in Shanghai[21] and the Garden of Look Forward (Zhan Yuan) in Nanjing. Drawing on the merits of both earth and stone mountains, most of the creations in the Jiangnan garden are diptychs espousing both soil and stone: using soil as the main material for the overall configuration and stone where peaks, cliffs, gullies, etc. are featured. This way, a mountain is created not only with feasible planting conditions for vegetal growth that promises a lush, woody mountain-forest effect, but also with a rich mountainscape of varying aspects.

[19]Little Cabin in the Woods.

[20]Literally "water rockery", as the floor of the grotto is covered with shallow waters that form an inlet.

[21]Planned by Zhang Nanyang during Ming Emperor Wanli's reign.

3.1.1.1.1.1 Basic Constituent of Man-Made Mound

The mountain, as a scenic image of the garden, is expressed by means of various basic features, namely a hillside, cliff, peak, grotto, ravine, stream, waterfall, and so on. The creation of a mountain is a process of selecting, rendering, and arranging these features according to the existing site conditions and building materials. Still there are other forms of features or treatments in the Jiangnan garden that insinuate the "mountain" notion, such as stone placement on the level ground that comes in the style of either a single stone, interspersed stones, or stone groupings. These are derivatives from the aforesaid basic features and a more abstract and epitomized expression of the mountain.

Foothill and Hillside

Build a low earthen grade with a few Mountain stones strategically set around and trees and flowers naturally disposed, these are the basic techniques of describing in space a hillside or the foot of an imaginary mountain range. This approach is both economical and effective in creating the sought-after result evocative of a mountain-and-forest environment. The location of such a construct usually abuts a garden wall or the border of an architectural structure such as a covered walkway. The mountain slope in the south of the Garden of Art (Yi Pu) in Suzhou had the partial remains prior to the Cultural Revolution (1966–1976), which used to be an excellent example of Zhang Nanyuan's "segmental landscape" precepts but regrettably lay waste during the Revolution era.[22] The extant examples of the foothill or hillside construction can be found at Veranda of Scenting the Sweet Osmanthus Fragrance (Wen Muxixiang Xuan) in the Garden that was preserved miraculously (Liu Yuan), the former imperial bureau garden of the Jiangnan textile superintendent of the Qing dynasty, the Sui Yuan garden, and the northeastern section of the Villa of beautiful surrounds (Huanxiu Shanzhuang), all of which are located in Suzhou (Fig. 3.1).

Cliff and Crag

Precipices are one of the most popular manifestations of stone mountains. In most cases, they are configured into a part of a mountain structure in the forms of either a perpendicular cliff (as with the yellow-stone mountains in the East Garden of Suzhou's Couple's Retreat Garden (Ou Yuan) and the Garden to delight one's parents (Yu Yuan) in Shanghai, and with the Lake-stone mountains in Nanjing's Garden of Look Forward (Zhan Yuan) and Hangzhou's Guo Zhuang garden at the West Lake) or an overhanging crag (as with the Lake-stone mountains in gardens like Villa of beautiful surrounds (Huanxiu Shanzhuang) in Suzhou and Schistose Mountain House (Pianshi Shanfang) in Yangzhou, created by Shi Tao) (Fig. 3.2).

[22](The modern-time renovation that extended the construction with a large quantity of rockeries has distorted the original naturalistic features beyond recognition.)

a

b

Fig. 3.1 (**a**) The foothill and hillside of the earthen mountain on the west bank of the lake in Zhan Yuan garden, Nanjing. (**b**) The foothill and hillside of Wen Muxuxiang Xuan in the Liu Yuan, Suzhou. (**c**) The remains of the foothill and hillside in the former imperial bureau garden of the Jiangnan textile superintendent of the Qing dynasty, Suzhou. (**d**) The foothill and hillside near the inscription "Bantan Qiushui Yifang Shan" (roughly "Half-filled pool of autumn waters by the house of the hill" created by Ge Yuliang) in Huanxiu Shanzhuang, Suzhou

There is another prototype of precipice making that is termed *qiaobi shan*, or "cliff mountain", in *The Craft of Gardens* (Yuan Ye). Raised against a wall surface, a *qiaobi shan* is conventionally used for masking garden walls or property boundary. Examples of this type of structure can be found in Garden of Look Forward (Zhan Yuan) in Nanjing; Little Lotus Manor (Xiaolian Zhuang) in Nanxun—a town in the city of Huzhou in Zhejiang Province; and a number of gardens in Yangzhou, such as Schistose Mountain House (Pianshi Shanfang), A Small Heaven and Earth Created

c

d

Fig. 3.1 (continued)

(Xiao Pangu), Villa of Reposing Roars (Jixiao Shanzhuang), Garden of Art (Yi Pu), and Garden to delight one's parents (Yu Yuan) (Fig. 3.3). Relieved by their inverted reflection, precipices constructed by the edge of water appear to be loftier and livelier. To take advantage of the waterfront, a landscape of a shoal or a few rock promontories can be created beneath the bluff, or a rock trail can be hewn winding up the bluff that allows the visitor a closer experience of touring the structure, which undoubtedly creates more interest than viewing the cliff at a distance. The waterfront circuitous footpath at the base of the bluff in Villa of beautiful surrounds (Huanxiu Shanzhuang) is one such structure that imparts the trekker an intimate sense of "stepping into the picture"(Fig. 3.4).

Fig. 3.2 (**a**) The yellow-stone cliff in the East Garden of Suzhou's Ou Yuan. (**b**) The Lake-stone cliff facing the water in Nanjing's Zhan Yuan. (**c**) The Lake-stone cliff in Hangzhou's Guo Zhuang garden at the West Lake. (**d**) The Lake-stone cliff in Huanxiu Shanzhuang in Suzhou (created by Ge Yuliang). (**e**) The Lake-stone cliff in garden like Pianshi Shanfang in Yangzhou, created by Shi Tao and abandoned

Fig. 3.2 (continued)

Fig. 3.3 (**a**) "Qiaobi shan" (or "cliff mountain" in Yuan Ye) in the northeast corner of Zhan Yuan, Nanjing. (**b**) "qiaobi shan" of Xiaolian Zhuang in Nanxun—a town in the city of Huzhou in Zhejiang Province (abandoned). (**c**) "qiaobi shan" in Xiao Pangu, Yangzhou

Fig. 3.4 (**a**) the waterfront circuitous footpath at the base of the bluff in Huanxiu Shanzhuang, Suzhou. (**b**) The waterfront circuitous footpath at the base of the yellow-rock mountain in Wangshi Yuan, Suzhou. (**c**) The waterfront circuitous footpath at the base of the rock mountain in Yi Pu, Suzhou

Peak and Luan

Creation of peaks stems from one's desire to obtain a distant perspective of the lie of the mountains and to intensify the steep atmosphere of a wooded mountaintop. Jagged mountains that undulate in the lie are called *luan*, or undulating mountain ranges. Among the surviving examples in the Jiangnan garden, *luan* are mostly created by stone setting, even for an earthen mountain. Stones used to effect *luan* need not be stacked up high but staggered around strategically. The earthen mountain placed in the western section of Suzhou's Garden that was preserved miraculously (Liu Yuan) provides an excellent exemplar of such execution (Fig. 3.5).

Mountaintops that jut up are called *feng*, or a peak, and the peak creation at the yellow-stone mountain in Couple's Retreat Garden (Ou Yuan)'s East Garden in Suzhou is one of the representative works of the kind (Fig. 3.6).

Usually, an entire large upright Lake stone—called "peak stone"—is used to affect a mountain peak, or sometimes multiple smaller pieces of stone are stacked to achieve the same result. The same approach is also applicable to creating earthen mountains peaks. Sizable *taihu* stones and stalagmites (Fig. 3.7) are often adopted in mountain making to symbolize nature-formed stone peaks like "Wanhu

Fig. 3.5 Stack stonepeaks on the earthen mountain in the west of Suzhou's Liu Yuan

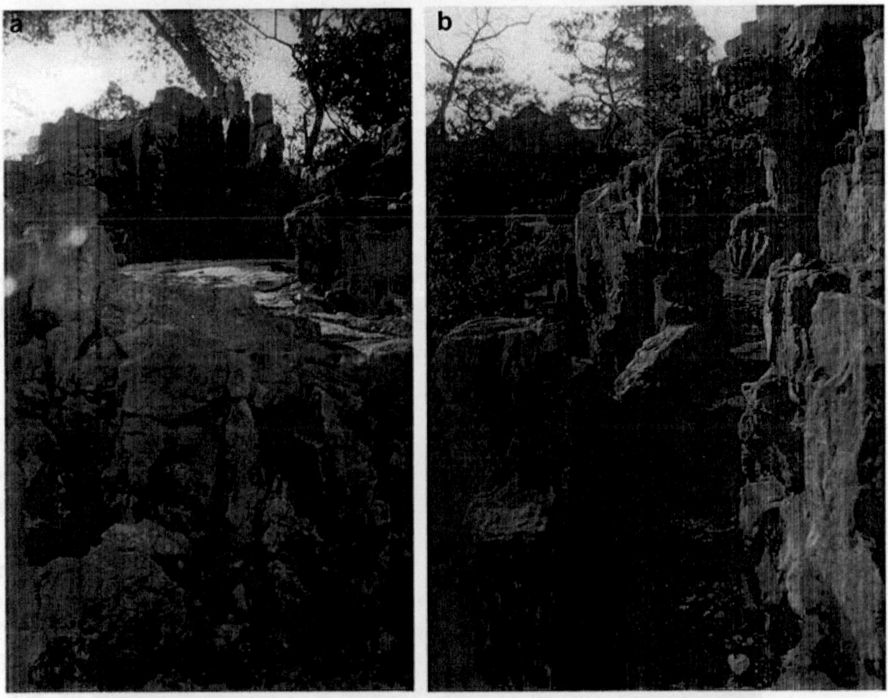

Fig. 3.6 (**a**) The stack stones at the peak of the yellow-stone mountain in Ou Yuan's East Garden in Suzhou. (**b**) The opening of the peak of the yellow-stone mountain in Ou Yuan's East Garden in Suzhou-The summit is made with a hollow interior for saving stones, increasing the magnitude of the mountain and forming a natural form of "pavilion" to shade rain

Fig. 3.10 (**a**) The stacked stone peaks on the lake-stone mountain of Suzhou's Huanxiu Shanzhuang. (**b**) The remained large stone peak made of lake stones on the earthen mountain in Suzhou's Xiao Lingyanshan Guan. (**c**) The five stacked stone peaks at the south yard of Wufengxian Guan in Suzhou's Liu Yuan. (**d**) Stone peaks piled up in front of the hall of Gu Wu Song Yuan in Shizi Lin, Suzhou

Fig. 3.11 (**a**) "Ting Qin Shi" symbolizing the old man under the window of Qin chamber in Yi Yuan, Suzhou. (**b**) A single Stone peak in the West Courtyard of Yi Feng Xuan in Liu Yuan, Suzhou

Fig. 3.12 (**a**) Ruiyun Peak (nicknamed "Xiao Xie Gu") in the western compound of the former Qing imperial bureau garden of the Jiangnan textile superintendent in Suzhou, exotic stones dedicated to Emperor Huizong of the Song dynasty. This peak was originally set in the pool, but now the water has dried up. (**b**) "Yu Hua" Stone Peak in Shanghai's Yu Yuan, also known as "Yu Linglong". Ming scholar Wang Shizhen marveled at this stone in his Yu Yuan ji for its "natural charms divulged through the sleek elegance, redolent of the legacy of the Sui and Tang classics". The peak is also said to be a relic from Emperor Huizong's exotic stone tribute. (**c**) Guanyun Feng in Liu Yuan, Suzhou. (**d**) Xiuyun Feng in Liu Yuan, Suzhou. (**e**) Ruiyun Feng in Liu Yuan, Suzhou. (**f**) Qunyu Feng in Zhan Yuan, Nanjing. (**g**) Meiren Feng in Wenlan Ge, Hangzhou. (**h**) A stone peak in Xu Yuan, Nanjing

Fig. 3.12 (continued)

The Nan Yuan garden in Yangzhou was known for its "Jiu Fengshi"[26] in the early Qing period; during his imperial inspection tour of the south, Emperor Qianlong favored the garden with the name "Jiufeng Yuan",[27] after which the emperor grew greatly attached to the nine rocks and picked two of the finest and took them back to his imperial palace in Beijing. Written accounts on the *taihu* stone traces to as early as the Tang period, when scholar-poet Bai Juyi was relegated to Hangzhou, where he "acquired one *tianzhu* stone and five *taihu* stones that he placed in the pond of his residential garden". In his *Taihu shi ji,*[28] Bai rendered an elaborate screed for the variety of the stones that he had observed at imperial prime minister Niu Sengru's mansion and villa. His figurative tribute to these stone formations reveals what could have been representative of the contemporaneous aesthetic views on the stone:

> (Of these stones), some are convolute in shape as if fresh clouds are curling upward from the Lingqiu hillocks; some stand upright majestically as a courtly minister, some are chiseled fine and smooth like fairy jade ware of ancient sacrificial rites, and some are sharp and pointed suggestive of a sword or halberd. Still some others evoking a dragon or phoenix that seems to be huddled up and yet about to stir, hovering over and yet ready to spring; neither ghost nor beast, it appears to be pacing and yet strikes to be galloping, baring fangs and brandishing talons. While at a dirty night when the gaping dark caverns (of the stones) seem to be gulping down clouds and rolling out thunders, the figures of the stones have metamorphosed to be all but forbidding. While in an otherwise fair morning when the stones are charmingly relieved under a cleared sky, the stones, as if brushing past the mountain mist and dabbing on its dark shadow, have become invitingly affable and intimately approachable. From dawn to dusk, the myriad transformations (of the stones) are beyond description. In a nutshell, all of the Three Hills and Five Mountains in the country and tens of thousands of the caverns and ravines from the four corners of the earth have been encompassed and embodied within this (mansion and villa garden).

—A portrayal redolent of the group mountain peaks at Lion Grove Garden (Shizi Lin) in Suzhou. The animated description reflects the general connoisseurship of *taihu* peak stones (including piled stones) that was prevalent through the late Qing period. The myriad manifestations of the stone via evocative metaphors as described in *Taihu shi ji* are ingeniously rendered between the "likeness" and the "unlikeness" of the allusive object, which allows the viewer the inkling of the object while it may not be. The *taihu* stone proper, in point of fact, is an artless and abstract formal beauty that is predisposed to quicken the imagination of its admirer. Take the Winter Hill at Yangzhou's Garden of Bamboo (Ge Yuan) as an example: the Hills is metaphorically named "Jiushi Tu"[29] and seemingly takes on the figures of a pack of wild lions; from an appreciative perspective, the elusive, subtle resemblance undoubtedly piques viewers' fascination (Fig. 3.13a).

Another such example is in the Master-of-Nets Garden (Wangshi Yuan) in Suzhou, where the pond-side piled stones that act as a trellis to keep up the climbing

[26] (Nine Peak Stones.)

[27] (Garden of Nine Peaks.)

[28] (Notes on the *taihu* stone.)

[29] (Image of Nine Lions.)

Fig. 3.13 (**a**) "Winter Hill"of Ge Yuan in Yangzhou—Jiushi Tu. (**b**) Lion-shaped stacked stone in Wangshi Yuan, Suzhou

wisteria also assume a sketchy shape of a lion, which leaves much space to viewers' interpretation (Fig. 3.13b).

Some stone layers, however, take the practice to such an extreme as to chisel out specific features of the similized objects in pursuit of the physical likeness. Besides lion, creatures modeled in figural stone setting also included leopard, elephant, and even the twelve animals representing the birth years. There are a number of motif compositions such as “Luotuo Feng”,[30] “Niu Chi Xie”,[31] and “Niu Chi Ha”.[32] Some of these works are worth certain appreciative value due to their implicit, nondescript quality while some others, as taking after reality too literally and too corporeally, have lost the subtlety to vulgarity. Lion Grove Garden (Shizi Lin) in Suzhou and the Garden of the Pavilion of Unity with Heaven (Tianyi Ge)[33] in Ningbo have in possession piled-stone lions of such nature (Fig. 3.14).

Grotto and Tunnel

Of many garden landscapes, the mountain cave, with its fantasy quality, is one that seldom fails to fascinate. There are two types: one allows the visitor to travel through and exist at the other end of the cavity, more or less a tunnel, as those built in the Lake-stone mountains at Hangzhou’s Wenlan Ge and Yangzhou’s Garden of Bamboo (Ge Yuan) (Fig. 3.15).

The other type is grotto, some of which afford modest pastime amenities to those who wish to make a short stop in the shelter for a drink or a game of *weiqi*, or simply to relax in the shade. There are Lake-stone grottos and yellow-stone grottos; examples of the former can be found in Suzhou’s Villa of beautiful surrounds (Huanxiu Shanzhuang) and Nanjing’s Garden of Look Forward (Zhan Yuan),[34] and those of the latter in Garden of Bamboo (Ge Yuan)’s Autumn Hill and Villa of Reposing Roars (Jixiao Shanzhuang)’s southwestern end, both gardens are located in Yangzhou (Fig. 3.16).

Creation of caverned spaces not only affords utilitarian functions that sophisticates the touring content, but also economizes the use of stone material that maximizes the scale of the mountain magnitude, killing two birds with one stone. From the structural point of view, the cave is built either on a continuous foundation or discrete stone base, and usually two types of roofing structures—beamed or vaulted—are employed. Although built on the same principle as that of a house, the cave is also a form of the plastic arts that aspires to embody the beauty of nature; therefore the construction is challenged to yield the natural-looking result for the cave. Mediocre designers often simply beam the cave roof with slabstones without any artistic manipulation, as if laying a lid over a coffin chamber. Ingenious designers, on the other hand, emulate nature’s creation and leave no trace of man-made effort by vaulting the roof with natural stones of different sizes and shapes, devoting much attention to securing the grip and lock between the ragged rocks by setting a concave surface of one stone against the convex surface of

[30] (Camel Peak.)

[31] (Ox Eats Crab.)

[32] (Ox Eats Frog.)

[33] (A private stack hall.)

[34] (The west side of the lake.)

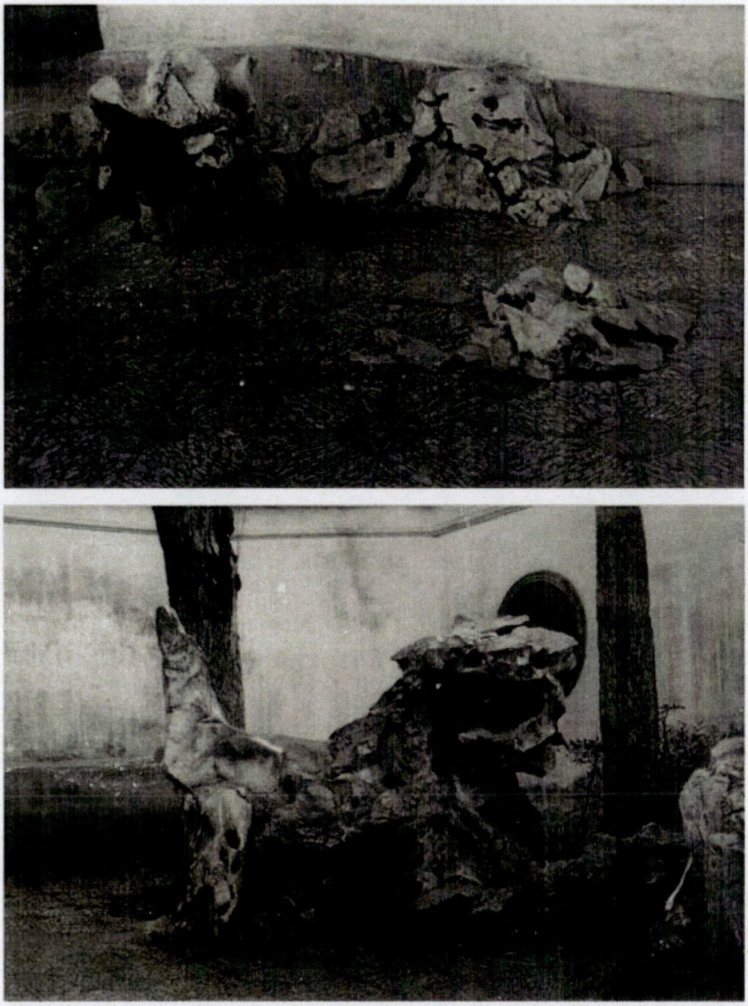

Fig. 3.14 (**a**) “Niu Chi Xie (Ox Eats Crab)” scattered lake stones in Shizi Lin of Suzhou, with the creation of “ox” and “crab” worthy certain appreciative value due to their implicit, nondescript quality. (**b**) Some stacked stones in Shizi Lin of Suzhou taking after reality too literally nd too corporeally, thus losing the subtlety to vulgarity. (**c**) Lion-shaped stacked stone in Tianyi Ge of Ningbo, which is also suspected of excessive pursuit of reality

another, a traditional technique termed “*goudai*” (clasping) by the Jiangnan local stonemasons. Grottos seen in gardens like Villa of beautiful surrounds (Huanxiu Shanzhuang) (Suzhou), Wenlan Ge (Hangzhou), and Garden of Bamboo (Ge Yuan) (Yangzhou) are prime exemplars in the category (Fig. 3.17).

Some creations go one step further as to attempt the scene of stalactites hanging reversely from the roof as seen in a natural *karst* cave, which has achieved a rather verisimilar effect. Examples include the caves of Xiao Lingyanshan Guan and

c

Fig. 3.14 (continued)

Qiayin Yuan's Xiao Lin Wu,[35] both in Suzhou, as well as those in Yangzhou's Di Yuan and Garden of Bamboo (Ge Yuan) (Fig. 3.18).

The cave tunnel of Xiao Lin Wu—the Little Cabin in the Woods—in Qiayin Yuan has a sinuous passage that leads to the mouth of the ulterior end of the tunnel; the length of the tunnel passage is congruent to the reach of the natural lighting emitting from both openings and, although the middle range of the tunnel catches the least illuminance, it is relatively short and expediently requires no artificial lighting while maintaining a desired dim atmosphere with looming visibility. This is a remarkable example of cave making in the Jiangnan garden but has suffered neglect, fallen into disrepair, and faces devastation (Fig. 3.19).

[35]Little Cabin in the Woods.

Fig. 3.15 (**a**) Lake stone tunnel at the entrance of the court garden in Wenlan Ge, Hangzhou. (**b**) Tunnel of Qushui Yuan in Qingpu, Shanghai, with stacked stools for rest. (**c**) Lake stone tunnel in Southwest of Jixiao Shanzhuang in Yangzhou. (**d**) Tunnel through the stone mountain in Wenlan Ge, Hangzhou. (**e**) Tunnel through "Summer Hill" of Ge Yuan in Yangzhou. (**f**) Tunnel digged through the rock on Xiling Yinshe, Hangzhou-Xiao Longhong Tunnel. (**g**) Tunnel with stepped mountain path in Huanxiu Shanzhuang, Suzhou-The entrance to the tunnel at the foot of the mountain. (**h**) Tunnel with stepped mountain path in Huanxiu Shanzhuang, Suzhou-The entrance to the tunnel on the mountain

Fig. 3.15 (continued)

The design for the grottos at gardens like Suzhou's Xiao Yuan, Xiao Lingyanshan Guan, and Villa of beautiful surrounds (Huanxiu Shanzhuang) also takes into consideration of using low-illuminant natural lighting. In addition to the entrance of the grotto, smaller openings are strategically positioned for lighting and aeration; an adequate vent is indispensable for ensuring a cool, dry, and fresh interior of the cave.

The cave at Villa of beautiful surrounds (Huanxiu Shanzhuang) has a small opening cut into the lower end of the cave wall on the far side, through which one can catch a glimpse of the water outside the cave (Fig. 3.20).

Besides the functions of ventilation and limited lighting—especially with the glisten from the water, the small opening affords a view of exuberant delights, sheerly a miniature portraiture of the large "Glistening Grotto" (Guang Yan) and other similar creations of nature, which are located on the shore of the Lijiang River in Guilin of Guangxi Province. The man-made production is precisely a result of thoughtful observation of nature.

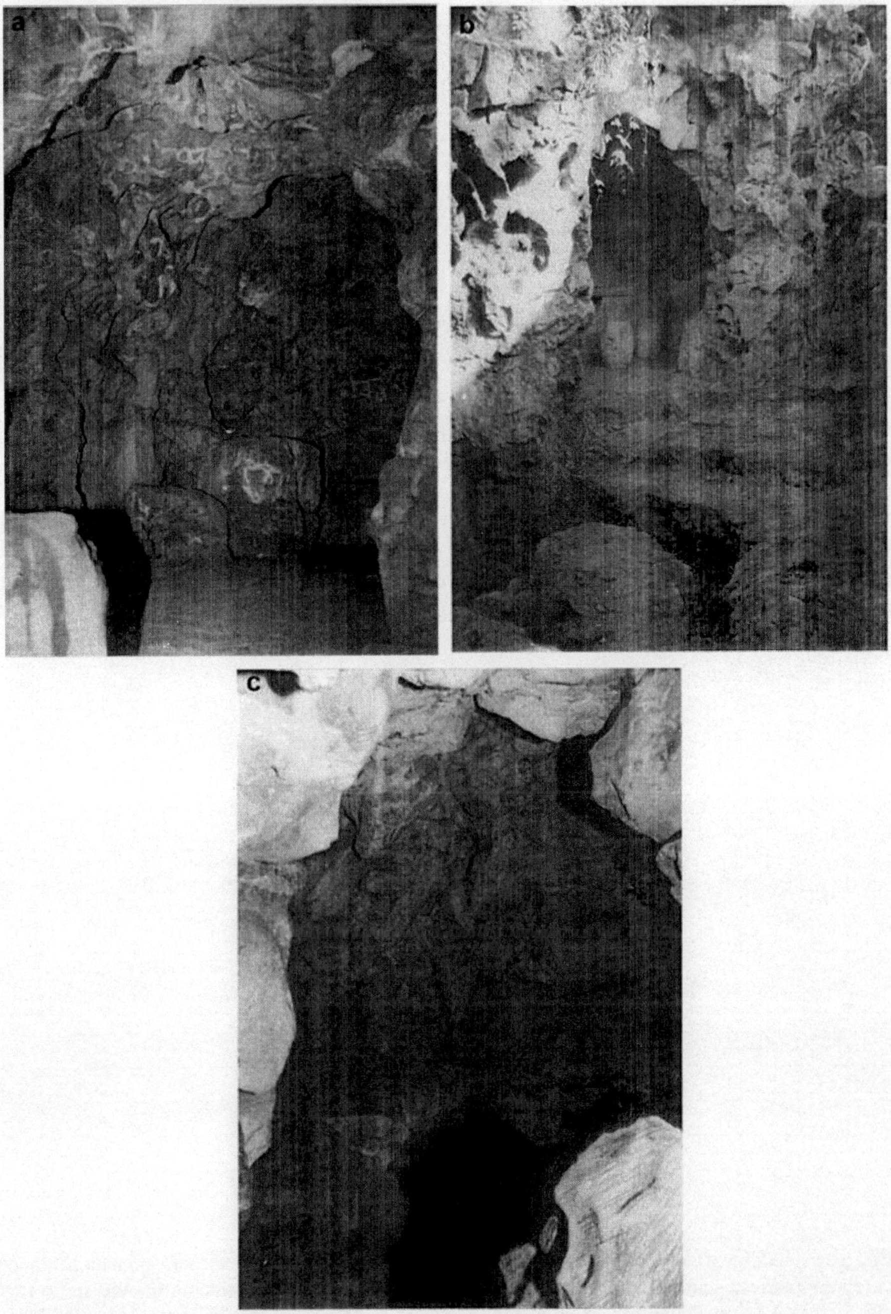

Fig. 3.16 (**a**) One of the caves on the stone mountain in Huanxiu Shanzhuang of Suzhou, inside which there is a karst scene with "tables" and "stools" made of natural stones. (**b**) Grotto No. 2 made of lake stones in Suzhou's Huanxiu Shanzhuang, where there are geometric-form walls, stone tables and stone drums inside. (**c**) Lake-stone grotto on the west side of the lake in Nanjing's Zhan Yuan—

Fig. 3.16 (continued) where there are stalactites hanging reversely from the roof, and natural stone ta (a long, narrow, and low bed) stacked in accordance to the shape of walls inside the grotto for visitors to take a rest. (**d**) The exterior appearance of yellow-stone grotto in "Autumn Hill" of Yangzhou's Ge Yuan-which was similar to the doors and windows of a building. (**e**) The interior of the grotto in "Autumn Hill" of Yangzhou's Ge Yuan-with the arrangement of natural stone tables and seats

Fig. 3.16 (continued)

a

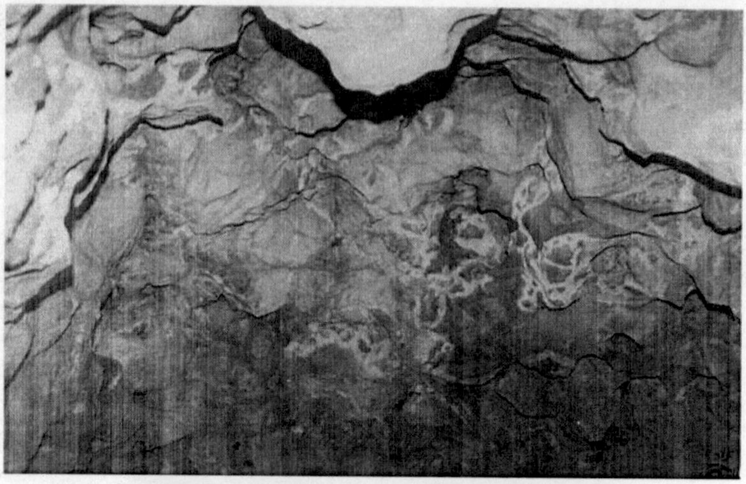

b

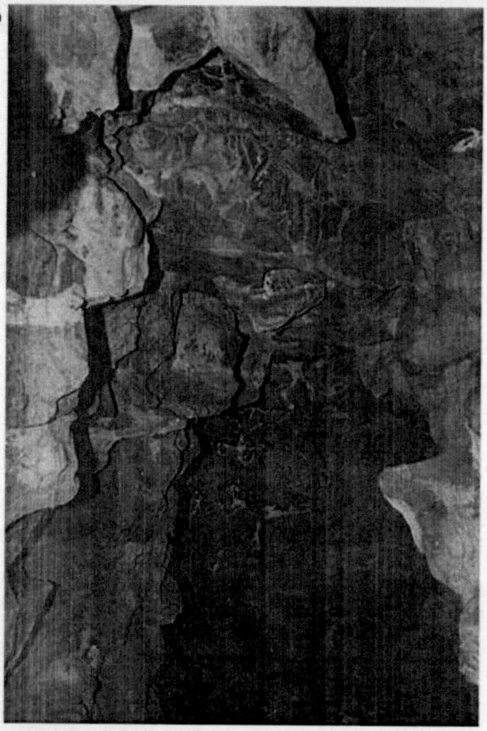

Fig. 3.17 (**a**) The piled stones on the roof of lake-stone grotto in Suzhou's Huanxiu Shanzhuang are the re-creation of the scene of stalactites in karst caves. (**b**) The treatment of the grotto roof in Nanjing's Zhan Yuan is a basic success, but there is a faulty stroke as well-a tilted "stalactite" that was set at a prominent location violates the fact that tips formed by dissolved limestone are downward vertically in nature

Fig. 3.18 (**a**) Karst cave in nature—stalactites in Shanjuan Dong of Jiangsu's Yixing Fig. 3.18. (**b**) The stalactites hanging reversely from the roof of the tunnel in Xiao Lingyanshan Guan, Suzhou

Ravine and Gully

To re-create, in a garden space, a naturalistic ravine or gully with a mountain brook flowing perennially, Jiangnan garden makers assume two differing approaches: *xie shi* and *xie yi*. Representative of the *xie shi* style is Villa of beautiful surrounds (Huanxiu Shanzhuang) in Suzhou, a paragon of the depiction of towering mountains

Fig. 3.19 The artistic interpretation of Lin Wu (Cabin in the Woods) Dong on the island of Taihu Lake—"Xiao Lin Wu" grotto in Qiayin Yuan, Suzhou

with swift-flowing rapids rushing down on the bottom of the deep valley. As they are created to a human scale, visitors are able to enter the landscapes and tour in the midst of the mountains and valleys and benefit from a profoundly personal experience. Alternatively, mountain brooks like those created in the east side of the Master-of-Nets Garden (Wangshi Yuan) and the west side of Crane Garden (He Yuan) and the one by Veranda of Scenting the Sweet Osmanthus Fragrance (Wen Muxixiang Xuan) in Garden that was preserved miraculously (Liu Yuan) are remarkable productions that offer visitors a footbridge above the waters so as to command a view of the purling water from a high vantage point, which adds a special *yi jing* appeal to the mountainscape (Fig. 3.21).

Among the *xie-yi* style creations, Gully of Eight Sounds (Bayin Jian) in Wuxi's Garden for Ease of Mind (Jichang Yuan) is an epitomized representation of a tinkling brooklet in nature that courses through a rocky ravine (Fig. 3.22). The

Fig. 3.20 The large daylight opening besides the natural stone table and small vent (visible to the water surface outside) at the lower part inside the grotto of Huanxiu Shanzhuang, Suzhou

depth of the valley is only as high as one person. The original work is capped by trees, rendering the atmosphere of the deep mountains and valleys with shades of greenery.

Due to landscape damages and disrepair over the time, however, the trees growing atop the hills no longer exist and the gully has since become an open-roofed "alley". It is hoped, though, that the garden conservancy reinstate the planting soon so as to restore the original landscape of a ingenious design. Ravines that are built with no presence of waters are most common in the Jiangnan garden; *Sui Gu*[36] in the East Garden of Suzhou's Couple's Retreat Garden (Ou Yuan) is one such typical example (Fig. 3.23).

Also, there is a type of passageway lined with rough natural stones and snaking between two raised vegetation platforms that is considered a variation of the ravine (Fig. 3.24).

Some ravines are accessible to visitors who may wish to track the tortuous mountain brooks as touring proceedings. *Haopu*, another manifestation of the

[36]Remote Valley.

Fig. 3.21 (**a**) The scene of ravine No.1 in Suzhou's Huanxiu Shanzhuang-with a natural flying bridge (geographical named "arch bridge") over it. (**b**) The scene of ravine No. 2 in Suzhou's Huanxiu Shanzhuang-with a stepstone footpath in it. (**c**) The stepstone footpath in the ravine of "Autumn Hill" of Ge Yuan, Yangzhou. (**d**) The stream in Yu Yuan, Shanghai. (**e**) Mountain brook in the east side of Wangshi Yuan (sluice at the end of the brook). (**f**) Mountain brook in the west side of Suzhou's He Yuan. (**g**) Mountain brook in Suzhou's Liu Yuan

Fig. 3.21 (continued)

ravine, is a gorge saturated with water. As water is the key factor in its making, *haopu* will be discussed in the section of "Creating Bodies of Water".

a

b

Fig. 3.22 (**a**) Bayin Jian in Wuxi's Jichang Yuan. (**b**) Brook and stepping stones of xie yi style of Bayin Jian in Wuxi's Jichang Yuan

Fig. 3.23 (a) One of the entrances of Sui Gu in the East Garden of Suzhou's Ou Yuan. (b) Sui Gu in the East Garden of Suzhou's Ou Yuan

Mountain Spring and Waterfall

Like ravine and *haopu*, mountain spring and waterfall are landscape features where mountain and waters are concurrent and interdependent. However, unlike water-suffused *haopu*, which is plausibly classified as a water feature rather than a mountain feature, spring and waterfall, as well as ravine, usually maintain a scant amount of, or often are completely free of, water. In classical Jiangnan gardens, the planning of the spring or waterfall mostly resorts to vertical composition foiled by piled-stone composition. The source of water relies mainly on the rainwater as the Jiangnan region sits on vast plains that lack a topography of the demanded head of water. In the absence of rainfall, however, garden visitors are but offered to admire the scene of rain-washed piled stones that manifest the scour of the spring or cascade. Stones used for making the spring and waterfall are mostly Lake stones. The crux of the construction lies in selecting and stacking the stones to shape the spout of the spring or the lip of the falls and in setting the water-hitting stones at descending stepped elevations. Jiangnan is a rainy country, and each time it pours, scenes of the splashing cliffside spring or gushing cascade transpire to feast the eyes of the beholders. To secure the source of water for the cascade, stones are usually piled high scaling the wall of a tall building in order to receive the confluent rainwater discharging off the building roof. This is why, when describing the techniques of making waterfall, Ji Cheng remarks in *The Craft of Gardens* (Yuan Ye) that "Waterfall looks like *qiaobi shan*".[37] To create a waterfall with a sizable drop from a substantial altitude, as recounted in *The Craft of Gardens* (Yuan Ye), the

[37] ("Cliff mountain" that is often built against the garden wall to mask the boundary of the garden.)

a

b

Fig. 3.24 (**a**) The stone-lined footpath snaking between two raised vegetation platforms of earth mountain in the East Garden of Suzhou's Ou Yuan—ravine in the style of xie yi, a canopy closure of big trees planted on the vegetation platforms is absent. (**b**) The stone-lined footpath that winds through the earthen hill in Suzhou's Liu Yuan—ravine in the style of xie yi

waterfall construct needs to command a certain height to get access to the roof water of the tall building, whose wall top serves as the gutter through which the rainwater will be amassed and transported to a small storing reservoir on the mountaintop; the falls transpire when the water buildup sluices out of the stone gullet and plummets along the side of the rockery. The previous waterfall in northwestern Villa of beautiful surrounds (Huanxiu Shanzhuang) was an example fashioned in this manner. It was formerly built scaling a rockery that featured a transverse structure spanning the connecting lane between the residential garden and the living area. As the transverse was removed, the waterfall has lost its flair of a sprayey cliffside spring—named *Feixue Quan*[38]—that once mesmericly hanged in the mist when viewed in the rain.[39] A similar structure nearby—a cascade built against the south end of the eastern wall—was made to the style of a mountain spring. Originally, a stone gully was set on the wall top to help generate the run of the cascade that descended the mountain slope when it rained. However, the stone gully—the *raison d'etre* of the running cascade—was dismantled during a modern renovation, which has virtually left this designated "main preserved national cultural relics" structurally bankrupt (Fig. 3.25).

Waterfalls and mountain springs that avail themselves of rainfalls are also discussed in *Zhangwu zhi*,[40] authored by Wen Zhenheng, a Ming garden-making theorist. In his writing, Wen introduces using a bamboo tube concealed in the crevice of the rocks as a device through which the rainwater is channeled from the roofing and spills down the piled stones to create "a stream of cliffside spring that purls gently down". Another technique of creating desired waterfall effect suggested in *Zhangwu zhi* is the use of *shuigui*, or water tank, which "holds up the water on the hilltop and, upon the arrival of the house guest, open the sluices, and the water will gush straight down", thus to effect the waterfall scene in honor of the guest. The waterfall in Suzhou's Lion Grove Garden (Shizi Lin) is an example built with this technique. However, Wen Zhenheng did not recommend this device without any reservation. He considered the contrivance to be "after all artificial and not as graceful as the method of running the roof water in the rain, which works more naturally". For natural-water generated waterfalls, one example is featured in Garden of Autumnal Glow (Qiuxia Pu) of Jiading County on the outskirts of Shanghai (Fig. 3.26).

Shoal and Promontory

The shoal and promontory, made to simulate the rocky rift or riverbed or riparian shoaly landscapes as they appear in nature (Fig. 3.27), are also a type of piled-stone structure that is closely associated with waters. In the creation of Jiangnan gardens,

---

[38] Flying-snow Spring.

[39] (Note: A replacement building was since built, which nevertheless has completely upset the original mountain-waterfall stone setting.)

[40] (A miscellanea on gardens and garden making.)

Fig. 3.25 (**a**) Remaining mountain-waterfall stone setting in southeastern stone mountain of Suzhou's Huanxiu Shanzhuang. (**b**) Part of the piled stones for collecting water of a cascade built against the south end of the eastern wall in Suzhou's Huanxiu Shanzhuang

they are mostly planned in combination with Lakes as a means to diversify the riparian scenery and relieve the water body. Examples of promontory creations include the rock promontory on the west of Prunus Mume Pavilion (Zhuwai Yizhi Xuan) in Suzhou's Master-of-Nets Garden (Wangshi Yuan); the promontory along the rocky shore beneath the mountain of Veranda of Scenting the Sweet Osmanthus Fragrance (Wen Muxixiang Xuan) in Garden that was preserved miraculously (Liu Yuan), also in Suzhou; and the lakefront promontory in front of the big yellow-stone rockery in Nanjing's Garden of Look Forward (Zhan Yuan) (Fig. 3.28).

Fig. 3.26 (**a**) Waterfall in nature—waterfall in Wu Yi mountain. (**b**) Waterfall beside Fei Pu Ting in Suzhou's Shizi Lin employed water tank—water spills down the piled stones at different height. (**c**) Waterfall in garden Qiuxia Pu of Jiading County—the piled stones are concave to create shade to set off the falling water

Hebu Tan[41] in Wu Xi's Garden for Ease of Mind (Jichang Yuan) is a creation of an earth-filled shoal speckled with natural stones; combined with plantings, the shoal scene takes on quite a delight of untamedness. Formerly, there was an accrete double-trunk pterocarya stenoptera tree that played a compositional role of setting

[41] Crane Steps Shoal.

Fig. 3.27 Rock promontory in nature—Dongting Dongshan island in the Taihu Lake

off the depth of the lake field, but it had regrettably died withered; although the tree was replanted, the "shoal of crane steps" was reinforced with disorderly arranged rocks that has tampered with the integrity of the landscape (Fig. 3.29).

As the water level fluctuates, riparian piled stones are often set at descending, irregularly stepped heights leading to the water so as to ensure for the lake waters a brimful effect even at a low water level. As a result of this stone arrangement, a promontoried landscape of a rocky lakebed has taken shape—a welcome derivative effect (Fig. 3.30).

Shoals and promontories are not only a form of stone composition that is interest provoking, they also have the sportful properties that afford the visitor the pleasure of "sitting on the rock by the flowing stream and feeling the clear water lipping the hands and feet".

The basic features discussed above are usually employed as compositional ingredients in making a mountain, which in most cases comprises two or more of the features. As a general rule in mountain making, the emphasis should be placed on the relationship between the mountain's overall configuration and the existing garden space and on the harmony the mountain enjoys with its surrounding features, including the water body, vegetation, architecture, and even ornamental animal life and poetic calligraphic epigraphs. To facilitate our discussion, the mountain will be divided into two types based on its landscape aspects: the earth and stone mountains, and will be studied respectively in the ensuing sections.

a

b

Fig. 3.28 (**a**) The rock promontory on the west of veranda "Zhuwai Yizhi Xuan" in Suzhou's Wangshi Yuan. (**b**) The promontory in front of the lake-stone mountain in Nanjing's Zhan Yuan is vivid. (**c**) Part of promontory in Nanjing's Zhan Yuan

c

Fig. 3.28 (continued)

3.1.1.1.1.2 Creating Earth Mound

In making a Jiangnan garden, earthen mountains are created to simulate the natural earthen mountains indigenous to Jiangnan landscapes, which usually have substantial overburden and significant vegetal cover. Frequently, these mountains have a rocky interior that tends to outcrop through the surface soil, an occurrence common in the areas susceptible to the eroding elements, areas like the top and foot of a mountain, the floor of a valley or grotto, and such; rocky outcrops are particularly common at the fractured precipices caused by natural or human forces. Denuded rocks are a favorite painting object in Chinese ink-wash landscape paintings as well. There are quite a number of well-executed earthen mountains in Jiangnan gardens that are either modeled directly after nature or copied from natural landscape paintings. The earthen mountains located in the western part of Garden that was preserved miraculously (Liu Yuan), in Garden of Surging Wave Pavilion (Canglang Ting), and in the lake of central the Humble Administrator's Garden (Zhuozheng Yuan)—all in Suzhou—are considered relatively large-scale creations representative of this kind. The largest earthen mountain extant in the Jiangnan garden is situated in Hui Long Tan in Jiading County on the outskirts of Shanghai. The mountain enjoys an open environment and affords visitors the pleasure of both viewing from the distance and visiting in the mountain itself (Fig. 3.31).

In Garden that was preserved miraculously (Liu Yuan) and Garden of Surging Wave Pavilion (Canglang Ting), the mountains do not provide a scenic guidance to distant viewing as in the case of Hui Long Tan but rather are created as *jinshan*[42] for intimate touring experience of the mountain. Making of *jinshan* lays much store on the spatial design for a woods-and-mountain environment. The earthen mountains, named "North Mountains", in the Humble Administrator's Garden (Zhuozheng Yuan) are placed in the middle of the lake as the principal scenery, designed to be

[42] Near mountain.

a

b

Fig. 3.29 (**a**) Earth-filled shoal in nature—shoal with trees in Tuochuan, Wuyuan County, Jiangxi Province. (**b**) Original configuration of "Hebu Tan" in Wu Xi's Jichang Yuan—the shoal's elevation is low, configured with double-trunk pterocarya stenoptera tree (sketch in 1956)

Fig. 3.30 Riparian piled stones are set at descending stepped heights to adapt to fluctuating water levels—a brimful effect of the lake water even at a low water level

viewed from Hall of Distant Fragrance (Yuanxiang Tang)—its key viewpoint—and from the surrounding banks of the lake. In this case, the mountain is created as *yuanshan*[43] for viewing from a distance, hence the execution pays more attention to the overall lie and form of the mountain and the varied depths and outlines of the vegetational disposition (Fig. 3.32).

Earthen mountains are prone to vegetal growth and afford plenty of natural rusticity. Prior to Ming, earthen mountains were particularly favored in the residential garden of the literati and officialdom, a fad that is reflected in the painting *Zhuozheng Yuan tu*[44] by Wen Zhengming during Ming Emperor Jiajing's reign. The earthen mountains in Garden of Surging Wave Pavilion (Canglang Ting) and the Humble Administrator's Garden (Zhuozheng Yuan) are the products of the incessant construction throughout the Five, Song, Yuan, and Ming dynasties. Although the earthen mountains we see today are the result of having been augmented with piled stones during later periods, the original flair of the untamedness and rusticity is still faintly palpable except for North Mountains of the Humble Administrator's Garden (Zhuozheng Yuan), whose former earthy appeal has been effaced with misled "embellishments" of excessive added rocks during a reconstruction in the 1960s. The former hummock in Suzhou's Garden of Art (Yi Pu) was also an earthen mountain with interspersed stone setting; created to its natural scale, they represented a mountain slope or the foot of a mountain. This masterwork of segmental realism was said to be the creation of famous realistic garden maker Zhang Nanyang during the late years of Ming Emperor Wanli's reign (early 17th century). It was a rare specimen of cultural relics but unfortunately suffered

[43]Distant mountain.

[44]Landscape of the Humble Administrator's Garden.

Fig. 3.31 (**a**) Earthen mountains in Hui Long Tan in Jiading County on the outskirts of Shanghai—scenery of stepstone footpath and winding paths. (**b**) Earthen mountains in Hui Long Tan in Jiading County on the outskirts of Shanghai—scenery of western foothill and hillside outcropping rocks

Fig. 3.32 (**a**) Space of earthen mountains in Suzhou's Canglang Ting. (**b**) The shape of the mid-lake earthen mountain in Zhuozheng Yuan—most of the central part of the mountain are covered stones and the rest are plants

structural damage during modern times. A scientifically researched restoration plan was rendered for recovering the original artistry of the mountain, but regrettably a complete rebuilding carried out in early 1980s did not follow closely the plan, as a result of which the mountain slope of the Ming legacy was entirely obliterated.

An earthen mountain laid with stones that simulate denuded rocks from a natural mountainscape is locally termed *tu bao shi*, or earth-covered rockery, made of either Mountain stones or Lake stones. Stone setting at the bottom of an earthen slope symbolizes natural rocks outcropping at the foot of a mountain, which not only aesthetically serves as a viewing object, but also practically functions as a retaining wall from the engineering point of view. Jiangnan is a region that abounds with rain. Laying stones around the foot of the hill helps prevent soil erosion by rainwater and defeature of the earthen configuration over the time. The principle and technique of such stone setting are parallel to those of stone mountains making (see next section), which means random arrangement of stones at varied heights. The aforementioned two earthen mountains in Garden of Surging Wave Pavilion (Canglang Ting) and Garden that was preserved miraculously (Liu Yuan) (the western part) are examples that employ yellow stones. The earthen mountain in Little Lotus Manor (Xiaolian Zhuang) of Zhejiang Province, on the other hand, is buttressed with Lake stones at the foot of the hill. The Lake stones here are piled to simulate the natural *karst* landscape of an earthen mountain whose bedrocks are exposed as a result of perennial rainwash; it is a successful execution that highlights the four desired qualities—*shou* (slender), *zhou* (wrinkly), *lou* (cavernous), and *tou* (perforated)—of the Lake stone (Fig. 3.33).

For earthen mountains made for the garden, stones are also frequently set in places like valley, gully, cave, stepstone mountain path, mountain ridge, and peak top. The mid-lake earthen mountain in the Humble Administrator's Garden (Zhuozheng Yuan) provides examples of such stone setting at its ravine and stepstone footpath (Fig. 3.34).

Stone setting on an earthen mountain that symbolizes peaks and *luan*[45] usually takes three approaches: first, the group disposition of piled stones that signifies *luan* as in the west of Garden that was preserved miraculously (Liu Yuan) in Suzhou; secondly, the multiple stone peaks jutting discretely as in Xiao Lingyanshan Guan in Suzhou; and lastly, the *xie yi*-style setting of a large, singular *taihu* peak stone as in the treatment in front of Hall of Illuminating the Way (Mingdao Tang) in Suzhou's Garden of Surging Wave Pavilion (Canglang Ting).

The extant earthen mountains are mostly post-Qing productions, especially after Taiping Rebellion, and were made with increasingly larger quantity of stone. Nevertheless, they are still considered earthen mountains as long as the basic landscape feature of an earthy mountain is not diminished by the presence of the stone. *Jiaxing xianzhi*[46] written during Emperor Kangxi's reign of early Qing has the following account: "In former days, mountains that were piled high with stones were

[45] Undulating mountain ranges.

[46] (Annals of Jiaxing County, Zhejiang Province.)

Fig. 3.33 (**a**) Yellow stone covering the foot of the mid-lake earthen mountain in Zhuozheng Yuan. (**b**) The foot of the earthen mountain and yellow stones along winding paths in Suzhou's Liu Yuan (the western part). (**c**) The earthen mountain in Xiaolian Zhuang of Zhejiang Province buttressed with Lake stones at the foot of the hill

considered the finest, and (people) were averse to the sight of soil". Here, the "former days" refer to the time earlier than the Kangxi period, which denotes that the trend of flaunting stone-made mountains probably began in the late Ming dynasty. Although such a propensity gave an impetus to the development of the stone mountain, it nonetheless impaired the creation and preservation of earthen mountains. In view of this situation, Qing scholar Li Yu, who was erudite in garden making as well, proposed "*yi tu dai shi*, *tu shi xiang jian*", an approach that tactically interspaces the earth with stones, accentuating the core presence of the earth and the

c

Fig. 3.33 (continued)

Fig. 3.34 Ravine and piled stones winding paths in the mid-lake earthen mountain in Zhuozheng Yuan

supplementary function of the stone.[47] Contemporaneous garden maker Zhang Nanyuan also remarked: "Lofty mountains that soar to the sky are not as good as small-scaled earth hillocks that are punctuated by stone arrangements".[48] If kept to

[47] (See *Xianqing ouji*—Casual notes on leisure matters.)

[48] (See *Zhang Nanyuan zhuan*, Biography of Zhang Nanyuan.)

the proper principle of the art of Jiangnan garden's mountain making, the superfluous stones loaded on some of the extant earthen mountains purely for wealth parade ought to be eliminated to achieve a better result for the landscape. However, it is obligatory that we categorically preserve the original productions, be they good or inferior, in their existing forms, for they are a living history that attests to the course and schools of the development of the Jiangnan garden art as well as provide a comparative platform for the study of the creative principles of the Jiangnan garden.

Aside from disposition of natural stones, the effect of the earthen mountain is largely achieved by use of vegetation that sets off the mountain-and-forest atmosphere. Particularly for the mountains intended to be viewed from a distance, the grandeur of the mountain range is actually effected by trees, which is why disposition of vegetation is an important means of complementing the lie of the mountain, especially the earthen mountain (Fig. 3.35).

For this reason, plantings on a distant mountain should follow the run of the mountain, and the heights and contours of the trees should accentuate the undulation of the mountain. Architecture is another complementary means to relieve the lie of the mountain. Placing a pavilion—foiled by arbores—on an earthen mountain that has a gentle lay often achieves an enhanced effect of a rolling mountain. Pavilions Embroidered Silk Pavilion (Xiuqi Ting) and Waiting for the Frost Pavilion (Daishuang Ting) in the Humble Administrator's Garden (Zhuozheng Yuan), Free Roaring Pavilion (Shuxiao Ting) in Garden that was preserved miraculously (Liu Yuan), and the small pavilions in Chan Yuan and Garden of Art (Yi Pu) are but a few good examples from the gardens in Suzhou (Fig. 3.36).

3.1.1.1.1.3 Creating Stone Mound

Creation of stone mountains in the Jiangnan garden is inspired by the rugged mountainscape in nature. Re-created through the art of garden, the entirety of an artificial stone mountain is not necessarily all stone-made. Based on the engineering techniques, the stone mountain can be generally divided into two types: the all-stone mountain, or otherwise called rockery, and the stone-clad mountain that is filled with earth interior. Judged by its building material, the all-stone mountain can further be divided into the Lake-stone mountain and the yellow-stone mountain.

The earth-filled stone mountain, also called *shi bao tu* by the Jiangnan native workmen, is an economical solution to making a large-scale stone mountain with only limited ready supply of affordable stone. To raise an earth-filled stone mountain, one needs to build the shell with the stone first and subsequently fill it in with the earth—a procedure that helps ensure a stable foundation of the piled stones. All cogitative creations follow the law of nature, which, in the case of making stone mountains, means never mixing the use of Mountain (yellow) stones with that of Lake stones. To tackle the material shortage that calls for use of both stone types, however, the Humble Administrator's Garden (Zhuozheng Yuan) in Suzhou provides an expedient solution. Take for example the yellow-stone mountain created to the south of Hall of Distant Fragrance (Yuanxiang Tang): the mountain has a built-in

a

b

Fig. 3.35 (**a**) Trees on the stone mountain in the Nanjing's Zhan Yuan enhanced the lie of the mountain. (**b**) Mountain-and-forest rusticity brought by lush disposition of vegetation in the mid-lake earthen mountain in Zhuozheng Yuan. (**c**) The newly built greening of earthen mountain in the former site of Gui yuan tian ju in Suzhou's Zhuozheng Yuan (eastern part)—trees compose higher skyline of the mountain

cave, whose interior is lined with Lake stones. Such a device has not only maintained the integrity of an all-yellow stone mountain when viewed from outside, but also realized karst cave scenery featured inside. Caves and tunnels built in an earth-filled stone mountain, incidentally, are also lined with stones (for the retaining purpose), in the same manner as in piling an all-stone mountain. Normally, under the circumstances when making all Lake-stone or all yellow-stone rockeries becomes impossible due to the scarcity of each type of the material, both stones can be used but for separate sections. Usually, the less-costly yellow stones are used for laying the

c

Fig. 3.35 (continued)

foundation—the base of the mountain, which is concealed beneath the earth, and the more-expensive Lake stones are put to best use from above the earth cover all the way up to the peak of the mountain. Gardens like Villa of beautiful surrounds (Huanxiu Shanzhuang) and Garden that was preserved miraculously (Liu Yuan) (central part) in Suzhou are prime exemplars of such techniques.

In the case of remaking an age-old, hardened earthen mountain into a stone mountain, stones are laid directly onto the surface of the solidified soil. In the area of the stepped mountain path, stones are not only paved on the surface of each step but more importantly also mounted onto all vertical planes of the steps and sidewalls. This execution is repeated in every aspect of the mountain and on every level of the mountain path that spirals up to the top of the mountain, hence to create a stone facade that conceals the presence of the earth and to complete the effect of a stone mountain. With a stone mountain, soil is usually exposed and reserved only as planting beds where disposition of vegetation is intended. Larger areas of soil on the mountaintop are often left unpaved for tree plantings (Fig. 3.37).

Combined with the configuration of the mountain path, peaks and ridges are arranged in the style of *feng hui lu zhuan*, a mountain making technique literally meaning "where the peak stands brings the next turn of the path". A peak, either a larger freestanding peak stone or multiple pieces of smaller stones stacking up, is set upright on the inside of the bend in the flexuous path (Fig. 3.38), so positioned to both intentionally obscure the sight of the road behind the peak and enticingly leads the curious visitor around the turn to continue the journey onto the path behind it. The caves and tunnels are made of rocky stones whose craftsmanship is the same as the pure stone mountain.

Making all-stone mountains consumes considerably more stone material. Limited by the individual financial resources, an all-stone mountain is usually made smaller

Fig. 3.36 (**a**) Architecture enhances the lie of mountain—Pavilion Xiuqi Ting in Zhuozheng Yuan in Suzhou. (**b**) A zigzag skyline formed by the Pavilion Feng Ting and hill-climbing wall on the stone mountain in Yangzhou's Xiao Pangu raise the mountain

Fig. 3.37 Wandering rocky trail and piled stones on the yellow-stone mountains in Yu Yuan in Shanghai

in scale than an earthen mountain or earth-filled stone mountain, hence there has been an old saying: "Use earth to make big mountains and stone small hills". Surely it is not categorically so; large rockeries are created where financial and material resources are available, the Summer Hill and the Autumn Hill in Yangzhou's Garden of Bamboo (Ge Yuan) are two such examples. To maximize the use of the given supply of stone, many stone mountains are made with a hollow interior, an ingenious solution that not only increases the magnitude of the mountain, but also allows creation of caves and tunnels. The Summer Hill in Garden of Bamboo (Ge Yuan) is piled with entire Lake stones, which is said to contain a dozen of caves; another large rockery in the same garden, the Autumn Hill, which is made of almost all yellow stones, also features cave construction; and in Hangzhou's Wenlan Ge, the Lake-stone mountain that was made in early Qing is, too, pierced with extensive cave configuration. As the stone mountain is more fictile than the earthen mountain and capable of variegated manifestations such as precipice, overhanging cliff, and grotto, its scenic structure takes on more complexity than its earthen counterpart and consequently its viewing route becomes more convoluted and more interesting. Making stone mountains in conjunction with water bodies that feature precipice,

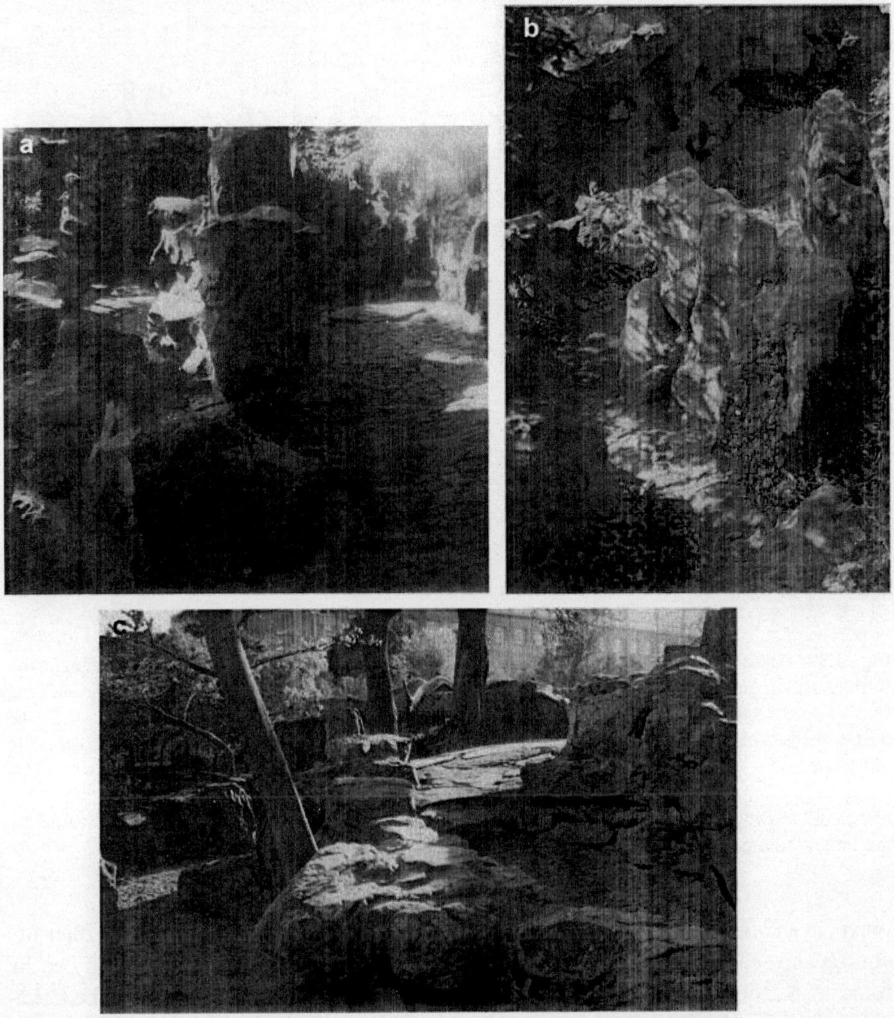

Fig. 3.38 (**a**) Stones stacking up in the style of "feng hui lu zhuan" on the yellow-stone mountains in Yu Yuan in Shanghai. (**b**) Stones stacking up in the style of "feng hui lu zhuan" on the lake-stone mountain in Nanjing's Zhan Yuan. (**c**) Low piled stones in the style of "feng hui lu zhuan" on the yellow-stone mountain in East Garden in Suzhou's Ou Yuan

ravine, and cave usually follows a set of formularized compositions and design treatments: firstly, create a shoal-and-promontory scene at the base of the cliff, then build a rocky trail that wanders into the ravine where a cave is tucked away that offers respite; from there, a stepstone footpath winds up the hill and leads to a stone footbridge overlooking the ravine—a setup intended to strike the visitor with the impression of a precipitous cliff and unfathomable vale; the circuitous mountain path then culminates on the top plane of the mountain or, in other cases, takes turns and

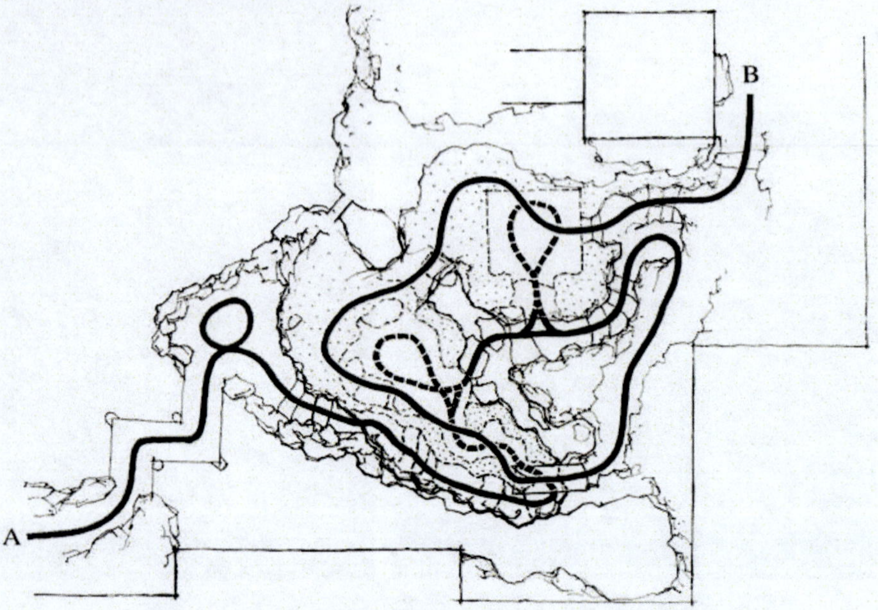

Fig. 3.39 Analysis chart of touring route of stone mountain in Suzhou's Huanxiu Shanzhuang (orthographic projection of three-dimensional route)—form A to B: promontory→zigzag bridge→promontory ground→the waterfront circuitous footpath at the base of the bluff→grotto→cave→stepstone in ravine→artificial cave→stepstone footpath in gully→hillside→stepstone footpath on the mountain→ridge→stone footbridge over the ravine→stepstone footpath→ground on the top of the mountain→paths in the style of "feng hui lu zhuan" leading to natural bridge in the north of mountain→stepstone footpath downwards the mountain→the pavilion with the inscription "Bantan Qiushui Yifang Shan"

twists in the *fenghui luzhuan* style until it hits the stepstone path that takes over the course and leads to descending the hill. Exemplary works of such creation can be found in Suzhou's Villa of beautiful surrounds (Huanxiu Shanzhuang) and the Lake-stone made valley "Yan Gu" in the southeastern part of the Yan Yuan garden in Changshu (Fig. 3.39).

The conspicuous achievement of the Jiangnan stone mountain making lies not only in its compositional arrangement of a scenic plane intended for static appreciation, but more significantly in the visitor's participatory sequential experiences in the kinetic viewing of the mountain inside the mountain itself. By way of the mountain path, the spatial features that deep and secluded mountains and valleys have—tortuous, rugged, remote, and labyrinthian—are being accentuated. Design of a mountain path sequence can be summarized into these seemingly paradoxical principles: descend first in order to ascend afterwards, lead left in order to proceed right, and vice versa, thus to increase the sense of depth and breadth of a limited space; additionally, *trompe-l'oeil* is often applied to trick the visitor's perception and enhance the interest of the kinetic experience as in a path that appears to lead ahead while it is in fact an impasse or another that seems to be lost in the maze while it springs open to an broad road at the end. In brief, the intent of the design is to create unanticipated situations that help enhance the viewing interest and enrich the touring experience (Fig. 3.39).

Making of artificial stone mountains initially aimed to simulate natural mountainscapes; however, over the course of the ongoing creative process (especially in the circumstance of using Lake stone), it has incrementally evolved into creation of abstract formalist beauty. The original four aesthetic criteria of the *taihu* stone—*shou* (slender), *zhou* (wrinkly), *lou* (cavernous), and *tou* (perforated), which were set forth by Song calligrapher Mi Yuanzhang for evaluating singular *taihu* Lake stones, have also gradually become the aesthetic criteria of stone mountains as well (Fig. 3.40).

Stone mountains of later periods were mostly made to pursue a craggy and cavernous fashion, featuring an intricate spatial configuration, which undoubtedly intensifies the sportful, labyrinthian interest of the mountain for the visitor. The rockeries in Suzhou's Lion Grove Garden are one such example. The creation of the Lion Grove Garden rockeries exhausted almost all the spatial compositional feats of stone-mountain making and provides the visitor with a convoluted space bristling with peaks, caves, and valley trails that circle up and down, left and right in a roundabout way, which would often disorient the visitor who, even though within a remarkably small area, could not find his way out for a while. The intricately-designed stone mountain in front of Veranda of Pointing to the Cypress (Zhibai Xuan), for instance, occupies merely 750 m^2 (less than 900 square yards), but one would easily spend 20 min to just tour round its main structure without a rest; frequently, one would hear his friend's voice just a stone's throw away yet could not

Fig. 3.40 Shou (slender), zhou (wrinkly), lou (cavernous), and tou (perforated) stone mountains (lake-stone mountain in Suzhou's Shizi Lin)

find a linking passage between them two, or he would see a passageway through the crevice in the rock and probe toward that direction but only find himself led to somewhere else. Interesting episodes also include that some out-of-town sightseers lost their way in the mazy rockery for a long time while about to miss their departing train and had to yell for help before they were "rescued". Whereas this is evincibly a successful production in terms of creating an environment of labyrinthian interest, it has nevertheless overdone its part in terms of re-creating a natural landscape representative of deep and remote mountains and valleys.

In terms of materials, Pure Stone Mountain also has two types: Hu Shi and Huang Shi. Using small natural stones, splicing them according to their shapes, and textures to form a natural mountain shape with overall momentum, this can actually be said to be a creation of sculpture. An extraordinary stone-mountain maker of the Jiangnan region, like an artist of landscape painting, will have first and foremost mentally mastered the image of the mountains and ravines as they exist in nature before he sets to work. This denotes that an elite rockery artist—distinct from commonplace rockery workmen—is not only perceptive and knowledgeable of the natural mountain but also has the aptitude to digest, ingest, refine, and epitomize the artistic nutrients culled from nature. Prior to the creative process begins then, the mountainscape to be created is already taken shape in the artist's mind and planned out in his design drawings or models. Eminent rockery maker Zhang Lian of the Qing dynasty, for instance, made the stone mountain in garden Crane Tending Islet (Fang He Zhou) for Zhu Maoshi, a man of wealth from Jiaxing near Shanghai, which was based on his drawing *Mo shi tu.*[49] Zhang's son, Ran, also a famous artificial-mountain maker,[50] painted *Yi Yuan shanshui tu,*[51] to which he created the rockery at Hall of Ten Thousand Willows (Wanliu Tang) for Feng Puyingin Beijing. In making a stone mountain, natural stones of moderate sizes are selected and pieced together according to their forms, textures, and veins in order to effect the consistency and integrity of the natural stone mountain—an undertaking bearing much affinity to sculpturing. A skillful rockery maker does not fall under the influence of the confusing shapes and sizes of the stones but rather maneuver the material with conversance: stones (usually on the smaller side) are seamlessly matched to amalgamate into a large, free-form ragged configuration, made to simulate the weathered and denudedrocky landscape (when Mountain stone is used) or the natural cavernous *karst* scenery (when Lake stone is used) (Fig. 3.41).

Unskilled rockery makers often stack the stones hit or miss that lacks strategic thoughtfulness and professional tactics; works as such are nothing more than a stockpile of haphazard stones (Fig. 3.42).

Design is crucial to the creation of an artificial mountain, and construction is a continuation of the design that merits equal attention. To satisfactorily carry out a production, the designer must partake in the construction and keep at the "on-the-

[49] Ink-wash sketch of the stones.

[50] Nicknamed "Rockery Zhang" in Beijing during early Qing.

[51] Picture of the mountains and waters in garden Yi Yuan.

Fig. 3.41 (**a**) Large, ragged configuration stacked with smaller pieces of lake stones (part of the lake-stone mountain in Suzhou's Huanxiu Shanzhuang). (**b**) Large, ragged configuration stacked with smaller pieces of yellow stones (Sui Gu in East Garden of Ou Yuan in Suzhou). (**c**) Landscape of denuded stone mountain in nature (Hu Qiu in Suzhou). (**d**) Landscape of denuded stone mountain in nature (Shangfang Mountain in Suzhou)

spot design" during the entire process until the construction is complete. Although the pre-construction design is conceived based on the building material, the actual dealing with the stones on the worksite often quickens one's impromptu ingenuity that deepens and substantializes the conception and even hones and improves the original design. It is true that refined workmanship depends very much on the artifice of the workmen, but the presence of the designer on the site enables him to provide

Fig. 3.42 (**a**) A stockpile of haphazard stones without entirety by the side of Jin Su Ting in Suzhou's Yi Yuan. (**b**) Constant pursuit of being shou (slender), zhou (wrinkly) of piled stones destroyed the integrity in Xi Yuan on the island of Dongting Dongshan in Suzhou

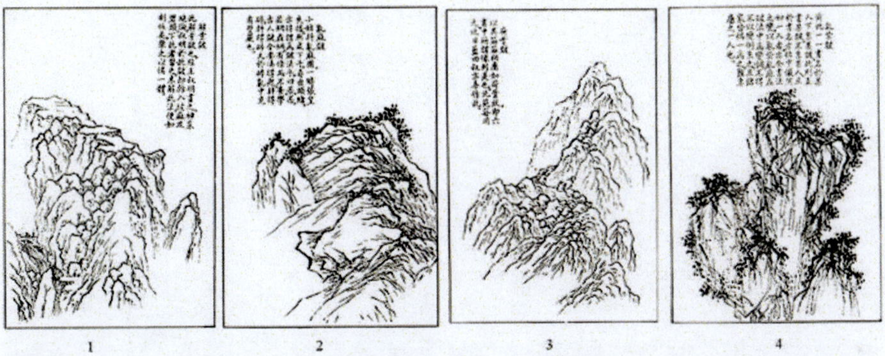

Fig. 3.43 Cun ink-stroke techniques of painting rocks and mountains (Excerpted from The Works of Paintings in Jiezi Yuan). (*1*) Jiesuo cun; (*2*) luanma cun; (*3*) heye cun (*4*) luanchai cun

immediate artistic direction and quality control throughout the entire creative process, which ensures the execution does not compromise at any stage. This is why, in *The Craft of Gardens* (Yuan Ye), Ji Cheng gives a ratio of nine to one to the role-play between the designer and the workmen. Artificial-mountain makers who has training in landscape painting or landscape painters who are engaged in garden making tend to apply their learning and artistry in mountain painting to mountain making. For instance, the art of composition in landscape brushwork helps configure mountainscapes in the art of garden, and the traditional *cun* ink-stroke techniques of painting rocks and mountains are also adopted in creating variegated shades and texture of a stone mountain (Fig. 3.43).

Rockery making influenced by painting artistry often emulates different *cun* brushwork styles, and the rockery in Villa of beautiful surrounds (Huanxiu Shanzhuang) is an example suggestive of the *jingguancun* style. Ming scholar Wang Shizhen commented in *Juyi lu*[52] that Zhang Nanyuanset artistic rendering as a rockery and resorted to the painting techniques of Yingqiu, Beiyuan, Dachi and Huanghe (Bu Fuming, 2005).

Finally, special mention should be made of a type of piled stone related to and derived from rockery arrangement—the planting bed of natural stones. Built in a similar fashion as with an artificial stone mountain, the planting bed is hemmed around in a free form with Lake or Mountain stones, the inside filled with soil for growing vegetation (Fig. 3.44).

Created to give a mountain-forest motif, planting beds are usually placed in a small *ting yuan* (court garden) or the courtyard of a large *zhai yuan* (residential garden), or in front or at the back of an architectural structure, such as a hall, where the space is inadequate for displaying large rockeries.

[52]Essays on Bai Juyi.

Fig. 3.44 (**a**) Natural planting bed in Zhilian Laowu in Suzhou's Ou Yuan. (**b**) Natural planting bed in Suzhou's Liu Yuan. (**c**) Natural planting bed in Suzhou's Tingfeng Yuan

3.1.1.1.2 Management of Water Surface

In nature, water has manifold functions. It regulates air humidity to ameliorate the microclimate and moistens the soil to propagate vegetation, to name but a few; additionally, the biogenic properties of water are vital to life of human beings, which all have made water a more congenial element in the natural landscape than mountains and form the material or utilitarian grounds for our appreciation of waterscapes. Water constitutes an important aspect of the garden's scenery. "Reflections (of the buildings) in the pool evoke (the feeling of) entering an underwater palace" (*The Craft of Gardens*),[53] demonstrating the spatial expansionary function of water (Fig. 3.45). In the classical Chinese garden, the water body—even with a relatively small amount of water—has distinctive expressional styles that are capable of accentuating the natural features of a waterscape, be it river, lake, sea, stream, waterfall, pool, or pond. In the history of Chinese garden making, creation of water bodies dates far back. *The Book of Songs*[54] recites: "The King comes to the 'Ling Zhao' pool, and the water teems with bouncing fish", which is the earliest known record of a garden's water scene, although it is unknown whether "Ling Zhao" was a

[53] Yuan Ye.

[54] Shi Jing.

Fig. 3.45 "Reflections (of the buildings) in the pool evoke (the feeling of) entering an underwater palace" (Yuan Ye)—Yu Yuan in Shanghai

natural water body that was artfully disposed with fauna and flora or a man-made pool created as a result of digging to raise "Ling Tai", the famous elevated terraced structure built by King Wen of the Western Zhou dynasty, which afforded a high vantage point for recreational and defensive purposes. "Ling Zhao" was created mainly for admiring ornamental fish, and the fishpond was accompanied by an animal park, "Ling You", also the property of King Wen, that kept exotic animals of ornamental value—"deer that are bare and sleek and white birds that are silky and pure" (*The Book of Songs*). Recorded in *Shuyi* ji,[55] "King Wu of the Eastern Zhou dynasty built Tien Chi (Heavenly Pool) in Suzhou, where he went boating in the blue-dragon boat", gives an example of a man-made water body used for recreational purposes. It is assumable that these pools were built in a free-form naturalistic style. Creators of water bodies in the early period had the liking for large expanse of water. During the reign of Emperor Wu of the Western Han dynasty, pools such as

[55]Notes on accounts of the extraordinary.

"Kunming" and "Kunling" were made to cover ten square kilometers (nearly four square miles). The ribbon-shaped water body, such as rivers and canals, were created to personify celestial formations. For example, during the Emperor Wu period, along a canal that ran north and south from the "Kunming" pool, two stone figures—representing Altair and Vega respectively, or the Cowherd and the Girl Weaver as in the traditional Chinese folklore—were placed facing each other as if gazing out at one another across the canal that symbolized the Milky Way—the callous deterrent that forcedly parted the longing couple—the Cowherd and the Girl Weaver, who could only cross the Milky Way to reunite once a year, as told by the folklore. Examples of the use of statuary in the water scenery during the Emperor Wu period also included the placement of stone whales in the "Kunming" pool as well as a chalcographic fountain structure mechanized to "draw water from the river and run it through a brass dragon that spits the water into the cup held in the mouth of a brass fairy where the water brims over and falls down". (*Hanguan dianzhi*). *Jisha wei zhouyu, jishui wei botao*, which is explained as "amass sands in the water to shape islets and break waters by arranging boulders in the midst of the dashing current to make waves", were some of the waterscaping techniques typically used in making large-scale gardens in the Western Han period. The private garden of wealthy gentry Yuan Guanghan from Maoling, built at the foot of Mount Beimang in the suburbs of today's Xi'an, was a paragon of such motional and acoustical waterscape that simulated the movement and sound of natural water scenery. Special mention should be made of that this creation took place in the second century BC, which set a rare precedent in the history of world's landscaping and garden making. This and the aforementioned examples of artificial water-body making illustrate an age-old tradition and diversed techniques of waterscape creation in classical Chinese garden making. Up in the Six dynasties (222–589 AD), the natural landscape theme in garden making saw significant development. The extant gardens of private residences in the Jiangnan region were mainly created in the tradition of the Six and later periods, which featured smaller scales and naturalistic sceneries, effacing the theme of immortal ideals common in the earlier period. While water reflections that are redolent of an "underwater palace" evokes the garden's spatial depth, the limpid pool waters form a reposeful counterpoint to the rhythmic built-up features surrounding the water—pavilion, tower, walkways, rockeries, flowers, and trees—and offers a contrasting pause in the busy space of the garden.

Water imparts different meanings and values to different subjects. For a chemist, water is a chemical composition—H_2O; for a hydraulician, the focus of attention is its hydraulic power; and for a landscape architect or garden maker, water is a landscape aspect in the form of river, lake, sea, stream, cascade, spring, pool, or pond. In the artistic structure of the garden, these landscape forms are precisely the basic features of water. Water has the distinct expressional styles unique to Chinese gardens, of which the Jiangnan garden is a prime exemplar. As with other landscape forms, creating bodies of water also needs to *xiao zhong jian da*[56] or *yi shao sheng*

[56]Evoke the large with the small.

duo,[57] which is analogical to the figurative acting in Beijing opera, where four walk-on actors marching in a simplified stage set can be effectively suggestive of thousands of troops fighting a battle. Among the existing Jiangnan gardens, a number of lake sceneries are created with small bodies of water. For instance, the Master-of-Nets Garden (Wangshi Yuan) in Suzhou features a sheet of water that covers merely 400 m^2 (under 500 square yards) but has successfully achieved the effect of rippling waters in an expansive lake. Another example of "achieve more with less" is Gully of Eight Sounds (Bayin Jian) at Garden for Ease of Mind (Jichang Yuan) in Wuxi, which is effected through a trickling stream of water that courses down through a small winding channel. The crux of making a lake, stream, or any other waterscape lies in the garden maker's superior understand of the scenic characters of different water bodies rather than being dictated by the given dimensions and volume of the waters. In the language of the garden art, larger waters do not necessarily make a larger form of water body (lake, for instance), and smaller waters a smaller form (pond, for instance). With different design techniques and treatments, to give an example, the waters in Xu Yuan (Xu's Garden) at the Slender West Lake in Yangzhou were crafted into a pond with a peripheral landscape styled in a pond-scape motif, even though this water body is actually larger in scale than the "lake" in the Master-of-Nets Garden (Wangshi Yuan). Therefore, defining the form of a water body should not be predicated on the magnitude of the waters.

The size of waters bears little significance in the principle of waterscape making, and to generalize all garden waters as "pool" also provokes intelligible ambiguity. From the perspective of the garden art, each water body in the garden carries its own scenic implication—the lake, pond, stream, spring, etc.—that articulates the qualitative difference of various waterscapes, and these qualitative differences are precisely the basis for classification of water bodies.

3.1.1.1.2.1 Basic Constituent of Water Surface

Water does not have a fixed shape. It takes the form of the container that holds it. In the natural world, water features are prescribed by the lay of the land, the contour of the shoreline, or other natural settings where water exists. Therefore, in creating a water body in a garden as an artistic undertaking, successfully describing varied characters of the water—in addition to its source, course, scouring, and other hydraulic disposition—largely depends on the treatment of the waters' shoreline and the backdrop of the water body.

Pond

Pond is a simple form of waterscape. It is a re-creation of either the regular-shaped pool usually found in the grounds of a temple, ancestral hall, guildhall, school, or private residence, or the natural lotus pond or fishpond commonly seen in the rustic

[57] Achieve more with less.

Fig. 3.46 Regular-form pond in Jiangnan's countryside—prototype in life of the regular pond of garden art

countryside (Fig. 3.46). At such, there are two types of pond making in the Jiangnan garden: one that is of the regular or geometric form, and the other the free or natural form. The geometric-form pond is mainly rectangle in shape and built with clean-edged bank revetment made mostly with slab stones, cut stones, flagstones, or other regular- or irregular-shaped stones. This type of pond is usually set on the central axis of the main hall of a building complex or within the entourage of a courtyard, such as the one in the court of the ancestral hall adjacent to Garden for Ease of Mind (Jichang Yuan) in Wuxi, as well as *Tianxia Di'er Quan*[58] at Mount Hui in Wuxi and "Jade Spring" (Yu Quan) in Hangzhou. The geometric-form pond is also found in naturalistic landscape gardens (Fig. 3.47).

Little Ink Pool (Xiao Mo Chi) opposite the study in the garden of Wan's residence on Wangxima Lane in Suzhou is one such example. The pool in Qu Yuan, the garden

[58]The Second Spring under Heaven.

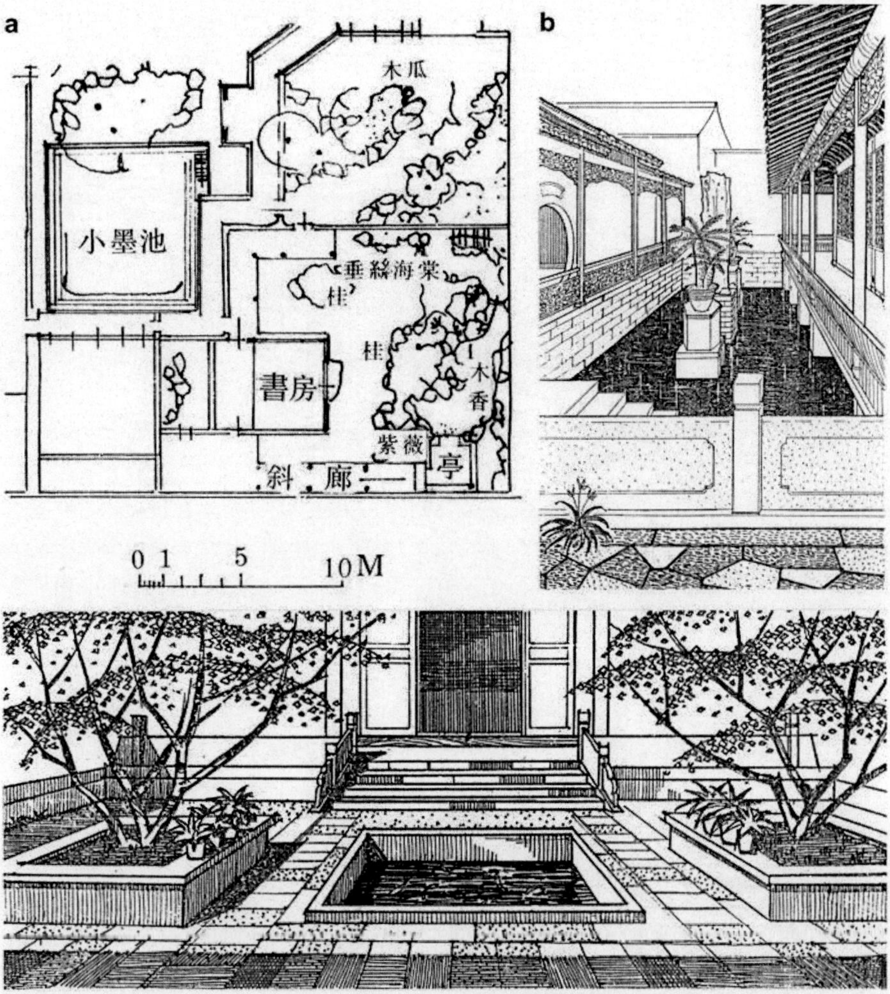

Fig. 3.47 (**a**) Regular-form pond (geometric-form) in the garden of Wan's residence on No. 7 Wangxima Lane in Suzhou—ichnograph of "Xiao Mo Chi"(Excerpted from Suzhou Classical Gardens). (**b**) Regular-form pond in a residential garden on Lu Xun Street in Shaoxing, Zhejiang Province (Excerpted from Popular Dwellings in Zhejiang). (**c**) Regular-form pond in Suzhou's Qu Yuan. (**d**) Regular-form pond in the court garden of Tianyi Ge in Ningbo, Zhejiang Province (Excerpted from Popular Dwellings in Zhejiang). (**e**) Regular-form pond in a residential garden in Wenzhou, Zhejiang Province (Excerpted from Popular Dwellings in Zhejiang). (**f**) Regular-form pond combined with regular-form planting bed and potted landscape in a residential garden of Fang Yiren's Drugstore in Linhai, Zhejiang Province (Excerpted from Popular Dwellings in Zhejiang)

d

e

Fig. 3.47 (continued)

of Yu's residence on Mayike Lane in Suzhou, and the pond south of the Thick Pines in Deep Mist (Fansong Aishen) building in Garden of Benefiting Descendants (Huiyin Yuan) all belong to this type. Ponds of the Jiangnan residential garden are mostly of the naturalistic free form. Compared with lake making, the natural-style pond has a relatively more foursquare plan with a shoreline smoother and less embellished such as with stone setting. The pond usually features no island or bridge but rather is grown with water lilies or stocked with algae and ornamental fish. The water body created in front of Hall of Listening to the Oriole (Tingli Guan) in Xu

f

Fig. 3.47 (continued)

Fig. 3.48 The pond in front of "Tingli Guan" in Xu Yuan at Yangzhou's Slender West Lake

Yuan at Yangzhou's Slender West Lake is an example of the natural-style pond (Fig. 3.48).

Lake

This is the most frequently used water feature in Jiangnan's *zhai yuan* (residential garden), or residential gardens. Whenever circumstances allow, the lake is employed as the core of the landscape composition for the entire garden; in other words, the garden's sceneries are mostly planned around a concentrated span of lake waters. A

Fig. 3.49 The earthen lake bank scattered with natural stones in Zhuozheng Yuan of Suzhou

lake is essentially a natural-form pond only more sophisticated in configuration. Generally, the lake has a more flexuous shoreline, intended to create a perspective of hidden inlets and branching streams. To simulate a natural lake, the bank of the garden lake is either laid entirely with natural stones or with earth alongside a few natural stones interspersed strategically. The bank's elevation is designed to near the level of the water so as to generate the effect of brimful waters lapping the shore of an expansive lake. To realize riparian scenery, the lake is often stylized with island, promontory, stepping-stone, towpath, bridge, dock, dike, or other landscape features characteristic of the Jiangnan region.

Examples of the earthen lake bank scattered with natural stones include the original creation of the "Embracing Beauty Lake" (Jin Hui Yi) lake embankment in Garden for Ease of Mind (Jichang Yuan) of Wuxi and the shore treatment with the earthen hill that banks the lake in the lake-and-mountain landscape sequence created in the Humble Administrator's Garden (Zhuozheng Yuan) of Suzhou (Fig. 3.49).

Lakes that are banked entirely with natural stones are most commonly found in the extant gardens from modern times, such as Little Lotus Manor (Xiaolian Zhuang) in Nanxun (Fig. 3.50).

Fig. 3.50 Naturally cavernous banks created with Lake Stones are piled in a natural style in Xiaolian Zhuang in Nanxun

Yi Yuan in Taicang, Xian Yuan in Mudu of Wu County, Villa of Reposing Roars (Jixiao Shanzhuang) in Yangzhou, Villa of Fenyang (Fenyang Bieshu) in Hangzhou, and Suzhou's Master-of-Nets Garden (Wangshi Yuan), Garden of Pleasure (Yi Yuan), and Garden that was preserved miraculously (Liu Yuan). For lakes that feature islands, the one in the Humble Administrator's Garden (Zhuozheng Yuan) boasts the largest island of its kind, created to simulate the lake-and-mountain scenery formed by the natural landmasses, named "Maji" and "Dongting", that jut from the waters of the Taihu Lake (Fig. 3.51).

In Yi Yuan of Nanxun, the succinct composition of the big lake and large island (Fig. 3.52) very much evokes the charm of "Garden of Mist and Rain Pavilion" (Yan Yu Lou) at the South Lake in Jiaxing.

Islands such as this in Nanxun's Yi Yuan—usually planted with vegetation—are in effect a manifestation of an oasis, a relief to the wide stretches of the waters. There are also lakes featuring small-scale islets, such as those in Suzhou's Garden that was preserved miraculously (Liu Yuan) (Fig. 3.53) and Jiading's Garden of Autumnal Glow (Qiuxia Pu) near Shanghai (Fig. 3.54). Promontories, another shoreline treatment, have been discussed in the section on "Shoal and Promontory" under "Basic Mountain Features".

Additionally, to set off lake scenery, bridges of various styles are often placed at tactical locations such as a narrow point of the waters, an inlet or bay, or where a river mouth is insinuated. In-depth discussion on the bridge will be conducted in "Placing Architectural Structures" in the proceeding segment.

Residential gardens in the Jiangnan region are restricted in space and do not have the acreage to luxuriate in great expanses of water. Garden makers are therefore

Fig. 3.51 Islands in nature—islands jutting from Dongting Dongshan island in the Taihu Lake

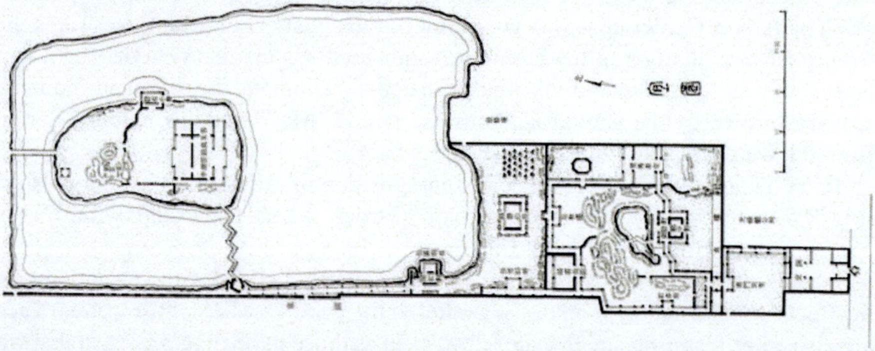

Fig. 3.52 Ichnograph of Yi Yuan of Nanxun, Huzhou, Zhejiang Province (Excerpted from A Miscellanea on Jiangnan's Gardens)

prodded to devise countermeasures to reclaim water coverage wherever possible. The straight embankment is one such solution: the upright revetment shortens the sloping section of the bank and allows accommodation of extra water coverage, resulting in an increase in the absolute dimensions of the water (Fig. 3.55).

In adopting this technique, however, integrated planning should be taken to include certain waterfront features such as a waterside pavilion or terrace. Besides the perpendicular bank, another bank treatment used to increase the span of water, or

Fig. 3.53 Hua lang combined with a bridge—wisteria trellises set atop the zigzag bridge leading to the "Little Penglai" island in Suzhou's Liu Yuan

Fig. 3.54 Islet in the lake in front of San Yin Tang in Jiading's Qiuxia Pu near Shanghai

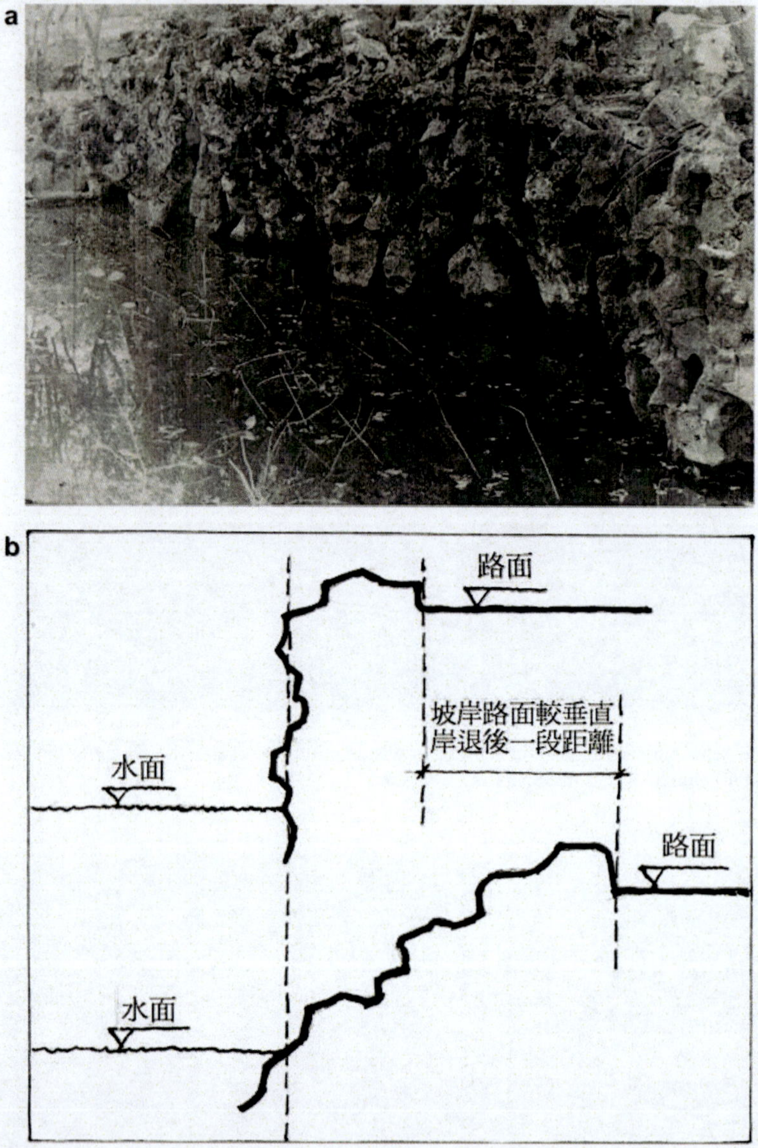

Fig. 3.55 (**a**) Straight embankment piled with Lake stones in Xiaolian Zhuang in Nanxun. (**b**) Comparative diagram for analysis between upright bank and sloping bank

rather the impression of such, is to arrange for natural stones to hang protrusive over the edge of the water (Fig. 3.56).

This is precisely an opposite approach to the perpendicular bank treatment. The protrusive stone bank that appears to obstruct partially the view of the water is

Fig. 3.56 Protrusive stone bank in Canli Yuan in Suzhou (Excerpted from Suzhou Classical Gardens)

intended to mask the edge of the water and create the illusion of deeper and further-reaching and yet self-effacing waters. The two very different treatments of the water's edge nevertheless achieve the same objective, which is to maximize the use of space and bring out the effect of greater water coverage in a restricted garden space.

River

Band-shaped water bodies are intended to delineate riverine scenery. They aim to simulate the natural watercourse in the Jiangnan country that teems with rivers and canals. The river created in Suzhou's Garden that was preserved miraculously (Liu Yuan) on the south of the fan-shaped pavilion "Vivacious Land (Huopopo Di)" and those in Garden of Autumnal Glow (Qiuxia Pu) and Huilong Tan Park, both gardens in Jiading, are a few examples of man-made riverscapes (Fig. 3.57).

This type of river is mainly earth-banked, with a few natural stones placed sporadically. The serpentine river course is designed to elongate the passage of the flow and animate the composition of the waterscape, evoking a long winding stream from a remote source. The band-shaped water feature in eastern Yi Yuan in

Fig. 3.57 (**a**) The river created on "Huopopo Di" (The Vivacious Land) in Suzhou's Liu Yuan. (**b**) River in Qiuxia Pu in Jiading. (**c**) River in Huilong Tan in Jiading

Nanxiang and eastern West Garden, located in the temple grounds of Jiechuang Si Temple (Monastery of Ordination Pennant) in Suzhou, are also this type of creation.

Mountain Stream

Among various water features of the Jiangnan garden, mountain stream is another manifestation of the ribbon-shaped water body. Like ravine and gully,[59] mountain stream is a landscape structure comprising interdependent mountain and water features. The difference, however, lies in that the ravine is stressed to express the spatial structural features of mountains and valleys and does not necessarily maintain the presence of waters, while in the case of the mountain stream, the focus is on the description of the waters in a mountainous region. Unlike making a river on a level land, the mountain stream is all laid with natural stones, creating an image of a rocky riverbed and rugged denuded stones scoured by the rushing torrent. The mountain stream created at Pavilion of Pagoda Reflection (Ta'ying Ting) in western the Humble Administrator's Garden (Zhuozheng Yuan) in Suzhou is one such example (Fig. 3.58).

Haopu

Haopu, the water-suffused gorge, is yet another waterscape hinged on the presence of mountain. As with the mountain stream, water is the focus of this landscape expression, hence it is discussed here as a water feature. In the circumstances where the water level is relatively low and the water body is of a long, narrow shape, *haopu* is often the landscaping solution. The water body is usually flanked by high rocky banks on both sides, forming a scene of steep-mountain sandwiched waters. In creating a *haopu*, the bank revetments are piled with natural stones, planted with vegetation that hangs down, climbs up, or crawls sideways, all lending itself to creating a secluded atmosphere of a deep gorge. A more dramatic effect of "towering mountains and deep waters" can be achieved by increasing the stature of the piled rocky banks in addition to capitalizing on the low water level. The eastern section of Couple's Retreat Garden (Ou Yuan) in Suzhou has a relatively high terrain and low water level. Exploiting the given site conditions, the waters are configured into a *haopu*-style water feature. This is a rather successful example also in terms of using vegetation to accentuate the characteristics of the *haopu* structure (Fig. 3.59).

The *haopu* also features a slabstone bridge set high above the water, lending a sense of deepness to the gorge. The elevated bridge, however, is not received well as it contradicts the view of "a small, low-lying bridge set close to the surface of the water", which is conventionally applied in making a Suzhou garden. Under this generalized, subjective conclusion that neglects the specific content of the water feature—the *haopu* in this case, the bridge was censured as a "faulty stroke" in the otherwise fine production and proposed to lower the height to conform to the general

[59] Discussed earlier in "Basic Mountain Features".

a

b

Fig. 3.58 (**a**) Natural stream in Si Ming Mountain in Yinxi, Zhejiang Province. (**b**) Rushing torrent on different levels of natural stream in Si Ming Mountain in Yinxi, Zhejiang Province. (**c**) The mountain stream created at Ta'ying Ting (Pavilion of Pagoda Reflection) in western Zhuozheng Yuan in Suzhou

c

Fig. 3.58 (continued)

aesthetic standard. It would have been amiss should that be done. Nevertheless, as the slabstone bridge was built in the form of a zigzag track bridge, the "faulty stroke", if any, should be the zigzag design rather than its elevated height as the zigzag line fragments the sense of unity of the scene (Fig. 3.60).

In the case of a higher water level, a *haopu* is also feasible by elevating the banks, the similar technique in raising mountains around waters.

Sunken Pool

A concentrated water body in a confined snug space is the key to creating the sunken pool. The deep-set water feature should be built in a shady rather than light spot, banked with stone rather than earth, and filled with water at a low elevation rather than flush with the bank. The pool in Suzhou's Garden of Listening to the Maple (Tingfeng Yuan) placed below Ban Ting (Half Pavilion) (Fig. 3.61a) and the one overlooked by the newly rebuilt "Bingli Tang Hall" in Wuxi's Garden for Ease of Mind (Jichang Yuan) are two such examples. Some sunken pools are built with stone walls that are jaggedly layered and set at descending stepped heights, making the pool waters accessible to the curious visitor, as in the case of Suzhou's Garden of Particles (Canli Yuan) (Fig. 3.61b) and Lin's residential garden on Xinmalu Road outside Suzhou's city gate—Chang Men. In Villa of beautiful surrounds (Huanxiu Shanzhuang), Suzhou, the little pool ensconced by the side of the pavilion with the inscription "Bantan Qiushui Yifang Shan"[60] is but a small creation of its kind that affords abundant delight (Fig. 3.61c).

In the western section of Garden of Surging Wave Pavilion (Canglang Ting), also in Suzhou, a depression in the rocky col below Pavilion of the Imperial Stele (Yubei

[60]Roughly "Half-filled pool of autumn waters by the house of the hill".

a

b

Fig. 3.59 (**a**) Haopu-style water feature in the eastern section of Ou Yuan in Suzhou. (**b**) Haopu-style water feature in Ban Yuan in Suzhou

Ting) is configured into a low-lying sunken pool surrounded by precipitous piled rocks, giving the desired impression of overlooking a deep chasm. Despite the stone arrangement that appears to be fragmentary, it is overall a successful creation of this water feature (Fig. 3.61d).

Some scholars take it as a "failed example" and hold that the banks of the pool are too high—an opinion based on the general idea of the "pond shore" design, neglecting the specific content of the sunken pool, a rather different water feature in concept from the pond. Other examples of the sunken pool include the deeply enchanting "Seal Spring" (Yin Quan) in the garden of Xiling Seal-engravers' Society (Xiling YinShe) garden in Hangzhou, which nestles in the shade beneath a

Fig. 3.60 It is appropriate that the haopu features a slabstone bridge set high above the water in the eastern section of Ou Yuan in Suzhou, however, the form of a zigzag track bridge fragments the sense of unity of the scene

clump of bamboo, and the *xie yi*-style water hole at the Gully of Eight Sounds (Bayin Jian) in Wuxi's Garden for Ease of Mind (Jichang Yuan) (Fig. 3.62).

Fountain Spring

There are two approaches in making fountain springs in the Jiangnan garden. One is to enhance the presence of an existing natural spring through artistic processing, and the other to create a naturalistic spring by means of artful emulation. The former is mostly seen in gardens located in natural scenic areas while the latter in *zhai yuan* (residential garden), residential gardens, in urban districts. For the artistically retrofitted natural spring, two different treatments are adopted: one resorts to architectural styling and creates a geometric pool with a stone-carved animal's head—a dragonhead most commonly—mounted on the pool's wall to let the spring water drop from the mouth of the animal (Fig. 3.63).

"Tianxia Di'er Quan"[61] at Mount Hui in Wuxi and "Jade Spring" (Yu Quan) in Hangzhou are two such examples. This is also a traditional style. On murals in the

[61] The Second Spring under Heaven.

Fig. 3.61 (a) Sunken Pool in Suzhou's Tingfeng Yuan. (b) Sunken Pool in Suzhou's Canli Yuan (Excerpted from Suzhou Classical Gardens). (c) The little pool ensconced by the side of the pavilion

Palace of Eternal Pleasure (Yongle Gong) from the Yuan dynasty, located in Yongji County, Shanxi Province,—a contrivance that can be traced to Emperor Wu's chalcographic fountain of "brass dragon that spits the water into the cup held in the mouth of a brass fairy where the water overbrims and falls down" during the Western Han dynasty. The other treatment assumes a more natural style. The fountain spring at Yunquan Jingshe[62] in Mount Tianping, Suzhou, for instance, adapted to the lie of the natural mountain, is made into a sunken spring pool embellished with natural stones and encircled by rustic stone railing. The famous fountain spring Hupaoquan at the West Lake in Hangzhou is another example that adopts this natural-flavored treatment.

Contrary to the relatively contained styles of the "processed" natural spring, the nature-inspired, manmade spring feature is created for urban residential gardens. A related subject on the mountain spring has been dealt with in the segment on "Mountain Spring and Waterfall" under "Basic Mountain Features", which discusses the need to command a certain altitude in order to effect substantial falls. Few springs made for Jiangnan gardens have the desired head of water, given the relatively plane terrain of the region, Gully of Eight Sounds (Bayin Jian), in Garden for Ease of Mind (Jichang Yuan) in Wuxi, becomes a rare example. Although there is only a small volume of the water, the chiming sound of Gully of Eight Sounds (Bayin Jian) vividly creates the image of "the clear spring runs over the rocks" (Fig. 3.64).

As a result of insufficient drop, the mountain spring is normally manifested by rock formation alone, only when the rain falls can the scene of a spring running down the hillside be viewed. The spring design in the existing Jiangnan gardens is mostly combined with the hillside rockery that features a stone cavity or mini pool at the foot of the hill, evoking an opening where the water seems to well up spring-like from the dark, unfathomable depth, as exemplified by "Flying-snow Spring" (Feixue Quan) in Villa of beautiful surrounds (Huanxiu Shanzhuang), Suzhou (Fig. 3.65).

"Flying-snow Spring" (Feixue Quan) at the foot of the Lake-stone rockery in the northwestern garden is a rare natural spring available in an urban residential garden, whose story is recounted in *Feixue Quan ji*—Notes on the Flying-snow Spring (Feixue Quan).[63] Other examples include the "Han Bi" spring in Suzhou's Master-of-Nets Garden (Wangshi Yuan), which is created in the fashion of the sunken pool, and the spring on the bank nearby the pavilion-capped bridge in Wuxi's Garden for Ease of Mind (Jichang Yuan), whose surroundings have unfortunately been ruined and lost the *yi jing* appeal.

Fig. 3.61 (continued) with the inscription "Bantan Qiushui Yifang Shan" (roughly "Half-filled pool of autumn waters by the house of the hill") in Huanxiu Shanzhuang, Suzhou (the pool is dry). **(d)** Sunken pool in the western section of Canglang Ting in Suzhou

[62]Exquisite House of Clouds Spring.

[63]The spring has been silted up since modern times though.

Fig. 3.62 (**a**) Little sunken pool in the Xiling Yinshe garden in Hangzhou—"Wen Quan" "Xian Quan". (**b**) Little sunken pool in the Xiling Yinshe garden in Hangzhou—"Yin Quan". (**c**) The xie yi-style water hole at Bayin Jian in Wuxi's Jichang Yuan. (**d**) Sunken pool in the Suzhou's Wangshi Yuan—"Leng Quan". (**e**) Little pool under the lake-stone mountain along the Moxiang Ge (Tower of fragrance of ink) in Suzhou's Tingfeng Yuan

c

d

Fig. 3.62 (continued)

3.1.1.1.2.2 Combination of Various Types of Water Surface

Generally speaking, given the small garden site typical to the Jiangnan *zhai yuan* (residential garden), creation of waterscape usually resorts to the single water-feature treatment: a pond, a lake, a sunken pool, or a mountain stream. Wherever there is space for maneuver, however, the combination of multiple water features is always resorted to enhance the scenery's variation. In *shanjing yuan*—gardens where mountain is the primary component and waters the secondary are mostly created into a band-shaped, circular water body such as a river and mountain stream, or as a small, concentrated water space such as a sunken pool. Examples include the stream

e

Fig. 3.62 (continued)

around the stone mountain in Suzhou's Villa of beautiful surrounds (Huanxiu Shanzhuang) (Fig. 3.66) and the sunken pool between the stone mountains in western Garden of Surging Wave Pavilion (Canglang Ting).

In *shuijing yuan*—gardens where waters are the primary component—or *shanshui yuan*—gardens where mountain and waters have equal presence, waters are mostly executed into a lake foiled by other water features. As with other garden

Fig. 3.63 A geometric pool in "Yu Quan" in Hangzhou, adopting a stone-carved animal's head as the mouth of spring

Fig. 3.64 Falling of spring in Bayin Jian, in Jichang Yuan in Wuxi

Fig. 3.65 "Feixue Quan" in Huanxiu Shanzhuang, Suzhou is built as a stone cavity

features, creating water bodies also has the tendency to fall in readily reproducible formulae. Typically, the principle water body is joined by one or more subordinate water bodies. For example, a large span of lake waters takes the center stage while one or more band-shaped rivulets or mountain streams are created as subordinate features. Here, the open space of the extensive lake is the principle view of the entire

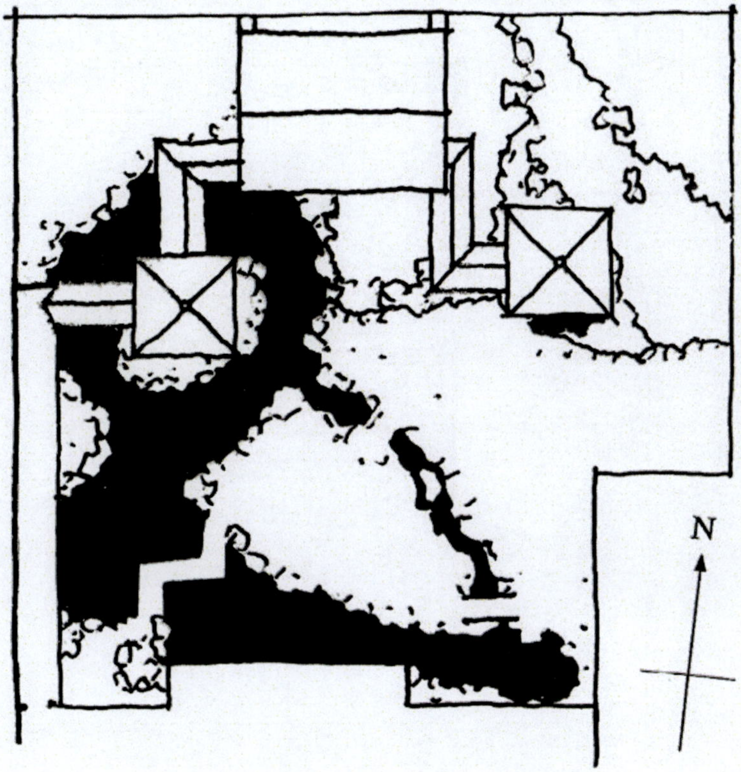

Fig. 3.66 Schematic plan of water feature in Suzhou's Huanxiu Shanzhuang

garden, and if the garden also features hills, the key scene should be arranged here in affinity with the waters. The principle hall of the garden, if any, should be placed here as well by the water (usually connected to the water by an ante-platform), facing the key mountain feature that is on the ulterior side of the water.

As for the lake itself, a more flexuous shoreline should be created to lend a perspective of hidden inlets and branching streams. If site conditions allow, extended water features, such as a rivulet or mountain stream, should be developed to establish additional scenic spaces. In Suzhou's Garden that was preserved miraculously (Liu Yuan) and the Humble Administrator's Garden (Zhuozheng Yuan), lake is the principle water feature of the garden and both lakes feature islands, which serve to increase the depth of field and evoke a sense of distance (Fig. 3.67).

In the case of Garden that was preserved miraculously (Liu Yuan), the subordinate water feature is a mountain stream, which is connected to the northwest of the principle lake. The offshoot stream not only insinuates the source and course of the lake but also, combined with piled stones, creates yet another garden space of a mountain-and-waters landscape. The two islands in the waters of the Humble Administrator's Garden (Zhuozheng Yuan) are significative of the lake-and-

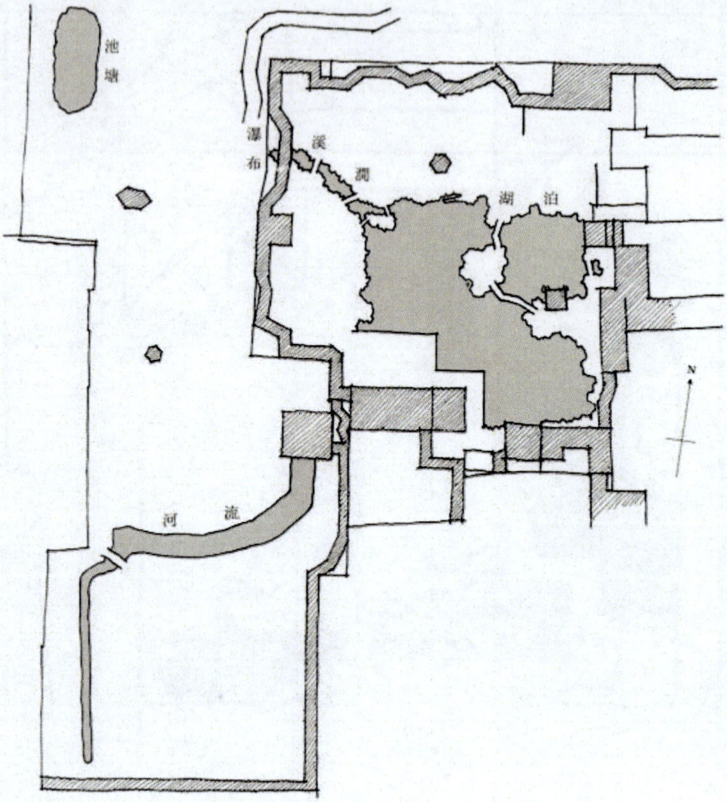

Fig. 3.67 Representative schematic plan of combination of water features—including lake, streams, waterfall, river and pond (Liu Yuan in Suzhou)

mountains scenic theme. Representing a mountain stream, the narrow channel formed by the two facing islands offsets the wide, open lake waters. In addition, a rivulet is created in the southeastern corner of the lake; it is a rather short rill and stops at "Crabapple Spring Hall" (Haitangchun Wu). At the mouth of the rivulet, a flagstone bridge is placed in imitation of the *lukuahe* bridge, a traverse road that crosses over the offshoot river on the lakeside promenade, a structure typically seen in regions teeming with rivers and lakes such as Jiangnan. The flagstone bridge also helps tone down the presence of the rivulet and creates an impression of a long stream coming from afar. The lake waters stretch further toward the south where more coves and inlets are formed. The lake in the Master-of-Nets Garden (Wangshi Yuan) of Suzhou also branches off a small stream in its southeastern corner, where a stone arch bridge, designed to accommodate sailboats, is placed at the entry of the lake, while the northwestern corner of the lake is configured into an inlet (Fig. 3.68a).

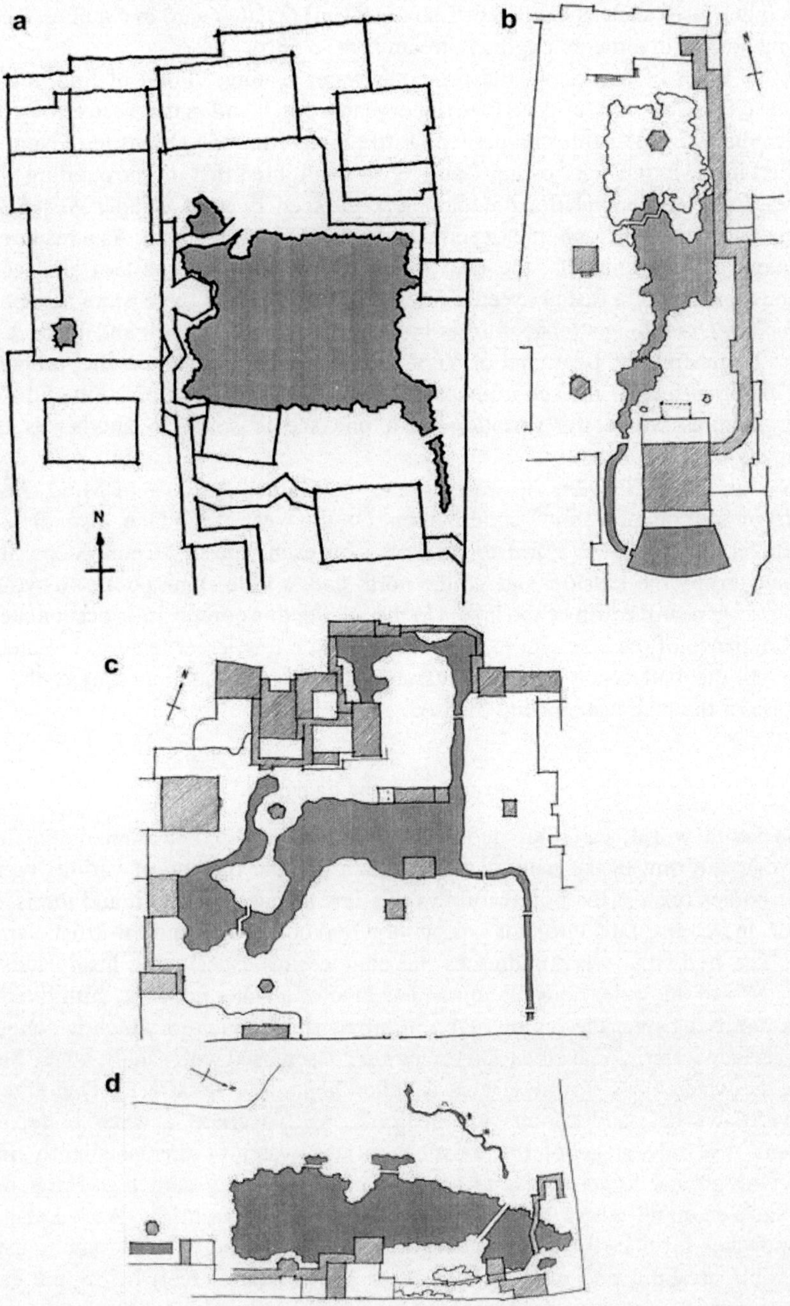

Fig. 3.68 (**a**) Schematic plan of water feature in Suzhou's Wangshi Yuan. (**b**) Layout plan of water feature in Nanjing's Zhan Yuan. (**c**) Layout plan of water feature in Nanxiang's Yi Yuan. (**d**) Layout plan of water feature in Jichang Yuan in Wuxi

In other cases such as Garden of Look Forward (Zhan Yuan) in Nanjing, the lake is combined with a much lengthier stream (Fig. 3.68b).

In Yi Yuan of Nanxiang, the principle water feature, “Pond of Sporting with Geese” (Xi’e Chi), is in effect a lake rather than a pond, and in the western part of the waters sits a sizable peninsula named “Little Pines Mound” (Xiaosong Gang). The “Little Pines Mound” (Xiaosong Gang) is so configured that its surrounding waters form a circuitous composition comparable to the scenery at the Slender West Lake of Yangzhou, imparting a unique experience for boating on the lake. As a result of the treatment of the peninsula, the lake waters are divided into eastern and western sections, each with a distinct scenic interest. On the east of the eastern section (the bigger side), two long streams further branch off, extending north and south respectively. In general, the treatment of Yi Yuan’s waterscape offers parallel preponderance to both principle and subordinate features, presenting a combination of different water features where the variation from one scenic space to another is rather conspicuous (Fig. 3.68c).

In contrast to Yi Yuan, the water scenery in Garden for Ease of Mind (Jichang Yuan) of Wuxi is of a subtle combination. On the west of the lake, a small *haopu*-style inlet is placed—as a foil to the lake—on each side of “Crane Steps Shoal” (Hebu Tan) by the hillside, and on the north end, a little stone pool or a symbolic spring is set near the arm of the lake. On the whole, this composition accentuates the predominance of the lake while, on a closer look, a few lesser features are incorporated into the waterscape to provide variations and break the monotony of the large expanse of the lake waters (Fig. 3.68d).

3.1.1.1.2.3 Expressing the Source and Flow of Water

In the natural world, waters are defined by the source and course as are mountains by the ridge and run. In the naturalistic Jiangnan garden, making of various types of water bodies (except the pond) follows this law of nature as well, and the issue of source and course falls into two aspects: the hydraulic force and the artistic expression. The hydraulic work addresses the engineering aspect—the headwaters and drainage—of the water body. Jiangnan is a land of a water network, with rivers and lakes ramifying over the region, even in urban areas, much of which is connected with surface water. Some cities and towns are crisscrossed with canals, where houses flank the watercourse, forming what is called hejie, a “river street”. There are also plenty of washes, and springs in rare locations. To create a water body, many gardens take advantage of the existing surface water or dredge slightly in the low-lying ground to get access to water. The Guo Zhuang garden in Hangzhou is one such example where the pool water is channeled from the West Lake. The groundwater level in the Jiangnan region is relatively high, and water is usually accessible at about one meter (about three and a quarter feet) below the earth’s surface, making it rather effortless to create a water body even by digging on a level ground. In making Jiangnan gardens, a wealth of hydraulic experience and techniques has been accumulated in maintaining both the quantity and quality of garden waters. A water body is usually made to be connected to surface running water outside the garden; this way, it not only helps generate the flow and scour of the water that maintains water quality, but also enables effective escape of excessive water in the garden during a rainy season.[64] For garden waters that cannot be connected to an external running watercourse, many of them are raised with fish to absorb microbes to prevent the water from going stale. To maintain certain level of cleanness of the water, another technique is applied: sinking a well in the bottom of the water. Termed "quanyan" (spring's mouth) by locals, the subaqueous well is able to draw water from the superficial groundwater; this way the surface water in the garden gets to be connected with the fluid subterranean water, hence helping ameliorate the water quality. According to a survey on ancient gardens conducted by Suzhou's department of cultural relics and garden management, two "quanyan" wells were discovered in the bottom of the pond in Garden of Pleasure (Yi Yuan); they are also found in the Humble Administrator's Garden (Zhuozheng Yuan), Lion Grove Garden (Shizi Lin), and Garden for Ease of Mind (Jichang Yuan). In Shanghai's Garden to delight one's parents (Yu Yuan), three such subaqueous wells are sitting in the floor of the pond, all of which are over three meters (about ten feet) deep. It is worth notice that natural wells in Jiangnan's residential gardens have become very rare nowadays.

From the perspective of landscaping, the source and course of various water bodies, with the exception of the pond, also need to be artistically expressed. Take the lake as an example: to signify where the water comes from and where it goes off, the lake—as aforesaid in the last segment of *Combing Water Features*—is often designed to be conjoined with a branching stream or water outlet, some of which are even extended to become a river or mountain stream. At such, the treatment of the end of a stream is key to the artistic creation of the source and course. There are two kinds of execution that expresses the source of water. One is to create a pseudomorphic intake by means of strategic stone setting that simulates the source of natural surface water; examples include the mountain stream next to Veranda of Scenting the Sweet Osmanthus Fragrance (Wen Muxixiang Xuan) in Suzhou's Garden that was preserved miraculously (Liu Yuan) and the stone arrangement at the river branch in Jiading's Garden of Autumnal Glow (Qiuxia Pu) (Fig. 3.69a). This execution is also often combined with the rockery design by the edge of a water body. As discussed earlier in "Fountain Spring" under the section on "Basic Water Features", the pseudomorphic intake takes the form of a fountain spring where the stone cavity signifies the mouth of a welling spring (Fig. 3.65).

The other execution is to create an actual intake mechanism, such as a water culvert, through which the outside water is channeled into the garden. A good example is the rockery tunnel in Hangzhou's Guo Zhuang garden that draws the West Lake water into the garden (Fig. 3.69b). To express the passage of water, three techniques are commonly adopted. One is through architectural device, such as

[64] Of late decades, however, many of such water passages have been silted up, dumped with waste, or even filled up and converted into roads. Subsequently, the pool and pond waters in many gardens have largely become stagnant.

a
b

Fig. 3.69 (**a**) Pseudomorphic intake by means of the stone arrangement at the river branch in Jiading's Qiuxia Pu. (**b**) The rockery tunnel in Hangzhou's Guo Zhuang that draws the West Lake water into the garden

placing at the end of a stream a waterside pavilion that has a cantilevered platform hanging suspended over the water (Fig. 3.70a) or a piled structure that rises above the water (Fig. 3.70b).

Such a contrivance allows the water to bite into the cavity under the building and conceals the very end of the water, creating an illusion of continuity of the stream underrunning the building to the ulterior side. Examples of such type of architectural device include "Crabapple Spring Hall" (Haitangchun Wu) in the Humble Administrator's Garden (Zhuozheng Yuan), the fan-shaped "Vivacious Land (Huopopo Di)" pavilion in Garden that was preserved miraculously (Liu Yuan), and the

a

b

Fig. 3.70 (**a**) The waterside pavilion's platform hanging over the water in Hu Yuan in Suzhou, creating an illusion of continuity of the stream. (**b**) Pavilion and corridor hanging over the water in Changyuan Ting in Suzhou, creating a sense of continuity of the stream. (**c**) A concave arch is set at the end of the garden stream, suggesting a continuing flow of water streaming through under the arch in Ban Yuan, Suzhou. (**d**) Using vegetation to obscure the terminal of the flow to create an illusion of continuity of the stream in Wenlan Ge in Hangzhou

c

d

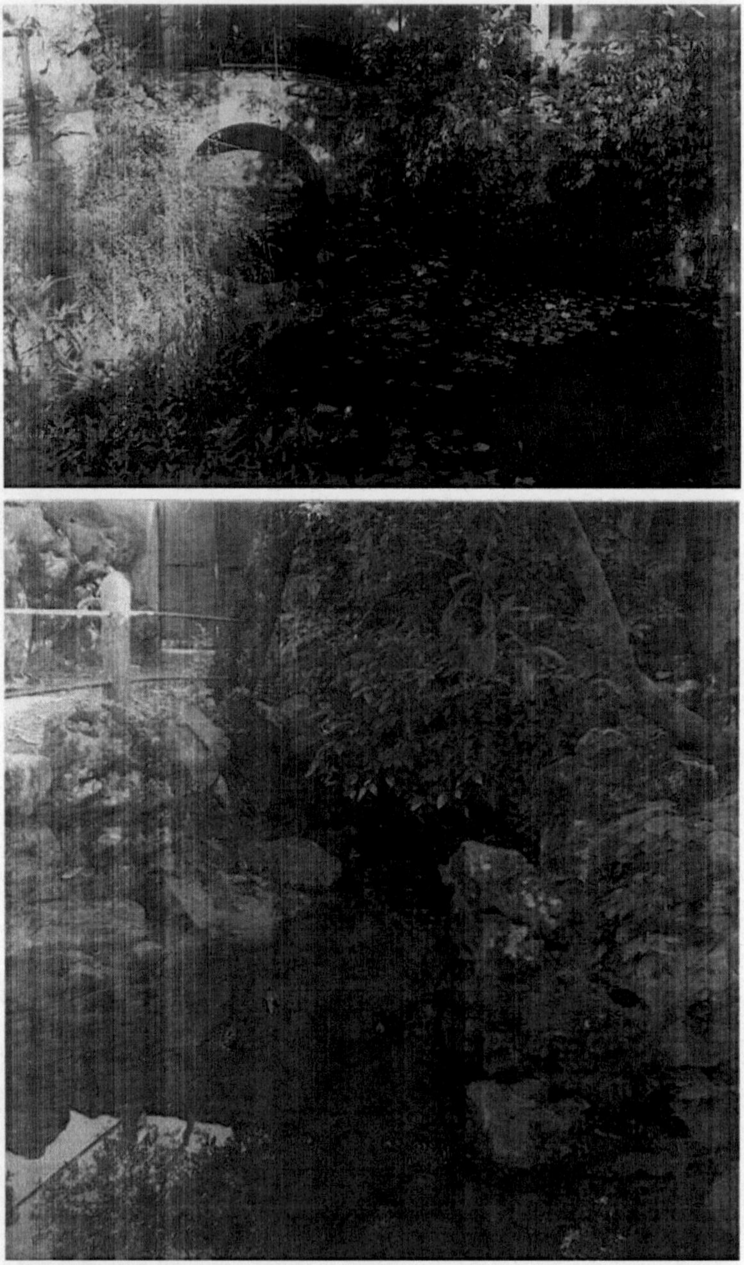

Fig. 3.70 (continued)

waterside pavilion in Garden of Miniature Landscape as if Contained in a Pot (Hu Yuan), all in Suzhou. On the same principle, arches can be used as a masking device as well, as in Half Garden (Ban Yuan) of Suzhou, where a concave arch is set at the end of the garden stream, suggesting a continuing flow of water streaming through under the arch (Fig. 3.70c).

Another commonly applied technique for the course's treatment is using vegetation to obscure the terminal of the flow. By letting it disappear amid the density of the plants, the intended arrangement evokes a lengthier watercourse that seems to stretch far beyond what the eye can see (Fig. 3.70d).

The use of piled stones is yet another technique often employed in the treatment of a course's end, where the stones are arranged into a stone culvert that symbolizes the passing, or the source, of a stream, depending on the terrain where the arrangement takes place. For instance, if arranged at the foot of a hill, the culvert represents the source (as demonstrated by the mountain stream at the foot of the hill by Veranda of Scenting the Sweet Osmanthus Fragrance (Wen Muxixiang Xuan) in Suzhou's Garden that was preserved miraculously (Liu Yuan)) or, if arranged in a low-lying flat landscape, it manifests the passing of the stream (such as the treatment of the rivulet's end near Ta'ying Ting—Pavilion of Pagoda Reflection—in western Humble Administrator's Garden (Zhuozheng Yuan)). Such stone arrangements are often accompanied by plantings of vegetation to add a shady and secluded effect.

Depending on the requirement of the composition, a stone culvert is also created as a part of the rockery formation in a waterscape that seeks to simulate natural rock formation as in the Guizhou and Guangxi regions. The culvert forms the visual focal point of the composition, reminiscent of the motion of the water passing through, which is yet another means of expressing source and course. Examples include 1) the aforementioned rockery water tunnel in Guo Zhuang garden of Hangzhou (Fig. 3.71); the rocky opening at "Little Red Cliff" (Xiao Chibi) (Fig. 3.72a) and the water cave opposite the *hehuating* (lotus hall) (Fig. 3.72b), both in Lion Grove Garden (Shizi Lin) of Suzhou; and the stone arch striding the water in Garden of Pleasure (Yi Yuan) of Suzhou (Fig. 3.72c).

For the same purpose, certain aquatic architectural structure such as *shui hua qiang*,[65] which often features an opening or entry such as a moon gate, is incorporated into the waterscape to evoke association with water or a passing boat. For instance, there are a couple of ponds in Shanghai's Garden to delight one's parents (Yu Yuan) that feature the whitewashed *shui hua qiang* striding across the water, each accented with a moon gate aptly called *shui yue dong*.[66] Such an architectural treatment not only lends a dynamic artistic effect to the aquatic composition but also creates the impression of a fluid and animated source and course that will otherwise be a stagnant water body.[67]

[65] Water ornamented wall.

[66] Water moon gate.

[67] *Shui hua qiang* is discussed more in detail in the next section, "Placing Architectural Structures".

Fig. 3.71 (**a**) Nanmu Ting from the Ming dynasty, located in Pianshi Shanfang in Yangzhou. (**b**) Poxian Qinguan in Yi Yuan, Suzhou

Fig. 3.72 (**a**) The stone arch striding the water in Shizi Lin in Suzhou—“Xiao Chibi”. (**b**) The stone arch striding the water opposite the hehuating in Shizi Lin in Suzhou. (**c**) The stone arch striding the water in Yi Yuan of Suzhou

c

Fig. 3.72 (continued)

Generally, lack of control of water volume in Jiangnan gardens results in fluctuating water levels: ponds and lakes overbrim during a rainy season while their water levels drop markedly as heat greatly evaporates the water during a dry season. This is much to the disadvantage of creating water bodies, especially pond, lake, and river. To solve this dilemma, Jiangnan garden makers have ingeniously invented the descending stepped stone setting along the bank of the water body.[68] This treatment allows the piled stones to slope to the water at descending stepped elevations and thus ensures a brimful effect of waters at any water level.

Water is indispensable to a Jiangnan garden. Even in a *ting yuan*, or court garden, which is usually a strip of narrow space shorn of conditions to place a water body, the least capacity of water is deployed however feasible to counteract its entire absence. The technique involves embedding a pot or vat in the ground of the courtyard and arranging a few natural stones around the edge of the container, thus to create a little atmospheric spot scene of "a bed of clear water floated with two or three duckweeds". Most of the occasions, though, simply a goldfish pot, a lotus vat, or a potted landscape is displayed in the courtyard for viewing, dispensing with the natural landscape treatment.[69] In a garden bare of water, such spot scenery or display is essential and, from the decorative point of view, plays a vital role as the finishing touch to the garden ensemble. If a garden can be likened to a symphony where the earth mound rumbles, rockery clangs, bamboo grove soughs, flowers muse, and veranda and hall resonate, a spot scene or display may well be a harmonic silver bell that chimes to perfect the symphonic structure of the garden.

[68] As discussed in the preceding section on "Shoal and Promontory" under "Basic Mountain Features".

[69] In-depth discussion on the subject of "Display" is included in the next section, "Placing Architectural Structures".

3.1.1.2 Planning of Architectural Elements

Architecture holds a special place in the creation of the Jiangnan garden. In defining the overall texture and artistic strain of a garden, architectural placement outweighs vegetational disposition, the latter to be discussed in the ensuing chapter. In the procedure of garden construction, architectural works are often scheduled ahead of the others. In light of the specific situation, though, vertical construction, such as molding the ground surface by digging a pond and piling a hill, can precede or coincide with the architectural construction provided it does not impede the undertaking of the latter. Nevertheless, if a relatively large building is designed to be raised on a rather short earthen hill, the foundation of the building then should be laid in the level ground, hence the architectural works should proceed first. Generally speaking, vegetational disposition project is subsequent to the basic completion of architectural and vertical construction. However, in the case where large nursery stocks are used for the garden planning, vegetal project then often precedes architectural works for the ease of transport and planting.

In their profuse variety, all landscape gardens contain, to a greater or lesser degree, some architectural element—a building, wall, bridge, pavement, and even small structures such as a stela or stone bench. The formal garden in the European tradition is even entirely designed and constructed based on the concept of "architecture", where trees and hedges are clipped into geometric shapes and flowers and turf are laid out in symmetric patterns. This geometric- or regular-style garden is also, so to speak, the "architectural-style" garden. On the other end of the spectrum, certain natural-style garden type is prone to total exclusion of architectural element. Except for the inevitable (a paved walk, for instance), the garden is freed of any aspect that marks architecture; even the minimal but necessary accoutrements, such as a resting facility, are obscured by plantings for fear of marring the naturalistic scenery effected mainly by vegetation. Divergent from both of these garden styles, the Jiangnan garden represents a dialectical understanding of architectural planning unique to the art of traditional Chinese garden making. It incarnates the opposite unity of man and nature in the garden art, predicated on which the organic marriage between architecture and natural landscape is constructed. The garden concept couched in the making of Jiangnan gardens reflects that garden is a naturalistic manmade environment that has a definitive functional property intended for everyday garden living. Based on this notion, Jiangnan garden making not only does not shun the artificial architectural treatment but moreover accentuates the architectural interest in the naturalistic creation, where an integral and harmonious artistic effect is attained by way of the contrast and balance between the natural and the artful. In tandem with the natural approach of garden making—shaping ground surface, disposing vegetation, and placing ornamental fauna, the Jiangnan garden maker also attaches great importance to the artificial approach—architectural planning, many principles and techniques of which are yet to be discovered and studied from the remaining garden examples. Architectural planning in garden making has a considerable bearing on the artistic appearance and functionality of the garden, and from the standpoint of garden construction, it holds a greater economic concern than vegetal plantings or animal placement. From this standpoint, architecture is an especially important component in the artistic fabric of the Jiangnan garden.

The architecture is a basic element contributing to the recreative function of the garden; it is an organic medium of the garden art. Structures such as pavilion, terrace, tower, hall, or roofed walkway are compositional ingredients of the garden and should not be isolated from the concept of garden. Classical gardens in Suzhou used to be classified into two types: "gardens containing architecture" and "gardens surrounded by architecture". This notion, nevertheless, isolates the architecture from the garden, viewing only hills, waters, and vegetation as garden and pavilion, corridor, tower, and hall as non-garden. This evidently does not speak accurately of the nature of the Jiangnan garden. The tradition of Chinese garden making has consistently set store by architectural planning. The artistic form of the garden that marks architecture as an important scenic content can be traced back to the Qin and Han dynasties when creation of immortal, otherworldly abode in the pompous palatial style was the rage in garden making. Fundamentally, the creation of such a garden form was rooted in man's longing and pursuit of functional dwelling in an ideal prospect affined with nature. Gardens with the natural and pastoral theme created in the Wei and Jin dynasties through Ming still featured architecture designed for carefree secluded living amid mountains, waters, trees, and flowers, only the architectural styles were different, mainly thatched huts and cottages in the rustic idyllic manner. The usage of the garden determines the primary scale and the functional nature of the architecture. The existing Jiangnan gardens, mostly built or rehabilitated during the Ming and Qing dynasties, see an escalated proportion in architectural construction consequent on the practical needs for pleasure-seeking garden dwelling that were intensified in this period. Garden owners during the Qing dynasty, largely court bureaucrats, scholar officials, and wealthy merchants, regarded their private residential garden as a component part of their everyday living space where family recreations and social functions usually performed. For this reason, the garden comprises not only the general "sight-seeing architecture" such as a pavilion or corridor but also the requisite functional architecture such as a banqueting hall. Such a functional structure is usually also strategically positioned to be the main vantage point to command the principal view of the garden, a compositional arrangement that almost becomes a set formula for zhai yuan (residential garden) design in the Jiangnan region. Take the following as examples: "Hall of Distant Fragrance (Yuanxiang Tang)" in the Humble Administrator's Garden (Zhuozheng Yuan) of Suzhou, "Hall of Hiding Beautiful Scenery (Jingmiao Tang)" in Garden of Look Forward (Zhan Yuan) of Nanjing, "Jinsha Zeyuan" in Jinxi Bieye of Hangzhou, "Clean Water and Lush Trees (Shuimu Mingse)" in Yi Yuan of Nanxiang, and "Banhu Yunjin Wanfurong" in Yi Yuan of Nanxun are all key architectural structures designed for gathering purposes that visually corresponds to the principal scenery of the garden. Additionally, these structures often take on special roles. For instance, the waterside "Hall of Thirty-six Mandarin Ducks and Eighteen Datura Blossoms" (Saliu Yuanyang Guan/Shiba Mantuohua Guan) in the western section of the Humble Administrator's Garden (Zhuozheng Yuan) (also called Subsidiary Garden (Bu Yuan)) has been concurrently used as a stage for dramatic performances and entertainments and, in Villa of Reposing Roars (Jixiao Shanzhuang) of Yangzhou, the quadrangle pavilion named "Little Square Pot (Xiao fang hu)" standing in the pond is also a small aquatic stage for theatrical festivities. Garden, as a means of pleasure seeking aspiring to natural interests, necessitates practical and functional paraphernalia that satisfy such pursuit. The presence of garden architecture is first and foremost ascribed to the functional nature of the garden art. From the artistic aspect of the landscape, the classical Jiangnan garden is deeply indebted to the influence of poetry, literature, and particularly landscape paintings, which are often embellished with architectural images—pavilion, bridge, bamboo fence, thatched hut, tower, or waterside veranda, with the intent to make the scenery more affable and more enticing. The inviting quality of a landscape brush-work—one that piques the viewer's desire "to stroll, to contemplate, to sport, or to dwell" in the depicted scenery—was greatly relished by Guo Xi, the classical painting theorist of the Song dynasty, who defined it as the criteria of the finest landscape paintings (*Linquan gaozhi ji*). In comparison, creation of the Jiangnan garden, in virtue of its functional nature, demands realistically that the scenery be not only affable and enticing but also practicable for "strolling", "contemplating", "sporting", and "dwelling". In other words, the inviting quality appreciated in the landscape painting is all the more essential to the Jiangnan garden and consequently, the architectural element in Jiangnan garden design is all the more indispensable, which goes so far as to place architecture in the core of the garden's scenic composition.

How the architectural content, necessitated by both aesthetic and functional needs, is to be assimilated in the garden scenery and in what artistic form the architectural content is to be expressed are focal issues in garden creation. From the examples of European garden making, the English Landscape style of the eighteenth century, designed to evoke a pastoral idyll, used farmhouse or cottage as architectural ornament so as to be in harmony with the surrounding landscape. In France, the Romantic landscape design of late eighteenth and early nineteenth centuries, excited by novelty and exoticism, featured peculiar, or strictly speaking, mutated Oriental (such as Turkish, Chinese, Japanese, or Indian) architectural structures, reflecting the contemporaneous romantic taste for the exotic and particularly the romantic fascination with the Orient, in which case garden aesthetics was taking on a socio-cultural flair. In some of the Romantic-style (termed "Sensibility" by the Western academia) gardens, the landscape was speckled with such architectural ornaments as ruins of an ancient castle or tomb or a rustic stone bridge or arches, resembling the strange and fanciful character of Medieval or Gothic romances and showing an affinity for melancholy and melodramatic subjects likely to evoke awe or passion for the past. The history of gardens reveals that the above-mentioned European landscape or natural garden schools are the result of the direct or indirect influence of Chinese gardens through Japan. In the realm of the classical Jiangnan garden, examples of architectural treatment from the remaining gardens demonstrate that buildings and landscape are a unified entity, as commanded by the basic principles of scenic construction in Jiangnan garden making. Life is the source of artistic creation, which is true of garden creation as well. The hilly country and canal-filled villages typical of the Jiangnan region serve as the principal wellspring for the creation of the artistic aspects of the Jiangnan garden, as a result of which the garden architecture takes largely on the style of the region's popular dwellings. Nevertheless, compared with its popular dwelling counterpart, garden architecture as an artistic re-creation is processed imagery and hence more elaborate in form, more varied in space, and more wondrous in ornamentation. Among the extant zhai yuan (residential garden) gardens, those built during the Qing dynasty, owing to the demands for utilitarian and solicitous garden living, exhibit an ornate residential character, which is conspicuously reflected in the largely increased proportion of the built-up areas in the garden. Its predecessor, the Ming garden, characterized by a tranquil, carefree, and stoic quality, had featured sparse architectural structures of rustic natural materials, such as a thatched pavilion with a bamboo fence, which since disappeared or were altered into the luxurious style of Qing with intense ornamentation, mirroring a time of the more extravagant taste of the Manchu rulers and Manchu love of decoration as well as a time of incremental demands for material comforts and indulgence among the privileged and leisured classes in the last phase of the imperial age.

The utility nature of the garden determines the existence, scale, and functionality of the garden architecture, whereas the artistic ideology of the garden prescribes its artistic aspects and relationship of the scenic structure. Furthermore, the ideological theme of the scenery dictates not only the selection of landscape type and vegetal variety but also architectural style. For instance, constructed on the theme of the reclusive living in Jiangnan's water-filled region, Garden of Art (Yi Pu) in Suzhou features the principal architecture by the waterside against a background of sparsely wooded flat hills, the structure itself, built in the horizontal composition, assumes a plain and simple style of the popular dwellings in the water-filled country. In Garden of Surging Wave Pavilion (Canglang Ting) of Suzhou, also known as a temple garden—to take another example, the thematic architecture in the mountain forest landscape—the garden's principal scenery—adopts the form of the pristine stone-columned quadrangle pavilion, a shelter structure typically seen off the roadside or at the ferry crossing in Jiangnan's rural areas, which hence accentuates the rustic country flavor of the garden. It becomes clear that the architecture is an integral part of the composite scenic imagery of the garden and that the creation thereof is subordinate to the entirety of the garden scenery. In the past, there was a classification of The Classical Gardens of Suzhou, namely: "architecture embraces gardens" and "gardens embraces architecture". This reflects the concept that only the flowers and trees in the mountains and ponds are regarded as gardens, and buildings, pavilions and other buildings are regarded as non-garden things, it is not in line with the reality of Jiangnan gardens.

As the artificial constituent in the garden's scenic construction, architecture exists in an opposite unity with the natural constituents, which are topography, vegetation, and animal life. The artificial beauty manifested through the architectural contrivances and the natural beauty embodied in the lay of the land, vegetational disposition, and ornamental animal placement are two antithetical aspects in the fabric of the garden. However, as the architectural structures in the Jiangnan garden are planned and executed to conform to natural landscapes, they are also impregnated with natural beauty; and on the other hand, as the hills, water bodies, and plant and animal life are condensed and refined into concentrated garden scenes, they are inevitably marked with processed artful beauty as well. In view of this, the garden's two antitheses are in fact mutually inclusive and therefore unified under the common thematic subject of the garden.

3.1.1.2.1 Name and Types of Building in the Garden

Judging by the names inscribed on the horizontal plaque traditionally hung on the frontal of the structure, there are nearly a score of architectural forms present in the Jiangnan garden. While the rough translation is merely for reference, the names include ting,[70] tang,[71] guan,[72] shi,[73] xuan,[74] xie,[75] ting,[76] lang,[77] lou,[78] ge,[79] zhai,[80] fang,[81] wu,[82] lu,[83] she,[84] chu,[85] suo,[86] and so on. They are mostly ancient structures, each originally standing for a specific architectural style with individual functions and since have been often used interchangeably in garden architecture. Ting (hall), according to Shiming: shigong,[87] "is used for hearing cases", referring to a structure where imperial routine work is administered. "Tang . . . is the central room that faces the sun", which originally purports to the central principal structure of a building complex that has a southern exposure, implies a bright and spacious hall. Ting and tang are similar in form and function, and the two words are since frequently used together as tingtang. Guan, according to one source, is "a place where one temporarily lodges; it can also mean an alternative dwelling place". Shi, explained in

[70] Hall.
[71] Principal hall.
[72] Studio.
[73] Chamber.
[74] Windowed veranda.
[75] House on a terrace.
[76] Pavilion.
[77] Roofed corridor.
[78] Tower.
[79] Multistoried pavilion.
[80] Den.
[81] House.
[82] Room.
[83] Cottage.
[84] Hut.
[85] Point.
[86] Place.
[87] A nomenclative dictionary: building names.

Shiming: shigong, is "a closed space that accommodates people or goods". This refers to the interior space of a building whose entire usable floor area was built above the ground versus the previous primordial structure of semi-cave dwelling that was built with the roofing above the ground and the living space underneath). Hence, "shi is house", as stated in Erya: shigong, or the "Dictionary of Terms" from before the first century. After the Shang and Zhou dynasties, shi, compared with tang—a bright and spacious hall, was a closed and subdued room mainly used for repose and sleep. Xuan, originally a high-fronted carriage with enclosing canopy used in ancient times, has since become the term for the architectural structure that has the similar configuration. The anteroom of a tingtang (hall), for instance, has been called xuan by the Jiangnan workmen to this day. A building with open-walled structure on three sides is also called xuan. Xie, according to Shuowen,[88] the first Chinese character dictionary, is "a house built on the terrace", alluding to a form of domicile structure constructed on a high terrace intended for dwelling and recreation during the Shang and Zhou dynasties; it was a favorite form of architecture among the regal rulers of the day. Ting (pavilion), as per Shiming: shigong, is a structure "where people stop off to gather or transfer (between journeys). People come to rest and move on again, and so on and so forth, hence (the property) does not have a permanent proprietor". Ting was originally a simple arbor-like structure intended for en route passengers to shelter or break, and it is one of the early structures adopted in garden architecture. Lang, whose ancient meaning is explained in Yupian as equivalent to wu, which are "houses that encircle the periphery of a tang" (Yan Shigu, ...). In the Han dynasty, the conjoined word langwu was already in use, referring to an architectural form that encompasses the courtyard. Langwu with an open-sided structure that was transformed into a covered corridor was later adopted as a garden structure intended for perambulatory use. Lou is "a multistoried building", as interpreted in Shuowen; ge was originally the name for a wooden doorstop member and later in Qin and Han dynasties also used to describe a type of covered passageway. In Yupian, ge is equated with lou; it further explains that a building with a simpler and more elongate shape is called lou and one with a more condensed and more complex form is called ge. In later periods, the distinction between lou and ge was blurred, which derived the conjoined term lounge to refer to multistoried buildings in general. Thus it can be seen that many of these terms are originally not the designations for garden structures, and those that have adopted these terms are not necessarily true to type or in keeping with their original architectural form. With the exception of ting (pavilion), lang (corridor), and lou (tower), which have maintained their distinct forms, most of the terms of the garden structures in Jiangnan gardens do not necessarily have a fixed style or architectural significance any more—the terms are taken on merely to lend a sense of antique elegance and an unsophisticated undertone to the garden architecture. Some of them are even used interchangeably; for example, tang (principal hall) and guan (studio) have been both used in naming the principal building of the garden, which usually has the most spacious interior, such as "Hall of Distant

[88] Explicating Chinese characters.

Fragrance" (Yuanxiang Tang) in the Humble Administrator's Garden (Zhuozheng Yuan) and "Hall of Forests and Streams for the Senior and Accomplished" (Linquan Qishuo Zhi Guan) in Garden that was preserved miraculously (Liu Yuan). It is also quite common for garden owners who are infatuated with novelty to make unconventional use of the terms in naming their garden structures. For instance, xie (house on a terrace) has since typically referred to an open-walled house built on a waterside platform in a garden (which is hence also translated as "waterside pavilion"); however in Nanxiang's Yi Yuan, the xie by the lotus pond is named with ge (storied house) instead, called "Tower of Floating Bamboo" (Fuyun Ge), whereas the house further away from the pond is named "Pavilion of the Fragrance of Lotus Roots" (Ouxiang Xie). Such examples are omnipresent: in Jinxi Bieye of Hangzhou, the quadrangle pavilion by the water's edge is entitled "Veranda of Redolent Snow" (Xiangxue Xuan), using xuan (windowed veranda) instead of the customary xie; and in Lion Grove Garden (Shizi Lin) of Suzhou, a multistoried building is named with shi (chamber)—"Woyun Shi" (Chamber of Reposing Clouds) rather than lou or ge—"Woyun Lou" or "Woyun Ge" (Tower of Reposing Clouds), which would otherwise be more befitting its architectural style. In view of such, distinguishing different forms of the Jiangnan garden architecture cannot be simply based on its apparent attributes like the inscribed plaque title that are artfully bestowed upon the building.

Structurally speaking, architectural design in the Jiangnan garden features open, semi-open, and enclosed constructions; there are also great variations in configuration and magnitude within each type of the construction. Roofing in the Jiangnan garden architecture poses a key factor that decides the overall shape of a construction. There are sundry designs in roof making, including the most basic gable roof, termed the *yingshan*[89] roof in the Jiangnan region; the *xianshan*[90] roof, which is an elaborated gable roof with added sideward eaves extended on the gable sides from below the pediments, a treatment that makes the pediment--the tip of the mountain ornamentally exhibitive (Figs. 3.73 and 3.74a); the cone-shaped *cuanjian*[91] roof in circular, square, or polygonal forms, most commonly seen with pavilion constructions; the rare *wudian*[92] roof, which has four sloping sides crested with a short range of ridge (Fig. 3.75b); and the double-eaved roof (Fig. 3.76a). In the creative process, the design of the shape, scale, and structural space of a construction is combined accordingly to support the thematic subject and functional requirement of the garden: some emphasize the ornamental value of the exterior structural figure while some others stress the flowing effect of the interior structural space.

At such, distinguishing different architectural forms in the Jiangnan garden should be predicated on the configurative and spatial characteristics of the building or the thematic role it plays in the scenic construction of a given garden space. It is

[89] Sharp gable.

[90] Revealed pediment.

[91] Bunched-up spire.

[92] Palatial.

Fig. 3.73 Yuanxiang Tang of Zhuozheng Yuan in Suzhou

Fig. 3.74 (a) Yuancui Ge in Liu Yuan. (b) Fucui Ge (on the mountain) in the western section of Zhuozheng Yuan (also called Bu Yuan)—lounge on the mountain. (c) Jianshan Lou in Zhuozheng Yuan—lounge by the water

Fig. 3.74 (continued)

Fig. 3.75 (**a**) An open-sided hexagonal pavilion with the cone-shaped cuanjian roof—Huxin Ting in Suzhou's Shizi Lin. (**b**) A hexagonal construction that is topped with a four-sided wudian-style roof—Zhile Ting in Suzhou's Liu Yuan

Fig. 3.76 (**a**) Hexagonal pavilion with the double-eaved roof—Huxin Ting in West Garden of the Jiechuang Si temple in Suzhou. (**b**) The double-eaved square pavilion in Shanghai's Yu Yuan

essential that the architecture, as a scenic component of the landscape, be tuned to the naturalistic environment of the Jiangnan garden. Garden architecture, whether taking after mountain village dwelling or canal-crisscrossed urban habitation or simply inspired by a landscape painting, is always planned and executed in tandem with those of natural constituents in the garden—hills, water bodies, plants, etc. It is through direct interfacing with natural constituents that the garden architecture (including minor structures and general architectural treatments) is incorporated in the scenic composition and becomes an integral part of the scenic imagery of the garden, taking the waterside pavilion as an example, where the ante-terrace of the structure borders on the water body that concurrently serves as the bank revetment of the water. At such, it is essential to examine the relationship between the architecture and its interfaced natural constituents, particularly the topography and vegetation, so as to recognize the different forms and characteristics of the architecture in the Jiangnan garden. In the ensuing section, the discussion will focus on the general form of some level-land constructions, including *tingtang* (hall), *lounge*, *ting* (pavilion), and *lang* (corridor), as well as on the architectural treatments of the ornamental contrivances, such as *dong men*,[93] *dong chuang*,[94] *lou chuang*,[95] and *feng chuang*,[96]

[93]Hole gate.

[94]Hole window.

[95]Openwork window.

[96]Air window.

followed by "Architectural Treatments Associated with Mountains Features" and "Architectural Treatments Associated with Water Features".

Although the scenery of the Jiangnan garden is essentially re-creation of natural mountain- and waterscapes, not every garden space is planned to bristle with hills and water bodies. Ina smooth area, a single rock with a few trees and flowers can be arranged to bring out the natural texture of the space. Architecture arranged in these circumstances often forms a separate space compartment of its own that is indirectly connected to mountains and waters by means of certain architectural treatments. Representative of the level-ground architecture is *tingtang* (hall).

3.1.1.2.1.1 Tingtang (Hall)

Tingtang (hall) is an indispensable placement in the Jiangnan garden. A space where the garden owner and the family gather for recreation or entertaining guests, it is typically large in space and scale and naturally becomes the principal structure of the architectural complex. For convenience, it is usually situated nearby the garden entrance that leads to the living quarters of the residence. According to *Yuan Ye*,[97] "Before breaking ground for a new garden, one should first position the *tingtang* (hall) and find it a view; most ideal is having a southern exposure", which alludes to that the *tingtang* (hall) is the first to build in the sequence of the garden construction. In terms of the landscape structure, the *tingtang* (hall) is the key viewpoint of the principal scenery—usually a lake-and-mountain composition—in the entire garden. By "southern exposure", the *tingtang* (hall) sits on a north-south axis with its facade facing south. This exposure is traditionally considered "most ideal" as it gives the interior space optimal sunlight and ventilation and from the *tingtang* (hall)'s standpoint, the corresponding principal scenery—the main scenic aspect thereof—can be viewed in favorable light as a result of this positioning. That is to say, in the case of a south-facing *tingtang* (hall), the principal scenery is positioned on its south with its main scenic aspect fronting north to greet the *tingtang* (hall) frontage; conversely, should the *tingtang* (hall) has a northern prospect, the principal scenery would be positioned on its north and its major scenic aspect would front south to counterfoil the north-facing *tingtang* (hall). The exposure of the structure or scenery is ultimately governed by the site conditions and the light requirement for the scenic aspect. The relations between a *tingtang* (hall) and its surrounding scenes also influence the design of the structure. There are three basic forms of *tingtang* (hall) commonly used in the Jiangnan garden, termed by the local craftsmen respectively *hehua ting*,[98] *yuanyang ting*,[99] and *simian ting*.[100]

[97] The Craft of Gardens.

[98] Lotus hall.

[99] Tandem hall.

[100] Four-sided hall.

Hehua Ting (Lotus Hall)

This is a simple form of *tingtang* (hall) with a single interior space and two open sides, front and back, and it normally has a width of three bays. In terms of the landscape scheme in relation to the building, usually suffice it to simply place the principal scenery—often a mountain-and-water composition—in front of the *hehua ting*. Typically, the *hehua ting* is adjacent to a pond where it looks out to the mountain scene from across the water. More often than not, the pond is grown with stands of lotus and popularly known as *hehua chi* (lotus pond), hence the name *hehua ting* (lotus hall). The hall has an open structure on the north and south sides, where floor-to-ceiling *chang chuang*[101] or the combination of *ban qiang*[102] and *ban chuang*[103] that span the upper space is installed to distinguish inside from out. On the east and west sides are blank gables, where, depending on the outside scenery, windows may be put in to take in the views. Examples of the *hehuating* include Verdure Mountain House (Hanbi Shanfang) in Garden that was preserved miraculously (Liu Yuan) (Fig. 3.77) and Pavilion of the Fragrance of Lotus Roots (Ouxiang Xie) in Garden of Pleasure (Yi Yuan) (Fig. 3.78), both in Suzhou.

Yuanyang Ting (Tandem Hall)

Extending three or five bays across, a *yuanyang ting* comprises two interior spaces—one behind the other—and two open sides (south and north, or front and back). The structure is usually roofed in the *yingshan*[104] style or the *xianshan*[105] style. Underneath the roofing, the ceiling comprises two or more arches of differing decorative styles, each of which is locally termed *xuan* and framed with ornamental rafters sealed by roof bricks. The rafters are wrought into different curvilinear forms and hence the arches are fashioned into various styles, such as the bow-shaped arch, *yizhixiang* arch, water-caltrop arch, crane-neck arch, boat-awning arch, teapot-handle arch, and crabapple-blossom arch (Fig. 3.79).

Between the multi-arched ceiling and the saddlebacked roofing forms an excellent thermal insulation interlayer, which makes the construction ideal for use in all seasons. The interior of the *yuanyang ting* is divided into two spaces—the south hall in the front and the north hall in the back—by means of latticework partition doors, folding screens, and/or ornamental casings. To diversify interest, the south and north halls differ from one another in arrangement and decoration as well as usage, forming two antithetical spaces, hence the name "*yuanyang*", which metaphrastically means "a mandarin drake and a mandarin duck" or otherwise often alludes to "an affectionate couple".

[101] Long windows, also explained as latticework partition doors.

[102] Short wall.

[103] Short windows.

[104] Hard gable.

[105] Revealed pediment.

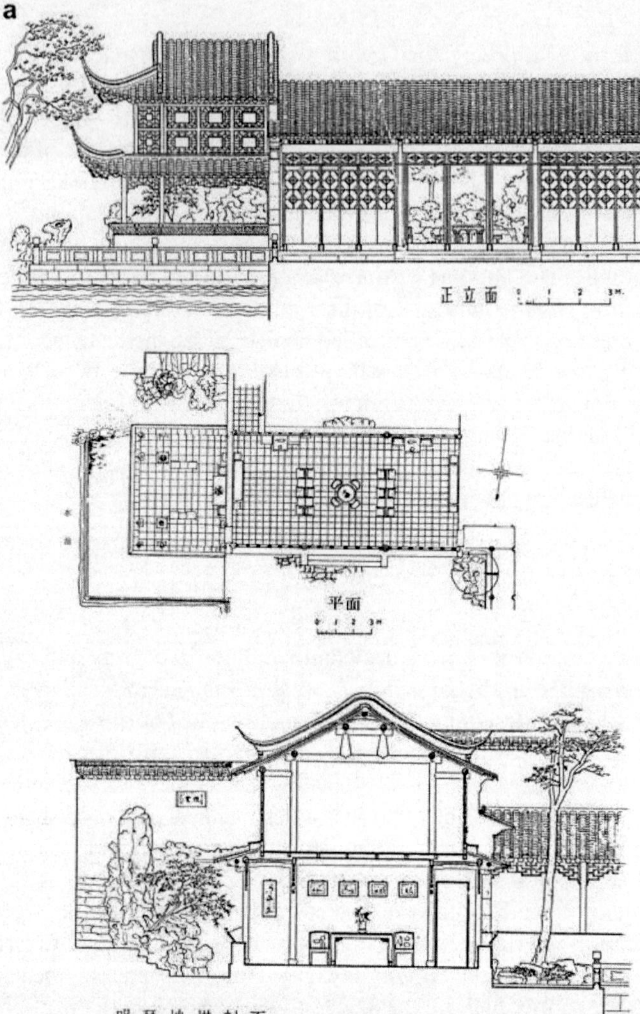

Fig. 3.77 (**a**) Layout, elevation and section plan of Hanbi Shanfang in Liu Yuan of Suzhou (Excerpted from Suzhou Classical Gardens). (**b**) The outside view of Hanbi Shanfang in Liu Yuan of Suzhou

b

Fig. 3.77 (continued)

For example, the *yuanyangting* in Garden that was preserved miraculously (Liu Yuan) of Suzhou, entitled "Hall of Forests and Streams for the Senior and Accomplished (Linquan Qishuo Zhi Guan)", features a double-arched ceiling, under which folding screens, lattice partition doors, and circular floor casings divide the space into south and north halls (Figs. 3.80 and 3.81d).

In western the Humble Administrator's Garden (Zhuozheng Yuan) of Suzhou, the verbosely-named *yuanyang ting* "Hall of Thirty-six Mandarin Ducks and Hall of Eighteen Datura Blossoms (Saliu Yuanyang Guan/Shiba Mantuohua Guan)" features a quadruple-arched ceiling, and its south and north halls are separated by lattice partition doors and floor casings (Fig. 3.82). Generally, the north hall fronts the principal scenery and takes in the mountain view from across the water via the moon-viewing terrace adjoining the house. This arrangement, which places the principal scenery to the north of the building, intends the principal prospect to be viewed in the sprightly frontal sunlight striking full from the south that tends to bring out the best color and relief of the landscape. Facing north, the north hall receives no direct sunlight, which keeps its interior cool, and its outer terrace, which usually lies in the shade of the building, offers a cool spot ideal for summertime relaxation. In contrast, the south hall is usually flooded with sunlight whereas its corresponding landscape—typically composed of woods and rockeries—is often bathed in shadow with backlighting, lending an implicit feel while being interest provoking. Naturally, the north hall is customarily put to use for estival and autumnal activities while the south hall is reserved for hiemal and vernal use. Take the *yuanyang ting* in western the Humble Administrator's Garden (Zhuozheng Yuan) as an example: north hall Hall of Thirty-six Mandarin Ducks (Saliu Yuanyang Guan) provides an affable environment to estivate where people can lounge in the shade during the summer

Fig. 3.78 (**a**) Ouxiang Xie in Yi Yuan in Suzhou. (**b**) The interior of Ouxiang Xie in Yi Yuan in Suzhou—the front are long windows and octangle windows are set on the blank gables

and autumn, catching a cool breeze on the terrace while watching mandarin ducks sporting in the pond. Conversely, for winter and spring, south hall Hall of Eighteen Datura Blossoms (Shiba Mantuoluo Guan) offers a sun-drenched indoor space where people can bask in the abundance of warmth while admiring datura blossoms outside the windows. Likewise in Garden that was preserved miraculously (Liu Yuan)'s *yuanyang ting*, north hall Qishishou Taigu is mainly used in summer and autumn while south Hall of Forests and Streams for the Senior and Accomplished (Linquan Qishuo Zhi Guan) is mainly for winter and spring.

Fig. 3.79 Different types of ceilings in the form of Xuan (Excerpted from Suzhou Classical Gardens). (*1*) Bow-shaped arch (Suzhou's Wangshi Yuan); (*2*) water-caltrop arch (Yi Yuan in Suzhou); (*3*) crane-neck arch (Zhuozheng Yuan in Suzhou); (*4*) boat-awning arch (No. 22 on Tieping Lane in Suzhou); (*5*) teapot-handle arch (Yi Yuan in Suzhou)

Fig. 3.79 (continued)

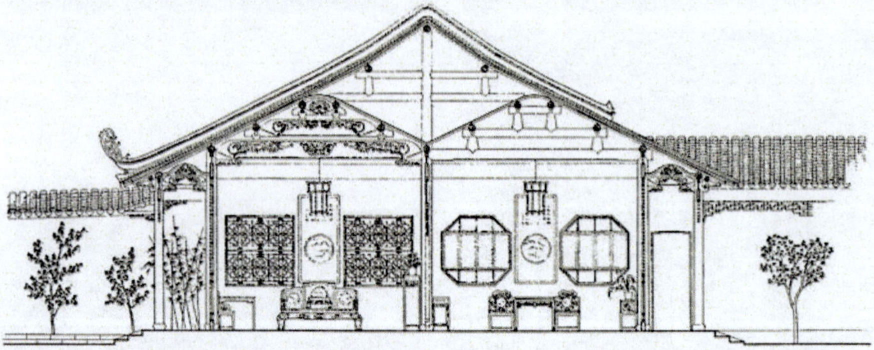

Fig. 3.80 Section of ceiling termed "xuan" in yuanyang ting ("Linquan Qishuo Zhi Guan") (Excerpted from Suzhou Classical Gardens)

Simian Ting (Four-Sided Hall)

Roofed in the *xianshan* style, a *simian ting* mostly spans three or five bays across. As the name suggests, a *simian ting* is a *tingtang* composed of a single interior space that is open on all four sides, tending to command a clearer view with minimal obstruction from all around. Commonly, the open interior is enclosed with lattice partition doors and encircled by open corridors on all four sides. Hall of Distant Fragrance (Yuanxiang Tang) of the Humble Administrator's Garden (Zhuozheng Yuan) (c) in Suzhou is one such example (Fig. 3.73), so are Mountain House of Green

Fig. 3.81 (**a**) Folding Screen of Linquan Qishuo Zhi Guan in Suzhou's Liuyuan, carved with ancient stone peak paintings on the south side and set with couch, flower-pattern beside, and chair and flower table in the hall. (**b**) Folding Screen in Yanyu Tang of Shizi Lin in Suzhou, with Records of Rebuilding Shizi Lin carved on the north side with calligraphy and painting scrolled decoration. Under the screen is placed a large tall narrow table in front of which are another table and chairs. (**c**) Gauze screen decorated with landscape painting in Chengqu Caotang in the east garden of Ou Yuan,

Fig. 3.81 (continued)

Fig. 3.81 (continued) Suzhou. (**d**) White panel screen in Zhilian Laowu of Ou Yuan in Suzhou, hung with nave banners and couplets. Under the screen is a long narrow tall table with a long table at the front and chairs at the left and right. (**e**) Six-partition screens in the Saliu Yuanyang Guan in Zhuozheng Yuan, Suzhou. (**f**) Rare five-partition gauze screens in Jixu Zhai of Wangshi Yuan, Suzhou. (**g**) Lattice-work screen in Qingfengchi Guan of Liu Yuan, Suzhou. (**h**) A single lattice-partition screen in "Wen Muxixiang Xuan" in Liu Yuan, Suzhou. (**i**) A small space partitioned by lattice partitions in Wangshi Yuan, Suzhou e

f

Fig. 3.81 (continued)

Tranquility (Lüjing Shanfang) of Yi Yuan in Nanxun and Veranda of Rain Delight (Yiyu Xuan) of Garden of Bamboo (Ge Yuan) in Yangzhou.

In Jiangnan gardens, many buildings are named as a *guan*, *xuan*, *fang*, *shi*, *lu*, or *she*, as in *huaguan* (painting studio), *shixuan* (poetry veranda), *shufang* (book room or study), or *qinshi* (music chamber), while in fact they are the same structures as the *tingtang* (hall), only smaller in scale. These scaled-down *tingtang* (hall) buildings

Fig. 3.81 (continued)

are generally termed *hua ting*,[106] most of which, like the *hehua ting* and *yuanyang ting*, feature a multi-arched ceiling and open sides on the front and the back where long windows or the combination of a short wall and short windows are installed.

The oldest surviving *hua ting* is Nanmu Ting[107] from the Ming dynasty, located in Schistose Mountain House (Pianshi Shanfang) in Yangzhou (Fig. 3.71a). Some other *hua ting* exemplars are Poxian Music Studio (Poxian Qinguan) in Yi Yuan,

[106]Garden hall.

[107]*Nanmu*-wood Hall.

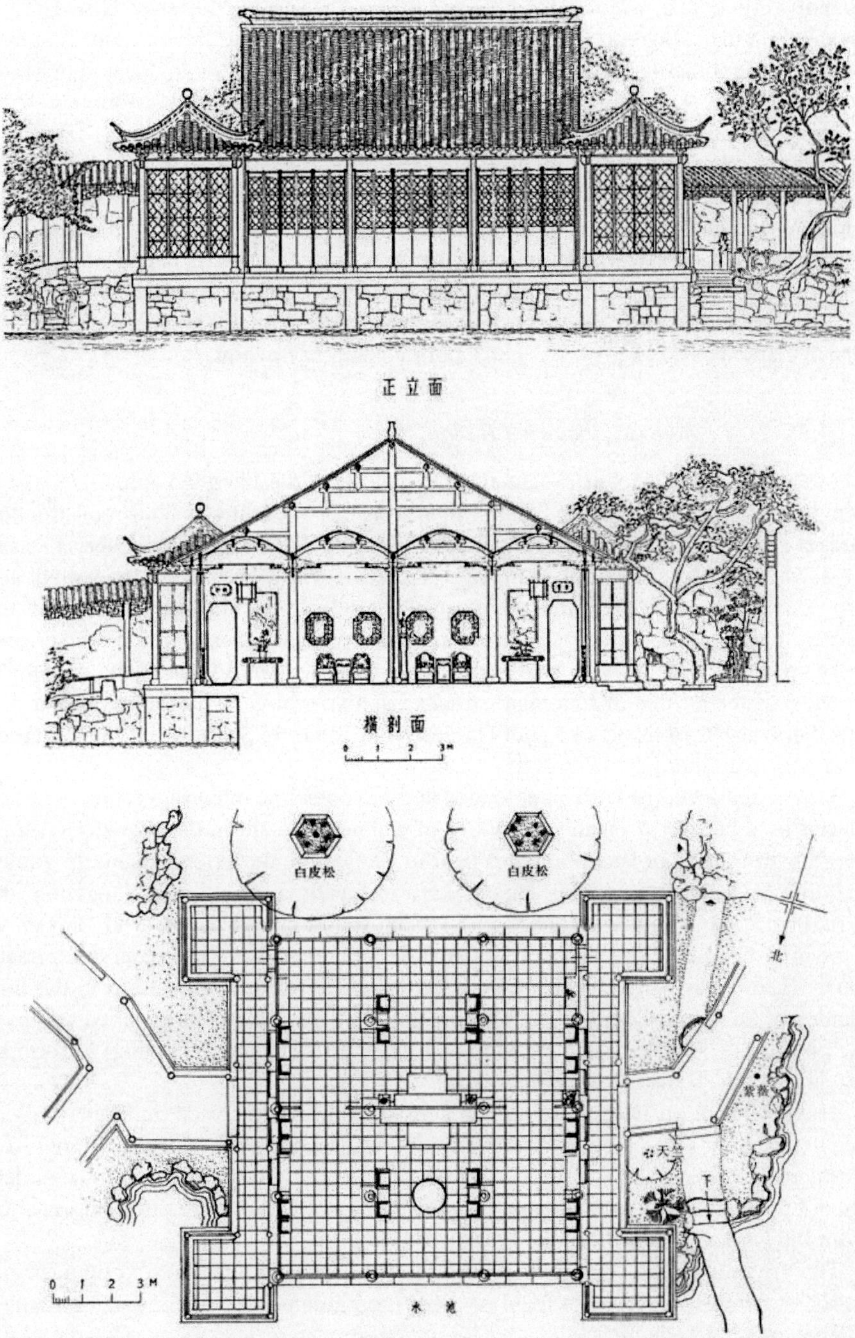

Fig. 3.82 "Saliu Yuanyang Guan/Shiba Mantuohua Guan" in western Zhuozheng Yuan of Suzhou (Excerpted from Suzhou Classical Gardens)

Suzhou (Fig. 3.71b); Peony Study (Dian Chun Yi) in the Master-of-Nets Garden (Wangshi Yuan), Suzhou (Fig. 4.24); Youyu Study (Youyu Shuwu) and Thatched Cottage of Embracing Mountains (Huanshan Caolu) in Xian Yuan, located in Mudu of Wu County; Lingering Veranda (Qiezhu Xuan) in Red Oak Mountain Villa (Hongli Shanzhuang), Hongzhou; and Hall of Splendid Clouds (Qixia Xianguan) in Yi Yuan, Nanxun. In the interior of the *hua ting*, the columns in the central area, which usually serves as the parlor, are transformed into inverted capitals topped with a lotus encarpus that hang down from the crossings of the roof beams. The *hua ting* whose inverted capital finials are carved into a flower basket is called *hualan ting* (flower-basket hall), as exemplified by the study in the residential garden at No. 7 Wangxima Lane (Fig. 3.83) and Flower-Basket House (Hualan Lou) in Lion Grove Garden (Shizi Lin) (Fig. 3.84), both located in Suzhou.

3.1.1.2.1.2 Lounge (Multistoried Building)

Another important structure essential to the making of the Jiangnan garden, *lounge* is comprised in every garden plan whenever site conditions and construction budget allow. Exhibiting both functional and ornamental values, a *lounge* is accessible for ascending high to command a distant view and, with its varied and expressive configurations, the architecture enriches the viewing content of the garden. Often the height of the *lounge* rises above the treetops, which embellishes the garden's skyline as well as affording the visitor an aerial view of the garden.

Most of the *lounge* in Jiangnan gardens rise two stories, with the exception of a few that consists of three, like Half Garden (Ban Yuan) in Suzhou and Inner Garden (Nei Yuan) in Shanghai (Fig. 3.85).

Larger-scale *lounge* with an extended and flexuous line of configuration are often placed as a backdrop on the periphery of a garden, such as the two-storied long building in Garden of Bamboo (Ge Yuan) in Yangzhou; the extended circum-garden building in Villa of Reposing Roars (Jixiao Shanzhuang), also in Yangzhou; the Winding Stream Tower (Quxi Lou), West Pavilion (Xi Lou), and Tower of Crowning Clouds (Guanyun Lou) in Suzhou's Garden that was preserved miraculously (Liu Yuan); and House of Double Reflection (Shuangzhao Lou) in the East Garden of Suzhou's Couple's Retreat Garden (Ou Yuan), which is attached square to the end of the Thatched Cottage at the City Corner (Chengqu Caotang) to form an extended L-shaped structure (Fig. 3.86).

However, not all *lounge* are garden construction. Multistoried buildings such as Wenlan Ge in Hangzhou and the Pavilion of Unity with Heaven (Tianyi Ge) (formerly a private stack hall) in Ningbo (Fig. 3.87), albeit affiliated to a garden, are not a component part of the garden composition and therefore different in nature from the garden *lounge*.

On the other hand, the smaller-scale *lounge*, usually with more compact and varied structures, serves as a focal point of the garden's scenic composition and is often arranged in combination with a mountain- or waterscape. Take as examples the following *lounge* placements in some of the Suzhou gardens: Yuancui Ge in Garden that was preserved miraculously (Liu Yuan), Fucui Ge (on the mountain) in western

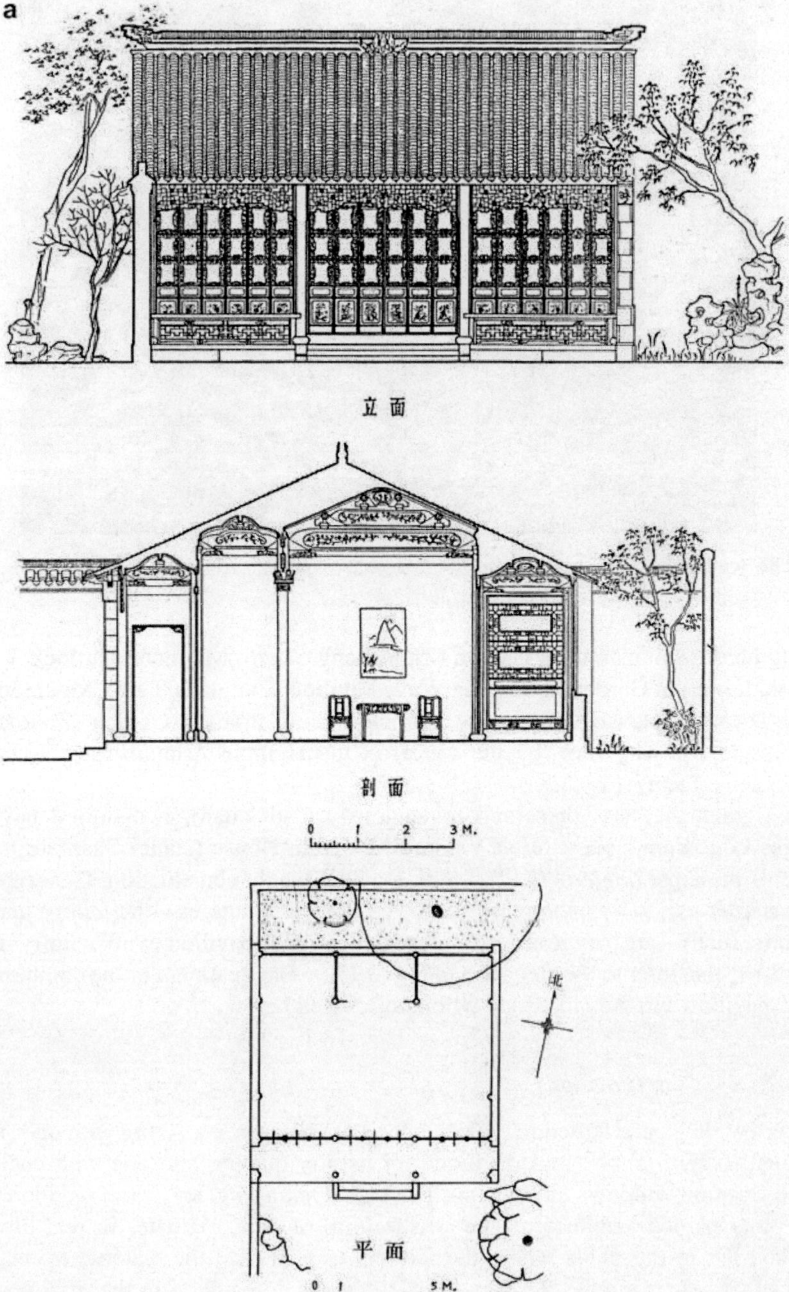

Fig. 3.83 (**a**) Hualan ting in the residential garden at No. 7 Wangxima Lane, Suzhou (Excerpted from Suzhou Classical Gardens). (**b**) Vertical flower basket in hualan ting at No. 7 Wangxima Lane, Suzhou

b

Fig. 3.83 (continued)

the Humble Administrator's Garden (Zhuozheng Yuan), Mountain Outlook Tower (Kanshan Lou) in Garden of Surging Wave Pavilion (Canglang Ting), Mountain-in-View Tower (Jianshan Lou) in the Humble Administrator's Garden (Zhuozheng Yuan), and Daoying Lou (by the water) in the Humble Administrator's Garden (Zhuozheng Yuan) (Fig. 3.74).

In Garden that was preserved miraculously (Liu Yuan), two-storied pavilion Mingse Lou adjoins one-storied Verdure Mountain House (Hanbi Shanfang)—the garden's principal *tingtang* (hall), forming a dual-height construction. Connected to the water by a moon-viewing terrace, the conjoined structures—the *lounge* and the *tingtang* (hall)—are suggestive of a stationary boat-pavilion with a lofty prow (Fig. 3.88; also refer to the elevation in Fig. 3.77a). The treatment of the combination of the pavilion and the landscape will be discussed below.

3.1.1.2.1.3 Ting (Pavilion)

Ting is a basic garden structure typical in the Jiangnan garden. A *ting* with open sides is called *liang ting*, or "cool pavilion" for its airy quality, and one with enclosed lattice partition windows might as well be called *nuan ting*, or "warm pavilion". As with other garden architecture, the artistic form of *ting* originates in real life: the thatched hut in the fields where the farmers take breaks, the waterwheel or well shacks that scatter around the farmstead, the roadside pavilion in the rural country that lends pedestrians a resting place and protection from the elements, and the ferry shelter along the river bank are precisely some of the prototypes for the garden *ting* (Fig. 3.89). In olden times, the roadside pavilion was customarily a setting where

Fig. 3.84 (**a**) Vertical flower basket in hualan ting in Shizi Lin of Suzhou. (**b**) Beam of hualan ting and vertical flower basket in a garden in Hangzhou

people gave a farewell repast to friends or loved ones before seeing them off to a long journey, and thus this type of roadside as well as riverside shelter structures historically became a landscape element redolent of parting, nostalgia, and a lone journey away from home that relished of much poetic sentimentality and somber aesthetics. The type of structure, including the farmer's resting hut in the fields, takes on a simple and open construction that affords a good field of vision, which, together with its aesthetic strain, makes it a favorable ornamental subject in the landscape painting and literature. Likewise, with its functional and ornamental properties, the structure has also long been assimilated into the art of garden making and

Fig. 3.85 (**a**) Lounge consists of three stories in Ban Yuan in Suzhou—a stack hall (occupied by a factory and decoration and environment changed (**b**) Guantao Lou with three stories borrow the scenes of Huangpu River in Nei Yuan of Shanghai

transfigured into an aesthetic object of viewing—*ting*—that provides a focal point for the garden and concurrently functions as a lookout point and resting spot in the garden. A *ting* can be set in any spot where perambulators can stop off for a break—on the hill, by the water, or simply amid a grove of plants. A product of the creative development of landscape and garden making over a long period of time in history, the *ting* of the Jiangnan garden assumes a rich variety of shapes and forms.

Generally speaking, the Jiangnan garden *ting* is a relatively small-scale construction, most of which are capped with a quadrangular or hexagonal tiled roof in the cone-shaped *cuanjian* (bunched-up spire) style or the *xianshan* (revealed pediment) style. Besides the tiled roof, the *ting* with a thatched roof that savors of rusticity and artlessness of the Ming garden and its predecessors is no long seen in the existing private residential gardens except in certain scenic areas of the Jiangnan region where occasional thatched pavilions as a simple decorative garden appurtenance can be spotted (Fig. 3.90).

Approximated to the style of the farmer's resting hut or field waterwheel shack, the quadrangular and hexagonal garden pavilions with the cone-shaped cuanjian roof are more common in private residential gardens than their octagonal counterpart. Usually, the geometric configuration of the pavilion body is consistent with that of its roof. For example, in Suzhou, open-sided Green Ripples Pavilion (Lüyi Ting) and enclosed Piny Wind Pavilion (Songfeng Ting) in the Humble Administrator's

Fig. 3.86 (**a**) Hutian Zichun Tower in Ge Yuan in Yangzhou. (**b**) A corner of circum-building in Jixiao Shanzhuang of Yangzhou. (**c**) Quxi Lou in Liu Yuan of Suzhou. (**d**) Xi Lou (West Pavilion), connecting with Quxi Lou in Suzhou's Liu Yuan. (**e**) Shuangzhao Lou in the East Garden of Suzhou's Ou Yuan. (**f**) Chengqu Caotang, connecting with Shuangzhao Lou in the East Garden of Suzhou's Ou Yuan. (**g**) Cangbao Lou in Shanghai's Yu Yuan (the upstairs is Qinghua Shi)

Fig. 3.86 (continued)

Fig. 3.86 (continued)

Fig. 3.87 Tianyi Ge (a stack hall) in Ningbo

Fig. 3.88 Mingse Lou in Suzhou's Liu Yuan

Fig. 3.89 (a) Well shacks and thatched hut for breaks in Jiangnan's fields. (b) The roadside pavilion in the rural country in Wuyuan, Jiangxi Province—leaning balustrades are set on the three sides of pavilion. (c) The roadside pavilion in the rural country in Wuyuan, Jiangxi Province. (d) The roadside pavilion on the outskirts of Quanzhou, Fujian Province. (e) The ferry shelter along the river bank in Huzhou, Zhejiang Province

Garden (Zhuozheng Yuan) are both quadrangular pavilions with a four-sided, spired cuanjian roof (Fig. 3.91); Mid-lake Pavilion (Huxin Ting) in Lion Grove Garden (Shizi Lin) is an open-sided hexagonal pavilion under a six-sided cuanjian roof culminated with a tall spire (Fig. 3.75a); and Pavilion of Pagoda Reflection (Ta'ying Ting) in western the Humble Administrator's Garden (Zhuozheng Yuan) is an enclosed octagonal pavilion covered by an eight-sided cuanjian roof (Fig. 3.92). There are exceptions, however, such as Pavilion of Supreme Delight (Zhile Ting) in Garden that was preserved miraculously (Liu Yuan), which is a hexagonal construction that is topped with a four-sided wudian-style roof (Fig. 3.75b). Four-sided

Fig. 3.89 (continued)

Fig. 3.89 (continued)

Fig. 3.90 Thatched pavilion in Tianping Mountain of Suzhou

Fig. 3.91 (**a**) An open-sided quadrangular pavilion with the cone-shaped cuanjian roof—Jinsu Ting in Suzhou's Yi Yuan. (**b**) An open-sided quadrangular pavilion with the cone-shaped cuanjian roof in He Yuan in Yangzhou—made for stage. (**c**) An enclosed quadrangular pavilion with the cone-shaped cuanjian roof—Songfeng Ting in Zhuozheng Yuan

c

Fig. 3.91 (continued)

garden pavilions that are roofed in the xianshan style resemble the rural roadside pavilion and riverside ferry shelter. Examples include the square pavilion in Suzhou's Garden of Surging Wave Pavilion (Canglang Ting) (Fig. 3.93a), the square pavilion in Shanghai's Garden of Various Fruits (Jiuguo Yuan), and the rectangular pavilion at Snow-Like Fragrant Prunus Mume Pavilion (Xuexiang Yunwei)", the lake-and-mountain landscape sequence in the Humble Administrator's Garden (Zhuozheng Yuan) (Fig. 3.93b). In Garden of High Virtue (Gaoyi Yuan) at Mount Tianping in Suzhou, Freedom Pavilion (Xiaoyao Ting), another example of the xianshan-roofed square pavilion, is built above the garden's lower-level entrance. A change in level creates interest, and perching on the ramp over the entryway, the construction demonstrates an interest of spatial composition that capitalizes on the height difference of the terrain (Fig. 3.93c). Pavilions with the double-eaved roof are rarely seen among private residential gardens. They are more frequently featured in gardens with a common nature, such as a temple garden. The large hexagonal pavilion located in the middle of Animal Release Pond (Fangsheng Chi) in West Garden of the Jiechuang Si Temple (Monastery of Ordination Pennant) in Suzhou is one such example (Fig. 3.94a) so are the one in the Wenlan Ge of Hangzhou and the double-eaved square pavilion in Shanghai's Garden to delight one's parents

Fig. 3.92 An enclosed octagonal pavilion—Ta'ying Ting in western Zhuozheng Yuan

(Yu Yuan), which once came into possession of a pecuniary trade association in the late imperial period (Fig. 3.76b). In keeping with the garden's naturalistic scenery, pavilion designers aspire to a succinct style, taking into account the configuring scale of the structure as well as its spatial relationship with the surrounding landscape. In this respect, Jiangnan gardens boast a good many successful creations. zhai yuan (residential garden) or ting yuan (court garden) are some of the urban gardens that are notoriously restricted in space, hence the expedient contrivance of the "half pavilion"—precisely a half pavilion built again a wall construction, seeking to economize the land and reclaim more garden space. Half pavilions, both open and enclosed, created in Half Garden (Ban Yuan) and Garden of Listening to the Maple (Tingfeng Yuan) in Suzhou, Pavilion of Unity with Heaven (Tianyi Ge) in Ningbo, and Yan Yuan in Changzhou are but a few representative works of the kind (Fig. 3.95). Exemplary pavilions of the Jiangnan garden bear a distinct resemblance

a

Fig. 3.93 (**a**) The square pavilion roofed in the xieshan style—Suzhou's Canglang Ting. (**b**) The rectangular pavilion roofed in the xieshan style—"Xuexiang Yunwei" in Suzhou's Zhuozheng Yuan. (**c**) Xiaoyao Ting at Gaoyi Yuan at Mount Tianping in Suzhou—taking the role of entrance. (**d**) The square pavilion with three sides open—Jiashi Ting in Suzhou's Zhuozheng Yuan.

to the image of their prototype—the rural roadside pavilion and riverside ferry shelter. Albeit artistically transfigured, they maintain an unaffected quality shared by their predecessor that tends to keep the artificial structure in tune with the overall naturalistic landscape of the garden. The harmony of a pavilion landscape composition is manifested through the relationship between the pavilion construction and the mountain- or waterscape arrangement that surrounds it, a subject that will be dealt with in greater detail in the subsequent segments. In contrast to the naturalistic style, there is yet a different trend in pavilion construction that attempts exotic interest and extraordinary designs, exemplified by the flabellate pavilion (termed "fan pavilion" in the Jiangnan region) in Lion Grove Garden (Shizi Lin); the circular pavilion roofed in the style of a conical bamboo hat in the Humble Administrator's

b

Fig. 3.93 (continued)

Garden (Zhuozheng Yuan); the quadrangular pavilion—styled "Recluse Abode of Bamboos and Chinese Parasol Trees (Wuzhu Youju)"—enclosed by four-sided whitewashed walls with a moon gate open in each enclosure, also in the Humble Administrator's Garden (Zhuozheng Yuan); "Square Triumph Pavilions" (Fang Sheng Ting), a brace of interlinked quadrangular pavilions in Xu Yuan of Nanjing, once the west garden of the self-proclaimed "King of Heaven" of the short-lived Heavenly Kingdom of Taiping in the mid-nineteenth century; the "plum pavilion" in Yi Yuan of Nanxiang near Shanghai, which is built on a plum-blossom-shaped base

c
d

Fig. 3.93 (continued)

Fig. 3.94 Nuan lang along the hill at Jinshan Si in Zhenjiang of Jiangsu Province

and enclosed by walls punctuated with plum-blossom-patterned dong chuang (hole windows); and the triangular pavilion in Meandering Water Park (Qushui Yuan) of Qingpu near Shanghai as well as Three Ponds Mirroring the Moon (Santan Yinyue) in Hangzhou (Fig. 3.96). Nevertheless, examples such as these do not represent the mainstream pavilion construction in the nature-aspiring landscape of the Jiangnan garden.

3.1.1.2.1.4 Lang (Roofed Corridor)

In the Jiangnan region, *lang* is originally a covered walkway that shelters pedestrians from the recurrent rain and sweltering sun, especially during the long sultry, rainy season. It is assimilated into garden architecture and serves to facilitate garden perambulation in the rain or snow or under a scorching sun. The colonnade and lintels of the roofed corridor form a procession of picture frames presenting a sequence of views, which increases the layer of the scenery and the depth of field and thus enhances the interest of the garden's perambulatory viewing. In the meantime, *lang*, with its open structure, also functions as a subtle division between garden spaces. Aesthetically, *lang* has significant ornamental value, which makes it an object of viewing as well. It is no wonder that *lang* is widely adopted in garden making and has been elaborated in its design and configuration. In respect to

Fig. 3.95 (**a**) Half pavilion in Suzhou's Tingfeng Yuan. (**b**) "Shiran" Half pavilion in Suzhou's Tingfeng Yuan (an enclosed pavilion roofed in the xieshan style). (**c**) Half pavilion in Tianyi Ge in Ningbo. (**d**) Half pavilion in Suzhou's Wangshi Yuan

c

d

Fig. 3.95 (continued)

Fig. 3.96 (**a**) The flabellate pavilion (termed "fan pavilion" in the Jiangnan region) in Shizi Lin of Suzhou. (**b**) The flabellate pavilion in western Zhuozheng Yuan—Yushui Tongzuo Xuan. (**c**) Hexagonal Shuxiao Ting with round roof in Liu Yuan in Suzhou. (**d**) "Wuzhu Youju" in Suzhou's

configuration, the Jiangnan garden *lang* can be categorized into the following five types: *kong lang*,[108] *ban lang*,[109] *fu lang*,[110] *nuan lang*,[111] and *hua lang*.[112]

Kong Lang (Open Corridor)

This is a roofed corridor with an open structure on both sides. Short walls or seating balustrades are usually raised between columns to lend perambulators a place to rest if need be (Fig. 3.97). The covered walkway that extends from the entry hall to the main hall in the rear of Crane Garden (He Yuan) in Suzhou is an example of a flexuous *kong lang*.

Ban Lang (Half Corridor)

A roofed corridor that is open only on one side, the *ban lang* is often built abutting a garden wall, serving to mask the enclosure and embellish the background. The covered walkway alongside the east garden wall in Garden of Look Forward (Zhan Yuan) of Nanjing and the one along the garden wall in Garden of Pleasure (Yi Yuan) of Suzhou (Fig. 3.98), and the "Wanchun Lang" gallery in Yi Yuan of Nanxun are but a few examples of the *ban lang*. To avoid monotony, *ban lang* and *kong lang* are often placed alternately, in which case the two structures are in effect one continuous, flexuous corridor that becomes a one-sided *ban lang* when it runs alongside a wall and, when it veers off the wall, turns into an unattached *kong lang* that opens out on both sides. The corridor next to Veranda of Scenting the Sweet Osmanthus Fragrance (Wen Muxixiang Xuan) in Garden that was preserved miraculously (Liu Yuan) (Fig. 3.99) and the one north of the "Mianbi" pavilion in Yi Yuan—both gardens in Suzhou—are two such examples. Sometimes, to "correct" the optical one-sidedness of the *ban lang* and construct a balanced view and lines of sight, mural epigraphy is showcased along the enclosed side of the corridor for perambulators' leisurely browse, as exhibited in the *ban lang* examples mentioned

Fig. 3.96 (continued) Zhuozheng Yuan—the quadrangular pavilion enclosed by four-sided whitewashed walls with a moon gate open in each enclosure. (**e**) An open-sided pavilion in the from of variable cross-section with Tongban tile and a round roof—the pavilion roofed in the style of a conical bamboo hat in Suzhou's Zhuozheng Yuan. (**f**) Compound roof—Yuanyang Ting in Nanjing's Xu Yuan. (**g**) The triangular pavilion—Kai Wang Ting in Santan Yinyue in Hangzhou. (**h**) Compound roof—Dezhen Ting in Suzhou's Zhuozheng Yuan. (**i**) An open-sided pavilion with ridge, a round roof and Hudie tile—Jinyue Ting in He Yuan of Yangzhou

[108] Open corridor.

[109] Half corridor.

[110] Double corridor.

[111] Enclosed corridor.

[112] Trelliswork walkway.

Fig. 3.96 (continued)

f

g

Fig. 3.96 (continued)

above (Figs. 3.98 and 3.99) and at the Wen Tianxiang "poetry tablet pavilion" in Lion Grove Garden (Shizi Lin). The use of literature inscriptions in garden making will be discussed in greater detail in "Exploiting Poetic Literature" in the following chapter. There is yet another type of *ban lang* that is virtually derived from a *kong lang* structure. It is formed by walling up the intercolumnar spaces on one of the two open sides of the *kong lang*, usually for the purpose of separating spaces. This type

Fig. 3.96 (continued)

a

b

Fig. 3.97 (**a**) A roofed corridor with an open structure on both sides and seating balustrades are raised between columns to lend perambulators a place—corridor at Zizai chu of Liu Yuan in Suzhou. (**b**) A roofed corridor with lou chuang on both sides—Xi lang in the East Garden of Suzhou's Ou Yuan

Fig. 3.98 Ban lang on one side of the wall—Canglang Ting in Suzhou

Fig. 3.99 Ban lang made of slab stones inscribed with poetry and kong lang are placed alternately—a roofed corridor at the side of Wen Muxixiang Xuan in Liu Yuan

Fig. 3.100 (**a**) Ban lang corridor with Louhua qiang in Suzhou's Ou Yuan. (**b**) Ban lang with dong chuang in Wuxi's Jichang Yuan. (**c**) Ban chuang in the ban lang in Suzhou's Ou Yuan

of *ban lang* often features a sequence of decorative openings in the wall, such as *lou chuang* (openwork windows) or *dong chuang* (hole windows), to create a more interpenetrative scenic environment, or in other words, establish an interaction between the two scenic spaces from each side of the wall; in addition, the wall windows also provide the anticipative visitor with a "sneak peek" of the ulterior garden scenes (Fig. 3.100a, b). Another variation of the *ban lang* is to have a short

c

Fig. 3.100 (continued)

wall raised on both sides of the corridor, with one side also featuring *ban chuang* (short windows) spanning the upper space to suggest a subtle division between garden spaces, the effect of which evokes another garden structure, *xuan* or *xie*, a windowed veranda, exemplified by the *ban lang* corridor in Suzhou's Couple's Retreat Garden (Ou Yuan) (Fig. 3.100c).

Fu Lang (Double Corridor)

Fu lang is a compound corridor composed of two parallel *ban lang* (half corridors) that share a party wall in the center (Fig. 3.101). Normally *lou chuang* (openwork windows) or *dong chuang* (hole windows) are put in in the party wall to create the depth of the scenic space and prefigure the landscape on the opposite side. In Lion Grove Garden (Shizi Lin) of Suzhou, the double corridor lying between the Xiuzhu Ge pavilion and Standing-in-the-Snow Hall (Lixue Tang) boasts a unique design in the construction: featuring openwork windows in the party wall, the *fu lang* actually has its both sides walled up, each side perforated with a sequence of decorative windows in circular and hexagonal shapes, intended to take in the garden views (Fig. 3.102). This design treatment of the *fu lang*, however, evokes a considerable exterior resemblance to a *nuan lang* corridor.

Fig. 3.101 Fu lang in Suzhou's Yi Yuan

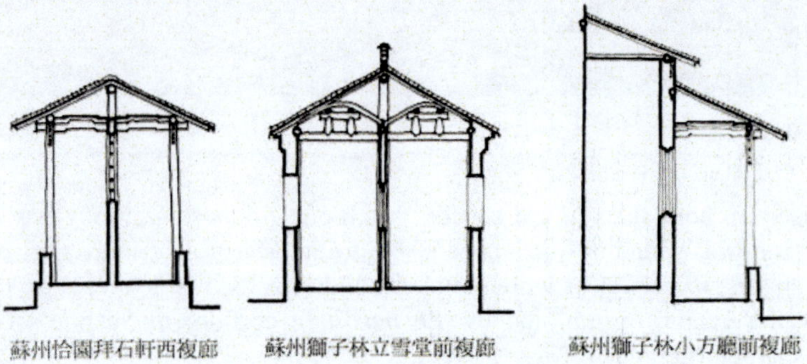

Fig. 3.102 Section of fu lang in Suzhou's Yi Yuan and Shizi Lin

Nuan Lang (Enclosed Corridor)

A derivative of an ancient form of indoor passageway, *nuan lang* is a roofed corridor with one or both sides installed with intercolumnar lattice doors or short walls topped with lattice windows, lending itself to protection from the cold and the wind, hence the name "*nuan lang*", literally "warm corridor". This type of enclosed corridors are often employed at monasteries in Jiangnan's mountain areas for its protective quality against the chill blowy climate, as seen in Temple of Golden Mountain (Jinshan Si) in Zhenjiang of Jiangsu Province (Fig. 3.94) and Tiantong Si in Yinxian County of Zhejiang Province (Fig. 3.103). Assimilated into the art of garden making, *nuan lang* has been transfigured into a complex form of garden construction, exemplified by the double-decked corridor adjacent to the waterfront Mountain-in-View Tower

Fig. 3.103 Nuan lang along the hill at Tiantong Si in Yinxian County of Zhejiang Province

(Jianshan Lou) in Suzhou's the Humble Administrator's Garden (Zhuozheng Yuan) (Fig. 3.100a). This structure was originally a two-storied *kong lang*—both the upper and lower corridors were open-sided—until recent times when the upper corridor was installed with sash windows and turned into a *nuan lang* (Fig. 3.104).

It is worth mentioning here that in ancient times (the Qin and Han dynasties) an overhead walkway was termed *fei ge* (flying pavilion) and a double-decked walkway "*fu dao*" (binary walk); the present-day double-decked corridor spanning the water in the Humble Administrator's Garden (Zhuozheng Yuan) is arguably a cross between the "flying pavilion" and "binary walk".

Hua Lang (Trelliswork Walkway)

In Jiangnan gardens, *hua lang* is an artistic rendering and creative elaboration of the everyday gourd trellis or leguminous plant lattice fence from one's backyard. Most commonly constructed in wood or bamboo, or occasionally stone for the refined-taste aspirer, *hua lang* somewhat resembles the roofed corridor in structure, which consists of parallel colonnades that support the roof, except that the *hua lang*, instead

a

b

Fig. 3.104 (**a**) The two-storied corridor in Jianshan Lou in Suzhou's Zhuozheng Yuan (the upper corridor is a nuan lang). (**b**) The two-storied corridor in Jianshan Lou in Suzhou's Zhuozheng Yuan (the upper corridor was a kong lang)

Fig. 3.105 (**a**) Hua lang combined with a hall—the Chinese wisteria trellises on the north of the principal hall, "Jingmiao Tang", in Zhan Yuan of Nanjing (**b**) Waterside hua lang combined with a hall—the Chinese wisteria trellises in front of "Sutai Chunman", which overlooks Fangsheng Chi in West Garden of the Jiechuang Si Temple in Suzhou

of a closed tiled roof as with the aforesaid corridor features, has an open roof of girders and cross rafters over which climbing plants such as rose, grape, or wisteria are trained. *Hua lang* that are simple and rustic in style recalls pastoralism while those elaborate and embellished urbanity. There are not many surviving examples of *hua lang* to this day; many are found in the drawings of ancient gardens of the Ming and early Qing periods. Among the few in existence, most are built adjoining a building such as a hall or pavilion, similar to a porch in form and function. The Chinese wisteria trellises on the north of the principal hall, "Hall of Hiding Beautiful Scenery (Jingmiao Tang)", in Garden of Look Forward (Zhan Yuan) of Nanjing

(Fig. 3.105a) provides one such example, so does the age-old-rattan-entwined colonnade abutting the front of the waterside principal hall, "Sutai Chunman", which overlooks Animal Release Pond (Fangsheng Chi) in West Garden of the Jiechuang Si Temple (Monastery of Ordination Pennant) in Suzhou (Fig. 3.105b). *Hua lang* are also built in combination with a path to form a climber-roofed walkway, such as the wisteria pergola walk near the entrance of Garden of Surging Wave Pavilion (Canglang Ting) in Suzhou (Fig. 3.106a) and the arched colonnaded stone pergola in Meandering Water Park (Qushui Yuan) in Qingpu, on the outskirts

Fig. 3.106 (**a**) Hua lang combined with a path—the wisteria pergola walk near the entrance of Canglang Ting in Suzhou (**b**) The arched colonnaded stone pergola in Qushui Yuan in Qingpu, on the outskirts of Shanghai

of Shanghai (Fig. 3.106b). There are also *hua lang* that are built on the top of a bridge, like a covered bridge, as seen in Garden that was preserved miraculously (Liu Yuan) where an ironwork wisteria trellis is set atop the zigzag bridge leading to the "Little Penglai" island (Fig. 3.53) and in the Subsidiary Garden (Bu Yuan) of western the Humble Administrator's Garden (Zhuozheng Yuan) where a similar structure is mounted on the ironwork bridge connecting the Stay and Listen Pavilion (Liuting Ge) and Hall of 36 Mandarin Ducks (Saliu Yuanyang Guan).

Lang, as a garden structure, has evolved significantly in the making of the Jiangnan garden. In addition to the various types discussed above, there is a variant of *lang* that combines the spatial quality of the *lang* with the structural features of a building like *lou*, *ge*, *xuan*, or *xie*. This treatment results in a unique architectural environment that enjoys both the openness of a corridor and the commodiousness of a house or veranda. In Suzhou's Garden that was preserved miraculously (Liu Yuan), the spatial arrangement at Crane's Place (He Suo) and the area surrounding the "Interwined Old Trees" (Gumu Jiaoke) courtyard on the east of the "Green Shade" (Lüyin) veranda are two prime exemplars (Fig. 3.107).

The figuration of a *lang*, while largely prescribed by the concurrent landscape structure, is most commonly seen in a flexuous shape, both meandering left and right on the horizontal plane and undulating up and down by following the natural contours of the land. Further development also sees more sophisticatedly built corridors that "climb hills" and "wade water", hence the names *pashan lang* (hill-climbing corridor) and *shui lang* (water corridor), both of which will be discussed in greater detail in the segments on "Architectural Treatments Associated with Mountain Features" and "Architectural Treatments Associated with Water Features".

3.1.1.2.1.5 Dong Men (Hole Gate), Dong Chuang (Hole Window), and Lou Chuang (Openwork Window)

As accessorial architectural ornaments, *dong men*, *dong chuang*, and *lou chuang* of variegated designs are undeniably extraordinary creations of the Jiangnan garden. In contrast to their counterparts in northern China and the Lingnan region (Guangdong and Guangxi Provinces), these ornamental contrivances maintain a distinctively delicate and ethereal style of its own. *Dong men*, *dong chuang*, and *lou chuang* originate from the simple-structured gate and window arrangement in the Jiangnan popular dwellings where tiling is utilized to construct the casing or the window lattice (Fig. 3.108). The tile-arranged windows can be traced back to the Shang and Zhou dynasties when roof tiles and potsherds were used to lattice windows, named "*wa you*" (tile windows), a simple and primitive form of its kind. The often-featured *lou chuang* ornamental window in the Jiangnan garden architecture is precisely an artistic re-creation of such latticed tile windows. Up to modern times, the creativity and artistry in making these ornamental architectural accessories had greatly matured, both in design and from the perspective of their function as a means of scenic construction, and there are ample valuable examples that have survived to lend themselves to the garden architectural study.

a
b
c

Fig. 3.107 (**a**) The east of the "Lüyin" veranda—the spatial arrangement surrounding the "Gumu Jiaoke" courtyard in Suzhou's Liu Yuan (**b**) He Suo in Suzhou's Liu Yuan (**c**) The interior of He Suo in Suzhou's Liu Yuan

a

b

Fig. 3.108 (**a**) Simple-structured window arrangement of tiling in the Jiangnan popular dwellings—prototype of lou chuang in garden art. (**b**) Simple tile-arranged windows in the Jiangnan popular dwellings—prototype of lou chuang in garden art. (**c**) One of lou ming chuang in the Jiangnan popular dwellings—prototype of lou chuang and hua qiang in garden art

Dong Men (Hole Gate)

Termed *men kong*[113] in *The Craft of Gardens* (Yuan Ye), *dong men* refers to a leafless door opening that is made in a wall of a compound or in a garden feature such as a pavilion, roofed corridor, or windowed veranda. Walls in the Jiangnan garden are generally coated with lime plaster and popularly called "whitewashed

[113] Gate cavity.

c

Fig. 3.108 (continued)

wall". When a *dong men* is set rimless in a whitewashed wall, it gives a natural and artless look, like the rimless moon gates in Suzhou's Garden of Art (Yi Pu) and Crane Garden (He Yuan), and the garden of Hangzhou's Xiling Seal-engravers' Society (Xiling Yinshe) (Fig. 3.109a, b). More often than not, the opening is framed with a rim of one or more inches wide and made of dark gray ground tiles, which, while foiled by the pristine white wall, appears quite classic and elegant but protects edges (Fig. 3.109c–e). A classic form of *dong men* is in the shape of a full moon, commonly called "moon gate". It is termed *yuechuang shi*[114] in *The Craft of Gardens* (Yuan Ye) and can be applied to both gate and window making. The geometric shape of the *dong men* tends to attract the sight line of the viewer and forms a focal point of viewing, which is the reason that a *dong men* is often used as a picture frame for an offsetting scenery. The cutout *dong men* is a rare form of architectural treatment and literally an eye-catching feature that adds interest and anticipation to a garden structure, and has a stronger guiding quality and more inviting nature than an otherwise regular doorway. The most common forms of *dong men* are the plain quadrangle (Fig. 3.110) and the quadrangles with slight variation such as an arched lintel, a teapot-handle shaped lintel, notched upper corners, or upper corner ornaments (Fig. 3.111). These styles, however, do not feature a border at the base of the opening or a raised threshold that would complete

[114]Moon-window style.

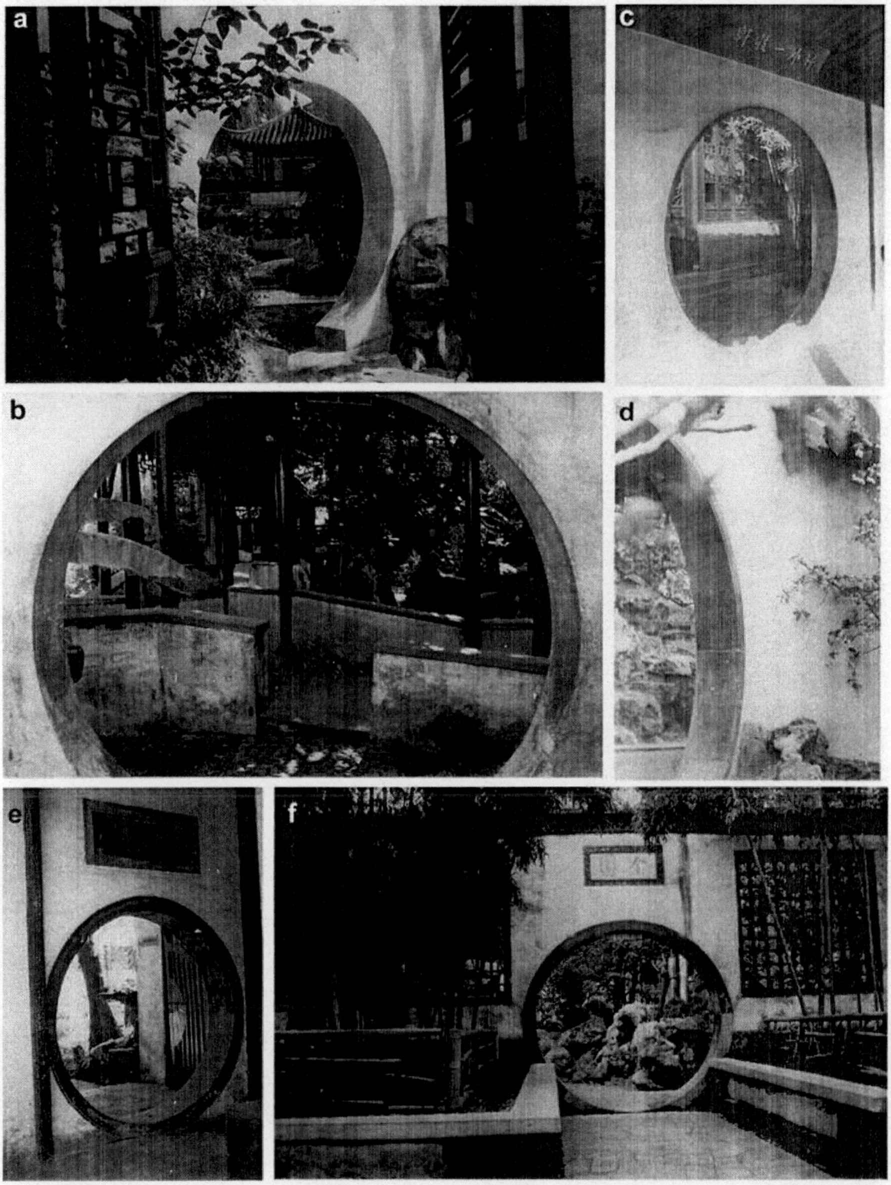

Fig. 3.109 (**a**) The rimless moon gate, with a natural and artless look, in Suzhou's Yi Pu. (**b**) The rimless moon gate, with a natural and artless look, in He Yuan. (**c**) The rimless moon gate with a rim made of dark gray ground tiles in the veranda "Zhuwai Yizhi Xuan" in Suzhou's Wangshi Yuan. (**d**) A part of the rim made of dark gray ground tiles. (**e**) Entrance of Yangzhou's Xiao Pangu—moon gate with a wide rim. (**f**) Entrance of Yangzhou's Ge Yuan—moon gate with a wide rim

Fig. 3.110 Dong men in the form of quadrangle (Suzhou's Wangshi Yuan) in Yuan Ye

the enclosure of a *dong men*, and therefore they lack an accentuated cutout effect that is characteristic of a "hole gate". The elongate octagon, another common style of the *dong men* design, forms a more complete *dong men* structure with its base border in the loop (Fig. 3.112). There are more than a score of different *dong men* styles in the Jiangnan garden, many of which, as identified in *The Craft of Gardens* (Yuan Ye),

Fig. 3.111 (**a**) Dong men in the form of quadrangle with notched upper corners next to Zhenyi (quadrangle) in Suzhou's Wangshi Yuan. (**b**) Dong men in the form of quadrangle with an arched lintel (Shizi Lin in Suzhou) in Yuan Ye. (**c**) Dong men in the form of quadrangle with a teapot-handle shaped lintel (Suzhou's Wangshi Yuan). (**d**) Dong men in the form of quadrangle with upper corner ornaments (Liu Yuan in Suzhou)

were available as early as in the Ming dynasty (Fig. 3.113). Ample examples can be found in the surviving Jiangnan gardens, including the "*tianyuan difang*" style (so termed by *The Craft of Gardens* (Yuan Ye), metaphrastically "circular sky and square earth"), which resembles the shape of a horseshoe or keyhole (Fig. 3.114a, b);

Fig. 3.112 (**a**) The elongate octagonal dong men with broad edges on the wall in Jixiao Shanzhuang, Yangzhou. (**b**) The elongate octagonal dong men with narrow edges on the pavilion in Zhan Yuan, Nanjing

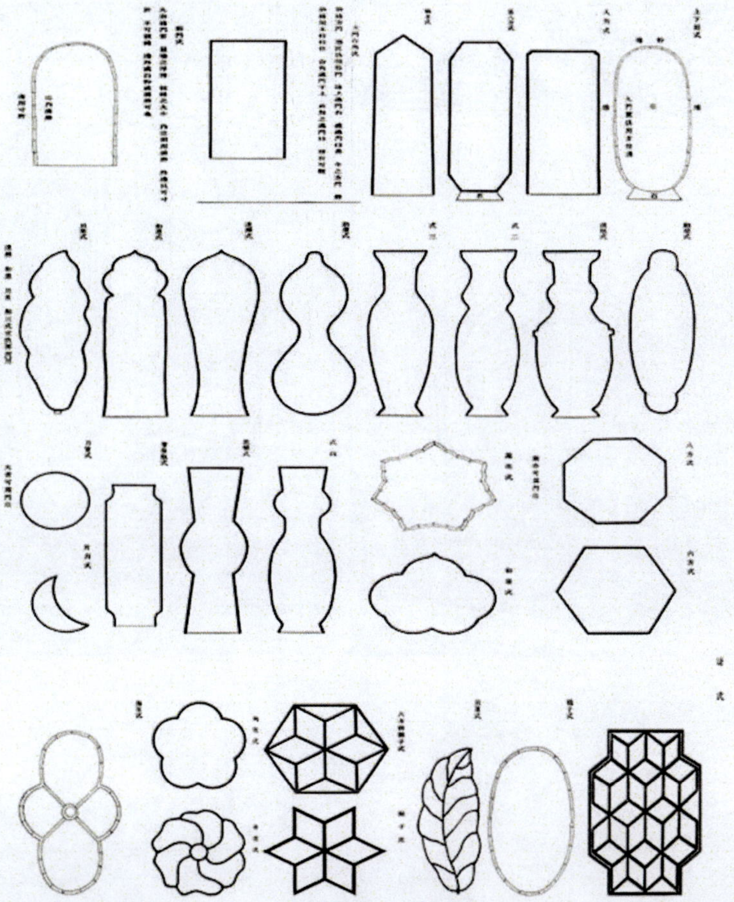

Fig. 3.113 Schema of dong men, dong chuang and special-shaped lou chuang (openwork window) in Yuan Ye of Ming Dynasty

the oval "crane egg" style; the "upper and lower arches" style, which is an oblong with an arched top and inverted arch at the bottom; and a variant "lotus petal" style (Fig. 3.114c–e). Other styles include the "shell leaf" (Fig. 3.115), "Han vase" (Fig. 3.116), "Chinese flowering crabapple blossom" (Fig. 3.117), "peach" (Fig. 3.118), "bottle gourd" (Fig. 3.119), and "silver ingot" (Fig. 3.120). There are a few variant designs of the "*ruyi*" style, including the "small *ruyi* head" (Fig. 3.121) and "great *ruyi* head", with the latter redolent of a wavy-edged mushroom cap (Fig. 3.122).

Fig. 3.114 (**a**) "Tianyuan difang" dong men of Yu Yuan, Shanghai (Drawn in accordance to the chart board of Jiangnan Yuanlin Zhi). (**b**) "Tianyuan difang" dong men in Qiuxia Pu, Jiading. (**c**) The dong men of "crane egg" style (so termed by Yuan Ye) in Fenyang Bieshu, Hangzhou (Drawn in accordance to the chart board of Jiangnan Yuanlin Zhi) (**d**) The dong men of "upper and lower arches" style (so termed by Yuan Ye) in Pang Zhai at Shijia Alley, Suzhou (Excerpted from Suzhou Classical Gardens). (**e**) The dong men of variant "lotus petal" style in Yu Yuan, Shanghai

d

e

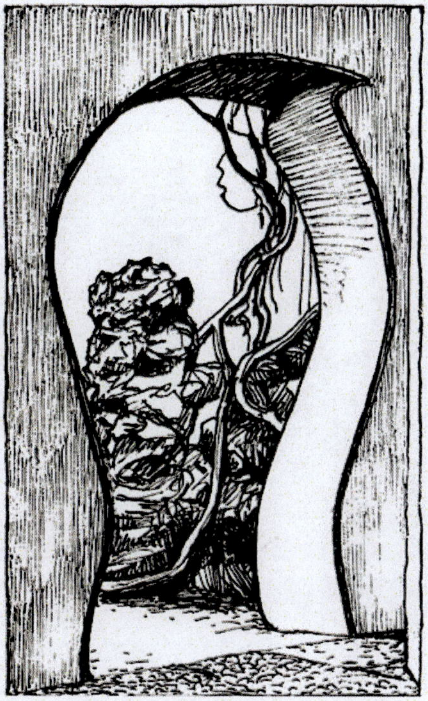

Fig. 3.114 (continued)

Dong Chuang (Hole Window)

Also termed *kuang chuang*[115] and *kong chuang*,[116] *dong chuang* is made on the same aesthetic principle as *dong men*, aiming to accentuate the ornamental interest of the wall by creating cutout spaces in strategic places. Like *dong men*, *dong chuang* also serves as a picture frame that crops the view of a single garden scene and turns it into a framed painting, the result of which is plainly described in *The Craft of Gardens* (Yuan Ye) as *kongxin hua*, or the "hollow painting" (Fig. 3.123b, c). A construction in eastern Garden that was preserved miraculously (Liu Yuan) of Suzhou provides a prime exemplar of an architectural space with garden interest effected by the *dong chuang* treatment, which links the flexuous indoor and outdoor spaces through a sequence of large cutout windows in the wall, resulting in an

[115]Case window.

[116]Cutout window.

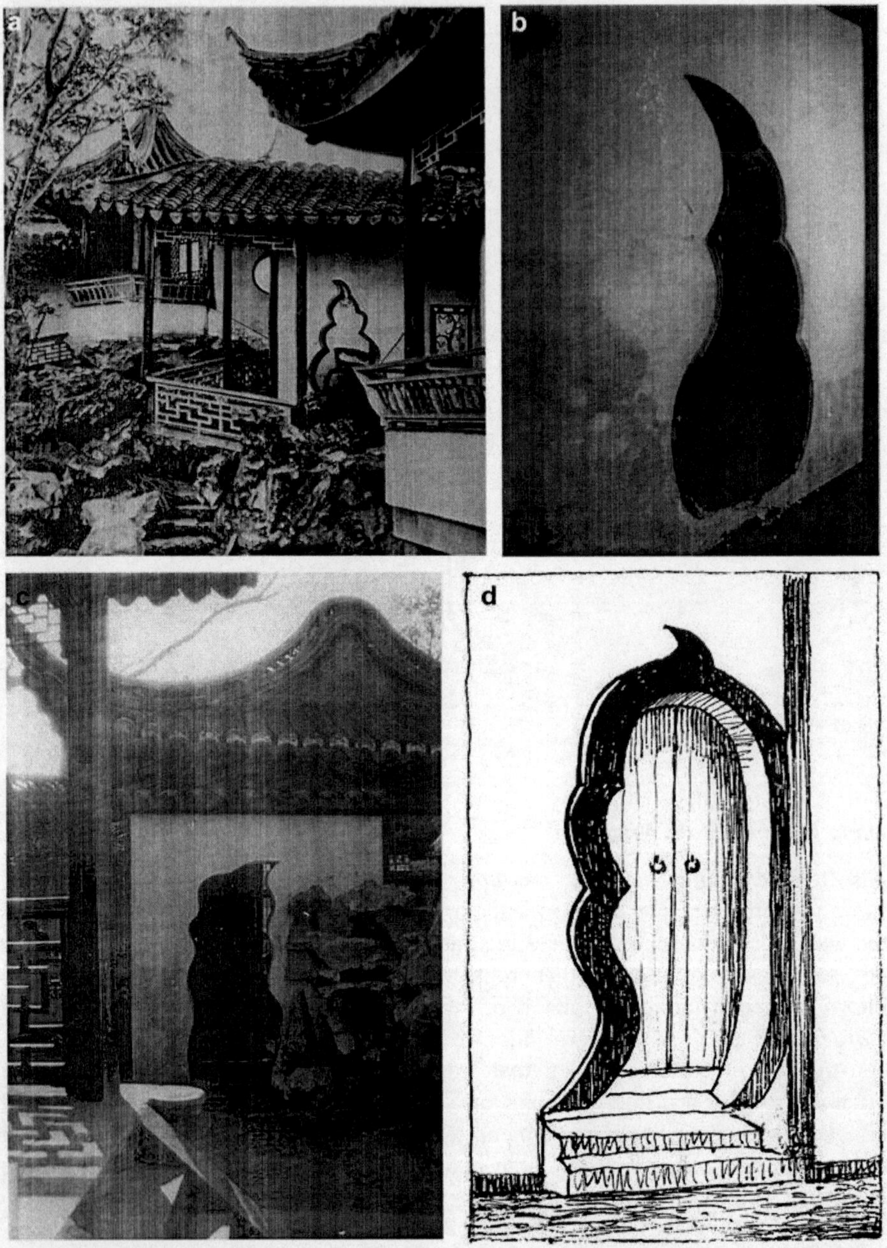

Fig. 3.115 (**a**) The dong men of "shell leaf" style in Chang Yuan, Suzhou. (**b**) The dong men of "shell leaf" style in the side of Yubei Ting (Pavilion of the Imperial Stele) in Shizi Lin, Suzhou. (**c**) The dong men of "shell leaf" style on the left of Yi Fang in Yu Yuan, Shanghai. (**d**) The dong men of "shell leaf" style in Xiling Yinshe, Hangzhou, the leaf stalk is up (Drawn in accordance to the chart board of Jiangnan Yuanlin Zhi)

Fig. 3.116 (**a**) The dong men of "Han vase" styled with perspective bottle in Shi Yuan, Nanxun (Drawn in accordance to the chart board of Jiangnan Yuanlin Zhi). (**b**) The dong men of "Han vase" style in Canglang Ting, Suzhou. (**c**) The dong men of "Han vase" style in front of the Weng's Residence, Changshu (Drawn in accordance to the chart board of Jiangnan Yuanlin Zhi). (**d**) The dong men of "Han vase" style in Qiuxia Pu, Jiading. (**e**) The dong men of "Han vase" style on the left of Yi Fang in Yu Yuan, Shanghai. (**f**) The dong men of "Han vase" style in Xiao Pangu, Yangzhou

f

Fig. 3.116 (continued)

interlaced interior and exterior environment where the vibrancy of the outside rocks and plants seems to infiltrate into the stillness of the room (Fig. 3.124). In respect to the *dong chuang* style, the variety is comparable to that of *dong men*, if not more so. The variegated geometric forms include upright quadrangle (as seen in the aforesaid Garden that was preserved miraculously (Liu Yuan) construction), horizontal quadrangle (Fig. 3.123a), upright and horizontal elongate octagon (Fig. 3.125), hexagon (Fig. 3.126a*)*, octagon and its variant with curved-in sides (Fig. 3.126b, c), round or "full moon" (Fig. 3.126d, e), as well as those with distinctive characters like the bottle gourd (Fig. 3.126f, g), Han vase (Fig. 3.126h), shell leaf (Fig. 3.126i), Chinese flowering crabapple blossom (Fig. 3.126j), plum blossom (Fig. 3.126k), fan (Fig. 3.126l), picture scroll (Fig. 3.126m), and oval (Fig. 3.126n). Some *dong chuang* styles have evolved to be more irregular and more intricate in configuration, as seen in Suzhou's Garden of Surging Wave Pavilion (Canglang Ting) and Nanxiang's Yi Yuan (Fig. 3.127). These windows have actually transformed and become borderline *lou chuang*, the openwork window, or alternatively, a free- or irregular-form *lou chuang*. The *dong chuang* with the "flowers and butterflies" motif in Nanjing's Yu Yuan perhaps represents the most vivacious and most liberal of its kind (Fig. 3.128), and the moon window in Taicang's Half Garden (Ban Yuan), with its lower space masked with decorative cloud openwork, suffices to be classified as a *lou chuang* (Fig. 3.129).

Additionally, in Garden of Bamboo (Ge Yuan) of Yangzhou, a grouping of small rimless moon windows are set at equal intervals in a whitewashed wall, forming a "light-seeping wall", which is a singular example of such creation. The moon-window perforated wall is entitled "*Touyue loufengqiang*",[117] situated on the south of a building given the same title and as the southern backdrop of the Winter Hill (also called Snow Hill). On the west, two small moon windows are made side by

[117] Wall of seeping moonlight and penetrating wind.

a

b

Fig. 3.117 (**a**) The dong men of crabapple blossom style in Shizi Lin, Suzhou. (**b**) The dong men of crabapple blossom style in Qiuxia Pu, Jiading

Fig. 3.118 The dong men of peach style in Yangzhou's Xiao Pangu--stalagmites laid out next to the tao men symbolizes the god of longevity that borrows the implied meaning of "Shouxing Xiantao (The god of longevity offers peaches)". This kind of mundane theme reflects the garden interest of merchants and citizens

side in the lower part of a different wall, which echo the grouped moon windows in the "*Touyue loufeng qiang*" and at the same time lend peephole views of the neighboring Spring Hill scenery on the other side. This is a sequence that shows great ingenuity in *dong chuang* creation (Fig. 3.130).

Fig. 3.119 The gate of calabash style in Yi Yuan, Nanxiang

Lou Chuang (Openwork Window)

A *lou chuang* is an ornamental window screened with decorative openwork. In *The Craft of Gardens* (Yuan Ye), walls that feature such windows are termed *louzhuan qiang*[118] and *louming qiang*.[119]

[118] Porous brick wall.

[119] Light-seeping wall.

Fig. 3.120 The gate of silver ingots style in He Yuan, Suzhou

The half-concealing, half-revealing nature of the *lou chuang* allows unobtrusive privacy and hence they are mostly made in a whitewashed wall that encloses a garden or garden space or a whitewashed wall of a semi-open structure, such as a pavilion, roofed corridor, and windowed veranda; they are sometimes also put in in the corner walls of the outer corridor of a building like a hall or house (Fig. 3.131). The multifarious decorative patterns of the *lou chuang* range from plain and simple to detailed and crafted, each of which is expressive of the interest and subtlety of its own. The most basic are those modeled directly after the prototype of the native popular dwelling, using roof tiles to arrange fish scale, ingot, or linked copper coin patterns, or simply using strip tiles to set window lattice. However, these patterns are not included in *The Craft of Gardens* (Yuan Ye) as garden's *lou chuang* design models for their "commonplace" nature, and, even occasionally adopted for garden use, they have mostly been elaborated and transfigured (Fig. 3.132). There are three types of *lou chuang* in terms of the materials and techniques employed: those set with tiles or wood strips, those carved out of brick, and those sculpted out of marl. Based on the patterning, though, *lou chuang* can be classified into three styles: the geometric style, free style, and mixed style of the two. Sundry geometric-style *lou*

Fig. 3.121 (**a**) The dong men of "small ruyi head" style in Jixiao Shanzhuang, Yangzhou. (**b**) The dong men of "small ruyi head" style in Yi Yuan, Nanxiang (Drawn in accordance to the chart board of Jiangnan Yuanlin Zhi). (**c**) The dong men of variant small ruyi head style (fusing teapothandle and small ruyi head) in Shizi Lin, Suzhou. (**d**) The dong men of kuiwen style in a courtyard at Fanmenqiao Lane, Suzhou (small ruyi head variant) (Excerpted from Suzhou Classical Gardens)

Fig. 3.122 The dong men of "great ruyi head" style of Yu Yuan, Shanghai (Drawn in accordance to the chart board of Jiangnan Yuanlin Zhi).

chuang designs are noted in *The Craft of Gardens* (Yuan Ye), including (in repeated patterns) cross, herringbone, hexagon, octagon, water-caltrop blossom, swastika (with the ends of the arms extended left), writing brush, silk braid knot, interlinked diamonds, high mallow blossom, waves, plum blossom, crabapple blossom, ice crackles, and conjoined petals. More often than not, two or more of these patterning elements are combined in one design, like crosses inwrought with crabapple blossoms, or ice chips with interludes of plum blossoms (Fig. 3.133). The free-style *lou chuang* design draws mostly on the subject of flora and fauna that embodies propitiousness or refinement, such as deer, crane, bat, phoenix, pine, bamboo, orchid, plum, chrysanthemum, plantain, lotus, pomegranate, and peach (Fig. 3.134a, b), and the "freeness" of the style is sometimes also echoed in the design of the window shape (Fig. 3.134c, d). Other favorable subjects often featured in the free-style openwork window design see Chinese characters, the four treasures of the study (i.e. writing brush, ink stick, inkstone, and paper), as well as historical figures and stories. Occasionally, a free-style and a geometric-style patterns blend into one design, where the geometric pattern usually borders the window or serves as a foil while the free-style pattern decorates the center. Take the *lou chuang* in Garden of Surging Wave Pavilion (Canglang Ting) of Suzhou as examples: one features repeated patterns of the conjoined swastika and square that set off the bamboo flute and palm-leaf fan design in the center, and the other curvilinear tracery inwrought in the middle with a "double happiness" character (Fig. 3.135). Generally, the opening of the *lou chuang* is fully filled with the ornamental openwork. In Couple's Retreat Garden (Ou Yuan)'s East Garden in Suzhou, however, the design of a circular *lou*

a b c

Fig. 3.123 (**a**) The view-cropping frame composed of the horizontal quadrangle dong chuang of ban lang in Jichang Yuan, Wuxi. (**b**) "Kongxin hua" (hollow painting) composed of the horizontal quadrangle dong chuang of Xiangyue Lang in Yi Pu, Suzhou. (**c**) "Kongxin hua" composed of the horizontal quadrangle dong chuang in Wangshi Yuan, Suzhou

Fig. 3.124 The scenic space with intersected dong chuang in the east part of Liu Yuan, Suzhou

chuang with the sunflower motif and a decorative periphery in relief shows a hollow in the center of the window (Fig. 3.136a). This evinces a configurative transition to a *feng chuang* (air window)-style *lou chuang*, the former being an interior wooden lattice window that tends to feature a void space in the center. The transitive style further evolves into the "hollow style" with the increasingly large unfilled space bordered by the receding decorative openwork, which, in this case, is made after the fashion of the *feng chuang*'s latticed woodwork (Fig. 3.136b, c). Analogous to a picture frame and intended for use as a view-cropping device, these *feng chuang*-style *lou chuang* also recall the function of the afore-discussed *dong chuang*, the hole window. The tiled *lou chuang* in the East Garden of Garden to delight one's parents (Yu Yuan) in Shanghai is another example that assumes the *feng chuang* style by simulating its wooden latticework (Fig. 3.136d). More discussion on *feng chuang* is resumed in the later section on "Architectural Ornament, Furniture, and Interior Display".

Fig. 3.125 (**a**) The dong chuang of horizontal elongate octagon style of "Zhuwai Yizhi Xuan" in Wangshi Yuan, Suzhou. (**b**) The dong chuang of upright elongate octagon style of "Lüyin" in Liu Yuan, Suzhou

3.1.1.2.2 Harmonic Union of the Architectural and Natural Elements

In making of the Jiangnan garden, placement of architectural structures—the artificial constituent of the garden—and disposition of topography, vegetation, and other natural constituents are closely coordinated. A reciprocal correlation of the two is a time-honored subject in the artistic creation of the Jiangnan garden, which has yielded a wealth of valuable experience throughout the generations. In Jiangnan gardens, architectural structures are mostly arranged surrounded by mountain and

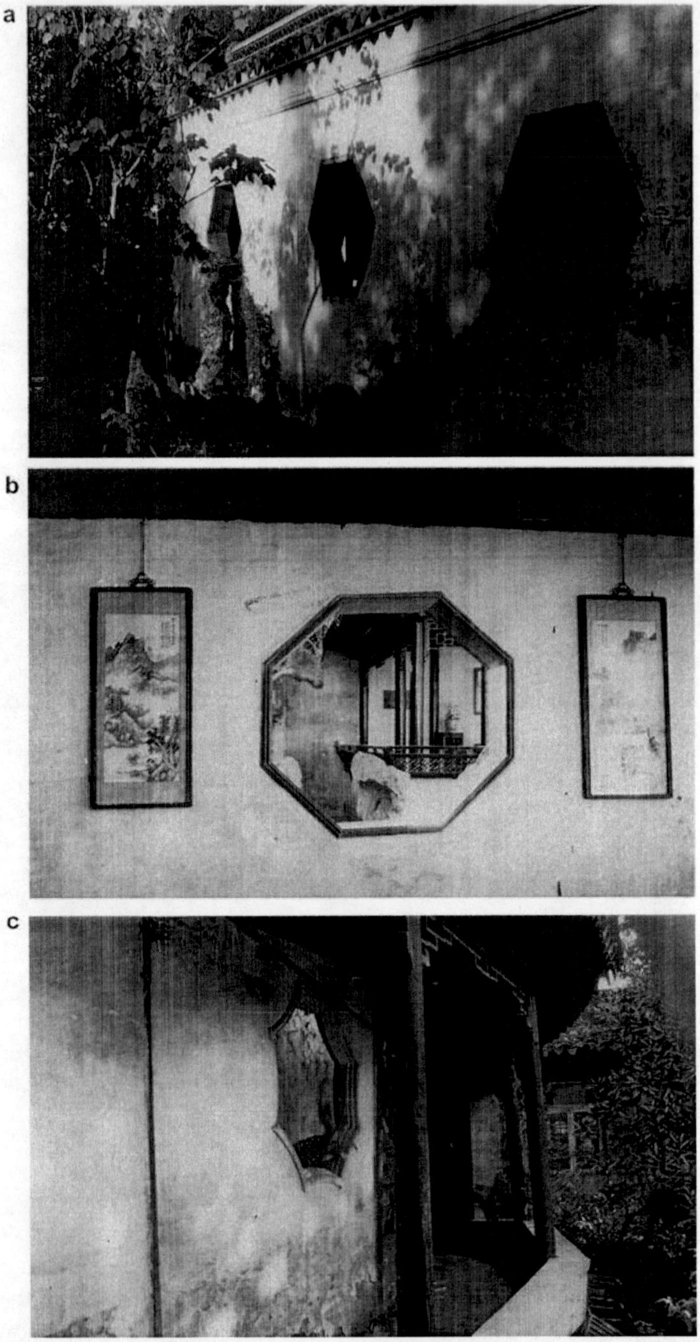

Fig. 3.126 (**a**) The dong chuang of "hexagon" style of Shizi Lin, Suzhou. (**b**) The dong chuang of octagon style of "Lüyin" in Liu Yuan, Suzhou. (**c**) The dong chuang of variant (arc) octagon style in Ban Yuan, Suzhou. (**d**) The dong chuang of moon-window style with no frame in the west of Letian

water features, which requires that the architectural planning be thoughtfully incorporated in the mountain-and-water context. In artistic epitome, the garden architecture takes largely after the fashion of the region's popular dwellings in mountain villages and canal-filled towns. In the subsequent segments, the discussion will focus on the association between some architectural placements and mountain and waters features.

3.1.1.2.2.1 Architectural Treatment of Building Adjoining Man Made Mound

Architectural treatments that are associated with mountain features can be classified into two types: those that serve to foil the mountain scenery and those that delineate and accentuate the architectural feature in a mountainous setting. In the former case, the mountain feature takes the center stage while the role of the architectural treatment is minimal but nevertheless often gives the finishing touches to the scenery. In the latter situation, the architecture is the predominant factor and takes a larger proportion in the construction, the key issue, however, lies in the architectural treatment implemented to manifest the character of a mountainous environment.

Stone Paving in Mountain Fields

As an architectural contrivance that works in with mountain scenery, ground paving is often completed by using fieldstone or flagstone to effect natural-looking pavage that maintains a rough surface with spontaneous tilts and dips of the stones, a realistic paving style intended to simulate a natural mountainous area. This touch of architectural treatment however, unlike the pavement in ice-crackle patterning, is actually a form of stone setting rather than an architectural execution in the full sense. On the hill of the Free Roaring Pavilion (Shuxiao Ting) in the western part of Garden that was preserved miraculously (Liu Yuan) in Suzhou, the rough stone pavement is a good example of the realistic-style paving that relishes of unkempt rusticity, so is the hilltop pavement that corresponds to the piled stones in the East Garden of Couple's Retreat Garden (Ou Yuan), also in Suzhou (Fig. 3.137). The ice-crackle pavement is a further symbolic presentation of the natural, irregularly formed stony ground surface in a mountainous area. It is usually a smoothly paved

Fig. 3.126 (continued) Lou in Tianping Shan Gaoyi Yuan, Suzhou. (**e**) The dong chuang of moon-window style with frame in the side of Lixue Tang in Shizi Lin, Suzhou. (**f**) The dong chuang of calabash style in Yi Yuan, Suzhou. (**g**) The dong chuang of calabash style in Chengbi Shanzhuang, Changshu (Drawn in accordance to the chart board of Jiangnan Yuanlin Zhi). (**h**) The dong chuang of Han vase style in Liu Yuan, Suzhou. (**i**) Dong chuang of shell leaf style in Yi Yuan, Suzhou. (**j**) The dong chuang of crabapple blossom style in Jixiao Shanzhuang, Yangzhou. (**k**) The dong chuang of plum blossom style in Jixiao Shanzhuang, Yangzhou. (**l**) The dong chuang of fan style of "Yushuitongzuo Xuan"—Shanmian Ting (Fan Pavilion) in Zhuozheng Yuan, Suzhou. (**m**) The dong chuang of picture scroll style in Xiao Pangu, Yangzhou. (**n**) The dong chuang of oval style in Xian Yuan, located in Mudu of Wu County, Suzhou (Excerpted from the atlas of Chinese Architecture)

Fig. 3.126 (continued)

patchwork of flagstones that are more trimly seamed and leveled than the rough stone-paving surface. To work in with the construction of a mountain feature, the ice-crackle patterning is the most frequently adopted paving form in stone ground pavement in the Jiangnan garden. A mountain-theme architectural treatment does not necessitate the physical presence of a mountain, nor does it have to be executed

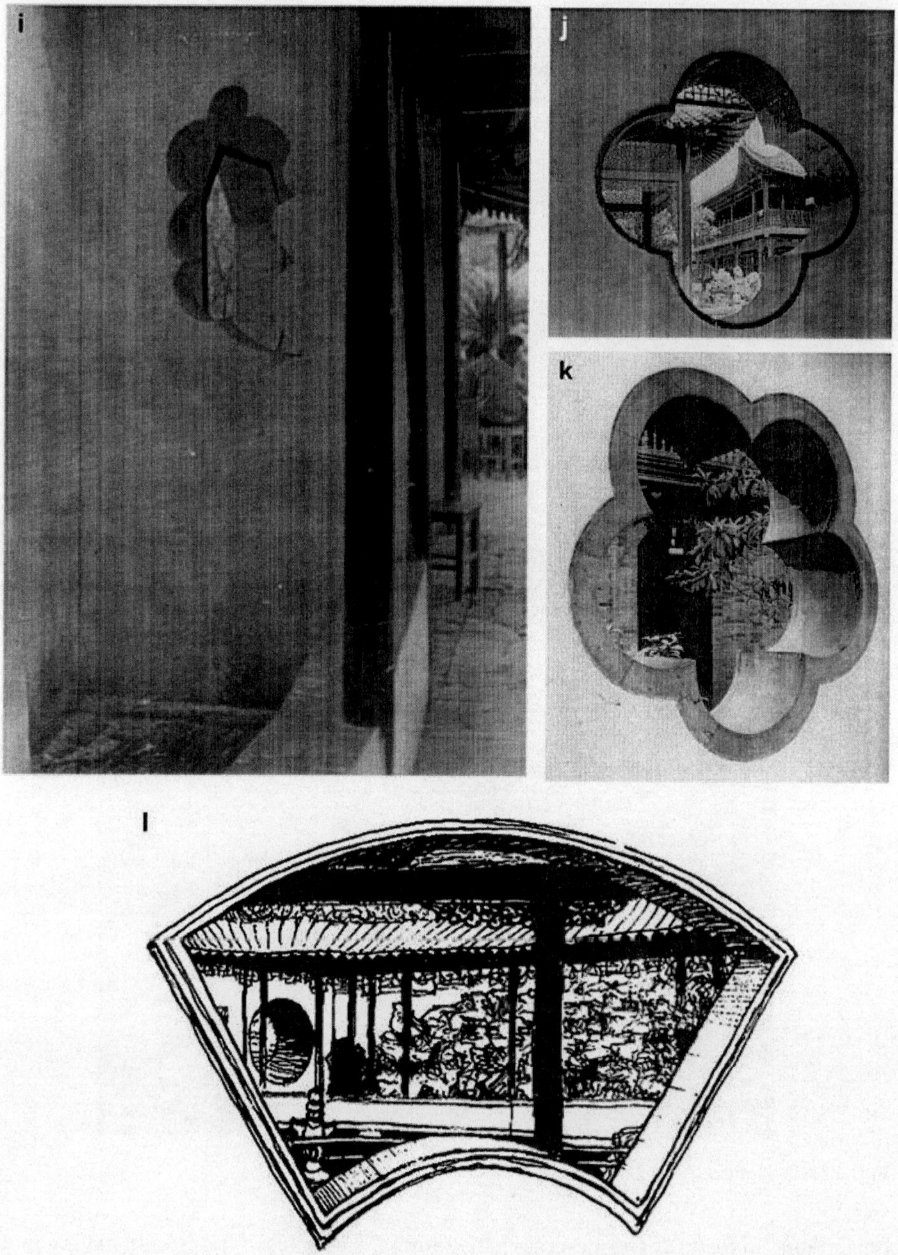

Fig. 3.126 (continued)

m

n

Fig. 3.126 (continued)

on a mountain feature. The ice-crackle patterning is often used for ground paving in a flat architectural environment as well where only simple stone setting is featured to echo the mountain theme, such as an upright freestanding solitary Lake stone that suggests a mountain peak or a group of strategically dispersed rocks that recall undulating mountain ranges. In Suzhou, the front yard of The Five Peaks Studio (Wufeng Shuwu) in the Master-of-Nets Garden (Wangshi Yuan) and the south yard of Pavilion of the Fragrance of Lotus Roots (Ouxiang Xie) as well as the front yard

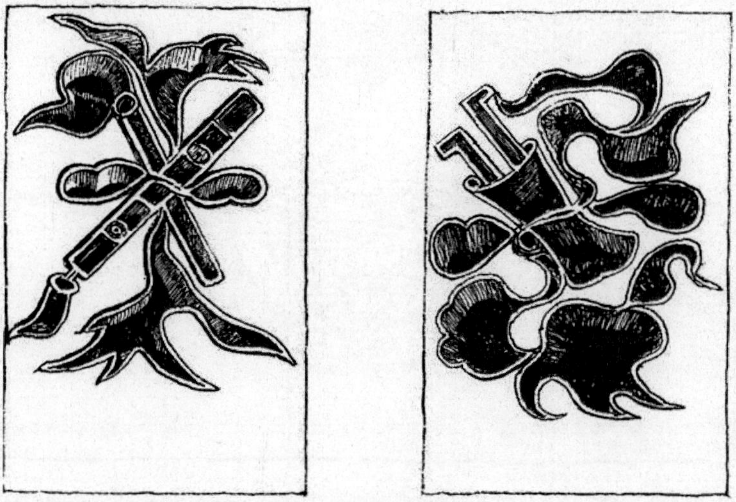

Fig. 3.127 The dong chuang or lou chuang of free style in Yi Yuan, Nanxiang (Drawn in accordance to the chart board of Jiangnan Yuanlin Zhi)

Fig. 3.128 The dong chuang of free style in Yu Yuan, Nanjing, also can be regarded as a lou chuang—butterflies and flowers (Drawn in accordance to the chart board of Jiangnan Yuanlin Zhi)

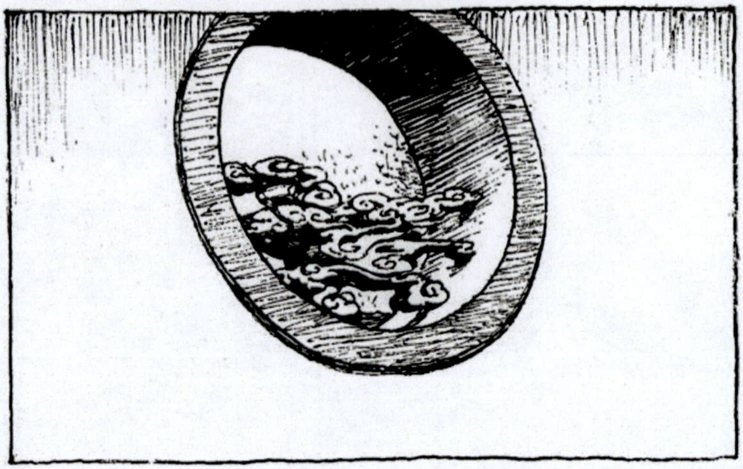

Fig. 3.129 The dong chuang of full moon style in Ban Yuan, Taicang, with its lower space masked with decorative cloud openwork, suffices to be classified as a lou chuang (Excerpted from Jiangnan Yuanlin Zhi)

of Thatched Hut of the Cold Season (Suihan Caolu) in Garden of Pleasure (Yi Yuan) all provide but a few such examples. Owing to its natural vein and decorative effect, the flagstone ice-crackle pavement is sometimes placed regardless of the absence of a mountain feature in the setting, as in the case of Mountain House of Lingering Clouds (Liuyun Shanfang) in Suzhou's Garden of Natural Delight (Chang Yuan), whose front terrace is paved in ice-crackled pattern with irregular flagstones, redolent of a mountainous environment that ties in with the theme of the architecture (Fig. 3.138). Ice-crackle paving materials are not confined to flagstone alone; occasionally, a combination of strip bricks, tiles, pebbles, or glazed pottery and porcelain shards is used to effect an ice-crackle patterning. However, the result of such ice-crackle paving no long bears any direct implication of the mountain theme, as its stylized patterning takes on certain luxuriance and has become a decorative treatment that sets off a gracious atmosphere of the architectural environment implicit of natural beauty. A special discussion on non-mountain feature related paving surface will be given in the segment on "Garden Paths" in a later section.

Natural Stone Bedding and Stepstone

In Jiangnan gardens, the architectural structure that is most commonly used to garnish a mountainscape is *ting*, the pavilion. In residential gardens of Suzhou, the lake-and-mountain composition is customarily the thematic subject of the garden's principal scenery, where a small-scale, open-sided square or hexagonal pavilion roofed in the cone-shaped *cuanjian* style is typically placed on the hilltop or hillside. Occasionally, a rectangular pavilion is placed instead, reminiscent of a rustic mountain village inn ensconced in a wooded hill. To help manifest the character of

Fig. 3.130 (a) "Touyue loufeng qiang" in Ge Yuan, Yangzhou—a special light-seeping wall. (b) Two circular dong chuang in the lower part of the wall in the west side of "Touyue loufeng qiang" of Ge Yuan, Yangzhou

mountain architecture, such as a hilltop pavilion, natural stone like Mountain stone or Lake stone is often arranged around the base or bedding of the structure, evoking a hilly environment with denuded bedrocks where the structure is situated. Garden of Surging Wave Pavilion (Canglang Ting) is a prime exemplar of such effecting that imparts an effective atmosphere (Fig. 3.139); others examples see Pavilion of Crowning Clouds (Guanyun Ting) in Garden that was preserved miraculously (Liu Yuan) and Cold Spring Pavilion (Lengquan Ting) in the Master-of-Nets Garden (Wangshi Yuan) (Fig. 3.140), all three gardens located in Suzhou. In some occasions, only a few pieces of natural stone are placed to convey the mountain theme, as in the case of Wufengxian Guan in Suzhou's Garden that was preserved

Fig. 3.131 (a) The lou chuang in the hill-climbing wall of Jixiao Shanzhuang, Yangzhou. (b) The lou chuang in the garden wall of Jixiao Shanzhuang, Yangzhou. (c) The lou chuang in the partition wall of Yi Yuan, Suzhou. (d) The lou chuang in the enclosed partition wall of Liu Yuan, Suzhou. (e) The lou chuang in the Qingfengchi Guan (xuan, xie) of Liu Yuan, Suzhou. (f) "Lingzhao Tingzhi" lou chuang in "Jingguan (static appreciation)" hall of Nei Yuan, Shanghai. (g) The lou chuang in the corner whitewashed walls of the outer corridor of the hall in Qushui Yuan, Qingpu

miraculously (Liu Yuan) and Veranda of Rain Delight (Yiyu Xuan) in Yangzhou's Garden of Bamboo (Ge Yuan), where the irregular stepstone is spontaneously set next to the raised ground of the house (Fig. 3.141). Such placement, especially when combined with the arrangement of a natural stone planting bed, conjures a mountainous atmosphere even in flat surroundings. Additionally, a fitting scale of landscaping together with a proportionate configuration of the architecture can also

Fig. 3.131 (continued)

render a unified effect between the building and the mountain-theme environment, like the construction of the hilltop Free Roaring Pavilion (Shuxiao Ting) and its surrounding landscape in Garden that was preserved miraculously (Liu Yuan) of Suzhou.

Pashan Lang (Hill-climbing Corridor)

To intensify the effect of mountain architecture, buildings, such as a mountain villa, are sometimes set on the top of a hill, which expediently gives the building a natural elevated base. With the hillside hewn into a staircase that leads up to the building, the construction is in fact analogous to a multistoried building in usage (Fig. 3.142). Mountain Outlook Tower (Kanshan Lou) is a two-storied pavilion compound perched atop a stone mountain in Suzhou's Garden of Surging Wave Pavilion (Canglang Ting). The stone mountain, as the foundation of the building, is hollowed out to accommodate two rooms, felicitously named "Yinxin Shiwu" (Heartstring Stone House), which renders the building three-storied in structure. To reinforce the correlation between the building and the mountain, the roughhewn rocky stairway that winds up the mountain slope and leads to the entrance of the pavilion compound is built in combination with a roofed corridor, forming a *pashan lang*, the hill-climbing corridor, whose progressional configuration conforms to the natural contours of the hill (Fig. 3.143). In the case of multistoried buildings that are built in a level-ground environment absent of desired topographical relief, stone features, such as a rockery, are often affixed to the setting to articulate the mountainous space. In some cases, natural stone is cut and laid in the fashion of a rough stepstone mountain path, replacing an otherwise regular wooden or brick staircase of the building (Fig. 3.144).

To accentuate the architectural interest and intensify the undulating effect of the lay of the land, the mountain passageway that winds up to a mountain-based construction is often built in the form of a *pashan lang*. A symbolic "hill-climbing"

Fig. 3.132 (**a**) The lou chuang made of dark gray ground tiles with geometric patterns (Jixiao Shanzhuang in Yangzhou). (**b**) The detail of lou chuang made of dark gray ground tiles. (**c**) The lou chuang made of delicate ground tiles in Ge Yuan, Yangzhou. (**d**) The lou chuang made of elegant whitewashed ground tiles in Ge Yuan, Yangzhou. (**e**) The lou chuang made of elegant whitewashed ground tiles with the layout of roses in Suzhou's Yi Yuan is more attractive. (**f**) The lou chuang assembled from whitewashed tiles in Yi Yuan, Suzhou. (**g**) The oval lou chuang of geomatric pattern assembled from tiles in Qushui Yuan, Qingpu. (**h**) The lou chuang of picture scroll style with regular ice crackle pattern in Xu Yuan, Nanjing. (**i**) The circular lou chuang with variant sunflower motif and hexagonal lou chuang with ice crackle pattern in Wangshi Yuan, Suzhou

pashan lang can be built on a level ground as well to lend a mountain architectural effect in a flat environment, provided that the side of the raised bedding that creates the grade of the corridor is strategically concealed with the arrangement of natural stones to give the illusion of a hillside. The form of *pashan lang* usually assumes two

Fig. 3.132 (continued)

basic fashions: the stepped style and the rolling style. The former runs at a descending (or ascending) stepped elevations (Fig. 3.145a) and the latter rolls in accord with the contours of the hill (Fig. 3.145d, e). In Xiao Pangu of Yangzhou, the *pashan lang* nevertheless demonstrates a unique roof style, which is built abutting a *pashan qiang* (hill-climbing wall) in the form of half pavilion roofs with upturned eaves (Fig. 3.145f, h).

Pashan Qiang (Hill-Climbing Wall)

Like *pashan lang*, *pashan qiang*, the hill-climbing wall, is an accessorial structure in the Jiangnan garden, used to heighten the heaving mountainous feel and accentuate the continuity of the mountain ridge. It is one of the most expressive architectural treatments in the Jiangnan garden. Taking after the enclosing wall in the local mountainous region, which is marked by a free-form curvilinear shape due to its conformity to the lie of the terrain, *pashan qiang* is an artistic interpretation (Fig. 3.144a) of its real-life prototype that gives a mountain living atmosphere to

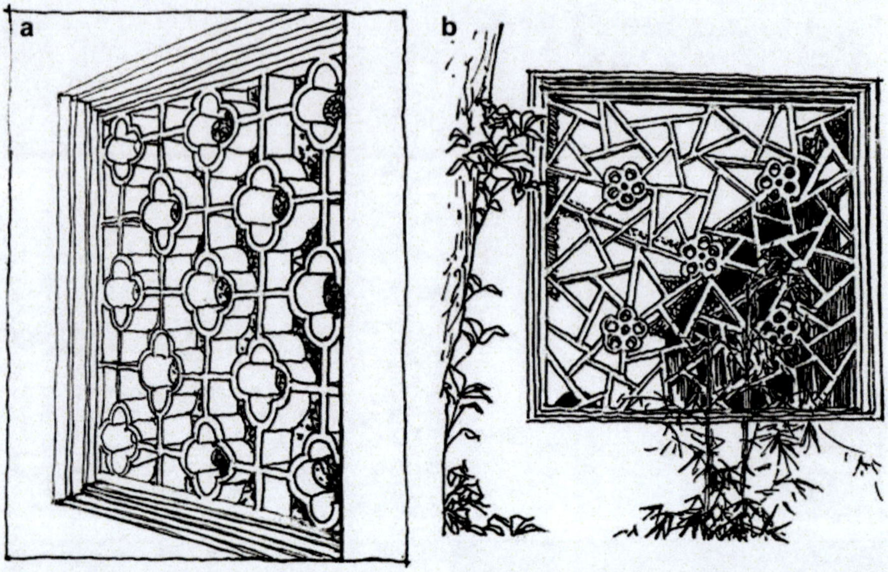

Fig. 3.133 (**a**) The lou chuang with the pattern of crosses inwrought with crabapple blossoms (Liu Yuan, Suzhou). (**b**) The lou chuang with the pattern of ice chips with interludes of plum blossoms (Wangshi Yuan, Suzhou)

an urban residential garden. Similar to *pashan lang*, the configuration of *pashan qiang* also takes on two symbolic styles: the stepped style that extends at descending stepped heights and the rolling style with a signature wavy coping that suggestively runs along the ridgeline of an undulating mountain. Because the curves in the wavy design resemble the curls of the cloud, the rolling-style *pashan qiang* is also called "*yun qiang*" (cloud wall). With its livelier and more natural conformation, the cloud wall is the more popular choice for Jiangnan garden makers than the abstract stepped-style *pashan qiang*. In the artistic execution, the sinuosity of the cloud wall coping does not necessarily conform literally to the lay of the land but rather is created for the symbolic and aesthetic effect attained, and the progression of the waves can be both free and regular in form. For this reason, the wavy coping is often licentiously implemented even when the wall is raised on a flat ground. Gardens like Yi Yuan in Nanxiang and Xiao Pangu in Yangzhou feature some of the few cloud walls with an irregular-style wavy coping that heaves freely with the landform and relishes of a naturalistic flavor (Fig. 3.146). The naturalistic leanings of the cloud wall are notwithstanding evident even when the wall is coped in the patternized regular wavy style, like the one in western Garden that was preserved miraculously (Liu Yuan) of Suzhou, which is built on the hill to serve as a foil to the garden's mountain-theme landscape (Fig. 3.147a, b). The key to a successful creation of the cloud wall is that the height of the wall is raised only to the eye level or lower in order to externalize the elevated sense of the mountain altitudes and impart the illusion of standing at a higher vantage point. In the case of the above-mentioned

Fig. 3.134 (**a**) The lou chuang with painted sculpture of red-crowned crane (Hui Long Tan of Jiading). (**b**) The lou chuang with painted sculpture of squirrel and grapes (Hui long Tan of Jiading). (**c**) The lou chuang with motif of entangled branches (Shizi Lin of Suzhou). (**d**) The lou chuang with motif of lotus (Canglang Ting, Suzhou)

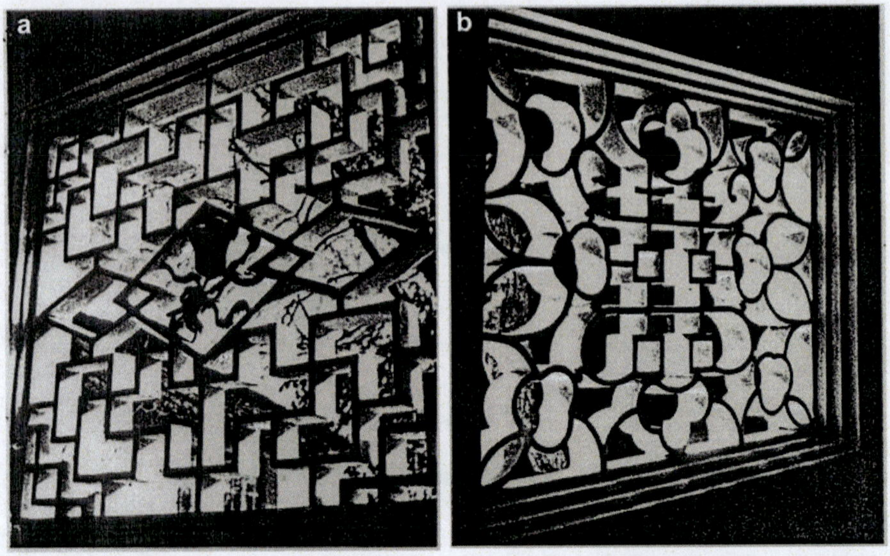

Fig. 3.135 (**a**) The lou chuang with patterns of conjoined swastika and square that set off the bamboo flute and palm-leaf fan in Canglang Ting, Suzhou. (**b**) The lou chuang with curvilinear tracery inwrought in the middle with a "double happiness" character in Canglang Ting, Suzhou

Garden that was preserved miraculously (Liu Yuan), the top of the cloud wall is barely to the eye level when perceived from either side of the wall. In Suzhou, aside from Garden that was preserved miraculously (Liu Yuan), gardens like Lion Grove Garden (Shizi Lin) and the Humble Administrator's Garden (Zhuozheng Yuan) also have masterful creations of the cloud wall (Fig. 3.176c, d). Low is the key. However, in a relatively open space, tall cloud walls are seen built by following the contour of the hill, such as in the garden of Tower of Rising Clouds (Yunqi Lou) near Mount Hui in Wuxi (Fig. 3.147e). When a cloud wall with the regular curvilinear coping is built on a flat land, the structure becomes more abstract in its symbolic capacity and requires coordinated arrangement of stone or earth features, such as a rockery or an earthen acclivity, to work in with the mountain theme that a cloud wall embodies. The cloud wall in the Loquat Garden (Pipa Yuan) inside the Humble Administrator's Garden (Zhuozheng Yuan) of Suzhou provides a fine exemplar of flatland cloud wall construction: built with the support of piled stones flexuously arranged in separate but coherent groupings that form a rolling range of hills, the wall is elevated to a height low enough that the rooftops of the houses on the ulterior side are visible, which not only increases the depth of field of the garden but more desirably creates the impression of a high vantage point on the hill (Fig. 3.148a). In Suzhou's Garden of Pleasure (Yi Yuan), the cloud wall next to the Hoeing-the-Moon Windowed Veranda (Chuyue Xuan) is another fine example of such a kind (Fig. 3.148b). The challenge of making the mountain-theme cloud wall on a flat ground, however, has yielded a few awkward productions as well, with the one in Xu Yuan of Nanjing

Fig. 3.136 (**a**) The lou chuang with the decorative periphery in relief (variant sunflower motif) in Ou Yuan's East Garden, Suzhou—shows a hollow in the center of the window, which evinces a configurative transition from lou chuang to the decorative periphery in relief. (**b**) The feng chuang (air window)-style lou chuang with decorative periphery in relief in the side of Yanyu Tang in Shizi Lin, Suzhou. (**c**) The feng chuang (air window)-style lou chuang with decorative periphery in relief in Zhan Yuan, Nanjing. (**d**) The feng chuang (air window)-style lou chuang with pattern of swastika beside the pavilion-capped bridge in Yu Yuan's East Garden, Shanghai

being the most stilted as the tall and stiff structure was built without the coordinative stone or earth features. The cloud wall in Meandering Water Park (Qushui Yuan) of Qingpu near Shanghai, to give another example, though raised low, also fails to bring out the desired result in the sheer absence of offsetting stone or earthwork

Fig. 3.137 The rough stone pavement (which is the natural form of ice-crackle pavement) on the hill of Ou Yuan's East Garden, Suzhou

Fig. 3.138 The ice-crackle pavement in the front moon-viewing terrace in Liuyun Shanfang in Chang Yuan, Suzhou

Fig. 3.139 The design with a unified effect between the Canglang Ting in Suzhou and the mountain-and-forest environment

(Fig. 3.149). In a relatively open environment, the cloud wall is often configured in curvilinear shape on both vertical and horizontal planes, forming a wavy-topped wall snaking through a garden space, which gives rise to the name "*long qiang*".[120] This figurative implication of the cloud wall, however, tends to be taken too literally by unwitting workmen. For instance, the *long qiang* in Shanghai's Garden to delight one's parents (Yu Yuan) was constructed with a carved dragon head attached to one end of the rolling wall top and a tail to the other, which turned the entire coping of the cloud wall into a wriggly monstrous reptile (Fig. 3.150), catering to the gaudy taste of the proprietor of the garden that had been in the possession of the Money Industry Trade Union since the Emperor Qianlong era of the Qing dynasty. This treatment, however, has completely deprived the structure of its original naturalistic connotation that a cloud wall is meant to represent—the undulating mountain motif. As with

[120]Dragon wall.

Fig. 3.140 The platform foundation of Lengquan Ting in Suzhou's Wangshi Yuan is concealed with the arrangement of natural stones

the roof or eaves embellishment of lion or bat carvings, it has become mere ornamentation symbolic of wealth and auspiciousness. In the making of the natural-landscape Jiangnan garden, where the aesthetic criterion of the garden scenery is its poetic and pictorial quality reminiscent of nature, such a garish style nevertheless reflects a creative proclivity for the grotesque appeal rather than natural interest, and a fang-baring dragon figure atop the cloud wall is indubitably non-simpatico with the creative principle of the Jiangnan garden.

3.1.1.2.2.2 Architectural Treatment of Building Adjoining Water

As far as the relative position between architecture and water is concerned, architecture can be found on the waterfront or on the water. If they are on the same water, buildings and bridges have different scenes. Therefore, this should not be the focus of the discussion. Here, according to the characteristics of the scene to choose its main discussed below.

Yue Tai: Moon-Viewing Platform

In a residential garden, the tingtang (hall) construction is the principle architecture, which usually takes the form of a hehua ting (lotus hall), yuanyang ting (tandem hall), or simian ting (four-sided hall). It is customarily positioned to front the garden's principle scenery, typically adjoining to a body of water. The interaction between the tingtang (hall) and water is effected mainly through the intermediary structure, a moon-viewing platform, which is an outdoor extension attached to the front of the building (Fig. 3.151). In the Dunhuang cave temples in northwestern

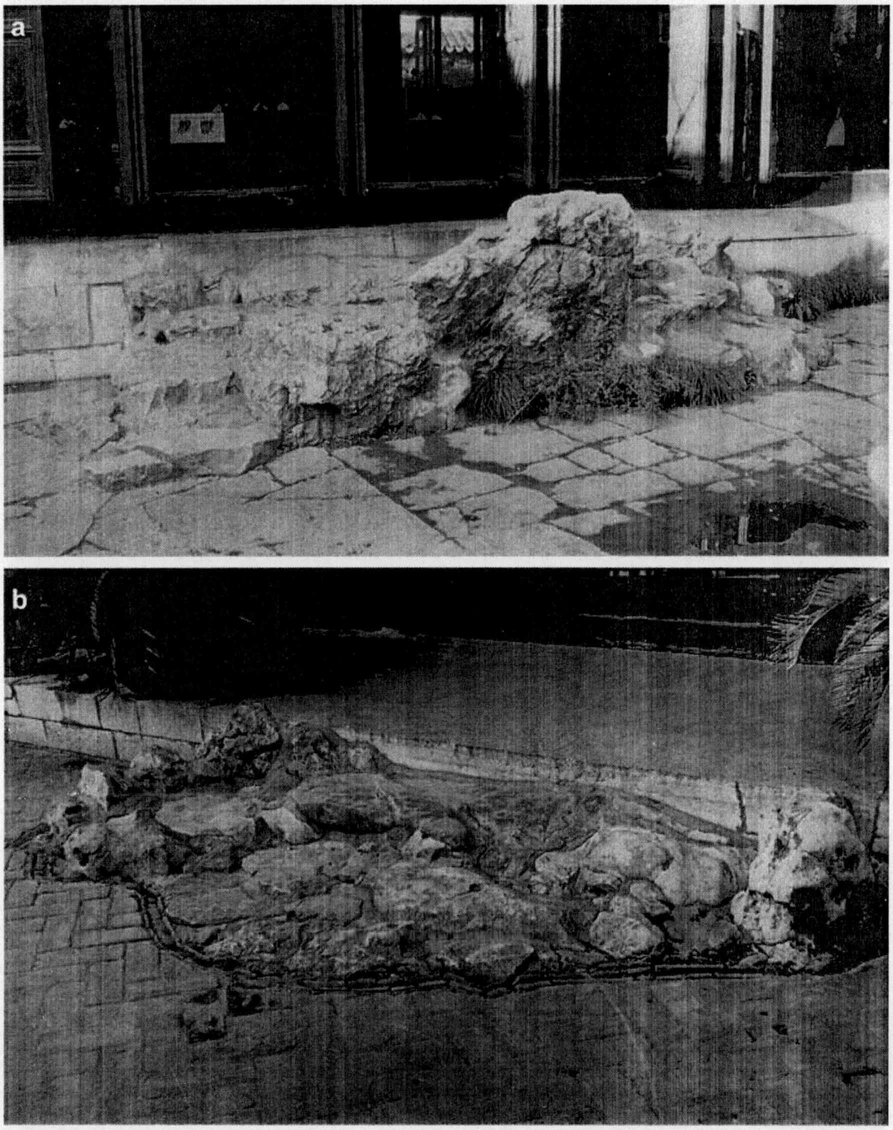

Fig. 3.141 (**a**) The stepstone made of natural stones in Wufengxian Guan of Liu Yuan, Suzhou. (**b**) The stepstone made of natural stones in Yiyu Xuan of Ge Yuan, Yangzhou

China, descriptions of the palace built with a yue tai by the lotus pond are painted in the Tang frescoes, demonstrating a long history of the yue tai as a garden structure.

Also called liang tai (cool platform) or lu tai (dew platform), the yue tai offers an ideal platform for people to develop a kinship with water and affords an idyllic open space for appreciating the waterscape and admiring a full moon. In small residential

Fig. 3.142 The scenery combined Wen Muxixiang Xuan with mountains in Liu Yuan, Suzhou

gardens where a full-scale moon-viewing platform cannot be accommodated, the vestibule of the tingtang (hall) that fronts the water is often opened up and configured into the form of a yue tai, as done for the front of Mountain House of Lingering Clouds (Liuyun Shanfang) in Garden of Natural Delight (Chang Yuan) of Suzhou (Fig. 3.152).

Shui Ting: Waterside Pavilion

The waterside pavilion is an artistic interpretation of the real-life ferry shelter in the rural country intended to lend the waiting passengers a resting place and protection from the elements. The ferry shelter is usually situated on the frontage of the water rather than above the water, as what the waterside pavilion in the Jiangnan garden has become in order to assume a closer affinity with water. Most of the existing *shui ting* constructions belong to the late Qing period and were constructed in this fashion. Formally, the waterside pavilion is identical with its land- or mountain-based counterpart in the garden, only differing from the latter in the base treatment. There are three techniques commonly employed in constructing the base of a waterside pavilion. One is to raise a solid geometric platform joined to a regular-shaped embankment or stone-lined bank, as in the case of Green Ripples Pavilion

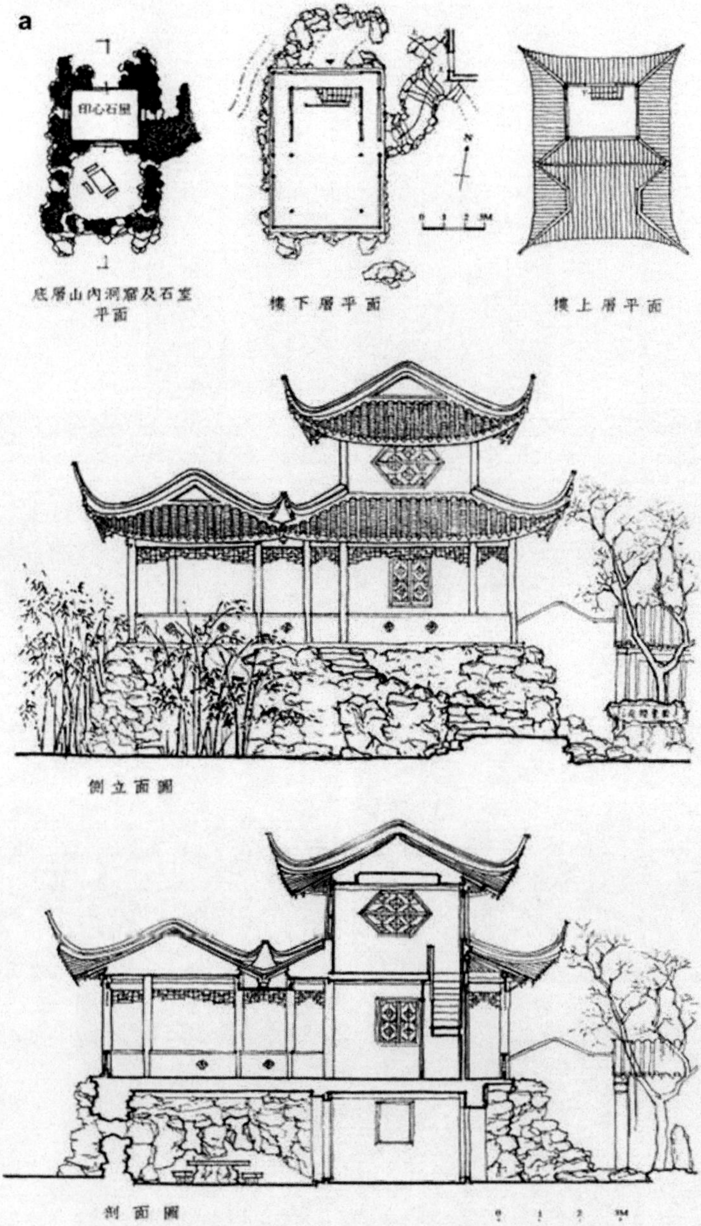

Fig. 3.143 (**a**) The combination of Kanshan Lou and mountains in Suzhou's Canglang Ting—plan, elevation and section (Excerpted from Suzhou Classical Gardens). (**b**) Kanshan Lou in Canglang Ting, Suzhou. (**c**) The staircase combined with rockery of Kanshan Lou in Suzhou's Canglang Ting—the form of pashan lang. (**d**) The ground floor of Kanshan Lou of Suzhou's Canglang Ting hidden inside the mountain—Yinxin Shiwu. (**e**) One of the rooms in Yinxin Shiwu of Canglang Ting, Suzhou

Fig. 3.143 (continued)

Fig. 3.144 (**a**) The rockery of Guanyun Lou in Suzhou's Liu Yuan was designed as a staircase. (**b**) The rockery made of lake stones of Duhua Lou in Suzhou's Wangshi Yuan was designed as a staircase. (**c**) The natural staircase of Mingse Lou in Suzhou's Liu Yuan, beside which is the stone peak named "Yi Ti Yun" (looking north from "Qiahang", the ground floor of Mingse Lou). (**d**) The rockery made of yellow stones in the pavilion of Qushui Yuan in Shanghai's Qingpu was designed as a staircase

(Lüyi Ting) in Suzhou's Humble Administrator's Garden (Zhuozheng Yuan), which also features at the bottom of the base a short staircase leading down to the water, reminiscent of a ferry crossing that the *shui ting* originates from (Fig. 3.153). The second technique takes after the stilted style of pile dwelling, with the intention to intensify the relationship between the structure and water (refer to Fig. 3.91c from

a

b

Fig. 3.145 (**a**) The pashan lang with the stepped style (Xihui Park, Wuxi). (**b**) The pashan lang with the rolling style (Xian Yuan in Mudu of Wu County, Suzhou) (photographed in the 1930s based on Jiangnan Yuanlin Zhi). (**c**) The pashan lang with therolling style of Kanshan Lou in Canglang Ting, Suzhou. (**d**) The lower part of pashan lang with the rolling style (Wangshi Yuan, Suzhou). (**e**) The pashan lang with a unique roof style (Xiao Pangu, Yangzhou). (**f**) The interior space of pashan lang (Liu Yuan, Suzhou). (**g**) The interior of pashan lang in Xiao Pangu, Yangzhou

the previous section). The third commonly adopted technique in making a *shui ting* foundation is using free-form natural stone, which involves three styles of execution. One is to stack stone piers in the water to form a trestle where the pavilion stands (Fig. 3.154), and another can be viewed as an elaborate form of the first, where the stone piers overarch the water surface and form a natural stone cave or tunnel beneath the pavilion (Fig. 3.155). Pavilions perched above the water on the stilted stones such as these are more expressive of an otherworldly sensation. The third style of execution is where the entire foundation of the pavilion is laid closely with

Fig. 3.145 (continued)

Fig. 3.146 The cloud wall in Xiao Pangu, Yangzhou

stones, as if building a solid natural-stone bank (Fig. 3.156, also refer to Fig. 3.96a and Fig. 4.39b). The natural stone overhead water pavilion can produce the effect above the Haopu river, which is very fantastic.

Shui Xie: Waterside House

Shui xie as a garden structure originates from the local pile dwelling, a riparian housing in the Jiangnan region, which abounds with rivers, lakes, canals, and swamps (Fig. 3.157). In making *shui xie* in the Jiangnan garden, stone is usually used as the piling material for its aesthetic quality as well as the anticorrosive property instead of timber, which is typically used for the popular dwelling. To minimize the magnitude of the construction, in most of the cases, only the waterfront overhang of the *shui xie* is supported with stone pillars piled in the water, as shown in the examples (Fig. 3.158).

A *shui xie* in multiple stories is in effect a waterside *lounge* (multistoried building), as demonstrated by the creation in Garden to delight one's parents (Yu Yuan) of Shanghai, where the veranda of the building sits above the water on piles (Fig. 3.159).

Depending on the theme of the garden, styles of the *shui xie* are adapted to take on different nuances to be tuned to the thematic landscape. For instance, to help effect peaceful scenery descriptive of a canal-filled village and acquire a sense of a close kinship with water, the *shui xie* in Garden of Art (Yi Pu) of Suzhou is piled low,

Fig. 3.147 (**a**) The cloud wall fusing the west side of principal scenery and the mountain scenery in Liu Yuan, Suzhou. (**b**) An overlook of the cloud wall from the hill in the west of Liu Yuan, Suzhou. (**c**) The cloud wall on the hill beside Woyun Shi in Shizi Lin, Suzhou. (**d**) The cloud wall on the hill beside Yiliang Ting of Bu Yuan in Zhuozheng Yuan, Suzhou. (**e**) The tall cloud wall built by following the contour of the hill in front of Yunqi Lou near Mount Hui, Wuxi

Fig. 3.147 (continued)

whereat the floor height is brought near to the surface of the water (Fig. 3.160). In a different setting, however, where the *shui xie* is perched high above a running mountain stream to adapt to a ravine-theme landscape, the floor then sits relatively further away from the water, as with "Hall Among the Mountains and Water (Shan Shui Jian)" in Suzhou's Couple's Retreat Garden (Ou Yuan) is one such example.

There is a variant construction of the *shui xie*. Instead of the piling style, the stone-laid aquatic foundation has an open structure and forms a beamed tunnel through which the water runs back and forth underneath the *shui xie*. In some cases, a railing fence is put in to guard the opening, as with the *shui xie* in front of the Weng's Residence in Changshu, Jiangsu Province. In Suzhou's Garden that was

Fig. 3.148 (**a**) The cloud wall with the flatland style in the Loquat Garden inside Zhuozheng Yuan of Suzhou—connected with the stone mountain in front of the Yuanxiang Tang and the earthen hill near Xiuqi Ting and built with the support of piled stones that form a rolling range of hills. (**b**) The cloud wall with free style in Ou Yuan's West Garden, Suzhou

preserved miraculously (Liu Yuan), "Vivacious Land (Huopopo Di)" features a similar structure except that its foundation is not open all the way through but rather forms a concavity where the flow of the water ends instead of passing through (Fig. 3.161). Contrivance such as this is merely intended to create the illusion of continuation of the stream.

Fig. 3.149 The cloud wall in Qushui Yuan of Qingpu near Shanghai, though raised low, also fails to bring out the desired result in the sheer absence of offsetting stone or earthwork

Fig. 3.150 The long qiang with the interest of seeking novelty in Shanghai's Yu Yuan, which has no connection with mountain wall already

Fig. 3.151 (a) The waterside yue tai in the north of Yuanxiang Tang in Zhuozheng Yuan, Suzhou. (b) The yue tai facing lake-and-mountain in the north of Hanbi Shanfang in Liu Yuan, Suzhou

Fang: Stationary Boat

Popularly known as "*han chuan*" (land boat) or "*chuan ting*" (boat hall), *fang* is a unique form of garden architecture commonly used to work in with a waterscape in the Jiangnan garden. The stationary boat stems from *hua fang* (decorated pleasure boat) or *lou chuan* (multistoried boat) (Fig. 3.162), which was a favorite tool of aquatic pastime for the wealthy in the old days, who frequented the region's famous

Fig. 3.152 The waterside yue tai in front of the Liuyun Shanfang in Chang Yuan, Suzhou

resorts of scenic beauty, such as Hangzhou's West Lake, by cruising down in one of these commodious contrivances. *Hua fang* afforded an ideal venue where leisure-time pursuits of the elitist class took place; an epicurean wine party was customarily given where a great variety of *chuan cai* (boat dishes)—dainty snacks and refreshments specially designed for the boat menu—was served and a convivial company of entertaining young women was usually provided. In a restricted residential garden where water coverage is limited, however, sport cruising in a *hua fang* is unlikely to be accommodated, hence the creation of *fang*. Intended to imitate *hua fang* physically or metaphysically, three forms of *fang* are used in the Jiangnan garden: stylistic, realistic, and abstract forms.

Fang in the stylistic form is made to station on the waterfront, which would otherwise resemble a regular waterside construction except for its stylistic structural and spatial configuration that unambiguously exhibits a close affinity to its mobile

Fig. 3.153 Lüyi Ting of Zhuozheng Yuan in Suzhou—the combination of regular-shaped bedding of shui ting and stepstone at the dock highlights the characteristic of riverside ferry shelter

Fig. 3.154 The shui ting in Qushui Yuan, Qingpu

a

b

Fig. 3.155 (**a**) "Yuedao Fenglai" Pavilion in Wangshi Yuan, Suzhou. (**b**) Xiuzhu Ge pavilion in Shizi Lin, Suzhou

Fig. 3.156 (**a**) The waterside pavilion in Xiao Pangu, Yangzhou. (**b**) The lake stone foundation of Duo Han Dou strengthens the connection of building and water, and creates the illusion of continuation of the stream. The foundation of adjacent shui ting fares likewise (Xiao Pangu, Yangzhou)

prototype, *hua fang*. The design of both its exterior appearance and interior spaces aims to create an authentic boat feel to the visitor. The base of this type of *fang*, however, does not follow the design of a *hua fang* to be fashioned in the shape of a shipboard; but rather, it is made into a slabstone-laid platform that resembles the

Fig. 3.157 (a) Pavilion number one supported with piles of riparian housing in the Jiangnan region. (b) Pavilion number two supported with piles of riparian housing in the Jiangnan region

foundation of a regular waterside construction (Fig. 3.163). To simulate the triplex cabin configuration of the *hua fang*, the stylistic stationary boat is comprised of three adjoining sections on the deck—the slabstone platform—that represent respectively the fore cabin located near the square-shaped bow, the middle cabin at the center,

a

b

Fig. 3.158 (**a**) The piling style aquatic foundation of "Shan Shui Jian" shui xie in Ou Yuan, Suzhou (photographed in 1956). (**b**) The piling style aquatic foundation of "Shan Shui Jian" shui xie in Ou Yuan, Suzhou (photographed in 2009, and has remained the original form since 1989). (**c**) The yue tai of piling style set on waterside house supported with deeply penetrated pillars, strengthening the sense of a close kinship between building and water (Jiushi Xuan in Shanghai's Yu Yuan)

c

Fig. 3.158 (continued)

and the aft cabin at the stern. The three joint "cabins" are in effect three individual structures, each of which has its own distinct architectural style: the fore cabin is built in the fashion of a *xuan* (windowed veranda) that is opened up to a small moon-viewing platform reminiscent of the open decking area at the bow; the middle cabin is analogous to a *shui xie* (waterside house), which sits low next to the fore cabin; and the aft cabin rises in the form of a *lounge* (multistoried house) that stands the most prominent of the three at the rear of the deck. The entire interior space of the stationary boat is also modeled after that of a boat to affect the look and feel of the *hua fang* cabins. As a form of architectural structure incorporated into its surrounding aquatic scenery, the stylistic *fang* is a remarkable creation with curious properties. At first glance, it is a peculiar alignment of a *xuan*, *shui xie*, and *lounge*, but a careful perusal of the coherent contour and flow of integral spaces would tell otherwise—a quaint pleasure boat impregnated in the image of a triplex building. This has not only resolved the dilemma of accommodating a cruising pleasure boat in a restricted residential garden but also given the construction itself an implicit quality of an artistic creation, where the subtlety lies between the likeness and non-likeness of its prototype.

Fig. 3.159 The waterside lounge of piling style in Yu Yuan, Shanghai

Fig. 3.160 After the renovation of the low piles in principal tingtang (hall) of Suzhou's Yi Pu, protective plates were added and the supporting structure of piles was covered, which lost its pristine interest.

The realistic *fang*, on the other hand, is stationed entirely on the water. This form of the stationary boat aspires to directly copy the real-life *hua fang*, including its platform foundation, which is modeled after the shipboard, rendering it a faithful replica of a *hua fang* except for its building material, which is stone instead of wood. Examples of the realistic *fang* include "Need-not-moor Boat (Buxi Zhou)" in Xu

Fig. 3.161 "Huopopo Di" shui xie of Suzhou's Liu Yuan—its foundation forms a concavity which creates the illusion of continuation of the stream (Its present situation is not implicit enough as the lack of the plant cover)

Fig. 3.162 Hua fang, lou chuan

Yuan of Nanjing (Fig. 3.164a, b) and one in Lion Grove Garden (Shizi Lin) of Suzhou, the style of which shows a distinct Western colonial flavor from the modern period (Fig. 3.164c).

The result of seeking after physical resemblance to a real-life boat, however, does not lend itself to enhancing the aesthetic appeal of the garden's aquatic scenery. On the contrary, an unwieldily proportioned stone watercraft, jamming in a narrow body of water, tends to upset the unity of the scenic imagery and distort the perceived scale

Fig. 3.163 (**a**) The fang (Stationary Boat) in Suzhou's Zhuozheng Yuan—"Xiangzhou, Chengguan Lou". (**b**) The fang in Suzhou's Zhuozheng Yuan—the interior space of "Xiangzhou, Chengguan Lou". (**c**) The hua fang zhai (decorated pleasure boat den) in Suzhou's Yi Yuan—"Fangzhai Laiyou Xiaoxishan". (**d**) The Buxi Zhou in Nanxiang's Yi Yuan

Fig. 3.163 (continued)

Fig. 3.164 (**a**) The Buxi Zhou in Nanjing's Xu Yuan—fore cabin. (**b**) The Buxi Zhou in Nanjing's Xu Yuan—the aft cabin is near the frame. (**c**) The realistic shi fang (stone boat) in Suzhou's Shizi Lin (with Western colonial flavor)

of the on-shore landscape and buildings. More significantly, fidelity to real-life objects stands against the grain of the creative principle of the Jiangnan garden, where the art of garden making rests on masterly epitome and artistic condensation of the beauty of the natural world. Placing a lifelike stationary boat in a garden space

Fig. 3.164 (continued)

is like using a real animal such as a horse or buffalo on the stage of a Beijing opera; the presence of livestock on stage falls completely out of accord with the symbolic style and artistic structure of the Beijing opera. Like the cloud wall coped with a fanged and clawed dragon, the verisimilar *fang* gives very little room for imagination or development of the *yi jing* appeal.

The third manifestation of the stationary boat is in the abstract *xie yi* style. Differing from the other two manifestations, the *xie-yi* style stationary boat aims to capture the spirit of the *hua fang* rather than mirror its outward appearance. There are two types of treatment in creating this abstract form of the stationary boat. In the first type of treatment, the stationary boat, though placed on the edge of water, does not follow the configuration of a ternary *hua fang* but is rather in the shape of a *shui xie*, and to effect an association with a boat, it features an elongated interior space fixed with sash windows or *ban chuang* (short windows) in the side to evoke a boat cabin with portholes. Additionally, the thematic inscription customarily given to the structure also helps conjure the image of a *hua fang* floating to and fro on the water. "Boat-That-Never-Sails Veranda" (Zhouerbuyou Xuan) in Garden of Autumnal Glow (Qiuxia Pu) in Jiading is one such example, which, compared with its comparables, bears a closer resemblance to the shape of a *hua fang* as it is composed of a raised structure on the front implicative of the fore cabin and an elongate lower structure behind it to represent the middle cabin, omitting only the upright multistoried aft cabin at the end (Fig. 3.165a). Stepping up the abstraction and symbolism, "Virtual Boat" (Xu Zhou) in Garden that is near enough to be called a garden (Jin Yuan) of Changzhou features only a span of eaves projecting from the

Fig. 3.165 (**a**) "Zhouerbuyou Xuan" in Qiuxia Pu, Jiading. (**b**) "Xu Zhou" in Jin Yuan, Changzhou. (**c**) The partial boat hall of "Di wo chenjin" in Chang Yuan, Suzhou. (**d**) The boat hall of "Di wo chenjin" in Chang Yuan, Suzhou (photographed in 2009)

gable side to form a waterside veranda whereby to suggest a brief presence of a fore cabin that is open to the lake (Fig. 3.165b). Further removed from the boat image is an even more conceptual creation located in Garden of Natural Delight (Chang Yuan) of Suzhou: entitled "Cleansing my earthly filth (Di wo chenjin)", the "boat hall" is installed with sash windows overlooking the water, the only structural element reminiscent of the image of a *hua fang* (Fig. 3.165c).

Some other constructions, such as "Buqiu Fang" in Villa of beautiful surrounds (Huanxiu Shanzhuang) of Suzhou, though titled "Fang" and situated by the water, do not possess any particular feature of a boat nor adopt any symbolic treatment of the boat imagery. The name was given merely because of the waterfront location of the building. To be in accord with its surrounding landscape, "Mountain House Complementing Autumn (Buqiu Shanfang)" might be a more appropriate name. In the second type of treatment of the abstract-style stationary boat, the construction is land-based and structurally removed from water; architecturally, it does not bear any resemblance of a *hua fang* either. The only association with the idea of a boat comes from certain topical architectural treatment or the literary inscription of the building. For instance, to impart the impression of a *hua fang*, the "Hong Luo" stationary boat in Shanghai's Garden of Various Fruits (Jiuguo Yuan) features an elongate space where floor-to-ceiling *chang chuang* (long windows or latticework partition doors) are mounted on the front end and a combination of *ban-chuang* (short windows) and *ban-qiang* (short wall) is installed on both sides (Fig. 3.166a). Though not exactly next to a body of water, the land-based *fang* sits in the vicinity of a lake and benefits from the waterscape that, combined with the literary inscription on the building, helps enhance the interest of a waterborne pleasure-boat. In Garden of Benefiting Descendants (Huiyin Yuan) of Suzhou, "Fishing Boat" (Yu Fang) presents a more metaphysical example of the land-based stationary boat. A regular small-scale building in the *tang* or *xuan* style, "Yu Fang" does not have a narrow, elongated space evocative of the shape of a *hua fang*; instead, it has an open front structure that suggests a fore cabin open to the bow and sash windows mounted in both sides to insinuate the look of a boat cabin. Furthermore, the impression of the *hua fang* is reinforced by the inscription of the building and the nearby waterscape that the building looks on (Fig. 3.166b). The abstraction of the land-based stationary boat is yet further pushed as seen in a fine piece situated in Shanghai's Garden to delight one's parents (Yu Yuan). Named "Yi Fang" (Boat Too), the structure is anchored in an enclosed courtyard, a setting where water is completely absent. However, the structure has kept certain traits of the *hua fang* manifested through a few architectural treatments, such as its stone-laid base that has the face carved in the shape of a shipboard, the sash windows that simulate the board windows in a boat, and its overall configuration and spatial division that subtly resemble those of a *hua fang*, all of which, along with the inspired name of the building—"Yi Fang", have contributed to the aquatic theme of this water-free environment and created by mental association an image of a *hua fang* floating on undulating waves (Fig. 3.167). It can be regarded as an excellent work. The east side of Suzhou Yiyuanpu Lvxuan, "Sit Alone Around the Huilu" and has the title "Stone Boat", can also be said to be a dry boat or boat hall. Its architectural form and environment have nothing to do with the

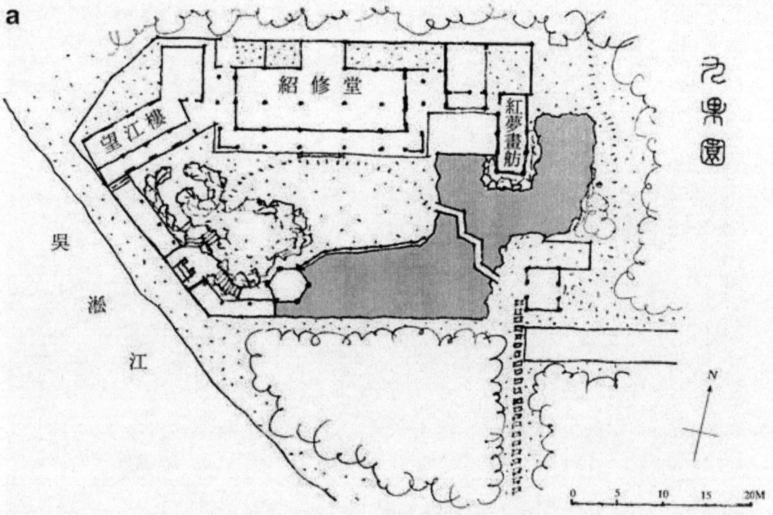

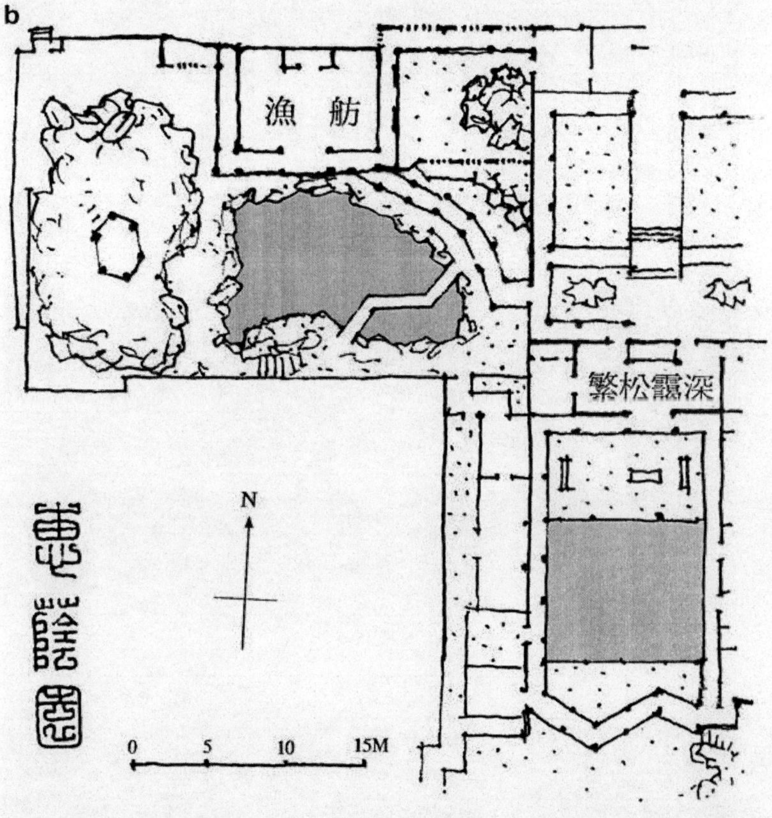

Fig. 3.166 (**a**) "Hong Luo" stationary boat in Shanghai's Jiuguo Yuan, not exactly next to a body of water (The plan of Jiuguo Yuan, drawn in accordance to Jiangnan Yuanlin Zhi). (**b**) The "Yu Fang" in Suzhou's Huiyin Yuan is not exactly next to the water and wasn't modeled after the shape of stationary boat (Partial plan of Huiyin Yuan, drawn in accordance to Jiangnan Yuanlin Zhi)

Fig. 3.167 (**a**) The base of "Yi Fang" in Shanghai's Yu Yuan was fashioned in the shape of a shipboard; Sash windows (balustrades, removable windows) can be evoked with portholes. (**b**) The bedding and stepstone of Yi Fang in Yu Yuan, Shanghai

image of anti-painting and waterscape. There is only an abstract connotation that is thought-provoking as shown by a surprisingly winning title.

Bridge, Beam, and Stepping-Stones

In principle, the Jiangnan garden is a re-creation of the natural mountain-and-water landscape of the Jiangnan region. Varigated types of water-crossings—bridges, beams, stepping-stones, and so forth—are a characteristic scenic component of the Jiangnan water-filled country landscape and have therefore long been assimilated in the art of the Jiangnan garden and become an indispensable embellishing element in making a water scene of the garden. Generally, bridges are mostly adopted in combination with water features such as lake or river while beams and stepping-stones are often paired with ravine or gully. Examples remaining in the extant gardens are mainly stone bridges—such as slabstone bridges in flat, trapezoidal, or zigzag shapes; stone arch bridges; covered (or sheltered) bridges; stone beams; and stepping-stones. Wooden and earthen bridges as depicted in the Ming garden painting have become increasingly rare nowadays (Fig. 3.168). The flat slabstone bridge is a common form of river crossing in rural areas of Jiangnan, and it is often built with low-rise slabstone railings along both sides of the bridge to provide some protection (Fig. 3.169). The style of the garden bridge is very close to that in real life, except for the smaller building scale. In the Humble Administrator's Garden (Zhuozheng Yuan), the stone bridge that sits to the north of Crabapple Spring Hall (Haitangchun Wu) and that serves as its *duijing*, an offsetting scene, is of this type. Said to be a Ming relic, it was built to the formal design of stone *shigou* railing and, over the centuries, has gained a patina of age and wear (Fig. 3.170).

Fig. 3.168 The earthen bridge in a garden of Jiangsu

Fig. 3.169 (**a**) The low-rise shi ban (slabstone) railings of flat slabstone bridge in Jiangnan's water-filled region. (**b**) The low-rise tiao shi (slabstone) railings of flat slabstone bridge in Jiangnan's water-filled region. (**c**) The slabstone bridge with no railing over the towpath of river branch in Jiangnan; A three-span bridge has its span raised in the middle to accommodate passing boats, which forms a trapezoidal shape. (**d**) The middle span of little slabstone bridge over the towpath of river branch in Jiangnan was elevated to give the needed clearance, and stairs were built for passersby

In real life, the slabstone bridge built across watercourses often has its span raised to accommodate passing boats. In the case of a three-span bridge, only the middle span is elevated to give the needed clearance, which forms a trapezoidal shape in the bridge (Fig. 3.170). In addition to its functionality, the heightened span also adds viewing value to the bridge and therefore has become a favorite bridge style of

Fig. 3.169 (continued)

garden markers. The stone bridge inside the garden entrance in Changzhou's Garden that is near enough to be called a garden (Jin Yuan) is one such example (Fig. 3.171).

In the art of garden making, the trapezoid-style bridge is often being transfigured to fit a specific space in the garden by reducing the scale and height of the bridge. The three-span arc-shaped stone footbridge in Suzhou's Garden of Art (Yi Pu) can be seen as a variant of the trapezoid-style bridge.

The most commonly adopted bridge style in the Jiangnan garden is the zigzag slabstone bridge that stretches low above the water surface, whose prototype comes from the towpath that is traveled by boat trackers. To parallel the course of the boat movement, the construction of a towpath usually follows a straight line; however, the path often bends and curves to meet the conditions of the waters or the shape of the shoreline (Fig. 3.172). In the Jiangnan garden, the slabstone bridge modeled after a rectilinear towpath is rather uncommon. Built approximately in the late Ming or early Qing dynasty, the "Seven Star Bridge" (Qixing Qiao) in Wuxi's Garden for Ease of Mind (Jichang Yuan), which diagonally spans the northeastern part of the lake, is one of the rare creations of the straight towpath-style garden bridge (Fig. 3.173). Although flat and straight, Seven Star Bridge (Qixing Qiao) is different from the common flat slabstone bridge and, judging by its scenic content, should be classified as a towpath-style bridge. The footbridge in the southwestern courtyard of Garden of Art (Yi Pu) in Suzhou, though short in length, is arranged in the flat,

Fig. 3.170 The stone bridge of Ming Dynasty that sits to the north of Haitangchun Wu in Zhuozheng Yuan, Suzhou

straight towpath style as well, which can also be considered as the style of a double-span stone beam (Fig. 3.174). Garden critics often have unfavorable opinions on Seven Star Bridge (Qixing Qiao), citing that it violates the *duan chu tong qiao*[121] creative principle or simply cold-shouldering its long, ramrod configuration, unwitting of that it is in fact a more realistic re-creation of a towpath scene. If anything derailed from the towpath concept, however, it is the addition of the railings to both sides of the bridge, which is nonetheless arguably necessary as they function as a protective feature befitting a garden structure. Among the existing towpath-style bridges, most of them take the form of a flat slabstone bridge in zigzags, so designed to meet the landscape compositional requirement as well as facilitate the perambulatory viewing proceeding intended to expose the visitor to different scenic aspects of the landscape. The most common zigzag bridges have two or four bends, which create three or five zigzag sections for the entire bridge (Figs. 3.175 and 3.176).

[121] Frame a zigzag slab bridge on a river.

Fig. 3.171 (**a**) The trapezoid-style stone bridge in Jiangnan's water-filled region (Huzhou, Zhejiang). (**b**) The trapezoid-style slabstone bridge facing the island mountain inside the garden entrance in Jin Yuan, Changzhou. (**c**) The stone bridge variant of trapezoidal shape in Suzhou's Yi Pu—three-span arc-shaped stone footbridge

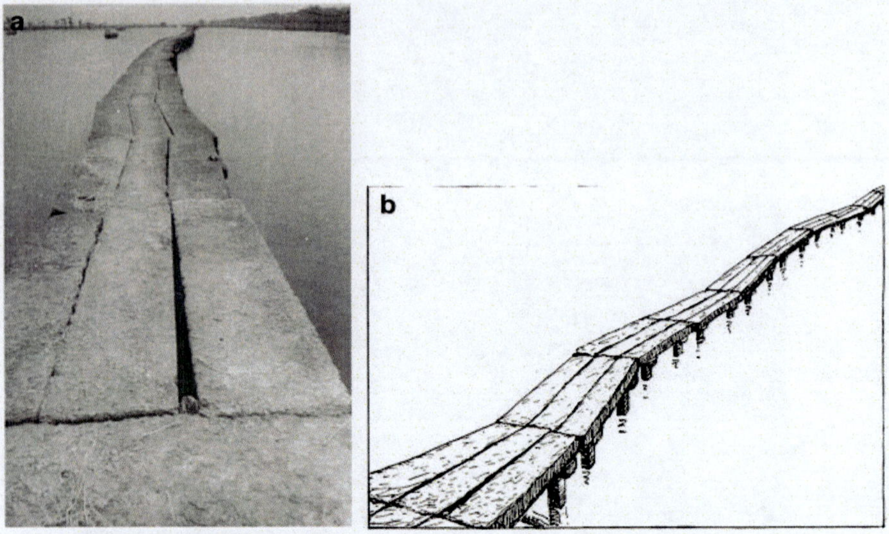

Fig. 3.172 (**a**) The towpath in Jiangnan's water-filled region (Shaoxing)—prototype for garden zigzag bridge. (**b**) Towpath in Jiangnan's water-filled region

Fig. 3.173 Rectilinear Qixing Qiao modeled after a towpath in Jichang Yuan, Wuxi

There are also those with one, three, or even eight bends, which are less common in the Jiangnan garden (Fig. 3.177). To facilitate natural perambulatory motion, the bends of the zigzag bridge usually take on an obtuse angle; rarely do they bend at a right angle. The eight-bend zigzag bridge at the Mid-lake Pavilion (Huxin Ting) in the south of Shanghai's Garden to delight one's parents (Yu Yuan) and the multi-bend zigzag bridge in the south of Garden of High Virtue (Gaoyi Yuan) in Suzhou

Fig. 3.174 Little two-span rectilinear towpath made of roughly processed slabstones sat low above the waters and was free of any railing in Suzhou's Yi Pu, which is closest to the prototype of towpath

Fig. 3.175 The two-bend zigzag bridge sat low above the waters and was free of any railing in Yi Pu, Suzhou

Fig. 3.176 The low slabstone railing of four-bend bridge in Suzhou's Zhuozheng Yuan endows the whole body of the bridge with a perfect shape, which is a very elegant example

are but a few examples of the zigzag bridge with right-angled bends (Fig. 3.178). Generally speaking, right angles render a zigzag bridge stiff and harsh looking, and therefore "slab bridges should avoid square curvature", remarked by Wen Zhenheng, a Ming dynasty scholar and landscape garden designer, in his *Zhang Wu Zhi* (Treatise on Superfluous Things), an encyclopedic book on garden architecture and interior design. In Jiangnan's water-filled rural country, the towing paths are built to sit low above the waters and free of any railing, so are the zigzag bridges in the garden that are most expressive of such rustic flavor (Figs. 3.175 and 3.177). In most of the residential gardens, however, railings are added to the bridge for the safety of the elderly, women, or youngsters of the family who often take a stroll in the garden or sport around the bridge, which nevertheless tends to dent the appeal of the scenery. To minimize the presence of such an addition, the railing is designed to be brief and simple. Although Wen Zhenheng does not explain in *Zhang Wu Zhi* the scenic connotation of the zigzag slabstone bridge as a garden structure, from the aesthetic perspective of the bridge, he suggests that "A slab bridge should have two bends and a rail of a single wooden bar, avoiding using a railing in vermilion made in the swastikas pattern; some use Taihu lake stone for the railing, which also renders [the bridge] gaudy".

Among the existing examples of Jiangnan gardens, railings are added to the bridges, as is customary for China's water towns. Railings typically consist of two horizontal wooden rails (in ancient paintings of the Ming and Qing Dynasty, the Yi Mu Wei Lan style is more common, which is a single-piece horizontal wooden handrail). They were made of wood in earlier periods, while iron is used more often in more modern examples. In the early days, only one railing was set on one side of

Fig. 3.177 (**a**) The roughly processed three-span zigzag bridge with one bend in Suzhou's Yi Pu, sat low above the waters and was free of any railing, modeled after a towpath in water-filled region, which is an example that is closest to the prototype of towpath. (**b**) The three-bend zigzag bridge at the foot of stone mountain in Zhan Yuan, Nanjing

Fig. 3.178 The multi-bend zigzag bridge with right-angled bends over Shijin Tang pond in the south of Gaoyi Yuan at Mount Tianping, Suzhou

the bridge to minimize their use, imitating the ordinary bridges found in water towns. This feature is reflected in such drawings as those found in the drawing collection *Qinggong Baimei Tu*. Moreover, drawings of ancient Jiangnan gardens, such as *Zhuozheng Yuan Tu* by artist Wang Yun of Qing Dynasty's Xianfeng Period (1851–1861), describe zigzag bridges with one railing. Likewise, the *County Records of Shanghai* record single-rail double-zigzag bridges and triple-zigzag bridges in Yeshi Garden (Yeshi Yuan)[122] during the Tongzhi Period (in 1871) of the Qing Dynasty (Fig. 3.179). However, bridges with only one railing on a single side are a rarity among existing gardens, where the iron railing of the stone beam bridge in the Garden of Miniature Landscape as if Contained in a Pot (Hu Yuan) of Suzhou and the one in the Villa of beautiful surrounds (Huanxiu Shanzhuang) of Suzhou are later additions (Fig. 3.180). The single-railing slab stone bridge in the East Garden of the Garden to delight one's parents (Yu Yuan)[123] in Shanghai and the Huilong Tan Park in Jiading are both rare and precious examples of integrated designs (Fig. 3.181). As for bridges with two horizontal railings, the one in Nanxiang's Yi Yuan serves as an example of low railings while that in the Garden for Ease of Mind (Jichang Yuan) in Wuxi is an example of high railings. These examples are all stone balusters with wooden horizontal bars (Figs. 3.182 and 3.173). Meanwhile, the zigzag bridges in Garden of Pleasure (Yi Yuan) in Suzhou

[122]The garden was destroyed in World War II, and now there are no remains.

[123]The East Garden was once an independent garden and was later absorbed into Garden to delight one's parents (Yu Yuan).

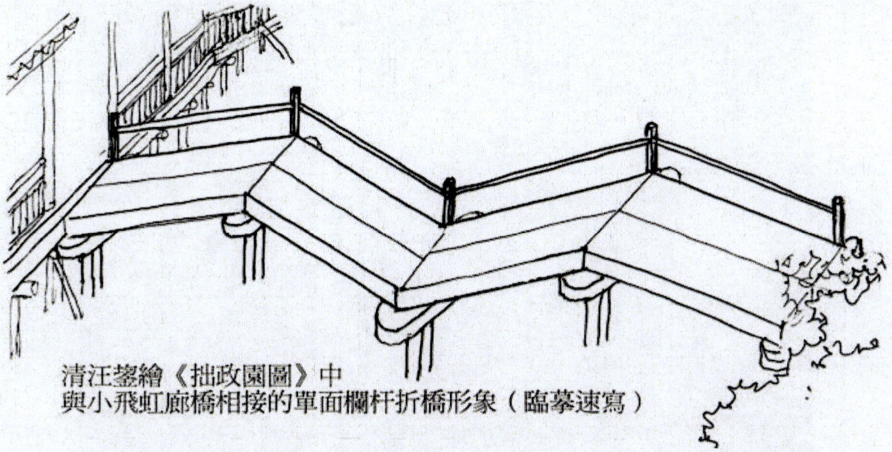

Fig. 3.179 The single-rail zigzag bridge demonstrated in Zhuozheng Yuan Tu by Wang Yun of the Qing Dynasty

and on the south of the Hall of Distant Fragrance (Yuanxiang Tang) of the Humble Administrator's Garden (Zhuozheng Yuan) in Suzhou are built with iron balusters with iron horizontal bars (Fig. 3.183). Iron bridge railings, in particular, can be made slender to help zigzag bridges keep their original shape. Slab stone bridges can also have stone-bar railings, with one example being found in the zigzag bridge around the Pavilion in the Lotus Breezes (Hefeng Simian Ting) in Suzhou's Humble Administrator's Garden (Zhuozheng Yuan) (Fig. 3.176). This type of stone-bar railing endows the whole body of the garden bridge with a perfect shape, allowing a rich level of creativity. The use of natural stone in railings of zigzag bridges, as seen in Suzhou's Mu Garden (Mu Yuan) and Qingpu's Meandering Water Park (Qushui Yuan) (Fig. 3.184), appeared as early as the Ming Dynasty according to Wen Zhenheng's *Zhang Wu Zhi*. This variant is a departure from the towpath style bridge, striving for novelty in form. In doing so, it becomes a type of scenery that though intriguing lacks specific usefulness, merely serving as a show of the harmonious landscapes of mountains and water. Overall, zigzag bridges boast an eclectic array of railing types in Jiangnan gardens. Furthermore, the variants of towpath-style bridges have given rise to garden bridges of various shapes that emphasize the ornamental value and appeal of their twists and turns rather than the direct significance of the bridge itself.

The stone arch bridge is also crafted in accordance with the real needs of daily life. As there are a great many rivers and branching streams in Jiangnan water towns, bridges are needed at the intersections of roads and waterways, forming a three-dimensional flow of traffic. For navigation under the bridge, in addition to the trapezoidal bridge mentioned above, stone arch bridges are built with high arches under which large sailboats are able to pass (Fig. 3.185). As an attractive feature of the beautiful Jiangnan water town, these bridges have been imitated in garden art

Fig. 3.180 The single-side railing on arc-shaped stone beam of Hu Yuan, Suzhou

(Fig. 3.186). The arched bridges in Suzhou's Lion Grove Garden (Shizi Lin), the Master-of-Nets Garden (Wangshi Yuan), the Half Garden (Ban Yuan), Qingpu's Meandering Water Park (Qushui Yuan), and the Nanjing's Xu Garden (Xu Yuan, once the west garden of the self-proclaimed "King of Heaven" of the short-lived Heavenly Kingdom of Taiping in the mid-nineteenth century) are all picturesque examples of navigable arched bridges in Jiangnan water towns (Fig. 3.187). Originally placed in inlets and branches in water towns, in garden art it is often placed as a backdrop to the main scenery. This backdrop often occupies the corners of inlets and branches of the water surface with a reduced scale, giving a profound sense of depth through the effect of perspective.

Commonly found in fields and mountains, the pavilion bridge (also called the covered bridge or sheltered bridge) is a combination of a pavilion and a bridge (Fig. 3.188). It serves to cross water, provide a place of shade and shelter, as well as a rest stop for pedestrians. With water burbling below, it boasts ornamental and interest value and is often reproduced artistically in Jiangnan gardens. The artistic versions require more vividness both in shape and in position. True pavilion bridges are so rare that the only existing example is found in Nanxiang's Yi Yuan in Jiading County, Shanghai (Fig. 3.189). The covered bridge in the Garden for Ease of Mind (Jichang Yuan) in Wuxi is the most similar to the original form found in water towns, while the Small Flying Rainbow Bridge (Xiao Feihong) in the Humble Administrator's Garden (Zhuozheng Yuan) in Suzhou is an arched covered bridge featuring a

a
b

Fig. 3.181 (**a**) The single-railing slabstone bridge in Yu Yuan's East Garden, Shanghai. (**b**) The single-railing slabstone bridge in Huilong Tan, Jiading

trapezoidal shape. (According to Wen Zhiming's *Zhuozheng Yuan Tu* painting, the Small Flying Rainbow Bridge (Xiao Feihong) was only a trapezoidal wooden bridge during the Ming Dynasty, and was reconstructed into a stone covered bridge at a later time) (Fig. 3.190).

Works imitating earthen bridges (i.e. a wooden structure with earthen surfaces) and wooden bridges are common in countryside gardens from the Ming Dynasty to the middle of the Qing Dynasty. An example of this can be seen in Dai Xi's painting *Zhuozheng Yuan Tu* from the 16th year of Daoguang's reign in the Qing Dynasty (1836). Other rare examples include a wooden bridge to the north of the Mountain-

Fig. 3.182 The zigzag bridge with horizontal lattice low railing in Yi Yuan, Nanxiang

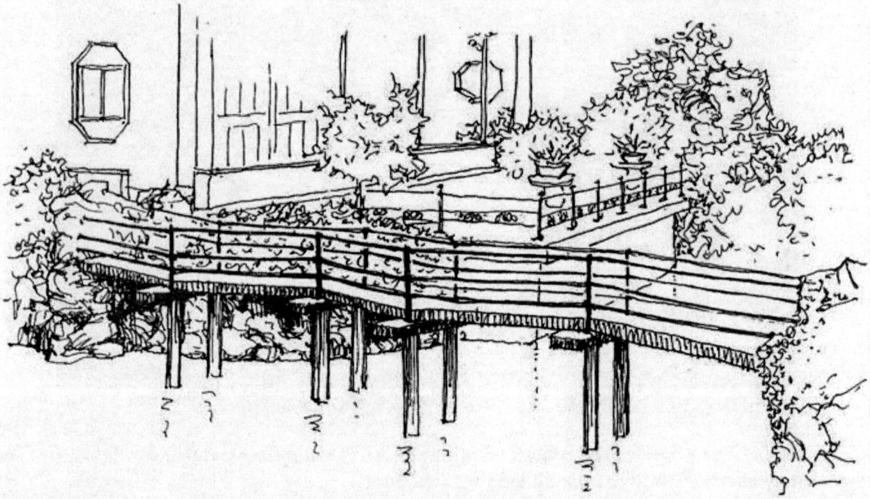

Fig. 3.183 The zigzag bridge with iron lattice railing in Yi Yuan, Suzhou

in-View Tower (Jianshan Lou) in the Humble Administrator's Garden (Zhuozheng Yuan) of Suzhou (note: it has recently been rebuilt into a stone zigzag bridge) and an earthen bridge built later in the Garden that is near enough to be called a garden (Jin Yuan, a.k.a. Yun's private garden) in Changzhou.

Additional features of such structures are small stone beams across streams and inlets. The so-called stone beam refers to a small-spanned bridge made of one to three stone slabs and no railing. In short, it is a simple bridge built across narrow

a

b

Fig. 3.184 (**a**) The zigzag bridge with Taihu stone railing in Mu Yuan, Suzhou. (**b**) The zigzag bridge with Taihu stone railing in Qushui Yuan in Qingpu, Shanghai

waters, similar to a single-plank bridge, offering a foothold for pedestrians to step on when traversing mountains and villages. They are not as formal as stone-slab bridges and have a more natural and rustic country style. Splendid examples of such a bridge are the natural stone beam on the east side of the Waterside Pavilion of Washing

1 2 3 4

Fig. 3.185 The stone arch bridges in Jiangnan's water-filled region.

Fig. 3.186 The three-arch stone bridge over the watercourse at the Shantang Street, Suzhou

Fig. 3.187 (**a**) The stone arch bridge in Shizi Lin, Suzhou. (**b**) The stone arch bridge in Wangshi Yuan, Suzhou. (**c**) The arch Bridge in Qushui Yuan, Qingpu. (**d**) The symmetrical stone arch bridge in the principal tingtang of Xu Yuan, Nanjing

d

Fig. 3.187 (continued)

Hat-Ribbons (Zhuoying Shuige) in the Master-of-Nets Garden (Wangshi Yuan) in Suzhou (Fig. 3.191) and the natural stone beam in Huilong Tan Park of Jiading, which intriguingly is broken off in the middle (Fig. 3.192). Another type of stone beam that imitates the low-rise slab stone bridge on the inlets of lakes or branches of rivers is shown in the Garden for Ease of Mind (Jichang Yuan) in Wuxi, the Tiger Hill (Hu Qiu) in Suzhou, the Garden that was preserved miraculously (Liu Yuan) in Suzhou, the Garden of Art (Yi Pu) in Suzhou, and the Garden of Look Forward (Zhan Yuan) in Nanjing (Fig. 3.193). In addition, in mountainous forests, stone beams are often used where woodsmen and travelers typically wade across narrow valley streams. This kind of stream stone beam is also reflected in garden art, such as the stone beams on the mountain streams of Villa of beautiful surrounds (Huanxiu Shanzhuang) and the Humble Administrator's Garden (Zhuozheng Yuan) in Suzhou and the Garden of Look Forward (Zhan Yuan) in Nanjing, all of which are masterpieces with complete scenery and touching artistic concepts (Fig. 3.194).

Some people have surmised that low-hanging bridges over water surfaces are a specific feature of Suzhou gardens. Therefore, examples of elevated bridges, such as the one in the Couple's Retreat Garden (Ou Yuan) in Suzhou, have been deemed as an architectural failure and the beams of which have received heavy criticism that they should be changed to lower beams so as to conform to their notion of what the bridges are supposed to be like. Such criticism is the result of judging the bridges by their physical structure alone, without any regard to context. However, if we analyze this issue using the creative principles of garden art, it is not difficult to see that whether a bridge hangs low near the water surface or is elevated overhead depends on the composition of the scene as a whole. For example, zigzag bridges that hang low over the surface of the water are undoubtedly successful creations that imitate wetland scenery, while highly-elevated stone beams imitate the depths of a ravine.

Fig. 3.188 (**a**) The sheltered bridge in the fields in Wuyuan, Jiangxi province (also called corridor bridge). (**b**) Benches for passers-by to rest on both sides of the corridor bridge in Wuyuan, Jiangxi province

Fig. 3.189 Pavilion Bridge in Yi Yuan, Nanxiang Town, Jiading County, Shanghai

Referring to a structure as having high-rise stone beams does not imply that they are high in absolute height, with even those built on a small scale still successfully imitating the scenery of high cliffs and deep ravines. This fact is shown in examples such as the small-scaled stone beam in front of the Winding Stream Tower (Quxi Lou) in the Garden that was preserved miraculously (Liu Yuan), Suzhou, which imitates cliff stones as well as the small stone beam in the valley near the Veranda of Scenting the Sweet Osmanthus Fragrance (Wen Muxixiang Xuan) (Fig. 3.195). Arched-shaped stone-beam bridges are another exquisite form of garden art as shown in Nanxun's Shi Garden (Shi Yuan, the garden that is comfortable) (Fig. 3.196).

Steppingstones that imitate the nature of streams, cliffs, and other places in the Jiangnan gardens. Excellent examples of such features include those in the valley of Villa of beautiful surrounds (Huanxiu Shanzhuang) (Fig. 3.197, the lower rocky beach of Dashi Mountain, the stream in the west side of the Garden of Look Forward (Zhan Yuan) in Nanjing (Fig. 3.198), and a freehand masterpiece in the Gully of Eight Sounds (Bayin Jian) in the Garden for Ease of Mind (Jichang Yuan) in Wuxi (see Fig. 3.22b).

Water Corridors and Water Walls

The so-called "water corridor" refers to a corridor overhanging water, which is a creation inspired by the covered (or sheltered) bridge, as displayed in the Garden for Ease of Mind (Jichang Yuan) in Wuxi, the Fenyang Villa (Fenyang Bieshu) in Hangzhou, and Changshu Xukuo Residence (Changshu Xukuo Ju) (Fig. 3.199). The Humble Administrator's Garden (Zhuozheng Yuan) also had a water corridor during the reign of Emperor Xianfeng in the Qing Dynasty (see Fig. 4.65a). In addition,

a
b
c

Fig. 3.190 (**a**) A rustic wooden porch bridge of Jichang Yuan in Wuxi. (**b**) Xiao Feihong in Zhuozheng Yuan in Suzhou. (**c**) Corridor bridge in the East Garden of Yu Yuan, Shanghai. (**d**) Corridor bridge of Huanxiu Shanzhuang in Suzhou. (**e**) Variation of corridor Bridge in Qushui Yuan in Qingpu, Shanghai

Fig. 3.190 (continued)

Fig. 3.191 Natural Stone Beam on the East Side of Zhuoying Shuige in Wangshi Yuan, Suzhou

Fig. 3.192 Broken Natural stone beam of Huilong Tan in Jiading

there is a type of water corridor built on the waterside that appears to be floating on the water, as seen in the western section (Subsidiary Garden (Bu Yuan)) of the Humble Administrator's Garden (Zhuozheng Yuan). In a broader sense, water corridors also include verandahs (Fig. 3.200).

The so-called water wall constitutes a whitewashed wall directly facing the water, decorated with *lou chuang* (openwork windows) or *dong chuang* (hole or light windows) on the surface or ribbon decorations on the top section, which renders

a

b

Fig. 3.193 (**a**) Stone beam in Liu Yuan, Suzhou. (**b**) Stone beam of Hu Qiu, Suzhou. (**c**) Stone Beam in Zhan Yuan, Nanjing

c

Fig. 3.193 (continued)

the wall more scenic with the reflection of the water. This is a creation that originated from the waterside courtyard walls of residential houses. The most realistic example of a water wall is in Jiaxing's Luofan Pavilion. Another example with more artistic characteristics is the water wall of the West Garden of the Jiechuang Si Temple (Monastery of Ordination Pennant) in Suzhou, which partially strides over the water, closely hugging its surface. The water wall of Hangzhou' Yi Yuan exhibits assorted hole or light windows decorated with lanterns, casting magical reflections across the scene at night (Fig. 3.201). On the water walls of the Garden to delight one's parents (Yu Yuan) and Yeshi Garden (Yeshi Yuan) in Shanghai, an open moon gate straddles the water. This intriguing creation not only reflects the scene on both sides of the water wall but also invokes imagery of boating (Fig. 3.202). Taicang Yi Yuan's dragon wall extends into the water at its end, creating a reflection of the courtyard wall of the water towns. Moreover, it also adds the playful element of a dragon playing in the water (Fig. 3.203).

Dikes

In order to allow for the planting of rice, water caltrop, and lotus root, as well as fish farming, dikes, and dams have been built in many rivers and lakes in Jiangnan water towns. In turn, these dams have also been depicted in their garden scenery. For example, in Suzhou's Humble Administrator's Garden (Zhuozheng Yuan), dikes connect the Pavilion in the Lotus Breezes (Hefeng Simian Ting) with the five-zigzag bridge in the western part of the peninsula, as well as the East Mountain lake with the three-zigzag bridge in the eastern part of the peninsula. Both of these examples are artistic representations of dikes (Fig. 3.204). Moreover, there is also a depiction of a

Fig. 3.194 (a) Elevated Stone Beam on the Stream of Huanxiu Shanzhuang in Suzhou. (b) Stone beam on the mountain streams of Zhuozheng Yuan in Suzhou. (c) Stone Beam on Huangshi Valley Stream in Zhan Yuan, Nanjing

Fig. 3.195 Small Stone beam at the Cliff Shore of Qingfengchi Hall in Liu Yuan, Suzhou

Fig. 3.196 Small arched stone beam in Shi Yuan, Nanxun

Fig. 3.197 Stepping stones in the valley of Huanxiu Shanzhuang, Suzhou

whole dike with the isolated river street long stone dike at the west side of Qingpu's Meandering Water Park (Qushui Yuan)'s lake (Fig. 3.205).

The above section offered a brief discussion on the relationship between architectural design and elements of mountains and water. The relationship between buildings and plants will be discussed in the next chapter: Planning of Vegetation.

3.1.1.2.3 Specific Architectural Treatment of Garden Buildings

Jiangnan garden architecture, in terms of artistic style, is different from the dignified northern garden architecture and the fiery Lingnan garden architecture. As far as the angle of the roof wing is concerned, regardless of the style of eaves employed, the curve of the roof ridge contour line is invariably larger than that of the northern roofing style but smaller than that of the Lingnan style and shows light, elegant temperament. Of course, these are all obvious features that we need not grace with

Fig. 3.198 Stream Stepping Stone on the West Side of Nanjing Zhan Yuan

further discussion. What this section will discuss is the special treatment of some principles in the creation of Jiangnan garden architecture.

Within the broad discipline of architecture, garden architecture has its own particular regularities and approaches. Nevertheless, it can also be said that some laws of architectural art are subordinate to the general laws of garden art. As is the case with modern opera, drama, and stage art, garden art is a depiction of natural scenery and reproduces the typical living environment. The design of modern stage scenery, using painting techniques that follow the principles of perspective, color and projection geometry, creates different space environments on the narrow stage and renders a specific atmosphere. The architectural design of Jiangnan Gardens also uses perspective, color, and light and shadow effects to create depth of field, a rendered atmosphere, and, like the stage, is treated as a fake version of scenery, as is sometimes necessary in order to highlight its space or shape.

3.1.1.2.3.1 Design of Architectural Dimensions

Through countless generations of architectural practice, a standard dimension has formed in general architectural design that takes its cue from the size of the human body. However, due to its positioning on narrow strips of land and the small landscape scale of Jiangnan gardens, it is necessary for buildings to be built on a smaller scale. Especially for scenes that mainly serve for viewing pleasure, more attention is paid to the design of scale. For example, within the main scene comprising a lake and a mountain at the Humble Administrator's Garden (Zhuozheng Yuan) in Suzhou, the mountain is small in scale in order to match the lake. This requires that buildings such as the Snow-Like Fragrant Prunus Mume Pavilion (Xuexiang Yunwei), which are positioned on the mountain, also be scaled down to

a

b

Fig. 3.199 (**a**) Water Corridor of Jichang Yuan in Wuxi. (**b**) Water Corridor of Xukuo Residence in Changshu (Sketch Based on Records of Jiangnan Gardens)

form the mountain and forest scene with suitable proportions. The Snow-Like Fragrant Prunus Mume Pavilion (Xuexiang Yunwei), which is an open pavilion that symbolizes a village cottage, is a rectangular pavilion resting on top of the mountain with its long side divided into three parts. According to the normal scale of classical architecture, its cornice should be over 300 cm high but is only about 220 cm. In contrast to this example, the open Hushan Zhenyi pavilion on the western hill of Wanshou Mountain in the Beijing Summer Palace was built within the natural scenery and has a normal architectural scale. As Hushan Zhenyi is similar to the

Fig. 3.200 (**a**) Water Corridor in Bu Yuan of Zhuozheng Yuan, Suzhou. (**b**) Water Corridor of Jiuqu Yuan in Changshu (according to Records of Jiangnan Gardens)

Snow-Like Fragrant Prunus Mume Pavilion (Xuexiang Yunwei) in architectural form, the comparison of the two reveals the reduction in the scale and volume of the Snow-Like Fragrant Prunus Mume Pavilion (Xuexiang Yunwei) (Fig. 3.206).

The Passable Pavilion (Ke Ting) in Suzhou's Garden that was preserved miraculously (Liu Yuan) is also an excellent example of maximizing scale reduction to achieve a good scenic effect (Fig. 3.207). The so-called maximization of scale reduction refers to the minimum scale that is the practical and immersive psychological limit of scale reduction. Taking height as an example, if the eaves are 220 cm high, the middle part of the eaves should be 195 cm above the ground. According to the average height of people in the Jiangnan area, they can go in and out of the structure without hitting their heads. Entering this small, elegant open space will not elicit feelings of depression. This is the meaning of minimum scale. If the scale is

Fig. 3.201 (a) Water wall of Luofan Pavilion in Jiaxing (Sketch Based on Records of Jiangnan Gardens). (b) Water wall of West Garden of the Jiechuang Si Temple in Suzhou. (c) Part of the water wall of West Garden of the Jiechuang Si Temple in Suzhou. (d) Water wall with assorted hole windows in Yi Yuan, Hangzhou (Sketch Based on Records of Jiangnan Gardens)

reduced further, it not only affects the practicality but also creates a psychological sense of model abnormality. Those structures which are built too small to accommodate people, such as the tiny pavilions, bridges, and archways (referred to as torii

c

Fig. 3.201 (continued)

in Japanese books) with small-scale landscapes in Lingnan gardens or Japanese gardens, belong to the bonsai vein of garden art.

In narrow space scenes, when a hall with functional requirements of large space volume cannot be reduced in size, the method usually employed is to break up the whole shape into parts. For example, the Hall of Thirty-six Mandarin Ducks and Hall

Fig. 3.202 (**a**) Flower Wall with an open moon gate in Front of Wanhua Building in Yu Yuan, Shanghai. (**b**) Flower Wall with an open moon gate on the East Side of Yuhua Hall in Yu Yuan, Shanghai. (**c**) Flower Wall Combined with Stone-Folding Water Culvert in the East of Biwu Xuan in Qiuxia Pu, Jiading. (**d**) April Wavy eadge of the moon gate wall echoed with the water (Yu Yuan, Shanghai)

d

Fig. 3.202 (continued)

of Eighteen Datura Blossoms (Saliu Yuanyang Guan/Shiba Mantuohua Guan) in the western section of the Humble Administrator's Garden (Zhuozheng Yuan) in Suzhou have attached pavilions to the four corners of the hall so that small buildings with four corners on the exterior wall are able to block out large halls. The roof also has four pyramidal sections correspondingly attached to the four corners of the main body, thus changing the impression of scale and reducing the appearance of its scale (Fig. 3.208).

This method of reducing scale is used not only for buildings, but also for ordinary bridges, hill-climbing corridors, and other architectural structures. The small arch bridges in the Lion Grove Garden (Shizi Lin) and the Master-of-Nets Garden (Wangshi Yuan) in Suzhou are typical examples. Compared with the Jade Belt Bridge (Yudai Qiao) in the Summer Palace, which is close to the original scale, the degree of their scale reduction is noticeable (Fig. 3.209). In the No. 31 garden of Beishan Street, Hangzhou, each step of the hill-climbing corridor is compressed to a height of 7–5 cm, which is only 1/6 of the normal height for stone steps. This is done in order to set off the rather short earthen hill with a height of no more than 150 cm

Fig. 3.203 The dragon wall of Yi Yuan in Taicang, Jiangsu extending into the water at its end (according to Records of Jiangnan Gardens)

Fig. 3.204 Dike near Hefeng Simian Ting in Zhuozheng Yuan, Suzhou

(the corridor was destroyed in the 1950s, leaving only stone steps, with the rest having been completely dismantled in recent years). Due to the reduction of the stone steps height, the elevation of a slope with a height of 55 cm, which can normally be traversed in less than two steps, will now require eleven steps to cover the same vertical distance. This increases the number of stone steps, and also the number of horizontal tread lines on the landscape surface so that the earthen hill appears to be higher. Tourists who make the gradual climb up the numerous

Fig. 3.205 Shaped dike of Qingpu Qushui Park

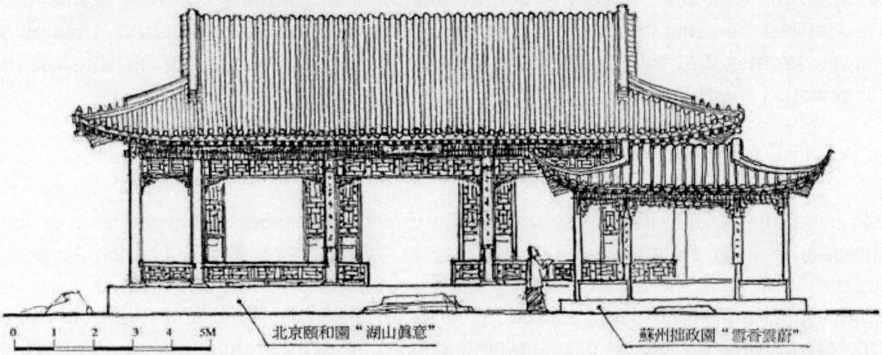

Fig. 3.206 Scale comparison of the open pavilion “Hushan Zhenyi” of Wanshou Mountain in the Beijing Summer Palace and “Xuexiang Yunwei” on the North mountain of Zhuozheng Yuan

Fig. 3.207 Scale Sense of Ke Ting in Liu Yuan, Suzhou

steps will thus have an experience similar to climbing a mountain. This small-scale treatment is also adopted for the Buqiushan Corridor in Suzhou's Villa of beautiful surrounds (Huanxiu Shanzhuang) (Fig. 3.210).

Scale reduction, therefore, serves not only to adapt to narrow spaces but also has a deeper significance in exploiting the depth of field by using the principle of perspective. The design of Jiangnan gardens often obeys the perspective principle that buildings close to the main viewing position appear large, while distant objects appear small. In general, main halls or other practical buildings are not scaled down, while accompanying buildings, as ornamental objects of these larger buildings, need to be scaled down so as to create an impression of a far-reaching scale. This creates the effect of expanding the space through perspective to achieve more ideal visual effects. This is precisely the same principle employed in ancient Greek and Roman architecture with the thickening and heightening of columns far from the central viewpoint to correct the reduction of the form caused by perspective. Jiangnan garden architecture adapts the principle of perspective to create an illusion of far-reaching depth.

3.1.1.2.3.2 Artistic Treatment of Orientation and Lighting of Buildings

The creation of Jiangnan garden architecture carries a special interpretation of the direction of light. Fundamentally speaking, this is because Jiangnan Garden Art is an art form that manages nature, making full use of natural conditions in the process of creating natural scenes. Its creation not only arranges the wind and rain in a scene (mainly through the means of plants) but also arranges daylight and moonlight in the

Fig. 3.208 (**a**) The roof of the Saliu Yuanyang Guan in Zhuozheng Yuan, Suzhou broken up into parts–the four pyramidal roofs on the corners of the main roof concealing the huge size of the main body. (**b**) Corner Pavilion of the Saliu Yuanyang Guan in Suzhou's Zhuozheng Yuan

scene. For example, the large main hall is used to create a shadow from the north platform of the hall and, while facing the platform and hall, the main viewing object is arranged in the south. In unison with a lake-mountain scene embellished with a pavilion and windowed veranda, a light and shadow effect is generated by side light from the morning until night, thus enhancing the three-dimensional spatial sense of the main viewing scene. This treatment of main viewing points and main viewing objects with sunlight operates on the same principle as stage lighting employed in theater productions.

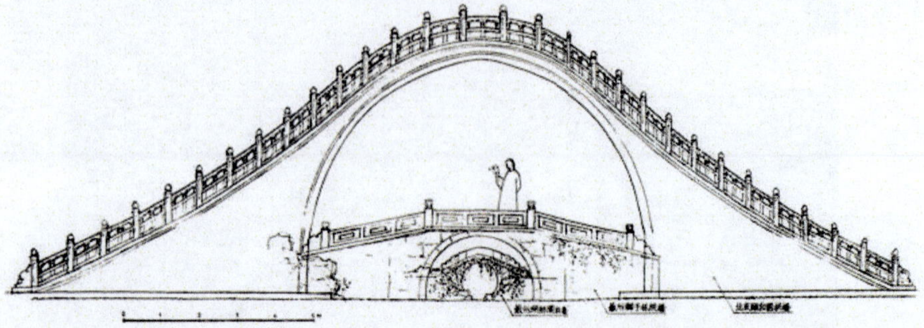

Fig. 3.209 Comparison of small arch bridges in Wangshi Yuan and Shizi Lin and large-scale arch bridge in royal gardens in the natural scenic area (Yudai Qiao in the Summer Palace)

In the creation of scenes, the treatment of sunlight and moonlight can be divided into direct lighting and indirect lighting.

Direct lighting primarily utilizes light and shadow effects and is commonly used in the lighting of whitewashed walls. The artistic creation of white walls, in contrast with the shapes and colors of mountains, water, flowers, trees, pavilions, and verandas, not only highlights the white color of the whole space but also highlights light and shadows. An apt description of the effect is given by a line from *The Story of the Western Wing*:[124] "The shadows of flowers dance about the wall". While using moonlight as the light source has its charm, using sunlight to project the beautiful shadows of the whirling flowers and trees provides a more poetic aura (Fig. 3.211). Moreover, whether using sunlight or moonlight, projecting bamboo trees from the front of windows into the interior can deepen the artistic conception of the scene. When combined with a light breeze, this treatment creates a mesmerizing, dreamlike scene (Fig. 3.212).

Indirect lighting mainly includes reflected light and filtered sunlight. Shen Fu (courtesy name San Bai) recorded the practice of using reflected sunlight to improve indoor illumination in his work *The Six Chapters of Floating Life*. In gardens, the reflected light from white walls, water surfaces, and plants is used as the source of illumination for the *shi* (chamber), *xuan* (windowed veranda), *xie* (house on a terrace), and *fang* (house), increasing the artistic flavor of internal scenery. The reflected light from the white wall's floating projection of flowers and trees and the reflected light from the water surface's glittering golden waves produces the effect of a smoky haze and blue clouds sweeping the room. Meanwhile, the reflected and filtered light from plants adds a quiet atmosphere to the interior as though it were resting in the shade of a grove. As for filtering light, it is best to use strong direct sunlight. In summer, sunbeams strike from a high angle so that the sunlight from the south cannot penetrate directly into the room. Therefore, only angled light from the

[124]The Story of the Western Wing (Xixiang Ji), by Wang Shifu, is a 13th-century adaptation of an epic romance of the twelfth century.

a b

Fig. 3.210 (**a**) Small-scale steps on the hill-climbing corridor of No. 31 garden, Beishan street, Hangzhou. (**b**) Small-scale steps of Buqiushan corridor in Huanxiu Shanzhuang, Suzhou

Fig. 3.211 (a) Light and Shadow Effect of white wall in Zhuozheng Yuan, Suzhou. (b) Light and Shadow Effect of Flower Leakage Wall in Suzhou Guangfu Situ Temple Garden. (c) Light and Shadow Effect of pergola in West Garden of the Jiechuang Si Temple in Suzhou. (d) The illusion of light and shadow effect of colored glass of Saliu Yuanyang Guan in Zhuozheng Yuan, Suzhou

east or west can fill the room, especially filtered light from the west in the afternoon. Generally, buildings in the Jiangnan area are designed for sunshine from the south, while the low-angle slanting sunlight during the summer will enter the room and increase the indoor temperature of the sections facing east or west. On the other hand, the favorable microclimate provided by plants and water in gardens allows for different orientations of sunlight to be considered. Strong, slanted light, like the surface light of a stage, is often needed for special architectural effects. The reason that "Chinese parasol and banana trees covering half of the window " in *Yuanye* (The Craft of Garden) looks elegant is mainly that it reflects green light into the room, allowing visitors to make the most of the shade on sunny days. However, indoor light can only maintain a higher illumination effect with ample direct light, making afternoon light an ideal light source. Being filtered by an appropriate plant such as paulownia, Chinese parasol, or bamboo clusters, light can be split into both a bright and elegant form. Therefore, the study and music chambers in gardens sometimes

Fig. 3.212 Shadows of trees (woods-and-mountain environment of Shuxiao Ting in Suzhou's Liu Yuan)

include an open window that faces west and has appropriate filter plants arranged to achieve light effects using the west sun. For example, The Locking Green Pavilion (Suolü Xuan) in the Garden of Pleasure (Yi Yuan) in Suzhou is positioned in a westward orientation, with bamboo forests arranged in the west of the veranda. With the strong afternoon sunshine and filters of foliage of banana, Chinese parasol or bamboo, green light fills the veranda and graces the environment with a poetic flavor of lock green as suggested by the garden's name (unfortunately, this aspect has been tampered with during renovation) (Fig. 3.213). Like the Locking Green Pavilion (Suolü Xuan), there are bamboo forest filters under the west window of the Veranda of Pointing to the Cypress (Zhibai Xuan) in the Lion Grove Garden (Shizi Lin), Suzhou.

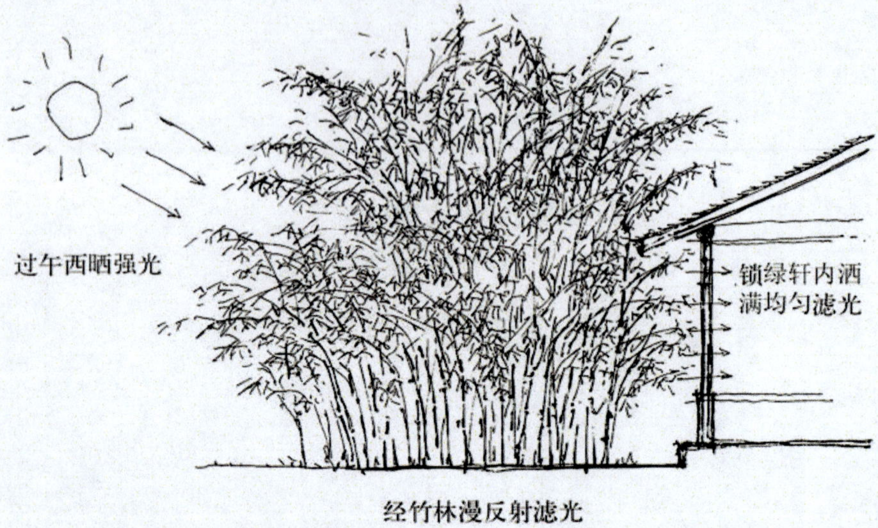

Fig. 3.213 Lighting treatment of Suolv Xuan in Yi Yuan, Suzhou

3.1.1.2.3.3 Deception: A Design Tool to Alter the Shape and Sizes of Buildings and Spaces

Architectural deception is another important feature of Jiangnan garden architecture design, with many examples often including fake mountains, fake water features, and fake buildings. The scenes crafted in theater stages and movie studios are fake versions of what they depict, but long as the lighting is combined effectively to give the audience a sense of realness, even a beautiful palace can be constructed of mere pictures. The function and form of general garden architecture are unified, and with the appearance of pavilions, corridors, mansions, terraces, and so on are consistent with their internal spaces, having their own practical value. The important difference between garden architecture and general architecture lies in its outstanding ornamental function, even to the extent that ornamental function is often the only function. Therefore, some artistic treatments made not out of utility but out of ornamental requirements are completely reasonable and necessary for garden architecture, even though these treatments are sometimes disconnected from practical functions. Sometimes deception is employed due to the needs of scene composition within the limitations of a particular location. In this situation, if the desired effect is achieved, then it is not deception but truth—a truth of art. For example, the Winding Stream Tower (Quxi Lou) in Suzhou's Garden that was preserved miraculously (Liu Yuan) was seemingly built for height in consideration of its view. Nevertheless, in reality, it is only half a tower with an interior that does not match the exterior, with the final design made in order to take up less space (Fig. 3.214). The deception doesn't hinder its role in garden art, but, on the contrary, is a necessary artistic treatment as is the fakeness involved with acting. One could say that it is these

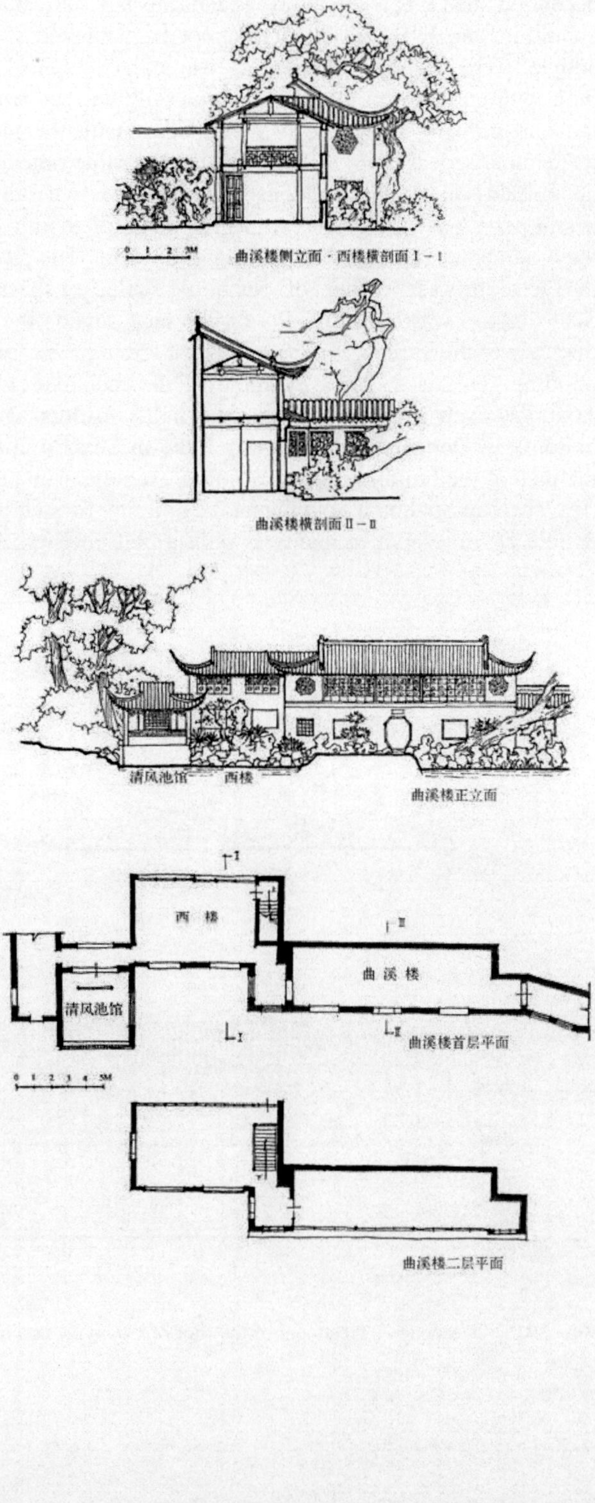

Fig. 3.214 "Deception" Treatment of Quxi Lou in Liu Yuan, Suzhou

deceptions that create the alluring landscapes of lake-side towers, where under such conditions, such scenic effects cannot be achieved if the deception is not fully adopted. The Tower of Crowning Clouds (Guanyun Lou) in the same garden is built with cornices on the side that can only be seen from inside the garden, while there are no wing corners or even cornices outside the garden (Fig. 3.215). Not all of its features appear in the field of view at the same time and those features which are not unified will not simultaneously come into view within the landscape. However, it is still perfect as a garden scene, as the incomplete unity does not in any way lower the ornamental value of the building as a whole. The pavilion on a ruined mountain in the northwest corner of Suzhou's Villa of beautiful surrounds (Huanxiu Shanzhuang) was built on stilts on the lane outside the garden (Fig. 3.216). From the view of the garden, it appears to be a scene in the forest, but in fact, it is a street building. This is a clever example of the reclamation of space. There are other examples such as half-pavilions and half-corridors against the wall. The carved building in Dongshan of Dongting Lake in Suzhou, a half pavilion hung high on the wall of the building, is also a fitting example. The pavilion in the Crane Garden (He Yuan) in Suzhou is actually a fake building formed by adding windows between double eaves, which in reality is only a hall porch without a floor and represents

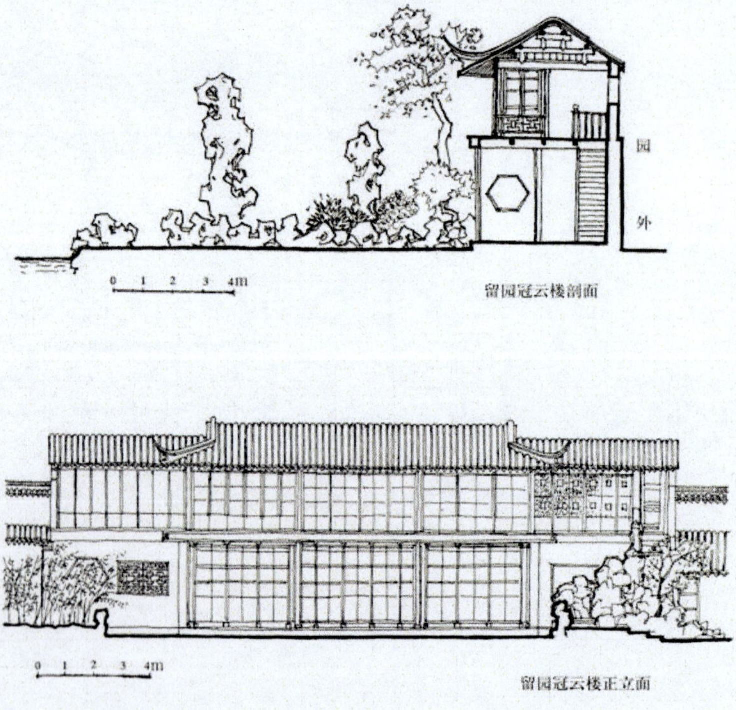

Fig. 3.215 "Deception" Treatment of the roof of Guanyun Lou in Suzhou Liu Yuan

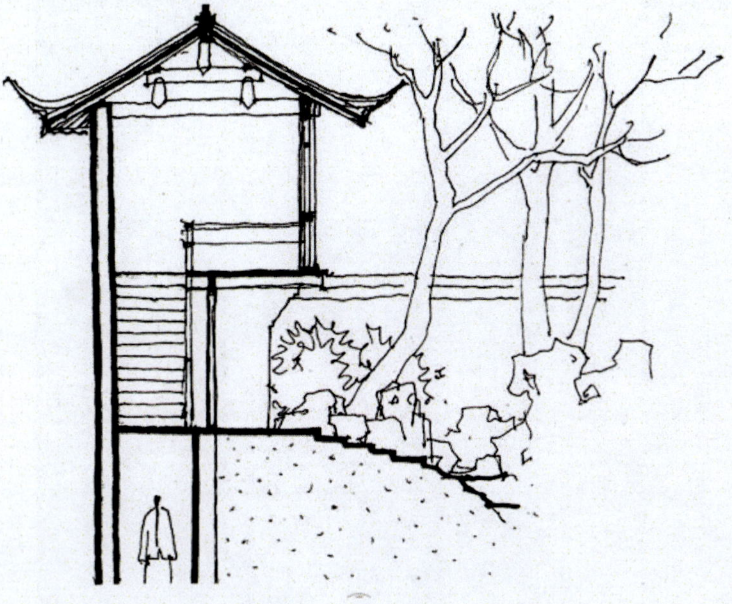

Fig. 3.216 Northwest mountain pavilion of Suzhou Huanxiu Shanzhuang (actually a street crossing building)

another form of the deception treatment (Fig. 3.217). It is akin to a carving artist who dismisses the garden wall, creating a sense of realism in the shape and space of art. This treatment can be applied to small details as well, such as the hollow painting (a special hole window for developing space) and even fake windows (such as the octagonal window of the Winding Stream Tower (Quxi Lou) in the Garden that was preserved miraculously (Liu Yuan), Suzhou), all of which serve as examples of real art achieved through deception.

3.1.1.2.4 Architectural Ornament, Furniture, and Interior Display

In general, the ornamentation, furniture, and interior display in Jiangnan gardens are similar to those found in local houses but are more decorative and intriguing due to their demanding viewing requirements. Furthermore, the supplementary decoration of the building and the modeling and arrangement of furniture and furnishings should be heavily coordinated and unified with the processed natural beauty environment of the whole garden.

Fig. 3.217 (**a**) The decorative half-pavilion shaped balcony of the "Carved Building" in Dongshan of Dongting Lake in Suzhou. (**b**) "pavilion style" hall porch of He Yuan in Suzhou

3.1.1.2.4.1 Architectural Ornament (Referred to as Zhuangzhe *in* Yuanye*)*

The concept of architectural ornaments in Chinese classical architecture differs from that of architectural decoration. Architectural decoration is a kind of beautification mode of architecture, which often plays a role in strengthening the artistic effect of the building's shape and space. It is the beautification of the structure combining decorations with utilitarian components that conveys a sense of beauty. It highlights the characteristics and utilization of the materials, the structural mechanics and building techniques, and the functional features and the practical efficiency of the components. These components are the basic principles of Chinese classical architectural decoration. The concept of architectural ornaments, on the other hand, refers to small wooden objects such as doors, windows, and lattice partition doors rather than large wooden structures such as beams and columns. The importance of these small wooden decorations cannot be overstated. Although they play an obvious decorative role and are referred to as architectural ornaments by craftsmen, their function lies far from the word ornament. They play a role not only in decoration and beautification but also in enriching the space and shape of the building. For example, lattice partitions of doors and windows are structures that enclose the space, hangings, and railings between colonnade and lintels enhance the spatial sense of the porch and corridor, and various kinds of folding screens and casings play the role of organizing the space. Therefore, various types of decoration form an essential part of Jiangnan garden architecture. Traditional artisans call the decoration on the outside of the building's outer-eaves decorations and the interior decoration inner-eaves decoration. The outer-eaves decoration includes railings, hanging, *chang chuang* (long windows or latticework partition doors), *ban-chuang* and *ban-qiang* (short windows and short wall), terrace windows (sill windows), Hehe windows (hung and removable windows), crosswind windows (horizontal drapes), brick frame windows, etc. The inner-eaves decoration includes folding screens, casings and gauze partitions (latticework partition doors), etc.

The architectural ornamentation of Jiangnan garden is as delicate and elegant as its architectural shape and space treatment. Wooden structures are rarely decorated with colored drawings (except for the golden decoration of the True Delight Pavilion (Zhenqu Ting) in the Lion Grove Garden (Shizi Lin) in Suzhou, which is a rare case in the Jiangnan area and is suspected to represent an introduction of the Lingnan style), while most embellishments come in the form of carvings. Carved decorations such as doors, windows, lattice partition doors, folding screens, and ornamental casings often use valuable furniture-making wood such as nanmu, mahogany, boxwood, red sandalwood, rosewood, and ginkgo. Treatment with oils and waxes allows the decorations to keep their original, woody color, just like home furniture. The use of pine, fir, and other general wood decoration are painted in chestnut color, ochre yellow and other colors, especially black-brown, imitating red sandalwood, mahogany, rosewood, boxwood, nanmu, etc. As a known traditional Jiangnan architecture wood carving skill, furniture decoration processing since the Ming Dynasty has used principles of furniture carpentry as a reference to produce works of such exceptional beauty that even the Qing Emperor Qianlong recognized the

Yangzhou works as having superior quality. Though they lack luxurious inlays of pearls and jade, they feature traditional delicate carvings of boxwood and ginkgo as well as other expensive wooden carving inlays that remain to this day.

Halls, both large and small, are the key ornamental buildings in gardens, utilizing nearly every architectural ornament method to serve as a suitable focus of study.

In the ensuing section, the discussion will focus on different types of architectural ornaments.

Railings

Wooden railings are a decorative barrier structure and can be divided into three types: balustrade,[125] half bar,[126] and gooseneck chair.[127] Balustrades are mostly used between the outer columns of the halls and pavilions, or under the terrace windows and sash windows to replace the half walls for ventilation. All wooden railing types are very ornate and decorative, with design varying vastly in complexity (see Fig. 3.218).

The half bar provides seats for convenient resting and is mostly used between the outer columns of the halls or between verandah columns. The simplest half bar is made of a dry or plastered low wall, or fair-faced brick wall showing a hollow space. Like balustrades, wooden half bars can also be designed in various grid patterns (Fig. 3.219).

Backrest railings are also a feature found in residential houses. Nowadays, due to the ongoing reconstruction of towns and villages, there are not many remaining examples of backrest railings. A rare example is under the ancient river street arcade in Tuochuan, Wuyuan, Jiangxi Province (Fig. 3.220). This backrest railing mainly used a gooseneck shape for the vertical bars and so is also called a gooseneck chair (Fig. 3.221). It boasts a rich, colorful style, which of course is not limited to the gooseneck. (Fig. 3.222) The gooseneck chair is convenient for sitting and resting and is mostly used in places where visitors linger for a period time, such as at waterside pavilions, windowed verandas, principal halls, houses on a terrace, and towers, as well as in buildings that are placed in fine scenes between mountains and flowers.

Hangings

This kind of decoration is installed between the outer corridors of halls or verandah eaves and under the lintels between columns of pavilions and corridors. It is made of thin wooden strips forming various grid patterns and represents an extension and development of the hanging of curtains in the Han and Tang dynasties. The delicate

[125] Handrail.

[126] Stool railing.

[127] Back railing or beauty rest.

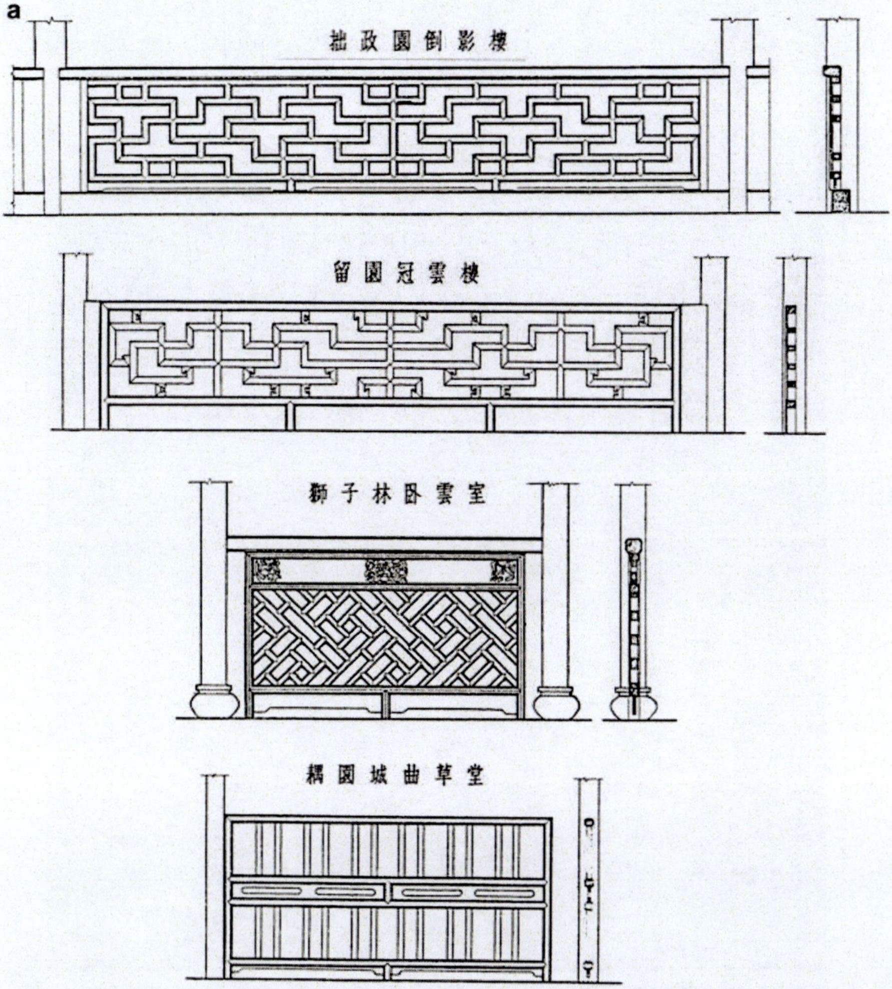

Fig. 3.218 (**a**) Actual measurement of four wooden hook bars in Suzhou classical gardens (extracted from Suzhou Classical Gardens). (**b**) Balustrade of Woyun Shi in Shizi Lin, Suzhou. (**c**) Balustrade of Jixu Zhai in Wangshi Yuan, Suzhou. (**d**) Balustrade of Yulan Tang in Zhuozheng Yuan, Suzhou. (**e**) Balustrade of Yanyu Tang, Shizi Lin, Suzhou. (**f**) Balustrade on the Ground Floor of Shuangzhao Lou in Suzhou's Ou Yuan. (**g**) Balustrade on the Upstairs of Jixiao Shanzhuang, Yangzhou. (**h**) Balustrade of Duhua Lou in Wangshi Yuan, Suzhou

wooden lattice is just like a curtain hanging under a lintel, covering the border and showing the division of space. Generally, the hanging and the railing are used alongside one another in alternating patterns, the echoes of which strengthen the spatial boundary and form a wonderful border for viewing the scenery (Fig. 3.223).

Fig. 3.218 (continued)

Fig. 3.218 (continued)

h

Fig. 3.218 (continued)

It is worth mentioning that *huayazi* (a decorative piece with hollow wood carvings derived from *queti*, the reinforcing component at the junction of a column and a lintel) replaces hangings on low architrave or secondary buildings (Fig. 3.224).

Casings

These were originally a tool used for hanging curtains and later developed into an interior decorative component that helps organize space. It gives space the appearance of being divided into different functional parts, which in turn gives the room an atmosphere of interconnectedness. Casings can be divided into ceiling casings and floor casings. Ceiling casings are the same as indoor hangings. Simple ceiling casings are composed of grid patterns with thin wood strips, while more complicated forms are mostly carved from wood boards and are similar to an arch with its two ends hanging down (Fig. 3.225).

Floor casing can be divided into three types: Free-type, gauze-partition-type (latticework-partition-doors-type), and hole-gate-type. The free-type casing is a ceiling casing with two ends rolled down, forming an irregular inner arch contour. The ornament patterns of free type casings are mostly festive or elegant objects related to plants and birds, such as *three durable plants of winter*—pine, bamboo, and plum blossom, *happiness appears on the eyebrows* (homonym for plum), *magpie climbing plum, squirrel, and grape,* interlocking branch pattern, and plantain pattern (Fig. 3.226). The gauze-partition-type casing is a regular form floor casing with a horizontal drape above and two screens on either side (Figs. 3.226, 3.227). The hole-gate-type casing is a further development of the floor casing, forming a hole gate by enclosing the two sides on the floor (even if the ends on the floor is not

Fig. 3.219 (**a**) Actual measurement of two wooden half-bars in Suzhou classical gardens (extracted from Suzhou Classical Gardens). (**b**) Half bar of the white sill wall in Yi Yuan, Suzhou. (**c**) Hollowed half bar made of polished terrazzo bricks in Wangshi Yuan, Suzhou. (**d**) Hollow Brick Carved Half Bar in Zhuozheng Yuan, Suzhou. (**e**) Wooden Half Bar of Jixiao Shanzhuang, Yangzhou. (**f**) Wooden Half Bar of South Corridor of Shuangzhao Lou in Suzhou's Ou Yuan

d
e
f

Fig. 3.219 (continued)

Fig. 3.220 Backrest Railings of Tuochuan River Street, Wuyuan, Jiangxi Province

enclosed, it can also form a hole-gate-type). The casing can take on various shapes such as an octagon, oblong octagon, or full-moon circular (Figs. 3.228 and 3.163b). The full-moon circular casing is a representative hole-gate-type, commonly known as the full moon casing or round light casing.

Folding Screens

Like other residential buildings, garden buildings often set folding screens (called screen door or drum door in *The Craft of Gardens* (Yuan Ye)) in the back of the hall

Fig. 3.221 Typical Goose Neck Chair at Ruyu Ting in Suzhou's Yi Yuan—Backrest with Curved Goose Neck Like Vertical Grills

to cover the rear entrance or stairs and facilitate the arrangement of furniture such as long narrow tables, dining tables, and chairs in the front of the hall. The folding screen is a flat panel screen comprising four, six or eight partition boards (similar to a latticework partition door without a frame), with six boards being the most common configuration. In *yuanyang ting* (tandem hall), the folding screen is placed in the middle of the hall, dividing the space into two secondary halls with casings. The screen is elaborately decorated, ornamented with carvings, paintings, or calligraphy in a unified style. For example, the screen in Yanyu Hall (Yangyu Tang) in the Lion Grove Garden (Shizi Lin) (of Suzhou features a carved map of the Lion Grove Garden (Shizi Lin) on the south side and an engraving of the account of *Rebuilding Shizi Lin* on the north side. In some halls, six latticework partition boards are placed in the center and can be regarded as a gauze-partition-style folding screen (Fig. 3.81).

Gauze Partitions

In the Jiangnan region, indoor lattice partition doors are called gauze partitions because it is often sandwiched with gauze and is used to divide the indoor space. Sometimes it is necessary to block the view and muffle noise to a certain extent, in which case lumber replaces the gauze, often including mounted or carved calligraphy, paintings, or antique rubbings (Figs. 3.228, 3.229, 3.230b, c). Decorative interior partition components can evoke an elegant atmosphere in halls and be an integral component of the architecture. As the assembling and disassembling of the gauze partition is convenient, it is so flexible that when large space is needed for festive banquets and other activities, the gauze partition can be completely disassembled and then completely re-assembled in no time.

a

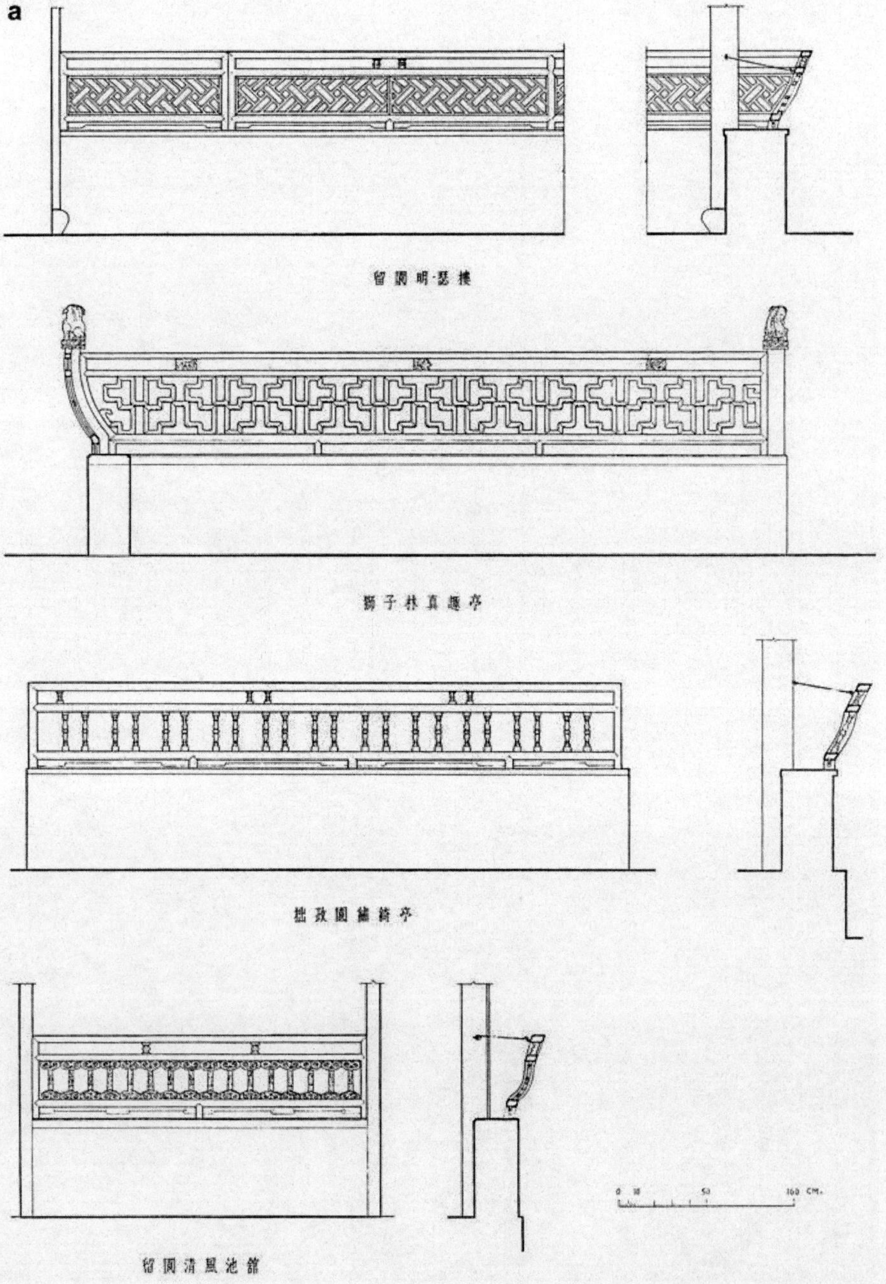

Fig. 3.222 (**a**) Measured Figures of Four Goose Neck Chairs in Suzhou Classical Gardens (From Suzhou Classical Gardens). (**b**) Goose Neck Chair at "Zhenqu" Ting in Shizi Lin, Suzhou. (**c**) Goose Neck Chair of "Lüyin" Xuan in Liu Yuan, Suzhou. (**d**) Goose Neck Chair with Seating Patterns in "Shan Shui Jian" in Ou Yuan, Suzhou. (**e**) Goose Neck Chair of Yi Yu Xuan in Ge Yuan, Yangzhou. (**f**) Details of Goose Neck Chair of Yiyu Xuan in Yangzhou's Ge Yuan. (**g**) Hollow Brick Stool and Gooseneck Chair with Wooden Backrest in Ban Yuan, Suzhou

Fig. 3.222 (continued)

Fig. 3.222 (continued)

Chang Chuang *(Long Windows)*

Chang Chuang (long windows), or long-lattice-style/Ge (lattice window) in *The Craft of Gardens* (Yuan Ye), are floor-to-ceiling windows and are a type of lattice-work partition used under outside eaves, usually for the front eaves, or just for the middle room. It is a door rather than a window and as an exterior structure, it is similar to the interior gauze partition in shape. Nevertheless, it has daylighting,

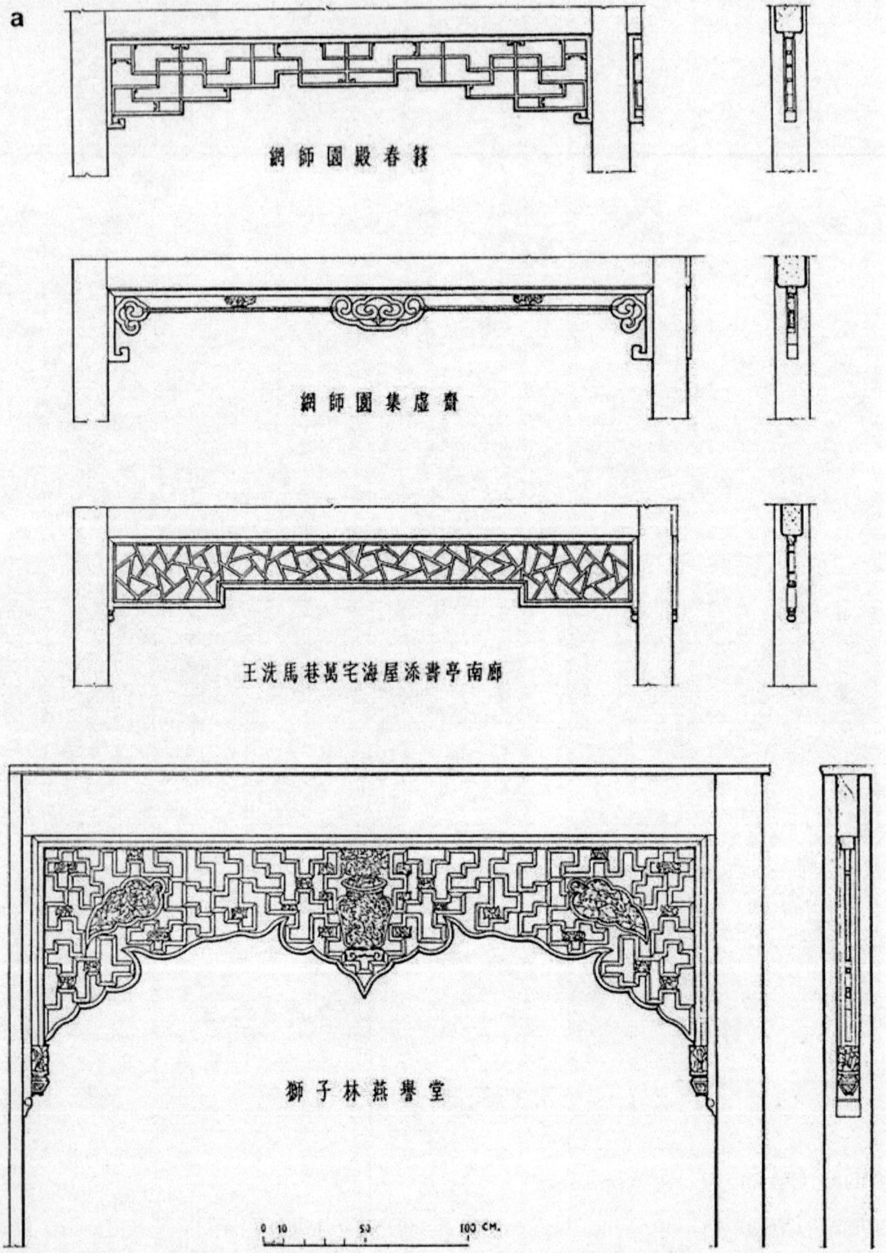

Fig. 3.223 (**a**) Four Cases of Hangings in Suzhou Classical Gardens (From Suzhou Classical Gardens). (**b**) Hangings of Zhilian Laowu in the West Garden of Ou Yuan, Suzhou. (**c**) Flower Basket Hanging Columns for Corner Hanging in the Outer Gallery of Flower Basket Hall in Shizi Lin, Suzhou. (**d**) Hangings and half-railings installed between the pillars of the pavilion in Zhuozheng Yuan, which makes the pavilion look exquisite and delicate and forms a framing border (taken in 2009, the water surface is full of lotus flowers for economic benefit, which violates the principle of water surface greening in classical gardens)

Fig. 3.223 (continued)

windproofing, and heat preservation requirements and in the past utilized *feng chuang* outside the *neixinzi* (latticework, called *ling kong* in *Yuanye*) for protection. *Feng chuang* is sandwiched with gauze or pasted with paper or installed with firmer light-transmitting materials like pieces of shell and mica. Having been replaced by glass in more recent times, it is closed at nighttime or during windy conditions to protect the long window. As mentioned in *The Craft of Gardens* (Yuan Ye), the ratio of the hollow frame at the upper part of the long window to the panel at the lower part (called shuyao (waist) and pingban (flat plate) in *The Craft of Gardens* (Yuan Ye)) is generally smaller than the 7:3 or 8:2, meaning that the lighting area is slightly smaller than that of the Ming-style. Moreover, the lattice works of existing long windows in Jiangnan gardens have various styles and complicated patterns, with a few of them exhibiting the plain wicker patterns of the Ming-style. As for panels on

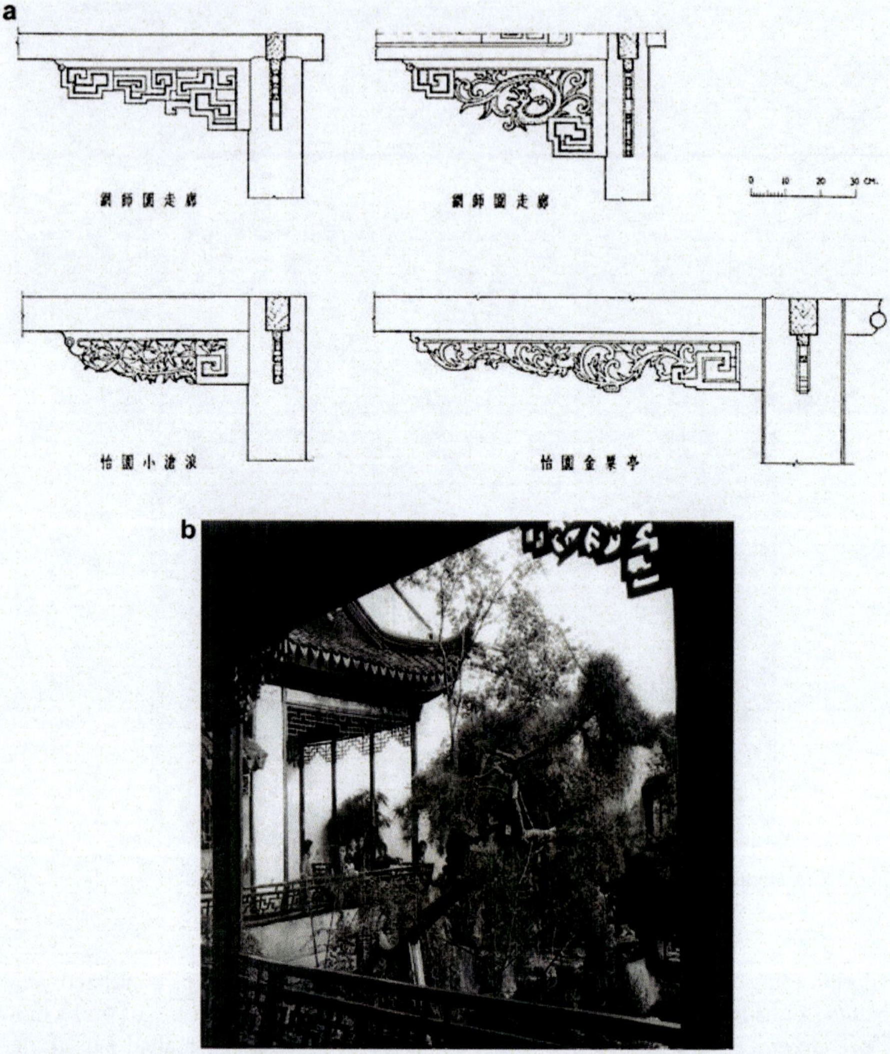

Fig. 3.224 (**a**) Measured Figures of Four Cases of huayazi in Suzhou Classical Gardens (From Suzhou Classical Gardens). (**b**) Huayazi of Zhuwai Yizhi Xuan of Wangshi Yuan in Suzhou, which makes the junction between the column and the fang exquisite and elegant. Hangings are used at the porch in front of the Sheya Lang in the distance

the lower part (jiatang or waist board and qunban or skirting board), there are both bare and carved versions (Fig. 3.231). There is another kind of long window called bright floor casing, which is made of latticework with its whole body exquisitely carved and is especially glittering and translucent when glass is used with modern

a

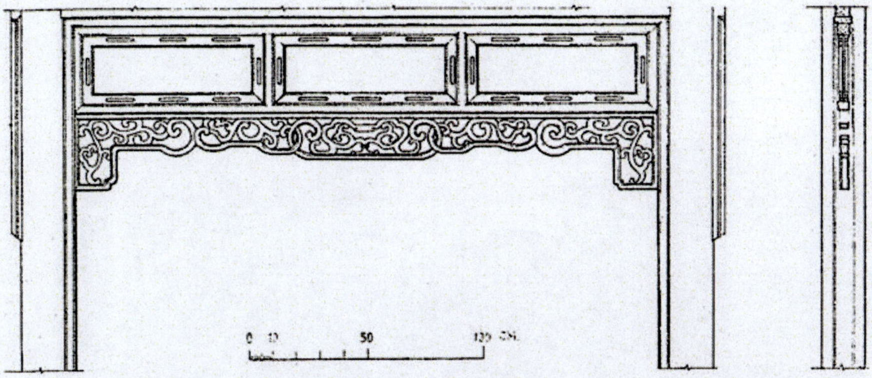

b

Fig. 3.225 (**a**) Actual measurement of ceiling casing of Dian Chun Yi of Wangshi Yuan, Suzhou (from Suzhou Classical Gardens). (**b**) "Xi Shang Mei Shao" ceiling casing of Liuting Ge of Zhuozheng Yuan, Suzhou (from Suzhou Classical Gardens). (**c**) Ceiling casing in Middle Room of Liuting Ge in Zhuozheng Yuan, Suzhou

versions. Owing to its four sides having bright floor casings, the Hall of Distant Fragrance (Yuanxiang Tang) of the Humble Administrator's Garden (Zhuozheng Yuan) in Suzhou has a bright and open indoor view with integrated interior and exterior scenes (Fig. 3.232).

Fig. 3.225 (continued)

Short Window

The decoration commonly called ban-chuang and ban-qiang (short windows and short wall) in the Jiangnan area corresponds to the short-lattice-style Ge (lattice window) in *The Craft of Gardens* (Yuan Ye). It is mainly used in the hall's Cijian (the room next to the middle one) under the eaves of columns in warm pavilions or warm corridors. The short window has upper and lower panels instead of the skirting board, the lattice pattern of which is similar to the long window (Figs. 3.231 and 3.233). It produces an excellent decorative effect together with the lower half wall which is either whitewashed or made of polished terrazzo bricks. As it is fully open, the internal and external scenes are integrated. The so-called short walls have different heights, some being equal to that of the long window skirting board, such as the one in the Cijian of Zhuyun An in the Garden that was preserved miraculously (Liu Yuan) in Suzhou and the one in the North Passage Hall of the Fry Pavilion (Ruyu Ting) in the Garden of Art (Yi Pu). On the other hand, others have a height of only 45 centimeters, like the one in the Cijian on the ground floor of the West Mansion in the Garden that was preserved miraculously (Liu Yuan) in Suzhou (Fig. 3.234). This low short wall of about 50 cm can serve the same seating function as the aforementioned half-bars when fitted with a sitting sill, and the addition of a backrest converts it into a gooseneck chair (Fig. 3.235).

Fig. 3.226 (**a**) Banana Floor Casing in the Hall of "Gu Wu Song Yuan" in Shizi Lin, Suzhou (**b**) "Sui Han San You" openwork floor casing with patterns of pine, bamboo and plum in Yiyu Xuan of Ge Yuan, Yangzhou. (**c**) Floor Casing with patterns of magpies on plum branches in Tiyun Shi of Wangshi Yuan, Suzhou. (**d**) Part of Floor Casing with patterns of magpies on plum branches in Tiyun Shi of Wangshi Yuan, Suzhou

Fig. 3.227 (**a**) Gauze-partition-type ceiling casing in Wufengxian Guan in Liu Yuan, Suzhou. (**b**) Part of the gauze-partition-type ceiling casing in Zhuoying Shuige of Wangshi Yuan, Suzhou. (**c**) Gauze-partition-type ceiling casing in "Shan Shui Jian" in the east garden of Ou Yuan, Suzhou

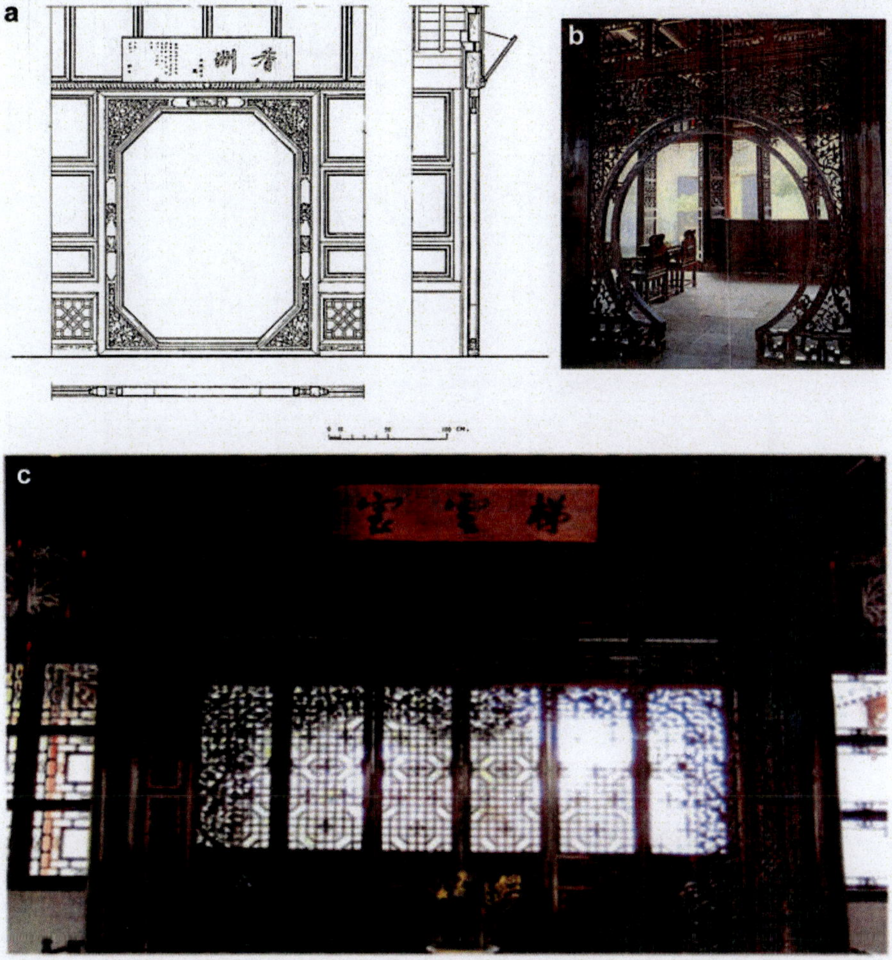

Fig. 3.228 (**a**) Measured Drawing of oblong octagon floor casing in Xiangzhou Fang of Zhuozheng Yuan, Suzhou (From Suzhou Classical Gardens). (**b**) Round light casing in Linquan Qishuo Zhi Guan in Liu Yuan, Suzhou. (**c**) Floor Casing in Tiyun Shi of Wangshi Yuan, Suzhou

Terrace Windows (Sill Windows)

The so-called terrace window is also a type of short window, with its lower part being a railing instead of a wall (Fig. 3.236). It is often used in the Cijian of the hall or under the outer eaves of mansions, pavilions, warm verandas, etc. (Fig. 3.237). The window can be opened and looked out of from the railing or closed in the winter to keep the warmth in. The lower balustrades are equipped with waterproof boards that are fixed with small rods and can be removed together with the windows when

a
b

Fig. 3.229 (**a**) Gauze Partition with Calligraphy and Painting Mounted on Inner Panel in Fucui Ge of Zhuozheng Yuan, Suzhou. (**b**) Gauze Partitions Carved Calligraphy and Painting on Inner Panel in Liu Yuan, Suzhou. (**c**) Gauze of Wufengxian Guan in Suzhou's Liuyuan—Interior Panel Mounted Antique Rubbings

Fig. 3.229 (continued)

necessary. Such flexible decorations of the upper window and lower board may be a variant of the ancient traditional Ta structure that has been preserved in Jiangnan residential buildings and serves only to change the upper hung windows into horizontal ones.

Hehe Window

The so-called Hehe window in the Jiangnan region is a type of short window that is opened vertically by propping and removing it just like a hung window. It is only the part lying in the visual field of viewers who are standing upright that can be opened, while the rest of the window still retains the decorative effect of a large area of plain latticework enclosure (Fig. 3.238).

Brick Frame Window

Currently known as the brick frame window, this is a type of window similar in style to those mentioned in *The Craft of Gardens* (Yuan Ye) such as fengchuang (wind window), shuchuang (book window), and xiuchuang (embroidered window), with a slight difference in that it is not an additional window for protection and may not be opened at will. By occupying a separate, dedicated hole in the wall, it has the same

a

b

Fig. 3.230 (**a**) Chairs and tea tables symmetrically arranged in the hall of Linquan Qishuo Zhi Guan in Liu Yuan, Suzhou, which divides the space into primary and smaller, secondary ones, providing a complete allocation of functions within the room dynasties. (**b**) The Wufengxian Guan

design concept as *lou chuang* (openwork windows), but uses a wooden lattice with sandwiched gauze, paste paper, or clamp oyster shell and mica sheets. One common style of the window only has ornamentation on the frame while leaving the center blank, with other styles including ordinary railings and latticework patterns. Most brick frame windows are rectangular or flat and some are shaped in geometric figures such as hexagonal, octagonal, and full moon (circular). For example, the window in the Veranda of Rain Delight (Yiyu Xuan) in Yangzhou's Garden of Bamboo (Ge Yuan) is a begonia-style window composed of four arcs (see Fig. 3.239). In addition, there are also freestyles, such as the brick frame window in Taicang's Yi Yuan, which is pomegranate-shaped. Glass is often present in existing examples of Jiangnan gardens in modern times, which serves to guard against wind and as insulation without obstructing the composition of the scenery.

3.1.1.2.4.2 Furniture and Interior Display

Furniture and displays inside and outside pavilions, halls, and mansions as well as among mountains, pools, flowers, and trees are an indispensable part of the garden landscape. An empty hall is not only unable to satisfy the practical needs of the garden but also unable to fulfill the expectations of an intriguing scenery. Therefore, furniture and displays are not merely accessories and decorations that are optionally or arbitrarily arranged in the architectural environment. This is the case not only inside the building but also in the outdoor natural scenic environment, where furniture and displays are an indispensable artifice of architectural management under the unified theme of the scene. Furniture and displays, as well as interior decoration such as folding screens, latticework partition doors, and casings, play an organizational role in the architectural space. For example, the tea table and chairs placed to the right and left in the hall divide the space into primary and smaller, secondary ones, providing a complete allocation of functions within the room (Fig. 3.230). The content, pattern, and style of furniture and displays, directly affect the emotional appeal of the environment, which is the case regardless of whether they are simple and natural or carved and luxurious. In Jiangnan gardens, the furniture and displays are based primarily on primitive simplicity, elegance, nature, or novelty. According to the residential functional requirements of a given garden, furniture can include beds, couches, tables, chairs, small tables, stools, cupboards, cabinets, shelves, etc. Displays can include candles, lights, flowerpots, bonsai, fish

Fig. 3.230 (continued) (Nanmu Hall) in Suzhou's Liu Yuan is separated by gauze screens. The south hall is recessed in the middle room, with a partition screen under which is a long narrow tall table with tables and chairs in front. Gauze screen is installed with mounted painting in the secondary room, and the tea table and chairs are symmetrically arranged in the center of the hall, completing the functional organization of the space. **(c)** Two secondary rooms of the North Hall of Wufengxian Guan in Suzhou Liu Yuan is recessed, and the space is also allocated with furniture. **(d)** Internal Perspective of Linquan Qishuo Zhi Guan in Suzhou's Liu Yuan—Decoration and Furniture as Supplementary Means of Space Organization (From The Catalogue of Jiangnan Gardens)

c

d

Fig. 3.230 (continued)

tanks, caged birds, tripod stoves, *qin*,[128] swords, antiques, calligraphy, paintings, the four treasures of the study (writing brush, ink stick, inkslab, and paper), etc. In modern times, items such as self-ringing clocks, full-length mirrors, and so on may also be present. As furniture and displays are movable objects, they are less likely to remain when compared to fixed architecture. The existing examples of Jiangnan gardens only contain partial and schematic restorations of some furniture and displays, most of which are not arranged in place, or with a layout that is not the

[128]The *qin* is a traditional Chinese stringed instrument in the zither family.

Fig. 3.231 (**a**) Measurement of Long Windows and Gauze Partitions in Suzhou Classical Gardens (From Suzhou Classical Gardens). (**b**) Long stained glass windows in western style of the Saliu Yuanyang Guan in Zhuozheng Yuan, Suzhou. (**c**) Long Window in Wufengxian Guan of Liu Yuan, Suzhou. (**d**) Jiatang and qunban made of boxwood carving in Tiyun Shi of Liu Yuan. (**e**) Long Window with Banana leaf pattern and "lü tian" (means green sky) pattern in Xiaolian Zhuang, Nanxun (with banana in front of the window)

Fig. 3.231 (continued)

original. In order to give readers a general understanding of the furniture and displays of Jiangnan classical private gardens, it is likely better to extract the relevant descriptions directly from the ancient texts of Chinese gardening. These descriptions not only introduce the layout of various garden buildings but also reflect the garden lifestyle and interests of the past. Of course, what is excerpted here reflects the requirements for the garden life of the specific owner of a garden under the historical conditions at the time. If examples of the ancient gardens could be exhibited today,

Fig. 3.231 (continued)

they would only be restored from a historical point of view that cannot ensure complete authenticity to the original. As far as this study on the subject is concerned, these materials are not only of great value to the understanding of classical gardens but also of great significance to the creation of new gardens.

The following is an excerpt from the *"Garden Displays"* section of Volume II of *Huajing* (short for *Michuan Huajing*, also named *Esoteric Garden Art*):

Tables in the Hall

In front of the hall, there is a large narrow table made of rosewood or nanmu, an ancient painting hanging on the wall above, a stone ornament on the table, and six polished or black varnished foldable chairs around the table. Inside the hall, there is also a narrow table, which should be on the left side and towards the east and not too close to the windowsill, where it would be at the mercy of the elements. On the table rests an old Duan Yan (inkstone), a pen container made of red sandalwood, rosewood or fragrant wood, a pair of compasses, a small bronze container of water for ink produced from an ancient kiln, and one mountain-shaped ink slab made of stone, crystal, or fragrant wood root. The ancients placed the ink slab on the left side, with an especially favorable position under the lamp, as light reflected from the ink is less offending to the eyes. There is also a small ink stick, a picture album, a paperweight, and a curio bottle. There is yet another small table, on which rest an ancient copper stove, an incense box (made of red sandalwood if it is not carved or varnished), a set of white copper incense tools, and a bottle for containing tools, which is either unearthed bronze, red sandalwood, or old tree root. A *qin* (a kind of Chinese zither) hangs on the left wall, while the right wall features a sword and a fly whisk. Gold and

Fig. 3.232 (**a**) Bright floor casings on all sides of Yuanxiang Tang of Zhuozheng Yuan in Suzhou. (**b**) Bright floor casing in the Middle Room of Yiyu Xuan in Yangzhou's Ge Yuan

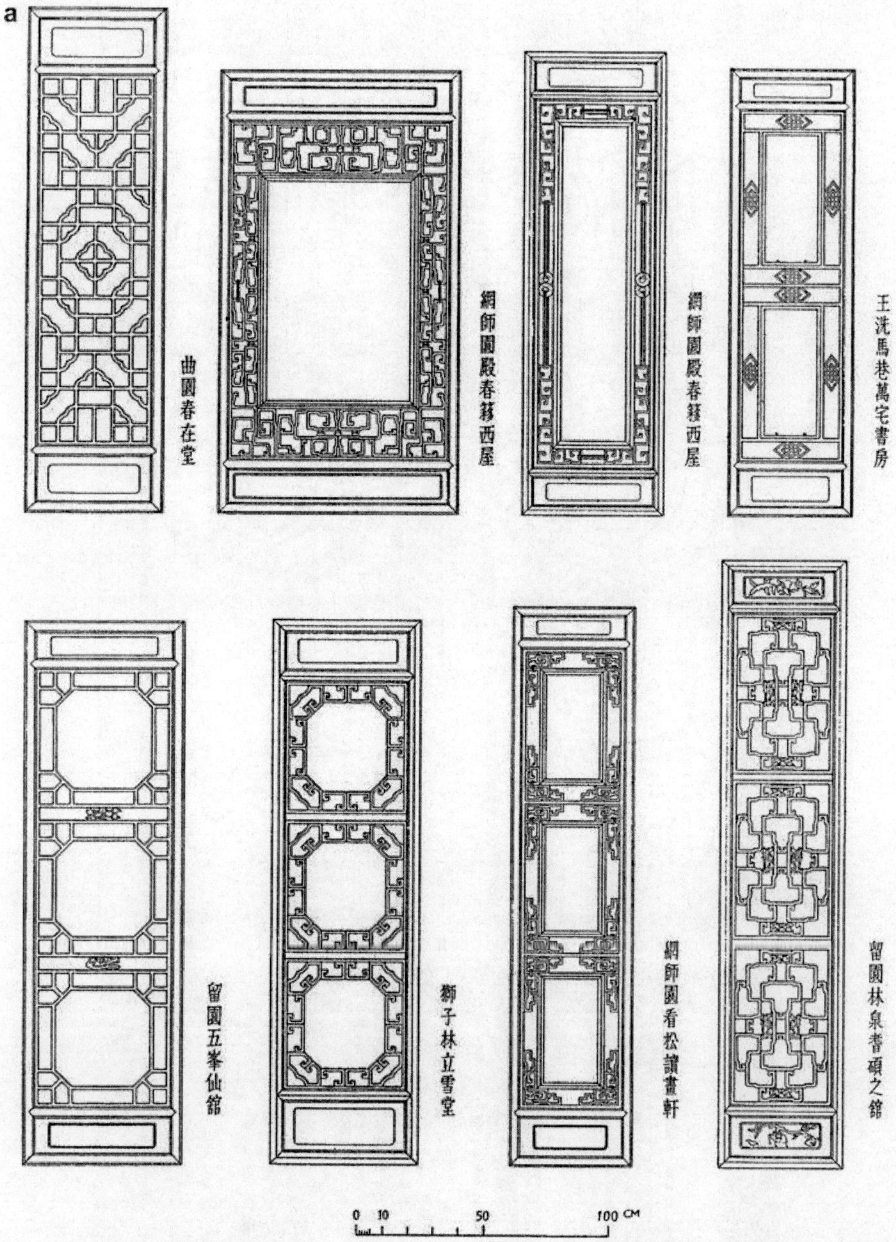

Fig. 3.233 (**a**) Measured Drawing of Half Window and Floor Window of Suzhou Classical Gardens (From Suzhou Classical Gardens). (**b**) Window lattice of antique characters and patterns in Xiaolian Zhuang, Nanxun. (**c**) Half Window of side rooms of Dian Chun Yi in Wangshi Yuan, Suzhou. (**d**) Half Window of Qingxiang Guan, Canglang Ting, Suzhou. (**e**) Half Window of Kansong Duhua Xuan in Wangshi Yuan, Suzhou. (**f**) Windows of seal character patterns in Xiaolian Zhuang, Nanxun

b

c

Fig. 3.233 (continued)

Fig. 3.233 (continued)

silver utensils must not be used in the garden, which, although viewed as a sign of wealth by foolish people, is regarded as vulgarity in the minds of more noble persons.

Couches in the Study Room

Only four chairs, two stools, one bed, and one sofa are allowed in the study. In the summertime, furniture made of mottled bamboo is the most comfortable, while in winter it should be decorated with brocade bedding or fur cushions. Other furniture, such as the ancient sumeru seat, short couch, pier table, long narrow table, big chair, and so on, can be set high, but lining up chairs against the wall should be avoided. At most only one folding screen can be placed there. Bookshelves and bookcases should be placed facing the sunny side so that books and historical records are readily visibleas in a bookstore. The bookcase also typically houses a ruler, a paper cutter, and an awl.

Fig. 3.234 (**a**) The ancient-style tall half-wall and half-window lattice of the north hall of Ruyu Ting in Yi Pu, Suzhou. (**b**) Low half wall and half window with swastikas patterns in Cui Ling Long, Canglang Ting, Suzhou

Displays in the Open Room

Built near the water with all the windows and railings removed, the *open room* is a pleasant dwelling place in the late summer. A Chinese parasol tree stands in front of the room, a bamboo forest extends behind it, and a pool of lotus flowers surrounds the exterior, complete with a waterside pavilion providing shade and a fragrant breeze. In the middle of the room is a particularly large wooden slab table, with a

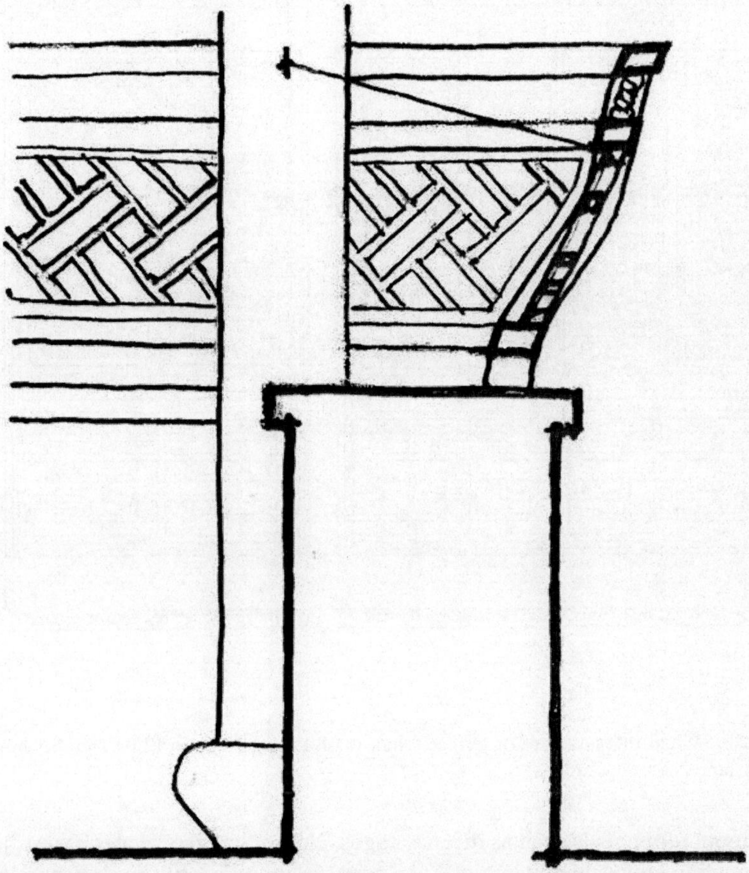

Fig. 3.235 Half Wall Backrest Bar

long narrow bed on each side. Valuable paintings are never hung in the open room, as they are vulnerable to cracking from the dry weather in summer and the back wall contains a gaping hole with no place to hang anything. The bed beneath the North window is covered with a bamboo mattress so that one may lie on it leisurely. On the table are a large inkstone, a green container of water, and a large wine vessel. Pots of *Cymbidium ensifolium* (L.) Sw., pearl orchid (ChlorantusSpicatusMak), and jasmine are placed on the table facing the wind. In addition to the ancient trees, enchanting peaks, and waterside lotus pavilions, curtains made of mottled bamboo cover the windows on all sides, conferring a delightfully cool appearance.

Furnishings in the Bedroom

Although decorative floor tiles and ceilings in the bedroom are inelegant, it is no harm using them without color paintings to keep the bedroom dry. In the bedroom, there is a bed facing south behind which there is a small room or a side room. This is

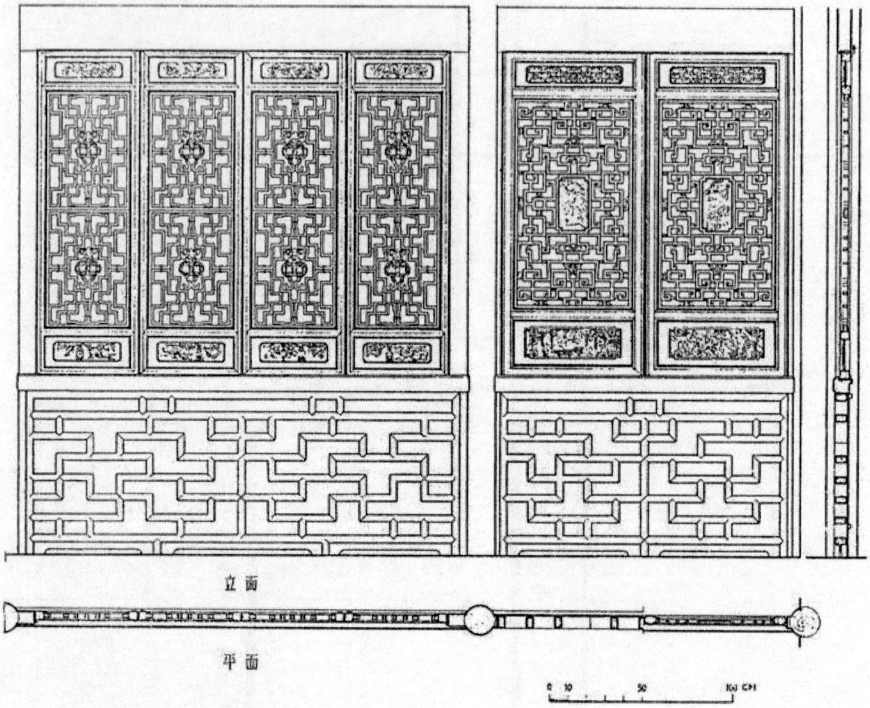

Fig. 3.236 Actual measurement of terrace window of Yifeng Xuan in Liu Yuan, Suzhou (taken from Suzhou Classical Gardens)

a storeroom for utensils such as incense cages, clothes hangers, water basins, jewelry cases, reading lamps, handkerchiefs, and soap cases and people don't generally enter it. The front of the bed features only a small table with nothing on it. To the sides are two small square stools and one small cabinet which is used to store incense and curios. The room is clean and elegant. Any gaudy or showy embellishment would give the appearance of a woman's boudoir, which breaks the elegance of sleeping in clouds and dreaming of the moon alone amongst the shadows of the trees. In addition, a bed is placed against an open window on the wall so that visiting friends can stay up all night talking. Under the window is a drawer with socks in it. Cheap trees cannot be planted in large numbers in the courtyard, but their highly ornamented variations can be taken into the room to be displayed as rare novel species, accompanied by decorative stones such as Yingshi and Kunshanshi. Two or three bonsai, imitating Chinese landscape paintings by Yunlin (Ni Zan, a painter in Yuan Dynasty) and Dachi (Huang Gongwang, a painter in Yuan Dynasty), fill out the narrow chamber.

Pavilion Furnishings

Generally, pavilions do not protect themselves from the wind and rain, so elegant furnishings are not placed in them. Tacky, inelegant furnishings are intolerable,

Fig. 3.237 Floor Window of Xi Lou of Liu Yuan, Suzhou

however. Furnishings featuring old paint, a square shape, thick feet, and primitive simplicity and nature will suffice. Open-air seating should consist of natural low flat lake stones, not circular stone stools or pottery stools. The use of wood shelters painted red on the brick road should be specifically avoided. Couplets on pillars must be inscribed on a board to avoid being defaced by wind and rain. If it is the couplet on a column of a hall or mansion, golden letters will be the first choice, cinnabar the second.

The Winding Corridor

There are two kinds of corridors, most of which wind around the house and have painted walls and columns. Reineckia carnea and Ophiopogon bodinieri Levl are planted beside the steps and there is a gauze lamp hanging overhead every dozen steps or so to provide light for the poets who walk in the night and appreciate the

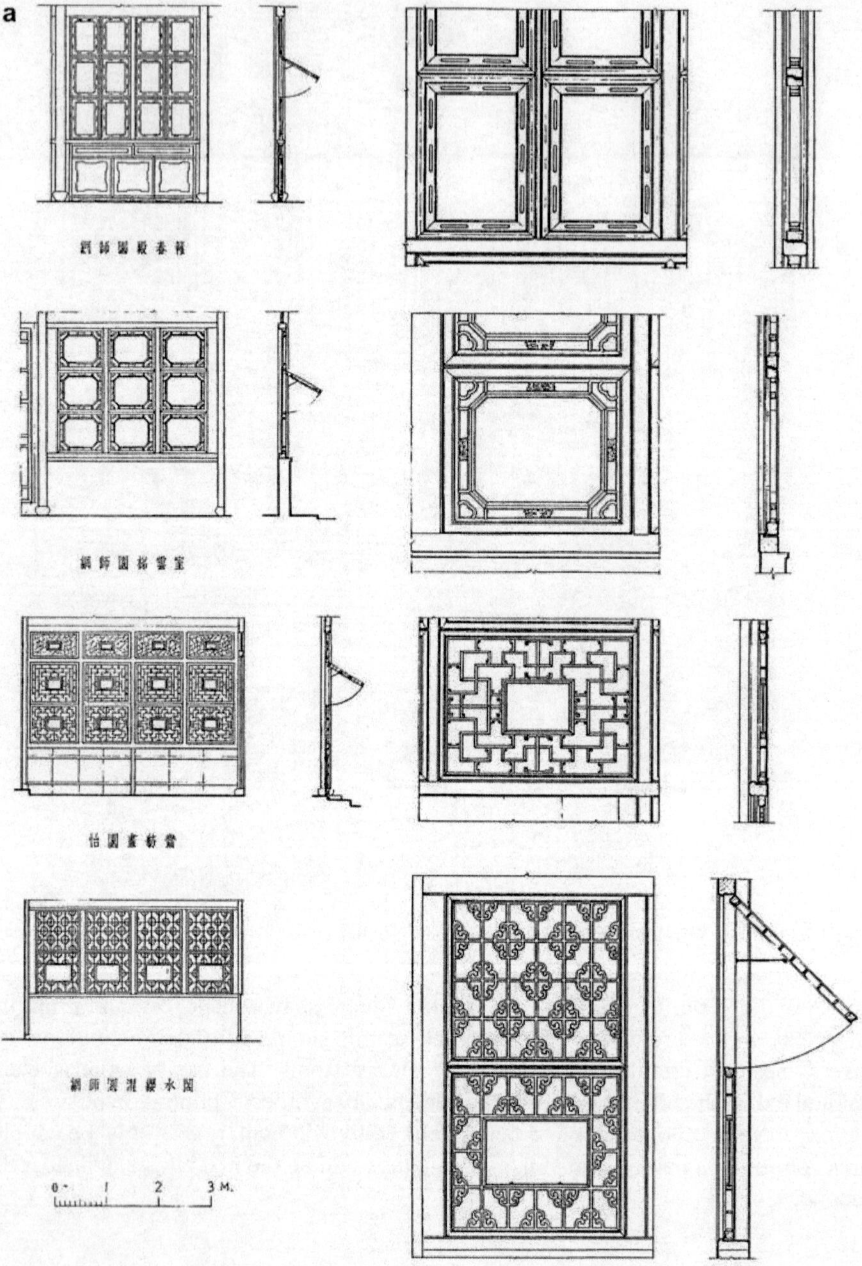

Fig. 3.238 (**a**) Measured Figure of Hehe Windows of Suzhou Classical Gardens (From Suzhou Classical Gardens). (**b**) Hehe Window of waterside pavilions on the east bank in Xu Yuan, Nanjing

Fig. 3.238 (continued)

fragrance of the flowers. Another kind of corridor built with bamboo rafters and without tiles is called the flower corridor. Trees such as hibiscus, camellia, sophora japonica, and cypress are used as walls, while plants such as banksia rose, rose, Chinese rose, Tangdi (Cerasus japonica), Rosa rubus, grape, and so on are used as sheds. Stone blocks and porcelain cylindrical benches are placed under them for rest.

The Chamber in the Overhead Corridor

Furniture should be as little as possible here. Only one ancient narrow desk is placed in this room, on which lie utensils like brushes and inkstone, incense boxes, incense stoves, and so on, all of them being small and elegant. Furthermore, there is a small stone table for storing tea and related items, as well as a small couch for reclining and sitting cross-legged when tired. Instead of hanging paintings, the chamber may be furnished with decorative ancient stones, a sandalwood figure of Lü Zu (ancestor of Quanzhen School, the mainstream of Taoism), or the golden Buddha figure.

In the Mansion

There are four square tables and more than ten round-backed chairs for the banquet. With windows open on all sides of the mansion, people can enjoy watching the distant mountains and water features outside. Playthings such as a chessboard, a Hushi[129] and a dice cup are readily available. As are writing tools such as pens, ink,

[129] A pot and arrows set for the pitch-pot game played at drinking parties, in which participants throw arrows into a pot.

Fig. 3.239 (**a**) Brick Frame Window (Wind Window) of Guanyun Lou in Suzhou's Liuyuan. (**b**) Brick Frame Window (Wind Window) of Lixue Tang in Shizi Lin, Suzhou. (**c**) Brick Frame Window (Wind Window) on the East Side of Wufengxian Guan in Liu Yuan, Suzhou. (**d**) Colorful Glass Flower Window (Wind Window) in Western style in Anxiang Shuying Lou of Shizi Lin in

inkstones, and paper for inscribing poetic prose. There are also several colored glaze gauze lamps used during the night for drinking festivities. A zither and a bamboo flute are also kept on site for entertaining guests.

Hanging Calligraphy and Painting

Ancient paintings should be hung in the elegant study room. Only one painting is allowed in one room at a time, as two paintings hung on opposing walls would be tacky and inelegant. Paintings must be replaced every so often. Long paintings can be hung high on the wall but must not be blocked or bent with bamboo sticks. On the drawing table, ornamental stone, blooming flowers, bonsai, and so on can be arranged, while vermilion painted shelves should be avoided. The main room should be hung with large drapes, while the secret chamber, on the contrary, should be hung with paintings of small scenes, flowers, and birds. Paintings in the shape of a vertical scroll, a fan, a square, a hanging screen, etc. do not offer an elegant viewing experience. There is a saying: the hanging painting had best be in harmony with its surroundings. The drawing table must be kept away from the painting to avoid staining.

Incense Stoves and Vase

The large, low square incense table should be on the daily sitting narrow table, on which is placed one incense stove, one large incense box filled with raw and cooked incense sticks, two small incense boxes containing Chinese eaglewood and ambergris, and one small bottle of incense tools. There can only be one stove in each room and it should not be placed on the table near the hanging paintings. In summer, it should be made of ceramics, while copper ones are used during winter. It must be of antique stock and cannot be used for heating or for bedding. All flower arrangements are made with vases placed on top of large or small low tables. In spring and winter, vases should be made of copper, or porcelain vases with a tin inner pot or with sulfur powder added in water. Porcelain is suitable for use in autumn and summer. Large vases are placed in the main room of high-rise buildings, while small ones are placed in the study room and the inner chamber. Copper and porcelain vases are preferred to gold and silver ones and viewing vases with ears or arranged in pairs are unsightly objects. Flowers in vases should be delicate rather than multifarious. If there is only one branch of flower, it should be peculiar and simple. Two flowers should have comparable features and more than two flowers, if not herbal ones, would be

Fig. 3.239 (continued) Suzhou. (**e**) Flat Square Wind Window with Back Patterns on Cui Ling Long East porch of Canglang Ting in Suzhou. (**f**) Flat Square Wind Window with ice crack patterns in Dian Chun Yi in Wangshi Yuan, Suzhou. (**g**) Wind window with moon and ice crack patterns of Xiaoshanconggui Xuan in Wangshi Yuan, Suzhou. (**h**) two-piece hexagonal wind window with ice crack patterns at Cui Ling Long porch of Canglang Ting, Suzhou. (**i**) Flower Window (Wind Window) of Ming-style in Wanjuan Tang of Wangshi Yuan, Suzhou. (**j**) Parallel wind windows of Dian Chun Yi in Wangshi Yuan, Suzhou (taken from The Catalogue of Jiangnan Gardens)

Fig. 3.239 (continued)

Fig. 3.239 (continued)

j

Fig. 3.239 (continued)

regarded as inelegant, like those found in bars and taverns. Windows should be open, and incense should not be used in the room where there are flower arrangements, for fear that the smoke of the incense will wither the flower, especially daffodils. It is also forbidden to put flower arrangements on painted tables, for fear of spillages damaging the painting.

The Prayer Room

Those who believe in Taoism may hang the painting of Qingniu and Lao Zi (Lao Zi was the founder of Taoism and his mount Qingniu was an ancient beast resembling a bull) or Chunyang and His Sword (Chunyang, also called Lü Dongbin, is the ancestor of Quanzhen school of Taoism) in the prayer room, both of which should be creations of famous painters from the Song and Yuan Dynasties. Those who believe in Buddhism may place a figure of Buddha in the room, such as a golden Tibetan Buddha with a graceful face or an Avalokitesvara figure made in the Song or Yuan Dynasties. Figures of Taoist gods such as Sanqing, Zitong, and Guandi are not suitable for the homes of Buddhist scholars. The location of this room should be in the Changsong Stone Cave and the addition of stone Buddhas and stone benches are preferred. The desk must be decorated with old porcelain vases, bowls filled with water, a stonetripod with incense, stone glazed lamps, ancient Japanese lacquer scripture cabinets on the left side to hold Buddhist scriptures or fairy chopsticks, an ornamental stone on the right side, as well as ornaments such as Dharani pillars, Ruyi scepter,[130] woven praying mat, chairs, and couches, all of which are designed

[130]The ruyi sceptor is a symbol of power and good fortune.

in a casual but delicate style. In the court, there are food tables, ancient stone seats, and a stone Dhanari column, under which is planted fragrant, colorful flowers.

The above introduced the different furniture and furnishings for enclosed halls or mansions, open pavilions and corridors, and outdoor areas. The choice and arrangement of furniture and furnishings will vary depending on the owner's activities and hobbies. The above description of *Huajing* can more or less represent the garden furniture and furnishings of literati and officialdom in the early Qing Dynasty. This furniture and furnishings are suitable for the practical and emotional appeal of garden life but are not exclusive to garden buildings. In addition, there is a category of small buildings, fittings, furniture, and furnishings with a natural artistic flavor, which are exclusive to gardens. These are the natural furnishings and natural papers listed in the *Garden Self-supply* section in *Huajing*. Natural utensils mainly refer to ornaments, furniture, or furnishings maintaining the natural form of plants. One type is made of living plants, evident in the saying "picking vines as curtains and bending bamboo to make fences", while the other type is made of bark logs or dead roots, such as cypress doors and pine tree seats. The shape of these natural furnishings, with the innate natural features of mountains and forests, are indeed symbols of the integration of life and nature and are creations that embody the garden philosophy. The natural furnishings listed in *Hua jing*, include:

Peach kernels, old vine canes, wooden pens, calamus leaf swords, pine whisks, lotus leaf cups, calabashes, grass book belts, banana leaf fans, coir rope, golden lamp shaped fruits, lotus bead, lotus leaf coats, cypress seed incense, weigela florida like a brocade ribbon, willow branches, hosta plantaginea like a jade hairpin, Vicia villosa, elm money, coconut shells, bamboo canes, cups carved from Nanmu root, lotus houses, seedling needles, coral beads, vervain, orchid accessories, sweetgums, and vine belts.

Huajing refers to plant leaves used for writing such as Tianran Jian (Natural paper) and are individually listed as follows:

- Red Leaf Jian, Banana Leaf Jian, Chinese parasol Leaf Jian, Persimmon Leaf Jian, Manchurian Catalpa Leaf Jian, Pattra Leaf Jian, Liyun Jian, Sanhua Jian, Tiao Jian, and Pu Leaf.
- It is considered elegant to inscribe improvised verses on these red, yellow, ochre, and brown leaves, which can serve as an excellent decoration for the study room of the garden.
- Furniture and furnishings closely related to garden life disappeared with the old lifestyle of days gone by. Therefore the remaining ones in the Jiangnan classical gardens, which are renovated and opened to the public, are not enough to fully reflect the atmosphere of life in the gardens, only providing a preliminary sense of the concept (see Fig. 3.230). The furniture currently on display mainly includes some tables, chairs, stools, beds, couches, basin stands, and the like used in the living room and there are also examples of the large long natural tables in front of the folding screen in the middle room. Additionally, there is the small tea table arranged beside the chairs, the slender high flower table with flowerpots, the long narrow altar table, the table for playing the zither (generally including the hollow

bricks of the eastern-Han-style, also called zither bricks in the Jiangnan area, which are ancient and elegant as well as playing a role of a resonance box). Also, there are square tables, round tables, various folding tables, various chairs, stools, and the like (Fig. 3.240). In Jiangnan gardens, there are still existing examples of natural furnishings, such as the pine seats described in *Huajing* as well as tables, chairs, and flower table displayed in the Garden that was preserved miraculously (Liu Yuan) and the Garden of Surging Wave Pavilion (Canglang Ting) in Suzhou, made of old vines and precious boxwood, nanmu, ginkgo, etc. roots (Fig. 3.241). Most indoor furnishings in the open Jiangnan classical gardens are blooming flowers, bonsai, calligraphy, painting, hanging screens, antiques, utensils, palace lanterns, etc., which still give an atmosphere of their own despite the lack of variety (Fig. 3.242).

- In addition to the indoor furniture and furnishings discussed above, there are also outdoor furniture and furnishings that can be left in the sun and rain, arranged in small weather-proof pavilions, and among moon terraces, courtyards, mountains, water, flowers, and trees. They are commonly made of stone, pottery, glaze, and porcelain. Examples include stone tables, stone stools, stone benches, stone couches, stone-carved sumeru seat, brick-carved sumeru seats, glaze sumeru seats, flower tables, stone drums, porcelain drums, glaze drum pier, etc. (Fig. 3.243). As for the mountainside, mountaintop, and in caves, the natural stone tables, stools, low tables, and couches combined with stacked stones are a kind of special furniture with more of a garden-style and organic connection with the natural scenery (Figs. 3.143e and 3.244). Outdoor furnishings generally include potted plants, bonsai, goldfish tanks, lotus pots, etc. (Fig. 3.245), as well as stone lamps, stone pagoda, stone Dhanari column, statues, etc.
- Large stone pagodas and stone Dhanari columns have gone beyond the concept of furnishings. As they are relatively rare in the private gardens of Jiangnan, here they are simply mentioned without further discussion. Cultural relics of value can add elegance to gardens, such as the stone Dhanari column built in the Tang Dynasty in front of the Huiyin Pavilion (Huiyin Ge) in Nanxiang's Yi Garden, the Pu Tong Pagoda (also known as the building with Buddhist scripture and an incense shape) built in the Song Dynasty in front of the Nanting Hall. There are also examples in modern times imitating ancient works such as the Huayan Sutra Pagoda in the garden of Hangzhou's Xiling Seal-engravers' Society (Xiling Yinshe) (Fig. 3.246).
- Small stone lamps, pagodas, and pillars are common small buildings in existing gardens, especially for those on water. Examples are the stupa-style stone lamp Three Ponds Mirroring the Moon (Santan Yinyue) in Hangzhou in Hangzhou, the small stone pagoda on the water beside the Winding Stream Tower in Suzhou's Garden that was preserved miraculously (Liu Yuan), the stone base on the water beside the Qingfengchi Hall, the small stone building on the water beside the north corridor of the Good-For-Both-Families Pavilion (Yiliang Ting) in Subsidiary Garden (Bu Yuan) of the Humble Administrator's Garden (Zhuozheng Yuan), and the small stone pagoda on the water near the Nanxue Pavilion (Nanxue Ting) in the Garden of Pleasure (Yi Yuan) in Suzhou (Fig. 3.247).

a

b

Fig. 3.240 (**a**) Wanjuan Tang with hanging paintings and long narrow table, Wangshi Yuan, Suzhou. (**b**) Bamboo Curtain Hanging from Window of Tingyu Xuan in Zhuozheng Yuan, Suzhou, with Qin Table and Chessboard at the font. (**c**) Redwood Round Table Drum Pier of Liuting Ge in Zhuozheng Yuan, Suzhou. (**d**) Qin Table and Flower Table of Xiaoshan Conggui Xuan in Wangshi Yuan, Suzhou. (**e**) Furniture and interior display of the "Qia Hang" Pavilion on the Downstairs of Mingse Lou in Liu Yuan, Suzhou. (**f**) Golden Brics Tea Table of Zhizaoshu Garden in Suzhou. (**g**) Round tables and stools in Poxian Qinguan of Yi Yuan in Suzhou. Qin table with Qin bricks (on which the ancient Qin was placed), ancient plate walls with long eight-square wind windows with ice cracks pattern, and hanging screens and other decorations

Fig. 3.240 (continued)

Fig. 3.240 (continued)

- The practice of placing statues in gardens can often be traced back to the Qin and Han Dynasties, but rarely to later periods. The garden of Hangzhou's Xiling Seal-engravers' Society (Xiling Yinshe) is an existing example in the Jiangnan area. In the example, the currently remaining statues such as the stone statues of Deng Shiru and Ding Jing, both seal carvings dressed as fishermen at the water's edge and the bronze figure of Wu Changshuo (a gift from Japan) in Shilong District (Fig. 3.248).

a
b

Fig. 3.241 (**a**) Tree root table, chairs, stools and tea table (Suzhou Canglang Ting). (**b**) Root Bed, Tea Table and Flower Table (Suzhou Canglang Ting). (**c**) Furniture made of rare tree roots, such as yellow poplar and nanmu, displayed in Mingdao Tang of Canglang Ting in Suzhou, such as square table, square stools, narrow table, tea table, chairs and full-length mirror. (**d**) Tree Root Furniture in the Saliu Yuanyang Guan in Zhuozheng Yuan in Suzhou, such as large narrow tall table, round table, round bench, imperial chair, Qin Table and full-length mirror

Fig. 3.241 (continued)

Fig. 3.242 (**a**) Nanmu Hanging Screen on Lotus Pavilion in Suzhou's Yi Yuan, carved with bamboo painting by Zheng Banqiao and filled with stone green, which is very elegant. (**b**) Palace lantern hanging under the porch of Yuanxiang Tang in Zhuozheng Yuan, Suzhou. (**c**) Palace lantern hanging in the hall in the inner garden of Yu Yuan in Shanghai

3.1.1.3 Planning of Vegetation

Plants constitute the main part of the natural ecological environment and are crucial to landscape resources. Using them in garden design creates a green natural environment full of vitality and beauty, with blooming flowers, in particular, providing a rejuvenating natural aesthetic. The salubrious properties of plants include the regulation of atmospheric humidity and temperature, de-dusting, and supplying oxygen, while the branches and leaves of trees provide shade, absorb heat, and contribute to water evaporation. This heat regulation effect of trees keeps forests around 1.7 °C cooler than buildings in the summer and 10 °C warmer in winter, providing a certain amount of comfort during hot summers and cold winters. Forests can also regulate humidity, with air humidity in pine forests being on average 13% higher than that of dry buildings. Forests can also reduce wind speed by acting as a wind buffer, helping to shelter people from a stiff winter's breeze and allow large particles in the air to settle. In addition, the pores, villi, and secretion of leaves can absorb and capture fine

Fig. 3.242 (continued)

dust particles through adhesion, thus purifying the air. For example, one hectare of a forest of pinus tabuliformis and dragon spruce can absorb over 30 tons of dust per year, which is very beneficial to reducing potential respiratory issues. Moreover, the coexistence of people and plants is a natural exchange, with people and animals

Fig. 3.243 (**a**) Waterproof stone tables and drums arranged in the small semi-open space of Yuyan Ting in Suzhou Yi Yuan. (**b**) Stone Table and Drum at Canglang Ting in Suzhou. (**c**) Stone Drum and Stone Flower in Suzhou Yi Yuan. (**d**) Stone Tables and Stools on the Hill of the East Garden of Ou Yuan in Suzhou. (**e**) Stone Bench Railings at Canglang Ting in Suzhou. (**f**) Bric Square Table and Porcelain Drum under the Window of the Inner Space of Shilin Courtyard in Suzhou's Liu Yuan

d

e

Fig. 3.243 (continued)

constantly inhaling oxygen and exhaling carbon dioxide, while plants, in contrast, absorb carbon dioxide in sunlight and release oxygen through photosynthesis. A hectare of forest absorbs about a ton of carbon dioxide a day and releases 700 kg of oxygen and a single person needs 0.7 kg of oxygen per day. If one were to rely solely on the oxygen supply of a green space in a small environment, 10 m^2 of trees, or 50 m^2 of flowers and grass, would be sufficient.

Another benefit of green plants is that they are exactly what the optic nerve needs for rest. In recent years, forests have been found to release a healthy substance that can be absorbed through the skin (this is the basis of so-called forest bathing). The above-mentioned salubrious properties represent the material basis for the feelings of euphoria that plants can provide. The flowers, leaves, and fruit of certain garden plants, on the other hand, are used for aesthetic enjoyment and eating, which is

f

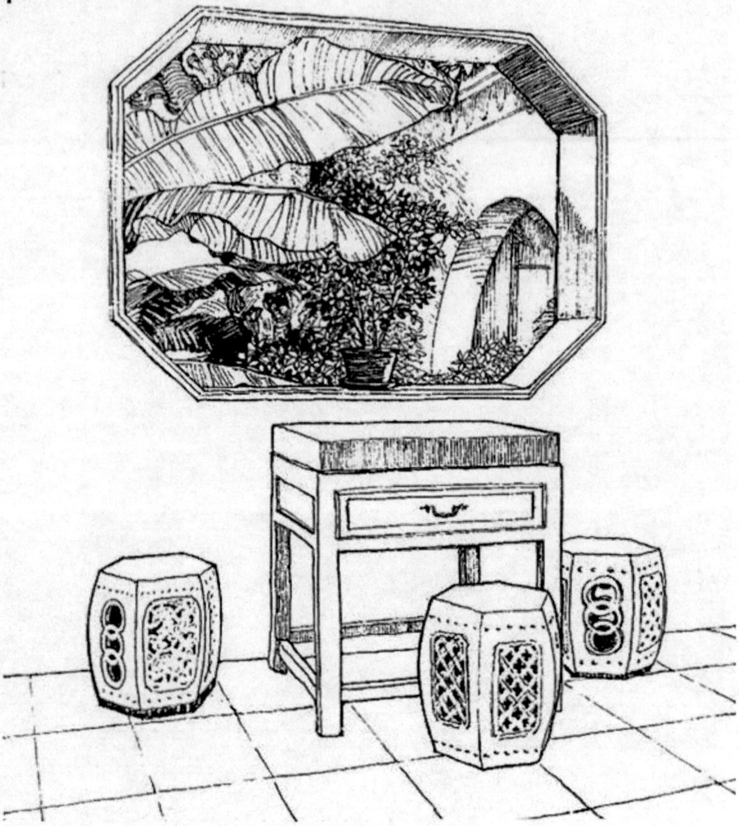

Fig. 3.243 (continued)

another material pleasure related to living in close proximity to a garden. Therefore, living plants serve as an essential complement to a non-living environment and gardens are arguably created to satisfy people's enjoyment of the plant environment. Plants are generally an indispensable ingredient for garden construction and, traditionally, Chinese and non-Chinese gardens alike have always been the objects of plant-based aesthetic enjoyment. European gardening is always based on plants, whether it be in the garden or the park. Although plants don't occupy a large proportion of the space in Chinese classical gardens, especially Jiangnan private gardens, they are still an essential element of the garden scene. Gardens can be free of mountains and water but they are only very rarely designed without plants. Admittedly, some of Japan's ku shan shui (dry landscape) court gardens are indeed special cases that lack plants. Nevertheless, these dry landscapes are often a part of a larger garden that has plants aplenty in the environment as a whole.

In terms of the artistic treatment of plants in gardens all over the world, the two general styles use either geometric or natural designs. The geometric style is also

Fig. 3.244 Natural Stone Tables and Stools in the East Garden of Liu Yuan, Suzhou

known as the integral or regular style. It includes Italian style gardens and European classical gardens, whose plant layout is based on the formal or architectural beauty of geometry, whether it be the overall layout or individual plants. Flowers are arranged in the same pattern as the so-called Parterre de la Broderie, laid out like a carpet. Trees are planted in rows, some are trimmed into geometrical shapes like squares and circles, while others are trimmed to represent a different object such as vases, pagodas, boats, people, and animals through the technical treatment of the garden. This is often referred to as green carving. According to Italian gardeners, these designs are based on the principle that the plant design of the garden must reflect the shape of the building to which it belongs. Moreover, the classical French garden designer André Le Nôtre stated that a major aspect of garden design is "forcing nature to accept the law of symmetry". The natural style of plant layout, also called freestyle or landscape style, is a free arrangement that imitates the natural growth of plants. In the early 17th century, the planting disposition advocated by the British empiricist philosopher Francis Bacon that he practiced in his own private garden was to adopt a "completely rustic charm, with native arbores and shrubs". Moreover, he added the importance of "making the garden as natural as the wilderness", which represents an extreme interpretation of the ideological trend of mechanical naturalism. In the 18th century, under the further influence of Chinese gardens, Europeans created landscape or picturesque gardens, which are called Chinese-style gardens (Jardin Chinois) or Anglo-Chinese gardens (Jardin Anglo Chinois). They imitate the plant layout of nature, striving for realistic truth in quantity, layout, and spatial scale, which also represent a type of mechanical naturalism. Chinese classical gardens, including their outstanding representatives—Jiangnan gardens—are also a natural

Fig. 3.245 (**a**) Potted plants in front of Qin Shi of Wangshi Yuan, Suzhou. (**b**) Stone Table with Natural Stone Bracket in Zhuozheng Yuan in Suzhou. (**c**) Lotus Pot Display in Suzhou Canglang Ting. (**d**) Landscape Bonsai Display (Suzhou Yi Pu). (**e**) Bonsai of Rocks in Suzhou Huanxiu

style that imitates nature in their plant layout. However, they represent the pursuit of artistic truth rather than realistic truth.

Among the existing Jiangnan garden works, those dating to earlier periods are about the same age as European natural gardens. The reason Jiangnan gardens have managed to be more encompassing and profound in artistic attainment in plant layout is that they benefit from a foundation of three thousand years of history in garden design. Early Chinese garden design, such as the management of emperors' and vassals' gardens during the pre-Qin period, paid great attention to the layout of plants and animals, with some gardens being dominated entirely by plants. The methods behind plant layout have been based on nature from the very beginning. In addition to maintaining natural forests, large-scale imperial gardens have adopted a natural method for the layout of certain precious species. Emperor Wu of the Western Han Dynasty further expanded the Shanglin Imperial Garden (Shanglin Yuan) of Emperor Qin Shihuang, planting 3,000 varieties of exotic flowers and trees presented from afar in the huge imperial garden that covered an area of over 300 Li (One Li is equivalent to 500 meters). with 70 individually detached palaces. According to ancient documents like *Records of the Imperial Architectural Constructions in the Han Capital and Its Environs* and *Compilatory Records of Han Institutions*, *Palatial Architecture and Gardens*, and *Imperial Anecdotage*, many palatial and private gardens of nobles and wealthy people in the Western Han Dynasty also featured layouts of precious flowers and trees in combination with mountain-and-water scenes. The Garden of Spring Trees (Fanglin Yuan) of Cao Wei in the Three Kingdoms is disposed of with high trees and hanging vines in combination with artificial mountains. In the Jin Dynasty, there were palatial gardens that were abundant in plants like the Garden of Immortals (Qiongpu Yuan), the Garden of Lucid Ganoderma (Lingzhi Yuan), the Garden of Mulberry and Catalpa Trees (Sangzi Yuan), the Garden of Grapes (Putao Yuan), etc. During the Northern and Southern Dynasties, small gardens with natural scenery began to appear and the arrangement of plants was planned so as to best utilize their natural beauty.

The Rabbit Garden (Tu Yuan) of King Liang, where "Lush forests with verdant leaves fill the garden, ... green grass, red flowers, water plants and liana, ... as well as flowing streams and mountain slopes with green bamboo forests ..." (*Ode to the Rabbit Garden*, Jiang Yan), and the Small Garden (Xiao Yuan), where "In the garden, there are several rows of elms and willows, and hundreds of pear trees and peach trees. Plants are so dense that one can view windows only by pushing aside lush branches and can reach paths only by walking on an incline, ... grass and trees grow closely, branches of plants intertwine with each other ..." (*Ode to the Small Garden*, Yu Xin), are both designed by making use of the natural and pastoral scenes of plants in small sites.

Fig. 3.245 (continued) Shanzhuang. (**f**) Display of Gold Fish Tank in Suzhou Yi Yuan. (**g**) Wells in Canglang Ting, Suzhou. (**h**) Lingbi Stone Display in Lengquan Ting of Wangshi Yuan, Suzhou (It is said to be a relic of Tang Bohu in Ming Dynasty). (**i**) Gold Fish Tank Display in Gu Zhai, Iron Bottle Lane, Suzhou

d

e

Fig. 3.245 (continued)

For the West Garden (Xi Yuan) of Emperor Yang in the Sui Dynasty, ribbons were cut into shapes of flowers and leaves to decorate branches during the withering season of flowers and trees. Moreover, the winter ice of ponds was broken and decorative plants, such as red algae, were arranged to decorate the pond. Although it seems extravagant and artificial, these approaches are man-made treatments of the seasonal characteristics of plants, which reflects the interest of humans in the natural beauty of plants. In the Tang Dynasty, whether it be imperial gardens or private gardens of dignitaries, precious flowers and trees were freely planted among mountain and river scenes and were also appreciated as independent, choice collections. The famous Imperial Garden of the Northern Song Dynasty, the Royal Garden of Northeast Longevity Mountain (Shoushan Genyue), included rare flowers, bamboos,

Fig. 3.245 (continued)

and trees from Zhejiang that were combined with animals to embellish the imagery of wild nature. Luoyang gardens, on the other hand, paid more attention to the appreciation of flowers. In addition to trees like pines, cypresses, junipers (Juniperus chinensis Linn.), catalpas (Calalpa ovata G. Don.), and Chinese parasol trees

Fig. 3.246 (**a**) Stone Pagoda of Hangzhou Xiling Yinshe. (**b**) Tang Toroni Sutra Stela in front of Huiyin Ge of Yi Yuan, Nanxiang

(Firmiana simplex (L.) W. Wight) as well as vines and bamboos, the gardens were planted with tree peonies (Paeonia Suffruticosa Andr.) and other flowers and trees between ponds and pavilions. Li Gefei wrote in *Famous Gardens of Luoyang*, when describing Li's Renfeng Garden (Renfeng Yuan): "The Minister Li Deyu has *Flora of the Pingquan*, where there are hundreds of varieties of plants". Nowadays, there are many skillful gardeners in Luoyang who can craft flowers and trees perfectly. This leads to an ever-increasing number of exotic species of flowers and trees, including dozens of varieties of peaches (Prunus persica (L.) Batsh), plums (Prunus salicina Lindley.), plums (Prunus mume Sieb. et Zucc.), apricots (Prunus armeniaca L.), lotus (Nelumbo nucifera Gaertn.), and chrysanthemums (Chrysanthemum sinense Sab.), and hundreds of varieties of tree peonies and herbaceous peonies (Paeonia lactiflora Pall.). Moreover, there are some exotic plants like orchid (Cymbidium virens Lindl.), jasmine (Jasminum sambac (L.) Aiton), Chinese snowball viburnum (Viburnum macrocephalum Fort. f. keteleeri (Carrière) Rehder), camellia (Camellia japonica L.), etc. They are difficult to grow in regions outside Luoyang, where they are a local specialty. There are therefore thousands of kinds of flowers and trees in the gardens of Luoyang". Jiangnan Gardens represent a special case in classical Chinese gardens. Regardless of the variety or natural method of the layout of its plants, it is traditionally representative. Up to even the Ming and Qing Dynasties, new varieties of ornamental flowers were being continuously cultivated. According to the records of Zhao Zhibi in *Illustrated Chronicle of the Hall Level with the Mountains*, "Herbaceous peonies in Yangzhou are the best in the world. There are as many as 39 kinds in old records, which have since become rare.

Fig. 3.247 (**a**) Stupa-style stone lamp "Santan Yinyue" of West Lake, Hangzhou. (**b**) Stone pagoda on the water beside Qingfengchi Guan in Suzhou Liu Yuan. (**c**) Stone pagoda on water surface in the North of Yiyu Ting in Zhuozheng Yuan, Suzhou. (**d**) Stone pagoda on water surface in Yi Yuan, Nanxiang

Fig. 3.248 Statue of Deng Shiru on the Bank of Xiaolonghong Dong of Hangzhou Xiling Yinshe

Shuangqi, binge, Zansan, and Jusi were not included in previous records, so people refer to these plants as 'Huarui', which means exotic flowers". Flowers for display are often planted expansively. For example, the herbaceous peony garden in the Garden of Thin Bamboo (Xiao Yuan) spans hundreds of mu.[131]

Plants are the primary means of describing the ecological environment and enhancing the seasonal features of scenery for Jiangnan Gardens to show a natural appeal. Their layout method mainly consists of integration into the mountain-and-water scene, which is a unique natural disposition that differs from that employed in European gardens. In addition, plants are often cultivated within an independent fence, nursery, pot, or vase, just like solitary stone peaks in mountain ranges. With pots and vases, flowers can grow indoors, which is the intermediary between architecture and nature. The natural landscape of Jiangnan Gardens gives people

[131] One mu is equivalent to about 0.16 acres.

the artistic feeling of lush vegetation. However, the number of plants used in these gardens is limited, the proportion of which is even smaller than that of architecture in the scenic structure. A group of just a few trees can be called a forest. This is not only because too many plants will obscure other scenery like mountains, rivers, pavilions and waterside pavilions but also because the unity of artistic style requires a small number of plants. Jiangnan Gardens are highly generalized artistic creations. Commentators often call them *xiao zhong jian da*,[132] which is not only shown in mountain buildings and water management but also in plant layout, which is sometimes called *yi shao sheng duo*.[133] If a garden is properly planned, even a single flower, blade of grass, tree, root, trunk, leaf, or fruit can reflect the charm of nature and can be enjoyed by people at all times (Figs. 3.249, 3.250). In the ancient environment free of modern industrial pollution, even the gardens in the city did not meet any strong demand for salubrious properties. The quantity of trees planted for mountain-and-water scenic imagery has already fulfilled the function of providing shade, which can reduce the radiant heat from bricks under the sun for the flowers and plants that serve as a foil to the landscape and yard. Therefore, in the designs of the classical gardens in the Jiangnan region, plant layout is mainly based on the artistic structure of its scenery.

Jiangnan is located between the 30th and 33rd northern latitudes and is near to the ocean. Its surface waters mainly comprise the Yangtze River and Taihu Lake. With abundant annual precipitation, the humidity and temperature of the atmosphere in Jiangnan are higher than that of North China, allowing a longer growth period for plants. There are many varieties of ornamental plants that can be grown in gardens there and horticultural technology has a long, rich history of development and practice in the area, which has made great achievements in plant layout. So far we have mainly discussed the art of plant layout as an element of scenic structure, without giving due attention to biological characteristics and cultivation techniques.

As plants and trees are living things, when they cannot fulfill the design intent at the beginning of garden construction, a growth process is needed to produce the expected scene effect. Therefore, in garden design, the plant layout is often treated differently in the near and long term. In the near future, when plants have not yet fulfilled the desired design effect, some fast-growing trees and herbs like poplars, willows, and bamboos are often planted first as a temporary, transitional fix. Because of the difficulty in achieving the desired effects with trees, garden design in Jiangnan Gardens particularly values the preservation and utilization of ancient trees. Ji Cheng said in *The Craft of Gardens* (Yuan Ye), "The new buildings are only suitable for planting poplars and bamboos, while old gardens with luxuriant ancient trees and blooming flowers can be rebuilt" and "Old trees obstruct the construction of eaves and walls, moving one step to the side can allow construction to begin, and cutting branches will not hinder the construction of roof. It is said that the construction of

[132]Evoking the large with the small.

[133]Achieving more with less.

a

b

Fig. 3.249 (**a**) The expressive roots of ancient trees in Zhuozheng Yuan, Suzhou. (**b**) The old twisted rattan in Suzhou's Huiyin Yuan shows the beauty of unrestrained natural forces

Fig. 3.250 (**a**) Autumn scenery of chrysanthemum in bloom in Suzhou's Zhuozheng Yuan. (**b**) Plants colors are obvious, highlighting seasonal features of scenery—autumn scenery in Yangzhou's Ge Yuan

beams and rafters is easy, but a shady tree is hard to grow". These insights offer a glimpse of Ji Cheng's gardening experience.

Plants are the most easily changed factors in gardens, while some trees may be withered and renewed in the use of gardens. The layout of flowers, especially herbaceous flowers, is more likely to evoke change. Changes in the supply of seeds, seedlings, nursery stocks, and gardeners as well as shifts in the interests of patrons and the owner of gardens are all typical causes for the change of plant layout. Therefore, the existing ancient gardens, especially the ancient gardens that were once abandoned and subsequently rebuilt, have many flowers and plants that have

Fig. 3.250 (continued)

been rearranged in modern times, with the exception of large trees. Some of them are replanted in the original pattern, while others are additions planned by managers. This undoubtedly increases the difficulty for us to discuss old original gardens. We strive to understand the original and to analyze examples with the help of the relevant ancient texts. Fortunately, the main garden workmen are traditional folk horticulturists whose techniques are not far removed from those of the past. In this way, even taking the current layout as the research object, the basic principle, and techniques of the originals can be generally understood.

3.1.1.3.1 Type of Commonly Use Plants and their Human Attributes

An extensive selection of plants is used in Jiangnan gardens. Common temperate trees and flowers are mainly used in the garden layout and some subtropical plants are also occasionally used. According to the current status of plant layout in existing ancient gardens and in reference to the records of ancient gardens, local chronicles, and ancient literature of horticulture like *The Book of Flowers*, *Treatise on Superfluous Things*, *Notes on All Various Herbs*, *The History of Flowers*, *The Esoteric Flower Mirror*, (Imperially Endorsed) *Extended Notes of the Peiwen Study on All Various Herbs*, *Essay on Various Plants*, *Illustrated Treatise on the Botany of China*, *An Illustrated Handbook on Various Flowers* etc., the plants used in Jiangnan gardens mainly include:

- Magaphanerophyte, Phanerophyte

Deciduous:

- Ginkgo biloba L.
- Liquidambar formosana
- Sapium sebiferum
- Salix babylonica
- Malus yunnanensis Schneid var. veitchii Rehd.
- Pterocarya stenoptera C. DC.
- Calalpa ovata G. Don.
- Firmiana simplex
- Sophora japonica (Schott) L.
- Ailanthus altissima (Mill.) Swingle
- Melia azedarach L.
- Ulmus pumila L.
- Celtis sinensis Pers.
- Maguolia denudata Desr.
- Zelkova schneideriana
- Alchornea davidii Franch.

Evergreen:

- Cinnamomum camphora (L.) Presl
- Eriobotrya japonica (Thunb.) Lindley.
- Pinus bungeana Zucc. ex Endl.
- Podocarpus macrophyllus (Thunb.) Sweet
- Pinus massoniana Lamb.
- Pinus thunbergii Parl.
- Juniperus chinensis Linn.
- Acer serrulatum Hayata.
- Magnolia grandiflora L.
- Michelia champaca Linn.
- Osmanthus fragrans (Thunb.) Lour.

- Ligustrum japonicum Thunb.
- Ilex chinensis Sims.
- Cryptomeria japonica D. Don.
- Trachycarpus fortunei H. Wendl.
- Nanophanerophyte, Flowering Shrub

Deciduous:

- Paeonia Suffruticosa Andr.
- Cercis Chineusis Bunge
- Lagerstroemia indica L.
- Syringa oblata Lindl.
- Hibiscus syriacus L.
- Jasminum undiflorum Lindl.
- Forsythia suspensa (Thunb.) Vahl
- Prunus triloba Lindley.
- Spiraea blumei G. Don.
- Prunus japonica Thunb.
- Prunus persica 'Duplex'.
- Malus micromalus Makina
- Chaenomeles speciosa (Sweet) Nakai
- Malus halliana Koehne
- Cydonia sinensis Thouin
- Punica granatum Linn.
- Acer palmatum Thunb.
- Prunus mume Sieb. et Zucc.
- Chimonanthus praecox (L.) Link
- Prunus persica (L.) Batsh
- Prunus salicina Lindley.
- Prunus armeniaca L.
- Pyrus serotina Rehder.
- Diospyros kaki L.f.
- Ziziphus jujuba Mill.
- Malus pumila Mill.
- Viburnum macrocephalum Fort.
- Hibiscus mutabilis L.
- Kerria japonica DC.
- Ficus carica L.
- Magnolia liliflora Desr.
- Tamarix chinensis Lour.
- Prunus pseudocerasus Lindl.
- Rosa rugosa Thunb.

Evergreen:

- Rosa chinensis Jacq.
- Camellia japonica L.

- Michelia fuscata (Andr.) Blume.
- Rhododendron simsii Planch.
- Nerium indicum Mill.
- Gardenia jasminoides Ellis.
- Fatsia japonica
- Buxus sempervirens L.
- Citrus tachibana (Makino) Tanaka.
- Myrica rubra Sieb. et Zucc.
- Serissa japonica (Thunb) Thunb.
- Jasminum officinale L.
- Citrus medica
- Nandina domestica Thunb.
- Damnacanthus indicus C.F. Gaertn.
- Jasminum sambac
- Viburnum odoratissmum
- Canarium album Raeusch.
- Litchi chinensis Sonn.
- Areca catechu
- Citrus medica L.var. sarcodactylis
- Yucca filamentosa L.
- Liana

Deciduous:

- Campsis grandiflora (Thunb.) Schum.
- Wisteria sinensis (Sims) Sweet
- Vitis vinifera
- Parthenocissus tricuspidata (Sieb.et Zucc.) Planch
- Rosa multiflora
- Rosa rubus
- Clematis florida Thunb.
- Evergreen:
- Rosa banksiae Ait.f.
- Lonicera japonica
- Ficus pumila L.
- Hedera helix L.
- Trachelospermum jasminoides (Lindl.) Lemaire.
- Akebia quinata
- Lycium chinense
- Polygonum orientale
- Rosa multiflora ‘Platyphylla’
- Quisqualis indica
- Rosa roxburghii Tratt.
- Bambusoideae
- Phyllostachys pubescens Mazel ex Houz Leh.
- Indocalamus tessellatus (Munro.) Kengf.

- Phyllostachys aurea
- Bambusa multiplex of auth.
- Phyllostachys nigra (Lodd.) Munro
- Chimonobambusa quadrangularis (Fenzi) Makino.
- Bambusa nana Roxb.
- Bambusa vulgaris 'Vittata'
- Phapis humilis Bl.
- Herbaceous Flower
- Dianthus chinensis L.
- Ophiopogon japonicus (Thunb.) Ker Gawl.
- Saxifraga stolonifera Meerb.
- Hemerocallis flava
- Commelina communis L.
- Musa basjoo Sieb. et Zucc.
- Paeonia lactiflora Pall.
- Chrysanthemum sinense Sab.
- Cymbidium virens Lindl.
- Iris tectorum Maxim.
- Hosta plantaginea Engl.
- Impatiens balsamina L.
- Celosia cristata L.
- Polygonum multiflorum Thunb.
- Althaea rosea (L.) Cav.
- Malva sylvestris L.
- Abelmoschus esculentus L. Moench.
- Bogonia evansiana Andr.
- Canna indica
- Aquatic Plant
- Nelumbo nucifera Gaertn.
- Nymphaea tetragona
- Plnagmites communis Trin.
- Myriophyllum spicatum L.
- Spirodela polyrhiza Schleid (Fig. 3.251).

The above plant species can all essentially be found in existing ancient gardens. Although it is not an exhaustive list of garden plants used in Jiangnan Gardens, it provides a general overview. Additionally, according to existing examples, there are dozens of plant species that can be used in small-and-medium residential gardens and over 100 species for large residential gardens.

While playing their role in landscape composition, plants are endowed with different attributes in addition to their general landscape significance as elements of the scenic structure. During their long history of viewing enjoyment, different thoughts and feelings have been attached to garden plants with some even becoming anthropomorphized. Such attributes are assigned according to ecological roles, physical features, meaning and homophonic sounds of the plant names, or because

Fig. 3.251 (**a**) Red-crowned cranes are raised in pairs, suitable to be arranged by waterside, on promontories or islets. (**b**) Peacocks' images are gorgeous, symbolizing auspiciousness. They should be raised in delicate courtyards or tang xuan and arranged with peony fences, peony flower bed, lake stones or planting beds. (**c**) Cats are usually kept indoors or in the courtyard. They sleep on couches or among rocks in the sun, adding a leisurely atmosphere to the scenery. It catches insects

of the influence of ancient literati and painters who expressed their feelings in relation to plants. For example, pines, bamboos, and plums are known as the "three friends of Winter" due to their cold-tolerant ecology, while pines and cypresses symbolize virtue and longevity as they are evergreen and have a long life. Meanwhile, bamboos have a sharp sense of integrity, lotuses “emerge from muddy water, yet are clean and pure” (*Ode to Lotus Flower*, Zhou Dunyi), and both are compared to Junzi (gentlemen). Plum blossoms are pure in nature with high-hearted flavor and grace, enjoy the sunshine, and are a companion to those in seclusion (such as in the story of the reclusive scholar Lin Bu who “took a plum as his wife and a stork as his son”). Tree peonies are graceful in shape and fragrance, and have been regarded as “exquisite beauty”, “the flower of wealth”, “the king of flowers”. Herbaceous peony, like tree peony, has a rich shape but is herbaceous and although both are considered as "the most noble among flowers", "tree peony is called the king of flowers", while "herbaceous peony is called the prime minister of flowers" (Treatise on Superfluous Things). Chrysanthemums are demure but not pretty or coquettish and bloom independently during the cold seasons. Praised by Tao Yuanming, “I pick fence side asters at will”, they have become synonymous with arrogance, elegance, and seclusion. Orchids have a "king's fragrance" and are endowed with attributes of supremacy and nobility. Lilacs (Syringa oblata Lindl.), osmanthus (Osmanthus fragrans <Thunb.>Lour.), banana shrub (Michelia fuscata <Andr.>Blume.), jasmines, with warm, gorgeous, sweet, and fragrant properties, are each compared to a beautiful maiden with her own special fragrance. Other varieties of flowers and plants like peach, plum (Prunus salicina Lindley.), apricot, and Chinese flowering crab-apple (Malus spectabilis <Ait.>Borkh) are regarded as fairies due to their appearances and colors. Feudal literati and scholars have always commented on ornamental plants, especially flowers, and there is no lack of monographs for such commentary. The Book of Flowers, written by Zhang Yi in the Song Dynasty, classified 71 kinds of flowers and other plants into nine categories ranked by value. Apart from the above-mentioned characters of plants, other examples include Chinese mahogany (Ailanthus altissima (Mill.) Swingle) and tawny daylily (Hemerocallis flava L.) being compared to parents, dwarf flowering cherry (Prunus japonica Thunb.) being compared to siblings, and phoenixes only perch on Chinese parasol trees, peach blossoms can exorcise evil spirits, pomegranate (Punica granatum Linn.) implies fertility, the word for beech (Zelkova schneideriana Hand. Mazz.) is a homonym for promotion in Chinese, etc. These also tend to be the basis for a varied selection in plant arrangement. Overall, the layout of plants in Jiangnan gardens often expressed different aspirations and artistic flavors in addition to natural

Fig. 3.251 (continued) and butterflies among plants, giving people natural pleasure. (**d**) Goldfish: small goldfish should be kept in delicate small waters. Large golden carps or other ornamental fishes should be raised in a large area of water. Goldfish and aquatic plants are laid out together, which produce a good effect. (**e**) Asian giant softshell turtle: The Yangtze giant softshell turtle in Fangsheng Chi in West Garden of the Jiechuang Si Temple in Suzhou.

tastes. This aspect also represents an ideological component that cannot be ignored in the artistic works of Jiangnan garden plants.

3.1.1.3.2 Artistic Functions of Vegetation in the Garden

In garden art, plants provide their visitors the enjoyment of appearance (the overall shape and the appearance of roots, stems, leaves, flowers, and fruits), color, and scent, all of which may delight the eyes, be delightful to touch, or please the nose. While exploring gardens, people can experience the intoxicating effect of nature through sight, touch, and smell. Of course, there is still much more to the function of plants in garden art. Unlike other garden materials, they also function as a unique scenic structure.

3.1.1.3.2.1 Expanding Space with a Shield Wall

Along the boundary wall of the garden or the back gable of the neighboring house, trees, shrubs, or climbing plants are planted to replace or decorate the dull brick, stone, ash, or soil background with the natural ecological shape of plants. It is written in *The Craft of Gardens* (Yuan Ye) that "the garden wall is faintly visible among the trailing plants", which not only produces a natural and lively viewing experience but also expands the sense of space of the garden as folded plants create a surprising depth illusion (Fig. 3.252).

Fig. 3.252 Vertical greening of wall surface (The garden wall is decorated with rosa banksiae in Wangshi Yuan)

3.1.1.3.2.2 *Shrouds and Shadows*

Under the broad background of the sky and no matter how well the landscape and architecture are treated with the arrangement of plants, the desired scenic effect cannot be achieved under the given scale and proportion if the flowers and trees are too short. In this case, tall trees are needed as the background and scope of the landscape to cover the whole scene in order to maintain emotional appeal. Tall trees not only form the background and border of the painstakingly-designed main scene but also sieve the broad daylight into a shady green, such as in the Fei's residential garden in the Peach Blossom Villa (Taohua Wu), Suzhou, where a century-old maple (Acer serrulatum Hayata.) beside the mountain shrouds the whole mountain in a rich, green shade. The green canopy of the trees is like a filter for stage lights, through which the sunlight casts and forms shadows that are necessary for creating an atmosphere of deep and remote mountains and valleys. Under the sunlight or moonlight, plant shadows creating such touching atmospheres as a flower shadow flickering on the wall, a plantain (Musa basjoo Sieb. et Zucc.) shadow shrouding the window, a Chinese parasol tree shadow enveloping the ground, or a locust tree (Sophora japonica <Schott>L.) shadow flowing down the courtyard (Fig. 3.212).

3.1.1.3.2.3 *Separated Yet Connected; Veiling the Depth of Field*

Plants also function as a type of spatial organization. When it is not appropriate to divide the scene by architectural means, plants such as tall trees, shrubs or bamboos can achieve the effect of completely blocking something from sight. In most cases, the use of plants can achieve a separate-yet-connected effect between adjacent scenes. Alternatively, the scene may be hidden in a sparser arrangement, thus making the scene look subtle and increasing the depth of field (Fig. 3.253), or the whole garden could be divided into landscapes and architectural scenes of several scenic spots. The connection between them can be strengthened through a common plant layout so that the artificial and natural constituents are unified in green scenery.

3.1.1.3.2.4 *Decorating the Landscape; Foiling Construction*

Among piled mountains and stacked stones, and the banks or surface of all types of water, plants are arranged to serve as a foil to the landscape. In the pictorial structure of scenic imagery, particularly the main viewing scenic plane of the principal scenery, trees, flowers, and bamboos need to be specially planned. In such cases, plants are often used as the key to the pictorial structure and play a role in supplementing and enhancing the atmosphere of the landscape (Fig. 3.254). The interior and exterior spaces of pavilions, corridors, halls, waterside pavilions, etc., also show their connection with nature through plants (Fig. 3.255). Garden building designs typically take into account the comprehensive compositional relationship between plants, shape, and space in both the courtyard space and the main scenic viewing side of the building. For example, there is an old wingnut tree (Pterocarya stenoptera C. DC.) leaning against the front of the Winding Stream Tower (Quxi

Fig. 3.253 Plants are used to separate the landscape, so as to make the scenery implicit and increase the depth of field and the level of scenery. (Shuxiao Ting in Suzhou's Liu Yuan)

Lou) in the Garden that was preserved miraculously (Liu Yuan), Suzhou (Fig. 3.256), where the function of plants as a foil to architecture is to integrate nature into artificial constituents.

3.1.1.3.2.5 Being Display and Appreciate; Bring out the Scene Theme

Ancient vines, trees, or rare flowers and plants can be regarded as a place of interest, which can constitute a scene's independent objects and are often displayed and appreciated in Jiangnan gardens. The term display appreciation refers to when plants are not directly organized in natural mountain-and-water scenic imagery but are displayed as a foil to tables, bases, rails, and fences in relatively independent environments such as yards, patios, platforms, intersections, etc. They are often coupled with close-view building settings like pavilions, corridors, verandas, water-side pavilions, etc., forming a scenic unit with plants as the theme. For example, the Intertwined Old Trees (Gumu Jiaoke) in the Garden that was preserved miraculously (Liu Yuan), Suzhou, and the dead lightning struck ginkgo tree (Ginkgo biloba L.) under the Jade Buddha Cave (Yu Fo Dong) in the Xu's Garden (Xu Yuan) in the Yangzhou's Slender West Lake Park (moved there in modern times) (Fig. 3.257).

a

b

Fig. 3.254 (**a**) Trees on the top of the mountain in Suzhou Ou Yuan make the pictorial structure of scenic imagery vivid and form a mountain-and-forest atmosphere. (**b**) The dense vegetation in Suzhou Zhuozheng Yuan forms a lake-and-mountain atmosphere, making the water vivid

On the contrary, the flowers and trees displayed in the vestibule of the hall and the courtyard of the study (not necessarily old trees and precious flowers) can also be used as the theme of scenes with natural appeal. Taking plants as the theme of the scene is certainly not limited to display and appreciation. Plant layouts that are integrated with the natural landscape can also play a role in the theme of scenery. Of course, it is more ideal to combine rare ancient trees with natural scenery, where they

Fig. 3.255 (**a**) The covered walkway in Ke Yuan is hidden among bamboo bushes, and the walkway and bamboo complement each other. (**b**) The canopy of towering trees on the stone mountain in Suzhou Liu Yuan, encircling the small pavilion–a unified pictorial structure of scenic imagery

Fig. 3.256 Boats could still been seen in Suzhou's Liu Yuan in 1930s. The pterocarya stenoptera tree leaning against the lake and the Quxi Lou form a perfect unified pictorial structure of scenic imagery. (It is drawn according to the photo in Catalogue of Jiangnan Gardens)

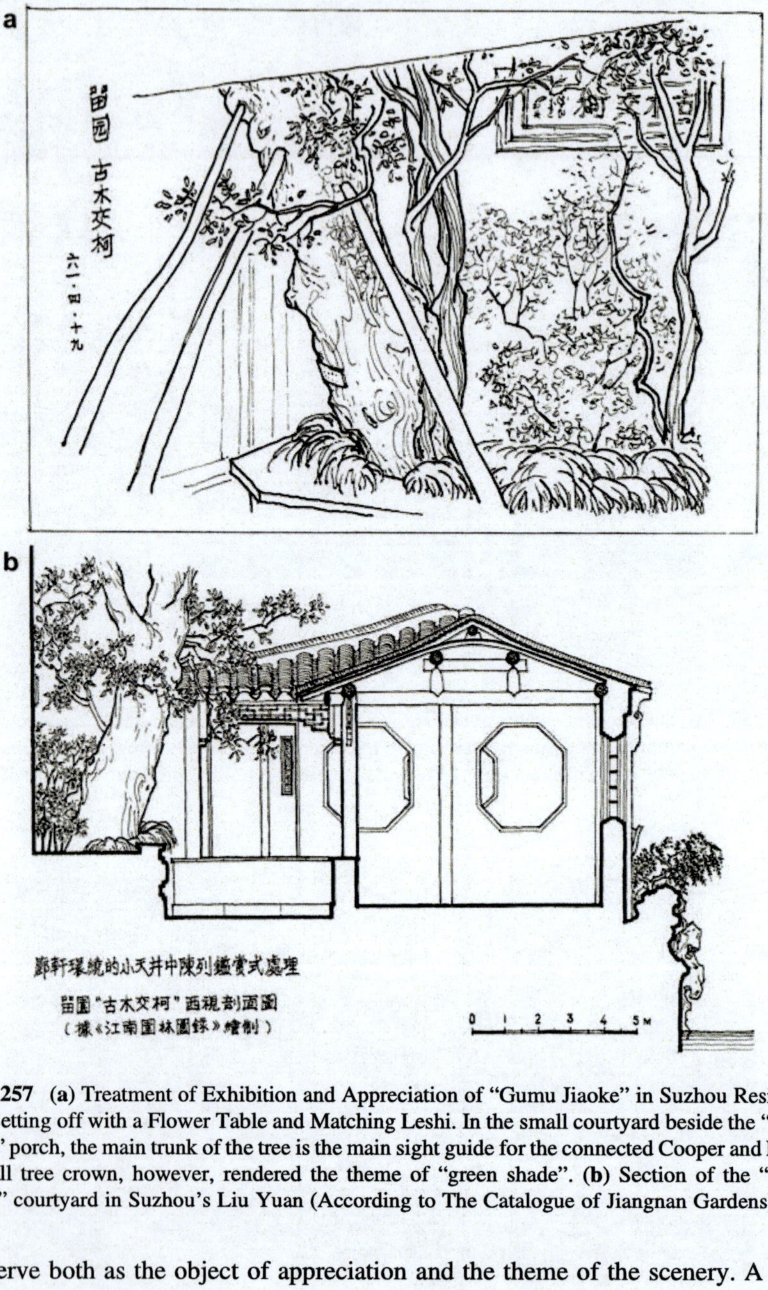

Fig. 3.257 (**a**) Treatment of Exhibition and Appreciation of "Gumu Jiaoke" in Suzhou Residence Park-Setting off with a Flower Table and Matching Leshi. In the small courtyard beside the "Green Shade" porch, the main trunk of the tree is the main sight guide for the connected Cooper and Privet. The tall tree crown, however, rendered the theme of "green shade". (**b**) Section of the "Gumu Jiaoke" courtyard in Suzhou's Liu Yuan (According to The Catalogue of Jiangnan Gardens)

can serve both as the object of appreciation and the theme of the scenery. A good example is the Watching Pines and Appreciating Paintings Studio (Kansong Duhua Xuan) in the Master-of-Nets Garden (Wangshi Yuan), Suzhou, which features scenery themed on ancient pines and cypresses (Fig. 3.258).

Fig. 3.258 (**a**) The plant layout in Suzhou's Yi Yuan—ordinary phlox peaches produce elegant and fresh effect in white architectural environment. (**b**) Suzhou Nethermaster Garden: The Theme Plant Ancient Pine and Cypress Combined with Rocks and Flower Terraces in front of Songdu Painting Gallery

3.1.1.3.2.6 Rich Landscape Color; Seasonal Highlights

In the creation of gardens, plants are not only the pigments of greening but also the means of rendering various colors. In the northern gardens, due to the short growing season, the vegetation withers in the long winter and leaves a stretch of gray-brown branches. Even if evergreen trees are planted, the color palette is quite plain. Therefore, much of the color is provided by the ornamentation of buildings, such as through decorative paintings. Jiangnan gardens, nevertheless, are different. They have superior natural conditions, longer growth periods, and numerous varieties of plants that can be used to meet the requirements of color and render bright, elegant, and natural scenes. The plain white walls, small green tiles, and tan wooden buildings in Jiangnan Gardens are created to set off the different kinds of green colors of vegetation and various colors of flowers. Garden scenes that depict nature should change with the seasons as nature does. It is the unique function of plants to show seasonal changes. In the Jiangnan region, each season has different blooming flowers, and trees are more or less distinctive in different seasons (Fig. 3.250). Trees blossom in spring fills out branches of leaves and seeds in summer, which in turn brings forth fruit in autumn. The appearance and color of deciduous trees constantly change with the turn of the seasons, from verdant in spring, to shady in summer, with fair leaves in autumn that mirror the blooms in spring, and withered branches in winter. The broad-leaved evergreen trees in Jiangnan are sufficient to maintain a spectacular garden scene and are well-suited to local conditions. Even a scene of a bleak forest with crows will not be desolate if given the appropriate disposition.

3.1.1.3.2.7 Wind, Rain, and Trees: Listening to the Sounds of Nature

In the beauty of the natural environment, wind and rain are natural phenomena that add a poetic touch to any scene. These natural phenomena are felt and appreciated not only through touch, vision, and hearing but are most appealing only through natural expression. The sights and sounds of rain falling on a lake and between rocks are vivid. A pool of water in spring ruffled by a light wind can be very touching, while it is plants that can even better show the charm of wind and rain. The spring breeze is invisible. The so-called graceful spring breeze is reflected by willows and poplars and the magical thing about the bamboo forest is that through it one can hear the wind. Bamboo forests are not only suitable for showcasing the sound of the wind but also the sound of rain. Planting bamboo near windows and listening to the "Night Rain in Xiaoxiang" has become a typical design feature of Jiangnan Gardens, for example, the Exquisite Emerald Bamboo House (Cui Ling Long) among the bamboo forest in the Garden of Surging Wave Pavilion (Canglang Ting), Suzhou (Fig. 3.259). Some broad-leaved plants are also great tools for hearing the rain and "Plantains and Chinese parasol trees are visible from a half-opened window" (*The Craft of Gardens*<Yuan Ye>). Plantains by the window of the Listening to the Rain Veranda (Tingyu Xuan) in the Humble Administrator's Garden (Zhuozheng Yuan), Suzhou, are not only for filtering sunlight but also for hearing the rain. "Leaving the remnant lotuses to help listen to the sound of rain", a verse written by Li Shangyin, pointed out the function of lotus leaves in hearing the rain (Fig. 3.260). Willows (Salix babylonica L.) Swaying in the Dawn Wind, and Plantains in the Night Rain,

Fig. 3.259 Borrowing bamboo as the window—Dian Chun Yi in Wangshi Yuan, Suzhou

Fig. 3.260 (**a**) Plantain planted under the window of Tingyu Xuan in Suzhou's Zhuozheng Yuan. (**b**) Plantain planted under the window of Dian Chun Yi in Wangshi Yuan, Suzhou. (**c**) Hearing Rain on the Banks of Xiangzhou and Chengguan Lou in Suzhou's Zhuozheng Yuan

c

Fig. 3.260 (continued)

the typical scenes mentioned in *The Craft of Gardens* (Yuan Ye), show that people already had an appreciation for the special use of presenting the wind and rain in garden making to listen to the sounds of nature as early as the Ming Dynasty.

3.1.1.3.2.8 Spreading Fragrance to Attract Bees and Butterflies

The appeal of garden art spaces is influenced by many factors, not only shape and color but also sound and smell. For visitors, experiencing garden works involves a comprehensive input from several senses, including vision, touch, audition, and smell. The olfactory effect of garden art is mainly provided by plants. The fragrance of vegetation makes the air in the garden fresher and stems, leaves, blooms, and fruits of some flowers are not only objects for viewing but also the source of fragrance. They also play the role of attracting butterflies and bees. The lush environment of the garden is also a special setting for attracting birds. Therefore, plants are also a means of organizing the animal scenery in the garden.

3.1.1.3.2.9 Seasonal Snacks: Roots, Leaves, Flowers, and Fruits

It is true that gardens are mainly objects of viewing, but the beauty of gardens lies not just in visual beauty. As mentioned above, the landscape is full of the fragrance and sounds of nature. It can be considered that the aesthetic enjoyment of gardens is an amalgamation of all the human senses. Being in a garden scene, one feels relaxed and pleased, not only because of the beauty of the scene but also the scene environment makes people comfortable. If you visit a garden tired with no place to sit or starving or thirsty with nothing to eat or drink, you will have no intention of enjoying the scenery no matter how beautiful it is. This is the same in both ancient and modern gardens. The main points of tourism activities in modern parks, scenic spots, and amusement parks can be summed up in five words: eat, live, play, view, and buy. That is to say that visitors must have food supplies, recreational facilities, something worth viewing, souvenirs to buy, and those visiting from afar need accommodation, all of which are very practical matters. Among these, we will turn our attention to food and beverages. Food and beverages with garden features, which are mainly provided by plants, are mainly related to natural flavors and seasons. Ancient private gardens were living spaces that were required to provide food and drink. In the face of beautiful and delicious flowers and fruits, the patrons who were seeking a life of seclusion would abide by the way of the countryside. That is the tasting of seasonal dried and fresh fruits and nuts like melons, pears, peaches, almonds, pine nuts, etc., picking roots, stems, leaves, flowers, and fruits to make various garden desserts and drinks like preserves, jams, juices, and fruit wines as well as purplevine cakes, osmanthus cakes, lotus-nuts sweet soup, marmalade, plum syrup, chrysanthemum drink, autumn pear grease, lotus root starch, dried persimmon, etc. The Hundreds of Flowers Brewage listed in the "garden self-supply" in The Esoteric Flower Mirror includes fruit wines and medicinal wines brewed with as many as 28 varieties of plants in gardens. Picking and tasting delicious fruit is a pleasure of garden living, while brewing flower and fruit food is another pastoral activity that has mental and physical health benefits. For plants, this is another important garden function that is often neglected by researchers. This function of garden plants plays a more important role in the spiritual aspect than the material aspect.

3.1.1.3.3 Unity of Scenic Imagery of Plant Layout

Plant layout should conform to the biological features of plants, such as whether a given plant is an sciophilous or heliophilous plant, hygrophilous or mesophilous plant, hekistothermal or thermophilous plant, has a preference for acidic soil or basic soil, etc. The planning of scenery art operates under this premise. In terms of art, it is necessary to consider the shape and color of plants, their assigned human attributes and the appeal and composition relationship with the environment. This aspect is represented by the experience summarized in *The Esoteric Flower Mirror*. The section titled Planting Location Method reads:

"......if the space of a garden is wide, it is suited to the planting of fruit trees, pines, and bamboos. If the space is narrow, it is suited for flowers, grasses, herbs, and seedlings. If there are luxuriant forests on the left, an open land must be on the right. If there are ponds in the front, terraced buildings must be built in the back. If there are winding paths outside, rare stones must be set inside to make the path look deep. Flowers that prefer sunshine should be planted in such a way that they face both the rising sun and setting sun, while those that require shade should be planted on the north side of the garden. Furthermore, the coordination of color and shape should also be considered. For example, tree peonies and herbaceous peonies are graceful in shape and are suited for planting near a marble terrace with the company of jagged rocks. Daffodils (Narcissus tazetta L.) and orchids should be planted in an earthenware pot with fine stones for decoration and be placed on the window of the bedroom, which can be enjoyed in the face of the setting sun. Peach blossoms are bright-colored and gorgeous, which are fit for places near villas at a mountain recess, or bridges and rivers with a mixture of willows, in the oblique sunglow of the setting sun. Apricot blossoms are prosperous and bright, which are fit for corners of houses and walls as well as places like sparse forests and broad pavilions. Charming pear trees (Pyrus serotina Rehder.) and pure white plum trees (Prunus salicina Lindley.) are suitable for broader courtyards and gardens, which can be made into strong wine or tea for entertaining guests. Garnet pomegranates and bright sunflowers are fit for white walls and green windows. On late July nights, with the charming moonlight and gentle breeze, an exotic fragrance fills the air as visitors wave off the summer heat with a feathered fan. Lotuses have beautiful appearance, which are suitable for water pavilions as their fragrance wafts and the surface of petals and leaves are full of dew. Chrysanthemums have personal integrity, which are fit for places like cottages and unadorned houses. Leave stamens float a nearby river where people collect dew to drink and fallen petals to eat. Chinese flowering crab-apple is tender and lovely, which is suited to carved walls and high houses covered with green window screens and lit with silver candles, where people lean against the railings or pillows. Sweet-scented osmanthus has a strong fragrance, which is suitable for high terraces and broad mansions. Chinese redbud (Cercis Chineusis Bunge) is glorious and long-lasting and is suitable for bamboo fences and sunken flowerbeds. Hibiscus (Hibiscus mutabilis L.) is a gorgeous and open plant and is suitable for places near rivers and marshes. Pines and cypresses are vigorous and are suitable for cliffs and peaks. Concealing wisteria (Wisteria sinensis<Sims>Sweet) and lofty Chinese parasol trees and bamboos are suitable for deep and serene yards and secluded pavilions brimming with the sound of birds chirping. Snow-white reed (Plnagmites communis Trin.) flowers and red maple (Liquidambar formosana Hance) leaves are suitable to appreciate from double eaves in a distance. Clumps of golden Japanese kerria (Kerria japonica DC.) and dense bright climbing rose (Rosa multiflora Thunb.) are suitable to be supported on racks. The rest of the exotic flowers cannot be described in detail but generalized like these. Because of their different qualities, flowering phases, and uneven colors, even herbs and wildflowers can be used for decoration with clever collocation to complement gardens..."

This is how the layout of plants is considered in the classical gardens in the Jiangnan region, from the perspective of the whole scene. Plant layout can be summarized into two major aspects, namely the relationship between plants and molding the shape of the ground surface and the relationship between plants and architectural design. Here, we outline their layout points in terms of the relationship between these two aspects.

3.1.1.3.3.1 Relationship Between Plant Layout and Molding of Ground Surface

Plants are planted on the earth's surface (including the ground and water). How plants are planted in Jiangnan gardens depends on the shape of the ground surface.

There are certain kinds of reasoning and methods gleaned from experience in the design of Jiangnan gardens, such as what kinds and sizes of plants to select and how to match them with the mountains, valleys, lakeside, or water surface features.

The layout of mountain plants is different in earthen mountains and stone mountains. As the name suggests, earthen mountains are dominated by earth and the foothill plants mainly include shrubs and indocalamus that are close to the ground's surface and match well with small trees. Using these plants to break up the line of sight helps to highlight the rising lie of the mountain at the foot of the traveler and make the stone and winding paths stand out more in the visual field, giving the visual effect of a deep and unpredictable winding mountain path. The lake-and-mountain scenery in the middle of the Humble Administrator's Garden (Zhuozheng Yuan), Suzhou, and the earthen mountain in the Garden of Surging Wave Pavilion (Canglang Ting) are good examples of this (Fig. 3.261). The hillside is treated as foothills except for planting large trees. At the top of the mountain, tall and large trees are planted and shrubs are appropriately matched so that layers of trunks can be seen in a head-up view, creating mountain forests with a certain depth of field. Looking up, branches intersect and shade the sun. Looking down, rocky bones are jagged and roots are scattered. The crown of the surrounding slope is low at the foot (mostly shrubs), which sets off the effect of the scene between the forest ridges on the mountains. The layout of the maple forest on the mountain top in the west of the Garden that was preserved miraculously (Liu Yuan), Suzhou, is quite a masterpiece of artistic conception. The earthen mountain in the Garden of Surging Wave Pavilion (Canglang Ting) is an example full of antiquity and wild flavor (Fig. 3.262). Earthen mountains are generally not the main object of viewing based on shape alone and their main function lies in the management of the forest space. Therefore, their design often doesn't focus on the guidance of the overall mountain

Fig. 3.261 Indocalamus in Stacked Stones at the Foothills of Canglang Ting in Suzhou

Fig. 3.262 (**a**) The maple wood with artistic conception on the west maintain top of Suzhou's Liu Yuan. (**b**) The scenery of the sun blotting out by the shade of tall trees in Suzhou's Canglang Ting is like a scene in a remote mountain

shape (making it difficult for tourists to see the overall mountain shape) but rather on the enhancement of the immersive effect of mountain forests. When earthen mountains are needed to be viewed as distant mountains, like the earthen mountains in the east part and lake of the Humble Administrator's Garden (Zhuozheng Yuan), Suzhou, the lie of mountains relies on tall trees to form the skyline (Fig. 3.263). In order to maintain the luxuriant vegetal cover in winter and exhibit seasonal features at the same time, a certain proportion between evergreen and deciduous tree species is maintained. Generally, the proportion of deciduous trees is slightly larger due to the need to highlight the seasonal characteristics.

The stone mountains, full of jagged rocks, feature fewer plants than earthen mountains, which mainly adapt to the features of natural stone mountains, namely less earth and less vegetal cover to highlight the beauty of the rocks. Therefore, the plant layout of stone mountains should be moderate, creating a mountain-and-forest atmosphere and highlighting the rocks at the same time. The reserved planting beds that are used for setting stones should be well designed. The yellow-stone mountain in the Garden to Delight One's Parents (Yu Yuan), Shanghai, created by Zhang Nanyang (a well-known master in mountain setting in the Ming Dynasty) is an example of well-placed flowers and plants. The Lake-stone mountain in the Garden of Look Forward (Zhan Yuan), Nanjing, not only highlights the rocks but also cleverly enhances the towering effect and the strong mountain-and-forest atmosphere by setting a two-layer platform behind the stone mountain and planting tall trees (Fig. 3.264). Planting beds reserved on stone mountains are not necessarily just

Fig. 3.263 Trees on the earthen mountain in the East Garden of Zhuozheng Yuan form its skyline

Fig. 3.264 (**a**) In Shanghai's Yu Yuan the yellow-stone mountain (Zhang Nanyang's creation in the Ming Dynasty) are disposed with plants sparsely so as to to fully show the beauty of the rocks of the stone mountain. (**b**) The main scenic plane of "Jingmiao Tang" on the south side of the lake—stone mountain in Zhan Yuan of Nanjing—showing the beauty of both rocks and vegetation. (**c**) A two-story terrace is piled up behind the peak of the stone mountain in Zhan Yuan, Nanjing, and trees are planted to form a rich vegetation background of the stone mountain

for planting flowers but more often than not are used for arranging trees, creeping fig (Ficus pumila L.), bamboos, and grasses, etc. Even if flowers are planted, they are just some small flowering trees, shrubs, or climbing plants like azalea (Rhododendron simsii Planch.), wild climbing rose, chrysanthemum, Chinese trumpet vine (Campsis grandiflora (Thunb.) Schum.), banksia rose (Rosa banksiae Ait.f.), wisteria, etc. Examples include the Chinese trumpet vines on the cave of the Lake-stone mountain below the Secret Fragrance and Dappled Shadows Tower (Anxiang Shuying Lou) in the Lion Grove Garden (Shizi Lin), Suzhou, the wisterias on the Lake-stone mountain below the Spiral Hairdo Pavilion (Luoji Ting) in the Garden of Pleasure (Yi Yuan), and the yellow stone Lion Mountain (Shizi Shan) in the Master-of-Nets Garden (Wangshi Yuan) (Figs. 3.13b and 3.265z). Generally speaking, there is a preference for plants on stone mountains to be delicate with vivid postures such as podocarpus (Podocarpus macrophyllus (Thunb.) Sweet), boxtree (Buxus sempervirens L.), white bark pine (Pinus bungeana Zucc. ex Endl.), Chinese hackberry (Celtis sinensis Pers.), crape myrtle (Lagerstroemia indica L.), procumbent juniper (Sabina procumbens (Endl.) lwata et Kusaka), etc. Curved and reclining trees are often planted among the precipices and mountain ranges which imitate the growth conditions between stone crevices under the strong wind of high mountains in nature. For example, the crape myrtles planted on the precipice of the Villa of beautiful surrounds (Huanxiu Shanzhuang), Suzhou, exhibit an exemplary design in its choice of proper proportion of small trees, special treatment of reclining posture, and the position. From a distance, the reclining crape myrtles and the lie of the mountain create a beautiful composition, while exploring the cliffside under the tree invokes an artist conception of a shady ancient path (Fig. 3.266). The sad thing is that the tree died and was felled and replaced with a spherical crown pine tree. Crape myrtles are moderate in shape and even in their younger age, inspiring artistic imagery of dragon-like crooked branches. It is easy to meet the design requirements with slight pruning and so are the most popular tree species used in the construction of Jiangnan gardens. Among pine trees, the white bark pines have characteristics similar to crape myrtles and they are often used in the arrangement of stacked stones. Stacked stones in ravines and gullies also rely on plants to create an atmosphere. The Gully of Eight Sounds (Bayin Jian) stacked with natural stones on the earthen mountain in the Garden for Ease of Mind (Jichang Yuan) in Wuxi mainly relies on the cover of wingnut trees on the earthen mountain by the ravine side to create a deep and serene effect. Unfortunately, due to poor management, these big trees have all withered and fallen away. Therefore, the Gully of Eight Sounds (Bayin Jian) has become a stone roadway completely losing the artistic conception of a secluded valley. The ravine and gully in the Villa of beautiful surrounds (Huanxiu Shanzhuang) were also exposed to the broad daylight for the same reason which greatly affected the effect of this masterpiece. From the gains and losses of the plant layout, the above-mentioned examples time and again prove the principles of the plant layout.

Solitary stone peaks are often planned with climbing plants like wisteria, climbing rose, Chinese trumpet vine, banksia rose, Chinese starjasmine (Trachelospermum jasminoides (Lindl.) Lemaire.), creeping fig, Japanese ivy

Fig. 3.265 (**a**) The old gingkoes in Suzhou Shizi Lin. (**b**) Weeping willows and lotus flowers are arranged around Hefeng Simian Ting in Suzhou's Zhuozheng Yuan. (**c**) The old pterocarya stenoptera tree in Jichang Yuan in Wuxi. (**d**) Phoenix trees/Firmiana simplex in the sunshine of autumn in Shanghai's Yu Yuan. (**e**) Camphor trees on the mountain of Canglang Ting in Suzhou. (**f**) The white bark pine (Pinus bungeana Zucc. ex Endl) leaning against the water in Suzhou's Yi Yuan. (**g**) Podocarpus macrophyllus beside the pavilion in Suzhou's Wangshi Yuan. (**h**) The black pines (Pinus thunbergii Parl) on stone mound of Jixiao Shanzhuang in Yangzhou. (**i**) The magnolia grandiflora in Suzhou's He Yuan. (**j**) The palm grove beside the corridor at the east part of Suzhou's Liu Yuan. (**k**) Lagerstroemia indica L on the mountain of East Garden of Suzhou's Ou Yuan. (**l**)

(Parthenocissus tricuspidata (Sieb.et Zucc.) Planch), etc. That is, as stated in *The Craft of Gardens* (Yuan Ye), "climbing roses are not trellised yet, might as well lean them on a stone". Stone peaks and stack stones are natural flower stands for plants. When arranged on stone peaks and stack stones, plants can embellish the stones with colorful flowers and vibrant leaves and can also serve to cover their imperfections. Plants also play a role of concealment and complementation, such as in the layout of the banksia roses on the Peak of Mountain in the Clouds (Xiuyun Feng) of the Garden that was preserved miraculously (Liu Yuan), Suzhou, and the wisterias on the solitary stone peaks in the vestibule of the Pavilion of Separate Peaks (Shufeng Guan) in the Xu's Garden (Xu Yuan) around the Slender West Lake, Yangzhou (Fig. 3.265c).

In areas near natural stone steps, stack stones, and ancient stumps, lichen and moss are planned to simulate antiquity. According to ancient texts on horticulture, mosses should be cultivated with water caltrop mud, horse dung, swill, and rice-water among others. The creation of moss marks the fineness of the plant layout of Jiangnan garden and its profound artistic accomplishment (Fig. 3.267).

Fig. 3.265 (continued) Malus halliana Koehne in Wuxi's Huishan Yuan. (**m**) Chaenomeles speciosa (Sweet) Nakai in Wuxi's Huishan Yuan. (**n**) Prunus mume Sieb. et Zucc in Wuxi's Huishan Yuan garden. (**o**) Red acer trees in autumn in Suzhou's Zhuozheng Yuan. (**p**) Hydrangeas (Viburnum macrocephalum Fort) beside Fucui Ge in Bu Yuan of Zhuozheng Yuan. (**q**) Magnolia liliflora Desr. in bud near the pavilion in Suzhou's Hu Qiu Yuan garden. (**r**) The white magnolia in front of Yulan Tang in Suzhou's Zhuozheng Yuan. (**s**) Cerasus yedoensis in Wuxi's Huishan Yuan. (**t**) Chinese redbud (Cercis Chinensis Bunge) in Wuxi's Huishan Yuan. (**u**) Oleanders (Nerium indicum Mill.) leaning over the water in Suzhou's Zhuozheng Yuan. (**v**) Fatsia japonica in front of the Ban lang (half corridor) in Suzhou's Tingfeng Yuan. (**w**) Winter jasmines beside the pool of Er Quan garden at Mount Hui in Wuxi. (**x**) Climbing wisteria on the lakestone cave in front of Biwu Qifeng building in Suzhou's Yi Yuan. (**y**) Yuccas beside Yanlü Xuan in garden of Qiuxia Pu in Jiading, Shanghai. (**z**) Campsis grandiflora on rock peaks of Suzhou's Wangshi Yuan. (**aa**) Wisteria on the "Little Penglai" island in Suzhou's Liu Yuan. (**ab**) Virginia creepers on rock peaks beside the corridor in Suzhou's Tingfeng Yuan; Iris and ophiopogon japonicus beside planting beds and stack stones. (**ac**) Rosa banksiae on the Xiuyun peak near Guanyun Lou in Suzhou's Liu Yuan. (**ad**) The Xiang Bamboo Forest in front of Cui Ling Long Building in Suzhou's Canglang Ting. (**ae**) Indocalamus tessellatus disposed with stack stones in Suzhou's Canglang Ting garden. (**af**) Ophiopogon japonicus disposed on the natural path in front of Wufeng Shuwu in Suzhou's Wangshi Yuan. (**ag**) Plantain trees planted at the moon gate of Suzhou's Canglang Ting. (**ah**) Red maples, the theme plant in Suzhou's Tingfeng Yuan. (**ai**) Chrysanthemum at the foothill of Xiuqi Ting on the artificial mountain in Suzhou's Zhuozheng Yuan. (**aj**) Aquatic plants beside watertown residential buildings in Suzhou's Yi Pu. (**ak**) Water lilies arranged in the goldfish pond on the north side of Linquan Qishuo Zhi Guan in Suzhou's Liu Yuan . (**al**) Scattered lake stones are arranged with hemerocallis in Suzhou's Zhuozheng Yuan. (from Suzhou Classical Gardens). (**am**) Cannas in front of Yifeng Xuan in Suzhou's Liu Yuan. (**an**) Reeds are disposed along the Yinbin Lake in the northern mountain area in Suzhou's Zhuozheng Yuan, which is full of wild interest. (from Suzhou Classical Gardens). (**ao**) Ficus pumila Linn and trachelospermum jasminoides on the arch bridge and the stack stones beside the stream of Suzhou's Wangshi Yuan. (**ap**) Climbing plants hung on the stone mountain—Yangzhou Xiao Pangu

Fig. 3.265 (continued)

As for waterscapes, plants are arranged on the shore, water surface, and in the water itself to help set off the scene. Most of the existing examples of Jiangnan gardens are disposed of in these two models, one of which is near the shore. Plants are arranged so as to break the monotony of the stack stones and such plants include

Fig. 3.265 (continued)

clusters of winter jasmine (Jasminum undiflorum Lindl.), Jasminum floridum Bunge, creeping fig, Chinese starjasmine, and dayflower (Commelina communis L.) which creep along stones and droop above the water surface. The second model consists of arranging plants such as Chinese pink herb (Dianthus chinensis L.), indocalamus (Bambusa multiplex of auth.), Ophiopogon japonicus (Thunb.) Ker Gawl., tawny daylily, iris (Iris tectorum Maxim.), fragrant plantain lily (Hosta plantaginea Engl.) and fringed iris (Iris japonica Thunb.) among the roads and stones near the shore to delineate the appeal of water towns, such as in the Master-of-Nets Garden (Wangshi Yuan), the Garden of Pleasure (Yi Yuan), and the Humble

Fig. 3.265 (continued)

Fig. 3.265 (continued)

Fig. 3.265 (continued)

Administrator's Garden (Zhuozheng Yuan) (Fig. 3.268). Serving as a background to the lake, the trees and shrubs are arranged in layers far from the lake itself. The key point is to leave space between elements of the scene and to leave an interconnected line of sight for viewing the water scenery. Plants emphatically disposed along lakes

u v w

Fig. 3.265 (continued)

Fig. 3.265 (continued)

Fig. 3.265 (continued)

and rivers are mostly trimmed to make them inclined to the water and adopt a graceful appearance, such as the flowering peaches (Prunus persica 'Duplex'.) near the pond in front of the Prunus Mume Pavilion (Zhuwai Yizhi Xuan) in the Master-of-Nets Garden (Wangshi Yuan), Suzhou, (alluding to the poetic verse "A single peach blossom stands out from the bamboo forest"), the Japanese black pines (Pinus thunbergii Parl.) planted nearby and the oleanders (Nerium indicum Mill.) by the waterside in the Humble Administrator's Garden (Zhuozheng Yuan) (Fig. 3.265u).

Fig. 3.265 (continued)

Willow is a proper plant to serve as a foil to the waterscape as its branches droop down to the water surface. In Jiangnan gardens, due to space limitations, willows are always trimmed to keep sparse and control body size. Generally, planting one or two willows with a few other trees along the bank will suffice (Fig. 3.104). Among the willow-themed scenes, for example, in the Zigzag Pathway with Willow Shade (Liuyin Luqu), an embankment landscape accompanying a waterscape in the Humble Administrator's Garden (Zhuozheng Yuan), the use of willow is very limited. Originally, three to five willows were planted and trimmed to keep the view

Fig. 3.265 (continued)

al

am

Fig. 3.265 (continued)

an

ao

Fig. 3.265 (continued)

transparent, but the view quality has since suffered from the addition of more trees. The principle of plant layout shown here is representative, which is also highly generalized under the principle of achieving more with less. There are few earthen banks in the existing waterscapes. The lake-and-mountain landscape in the Humble

ap

Fig. 3.265 (continued)

Administrator's Garden (Zhuozheng Yuan), Suzhou, is a successful example of this. Earthen banks are rich in wild artistic flavor. For the soil and water conservation, grasses are planted for slope protection. Vegetation is lush and the soil is less exposed, an imitation of nature. Aquatic plants can be planted along the bank, which is the consideration of the economy. In the above-mentioned example, if reeds are planted too abundantly, the aesthetic feeling of wildness and leisure will be broken, and the desolate scene of an overgrown wasteland will take its place (Fig. 3.265al).

The disposition of plants on the water surface must abide by the rule that the necessary scenery of the lake and sky be maintained. In general, duckweeds (Spirodela polyrhiza Schleid) and lotus leaves are always planted in places that aren't already covered by beautiful reflections. Due to scenic requirements, on the water surface near the Hall of Distant Fragrance (Yuanxiang Tang) and the Stay and Listen Pavilion (Liuting Ge) in the Humble Administrator's Garden (Zhuozheng Yuan), Suzhou, main plants related to the theme are disposed at specific positions with limited space to prevent them from spreading to completely cover the water surface. On a more technical level, in order to control the position and scope of the plants, lotus pots and stone plates are placed at the bottom of the water, or holes are left at places where lotuses are needed. The easiest and most commonly used method is to put lotus pots at the bottom of the water. The Humble Administrator's Garden (Zhuozheng Yuan) in Suzhou and the East Garden (Dong Yuan) of the Xu's Family in the Ming Dynasty in Nanjing both use this method (Fig. 3.269). The laying of stone plates underwater and holes left to control the scope of lotus growth above water was used in the Gu's Garden (Gushi Yuan) in Kunshan among others. Water lilies (Nymphaea tetragona Georgi) are delicate and exquisite and are often used in small pools, like the pond at the north of the Hall of Forests and Streams for the

Fig. 3.266 (**a**) Stone mountains and planting beds are created to plant flowering shrubs in Suzhou's Hu Qiu scenic area. (**b**) Planting beds are built to plant trees on the mountain in Suzhou's Yi Yuan (Contented garden). (**c**) The slanting lagerstroemia oblique on the lake-stone mountain of Huanxiu Shanzhuang in Suzhou

Senior and Accomplished (Linquan Qishuo Zhi Guan) in the Garden that was preserved miraculously (Liu Yuan), Suzhou and the pond in the Garden of Natural Delight (Chang Yuan), receiving the beauty of seclusion accompanied by goldfish and algae (Myriophyllum spicatum L.) (Fig. 3.265ak).

Plant layout is also important in other water environments, such as the shade of bamboo and woods enveloping a pond to give it a sense of depth or pines, maples and vines placed around waterfalls and cliffs symbolizing high mountains. While

c

Fig. 3.266 (continued)

drawing on natural landscapes or simulating landscape paintings, the creations follow the principle of *xieyi*[134] in order to achieve more with less.

The disposition of algae in water is an effective means to achieve seclusion and quietness and is often combined with goldfish cultivation, resulting in a comprehensive effect. Examples include the disposition of the Garden of Listening to the Maple (Tingfeng Yuan) and the Villa of beautiful surrounds (Huanxiu Shanzhuang) in Suzhou.

[134] Aiming to capture the spirit of the object rather than its physical likeness.

Fig. 3.267 Mosses growing naturally on the roots of ancient trees in Canglang Ting, Suzhou

3.1.1.3.3.2 Relationship Between Plant Layout and Architectural Design

The combination of architecture and nature takes into account coordination with plants as well as with the landform of the site. In Jiangnan gardens, architectural design and plant layout are coordinated. Some small buildings related to planting are also created according to the appeal of the landscape environment. The likes of the herbaceous peony bar, tree peony terrace (e.g. the Lion Grove Garden (Shizi Lin) in Suzhou), the parterre fences (e.g. the Garden that was preserved miraculously (Liu Yuan), the Garden of Pleasure (Yi Yuan), the Garden of Surging Wave Pavilion (Canglang Ting), etc.) (Fig. 3.270), and climbing plants like wisterias, banksia roses (e.g. the Swallow Garden (Yan Yuan) in Changshu), etc., combined with flowers, trees, and other plants together form garden creations of a pastoral sentiment, such as the Heavenly Scented Courtyard (Tianxiang Tingyuan), the Garden of Riches and Honour (Fugui Huayuan), the Trellis of Melons and Beans (Guapeng Doujia). Furthermore, movable pedestals of various flowers in potted landscapes are decorative architectural designs that show off plants.

The deeper relationship between plants and architecture lies in the combination of space and shape. Plants are the most flexible and vivid means of integrating natural and architectural spaces. The design of Jiangnan gardens mostly uses plants to harmonize the architecture with the landscape environment in which they are located. This involves disposing of flowers, grasses, and trees in architectural and landscape spaces, thus unifying the whole garden scene in a beautiful plant space. Jiangnan gardens attach great importance to older trees as they can be regarded as a

Fig. 3.268 (**a**) The plants near the lake on the side of earthen mountain in Suzhou's Zhuozheng Yuan are mainly luxuriant shrubs and indocalamus. (**b**) Trachelospermum jasminoides are arranged on stack stones beside the stream in Suzhou's Wangshi Yuan. (**c**) Ficus pumila and Trachelospermum are arranged on the stack stones on the lakeshore in Qushui Yuan of Qingpu near Shanghai. (**d**) Disposition of Mosses, Ferns, Ficus pumila, Trachelospermum and Calligonum on the stack stones beside the lakeshore of Suzhou's Yi Yuan

symbol of the ecological environment with landscape value. As mentioned above, outstanding garden makers in ancient Jiangnan took older trees into consideration of the established conditions at the beginning of the landscape planning and design process. If the layout of the architecture conflicts with the position of the trees, the gardener would modify the architectural plan by "rang yi bu (meeting halfway)" (*The Craft of Gardens* (Yuan Ye)). This is because trees grow for a long time. This is the so-called "trees take ten years to grow" analogy which has a slowing effect on planting design. The use of original trees can, therefore, be effective. For the relationship between scenes, what is more important is that the architecture "meeting halfway" embodies the thought of respecting and conforming to nature. In garden creation, buildings are placed to set off the shady trees that have been growing for

Fig. 3.269 (**a**) A lotus vat is placed at the bottom of the pool to control the position and range of plant on the water surface. The water surface cannot be covered with lotus leaves. (**b**) Before the "Cultural Revolution" in Suzhou Zhuozheng Yuan, lotus pots were once used to control the arrangement of plants on the lake surface. However, there were only two clusters, which were too stiff. This arrangement did not consider the composition of the water surface and the effect was not good. After the "reform and opening up", the lake was used to generate income from lotus roots. Lotus plants were planted all over the lake and the lake was completely lost

many years, especially the trees that are hundreds of years old. For example, the Intertwined Old Trees (Gumu Jiaoke) in the Garden that was preserved miraculously (Liu Yuan), Suzhou, features open corridors and verandas on the east of the shade surrounding ancient trees. Thus, this forming an interpenetrating relationship between the plant space (patio) with the theme of ancient trees with the architectural

a

b

Fig. 3.270 (**a**) Peony flower bed in Suzhou's Shizi Lin. (**b**) Bamboo fence in Suzhou's Liu Yuan. (**c**) Bamboo fence and doors in Hu Qiu Mountain, Suzhou

c

Fig. 3.270 (continued)

space (corridors and verandas) as the background (Figs. 3.257, 3.271). Ancient trees are not only a symbol of the ancient ecological environment but also a symbol of the culture of historical gardens. It is a pity that in the Intertwined Old Trees (Gumu Jiaoke) the great cypress and privet (Ligustrum japonicum Thunb.), which were close during life, also died together. Of course, life and death are decreed by fate. Perhaps they suffered from an incurable disease. It is hard to say that their death must be a dereliction of duty on part of those responsible for their safekeeping. However, one point should be brought to the attention of custodians, namely to never cut down these huge dead trees, particularly in the Garden that was preserved miraculously (Liu Yuan), one of the National Key Cultural Relics Protection Units. No one is authorized to take such action. In the courtyard of Intertwined Old Trees (Gumu Jiaoke), people still appreciate the trunks of these deceased trees, maintaining the emotional appeal of the scene's space. Although the trees are dead, they can be revived with climbing plants. The urgent course of action for these ancient dead trees is not to clear them and let the little courtyard be open and clear (losing the objects that it serves to highlight), but to properly embalm the dead trees so that they can decorate this historic garden as long as possible. We mention this regretful incident

Fig. 3.271 "Gumu Jiaoke" in the small courtyard of the "green shade" dongxuan gallery in Suzhou's Liu Yuan—privet and cooper (taken in 1956)

here so as to learn a lesson and improve the preservation of ancient gardens and to explain the value of ancient trees as a theme of the architectural environment.

Another type of relationship between architecture and plant spaces can be exemplified by the Veranda of Scenting the Sweet Osmanthus Fragrance (Wen Muxixiang Xuan) that is among the mountain forest on the west of the principal scenery of the Garden that was preserved miraculously (Liu Yuan) as well as the Pavilion of Listening to the Billow (Tingtao Ting) that is among the mountain forest on the west of the Lion Grove Garden (Shizi Lin), where the buildings are treated as a background for the forest. For pavilions, verandas, corridors, or hole door and hole window in open spaces, trees form a natural green curtain on the open side. Therefore, trees complement architectural design.

Plants are a powerful means of communicating architecture and natural space. The indoor display of potted flowers and vases can show the meaning of natural scenery permeating into the room. Nevertheless, this is not as vivid and interesting as outdoor branches and leaves of flowers and trees poking into the room from a hole door and hole window or climbing plants winding through openwork window. Can one deny that the bamboo branches poking through the hexagonal hole window in front of the Worshipping Peak Pavilion (Yifeng Xuan) in the Garden that was preserved miraculously (Liu Yuan), Suzhou, is a vivid example of natural space flowing into architectural space? Even if plants don't extend indoors, the flowers and trees leaning against the doors and windows can produce the same effect if they appear repeatedly and are strewn randomly in the architectural space. In the eastern part of the Garden that was preserved miraculously (Liu Yuan) in Suzhou, some courtyards were arranged among halls, verandas, and corridors near the Worshipping Peak Pavilion (Yifeng Xuan) with the layout of flowers and stones. Indoors were connected to the outside via hole door and hole window giving the feeling that the buildings, plants, and stones are interwoven, hardly distinguishing where the indoor space ends and the outdoor patio begins. This is an excellent example of interweaving plants into architectural space to show the integration of natural space and architectural space (Figs. 3.124 and 3.243f).

Another type of design is based on coordinating buildings to match the plants. Trees planted separately in Jiangnan gardens represent importance attached to posture and ancient trees with vivid postures can be arranged with large or simple-shaped buildings uniformly. For example, the land boat the Fragrant Islet (Xiangzhou), the Watching-Clear-Water Building (Chengguan Lou) in the Humble Administrator's Garden (Zhuozheng Yuan) and the Winding Stream Tower (Quxi Lou) in the Garden that was preserved miraculously (Liu Yuan), both of which have a great body size, are set off compositionally by giant enclosed ancient trees which lean onto the lake to make the landscape composition more vivid. Even saplings can also form a compositional relation with buildings through selection, trimming, or making trees with vivid postures. Planting trees and shrubs around buildings can partially overlap the landscape, which in itself acts to cover the building. The interweaving of buildings and plants creates the effect of building snares in natural scenes. Among them, the integration of architectural space and plant space is mainly to leave space for an open architectural environment and dispose of plants, or to insert small green patios among the architectural space. Examples of such treatment include the design of the long lane at the entrance of the Garden that was preserved miraculously (Liu Yuan) in Suzhou and the Worshipping Peak Pavilion (Yifeng Xuan) in the eastern part of the same garden.

3.1.1.4 The Embellishment of Gardens with Animals

Like plants, animals are one of the main components of natural ecology. In nature, animals maintain a certain ecological balance with plants and among their own species. Therefore, animals and plants as a whole ecological environment—an ecological chain—have been studied in modern environmental science. Nowadays, many wild animals and primitive plant covers are protected and utilized as scenic resources for tourism. Due to the close relationship between animals and plants, animals are often used in artificial gardens (especially natural-style gardens) as embellishments when arranging plants to ornament a garden. In recent times, animal gardens—public gardens with animals as the main ornamental objects have emerged. Generally speaking, zoos whose primary purpose is attracting visitors rather than scientific research also place an emphasis on science education, seeking a more complete collection of animal species and displaying them in evolutionary order, with less of an emphasis on natural scenery ornamentation.

Modern natural eco-tourism areas in Africa, Australia, and other regions have changed previous methods of zoo exhibition and viewing through safe tourism tools for people to observe wildlife in its natural habitat. This is pioneering work that allows people to observe animals in their natural ecological state, however, prevents visitors from having contact with nature. It is a special kind of park that cannot be freely enjoyed by visitors. In fact, gardens where visitors can enjoy themselves freely and animals live a free-range life have already been around for a long time. Of course, the animals can only be livestock, poultry, or other domesticated species. Due to the influence of Chinese gardens, some non-Chinese gardens also make use of free-range animals to decorate the natural scenery of gardens, such as deer in the classical gardens of Nara, Japan, and cows and sheep in the 18th century natural gardens in Europe.

China boasts a long history of embellishing gardens with animals. As early as 3,000 years ago in the Western Zhou Dynasty (1046–771 BC), King Wen's (1152–1056 BC) large-scale natural-style garden covered an area of almost 3,850 square kilometers and included wild pheasants and hares, deer, waterfowl, and fishes for ornamental purposes. In the *Book of Songs* (Shi Jing), it is recounted that "In Wondrous Park the king saw the deer in the ring lie at his left and right. How sweet sang the birds white. The king by Wondrous Pond saw fishes leap and bound", reflecting the scenery from that time. In the Warring States Period (475–221 BC), birds and beasts were kept in the gardens of the rulers of various countries. Patterns of gardens with birds, fish, or other animals are often found on the bronze wares unearthed from the tombs of the Warring States Period (Fig. 3.272). According to the Zhou Dynasty's regulations, officials such as You Ren were in charge of raising and managing animals in imperial gardens. Moreover, during the Qin (221–207 BC) and Han (206 BC–220AD) dynasties, evidence shows that the embellishment of gardens with animals was still widely used.

The East Garden (Dong Yuan, also called Liang Yuan, Tu Yuan), built by King Liang Xiao (?–144 BC) (the second son of Emperor Wen [203–157 BC] of the Western Han Dynasty [202 BC—8 AD]), covered an area of 122,500 km^2 and included such scenery features as rocks for apes and islets for cranes and wild ducks. It can be seen that animal viewing was combined with natural scenery. According to *Compilatory Records of Han Institutions, Palatial Architecture and Gardens, and Imperial Anecdotage*, in the Palatial Garden of Sweet Spring (Ganquan Gongyuan) Emperor Wu (156–87 BC) of the Western Han Dynasty kept deer, birds, and other various animals among the natural scenery. In Senior General Huo Guang's (?–68 BC) private garden, there were 36 pairs of free-range mandarin ducks in a large lotus pond, swimming about like colorful brocades. The private garden of a wealthy gentry Yuan Guanghan from Maoling, built at the foot of Mount Beimang in the suburbs of today's Xi'an, was also embellished with rare birds and beasts such as

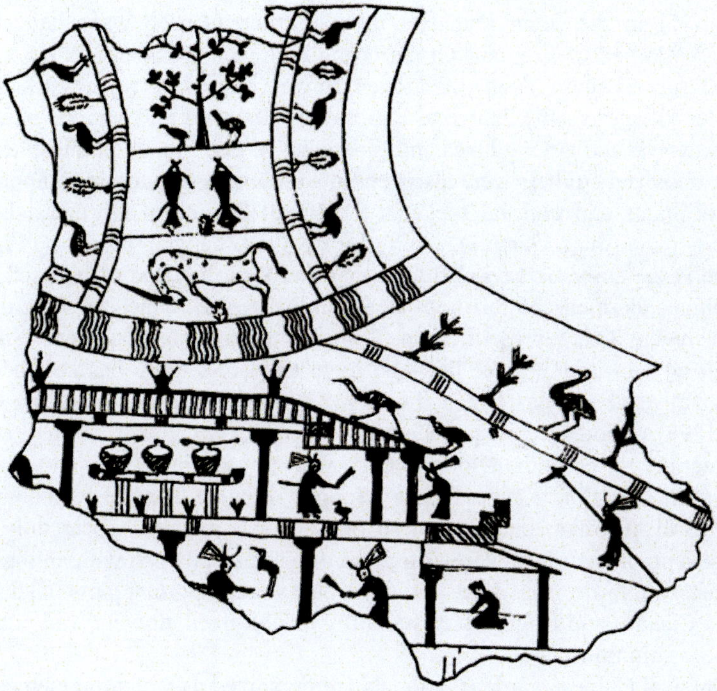

Fig. 3.272 On the bronze wares of the Warring States Period (475–221BC), animal figures are on patterns of gardens. In the garden, there are broad rivers and small streams, fish ponds, broad-leaved trees and coniferous trees, and free-range waders like ostriches, storks and cranes, etc. There are also trained animals with long horns like antelope and long tails like cattle.

white parrots, purple mandarin ducks, cows, and green rhinoceros among the artificial mountain-and-water scenic imagery.

The natural scenery of streams, sandbanks, and pampas grass of such garden even attracted some wild birds to inhabit and breed there, which added to the natural wild atmosphere. In the period of the Jin Dynasty (266–420), the Fragrant Grove Garden (Xianglin Yuan) of Cao-Wei political power (213–266) in the Three Kingdoms (220–280) was rebuilt into the Floral Grove Garden (Hualin Yuan). In its natural scenery, it also kept birds, animals, and fish. During the reign of Emperor Xuanzong (685–762) of the Tang Dynasty (618–907), Wang Wei, a poet, painter, and Minister of State, raised deer by mountain streams in Luzhai and cranes in the Nantuo scenery (both Luzhai and Nantuo are in Shaanxi province) in his villa garden in Wangchuan, Lantian County. The great poet Bai Juyi (772–846) also kept cranes[135] between hills

[135] In ancient China, cranes were considered to be the most noble birds, representing longevity and fidelity. At the same time, it was believed that the crane was "under one bird, above ten thousand

and pools in his residential garden in Luoyang. During the time of Emperor Huizong (1082–1135) in the Song Dynasty, the Fengzhen Garden was changed to the Forbidden Garden of Fragrant Grove (Fanglin Jinyuan, this one is not the Floral Grove Garden (Hualin Yuan) mentioned above), where the scenery of mountains and water villages in the Jiangnan region was designed and various animals and birds were arranged among lakes and mountains to add the wild atmosphere.

Rare birds and animals were also kept in a mountain-and-forest composition full of exotic plants and unusual rocks in the Royal Garden of Northeast Longevity Mountain (Shoushan Genyue), built during the Zhenghe period (1111–1118), Zhenghe is the emperor's reign title at that time). In the dead of night, the roar of wild animals alarmed and terrified people in the capital, which was considered an ominous omen. This represented a peak of raising animals in gardens. Since then, although no works quite like this one have appeared, there has still been much embellishment of gardens with animals in Chinese gardens. During the Qianlong (1711–1799) period of the Qing Dynasty (1636–1912), swans, cranes, and geese were naturally raised in the Old Summer Palace (Yuanming Yuan) and cranes were also kept in the Pine Crane Palace (Songhe Zhai), Chengde Mountain Resort (Chengde Bishu Shanzhuang), and so on. Animals did not appear only in royal gardens as private gardens were also more or less embellished with animals. At the very least, fish, birds, cats, dogs, and other pets were important parts of such private gardens. Cicadas and frogs, which naturally inhabit trees, flowers, and ponds, were even more common.

In classical Chinese gardens, animals are organized as structural elements of the landscape in the garden scenery. Some animals that were rare or highly prized for entertainment at that time served as independent objects of appreciation and entertainment. Let's not discuss the early creation of palatial gardens. Here only in terms of private gardens in the Jiangnan region, this point of view and the principle of design are clearly displayed in these gardens as well as in contemporaneous books about gardens. Animals and plants are both indispensable ecological elements in the scenic structure of gardens. In *The Craft of Gardens* (Yuan Ye), a famous garden art monograph published in the late Ming Dynasty (1368–1644), there is no special discussion given on plant layout or animal disposition, which is a shortcoming of the text. Fortunately, the book *The Esoteric Flower Mirror*, a monograph on garden plants that was written somewhat later than *The Craft of Gardens* (Yuan Ye), touches on the raising of animals for gardens. It can be regarded as work on the ecological elements of gardens and as a coincidental supplement to *The Craft of Gardens* (Yuan Ye). In cases where there are insufficient examples, this book will serve as an important reference for our study on animal layout in Jiangnan gardens.

---

birds" in ancient times, second only to phoenixes, and symbolizing the first. Therefore, the pattern of the official clothing of the Yi Pin (first-grade) officials in the Ming and Qing dynasties consisted of cranes. The "first-grade" is the name of the highest official rank in ancient times. Under the emperor, officials are divided into nine grades, the first being the highest. Therefore, ancient Chinese loved raising cranes in gardens.

3.1.1.4.1 Types of Commonly Used Animals and Methods of Appreciation

Limited by their lifespan, death from diseases, or with the change of ownership of gardens, animals considered as living scenery elements in gardens are easier to change than plants. Therefore, ancient trees more than 100 years old can often be seen in classical gardens but, generally speaking, the animals kept in gardens are unable to survive for such long time periods. In very rare cases, fish and shellfish can be kept for a long time in some temple gardens, which have experienced relatively little change. For example, five-colored carps that were over 100 centimeters in length lived almost 100 years in the pond of Fish Jumping at Jade Spring (Yuquan Yuyue) of Hangzhou's West Lake until the 1930s.

Due to a lack of original embellished animals in the classical gardens in the Jiangnan region, some researchers often neglect this important aspect of such gardens. In fact, since 1949, some gardens have made efforts to restore the embellishment of gardens with animals. For example, the Humble Administrator's Garden (Zhuozheng Yuan) in Suzhou once kept mandarin ducks in the pond in front of the Thirty-Six Pairs of Mandarin Ducks Hall (Saliu Yuanyang Guan) and the Garden of Art (Yi Pu) once raised rabbits in the mountain foothills. Unfortunately, such practices did not last long and they were not further enriched. Some institutions or residents occupy gardens that have not been officially opened to the public and kept some animals for living needs rather than for the restoration of garden scenery. For example, ducks were bred in the pool of the Garden of He Family (He Yuan), Yangzhou, in the 1950s. Although this could not be regarded as the original design of embellishing the garden with animals, it gives an impression of the scenery of animal embellishments and garden living in days past.

This is of great benefit to the study of our topic. Our discussion will largely rely on ancient books about animal raising in the classical gardens in the Jiangnan region. These materials not only reflect the views on garden animals at that time but also shared experience about animal breeding. This is only in reference to the material of *The Esoteric Flower Mirror*, the book I mentioned earlier, written by Chen Fuyao[136] in Hangzhou during the Kangxi period (1661-1722) of the early Qing Dynasty, which provides a guide about the animal species used in the classical gardens in the Jiangnan region and methods of appreciation.

In Volume 6 of *The Esoteric Flower Mirror*, garden animals are classified into four categories. These are birds, domestic animals, aquatic animals, and insects, and their habits and feeding methods are introduced respectively. Here, only the different categories and their points of appreciation will be introduced.

[136] A horticulturist in the Qing Dynasty, whose book, *The Esoteric Flower Mirror*, is an important ancient book of horticulture in China. He was born in about 1612, while the date of death is unknown.

3.1.1.4.1.1 Birds

Chen wrote: "The good birds on the trees and other precious animals under the trees are all enough to outshine those famous gardens and help discover the shortcomings of those gardens". To sum up Chen's views on bird appreciation, there are four specific aspects that are mentioned. These are beautiful feathers with rich ornamental value, charming sounds that are the voice of nature, fierceness and fighting ability to entertain people, and embellishing lake scenery with swimming birds. The species listed in the book are:

- Crane (Red-crowned crane), Grus japonensis.
- Luan—unknown species.
- Peacock, Pavo cristatus.
- Lu Si, Ardea garzetta Linn.
- Parrot, Psittacus
- Qin Ji liao (hill myna)—originally from southern China, it is clever and can imitate the way in which people speak and laugh. No Latin name.
- Niao Feng—Unknown Latin name.
- Qu Yu (crested myna), Acridotheres Cris tatellus.
- Hawk (or falcon), Common gos hawk.
- Eagle (or vulture), Aquila chrysaetus Linn.
- Harrier, Accipiter nisus
- Pheasant, Phasianus torquatus Gm.
- Chicken, Gallus domesticus Briss.
- Bamboo-partridge, Bambusicold Gould.
- Yellow-bellied tragopan, Meleagris gallopavo L.
- Mandarin duck, Aix galericuiata L.
- Purple mandarin duck, Nycticorax prasinosceles.
- Pigeon, Columba.
- Quail, Coturnix communis Bonn.
- Common blackbird—Unknown Latin name.
- Swallow, Hirundo rustica gutturaiis Scop.
- Chinese Hwamei, Garrulax canorus.
- Wren, Troglodytesfumigatus Temm.
- Flower guard bird—Unknown Latin name.

3.1.1.4.1.2 Domestic animals

Chen briefly recorded several kinds of livestock and animals that could be domesticated which he believed were also "suitable enough for garden entertainment".

- Deer, Cervus sika Temm.
- Rabbit, Lepus.
- Monkey, Simi Mensehenaffen.
- Dog, Canis familiaris L.

- Cat, Felis domestica Briss.
- Squirrel, Sciurus lis Temm.

3.1.1.4.1.3 Aquatic animals

Chen also included amphibians into this category. On his views regarding appreciation, he wrote: "Colorful swimming fish play among algae. Frogs call in the morning and at night. They are a pleasure in the life of the water towns and an essential part of the gardens".

- Goldfish, Carassius auratus var.
- Rumble/Fighting Fish, Zacco platypus.
- Mossback, Thecophora.
- Toad (frog), Bufo vulgaris Laur Toad.

3.1.1.4.1.4 Insects

Chen's arguments on insects include: "If butterflies and bees were not busy under flowers, there would never be any fun to be had. When mocking birds stops chirping and the autumn wind soughs, if there are no cicadas and crickets singing, the garden will be lonely and quiet. Where, then, will the fun of autumn come from?"

- Bee, Apis chinensis.
- Butterfly, Rhopalocera P apilio.
- Cricket, Cryllodes berthellus, Sauss.
- Cicada
- Japanese crickets, Homoeogryllus joponicus.
- Katydid, Mecopoda niponensis.
- Firefly, Luciola vitticollis, Kies.

Although the garden animals mentioned above from the *The Esoteric Flower Mirror* do not include all those used in the classical gardens of the Jiangnan region, birds, beasts, insects and fish are all mentioned. It is a detailed and accurate description of gardening practices from the past. This book makes up for the lack of examples of embellished animals from times past and there is immense gratitude to the author for creating such a useful resource.

In Jiangnan gardens, the animals that laid out freely among the mountain-and-water scenic imagery are all selected according to the theme of the scenery, regardless of whether they are livestock, poultry, domesticated wild animals, insects, or fishes. Therefore, the arrangement of these animals is conducive to the formation of the atmosphere of the design. The effect of embellished animals lies not only in their natural and interesting activities which enhance the wild and dynamic feel of the scenery but also in the sounds of animals among forests and springs, like the call of deer or cranes, twitter of birds or singing of cicadas. The sounds of crickets' chirping in cool nights, bees humming among flowers, or even fish-eating fallen flowers are all music with natural aesthetic pleasure that enhances the appeal of garden art.

The artistic creation of Jiangnan gardens is based on the principle of borrowing. For seemingly accidental and passive elements, such as birds and insects that appear with the changing of the seasons, people can understand and grasp the laws of their behavior to use them as a natural element of the scenic structure, so they will serve as objects of viewing appreciation. *The Esoteric Flower Mirror* has clearly stated this view: "The lively music and singing are not comparable to the beautiful birds on the branches. The sound of parrots in the home is not as good as that of swallows and orioles. There are a great many birds that can sing. If you don't catch them and destroy their eggs, you will not find it bothersome when these birds come to eat, drink aŋd sing". *The Craft of Gardens* (Yuan Ye) also mentions the "borrowed scenery" related to freely inhabiting animals, for example, when "curtains are rolled to attract swallows" and "the first calls of the phoenix echo from the shadow of the woods".

In terms of the appreciation of garden animals, what is significant is their sound effects. Plants, on the other hand, merely produce sounds with the help of the wind and rain and serve as a foil for quiet surroundings. Sounds emitted by the animals themselves allow the deep and quiet atmosphere of the natural scenery to stand out better, a phenomenon aptly described in one of China's ancient poems: "the forest is more peaceful while cicadas are chirping, the mountain is more secluded while birds are singing". As garden art, "the chirping of the birds and voice of the beasts" can even become objects of natural aesthetics.

The second volume of *The Esoteric Flower Mirror* titled "The Garden that Supplies Itself", lists sounds from nature and animals. Those animal sounds are from cranes, swallows, frogs, earthworms, fish, crickets, horses, wild geese, deer, magpies, dogs, chicken, cicadas, katydids, etc. Classical poetry related to the aesthetic enjoyment of gardens includes the verse "There is no limit to the rich sounds found in nature. It is a great thing that I have an ear for enjoying them". The section titled "Hundred Bird Officers" lists the interesting linguistic features of sounds originating from people imitating the sounds of birds over the ages.

The book lists the homophonic sounds of various birds, such as "Xing Bu De Ye Ge Ge (The journey is difficult)", "Feng Huang Bu Ru Wo (even the phoenix is inferior to me)", "Ni Hua Hua (The mud is slippery)", etc.

These all refer to the "words" of a certain kind of bird. For example, it is believed that the partridge's cry is saying "Xing Bu De Ye Ge Ge (The journey is difficult)", the bamboo partridge's cry says "Ni Hua Hua (The mud is slippery)", and the cry of the flower guard bird is "Mo Sun Hua (Don't damage flowers)". There are examples of vivid and interesting legends among all Chinese people, whose phonetics and dialects differ from north to south and they all have adopted different homophonic expressions of bird songs. What is recorded here mainly pertains to the Jiangnan region. For example, when cuckoos cry, northerners believe they are saying, "Singles carry garden hoes" (another version reads "Singles are very bitter"), while southerners think they are saying, "Reap the wheat and pluck the rice seedlings". Both of these phrases mean encouraging farming. According to the northern homophonic tones and the seasonal time when this migratory birds crow, people think the birds are reminding farmers to hoe the fields. In the south, when cuckoos call, it is time to cut wheat and plant rice seedlings, which corresponds to homophonic sounds of the local language.

This way of giving bird cries an interesting feature has a long history. *The Esoteric Flower Mirror* was completed in 1688 and so contains a history of at least 300 years, and almost certainly more than that. For example, Xin Qiji, a patriotic poet in the Southern Song Dynasty (1127–1279), wrote in *Buddhist Dancers—Written on the Wall of Zaokou, Jiangxi*, "Blue mountains can't stop water flowing, eastward the river keeps on going. At dusk it makes me sad and weep, to hear the partridge in mountains deep". The poem implies that the partridge's cry is saying "The journey is tough". Therefore, it can be speculated that believing that the personification of partridges saying "The journey is difficult" dates back to at least more than 800 years ago during the Southern Song Dynasty.

Chinese people love nature, which not only manifests in their enjoyment of mountains, rivers, flowers, and trees but also in their deep appreciation of animals. They have always projected their thoughts and feelings onto the animals they are interested in, as they do with plants. Namely, the Chinese personify animals, holding that dogs and horses know loyalty, crows understand filial piety, swan geese are faithful to their partners, deer, cranes, and turtles symbolize longevity, lions can protect people from evil, elephants represent auspiciousness, peacocks stand for happiness and wealth, mandarin ducks signify a happy and loving marriage, bees and butterflies infatuated with flowers symbolize love, etc. All these directly affect the creation and appreciation of animal scenery in gardens.

3.1.1.4.2 Unity Between Embellished Animals and Other Elements of Scenic Imagery

China's 3000-year garden construction tradition was inherited when the classical gardens in the Jiangnan region were built. In the design of natural scenery, there is not only the layout of hills, water, trees, flowers, and other plants but also many embellished animals. Some of the animals in gardens gather freely into the landscape environment, while others are specially raised by humans. There are also different ways of keeping animals, including free-range and cage-keeping. Animals that gather in gardens by themselves are mainly insects and birds. If a body of water in a garden connects to the surface water outside the garden, fishes and frogs will also cluster in the garden. Natural gathering requires time. Whenever gardens were in the initial stages of constructions, animals would be arranged manually according to the designed scenery, which imitated their ecological environment.

These embellished animals which are gathered in gardens by humans means like those that gather freely on their own can best demonstrate inherent ecological characteristics. Birds and cicadas on trees, crickets, grasshoppers, katydids, butterflies, dragonflies and fireflies in grass or crevices, under walls or steps and by ponds, and fishes and frogs in pools, etc. are all in harmony with the landscape and architecture and as a whole constitute a completely natural landscape. Other animals that are raised by humans are also selected according to the theme of the natural

Fig. 3.273 Caged Birds are under the eaves of the roofed corridor in Suzhou's Tingfeng Yuan

scenery so that they are unified with other scenic constituents. For example, deer and hares should be freely kept in mountain forests or by rivers. Egrets, mandarin ducks, cranes, and other waterfowl should be placed out in lake or sandbank scenery. Peacocks are raised in courtyards and are often arranged with Taihu lake stone pedestals and peony flower fences to form a rich and prosperous scene.[137] Parrots, mynas, larks, thrushes, and other songbirds are kept in cages or on racks and arranged under the roller shutters of the garden residence room, or at the front eaves of the porch (Fig. 3.273). The restoration of animal embellishments is generally neglected in the classical gardens that have been renovated and are now open. Nowadays, only a few goldfish are kept in garden ponds.

Chinese garden creation has always regarded animals as an element of the scenic structure. In scenery composition, the ground surface appearance, plant layout,

[137]As mentioned earlier, Chinese have always believed that peacocks symbolize happiness and wealth. In ancient China, Taihu Lake stone is very precious. Generally, it can only be owned and collected by the wealthy. Peony flowers are gorgeous when in full bloom, symbolizing wealth and happiness in life in China.

architectural design, and other elements of the landscape are all unified under an established theme. This principle has been mentioned by Cao Xueqin (about 1715–1763), a great writer in the early Qing Dynasty, in his masterpiece *A Dream of Red Mansions*. Arguments about garden art in this book can represent the dominant concept of landscape design and design principles in the Qianlong period (1736–1795), another peak of garden design in late feudal China. In his book, Cao clearly expounded the concept that animals should be unified with the emotional appeal of scenery.

For example, in describing the scenery design of the Paddy-Sweet Cottage in Grand View Garden, the book reads, "As they walked on talking, their eyes fell on some green hills barring their way. Skirting these they caught sight of brown adobe walls with paddy-stalk copings and hundreds of apricot-trees, their blossoms bright as spurting flames or sunlit clouds. Inside this enclosure stood several thatched cottages. Outside grew saplings of mulberry, elm, hibiscus and silkworm thorn trees, whose branches had been intertwined to form a double green hedge. Beyond this hedge, at the foot of the slope, was a rustic well complete with windlass and wells weep. Below, neat plots of fine vegetables and rape-flowers stretched as far as eye could see".[138] (The 17th chapter of *A Dream of Red Mansions*)

The author explained the embellished animals in this mountain village through the word of the story's character Jia Zhen, saying, "Other birds would be out of place here, but we ought to have some geese, ducks, hens and so on".[139] The eighteenth chapter of the book mentions that "the pleasure grounds were well-stocked with cranes, peacocks, deer, rabbits, chicken and geese to be reared in appropriate places".[140] In the later chapters, it mentions that the Bamboo Lodge with babbling streams and bamboo forests was embellished with parrots. These lines reflect the author's regard for animals as necessary embellishments in gardens. In other words, animals are a part of the scenery. This idea reflected in the book is based on the gardening practices of Jiangnan gardens.

Although in *The Craft of Gardens* (Yuan Ye), which summarizes the experience of garden construction in Jiangnan, there is no chapter devoted to the layout of animals in gardens, many viewpoints of animal appreciation and examples of the unified emotional appeal between animals and scenery in the narration of typical scenery do appear. Although these lines are not discussing theories of any sort, they can still show that in the aesthetic enjoyment of garden art, animal elements exist. These can serve as a reference for us to study the design of classical gardens.

[138] Quoted from *A Dream of Red Mansions* translated by Yang Xianyi and Gladys Yang.

[139] Idem.

[140] Idem.

3.1.2 *The Guiding Component of Scenic Imagery: The Relationship Between Scenery and Structure*

Guidance is a decisive factor in garden scenery that is key to the garden design. As an abstract concept, guidance integrates with specific scenic constituents to embody the whole thought and functionality of a garden. It is the organizing element of a garden and the organization of all scenic constituents depends on it. In fact, the scenic constituents mentioned above already contain a local aspect of guidance, which determines the relationships of various scenic spaces, organizes the change of front views, and arranges the invisible duration, visible direction, and viewing distance through which each front view is intended to be revealed to the viewer.

Generally speaking, it plays the role of editing the scenery, thus representing the relationship between scene and structure in a garden. This relationship, represented by guidance is not only a spatial relationship but also a temporal or even aspatial-temporal one. Some garden scenery effects, such as "a winding path leading to a secluded spot", "the position of the peak flows into the next turn of the mountain path", "a mountain springing to view as the door opens", "suddenly see the light", and "Mountains multiply, streams double back—I doubt there's even a road, willows cluster darkly, blossoms shine—another village ahead!" are a few examples achieved by means of scenic guidance. At the same time, guidance is also a medium through which to connect gardens and visitors. It is because of guidance that visitors can be personally present to visit gardens and thus gain exposure to such infectious garden art. Guidance, as the most active factor in gardens and the soul of gardens, determines the success or failure of garden design on the premise of an established material basis and is the key to understanding Jiangnan gardens. Only by understanding the nature and mechanism of guidance can we truly understand the composition of the scenic constituents mentioned above.

Guidance includes two basic aspects: *route* and *concealing and revealing*. Route is an active and flexible element that leads the aspect of guidance. The reason guidance becomes the most active element is due to the role of route. For the garden itself, route connects all scenic spaces that are fixed by concealing and revealing and forms the guidance together with concealing and revealing to arrange the landscape scheme. In regards to visitors, route directly links them to a path for visiting the garden and guides viewing, specifically arranging the route of the viewing and the location of garden living activities. It also matches sequential and non-sequential garden viewing (dynamic viewing and static viewing) to plan garden living proceedings and enhance the viewing experience.

Concealing and revealing is the environment of route, forming the composition of a viewing space. Locations and relationships of various scenic constituents, division of scenic space, and disposition of principal and subordinate scenery all depend on concealing and revealing. Compared with route, concealing and revealing is not so much an organizing factor as a segmenting factor. It is diametrically opposed to the connecting factor—route and divides scenery into complex groups and unifies them to form guidance.

In the scene arrangement, the density, distance, height, visibility and invisibility as well as the static, fixed or zigzaging disposition of the groups of scenic imagery determined by concealing and revealing can only succeed in displaying the ideological content, the changes in emotional appeal and guiding the viewing experience through the functioning of the active element, namely, route. It is therefore said that route enlivens the ideas of a garden and gives concealing and revealing its appeal.

In addition, these paved paths, courtyards, sites, winding paths, stepping stones, grotto and tunnels, valleys, bridges, covered walkways, halls, pavilions, pavilions on terraces, windowed verandas, towers, belvederes, are all the routes that visitors go through when visiting a garden, which themselves are partly hidden and partly visible. This is achieved through concealing and revealing, which is therefore based on these objects, and they contain various forms of routes. Dynamic routes and static concealing and revealing are opposite and dependent on each other, constituting inseparable whole—scene guidance, i.e. the relationship between scenery and structure.

3.1.2.1 Route

Route is a design that exists to provide convenience in visiting a garden. In a practical sense, it can be used for passing through and other garden living related activities. It can be a passageway, a place for garden living, and a means of traveling (cars, sedan chairs, boats, donkeys, mules, and horses, etc.). Due to the narrow space in private gardens in Jiangnan, small boats called grasshopper boats by (Li Qingzhao, 1084–1155, a famous female poet), are mainly used. Like paths, bridges, corridors and so on, it is both what visitors travel along and the object of visiting.

In the 1930s, Professor Tong Jun (1900-1983, a Chinese architect) took a picture of a small boat in the Garden that was preserved miraculously (Liu Yuan)'s lake, Suzhou (Fig. 3.256). This highlights the importance that visitors are guided through the sightseeing journey by various forms of passageways and are given the opportunity to sit, lie, or engage in some garden living activities by rest facilities, rest stops, and places of entertainment scattered throughout the garden by design. From the aspect of landscape viewing, route clips through various front views, organizes the emotional appeal of changeable scenery, enables people to continuously, and/or intermittently visit the changing scenic planes during the visit, thus organizing the feel of the visit and enabling people to enjoy the garden art.

A route directly linking the garden and visitors includes two components: garden path and view stop. Garden paths provide sightseeing traffic and organize continuous viewing (dynamic viewing). View Stops arrange garden living and organize fixed-point viewing (static viewing).

3.1.2.1.1 Garden Path: The Organizer of the Vein of Scenic Imagery and the Operator of Viewing Points

The unity of opposites between the continuity of tour guiding and the changeability of forms constitutes the essence of garden paths. Garden paths are vital to garden design as the active element of route composition. If the viewing route of a garden is not well organized, it is impossible to give full play to its ornamental role even if the scenery is beautiful. Garden paths cannot be separated from the viewing objects. However, in the design of some new parks, paths are not arranged as an organic part of the scenery, whether it be straight or curved but are somewhat separated from the viewing objects. This kind of isolated path planning does not allow paths to pass between scenery but instead separates paths from the scenery. This method of dividing scenic spots by paths originates from the common practice of urban and rural traffic road planning, which does not conform to the established design rules of traditional gardens.

The classical gardens in the Jiangnan region inspire us to establish a different concept for learning about paths in gardens. Garden paths not only provide for the need for passage but also introduce visitors into the arranged scenery. Whether a pictorial structure of scenic imagery is perfect or not actually depends on the various front views that are relative to garden paths. Conversely, the creation of the scenery cannot be separated from the garden path. This, in fact, has been determined by the principle that the garden path belongs to route and it is route that organizes scenery.

3.1.2.1.1.1 Characteristics of Garden Paths

Guiding Characteristic of Visiting Gardens

The path in Jiangnan gardens has the function of guiding a visitor. Although this is taken for granted, it can be understood on a deeper level through comparison. Some new parks, such as the Viewing Fish at Flower Pond (Huagang Guangyu), a key park in Hangzhou reconstructed in the 1950s, has its main garden path leading from one garden entrance to another. Visitors often set foot on this broad garden path expectantly, hoping that it will lead them into a beautiful place and proceed cheerfully along the path, only to arrive at another garden entrance and walk out of the garden without enjoying any of it. This is a terrible disappointment. The rock garden set of key sights carefully decorated with precious colorful stones with patterns cannot be reached only by relying on the guidance of its main road, which depends solely on blind luck, guidance from other people, or road signs.

As can be seen from the general tour map, the rock garden is a scenic spot far away from the main road. Chinese traditional gardens in Jiangnan are different. The viewing objects in Jiangnan gardens are arranged on a viewing route with the main objects on the corresponding route. In other words, the viewing route is contained in the scenery. In fact, the form of the viewing route itself is among the objects of viewing-scenery. This is why the garden path has become the best guide for visiting gardens. People visit gardens along the path and can see all the arranged scenery, just as though it were a painting. What is not worth viewing, such as the corners of the garden wall, has no clear path leading to it. It is only for the needs of management that unnoticeable paths are set up.

Correspondence with Scenery

The so-called "yinjing shelu; yinlu dejing" (build a path by following the scenes and capture scenes by following the path) in Jiangnan gardens illustrates the close correspondence between viewing route and viewing objects. Every viewing object arranged in Jiangnan gardens, ranging from a lake-and-mountain scenic theme to the embellishment of a few bamboos and rocks, must have a corresponding viewing position to view the front view. The main scenery is often seen from both afar and near in all directions. As viewing points move, a series of perspective front views continuously come into sight when one travels along the viewing route. Thus, an infinite number of front views will be seen from a limited amount of scenic imagery. This is the reason that people praise the fact that scenic planes are progressively revealed to viewers as they move and their perspective shifts in Jiangnan gardens.

There are two kinds of corresponding relations between viewing routes and viewing objects. One is that the travel route is consistent with the viewing route, which is gradual and fascinating, allowing people to reach the expected scenic spaces as desired. The other is that the travel route is inconsistent with the viewing route, which also can be divided into two types. The first is viewing routes that are connected directly while travel routes are not connected directly to create an unattainable suspense-alluring effect. The second is viewing routes that are not connected directly while travel routes are connected directly to create an effect of unexpectedness, or excitement and amazement.

Whatever the relationship is, the viewing route and the travel route are unified on the premise of enhancing the garden effect. It is the correspondence, or unity, between path and scenery. This brings us back to the basic principle that the path is scenery: A covered walkway or a bridge is a passageway for traveling as well as an object of viewing.

The Curvature of Tour Route

In real life, urban and rural traffic roads try to take a straight-path shortcut when possible, while paths in Jiangnan gardens are deliberately arranged as a line winding about and coordinating with the landscape environment. This form contains the content of nature. Compared with western geometric garden paths, the winding or freestyle garden path is different from that in modern Europe. In Jiangnan gardens, such twists and turns of paths follow fixed patterns, while the twists and turns of freestyle road in some early European gardens don't follow any rules.

This contingency as the 18th century European garden critics called was grotesque, but it was not Oriental or Chinese as they mistakenly believed. On the contrary, it derives solely from the Baroque and Rococo tastes of Europe. In Jiangnan gardens, paths are indeed winding and are not only two-dimensional but also three-dimensional. This has nothing in common with the early freestyle paths in Europe, which imitated paths of Chinese gardens out of curiosity, making unnecessary twists and turns on a flat surface without rhyme or reason, leaving a rather whimsical impression. The zigzags of viewing routes of Jiangnan gardens have their specific contents and certain principles that render them more natural.

The amount of winding in paths in Jiangnan gardens is determined by scattered scenes. The principle is that due to the twists and turns of the plane and the changes of the elevation, the path backgrounds can be covered up by each other in perspective. As the viewing progresses, scenes unfold gradually, thus avoiding dullness and tastelessness due to the fact that one can see everything in a glance. It can make the scenery implicit, increase the layering of the scenic imagery, expand the front view, enhance the viewing effect, and stimulate the artistic flavor of the viewing experience. Due to twists and turns, garden paths increase the journey distance and extend the viewing time. In regard to viewing, garden paths play the role of expanding space, making the limited garden area create the illusion of infinite scenery.

In terms of composition, single means monotonous. A straight line is more monotonous than a curve and the curves repetition following certain rules is more monotonous than a free curve. On the contrary, that which contains opposing factors or contradictions is vivid and rich. Therefore, the layout of tour routes in Jiangnan gardens is "left then right, up then down, understanding then confusion, and reveal then hide". An example can be seen in the stone mountain in the Garden of Look Forward (Zhan Yuan), Nanjing, the Lion Grove Garden (Shizi Lin), Suzhou (Fig. 3.274).

Circularity: Repeat Again and Then Again

In small Jiangnan gardens, in order to make full use of the space, viewing routes strive to spread throughout the whole garden to the greatest extent possible. Routes in the whole garden run through one another and are connected from beginning to end. Seen from a diagram, the path appears circular. Each part of the garden path appears to be a small circle and the whole garden path appears to be a large circle. This route system connects all the scenic spots in a garden and there are more than two routes leading to main scenic spots so that visitors do not have to turn around.

3.1.2.1.1.2 The Change of Route Form

While continuity is required by the route as a sort of tour guiding, it requires rich and varied changes when it becomes an object of viewing. When observing a general map of a Jiangnan garden, it seems at first glance that the paths are incoherent. Even if you are within the scenery, you often feel that the roads are indistinct. Sometimes it is a covered walkway, sometimes a path to a pavilion on a terrace, sometimes to a valley. The form of garden paths is ever changeable, making them interesting. Probably due to the habit of reading maps of modern parks, there was once an expert in Suzhou Classical Gardens who said, "In the general plan of Suzhou Gardens, there is no path".

Fig. 3.274 Layout of "Suspected Road" at the stone mountain in Nanjing's Zhan Yuan: It is shown as a downhill path (In the tunnel culvert there seems to have a downhill stepstone mountain path.), but it is actually impassable. The left side of the tunnel culvert is shown as a uphill stepstone mountain path (upper two levels), but it is actually a downhill path.

This rather misguided remark reflects how confusing the changing paths in Jiangnan gardens are. If the form of a garden tour route remains unchanged from beginning to end, it will be as monotonous as a highway. Even if there are constantly changing scenes along the road, it is just like looking at the exhibits in a gallery, where viewers are unable to delve into the painting. Touring Jiangnan gardens gives people a feeling of being able to walk about *in* a painting because the form of the route itself is a rich and changeable scene. The changeability of route form also shows that paths and scenery are an organic whole that cannot be separated distinctly.

Fig. 3.275 (**a**) Yong Lu in Jiangnan village—Life prototype of Yong Lu in garden art. (**b**) Yong Lu in Wangshi Yuan—Basic form of Yong Lu

Transformation of the Form of the Garden Path

Route is an abstract concept whose concrete form is provided by concealing and revealing. In order to establish the overall concept of the garden path, the route form is discussed together with it here.

- Basic form—Yong Lu[141] (Figs. 3.275 and 3.265ad)
- Simplified form of Yong Lu—steppingstone[142] (Fig. 3.276).
- Spreading form—courtyard, site (Fig. 3.277).

Forms combined with architecture–covered walkway, hallway, entrance-hall (see illustrations in the walkway section)

Transformation forms when garden paths lead to mountains—winding paths, stepstone mountain paths, ravine and gully, tunnels, and flying bridges (Figs. 3.278 and 3.279 and the illustrations in the Shaping of Ground Surface and Planning of Architectural Elements chapters).

Transformation takes place when garden paths lead to water—dikes, stepping stones, bridges (illustrations in the Shaping of Ground Surface and Planning of Architectural Elements chapters).

[141] Brick or stone-paved paths in gardens.

[142] Discontinuous pavement.

Fig. 3.276 Stepping stone is the simplified form of Yong Lu. This is stepping stones in fields in Wujin district, Changzhou city, Jiangsu province. This kind of stones are more used in Japanese gardens. There are few examples of this stepping stone in existing gardens in Jiangnan gardens.

Rationality of Paving

Path pavement types in Jiangnan gardens vary greatly. For the design of a garden, the choice of pavement still follows a unified idea rather than considering in isolation and arbitrarily choosing the paving form. According to original works that have not been renovated or reconstructed by later generations, the ground pavement is always subject to the surrounding scenery. Visitors can feel the unique emotional appeal of a given scenic space through vision and touch. Its rationality from an ornamental point of view lies in using materials appropriate for local conditions, as can be seen in woods-and-mountain environment, winding mountain paths, mountain-climbing stone paths and mountain floors that are usually paved by rubble, block stones, and slab stones and gravel, (here the materials are stones from mountain cutting or gravel from a mountainous rocky beach) (Fig. 3.280).

If bricks or other artificial materials are used, improper handling will damage the sense of unity. For example, even if the ground is flat over two groups of stack stones that correspond to one another, it is still paved with stones to strengthen the integrity of the mountain space. Block stones are mostly used for pavement in order to correspond to stack stones. If the paving method is coarse, the path will appear messier, and if it is slightly processed, it will be shaped like cracks in the ice. Finely processed ice crack patterns are paved with stone plates. After the irregular stone plates are cracked, the joints are then spliced together. The width of the joints is generally uniform (Fig. 3.281). In some relatively independent courtyards,

Fig. 3.277 (**a**) Haitang Chun Wu in Zhuozheng Yuan in Suzhou—The ground is paved with begonia pattern. (**b**) Decorative paving of the Pipa Yuan in Zhuozheng Yuan in Suzhou. (**c**) Decorative paving of the inside site of Suzhou's Yi Yuan after the reconstruction of the entrance

Fig. 3.278 (**a**) Scenes that conceal of stone mountains at the entrance of Zhuozheng Yuan in Suzhou. (**b**) The stone mountains at the entrance to Suzhou's Zhuozheng Yuan is a scene in contraposition to the south of Yuanxiang Tang

decorative paving is usually done to match the elegant layout of windowed verandas and halls.

For example, in the Loquat Garden (Pipa Yuan) and the Courtyard of Spring Crabapple Flower (Haitang Chun Wu) in the Humble Administrator's Garden (Zhuozheng Yuan), Suzhou, the ground is paved with decorative brocade-like mosaic patterns that produce rich and luxurious effects (Fig. 3.277). The two kinds of patterns for decorating the ground are geometric and freestyle. Geometric forms include various patterns composed of straight lines and curves (Fig. 3.282). Freestyle patterns mainly include those of animals and Bo Gu (decorative handicrafts with

Fig. 3.279 (**a**) Stepstone mountain path made of rectangular stones in natural mountain forest (Guanyin Shan, Yangzhou). (**b**) One of the forms of combination of path and mountain—winding path in mountains (Nanjing's Zhan Yuan). (**c**) The stepstone mountain path chiselld out of the rock of

patterns on their surfaces) symbolizing auspiciousness and erudition (Fig. 3.283). It is rational and practical to prevent the ground from being muddy due to rain and snow in order to facilitate passage and prevent dust from being blown up by the wind from pollution so as to protect cleanliness. Furthermore, gardens are for people to visit, so the ground pavement should be decorative. How smooth or rough or even bumpy a road surface affects how convenient the road is for walking, which in turn controls the traveling speed of visitors to move through the surrounding scenery. This is related to the organization of visiting time.

3.1.2.1.2 View Stop: Structure of Garden Living and Static Viewpoints

View stops play a role in providing perspective base point for key viewing front views as well as positions for garden living. As they mainly serve a practical function, most view stops are architectural objects such as halls, windowed verandas, pavilions, terraces, towers, and belvederes. In terms of viewing, view stops are well-matched with visitors' feelings. They are footholds for visitors to carefully observe key front views as well as components of the viewing objects and often serve as the center of the pictorial structure of scenic imagery. As a foothold for viewing, they also constitute a close shot of a scene or a picture frame of a front view. Main view stops always face main front views of principal scenery, which is a principle generally followed in Jiangnan garden design.

The foundation of the layout of a whole garden is established if the relationship between main front views and major view stops is determined. Locating a hall, the major view stop, is emphasized particularly in *The Craft of Gardens* (Yuan Ye), which writes: "Any determination of garden foundation design should be based on determining the location of the hall. The first thing to do is to design its scenery layout, preferably facing south". It is correct to first determine the location of the main view stop when considering view stops for viewing key scenery on the viewing route. This view stop has the most prominent practical function and is generally used for gathering activities. Food for feasting, playing, and other activities are supplied by the service part of the residence. Therefore, this base point should be arranged according to the requirements of convenient contact with the residence. Other view stops are flexible enough to adapt to the requirements of the established main view stop throughout the organized scenery. However, when it comes to the positioning of major view stops, the overall layout of a garden, especially the positioning of

Fig. 3.279 (continued) Gushan Mountain in Xiling Yinshe of Hangzhou. It has the charm of seal cutting on stones. **(d)** Lakeside winding path at the foot of the mountain (Qushui Yuan in Qingpu, Shanghai). **(e)** Gentle stepstone mountain path outside the west wall of the Pipa Yuan in Suzhou's Zhuozheng Yuan. **(f)** Natural stepstone mountain path on the lake-stone mountain, Wenlan Ge, Hangzhou. **(g)** semi-natural and semi-artificial stepstone mountain path in Hu Qiu Mountain, Suzhou. **(h)** The low stepstone mountain path of planting beds and stack stones in Suzhou's Tingfeng Yuan

f

g

Fig. 3.279 (continued)

h

Fig. 3.279 (continued)

principal scenery, and its requirements for major view stops should both be taken into account.

Therefore, to be precise, major view stops and their corresponding principal scenery should be arranged together as a central scenery group in a garden. As previously mentioned in the chapter on architecture, "The first thing to do is to design its scenery layout" in *The Craft of Gardens* (Yuan Ye) refers to the procedure for construction. As for the orientation of the hall (major view stop), it should have a good direction of sunlight while satisfying the landscape scheme (this is speaking from the salubrious property of the hall building). From the perspective of the external scenery of the garden, it is better facing north rather than south. This is because when being viewed from the north, the main front view of the principal scenery is facing south, which can make the main front view of the principal scenery receive the longest duration of sunshine. In this way, the layering of the scenic imagery can be highlighted, and the sense of space can be strengthened, as sculptures rely on the effects of light and shadow. Therefore, halls, the main view stops in Jiangnan gardens, are treated on both the north and the south main facades whenever conditions permit. This principle is also applied to the relationship between other secondary view stops and the corresponding front views.

3.1.2.2 Concealing and Revealing

The environment is the form and background on which routes depend and the environment is composed of basic scenes. Concealing and revealing fixes the relationships among these basic elements. Guidance can only reflect the significance

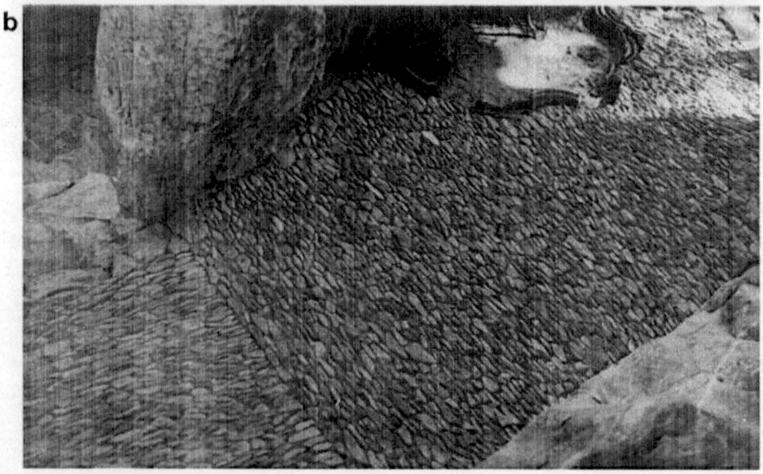

Fig. 3.280 (**a**) Rubbles and block stones are used to pave the land in mountain areas. (East Garden of Suzhou's Ou Yuan). (**b**) Mountainous areas are paved with gravels (Suzhou's Yi Pu)

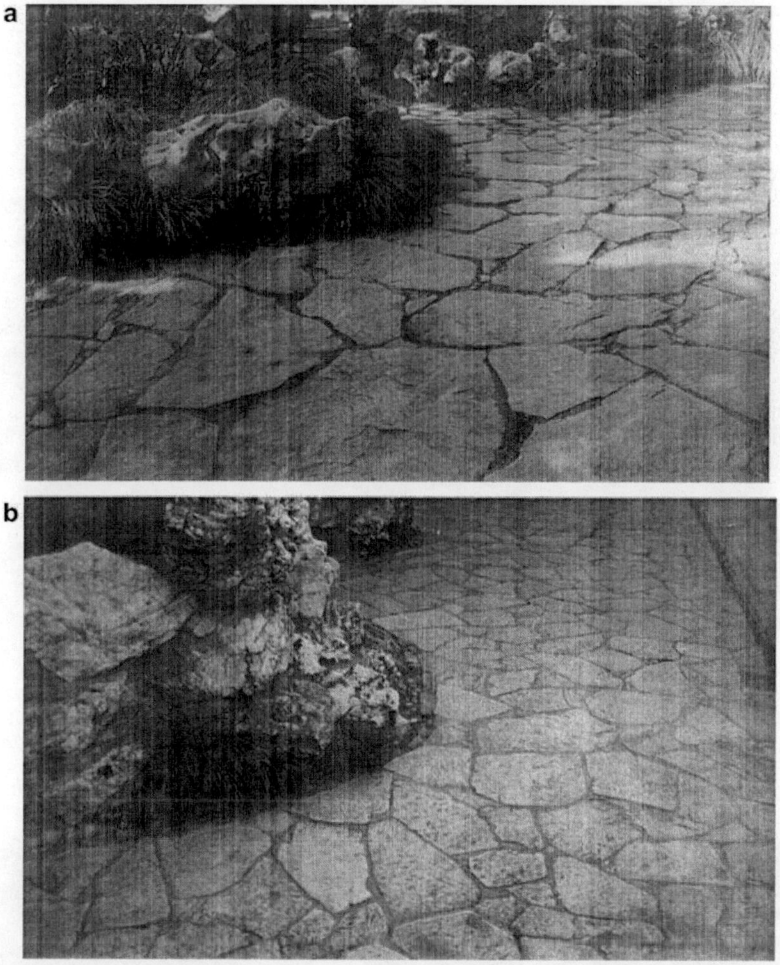

Fig. 3.281 (a) The ground with rough ice crack (Suzhou's Yi Yuan). (b) The ground with delicate ice crack (Yangzhou's Ge Yuan)

of route through concealing and revealing. The two, concealing and revealing and route, which take scenic constituents as the means and the design concept as the guidance and are inseparable and combine to form the landscape scheme. The type of landscape scheme should be considered in the design of Jiangnan gardens. The arrangement of viewing points and viewing procedures (including the viewing of this group of scenes in person and viewing from another scenic space) and their formalization should be considered accordingly while scenic constituents are organized as viewing forms.

In fact, the organization of scenic constituents and the formalization of viewing modes are two sides of the same problem. If the function of the former lies mainly in

a

Fig. 3.282 (**a**) Measure drawing of pavement pattern of Suzhou's classical gardens. (from Suzhou Classical Gardens). (**b**) Ground paving with geometric pattern-ice crack pattern paved with gravels and tiles (Courtyard of Jixu Zhai in Suzhou's Wangshi Yuan). (**c**) Ground paving of geometrical pattern—begonia pattern paved with gravels and tiles (Courtyard of "Sutai Chunman" in Jiechuang Si Temple in Suzhou). (**d**) Plaid pattern paved with small cut stones in Suzhou's Yi Yuan. (**e**) Diamond pattern paved with bricks in Xiangyue Corridor in Suzhou's Yi Pu. (**f**) Pattern of cross and octagonal inlaid with strip bricks and gravels in Suzhou's Huanxiu Shanzhuang. (**g**) Ground pattern of twisted branches in Yangzhou's Jixiao Shanzhuang—just like gorgeous brocade. (**h**) Ground paving of geometrical pattern—pebbles, gravels and tiles are paved as the background of warm gray color and begonia pattern of cold gray color. (Courtyard of Dian Chun Yi in Wangshi Yuan). (**i**) Paving pattern of diamond flowers in Suzhou's Liu Yuan. (**j**) Paving pattern of sunflowers in Suzhou's Liu Yuan

b

c

d

Fig. 3.282 (continued)

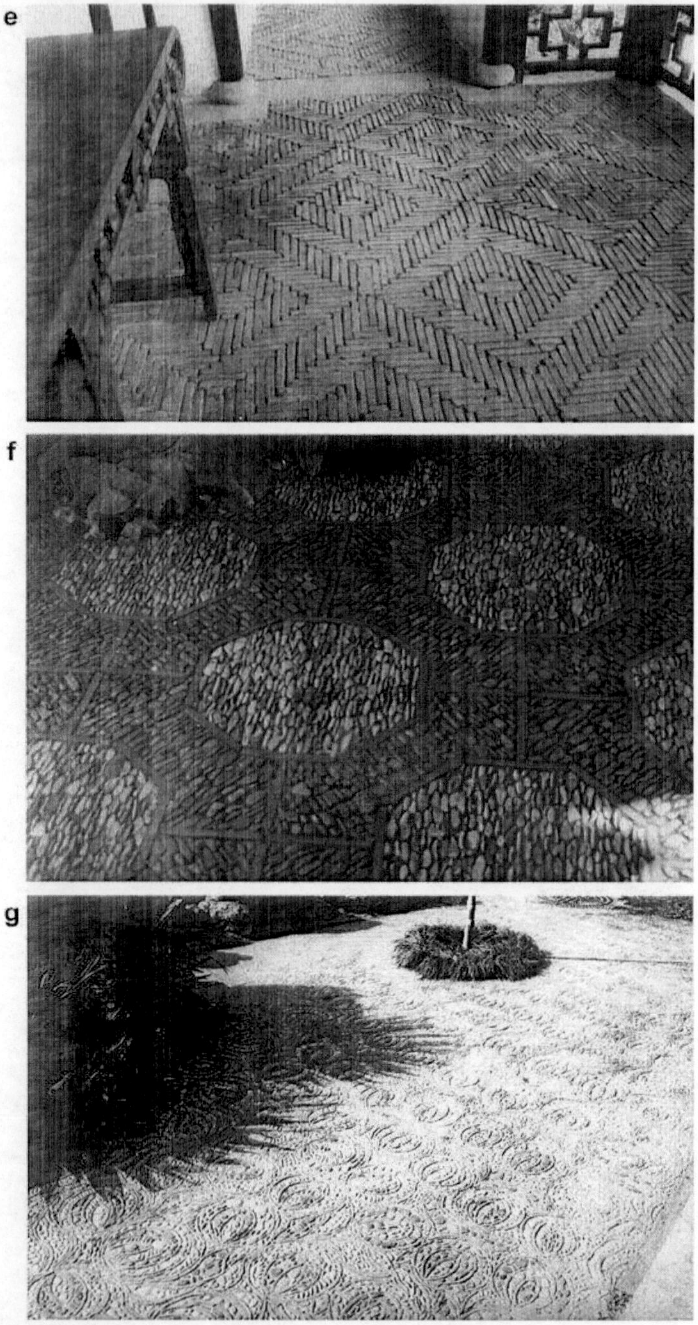

Fig. 3.282 (continued)

Fig. 3.282 (continued)

route, then the function of the latter lies mainly in concealing and revealing. For the sake of simplicity, the two are discussed separately. The previous section focuses on route, while this section focuses on concealing and revealing, which can also be understood as the pictorial structure of scenic imagery of route. Scenery exists according to fixed spatial structure. The effects of visiting gardens are overlapping of the continuous impressions of scenic planes on the retina of visitors. From this, we

a

b

Fig. 3.283 (**a**) Ground paving of animal pattern—Crane pattern paved with gravels, tiles, porcelain pieces, etc. (**b**) Ground decoration of Cornucopia pattern

can see that scenic structure or scenic composition is different from the front view and its sequence that visitors have observed.

3.1.2.2.1 Scenic Front Views and the Pictorial Structure of Scenic Imagery

Scenery is a concept of space and is composed of natural and artificial forms. Therefore, it can be abstracted as two opposite yet unified aspects of form and space. The beauty of scenery includes both form beauty and spatial beauty. Decorative objects displayed in gardens, such as caged birds, potted flowers, fish tanks, stone seats, and local ornaments of landscapes, flowers, trees, and buildings all emphasize form. The spatiality of a garden enables it to have the function of sightseeing. From the perspective of visual enjoyment, a garden is suited for viewing because its landscapes, flowers, trees, pavilions, and other forms are well-spaced. At the same time, even if the primary concern is visual enjoyment, space art is still needed to bring about an enchanting experience.

The so-called pictorial structure of scenic imagery (or structure) is mainly the composition of space. The spatial relationship between the front and back, the left and right, and the top and bottom shaped by specific forms of scenic constituents is based on the viewing route. Therefore, although the location of scenic constituents is fixed, the composition is relative. For a given spatial composition, the perspective effect of front views is different based on different viewing point positions. As the viewing point changes, so do the perspective front views, which is described as "scenic planes are to be progressively revealed to viewers as they move and their perspective shifts". When the visitors' viewing points are moving, the impression of scenery through visitors' eyes forms a continuous sequence of front views. This is concealing and revealing, which constitutes the forms or background of the route.

Therefore, to understand the nature of concealing and revealing, it is necessary to introduce the concept of front view, which is different from scenic imagery. Scenic imagery is three-dimensional, while front views are two-dimensional. What visitors actually see while they are in scenic imagery is front views. Scenic constituents are finite, while front views are infinite. Front views originate from scenic imagery but only under specific guidance conditions, i.e. only through the special interaction between route and concealing and revealing can front views be obtained. The key to the design of garden art lies in the management of scenic structure or the pictorial structure of scenic imagery. The pictorial structure of scenic imagery is relative to the viewing point and its focus is to achieve the design effect of the sequence of front views. To achieve this effect, the path of the corresponding viewing point (the route) should be arranged.

From this, we can see that the key to the creation of a route lies in how to obtain a sequence of front views with artistic effects. That is the selection of the form and background of the route—concealing and revealing, on which the route depends. This is to be realized through the pictorial structure of scenic imagery. The creation of garden art is indeed a process of repeated deliberation. The art of Jiangnan gardens reminds us that we should not consider the arrangement of scenery and route planning in isolation.

3.1.2.2.2 The Montage of Groups of Scenic Imagery: Orderly Arrangement of Static and Dynamic Views

A garden is composed of several groups of scenes, which have a certain structural relationship within themselves and with each other. This relationship is manifested through sightseeing activities and so this relationship is reflected in a series of visual images that visitors see during sightseeing. Therefore, although the positions of the foundation of various scenery in gardens are fixed, their structural relationships are changeable and fluid. In this way, the design of scenic spaces is actually the arrangement of viewing front views. The pictorial structure of scenic imagery is always considered relative to the viewing. The feeling imparted to visitors of garden art from the scenery of a given composition is the sum of a series of overlapping impressions made by front views.

Artistic garden effects like *qu jing tong you*[143] and *huo ran kai lang*[144] cannot be achieved by a single front view. Front views are like scenes in films, the organization of which is like editing movie scenes or creating a montage. Garden art can only display thoughts and interests and produce vivid appeal through a montage of front views. For example, when visiting the Master-of-Nets Garden (Wangshi Yuan) in Suzhou in days gone by, people followed the main viewing route. Visitors enter the residential building through the front gate—the brick gatehouse carved with the Chinese characters "Zaoyao Gaoxiang[145]"—pass through the small courtyard and arrive at the sedan chair hall where the sedan chair is placed. Upon reaching this point, visitors perceive a house space with a rigorous central axis that forms the basis of the environmental impression before you visit the garden.

With this impression, visitors enter the garden through the small door (on the top of the door are four Chinese characters, "Wang Shi Xiao Zhu", or The Fisherman's Small Abode) on the west side of the sedan chair hall. Entering the garden entrance, front views in the inner space of the covered walkway are concealed and revealed by rocks and flowers. Naturally, this refreshes the spirit. Then one enters the main hall, the Small Hill and Osmanthus Fragrans Pavilion (Xiaoshan Conggui Xuan) or its left and right corridors, the overlapping impression of similar front views is still present. In short, the comprehensive feeling of the series of front views after entering the garden reminds visitors of a secluded valley scene. After *feng hui lu zhuan*,[146] a bright front view of a lake appears as the theme scenery of the whole garden. At this time, the spirit of the visitors is again refreshed (Fig. 3.284). The pleasure and the

[143] A winding path leading to a secluded spot.

[144] Suddenly seeing the light.

[145] The name of the gate wishes people a bright and successful future.

[146] Where the peak stands brings the next turn of the mountain path.

Fig. 3.284 (**a**) Carved brick gate in the residence of Suzhou's Wangshi Yuan (Over the gate is inscribed Chinese characters "Zaoyao Gaoxiang"). (**b**) Sedan chair hall of Wangshi Yuan (with the plaque "Qing neng zao da" hanging). The main entrance to the garden is on the right (west) side of the sedan chair hall, and the gate of the garden is inscribed with the Chinese characters "Wang Shi Xiao Zhu". (**c**) The corridoor on the west side of Xiaoshan Conggui Xuan in Suzhou's Wangshi Yuan—after the entrance of the garden gate "Wang Shi Xiao Zhu", there is scenes that conceal of Yungang in the north, which block the principal scenery of the lake. (**d**) Scenery of Suzhou's Wangshi Yuan looking at the lake from southeast

sense of lightness of seeing a series of front views (scenery on the lake) are based on the impression of a secluded valley.

Although the comprehensive treatment provided by the natural shore and background has played a role, the main reason for the rippling and diffusing effect of the lake in a small and closed space is the successful use of montage front views between the garden door (The Fisherman's Small Abode (Wang Shi Xiao Zhu)) and the scenery on the lake. On the contrary, if visitors directly observe the water surface on the basis of the impression of open space outside the garden, they will feel congested and constrained instead of being aware of the open lake and sky. Following the principle of special relativity, applying the contrastive method to organize changes in front views is one of the basic experiences of Jiangnan garden design, which is the concept of *xiao zhong jian da*.[147] The area of Jiangnan gardens is not large. In order to achieve a deep and distant effect in scenic spaces, apart from reducing the scale of the scenes, special emphasis is placed on increasing the layering of the scenic imagery and expanding the depth of field.

The design of front view sequence is a crucial means of completing the process described above. The design of a front view sequence is the combination of the orderly arrangement of static and dynamic views or the arrangement of viewing sequence. From the perspective of the principle of garden art, it is restricted by the relationship of scenic structure, i.e. scenic guidance. Therefore, guidance organizes the orderly arrangement of static and dynamic views in order to change the fixed scenic constituents. The principle of guidance first lies in the route arrangement of the viewing point: up-down, left-right, front-back, far-near, and static-motion. The second lies in the coordination of the concealing and revealing of the background.

The classical gardens in the Jiangnan region are billed for presenting the scenic structure in a zigzagging manner with scenery performing primary and secondary disposition, which used to be seen in writings and paintings. In terms of the primary and secondary relationship, it is typical of Jiangnan gardens to have principal scenery which usually embodies the theme such as the Rocky Stone of the Villa of beautiful surrounds (Huanxiu Shanzhuang) and the Master-of-Nets Garden (Wangshi Yuan) with a lake at its center. The principal scenery of gardens in Jiangnan is usually the landscape in front of the main hall. The above-mentioned chapter has delved into this by detailing the meaning of the main hall serving as a view stop from the perspective of route. As a matter of fact, the primary mountain-and-water scenic imagery itself has view stops. Examples of this include the Snow-Like Fragrant Prunus Mume Pavilion (Xuexiang Yunwei) in the Humble Administrator's Garden (Zhuozheng Yuan) of Suzhou, the Angle Missing Pavilion (Buque Ting) and the Floating Bamboo Pavilion (Fuyun Ge) in the main lake-and-mountain scenery of Yi Yuan in Nanxiang, Tens of Thousands of Lotus Like Half Lake of Brocade (Banhu Yunjin Wanfurong) and the Small Waterside Pavilion of Rain-Listening (Tingyu Xiaoxie) in the main lake-and-mountain scenery of Yi Yuan in Nanxun.

[147] Evoking the large with the small.

Even without pavilions, base points such as stone tables and stools or naturally built stones for those who want to have a rest or sit quietly feasting their eyes are also available. All these view stops are manifestations of the scenic route. From the aspect of concealing and revealing, which is opposite and complementary to the route, visitors can have a view of the combination of front views as the backdrop of the route. Appreciation of a certain scenery from different angles and distances can produce endless front views. The difficult nature of designing each front view well usually boils down to ensuring the good composition of the main front view. In order to ensure the artistic effect of gardens, first of all, the composition of the principal scenery must be perfect in creation. In the principal scenery, perfection should be guaranteed in the composition of the main front view in front of the main view stop.

From the example of Jiangnan gardens, gardens where mountain is the principle structure, as exemplified by the Villa of beautiful surrounds (Huanxiu Shanzhuang) in Suzhou, whose principal scenery is the central stone mountain, its major front view is facing southwest and the major viewing view stop—the hall of Villa of beautiful surrounds (Huanxiu Shanzhuang) and the western veranda—are opposite to the stone mountain by the waterfront corridor. Therefore, priority is given to the composition of the south and west sides of the shape of the stone mountain and plant layout, which features small trees such as crape myrtle that grow obliquely under the mountain peak (Fig. 3.285). Gardens where waters forms the principal view of the landscape, as seen in the Master-of-Nets Garden (Wangshi Yuan) in Suzhou, where the principal scenery is the central lake surface, whose major front view is facing north, with the Watching Pines and Appreciating Paintings Studio (Kansong Duhua Xuan), along with the Prunus Mume Pavilion (Zhuwai Yizhi Xuan), upstairs of the Meditation Study (Jixu Zhai) and the Duck-Shooting Veranda (Sheya Lang) at the opposite side.

From the north to the south, the nearby front view is the rock, the middle plane is the cliff pavilion, and the distant plane is the pavilion on water. Especially the compressed scale of the stone arch bridge at the mouth of lakeside harbour branch, all of these form a complete and far-reaching picture of water village scenery. Mountain-and-water gardens (gardens where mountain and waters have comparable presence), such as the Humble Administrator's Garden (Zhuozheng Yuan), where the principal scenery is a lake-and-mountain scenery embellished with the Snow-Like Fragrant Prunus Mume Pavilion (Xuexiang Yunwei), the lake-and-mountain scenery has a south-facing major front view and a perfect front view with outstanding composition is formed with the Hall of Distant Fragrance (Yuanxiang Tang) at its opposite. Whether the principal scenery is designed successfully influences the artistic effect of the whole garden. A successful work, principally, should have perfect principal scenery.

The principal scenery in mountain-and-water Jiangnan gardens have formed a typical pattern, namely a water surface (usually a lake) is arranged in front of the main view stop (usually a hall). Moreover, mountains and forests are arranged on the other side of the water surface, small pavilions are embellished among the forests, and islands in the water can be set up while rocky beaches are arranged near the shore so as to form the middle front view. Examples include the scenery of the

a

b

Fig. 3.285 (**a**) Huanxiu Shanzhuang, Suzhou, looking south at the main hall from Liang Shi on lake stone mountain—"Huanxiu Shanzhuang" (newly built after restoration). (**b**) The main view of the main hall of Suzhou Huanxiu Shanzhuang, looking at the lake stone mountain from north, has not been restored, so the view is empty and loose. Lagerstroemia indica, which inclines to the west of the mountain and forms an integral composition, has died down. Now, pine trees with a thick crown have been replanted in situ. The required posture and height for the composition and artistic conception of the scene are gone

Snow-Like Fragrant Prunus Mume Pavilion (Xuexiang Yunwei) facing the Hall of Distant Fragrance (Yuanxiang Tang) in the Humble Administrator's Garden (Zhuozheng Yuan) in Suzhou, the Passable Pavilion (Ke Ting) facing the Verdure Mountain House (Hanbi Shanfang) in the Garden that was preserved miraculously (Liu Yuan) of Suzhou, Pavilion of the Fragrance of Lotus Roots (Ouxiang Xie) in

Suzhou's Garden of Pleasure (Yi Yuan) opposite Minor Surging Waves Pavilion (Xiao Canglang), the Surging Waves Studio (Canglang Shuwu) in Hangzhou's Gao Garden (Gao Yuan) opposite Xiang Pavilion (Xiang Ting), Hall of Hiding Beautiful Scenery (Jingmiao Tang) in the Garden of Look Forward (Zhan Yuan) in Nanjing facing the Mountain Pavilion (Shan Ting), and Clean Water and Lush Trees (Shuimu Mingse) in Yi Yuan in Nanxiang facing the Angle Missing Pavilion (Buque Ting), etc.(Fig. 3.286).

This is a basic format. If the depth is limited, the landscape in front of the main view stop will be arranged transversely, such as the arrangement in the Pavilion of Unity with Heaven (Tianyi Ge) in Ningbo, the arrangement of the scenic imagery group of the Fishing Boat (Yu Fang) in the Garden of Benefiting Descendants (Huiyin Yuan) in Suzhou, the arrangement of the scenic imagery group of the Self-Cultivation Hall (Shaoxiu Tang) in Shanghai's Garden of Various Fruits (Jiuguo Yuan), and the arrangement of the scenic imagery group of the Hall of Serenity (Qingxue Tang or Jingguan Ting) and Pavilion of Tall Green Trees (Songcui Ting) in the Inner Garden (Nei Yuan) of Shanghai (Fig. 3.287). This can produce an effect of static viewing based on the major view stop. It is not only the principal-subordinate relationships that can highlight principal scenery but also the different structural relations between the subordinates. In terms of route, the secondary view stops connected in series on the viewing route also correspond to their respective key viewing front views. In the organic structure of scenic imagery, each corresponding relationship is like a key node, through which all scenery is connected into a whole. Therefore, it is very important to grasp this key node of correspondence in garden creation.

The structural properties of key nodes can be divided into two types, which show two corresponding relationships, namely, obstructive scenery and contraposed scenery. As far as design is concerned, they can be understood as two methods of dealing with the separation and connection of scenic spaces, which are opposite and complementary to each other. Obstructive scenery and contraposed scenery are not fixed. For example, the stone mound at the entrance to the Humble Administrator's Garden (Zhuozheng Yuan) is like a screen wall (facing the gate inside a traditional Chinese courtyard) that blocks the view inside the garden and prevents you from having a clear view when you first arrive at the garden. From the perspective of the relationship between the entrance and the scenery in the garden, it is obstructive scenery. As far as this entrance of the garden is concerned, this group of stone mountain scenery is contraposed scenery of the entrance.

From the perspective of the view from the Hall of Distant Fragrance (Yuanxiang Tang), the building on the opposite of the entrance, this stone mountain as the end on the south viewing route is also the contraposed scenery. In the sense that the stone mountain conceals the entrance (the entrance cannot be seen from the Hall of Distant Fragrance (Yuanxiang Tang)), the stone mountain is obstructive scenery (Fig. 3.278). Therefore, obstructive scenery and contraposed scenery are mutually transformed with a change of base point for viewing. However, under certain preconditions, the meaning of obstructive scenery and contraposed scenery is fixed. Therefore, it is necessary to bear in mind these two opposite and complementary

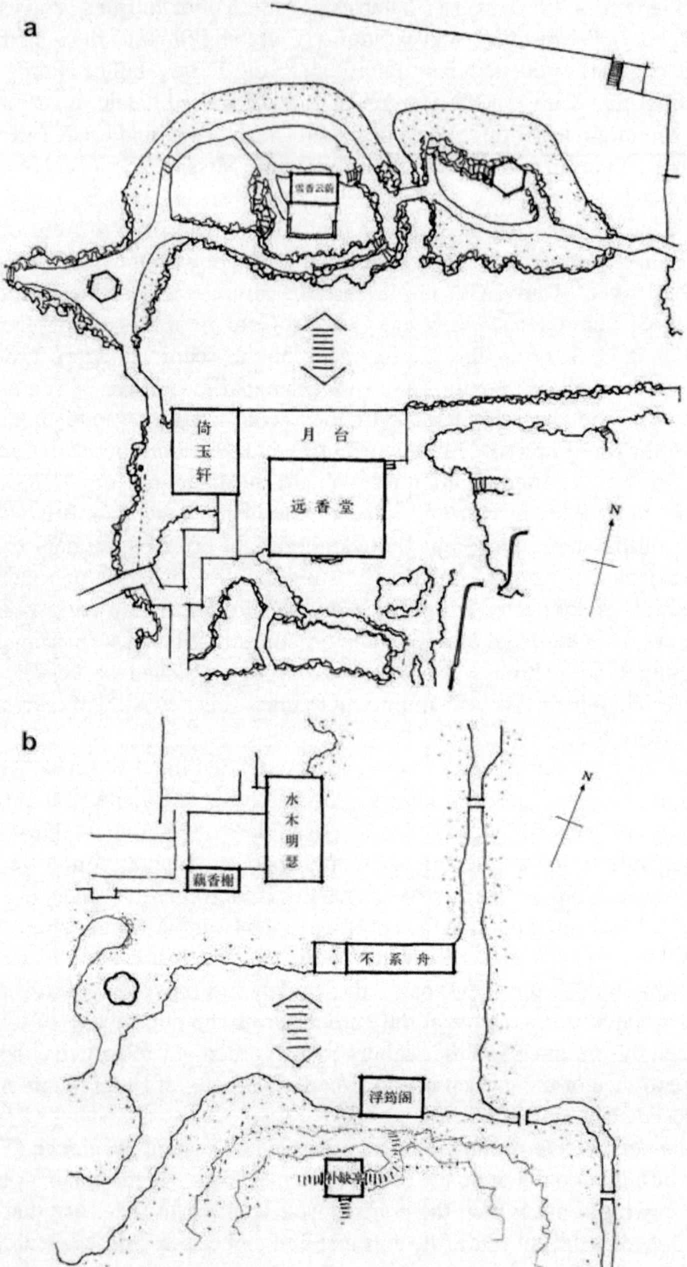

Fig. 3.286 (**a**) The plan of Yuanxiang Tang in Suzhou's Zhuozheng Yuan facing Xuexiang Yunwei. (**b**) The plan of Shuimu Mingse in Yi Yuan of Nanxiang facing Buque Ting

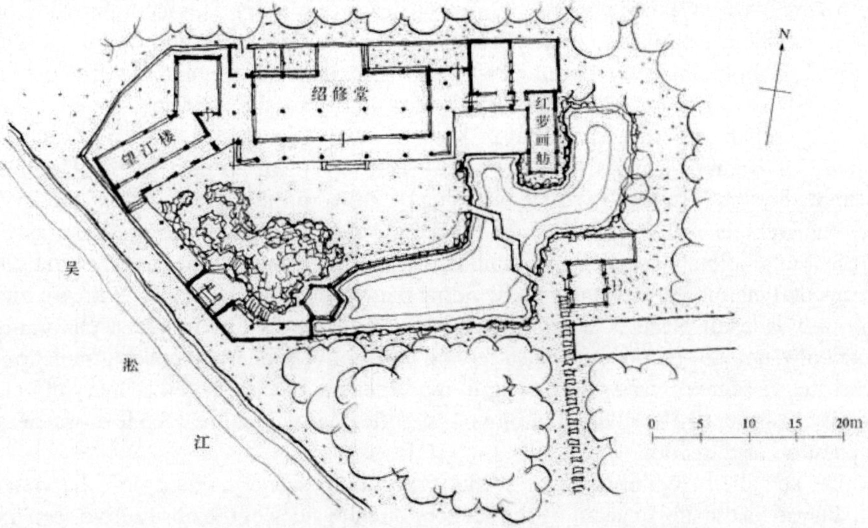

Fig. 3.287 Water and mountain is the principal scenery faced by Shaoxiu Tang of Jiuguo Yuan in Shanghai. (Copied from the illustration of The Catalogue of Jiangnan Gardens)

concepts for design or analysis of garden works. Through the above-mentioned example of stone mountain at the entrance of the Humble Administrator's Garden (Zhuozheng Yuan), it can be realized that the principle that obstructive scenery or contraposed scenery is the key node connecting adjacent scenery (such as the scenery in front and at the back of the stone mountain). In a garden, it is through this key node that all the scenery form an inseparable whole.

3.1.2.2.2.1 Obstructive Scenery

The separation of scenery depends on obstructive scenery. Meanwhile, the transition of scenery (orderly arrangements of dynamic views) often relies on obstructive scenery to "draw in". "Garden in a garden" relates to several groups of scenes with different styles or small gardens which are relatively independent separated by obstructive scenery. In real-life examples of Jiangnan gardens, some obstructive scenery cuts off the line of sight by using the undulation of earth mounds or stone mounds. The examples can be found in the earth mounds between the Mountain-in-View Tower (Jianshan Lou) and Recluse Abode of Bamboos and Chinese Parasol Trees (Wuzhu Youju) in the Humble Administrator's Garden (Zhuozheng Yuan), and the stone mounds between the Thatched Cottage at the City Corner (Chengqu Caotang) and Hall Among the Mountains and Water (Shan Shui Jian) in the Couple's Retreat Garden (Ou Yuan). On the other hand, some obstructive scenery relies on the architecture, namely the halls and other buildings, half corridors, overlapping passageways, or walls, to cut off the line of sight. To be specific, the principal scenery of the Garden of Art (Yi Pu) in Suzhou is separated from the scenery in the southwest corner by a whitewashed wall. Generally, obstructive scenery uses a combination of the ground surface, plants, and/or buildings. For example, trees and buildings (the Snow-Like Fragrant Prunus Mume Pavilion—Xuexiang Yunwei and Pavilion on the North of the Mountain—Beishan Ting) are a part of the aforementioned earth mounds separating the Mountain-in-View Tower (Jianshan Lou) and Recluse Abode of Bamboos and Chinese Parasol Trees (Wuzhu Youju) in the Humble Administrator's Garden (Zhuozheng Yuan). In addition to separating scenery, obstructive scenery is also used for masking borders of gates and walls so as to present unpredictable scenes and limitless interest for visitors. The orderly arrangements of dynamic views, such as "a winding path leads to a seclusion", "enticing one into a wonderful place", "winding path along high peaks", and "a vast panorama suddenly stretches before one", also rely on obstructive scenery to make a transition. Obstructive scenery varies according to requirements of the desired scenery effects and it can generally be divided into two categories: virtual and real, be it for scenery separation or transition.

The so-called "real obstructive scenery" partitions scenery, cutting off the viewing line so that visitors can enjoy the scenery on both sides of the obstructive scenery respectively. This method is commonly used between two groups of scenes that present different tastes and artistic flavors. For instance, mountains and forests partition the principal scenery—the lake area in the Garden for Ease of Mind (Jichang Yuan) in Wuxi from the subordinate scenery—Gully of Eight Sounds (Bayin Jian). In existing examples of Jiangnan Garden, partitioned obstructive scenery is much like the "cutting off without encircling" and "encircling with a gap" described in art of war. Partition can be regarded as a special connection. Therefore, it is reasonable to partition or cut off the viewing line and route in one direction instead of encircling them without a gap. Even if a scenic space needs to be enclosed, there must be a gap which is a hint about the scenery. In other words, in partitioned obstructive scenery, it is necessary to leave an inducing point to connect the scenes, be it from the perspective of a section or a single point, which is similar to the device of foreshadowing used in literature. The fragmented whitewashed wall of Three Ponds Mirroring the Moon (Santan Yinyue) of Hangzhou's West Lake is an approximate diagrammatic example of "cutting off without encircling", with an openwork window serving as an inducing point (Fig. 3.288). The fragmented water ornamented wall in the Garden to Delight One's Parents (Yu Yuan) in Shanghai is also designed in this way. There, the arched hole under the water ornamented wall, together with the hole's reflection on water, forms a moon shape (which on a scenic level can give an impression of journeying through on a boat). The moon-shaped hole serves as an inducing point, which foreshadows the scenery on the other side of the wall (Fig. 3.202a, b). A space, facing with the principal scenery—lake-and-mountain composition, is enclosed by cloud hill-climbing wall and earth mounds under the Embroidered Silk Pavilion (Xiuqi Ting) in the Loquat Garden (Pipa Yuan) of the Humble Administrator' s Garden (Zhuozheng Yuan) in Suzhou. A moon gate on the cloud hill-climbing wall, providing a fan-shaped view on both sides of the gate, serves as an inducing point (Fig. 3.289). Meanwhile, the Embroidered Silk Pavilion (Xiuqi Ting) on earth mounds, connecting the interior

Fig. 3.288 Whitewashed wall in Santan Yinyue in Hangzhou's West Lake

Fig. 3.289 The Pipa Yuan of Suzhou's Zhuozheng Yuan is parted by yunqiang, leaving the moon gates for walking and viewing

and exterior of the Loquat Garden (Pipa Yuan) by mountain paths, serves as another interesting inducing point. The Loquat Garden (Pipa Yuan) is created relatively independent from the whole garden by partitioned obstructive scenery, while connected to the whole garden by means of the two aforementioned inducing points. At the same time, its low cloud hill-climbing wall, opposite buildings as well as separating walls and forest crown are also types of inducing treatment that foreshadow the adjacent scenery.

The so-called virtual obstructive scenery enables permeability between scenery. This is an arrangement that appears to be separate and yet permeable, leaving adjacent scenes with a vague connection. The arrangement is mostly used to divide two groups of scenes with a similar appeal, or scenes that serve as a background for the opposite one without direct access. If the obstructive scenery is formed by architecture, continuously arranged openwork windows are often used. The permeability of scenery is controlled by the number of openwork windows and complexity of patterns. In Suzhou's Garden that was preserved miraculously (Liu Yuan), the arrangement between lake-and-mountain scenery and the area from the Interwined Old Trees (Gumu Jiaoke) to the Winding Stream Tower (Quxi Lou) is a prime example (Fig. 3.107a). When virtual obstructive scenery with architecture emphasizes more on permeability, half corridor with continuous hole windows is used to separate space, as is shown in the Garden for Ease of Mind (Jichang Yuan) in Wuxi. Sparse forests and bamboo curtains are often used in virtual obstructive scenery with vegetation, such as the bamboo curtains between the Exquisite Emerald Bamboo House (Cui Ling Long) and the Five-hundred-Sages Temple (Wubai Mingxian Ci), as well as between the Bamboo House and Mountain Outlook Tower (Kanshan Lou) in Suzhou's Garden of Surging Wave Pavilion (Canglang Ting) (Fig. 3.290). However, when there is only one or two openwork windows (leading to an increase in wall surface) in the above-mentioned architectural arrangement, namely when a lower limit of the permeability of obstructive scenery is reached, it actually becomes an inducement to partitioned obstructive scenery. Alternatively, if arranged with densely planted trees and shrubs, the permeable obstructive scenery is similar to partitioned obstructive scenery. When a half corridor with an openwork window on its wall side functions as obstructive scenery, its increasing permeability will surpass that of hole windows until it becomes an open corridor, namely an upper limit of the permeability of obstructive scenery is reached. In that case, virtual obstructive scenery has transformed into a connecting factor. As such, the colonnades and lintels of open corridors form a series of picture frames, thus strengthening the visual connection between scenes on both sides of the corridor. In terms of vegetation, an upper limit of permeability is reached when permeability is increased by arranging only a few bamboos and trees, which actually merges plants into two adjacent groups of scenes and forms a front or medium scene.

3.1.2.2.2.2 Contraposed Scenery

Obstructive scenery and contraposed scenery are opposite and complementary. The former relates to concealing and can be divided into real and virtual categories as an

a

b

Fig. 3.290 (**a**) Continuous hole windows in the half corridor of the Jichang Yuan in Wuxi, which produces the effect of separation and connection—virtual barrier. (**b**) Bamboo curtain barrier in Cui Ling Long, Canglang Ting, Suzhou element of separation, while the latter relates to revealing and can be classified into two categories of loose correspondence and close correspondence as an element of connection. The contraposed scenery from fixed viewing angles form a close correspondence. We begin with the principal scenery which corresponds to a principal hall. A close correspondence refers to a clear relationship (also known as the axial-line relationship) between a viewing point and scenery in sight. Generally, a complete composition from the angle of a hall is a prime example. This contraposed scenery is always found within the frames of colonnades, lintels, lattice partition doors, hangings, and railings of a building, which enables the contraposed scenery to be more clear and concentrated. For example, contraposed scenery can be found in the lake-and-mountain composition of the Snow-Like Fragrant Prunus Mume Pavilion (Xuexiang Yunwei) to the north of the Hall of Distant Fragrance (Yuanxiang Tang) in Suzhou's Humble Administrator's Garden (Zhuozheng Yuan), the Passable Pavilion (Ke Ting) to the north of Verdure Mountain House (Hanbi Shanfang) in Suzhou's Garden that was preserved miraculously (Liu Yuan), and Minor Surging Waves Pavilion (Xiao Canglang) to the north of Pavilion of the Fragrance of Lotus Roots (Ouxiang Xie) in Suzhou's Garden of Pleasure (Yi Yuan). Contraposed scenery matches a painting and varies in size. As such, the contraposed lake-and-mountain scenery of the aforementioned principal halls is relatively large. Meanwhile, subordinate halls also have correspondence to the principal scenery from different angles. For example, the contraposed scenery of Boat Shaped Studio (Huafang Zhai) can also be found between Spiral Hairdo Pavilion (Luo Ji Ting) and the west of principal scenery of lake-and-mountain composition in the Garden of Pleasure (Yi Yuan) (Fig. 3.291). However, most of the contraposed scenery is small-scale with some even being stone peaks or planting beds, such as the contraposed scenery of the southern and northern yard of the Five Peaks Library (Wufeng Shuwu) and scenery of the backyard of the Cloud Stairway Room (Tiyun Shi) in Suzhou's Master-of-Nets Garden (Wangshi Yuan) (Fig. 3.292).

In a general way, doors and windows of a building can be viewed as picture frames, thus, contraposed scenery is arranged in their corresponding positions, be it far or near. The White Pagoda (Bai Ta) and Five Pavilions Bridge (Wuting Qiao) in the distance can be seen from the hole doors and hole windows of the Mid-lake Pavilion (Huxin Ting) of the Slender West Lake in Yangzhou, which is an intentionally designed contraposed scenery. Small-scale contraposed scenery includes the leaning black pine to the south of the Prunus Mume Pavilion (Zhuwai Yizhi Xuan)'s moon gate in the Master-of-Nets Garden (Wangshi Yuan), the stone peaks stacked by Taihu stones in contraposition to the right door of the Knowing-Fish Waterside Pavilion (Zhiyu Jian) in Wuxi's Garden for Ease of Mind (Jichang Yuan), the opposite moon gates in Suzhou's Garden of Art (Yi Pu), as well as the plantain trees, Chinese parasol, and stone peaks outside the hole doors and hole windows in the east of Garden that was preserved miraculously (Liu Yuan) (Fig. 3.293).

Loose contraposed scenery refers to the objects of viewing from base points like pavilions or corridors (with seats), where visitors can stop to rest. These scenes, be it far or near, have their own particular appeal. But from the perspective of the viewing point, a complete scene with clear correspondence is not necessarily formed. This

a

b

Fig. 3.291 (**a**) Huafang Zhai in Suzhou's Yi Yuan—a base point of static viewing. (**b**) The scene in contraposition of Huafang Zhai in Suzhou's Yi Yuan—Luoji Ting set against by the cliff and crag beside the lake

a

b

Fig. 3.292 (**a**) The scene in contraposition of planting beds and stone peaks in the back yard of Tiyun Shi in Suzhou's Wangshi Yuan (Current Situation). (**b**) The scene in contraposition of the back yard of Tiyun Shi in Suzhou's Wangshi Yuan in1960s

type of contraposed scenery is similar to a handscroll, more often than not presenting well-designed compositions. Visitors are free to choose a viewing position in these loose contraposed scenes and can always enjoy satisfactory scenic planes in a fixed field of vision.

From the perspective of in-motion viewpoints, contraposed scenery, as the head-on sight on a main route, is often arranged at the turn of the route. Such contraposed scenery may be a pavilion, a peak, a tree, or a clump of bamboos, which often has a clear and close correspondence, such as the flowering shrubs planted at the turning point of the covered walkway in Suzhou's Crane Garden (He Yuan) (Fig. 3.294).

When you walk in the covered walkway and gaze around the garden, the hole windows are often applied to guide viewing lines and organize contraposed scenery as accompanies by hole doors.

Like obstructive scenery, contraposed scenery is also a relative concept. On one hand, these two scenes can be converted into each other based on viewing positions. On the other hand, contraposed scenery and its corresponding viewpoint can also be converted. However, a principal viewpoint always predominates in this reversible relationship. For example, the Hall of Distant Fragrance (Yuanxiang Tang) in Suzhou's Humble Administrator's Garden (Zhuozheng Yuan) is regarded as a main viewpoint while the mountain-and-forest composition where the Snow-Like Fragrant Prunus Mume Pavilion (Xuexiang Yunwei) is located is treated as contraposed scenery. In garden designs, the contraposed scenery with a dominant party is crucial for the overall layout.

3.1.2.2.2.3 Decorative Scenery

Obstructive and contraposed scenery are two basic parts of concealing and revealing which determine the guidance, while decorative scenery is embellishment in a fixed

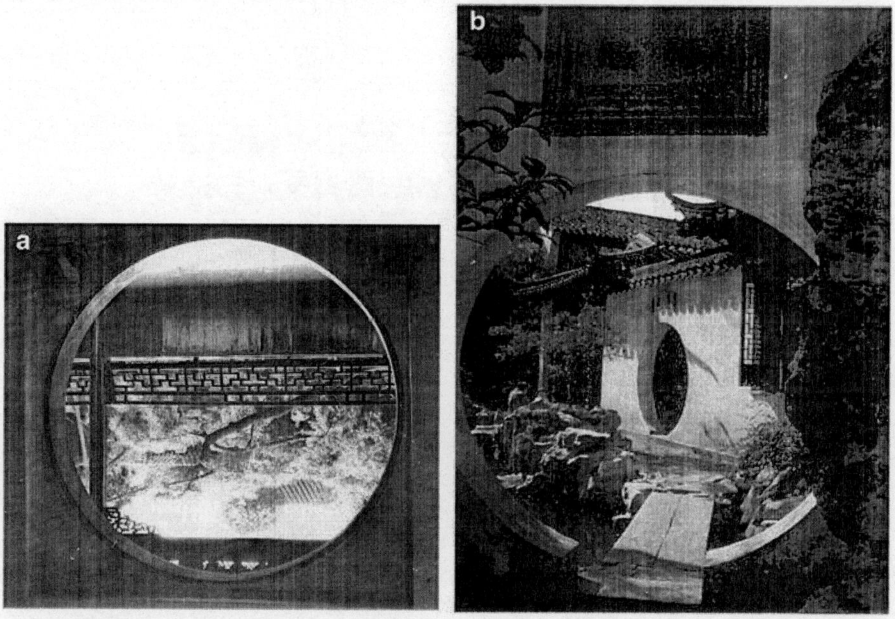

Fig. 3.293 (**a**) A black pine leaning south of the Xuanyue Cave in the Bamboo Garden of the Master of the Nets in Suzhou-the combination of the scene and the title of the Xuan, makes the scene full of poetry like "the thousand trees at the head of the river want to be dark in spring, and a branch outside the bamboo is better". (**b**) The moon gate of "Yuou" facing the moon gate of "Qinlu" in Suzhou's Yi Pu—scenes in contraposition. (**c**) The door of Yujie at the rightside of the entrance hall in the Jichang Yuan garden in Wuxi facing the stone peaks in Bingli Tang

Fig. 3.293 (continued)

overall guidance. In other words, decorative scenery is icing on the cake, playing a role in making up for empty space and enriching dull, inanimate corners of room or space. Decorative scenery, most of which is for close viewing, decorates a partial space and generally avoids affecting the overall picture. It usually consists of flowers, trees, stalagmites, stack stones, stone peaks, and carvings, as well as the small-scale shaping of the ground surface. As for its site, it is typically placed in narrow lanes around houses, small courtyards inside a house, or vacant spaces at the roadside and in the corners (Fig. 3.295). Some decorative scenery relies solely on vegetation. For example, palms and banksian roses are used to fill the small space at the turn of the half corridor near Pavilion of Pagoda Reflection (Ta'ying Ting) in the west of Suzhou's Humble Administrator's Garden (Zhuozheng Yuan). It sets off the space and distance between the half corridor and the wall, creating a delusive depth of field which is larger than that in reality. The delusion that you are walking through an open corridor surrounded by flowers and trees breaks the imbalance of the half corridor (Fig. 3.296). Some decorative scenery relies on plant layout together with

Fig. 3.294 Flowering Shrubs are arranged at the turn of the covered walkway in Suzhou's He Yuan as a scene in contraposition.

planting beds of lake stones or stalagmites. For example, in the empty space at the turn of the half corridor beside Yam Pavilion (Yuyan Ting) in Suzhou's Garden of Pleasure (Yi Yuan), only bamboos and moss covering the gravel are used, which reveal a simple, elegant, and rustic appeal (Fig. 3.297). Some decorative scenery relies on vegetation together with stack stones and a slight undulation, such as the scenery between right curved corridor of the Distant Green Tower (Yuancui Ge) and the garden wall, and the scenery at the turning point of the right corridor of Veranda of Scenting the Sweet Osmanthus Fragrance (Wen Muxixiang Xuan) and cloud hill-climbing wall in Suzhou's Garden that was preserved miraculously (Liu Yuan) (Fig. 3.298).

The composition of decorative scenery can be divided into both two- and three-dimensions. As is noted in *The Craft of Gardens* (Yuan Ye), "The whitewashed surface acts as paper and the rocks as the painting upon it" (Quoted from *The Craft of Gardens* (Yuan Ye) by A.M.H. p. 109.). Flowers and stones are set close to walls, just like two-dimensional painting, and viewing in one direction is limited when

Fig. 3.295 The scene that expose at the backyard corner of Jixu Ge in Suzhou's Wangshi Yuan—pictorial structure of scenic imagery is formed by high and low stalagmites and wall folding lines.

necessary. A typical example is walls in the decorative scenery adjacent to the Yam Pavilion (Yuyan Ting) that encapsulate framed scenery. What's more, two-dimensional decorative scenery can be found in scenery from the hole windows on curved corridors which stretch along a garden wall in Suzhou's Garden of Natural Delight (Chang Yuan) and Half Garden (Ban Yuan) (Fig. 3.299). Three-dimensional decorative scenery presents a spatial composition from different angles. For example, A Scene in the Garden that was preserved miraculously (Liu Yuan) (Hua Bu Xiao Zhu) provides 180 degrees of viewing range from the left, right, and center planes of view. Therefore, composition should take spatial relationship into consideration for integration of multiple scenic planes (Fig. 3.300). Parterre set with stones and flowers is used to set off and decorate the deadwood (which is similar to a scene of tree stump) in Wuxi's Temple of Mount Hui (Huishan Si). This is an example of three-dimensional decorative scenery as it can be appreciated following a route that circles the parterre.

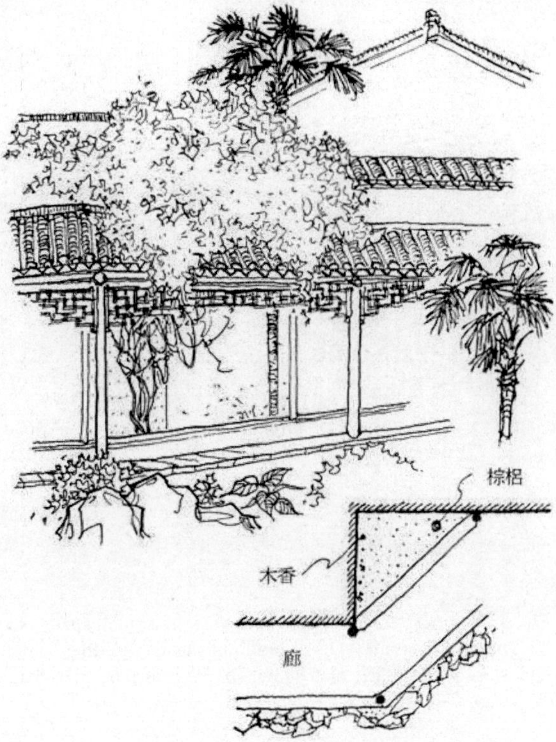

Fig. 3.296 Scenes that expose formed by only plants—the scene that expose of palm and radix aucklamdias in the half corridor near Ta'ying Ting of Bu Yuan in Suzhou's Zhuozheng Yuan

From the perspective of contents of decorative scenery, it can be divided into impromptu and titled decorative scenery. The former relates to garden owner's decoration on a whim. This type of decorative scenery includes many flowers, trees, bamboos, and stones in Yard of Rocks (Shilin Xiaoyuan) in Suzhou's Garden that was preserved miraculously (Liu Yuan), the flower bed for the deadwood in Wuxi's Temple of Mount Hui (Huishan Si), as well as scenery adjacent to Yam Pavilion (Yuyan Ting) in Suzhou's Garden of Pleasure (Yi Yuan). With owners' tastes changing constantly, impromptu decorative scenery has great variability, especially in terms of plants. For instance, in Suzhou's Master-of-Nets Garden (Wangshi Yuan), a flowering peach was planted after the 1960s under a hole window in the west of the Prunus Mume Pavilion (Zhuwai Yizhi Xuan) (this originally thematic plant scattered in the south of the pavilion). Titled decorative scenery has attributes of an independent scene and consists of a group of small-scale scenes with inscriptions of titles or names. Aforementioned A Scene in the Garden that was preserved miraculously (Liu Yuan) is an example. The Interwined Old Trees (Gumu Jiaoke) in the garden can be regarded as a three-dimensional titled decorative scenery. The nature of scenic spots and historic sites makes it an important constituent that affects the garden structure. Therefore, it is more appropriate to treat it as an independent scenic space in whole garden structure. The example of the Interwined

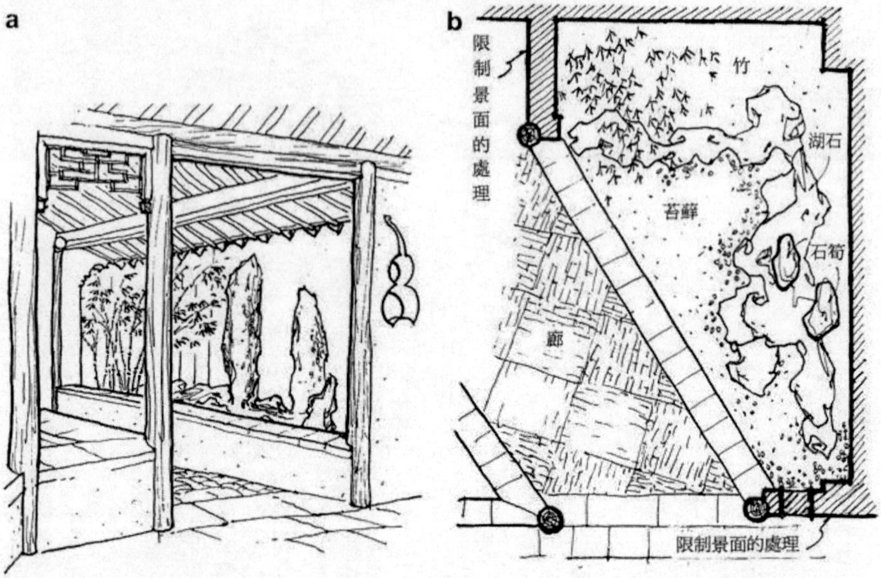

Fig. 3.297 (**a**) Second-degree spot view of a small space near Yuyan Ting half gallery in Suzhou's Yi Yuan (with small walls at both ends as the guiding limitation), with bamboo and gravel paving and moss forming the spot view. (**b**) Plan of the scene that expose near Yuyan Ting in Suzhou's Yi Yuan

Old Trees (Gumu Jiaoke) reveals that once the upper limit of decorative scenery is exceeded, the scenery can exert an influence on garden structure. If handled skillfully, a tiny decorative scene will become indispensable to the overall picture. In creation and management, an outstanding decorative scene, even if designed on a whim, should be consistent with the appeal of the scenery where it is located. Only in this way can the unity between decorative scenery and environment be maintained.

3.1.2.2.3 Intensification of the Depth of Scenes: Borrowed Scenery

In his treatise on garden making, *The Craft of Gardens* (Yuan Ye), a famous garden maker Ji Cheng of the late Ming dynasty put forward "borrowed scenery". In *A Philosophy of One's Own*, Li Yu of the early Qing Dynasty, who was erudite in garden-making as well, suggested that "the essence of finding a scene lies in borrowing of scenery", which contributed a great deal to the theories of classical garden making. Borrowed scenery is an important principle for intensifying the depth of a scene. Through the borrowing of scenery, the garden design, confined to a fixed space, is going to enhance the artistic effect of the garden based on the surrounding environment of its site as well as the weather conditions whose favorable factors are made full use of.

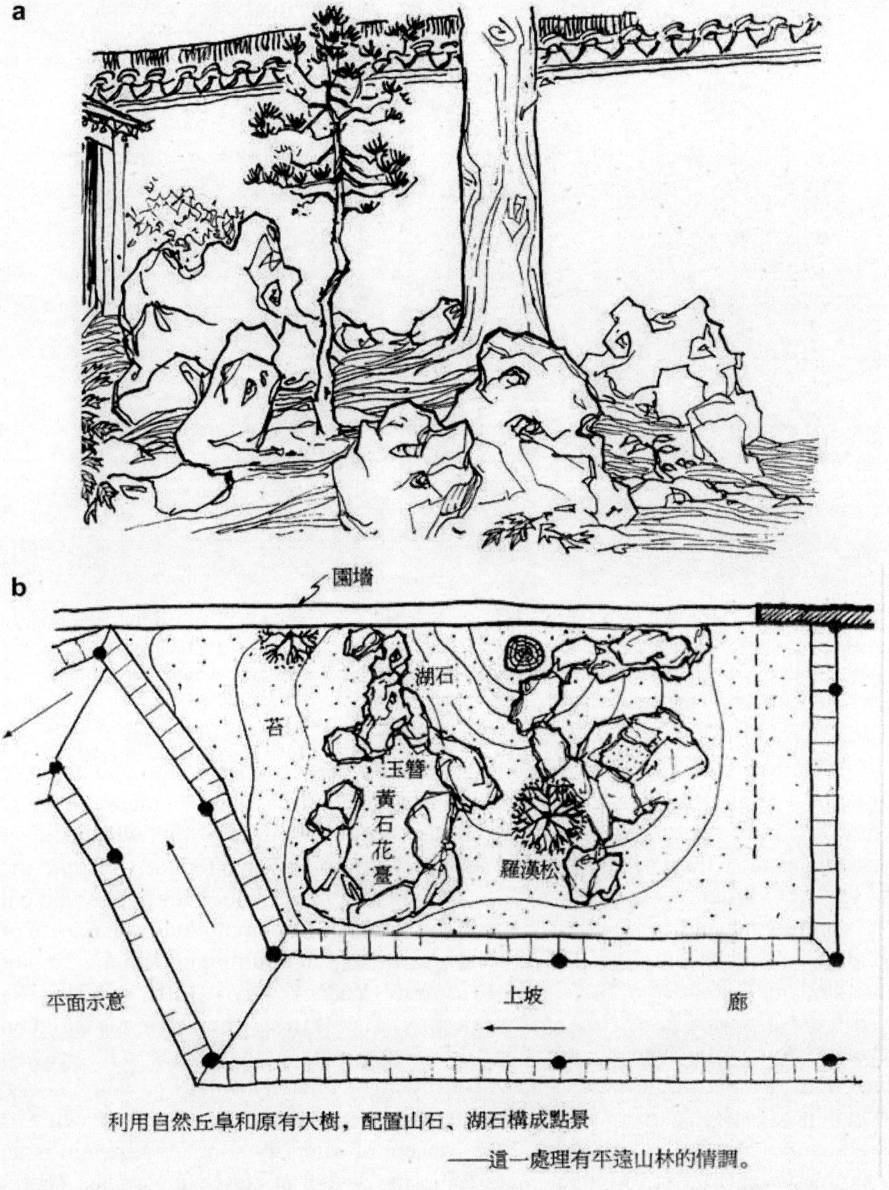

Fig. 3.298 (**a**) The scene that expose between the right curved corridor and garden wall of Yuancui Ge in Suzhou's Liu Yuan. (**b**) Plan of the scene that expose between the right curved corridor and garden wall of Yuancui Ge in Suzhou's Liu Yuan

Fig. 3.299 (**a**) The space between the half corridor and the garden wall is mainly composed of stacked stones to form a scene that expose (Ban Yuan of Suzhou). (**b**) The scene that expose composed of a single plant—sparse bamboo. The space of acute angle defines the guidance for viewing the scene that expose. (Suzhou's Yi Yuan)

According to Ji Cheng, "To borrow from the scenery means that although the interior of a garden is distinct from what lies outside it, as long as there is a good view you need not be concerned whether this is close by or far away, whether clear mountains raise their beauty in the distance or a purple-walled temple rises into the sky nearby. Wherever the view within your sight is vulgar, block it off, but where it is beautiful, take advantage of it; never mind if it is of just empty fields, make use of it all as a misty background. This is what known as skill in fitting in with the form of the land". (Quoted from *The Craft of Gardens* (Yuan Ye) by A.M.H. p. 121) (*The Craft of Gardens, The Theory of Construction*) That is to say, borrowing is based on the site of a garden. Meanwhile, favorable factors outside the garden are borrowed and used as a part of interior scenery. Borrowing is an extension of garden scenery, which is to expose gardens (an artificial environment simulating nature) to nature as well as the artificial environment. The concept of integrity contained therein is an extremely valuable and fundamental idea in the design of classical gardens. Therefore, Ji Cheng concluded that "Making use of the natural scenery is the most vital part of garden design" (Quoted from *The Craft of Gardens* (Yuan Ye) by A.M.H. p. 121).

What is borrowed is scenery outside a garden in the process of designing the interior scenery. Therefore, the connection between interior scenery cannot be designated as borrowing. Ji Cheng's explanations of borrowed scenery are limited

Fig. 3.300 (**a**) A scene that expose—a scenic plane viewing from the side of "Hua Bu Xiao Zhu" in Suzhou's Liu Yuan (current situation). (**b**) The spot scene with stalagmites and sparse bamboos as main factor (Suzhou's Yi Yuan). (**c**) The spot scene near Zhuoying Shuige in Suzhou's Wangshi Yuan

to a rhythmical prose style and yet consistently vague in content. The "Everywhere float drifting petals and the drowsy threads of willows" (Quoted from *The Craft of Gardens* (Yuan Ye) by A.M.H. p. 120), "You can gaze at the bamboo by a bend in the stream, and watch a fish from the banks of the Hao" (Quoted from *The Craft of Gardens* (Yuan Ye) by A.M.H. p. 121) and other descriptions probably refer to scenes in a garden. The so-called "Beyond the half-open window lies the emerald shade of plantains and paulownias, and vines and creepers spread their turquoise over the surrounding wall" (Quoted from *The Craft of Gardens* (Yuan Ye) by A.M.H. p. 121] evidently has nothing to do with borrowing of scenery. Therefore, it is necessary to scientifically clarify borrowing of scenery proposed by Ji Cheng and define the important concept.

According to Ji Cheng, there are five categories such as borrowing from afar, from nearby, from above, from below, and in relation to time. The so-called "there comes into view a skein of white egrets" (Quoted from *The Craft of Gardens* (Yuan Ye) by A.M.H. p. 121) and "you raise your glass and hail the bright moon" (Quoted from *The Craft of Gardens* (Yuan Ye) by A.M.H. p. 121) are not so much "borrowing from above" as "borrowing in relation to time". Meanwhile, it is better to treat "lean over the stream and enjoy the moon" (Quoted from *The Craft of Gardens* (Yuan Ye) by A.M.H. p. 121) and "the insects cry, hidden in the grass" (Quoted from *The Craft of Gardens* (Yuan Ye) by A.M.H. p. 121) as "borrowing in relation to time" rather than "borrowing from below". Therefore, these five kinds of borrowing can be summarized into three types, which are borrowing from afar, from nearby, and in relation to time. Borrowed scenery serves as a principle of structural relations of scenes, therefore, its classifications should be based on the relations. For example, borrowing from afar or nearby can be distinguished from the perspective of spatial relations. As such, borrowing scenery in relation to time is based on scenic appeals caused by time and weather conditions, such as the change of seasons, passing of time, or variations on climate which are also objective laws. Borrowing from above and below are distinguished from visitors' subjective actions. A row of flying egrets as a result of changing of seasons or a bright moon shown due to celestial bodies can constitute a background of garden scenery, adding to the pleasure of natural aesthetics. Therefore, it can be viewed as the coloring of scene by time and weather conditions. In this regard, borrowing scenery in relation to time can better disclose the principles than borrowing from above. By the same token, looking down at the moon shadow in the water or chirping insects in the grass should be viewed as "borrowing scenery in relation to time" instead of "borrowing from below". Borrowing scenery in relation to time is a supplement and foil of time and weather conditions to scenery as well as a factor to strengthen artistic conception. This problem will be discussed in "On the Creation of Scenic Imagery". Borrowing from afar and nearby are to arrange a guidance inside the garden to enjoy borrowed objects outside the garden. In other words, these methods manage a guidance to intensify the depth of field. In this way, it can be made clear that intensifying the depth of field serves as a function of borrowed scenery, and borrowing from afar and nearby as methods.

Generally, borrowed objects, be it in borrowing from afar or nearby, include mountains, water, and all natural and pastoral scenery, or scenic spots such as temples and pagodas, or other artificial garden scenery.

3.1.2.2.3.1 Borrowing From Nearby

Borrowing from nearby is to borrow scenery adjacent or very close to garden site. If the borrowed scenery is relatively high, garden wall is commonly masked so that exterior scenery appears to be involved in the garden. Concealment includes dense trees, piled mountains, stack stones, or architecture such as half corridors stretching along walls. Take Mount Hui outside the Garden for Ease of Mind (Jichang Yuan) in Wuxi as an example, the mountain-and-forest composition involving Gully of Eight Sounds (Bayin Jian) is used to hide the garden walls. In that case, Mount Hui seems to be in the garden from the viewpoint of the Knowing-Fish Waterside Pavilion (Zhiyu Jian) in the principal place of the lake, and the mountain-and-forest composition reaches the lakeshore directly, giving the appearance of being a lower part of Mount Hui (Fig. 3.301). Another example is the Second Garden (Er Yuan), which

Fig. 3.301 Wuxi's Jichang Yuan borrowing from near-by Huishan Mountain, in which earth mounds are built, stones are stacked, and plants are disposed to shield the garden wall, so as to connect Huishan Mountain outside the garden with the mountain forest inside the garden.

abuts on the western part of Suzhou's Humble Administrator's Garden (Zhuozheng Yuan)—the Subsidiary Garden (Bu Yuan) during the late Qing dynasty. Arrangements in the Second Garden (Er Yuan) borrow scenery from its adjacent garden. Pavilions and corridors are used to mask property line walls in all directions (thus, the existing property line walls' thickness is twice its original one and the "Moon Gate" is like a pipeline, which unexpectedly creates an effect of deep seclusion), thus tall pavilions, halls and trees in the adjacent garden appear as though they are another scenic spots inside the Second Garden (Er Yuan). In addition, another method of borrowing from nearby relies on pavilions in a forest-and-mountain composition to allow the appreciation of more scenic planes in neighboring gardens. For example, the Good-For-Both-Families Pavilion (Yiliang Ting) in the Subsidiary Garden (Bu Yuan) is designed to overlook the scenery in the Humble Administrator's Garden (Zhuozheng Yuan) (Fig. 3.302). "Yi Liang" is adapted from Bai Juyi's poem, "As neighbors, our courtyards enjoy the same bright moonlight and same spring trees", which reflects precisely "borrowed scenery".

For borrowing lower neighboring scenes, such as water, guidance along the water is required. The Garden of Surging Wave Pavilion (Canglang Ting) in Suzhou, for example, uses waterside pavilions and corridors to borrow waterscape outside the garden (Fig. 3.303). Garden of Fan Li (Li Yuan) and Garden of Fish Pond (Yu Zhuang), which were built along the bank of Taihu Lake in modern times in

Fig. 3.302 Viewing the Yiliang Ting in Bu Yuan from Suzhou's Zhuozheng Yuan

Wuxi, employ dikes, bridges, pavilions, and corridors that extend into Taihu Lake to integrate the scenery. Adjacent to West Inner Lake (Li Xi Hu), the Villa of Fenyang (Fenyang Bieshu) overlooks scenery of the lake by Tower for Admiring Su Dongpo (Jingsu Ge), and wharves and docks are set up along the lake to provide with recreational activities in the borrowed scenery. Another example is Mountain Villa of Red Oak (Hongli Shanzhuang) which abuts on Viewing Fish at Flower Pond

Fig. 3.303 Pavilion, roofed corridor, windowed veranda are built in Canglang Ting of Suzhou, borrowing from the near-by waterscape outside the garden.

(Huagang Guanyu) of the West Lake. Waterside pavilions and curved corridors are used to borrow scenery of the lake. In Suzhou, the East Garden of the Couple's Retreat Garden (Ou Yuan) employs a different method to borrow waterscape of the inner moat outside the garden. With the riverway playing a role of a street, the ground floor is not accessible to the river to prevent from noise and danger. Thus, Tower for Listening to Splashing Oars (Tinglu Lou) is built in the southeast corner of the garden to listen to the sculling sound of ships that pass by and overlook the river. In Shanghai, Viewing Waves Tower (Guantao Lou) in the Inner Garden (Nei Yuan) and River-Viewing Tower (Wangjiang Lou) in Garden of Various Fruits (Jiuguo Yuan) are designed with the same approach, with an original purpose of overlooking the Huangpu River (in modern times, high buildings obstruct the view of the Huangpu River, even if one is standing on the third story of Viewing Waves Tower (Guantao Lou) (Figs. 3.85b and 3.304).

3.1.2.2.3.2 Borrowing From Afar

Borrowing from afar is to enjoy exterior scenery that is far away from the garden. The specific methods are either to leave a clear viewing line in the corresponding positions, or to set pavilions or halls with high viewing points. In terms of the former, for instance, in Wuxi's Garden for Ease of Mind (Jichang Yuan), as the contraposed scenery of original principal hall—the Hall of Tea (Jiashu Tang) (a teahouse now

Fig. 3.304 Building of Tinglu in East Garden of Suzhou's Ou Yuan, borrowing from near-by water senery of city moat and boats

occupies the historic site of this hall), Embracing Beauty Lake (Jin Hui Yi) and its opposite pavilions and corridors are surrounded by sparsely-planted trees. Thus, the distant pagoda (Longguang Fota) on the Mount Xi is integrated into the scenery as a background. In fact, this borrowed scenery can be seen from many places in the garden (Fig. 3.305). Another example is in Suzhou's Humble Administrator's Garden (Zhuozheng Yuan). The contraposed scenery to the west of the Embroidered Silk Pavilion (Xiuqi Ting) and Recluse Abode of Bamboos and Chinese Parasol Trees (Wuzhu Youju) is permeable enough to connect distant North Temple Pagoda (Beisi Ta), thus forming a far-reaching background of a Buddhist temple (Fig. 3.306). In Changshu, Garden for the Vacant Mind (Xu Guo Ju) and Garden of Zhao (Zhao Yuan) adopt low corridors to connect scenes of Mount Yu. In terms of the latter, outstanding examples of borrowing from afar can be found in the following gardens. Tiger Hill (Hu Qiu), Jiechuang Si Temple (Monastery of Ordination Pennant) and southwestern mountains can be viewed from the Mountain-in-View Tower (Jianshan Lou) in the Humble Administrator's Garden (Zhuozheng Yuan), Pavilion of Supreme Delight (Zhile Ting) and the Free Roaring Pavilion (Shuxiao Ting) in the Garden that was preserved miraculously (Liu Yuan), and Mountain Outlook Tower (Kanshan Lou) in the Garden of Surging Wave Pavilion (Canglang Ting). Huangpu River could be appreciated from River Outlook Pavilion (Wangjiang Ting) in Shanghai's Garden to Delight One's Parents (Yu Yuan) in days past. In addition, a hall in Garden of Xiling Seal-engravers' Society (Xiling Yinshe) on Hangzhou's Gushan Mountain can overlook the West Lake (Fig. 3.307).

Fig. 3.305 Wuxi Jichang Yuan borrowing from afar Longguang Tower in Xishan Mountain

3.2 On the Creation of Scenic Imagery: Scenic Imagery Beyond the Bounds of Perception

The previous section discusses the formation of scenery, which is of great significance. However, the construction of specific scenery doesn't entail the completion of the design of Jiangnan gardens which applies the elements of shaping of ground surface, vegetational disposition, architectural arrangement, and animal embellishment. Only when the garden or the scenery is given the poetic and picturesque quality and artistic profundity to express the inspiration of nature and life as well as integrated with form of garden living does it come into possession of the quintessential artistry and arrive at the highest precinct of the garden art—*yi jing*. (Quoted from *A Treatise on the Garden of Jiangnan* by Yang Hongxun, p. 7.) That is to say, scenery alone, which belongs to form, is not the entirety of garden art. Therefore, strictly speaking, scenery itself is not a complete and independent art, furthermore, it

Fig. 3.306 Suzhou's Zhuozheng Yuan borrowing from afar the Beisi Ta—Viewing from "Wuzhu Youju" to the west

is not the highest aesthetic object of garden art. A perfect garden includes unity of ideas and interest of the garden and scenery, as well as unity of the scenery and garden living style, whose effects present the highest level of garden art, which is called artistic conception.

Artistic conception of a garden is aesthetically profounder than garden scenery that is appreciated intuitively. Thus, it is the ultimate criterion of a garden creation. Artistic conception of a garden is based on scenery which provides visitors with pleasuring experience. When specific, limited scenery integrates the utilitarian content of visiting experience as well as the spiritual content of poetic sentiment and picturesque overtone, ideas, and philosophies, it sublimates into an essential, infinite, unified, and perfect aesthetic object that presents a deeper and broader aesthetic experience. The appreciation of music lies not merely in the scales, timbres, volume, voices, rhythms, and melodies of the instruments involved but rather in the thoughts and feelings embodied through that which lingers on beyond the individual sounds, which is similar to "scenic imagery lying beyond the bounds of perception" that Sikong Tu wrote of.

Artistic conception of a garden hints at more charming seclusion through the physical scenery, which can be summed up as an endless meaning in limited scenery. From the perspective of visitors, a feeling, which stems from the pleasant scenery and deepens through the visitors' association, seems to be mysterious and enigmatic. However, from the standpoint of creation, the vivid and impressive artistic conception, which results from garden makers' Fsubjective ideals, feelings, and appeals, is knowable, and therefore its composition can be understood and mastered through rational analysis as well as being regarded as a guidance for creation.

a

b

Fig. 3.307 (**a**) Sizhao Ge in Xiling Yinshe at Gushan in Hangzhou borrowed the distant scenery of West Lake. (**b**) West Lake from the perspective of Sizhao Ge

This section will briefly discuss basic principles and formation of artistic conception.

3.2.1 *Unity of Function and Scenic Imagery*

In the second chapter, the functional content of Jiangnan Gardens was discussed specifically, which can be divided into two aspects: aesthetic and utilitarian. The former refers to ideas and interest of a garden while the latter relates to the fulfillment of the garden living and visiting proceedings. Both types of content are contained in scenery. Moreover, the unity of content and form (scenery) is an inseparable component of an outstanding garden and only in that case can artistic conception be created.

3.2.1.1 Unity of Scenic Imagery and Human Thoughts with the Intrigues of Life

According to *The Craft of Gardens* (Yuan Ye), "the essence which touches the heart must be fully imagined in your mind before you put pen to paper" (Quoted from *The Craft of Gardens* (Yuan Ye) by A.M.H. p. 121). That is, there must be a theme in mind at the outset of garden creation. As a matter of fact, apart from a general theme for the whole garden, themes with different artistic flavors for each part are also needed, which constitute the basis of scenic elements and relations. For instance, the Master-of-Nets Garden (Wangshi Yuan) in Suzhou, with a theme of "fisherman in seclusion", takes water as its principal scenery. As a place where "the master of nets" or "fisherman" resides, the water will be shaped into an ideal fishing ground—a lake, instead of a ravine or deep pool (Fig. 3.284d). Management of water surfaces—a misty lake, stones and plant layout are all conducted in accordance with the theme of "fisherman in seclusion". On one hand, stones are stacked to highlight shoals and promontories (a feature of lake) or serve as a backdrop to the water. On the other hand, in terms of plant use, to highlight rippling waves, lotus is not preferred here as lake is different from a lotus pond in canal-filled villages. The lake (subordinate scenery) sets off the lake-and-mountain composition (principal scenery) with a background of mountains and forests, which is exactly consistent with "fisherman in seclusion". On the basis of consistency, the lake presents scenic space with different tastes to enrich the garden, which exerts no negative effects on "fisherman in seclusion" and meanwhile highlights lake-and-mountain composition. Such scenic space with themes can leave a complete and focused impression on visitors because of the unity between scenic form and thematic content, which is also known as "harmony of the scenery depicted and the emotions expressed". Thus, intriguing scenery is created. If, on the contrary, the design of scenery fails to conform to the theme, this unity will be destroyed. The principal scenery of Suzhou's Garden that was preserved miraculously (Liu Yuan) is the lake-and-mountain composition which exhibits islands in middle of the lake and bridges across the islands. The scenery and ideas and interest of the garden are both complete and unified. However, wisteria trellises on the bridge destroys the scale of the lake and islands, narrowing the lake space. Furthermore, wisteria trellises arranged in a landscape of lake and islands fundamentally fail to conform to the thematic appeals (these trellises were already in place when the East Garden (Dong Yuan) of the Xu's Family was rebuilt during the reign of Jiaqing in Qing dynasty—1796-1820, which is shown in Liu Maogong's painting) (Fig. 3.53). Therefore, due to the messy effects, the scenery cannot create impressive and profound artistic conception. In addition, an effect of "towering mountains and deep water" can be achieved by elevated stone beams between the ravine and gully. A small span of stone beams can arouse excitement from trekking without compromising visitors' safety, which can be found in Suzhou's Humble Administrator's Garden (Zhuozheng Yuan). However, the arrangement in East Garden of the Couple's Retreat Garden (Ou Yuan) fails on this count. Although the slabstone bridge is elevated, however, it adopts the form of a zigzag bridge which is derived from the slab stone bridge found on lakes in water towns, thus destroying the sense of unity (Fig. 3.60), which confuses visitors and cannot impart a moving artistic conception. It can be concluded that the production of artistic conception on a high level is premised on the unity of scenery and ideas and interest of a garden.

3.2.1.2 Unity of Scenic Imagery and Form of Garden Living

Garden art features functionality and Jiangnan classical private gardens serve as space for everyday living, be it a court garden, residential garden or a villa garden. Functionality is the premise of functional art and restricts directly its aesthetic value. A functional artwork without any functionality cannot show complete aesthetic experience even if its appearance is beautiful. For example, as a practical vessel, an exquisite openwork glass will never arouse people's wholehearted affections, no matter how beautifully decorated, because it lacks functionality for holding wine. Beauty lies on utility. The beauty of functional art has a closer connection with utility, which is especially true of garden art that has a higher aesthetic significance in utility. A court garden, residential garden or villa garden without utilitarian functions cannot provide aesthetic enjoyment even though great pains are taken in scenic composition, First and foremost, a garden is a place for rest and recreation. Therefore, esthetic enjoyment of a garden is obtained by living or touring in a comfortable way. In other words, visitors have no intention of appreciating secluded and beautiful scenery of a garden which lacks facilities for recreational activities, such as areas for sitting and resting, shelters from sun and rain, food supplies, etc., let alone appreciating artistic conception. From the outstanding examples of Jiangnan gardens, it can be concluded that all the conveniences that meet the needs of garden living are organized into the scenery, from the pavilions for sightseeing and resting to the halls for banquets watching and drama performances. That is to say, conveniences are also viewing objects. Functionality is skillfully arranged in the scenic design which follows ideas and interest. For example, grottoes are set inside mountains to conform to the appeals of deep and remote mountains and valleys. In addition to the entrance of the grotto, smaller openings are strategically positioned for lighting and aeration. Moreover, natural stones in the grotto serve as tables and seats for recreational activities such as resting, playing chess, and drinking wine. In this way, the functionality of providing rest areas, entertainments and shelters from sun and rain is unified with appreciation of natural scenes (Fig. 3.16). For the creation of classical gardens in the Jiangnan region, it is necessary to arrange scenery in order to stay with the themes as well as some functional contents. For instance, the large-scale Hall of Thirty-six Mandarin Ducks and Hall of Eighteen Datura Blossoms (Saliu Yuanyang Guan/Shiba Mantuohua Guan) in the Subsidiary garden (Bu Yuan) in the western section of the Humble Administrator's Garden (Zhuozheng Yuan) was designed for banquets, dramatic performances, and other activities at the time. Without enough space, a garden cannot provide such recreational activities, nor can it reflect its artistic value. On the premise that setting up a large hall is a must, scenic management is carried out intentionally. For example, the structure is divided by combining concave and convex of guardrails and roof, thereby reducing its scale and coordinating it with the mountain-and-water scenery as much as possible. The integration of the architecture and its surroundings can be achieved by means of waterside railings, animals and plants such as mandarin ducks and camellias (Fig. 3.208). Therefore, on the whole, the architecture can be regarded as a success. However, some criticisms (namely, that the hall is too large) only focus on scenic form but ignore restrictions of functional content on the composition. In other words, the unity between scenery and garden living style is neglected, which is precisely the essence of a garden. That is to say, unlike the appreciation of paintings, gardens should be enjoyed in touring or daily life. The aesthetic enjoyment lies not only in visual sensations, but in auditory, olfactory, tactile sensations, and even the comfort of body and mind. Only in this way can artistic conception, the ultimate aesthetic object, come into being.

3.2.2 *Recreation of Nature with Scenic Imagery: The Generality of Scenic Imagery*

The artistic conception of a garden is shown in specific scenery, which begs the question: What kind of garden scenery is capable of effecting artistic conception? Creation of the garden's natural landscapes, including those of a pastoral interest as manifested in the mountain village or canal-filled country scenery, should follow the laws and principles of the natural ecologic world, namely the natural structural relationship among the mountains, water bodies, vegetation, animal life, etc., and approximate the garden landscape features to their natural prototypes. There is a significant similarity in the creative principle of painting and that of garden making. In Chinese classical painting theory, two antithetical concepts are used in defining the artistry of a landscape painting: xing si (formal likeness) and shen si (essential likeness). Essential likeness is premised on formal likeness but is nonetheless regarded as the highest aesthetic criterion in judging the artistry of a natural landscape painting. (Quoted from *A Treatise on the Garden of Jiangnan* by Yang Hongxun, p. 8.) Renowned poet Bai Juyi of the Tang dynasty once remarked: "There are no fixed rules in painting, and the rules lie in similarity. There are no set examples in learning, and the examples are truth itself". Master painter Qi Baishi of the Qing dynasty commented that the artistry of a painting lay "between likeness and non-likeness" of the painted subject, which resonates with the observation of great German poet J.W. Goethe that beauty exists "between truth and non-truth", both of which serve as an excellent footnote to Bai Juyi's argument. The "likeness and non-likeness" or "truth and non-truth" of an artistic image is precisely the knack to the artistic creation of garden. (Quoted from *A Treatise on the Garden of Jiangnan* by Yang Hongxun, p. 8.) Likeness, be it physical or essential, does not equal identicality. As a matter of fact, it is inevitable that gardens differ from reality or nature. Likeness is relative to identicality in artistic expression, and identicality refers to mechanical reproduction of naturalism. Traditional Chinese art invariably opposes naturalism and advocates artistic creation that reflects life and reveals truth. In other words, the essential content of life (i.e. the reality of life) should be revealed through physical characters, which is similar to essential likeness in classical art theory. Intuitive appearance in life often occurs by happenstance, even involving illusions that are incompatible with its essence. In artistic creation, recapitulatory typification is by no means conceptualization, indeed it is a concentrated way that reveals the discovered inner spirit and quality, which equals the generality of art. The artistic image that has been refined and processed (i.e. artistically epitomized) is, therefore, different from the appearance of life. It is worth mentioning that the secret of typification in artistic creation, on the contrary, lies in its individuality. Outstanding works reveal the true meaning of life through the personality that includes happenstance and illusion. These works highlight the spiritual essence from the midst of the external artistic image, leaving a deep impression on visitors with intuitive artistic feelings. The generality of traditional Chinese art does not count time and space and yet integrates them, presenting its own uniqueness. The ink-wash painting, which mainly utilizes Chinese impressionistic-style skill of brushwork, emphatically render the most essential characteristics of the depicted object in the image. In order to express certain content, the image is often subjected to sublation and exaggeration and even though it sometimes undergoes considerable deformation, it still embodies likeness. The specious formal likeness is vital for achieving essential likeness. In order to further understand the common connotations of traditional Chinese artistic styles, first, we are expected to take a glance at the fundamentals of essential likeness in Beijing opera. Like Chinese gardens, Beijing opera can be regarded as a representative of Chinese artistic styles. Its profound artistic principles has been recognized and appreciated by the realm of art in the world. Since the mid-1960s, European drama critics have made a comment that Beijing opera, performance systems of Stanislavski and Brecht are listed as the "three systems" in the world. Therefore, the generality of images in Beijing opera contributes to a better understanding of the generality garden scenery. It is prominent that sublation and exaggeration in images of Beijing opera disregards time and space. For example, Zhuge Liang, an outstanding strategist and politician famous for his strategy in the late Han Dynasty and the Three Kingdoms (3rd century AD), wears a half-meter long black beard (artificial whiskers worn by actors in Beijing opera) to show his loyalty, calmness, erudition and talents even though he is at the beginning of his career. The long beard is hung out of his cheek instead of sticking on the face, which enriches the stage effect of the character. Such formal likeness, which exaggerates the character's beard and age, can reveal the character's essence in a better way, thus, it is more real and vivid. In that case, essential likeness is achieved. However, if the role is depicted as an inexperienced young scholar in a naturalistic way, it may be far cry from Zhuge Liang in people's minds who has made brilliant achievements. The generality of gardening art in Jiangnan is the same as that of other arts. Scenery of the classical gardens in the Jiangnan region is chosen and designed by the garden maker subjectively, which involves his thoughts, feelings and tastes. It is a humanized nature with interest rather than a simple imitation of the natural objects' external forms. Thus, scenery will be exaggerated and transformed to express certain thoughts and tastes. Taking the description of mountain peaks as an example, peaks are parts of mountains, but in the design of stone peaks in garden art, the characteristics of lofty, weird, and steep are often exaggerated whereas discarding their lower transitional part, which has no direct relationship with these features. Stone peaks are often created with freestanding Taihu stones. There is no need to explain the relationship between stone peaks set on the mountain and foothills or hillsides. Stone peaks and stalagmites can also be set on level ground (Figs. 3.7 and 3.12), which is evocative of the imagery of "green looms up in stacking peaks", "rising peaks", and "an isolated peak springs up" mentioned in classical Chinese literature. Solitary stone peaks, when properly laid out, can capture the essential likeness in the rising stone mountains or the stone peaks in the approximate scene. The same is true of water scenery, such as sunken pools in nature, where, regardless of the water level, the water surface is relatively small and it is generally deep. The creation of sunken pool involves grasping the essence of its deep-set features and highlighting them on the image. It is therefore built in a confined and snug space with shade, and filled with water at a low water level (not brimming to the bank) so that visitors conjure up the image of the deep chasm of a well. As for the invisible bottom, it is not required to be deep, which means a sublation and exaggeration of the real sunken pools found in nature (Fig. 3.61). Another example is the slab stone bridge in a Jiangnan riverside town, the painting performance of which usually follows a straight line. However, the bridge can also be made to bend and curve in order to match the surface conditions. Slab stone bridges in gardens should match the conditions of the water surface or the shape of the shoreline rather than following a shortcut. These winding bridges are built to sit low above the water to set off the scenery of a rippling lake (Figs. 3.175 and 3.184). It is like the cloud hill-climbing wall, which depicts an undulating wall on the mountain, both of which maintain similarity with the characteristics of the original version and highlight and exaggerate the artistic images that reflect its essence (Fig. 3.147). This generality is just like the way that Beijing opera art reveals the characters through face painting, which not only reflects the inner spirit but also enhances the decorative formal beauty. This is the beauty of truth and non-truth. In the visual scene creation, generality through sublation and processing is the key to artistic conception in garden art.

3.2.3 Development of Scenic Imagery in Depth

Landscape garden work that reflects nature and real life is the result of the garden maker's subjective creation based on their understanding of nature and is therefore the product of thinking and labor. A landscape garden work embodies the logical thinking of the garden maker that is formed with his aesthetic judgment and outlooks on nature and gardens; it is also a product of his imaginative thinking fostered by his fascination with love of beauty and life. As such, the accomplishment and profundity of a garden—whether it is a living epitome of nature or a relishing of rich and deep artistic conception—rests largely with a rich and broad life and creative experience as well as the artistic erudition of the garden maker. In other words, the makings of a garden maker have an immediate bearing on his logical and imaginative thinking, which exerts a direct impact on the outcome of the garden creation.

From the perspective of creation, artistic conception is a reflection of real life as well as an expression of the garden maker's thoughts and feelings. From the perspective of garden appreciation, artistic conception exists objectively as well as the aesthetic enjoyment of garden maker's imagination. The appreciation of a poetic garden's artistic conception is the process of visitors' subjective thoughts and emotional activities. The empathetic experience of artistic conception is predicated upon the visitor's knowledge of nature and life, cultural attainments, aesthetic capabilities, and understanding of the garden art "language". The feelings about artistic conception of children differ from those of adults. The depth and extent of one's appreciation of artistic conception is in direct proportion to those of one's life experience, cultural sophistication, and artistic aptitude. Therefore, for a visitor, the deeper appreciation of a garden's artistic conception hinges upon the enhancement of visitor's cultural and artistic knowledge.

Artistic conception—"scenery beyond perception" is evoked by garden scenery which lies "between likeness and non-likeness, between truth and non-truth". Its profundity lies not only in changing scenic spaces, but also in the visiting process, change of seasons, passing of time, or variations in weather and climate such as rain, snow, sunshine, or clouds as well as expansion of scenery through poems and prose such as poetic couplets on plaques and inscriptions on tablet.

3.2.3.1 Continuation in Time Extension of Space

Reflecting the deep and wide nature of life by limited artistic images, such as poems, paintings and stages, requires not only the sublation of images as well as the generality and typicality of exaggeration, but also the intriguing implication of plot. Only in this way can the feelings of the readers and the viewers be stimulated and their mental associations inspired, thus deepening the artistic conception. Gardens are also a limited art form, especially classical Jiangnan private gardens. Producing a profound artistic conception from limited scenery requires not only generality and typicality, but also implicitness. Deep and wide artistic conception in scenery requires not only generality, typicality, but also implicitness. A mask of scenic imagery hinges on flexuous shaping of space. Orderly arrangement of scenes featuring in "hiding before revealing" is not only required to intensify the visiting experience, but also to mask scenes and deepen the artistic conception.

What determines the sequence of scenes is the guidance, which reflects the structural relationship of scenes not only in space and time respectively, but also between space and time. The spatiality of the scene is inseparable from its timeliness. In terms of the orderly arrangement of dynamic views, circuitous paths increase the sightseeing distance, while complex arrangement of concealing and revealing multiplies the layers of the depth of field, thus increasing the duration of the visit, which fulfills the purpose of expanding the space. It can be concluded that the curvature of scenes expands the depth of space and extends the time for aesthetic enjoyment. The time and space extended by a winding route compared with that of a straight line can be measured. However, the far-reaching artistic conception increased by expanding and implicating cannot be measured in numbers. For example, in the Gully of Eight Sounds (Bayin Jian) in Wuxi's Garden for Ease of Mind (Jichang Yuan), the straight-line distance between the entrance and exit at both ends of the gully is 41 m, while the actual distance of the meandering gully is 50 m, which is 9 meters longer (Fig. 3.308). It takes about 1 min to walk along a straight line, and about 1.5 min to travel through the winding valley, which implies a half-min increase in the sightseeing duration. This mechanical number reveals an extension of space and time, and the latter is calculated without stopping. However, the fact of the matter is

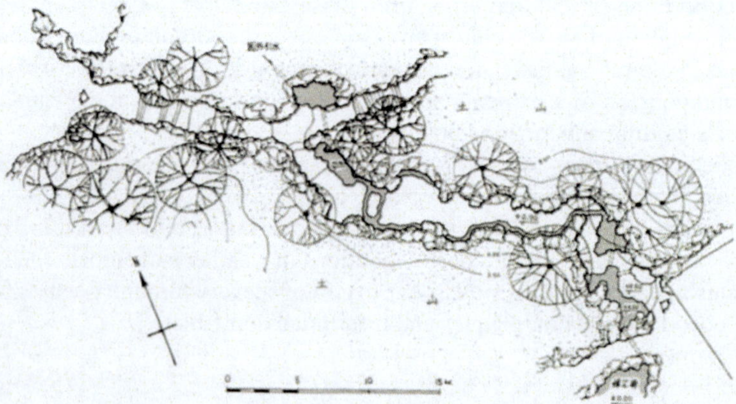

Fig. 3.308 Plan of Bayin Jian (Xuancong Jian) in Wuxi's Jichang Yuan (drawn according to illustrations in History of Ancient Chinese Architecture Technology).

that sightseeing is not a continuous process. The opposite part of the zigzagging path—rocky ground, fractured precipices, gurgling streams and cliff inscriptions as well as the dense shade that covered the original valley are unified with the path itself. Staggered scenic planes leave an impression of infinity, and ideas and interest contained in sequence of infinite scenic planes are immeasurable. The amount of space and time extended by flexuous scenes varies with the degree of flex, the number of base points for static views, activity setting and the ability to retain the artistic tastes. The space and time of artistic conception felt by visitors are much more than the actual number, which is precisely due to its implication. The implication caused by convolution lies in travelling routes as well as in concealing and revealing for viewing, such as decorative scenery or hollow paintings in hole door or hole window, walls, sources and courses of streams, blind windows and attics. Limited scenes lead to infinity and a feeling of experiencing "scenic imagery beyond the bounds of perception".

In addition to scenic space, scenic time is also an object of aesthetic enjoyment in garden art. Unlike reading, garden viewing enables visitors to immerse themselves in the landscape of garden, to be specific, both in the scenic space and time. The continuation for aesthetic enjoyment of the scenic space can deepen artistic conception, which is also true for scenic time. For example, music can expand the aesthetic enjoyment of artistic conception through the passing of time. The shape and space of the landscape art scene are organized in accordance with time, which is the so-called sequence. The sequence of garden scenery is actually a series of clips of scenic planes, which are displayed with the passing of time. Melodies and rhythms of music can also be experienced in the sentimental changes. The appreciation of garden art through "sightseeing" is precisely the appreciation of the organization of musical time. Scenes of mountains and waters, flowers and trees and so on will take on artistic plots, such as (for example) "where the peak stands to welcome the next turn of the mountain path", "a winding path leading to a secluded spot" and "a sudden view" as the time passes, which all have melody and rhythm. This will arouse viewers' feelings, mental associations and illusions, evoking a more profound artistic conception. The route of stone mountains in Suzhou's Villa of beautiful surrounds (Huanxiu Shanzhuang) is convoluted, winding through mountainsides, valleys and the foot of the mountain. The unfolded length of the route is much longer than that of its straight line. Continuation in time and extension of space can be embodied in this route. Grottoes, stone rooms and other base points on the route can hold visitors. These are about the routes. Furthermore, scenic planes of concealing and revealing, such as cliffs, shady ravines and green peaks and ridges, can also provide fantastical views. The promontory, cliffs, stone peaks, forests, ravines and grottoes boast the melody and rhythm of shape and space, in which you will be integrated if you immerse yourself into the scenes. As the tour goes on, visitors are overwhelmed by the flowing scenery in the process of exploring valleys and caves, crossing streams, climbing mountains, etc.. Due to the treatment of space art, the simple "stone pile" is pleasing to eyes, convoluted and deep, and full of artistic taste. It also serves to draw in the visitor to the infinite artistic conception of deep and remote mountains and valleys through the treatment of time art.

3.2.3.2 Borrowing of Scene in Relation to Time: The Coloring of Scenery with Time and the Natural Elements

Like the art of spatial planning, art of temporal planning can also generate a profound artistic conception, which is the result of integrating these two arrangements. In terms of scene sequence of garden art, temporal planning is of special significance, which we discussed briefly in the previous section. However, the temporality of the scene is not only manifested in duration and methods of the scenes, but also in the elements of nature: the changing of seasons, passing of time in a day, or variations in climate and weather such as rain, snow, sunshine, or clouds (Fig. 3.309). The

a

b

Fig. 3.309 (**a**) Snow in Zhuozheng Yuan of Suzhou (From Suzhou Classical Gardens). (**b**) The garden near Mount Hui in Wuxi with spring in the air. (**c**) Winter mountain features around "Xuexiang Yunwei" in Zhuozheng Yuan of Suzhou. (**d**) Canglang Ting of Suzhou at Sunset

Fig. 3.309 (continued)

"different scenery caused by moving" throughout the visit includes the factor of time. That is, the scenic planes change as viewers' perspectives shifts, while time passes as well. In this process, the pace of the tour (including rest) corresponding to the duration of the scenic planes (including dynamic and static effect) undoubtedly affects the artistic conception of the garden art. However, the coloring of scene with time and change in elemental conditions with the passage of time is of greater significance to the deepening of artistic conception. First of all, the shade and hue of the scene are rendered by the sky and cloud. During the day, from morning to noon, dusk, and then to night, sunshine, clouds, starlight and moonlight, like stage lights, constantly change and cast themselves onto the scene. In daytime alone, there is a spectrum of different lights including broad daylight, morning sunlight, and sunset. These lights shine onto scenes alternately, changing the shade and hue. The dust, clouds, rain and snow brought by the weather also envelope the scene with different sentiments. The four seasons are more directly reflected in the scene. In terms of vegetation, flowers bloom and wither away with the passage of time, with some flowers blooming for the whole year. Various flowers indicate different seasons, with their leaves changing shapes and colors with the passage of time. In terms of animals, migratory birds and insects also change their habitats, appearing and disappearing in accordance with time. Water also takes on different seasonal features. The change of nature, seasons and sentiment in the above-mentioned scenes is another manifestation of the temporality of the scene. Borrowing of scenery in relation to time in The craft of gardens refers to deepening the artistic conception through the coloring of scenery with time and the natural elements, which is an outstanding and important idea. The underlying principle is to master the planning of time and utilization of the elements of nature (clouds, rain, flowers, etc.) and seemingly accidental natural elements (swallows in spring and insects in autumn) into scenes, enriching the natural features of the garden.

The time spent appreciating scenery is that for coloring scenery with time and the natural elements. In other words, the time of scenic sequence is in line with the seasons and the elements of nature. Effects of "different scenery brought into view by moving" is combined with that of "different scenes brought into view by the passage of time and change in the elements", strengthening the appeal of time. As the garden scenery lies in the natural space and changes with natural elements and passage of time, it can leave a profound impression of artistic conception by means of nature.

3.2.3.3 Expansion of Scenery Through Poetry and Prose: Inscriptions and Poetic Couplets

Chinese paintings usually include side carvings and inscriptions that when stamped with a seal not only indicate the title, author, and time of creation but also often include the purport, associated feelings, or reasons of creating the painting. The characters written on the painting and the various seals stamped on it are integrated into the picture not only in form but also in content. The characters and seals on the painting have become indispensable components of Chinese paintings, whether the subject matter of the painting is flowers, feathers, fair ladies, or landscapes. It is essentially a collaboration between poetry and painting. Therefore, it is believed that Chinese painting is an integration of painting, calligraphy, and seal carving, which is not only in terms of form, but also from the perspective of content—the poetry and prose expressed by calligraphy and seal carving are also related to the theme of the painting and the author's aspiration and, therefore, enhance the aesthetic appeal.

In "The tumbler" painting, a work by famous Chinese painter Qi Baishi, stands a tumbler with a black gauze cap and a uniform robe (both are typical attire of Chinese officials), which, without any inscriptions, simply appears to be a toy for viewing. Only after viewers have read the author's unique inscriptions on the painting can they realize the painting's implication of social acrimony and sarcasm directed at corrupt officials who neglect their official duties day in and day out. Just as side carvings, inscriptions, and seals on Chinese paintings, similar expressions of poetry and prose also exist in gardens. There are not only names for gardens but also names for scenes in gardens as well as poems, prose, and inscriptions to express the sentiment. These names, poems, and inscriptions, when integrated into the scenery and combined with some structural elements of the scenery, become an integral part of garden art that serves sometimes for clarifying the motif and artistic taste of the scene or for becoming an object of appreciation as classic cultural relics. In this sense, it is safe to believe that poetry and prose are one of the means for the comprehensive creation of Chinese classical gardens, including Jiangnan gardens. This being the case, the importance of poetic couplets and inscriptions in gardens cannot be overstated. Otherwise, not only will the style of the individual scene be spoiled but also the whole scenery of the garden will be ruined.

3.2.3.3.1 The Role of Poetry and Prose in the Art of Landscape Garden

The roles of poetry and prose in gardens can be categorized as follows.

Firstly, poetry and prose can improve the style of a garden. Beautiful poetry and prose can endow a garden with a so-called "scholar's style" to turn it into an elegant garden, without which, the garden would be vulgar and lacking in cultural accomplishment. It is for this reason that not only names and inscriptions are an integral part in gardens but also some ancient calligraphy, books of model paintings or drawings, letters, and classical texts not directly related to the contents of the scenery are displayed in the scene as independent antique cultural relics for appreciation.

Moreover, poetry and prose add the finishing touch to garden scenery. In *A Dream of Red Mansions* (one of the four great Chinese classic novels) by Cao Xueqin, Cao expressed such a viewpoint in the words of Jia Zheng (one of the main characters): "(Grand View Garden) ... All those prospects and pavilions-even the rocks and trees and flowers will seem somehow incomplete without that touch of poetry which only the written word can lend a scene". (Quoted from the translation of *A Dream of Red Mansions* by David Hawkes.) (Chapter 17). Plain as it is, it is true. Moreover, the reason behind it lies in that poetry and prose play a role in highlighting the motifs and artistic tastes of a garden.

All Chinese classical gardens can be viewed as the so-called "subject garden". The names of gardens all emphasize ideas guiding creation as well as motifs and artistic tastes whether its content be about things, scenery, aspiration, or feelings. Names focusing on things include the Subsidiary Garden (Bu Yuan) and the Garden that was preserved miraculously (Liu Yuan) in Suzhou. Names focusing on scenery include Little Lotus Manor (Xiaolian Zhuang) in Nanxun and the Garden of Mist and Rain Pavilion (Yan Yu Lou) in Jiaxing. Names focusing on aspiration include the Garden of Returning to Native Village (Guitian Yuan) and the Humble Administrator's Garden (Zhuozheng Yuan) in Suzhou. Names focusing on feelings include the Garden of Natural Delight (Chang Yuan) and the Garden of Pleasure (Yi Yuan) in Suzhou. That poetry and prose are used to highlight the human thoughts with intrigues of life of a garden is not only reflected in the name of the whole garden but also in the names of scenes in the garden as well as the inscriptions that accompany them. Therefore, names and inscriptions of different scenes are to be found everywhere in Jiangnan gardens. Halls, multi-storied buildings, pavilions, and even stones (mountain stones or lake stones) are often inscribed with names or inscriptions whose content can also be roughly divided into things, scenery, aspirations, and feelings. Take the names of buildings as an example, names of buildings focusing on things include the Book-basking Pavilion (Pushu Ting, same as the name of the garden) in Jiaxing; names of buildings focusing on scenery include the River-Viewing Tower (Wangjiang Lou) in Shanghai's Garden of Various Fruits (Jiuguo Yuan) and the Clean Water and Lush Trees (Shui Mu Ming Se) in Nanxiang's Yi Yuan; names of buildings focusing on aspiration include the Self-Cultivation Hall (Shaoxiu Tang) in Shanghai's Garden of Various Fruits (Jiuguo Yuan) and the Washing My Ribbon Pavilion (Ke Zhuo Wo Ying) in the Mountain Villa of Red Oak (Hongli Shanzhuang) in Hangzhou; names of buildings focusing on feelings include The Den of Pleasure (Ke Zi Yi Zhai) in the Garden of Pleasure (Yi Yuan) of Suzhou and the Think Deep Aim High Hall (Zhi Qing Yi Yuan) in the Humble Administrator's Garden (Zhuozheng Yuan) of Suzhou. It is the same case with inscriptions, except that they are often fusions of feelings with the natural setting and are usually written as couplets on plaques combined with buildings.

The ultimate function of poetry and prose in garden art is to manifest the spirit of the scene, that is, to enhance the aesthetic appeal of the garden. It is the poetic names and inscriptions that arouse visitors' imagination and feelings spontaneously. For example, there are plum trees on the lake mountain in the Humble Administrator's Garden (Zhuozheng Yuan) of Suzhou. However, if it were not for the name of the building—Snow-Like Fragrant Prunus Mume Pavilion (Xuexiang Yunwei)—people would not feel the aesthetic appeal of stepping out onto the snow in search of prunus mume flowers. Originally, the horizontal scroll of the pavilion bearing the inscription "Shanhua Yeniao Zhijian",[148] written by Ni Yuanlu (a painter of Ming dynasty),

[148] Among Mountain Flowers and Wild Bird.

and the inscription on the plaque couplets reading an excerpt from a poem of the Tang Dynasty—The forest's more tranquil with cicadas' chirps; The mountain's more secluded with singing birds—were coordinated to enhance the aesthetic appeal of the wild mountain forests. Moreover, they were written by Wen Zhengming, one of the four gifted scholars in the Jiangnan region during the Ming Dynasty, adding to the literary charm of the scene.

The building in the bamboo forest of the Garden of Surging Wave Pavilion (Canglang Ting) is named Exquisite Emerald Bamboo House (Cui Ling Long) and inscribed with couplets reading "The windy bamboo is the flute; The running water plays the lute", which further manifest the secluded atmosphere that transcends the scenery of the bamboo forest. The Beauty Within Reach Tower (Xiexiu Lou) in the Master-of-Nets Garden (Wangshi Yuan) in Suzhou is inscribed with the lines "Enjoying clouds from rocks, I lean against a cane; Playing a zither under pines, I wait for cranes". The Standing-in-the-Snow Hall (Lixue Tang) in the Lion Grove Garden (Shizi Lin) is inscribed with the lines "The green bamboo and pine are truly guests; The bright moon and the breeze are always friends" and "To feel the interest of the pines and stones; To view the person from the wonderland" among others. These poems all play a role in further personifying and rationalizing the natural scenery, thus enhancing its aesthetic appeal.

The aforementioned examples are expressions of reclusive thoughts enjoyed by Confucian scholars or officials and other examples express indulgence in creature comforts such as "The birds prepare their nesting when air of spring fills; The plantains shelter windows as rain of night chills"(couplets on a plaque in the Waterside Pavilion of Washing Hat-Ribbons (Zhuoying Shuige) in the Master-of-Nets Garden (Wangshi Yuan). Both of these themes are mainstays of poetry and prose in Jiangnan gardens. The remainder of poetry and prose in gardens draw from commentary on the Confucian classics, such as mottos on cultivating one's morality, which is not directly related to natural scenery. Therefore, they may be used to furnish halls and study rooms but not the pavilions in natural landscapes as they are divorced from the appeal of the scenery and would be inappropriate and rather pedantic. The couplet attached to the renovated Embroidered Silk Pavilion (Xiuqi Ting) in the Humble Administrator's Garden (Zhuozheng Yuan) of Suzhou, that reads "To be a gentle and generous man; To live an honest and diligent life" is an example of disharmony. If the couplet were moved to the bamboo forest then it might be able to appropriately combine emotion with the scene, which is what will be discussed below.

3.2.3.3.2 Unity of Poetry, Prose and Scenic Imagery

As artistic literary forms, poems, ci[149] and articles (or essays) all have their independent ways of expression and appeal, whether their content mainly deals with

[149] A type of lyric poetry in Classical Chinese poetry.

things, scenery, aspiration, or feelings. They could have been created on a whim, but to play its unique role in gardens they need to conform to the purport of the garden scenes and be unified with the background. Therefore, names and inscriptions, which mostly cite beautiful lines in Chinese traditional poems, ci, and articles, and abridged or adapted sentences that serve as allusions to the originals are usually related to the scenes they belong to.

The pavilions in the Mountain Villa of Red Oak (Hongli Shanzhuang) in Hangzhou and the Garden of Pleasure (Yi Yuan) of Suzhou, as combinations of waterscape, both allude to the poetic line "If the water of the Surging Waves River is clean, I can wash the ribbon of my hat; if the water of the Surging Waves River is dirty, I can wash my feet", from *Mengzi: Lilou* as their title. Of the two names, the former basically records the original sentence as "the Washing My Ribbon Pavilion (Ke Zhuo Wo Ying)", while the latter reads "the Waterside Pavilion of Washing Hat-Ribbons (Zhuoying Shuige)". The water-front building of the Garden for Ease of Mind (Jichang Yuan) in Wuxi alludes to the debate between Zhuangzi (Zhuang Zhou) and Huizi (Hui Shi) (Both Zhuangzi and Huizi are famous Chinese philosopher around the 4th century BC during the Warring States period) on the philosophical meaning of a "fish's joy" where Hui Shi questions Zhuang Zhou: "You are not a fish; how do you know what constitutes the enjoyment of fish?" to which Zhuang Zhou answers "You are not me, how do you know that I don't know what constitutes the enjoyment of fish!", and names itself as, Knowing-Fish Waterside Pavilion (Zhiyu Jian). There are also scenes with self-created names. For example, the main hall on the island in the middle of the lake in Yi Yuan of Nanxun is named Tens of Thousands of Lotus Like Half Lake of Brocade (Banhu Yunjin Wan Furong) and the veranda in the osmanthus forest on half-way up the mountain in the Garden that was preserved miraculously (Liu Yuan) of Suzhou is named Veranda of Scenting the Sweet Osmanthus Fragrance (Wen Muxixiang Xuan). The pavilion beside the Flying-Snow Spring (Feixue Quan) in Suzhou's Villa of beautiful surrounds (Huanxiu Shanzhuang) is named Putting a Question to the Spring Pavilion (Wenquan Ting), while the pavilion borrowing scenery from sails on the mountain where Jiaxing's Lower a sail Pavilion (Luofan Ting) locates itself, is named "Shadow of Sailing (Fanying)". The same type of scene can have different poetic descriptions and different moral meanings, which gives the scene different artistic conceptions. For example, the theme of Hall of Distant Fragrance (Yuanxiang Tang) in the Humble Administrator's Garden (Zhuozheng Yuan) is lotus flowers and, therefore, its name borrows the meaning of "its subtle perfume is wafted far and wide (Xiangyuan Yiqing)"from Zhou Dunyi's *Ode to the Lotus Flower*. Similarly, in combination with the theme of lotus flowers, the Garden that was preserved miraculously (Liu Yuan) names one of its building as Stay and Listen Pavilion (Liuting Ge), which is in line with "Hearing the Raindrop Pitter-pattering on the Remains of Lotus (Liu De Can He Ting Yu Sheng)" by poet Li Shangyin. The former refers to a gentleman who "is not dirtied even though he grows from a muddy pond and does not become seductive even after having washed off in clear water for years", which means one who preserves his purity. It enables visitors to appreciate the artistic conception of purity and noble sentiment when watching lotus flowers. The latter leads people to the free and elegant artistic conception of enjoying listening to the heavenly sounds while viewing the scenery. It is also easy to change the names and inscriptions of scenes because they are usually attached to buildings in the form of Bian'e (inscribed board) and couplets. The owner of the garden can change their old work if their mood changes or they find and fancy other good lines. Sometimes, if a garden was deserted, the poetic couplets on plaques were easily lost and when they were rebuilt the plaques were replaced with new ones. Furthermore, when the owner of a garden changed, more changes were typically made in addition to names and inscriptions. For example, when the owner of the Swallow Garden (Yan Yuan) in Changshu, which possessed the famous masterpiece the Valley for Swallows (Yangu) created by Ge Yuliang (a famous rockery piling master), was changed in Guangxu's reign in the Qing Dynasty, the original owner removed all the inscriptions and couplets and took them with him, leaving the garden empty (Tong Jun: *Records of Jiangnan Gardens*). As a result, many of the original names and inscriptions inside or outside the buildings in the existing Jiangnan gardens have been lost due to a lack of preservation.

Even for the same scene in the same garden, the names and inscriptions may also have undergone massive changes. According to Professor Tong Jun's investigation before the War of Resistance Against Japan, several names and inscriptions of scenes in the Garden of Pleasure (Yi Yuan) in Suzhou were different from what they are now. For example, the main hall "Pavilion of the Fragrance of Lotus Roots, Hoeing-the-Moon Windowed Veranda (Ouxiang Xie, Chuyue Xuan)" was originally named The Den of Pleasure (Keziyi Zhai), the original name of Golden Grain Pavilion (Jinsu Ting) was "Here Plants the Osmanthus Dancing High in the Wind (Yun Wai Zhu Po Suo)", the original name of Stone Boat (Shi Fang) to the east of Locking Green Pavilion (Suolü Xuan) was "Sitting Alone After Going Around Cloisters (Raobian Huilang Huan Duzuo)", the original name of Veranda of Stone Worship (Baishi Xuan) was Thatched Hut of the Cold Season (Suihan Caolu), and the original name of Heavy Dew Hall (Zhanlu Tang) was "Here Comes the Beautiful Island (Qiongdao Feilai)" It is the same with the Humble Administrator's Garden (Zhuozheng Yuan) in Suzhou, inside which Veranda of Rain Delight (Yiyu Xuan) was originally named "Listen to the aroma in seclusion (Ting Xiang Shen Chu)", Minor Surging Waves Pavilion (Xiao Canglang) was originally named The Beautiful Pavilion (Qinghua Ge), The Magnolia Hall (Yulan Tang) was originally named The Hall of Painting (Bihua Tang), The Courtyard of Spring Crabapple Flower (Haitang Chunwu) was originally named "The Shadow of Plum Blossom on Half Window (Banchuang Meiying)", and the hexangular pavilion on the lakeside mountain was originally named "Encouraging Farming (Quangeng)"and is now simply called "The Pavilion on the North of the Mountain (Beishan Ting)"after losing its name and with no further investigation.

3.2.3.3.3 Scenes Organized with Poetry and Prose

Regarding the forms of poetry and prose, apart from calligraphy and paintings displayed indoors or hanging screens and the like, the subject-oriented poetry and prose, as well as independently-appreciated spot scenes, can take the forms of Bian'e, couplets on pillars, cliffside inscriptions, and poem-engraved stone slabs, among others. Mainly combined with buildings and rocks, these poetic forms not only enhance the aesthetic appeal of the garden with its literary content but also enrich the scene as ornamentation or decoration.

There are nameplates of scenes hanging on buildings in gardens, the horizontal ones of which are called "Bian" and the vertical ones "E" (Pronounced somewhat similar to the English filler word "uh"). These nameplates are usually hung under the eaves of the middle room of buildings and sometimes in the center of a room. Couplets on pillars are often used together with Bian'e, hanging symmetrically under them. These Bian'e are mostly made of wood, while couplets on pillars are sometimes made of bamboo and in-door ones may be made of paper or tough silk. In addition, there are also Bian'e on various moon gates (a circular opening in a garden wall that acts as a pedestrian passageway) which are usually made by brick carving or sometimes stone cutting. The basic shape for Bian'e is a rectangle but it may also be made into various shapes to highlight the artistic taste and ornamental effect of the garden.

The design of Bian'e generally adopts a simple and elegant style, with the exceptions of the Bian "Garden of High Virtue (Gaoyi Yuan)" in Tianping Mountain and the Bian "True Delight (Zhenqu)" in the Lion Grove Garden (Shizi Lin) in Suzhou that are designed with solemnity and luxury because they are treasures inscribed by Emperor Qianlong of the Qing Dynasty. The characters on these two Bians are made of gold, with the former one surrounded by a gold frame with a pattern of two dragons frolicking with a pearl and the latter one surrounded by a gold frame with a pattern of clouds (Fig. 3.310). *A Philosophy of One's Own: Room Part* by Liyu listed 8 different styles of Bian'e, namely (1) Banana Leaf Bian characterized by the shape of banana leaves with a green body, ink rib, and yellow characters, (2) Bamboo Bian made of half of the polished bamboo tube filled with azurite, turquoise, or ink characters, (3) Inscription E engraved with wooden boards in the same fashion as a stone tablet and with white characters on the black background, (4) Handscroll E taking the form of a hand scroll (banner scroll), (5) Album Bian taking the form of an album (folded album), (6) Xubai Bian, hollowed-out characters on thin boards with stencil tissue paper back cover, frosted black lacquer background, hung inside against the light (7) Shiguang Bian, similar to Xubai Bian and used in combination with stone mountains (8) Autumn Leaf Bian alluding to "Ode to Autumn Leaf created in the riverside of the palace (Yugou Tihong)".

In addition to horizontal and vertical rectangle Bian'e, there are also bamboo, wooden, or wooden zither-like couplets on pillars, as well as brick carving handscroll E (such as the E inscribed with "Qinlu"[150] on the moon gate in the Garden of Art (Yi Pu) of Suzhou and the E of Hezi[151] moon gate in the Villa of Fenyang (Fenyang Bieshu) in Hangzhou) (Fig. 3.311). The characters, mostly written in ancient seal script, official script, and sometimes lively cursive script but

[150]The Courtyard of Cress.

[151]A shape similar to oval.

Fig. 3.310 (**a**) "Lin Quan Qi Shuo Zhi Guan" Plaque in Liu Yuan of Suzhou—natural Nanmu bottom with mineral green Characters. (**b**) "Han Bi Shan Fang" E in Liu Yuan of Suzhou—black characters on a white background. (**c**) "Yueguan" Plaque in Xiaojinshan in Slender West Lake of Yangzhou—black characters on a white background like a horizontal scroll made of rice paper. (**d**)

h

Fig. 3.310 (continued)

rarely formal regular script, can be carved in two ways. These are Yin carving, which means the characters are carved sunken, and Yang carving, which means the characters are carved prominently. For example, there is a wooden couplet on pillars that reads "The moonlight forms a chessboard in groves of bamboo. The wind is playing strings in the forest of pines" in Minor Surging Waves Pavilion (Xiao Canglang) in the Garden of Pleasure (Yi Yuan) which references the Garden of Surging Wave Pavilion (Canglang Ting) and was rendered in cursive script written by Zhu Zhishan[152] that set off the natural and unrestrained artistic conception of the scene. Another example is the wooden screen in the Humble Administrator's Garden (Zhuozheng Yuan) in Suzhou, with the inscription written by Wen Zhengming,[153] the screen boasts an elegant appearance with white letters on the original wooden

---

Fig. 3.310 (continued) "Zhuwaiyizhi Xuan" Plaque in Wangshi yuan, Suzhou—natural Nanmu bottom with pinky white characters. **(e)** "Xiuqi Ting" handroll Plaque in Zhuozheng Yuan of Suzhou (so-called "Handroll E" in A Philosophy of One's Own). **(f)** "Hu Tian Zi Chun" Plaque hanging on the eaves of the pavilion in Ge Yuan of Yangzhou **(g)** "Gaoyi Yuan" Plaque in Tianping Mountain, Suzhou—it is designed in a palace style with glitter of gold and of jade because it is the treasure inscribed by Emperor Qianlong **(h)** "Zhenqu" Plaque in Shizi Lin in Suzhou—it is designed in a palace style because it is the treasure inscribed by Emperor Qianlong

[152] One of the four gifted scholars in the Jiangnan region in the Ming Dynasty.

[153] Another one of the four gifted scholars in the Jiangnan region in the Ming Dynasty.

Fig. 3.311 (**a**) The bamboo couplet attached to Xiuqi Ting in Zhuozheng Yuan of Suzhou—"Be a gentle and generous person, live an honest and diligent life". (**b**) The bamboo couplet "Shan Shui Jian" hanging on the columns of the waterside house in the east garden of Ou yuan in Suzhou (made of half bamboo tube with bamboo joint scrapped off). (**c**) Couplets made of board hanging on both sides of the hole window in Wangshi Yuan,Suzhou. (**d**) The "Bie You Dong Tian" brick carving handroll plaque (or gate E) in Liu Yuan of Suzhou. (**e**) The "Yunku" brick carving handroll plaque in Wangshi Yuan,Suzhou. (**f**) The "Zhenyi" brick carving plaque in Wangshi Yuan, Suzhou. (**g**) The brick carving plaque in the east garden of Ou Yuan in Suzhou

background. The inscriptions in the gardens are mostly written by famous masters, increasing the allure of the gardens (Fig. 3.312).

Cliffside inscriptions are common in famous mountains and scenic spots (Fig. 3.313). In Suzhou alone, there are several cliffside inscriptions on Tiger Hill (Hu Qiu), such as Sword Pond (Jianchi), Sword-Testing Rock (Shijian Shi), White Lotus Blossom (Bailian Kai), Spring of Simplicity and Honesty (Hanhan Quan), and Peach Shaped Stone (Shitao) (Fig. 3.314). Jiangnan gardens are the reproduction of natural scenic spots and, therefore, cliffside inscriptions are also made on piled

Fig. 3.311 (continued)

stones or solitary stone peaks inside them. Examples of such phrases include "Pingfeng Sandie"[154] inscribed on the peak stone of stone screen in the Garden of Pleasure (Yi Yuan) in Suzhou, "Fei Xue"[155] inscribed on the lake-stone mountain beside Flying-Snow Spring (Feixue Quan) in the Villa of beautiful surrounds (Huanxiu Shanzhuang), "Panjian"[156] inscribed on the waterside piled stone in the Master-of-Nets Garden (Wangshi Yuan), the large seal inscription "Xiling Yinshe" inscribed on the mountain stone of Xiling Seal-engravers' Society (Xiling Yinshe) in Hangzhou, and "Bayin Jian"[157] inscribed in the Garden for Ease of Mind (Jichang Yuan) in Wuxi (Fig. 3.315).

Poem-engraved stone slabs are mainly inlaid in the walls of half corridors. The half corridors beside the garden boundary don't have ornamental perforated windows on the garden boundary wall, so there is only one side on which to enjoy the view on such routes. Therefore, to balance the two sides and break the baldness of the garden boundary wall, inlaid poem-engraved stone slabs are often displayed on the wall for appreciation. Poem-engraved stone slabs are mostly written by famous

[154] Trifold of Screen.

[155] Flying-Snow.

[156] The Winding Stream.

[157] Gully of Eight Sounds.

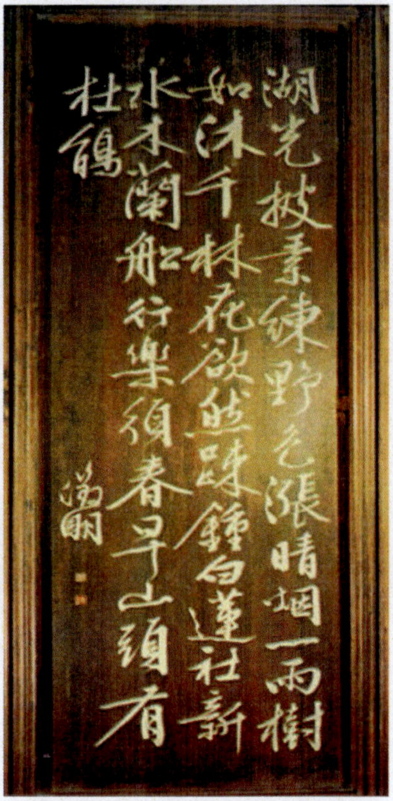

Fig. 3.312 The inscription on the wooden screen in Zhuozheng Yuan of Suzhou written by Wen Zhengming, one of the four gifted scholars in Jiangnan region in the Ming Dynasty

historical artists. The content of the engravings is not limited to poems and can include letters, classic literary works, or essays. For example, the half corridor in the west side of the Verdure Mountain House (Hanbi Shanfang) of the Garden that was preserved miraculously (Liu Yuan) in Suzhou stretches for more than 200 meters and houses more than 300 pieces of poem-engraved stone slabs embedded in the wall, which, known as *The Model Calligraphy of the Garden that was preserved miraculously (Liu Yuan)* (Fig. 3.316), includes famous calligraphy work from Jin, Tang, Song, Yuan, Ming, and Qing Dynasties. Other examples include the Garden of High Virtue (Gaoyi Yuan) (previously the Clan Hall of Fan Zhongyan) in Tianping Mountain of Suzhou, which exhibits the engravings of famous poems by Bai Letian, Su Zimei, Wang Junyu, the half corridors at both sides of the Pavilion of the Imperial Stele (Yubei Ting) in the Lion Grove Garden (Shizi Lin) and the half corridor in the west of the Waterside Pavilion of Washing Hat-Ribbons (Zhuoying Shuige) in the Master-of-Nets Garden (Wangshi Yuan).

Fig. 3.313 The cliffside inscription of the Ming Dynasty on Mount Wuyi—"Jingtai"

Fig. 3.314 (**a**) The inscription on a natural stone on Hu Qiu Mountain—"Shitao". (**b**) The cliffside inscription on Hu Qiu Mountain—"Jianchi"

a

b

Fig. 3.315 (**a**) The natural stone screen on the mountain in Yi Yuan of Suzhou—"Trifold of Screen" Inscription. (**b**) The inscription on a natural stone in Wangshi Yuan, Suzhou—"Panjian". (**c**) The inscription on a natural stone in Xiling Yinshe of Hangzhou—"Xiling Yin she". (**d**) The inscription on a natural stone in Wuxi's Jichang yuan—"Bayin Jian". (**e**) The inscription on a natural stone in Zhan Yuan, Nanjing—"Nüwa Shi". (**f**) The inscription on a natural stone in Zhan Yuan, Nanjing—"Shuijing". (**g**) The inscription on a natural brookside stone in Wangshi Yuan, Suzhou—"Daichao"

Fig. 3.315 (continued)

g

Fig. 3.315 (continued)

Fig. 3.316 "Model Calligraphy of Liu Yuan" embedded in the wall of the half corridor in Liu Yuan of Suzhou

Chapter 4
Critique and Discussion on Some Typical Gardens

In order to render a complete concept of Jiangnan Gardens after analyzing their structural principles, the final chapter of this book will introduce and discuss some representative examples of existing gardens in different environments, including urban sites, sites beside a mansion house, sites among mountain forests, riverside and lakeside sites and village sites according to the classification presented in *The Craft of Gardens* (Yuan Ye). Gardens with different functions are also represented, such as private residential gardens, villa gardens, and public gardens owned by clans, guilds, monasteries, or academic groups, as well as several public gardens in scenic spots of municipal projects. Among them, gardens in the city are mostly affiliated with residences, which means they are built on sites beside a mansion house. These examples also represent gardens on different scales. Scale, though a quantitative measure, will lead to qualitative discrepancies. There are no strict boundaries between the size of a garden or the size of its site, but we can make a rough distinction as a reference. That is that gardens covering an area of hundreds of square meters but less than 1000 m^2 can be classified as small gardens, gardens covering an area of about 1000–3000 m^2 can be classified as medium gardens, and gardens covering an area greater than 3000 m^2 can be classified as large gardens.

Next, some examples of gardens will be briefly introduced and reviewed. The creation of classical Jiangnan gardens is based on the layout of the main front view of the principal scenery (mountain-dominated, water-dominated, or mixed landscape) and its corresponding main view stop (main hall). Therefore, grasping the corresponding relations that determine the overall scheme of the garden is a high priority when analyzing and evaluating a garden. In this way, we can get an in-depth understanding of the foundation of the whole scenery and the relationships between scenes of a garden.

© Springer Nature Singapore Pte Ltd. 2022

H. Yang, *A Treatise on the Garden of Jiangnan*,
https://doi.org/10.1007/978-981-16-6924-8_4

4.1 Examples of Gardens in Cities

4.1.1 Garden of the Study at Southeastern Corner of House No 7, Wangxima Street in the City of Suzhou Jiangsu Province (Small Garden of Mountain Scape)

This is a court garden created on the vestibule of the house's study, covering only 200 m^2, and together with the small courtyard on the north side covers less than 300 m^2 (Fig. 4.1). The garden was later expanded and has become relatively independent. Therefore, it can generally be regarded as a residential garden.

According to *Suzhou Classical Gardens* written by Professor Liu Dunzhen, this small court garden of the study was the relic of Ren Daorong's residence in Tieping Lane during the reign of Emperor Guangxu in the Qing Dynasty, while its northern garden was expanded by its owner, Mr. Wan, during the period of the Republic of China.

The structure of this garden is simple with the principle scenery being arranged in front of the main building, which is the general layout of the court garden for flower halls and studies (Fig. 4.2). The main hall of this garden—the study—has its main axis stretching from the east to the west and its main facade facing east (generally, the main axis stretches from the south to the north and the main facade faces south). To the east (facade) of the study arranged the principal scenery, a mountainscape flanked by pavilions and corridors. The mountain was built with stacked lake stones (stone-covered earth) and planted with luxuriant Rosa banksiae, osmanthus, crape myrtle, begonia, and palm among others, highlighting seasonal features, and the garden wall behind the mountain is partially covered by climbing plants (Fig. 4.3). The main front view faces west, opposite the study. The study is a four-sided hall with long windows on its west and east sides and short windows on its south and north sides. The small patio with a skylight at its back is a subordinate scenery in which lake stone peaks matched with osmanthus are stacked. The four-side-open hall allows the principal and subordinate scenery to penetrate each other, thus extending the depth of field to the limits of the garden site in four directions.

The buildings, especially the pavilions and corridors in the principal scenery, are scaled-down while their layout still resembles that of a residence. For example, two corridors stretch out on both sides of the study, but in order to avoid symmetrical rigidity, a balanced asymmetry with a slight central axis is adopted. That is, the study is slightly closer to the right corridor than the left corridor, the right corridor, therefore, is slightly longer with square pavilions at its end, while the left corridor is slightly shorter with corner doors to the side courtyard at its end. The corridor on the right side, echoing the mountain, climbs up to the square pavilion on the mountain. The corridor on the left is horizontal and leads to the small courtyard on the north side (thus connecting with the expanded Small Ink Pond (Xiaomo Chi), presenting a relatively low-lying trend. It is a naturally appealing scene, as the relationship between the expanded Small Ink Pond (Xiaomo Chi) and the original garden is in line with the common logic of scenery). Going further north, the

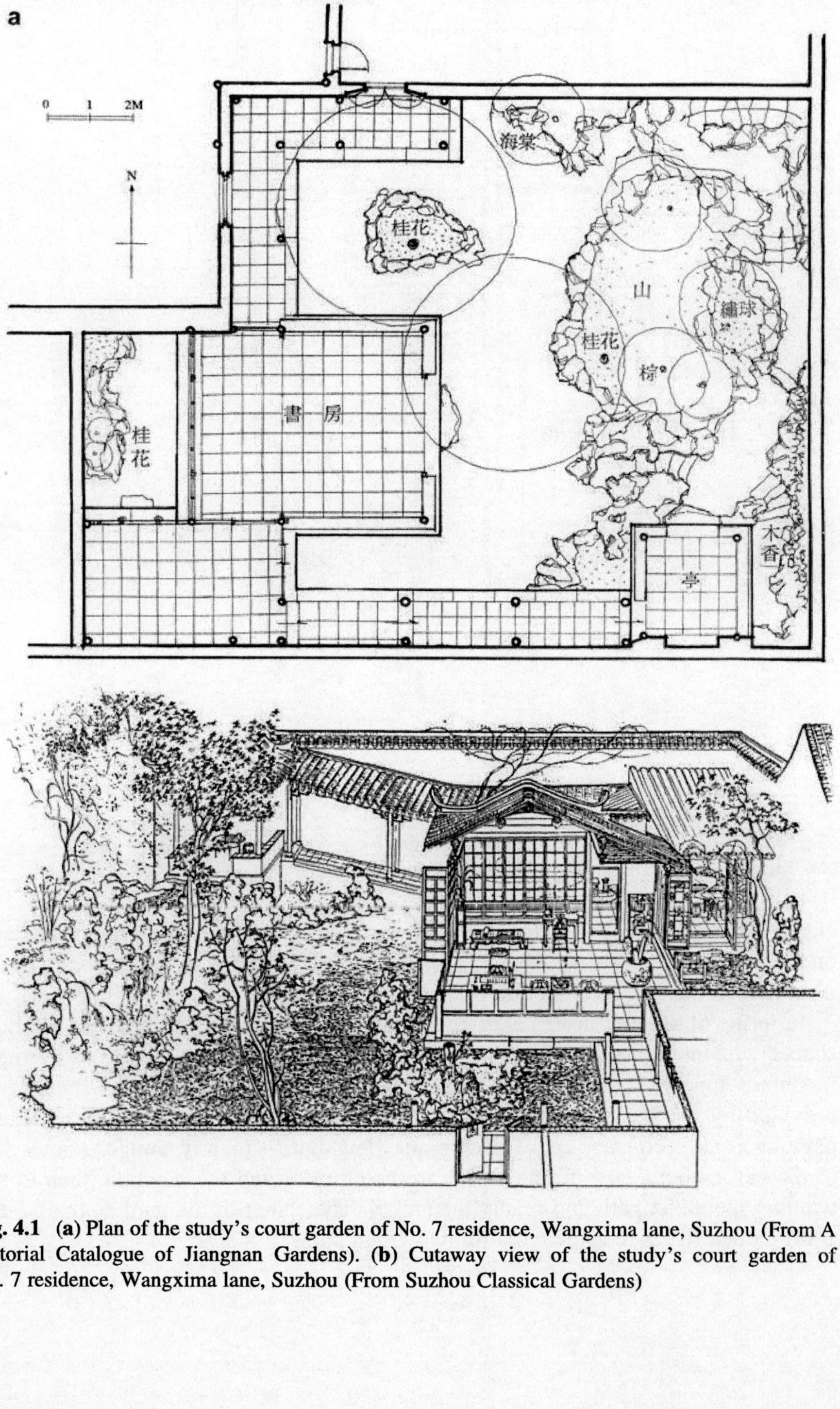

Fig. 4.1 (**a**) Plan of the study's court garden of No. 7 residence, Wangxima lane, Suzhou (From A Pictorial Catalogue of Jiangnan Gardens). (**b**) Cutaway view of the study's court garden of No. 7 residence, Wangxima lane, Suzhou (From Suzhou Classical Gardens)

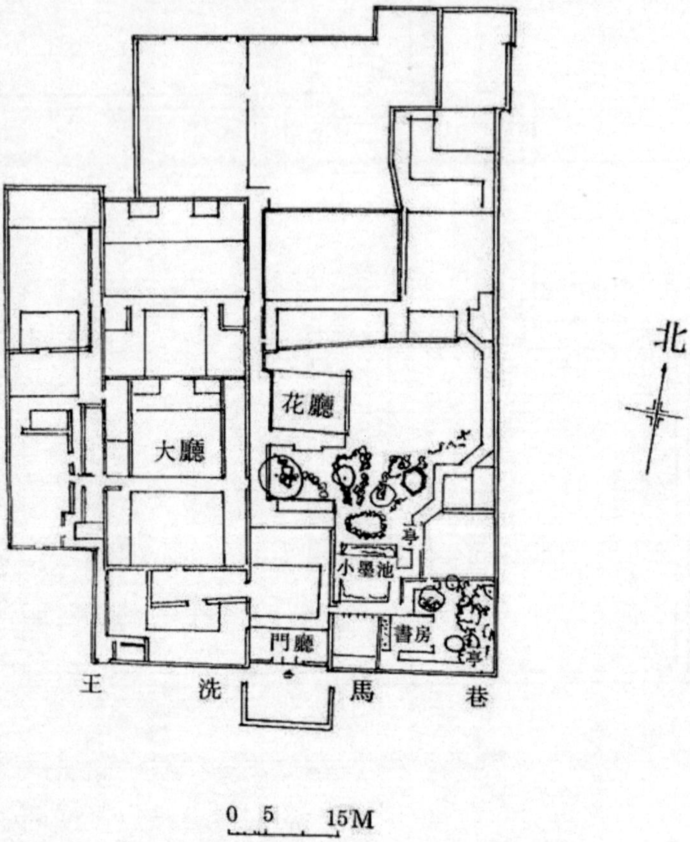

Fig. 4.2 The overall plan of the relationship between the study's court garden and the No. 7 residence in Wangxima lane, Suzhou (From Suzhou Classical Gardens)

expanded corridor leads to the back yard of the inner chamber for womenfolk, and on the left side of the study, there is a corridor leading directly to the courtyard of the front hall so that guests can enter the garden directly.

In terms of structural elements, the small court garden in front of the study is shaped with mountains, but without water, on its ground surface, presenting a simple mountain appeal. In terms of guidance, there are only two points of static view provided by the buildings, namely the study as the main one and the mountain pavilion as the secondary one. The dynamic view route is mainly along the outer ring of the garden boundary: from the hall to the corridor and the pavilion, then to the winding mountain path and finally the tunnel. The center of the garden, that is, the court garden site is the spreading of the garden path.

Fig. 4.3 Scenery of the pavilion and corridor as well as rocks and plants in the study's court garden of the No. 7 residence, Wangxima lane, Suzhou

4.1.2 Garden of the Study in Mansion of the Purple Cloud of Prosperity (Laizi Lou) in the County of Tiantai, Zhejiang Province (Small Garden of Mountain Scape)

The Mansion of the Purple Cloud of Prosperity (Laizi Lou) in Tiantai County is one of the 18 buildings of the ancient mansions that remains today. Most of these mansions were renovated or rebuilt in the Qing Dynasty with the date of their initial construction unknown. Nevertheless, judging from their architectural form and engineering practices, they preserve a number of customs from the Song Dynasty. The main building of the Mansion of the Purple Cloud of Prosperity (Laizi Lou) was built after the ownership of the mansion was changed during the reign of Jiaqing in the Qing Dynasty (1818).

The court garden is a side courtyard on the right of the central courtyard, therefore it is attached to the right wing-room, which is also the study of the house. The court garden covers an area of only 55 m^2 (Fig. 4.4), with the principal scenery, a mountainscape, positioned opposite the main view stop, the study. The mountainscape is relatively simple with lake stones stacked against three sides of the garden wall and the house. At its facade, there are rising mountain peaks and a rock overhang on the cliff, under which a stone teapot table is positioned. This is an echo of the stone planting beds against the wall on the central axis of the courtyard and if a stone ta (a long, narrow and low bed) is positioned under the rock overhang, it will futher increase the sense of seclusion. It is worth noting that this mountain is not made into a cliff mountain embedded in the surface of the garden wall. Instead, the folded peaks create a very thin shadow wall, leaving the space between the mountain peaks and the garden wall as a planting bed. The rock overhang stretches forward to enlarge the top area and reduce the floor area, thus increasing the depth of

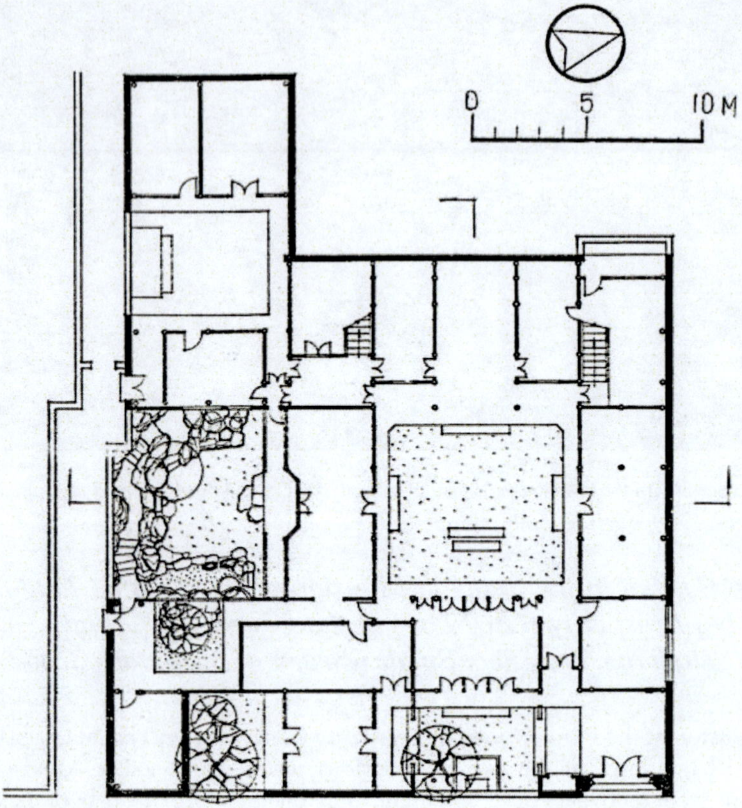

Fig. 4.4 Plan of the court garden attached to the study of Laizi Lou in Tiantai, Zhejiang (From Chinese Vernacular House)

field (Fig. 4.5). The mountain vegetation is densely configured with a sense of quiet seclusion created by forest vegetation. The scales and proportion allow for sprawling trees to grow windward amidst the peak stone crevices, and their vitality forms a spectacle to behold.

Two stalagmites placed on the platform near the left side of the study form a stone forest of sorts, under which a crossing way runs, turning right to the summit. A forked road on the mountain leads to side garden door and the mountain steps wind across and down the hill. Small as the mountain is, it is different from penjing (Penjing is a Chinese garden art based on miniaturized trees and landscape models that are generally grown and arranged inside pots) in Lingnan Garden Art in that visitors can immerse themselves in this mountain.

Serving as the background, the high garden wall is a continuous leaking wall with a connecting pattern, which weakens the artistic conception of the scene. It would have been much preferable to adopt the hill-climbing wall for such purposes.

a

b

Fig. 4.5 (**a**) Perspective view of the principal scenery of the court garden attached to the study of Laizi Lou in Tiantai, Zhejiang (drawn according to the chart in Chinese Vernacular House); (**b**) Elevation of the principal scenery of the court garden attached to the study of Laizi Lou in Tiantai, Zhejiang (drawn according to the chart in Chinese Vernacular House); (**c**) Cutaway view of the principal scenery of the court garden attached to the study of Laizi Lou in Tiantai, Zhejiang (drawn according to the chart in Chinese Vernacular House)

4.1.3 Garden of Miniature Landscape as if Contained in a Pot (Hu Yuan) in Suzhou, Jiangsu Province (Small Garden of Water Scape)

The Garden of Miniature Landscape as if Contained in a Pot (Hu Yuan) is located in the western part of house No. 7, Miaotang Lane in the old city of Suzhou, covering

Fig. 4.5 (continued)

an area of only 300 m^2 (Fig. 4.6). The garden is connected to the residence by a moon gate. After entering the door, the principal scenery, a space with water in its center (Fig. 4.7), immediately leaps into the eyes. The type of water surface is mainly lake and becomes narrow in the south, gradually forming a stream shape. The main hall is set to the north of the lake. In front of the hall, a narrow moon-viewing platform is set up near the water. This, together with the half pavilion overhead the water, which is connected to the left half corridor along the garden wall, plays the role of obscuring the edge of the lake, thus giving the appearance of a continuous flow of water.

Looking south from the main view stop—the hall—different views of the lake come into sight. The close views include a two-fold slab stone towpath-style bridge on the right, the promontory of the peninsula, the Malus halliana Koehne and the bamboo grove that slant into the lake on the left. The middle views include an inlet shaded by trees such as pinus bungeana and a stone footbridge, which not only serves as a garden path across the river but also obscures the stream to further increase the far-reaching effect. The distant views include the pavilion and the

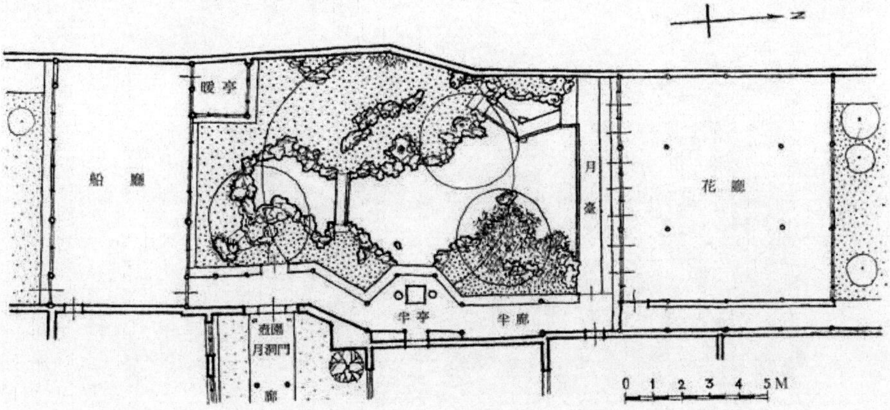

Fig. 4.6 Plan of Hu Yuan in Suzhou

windowed veranda shaded by bamboos at the vanishing point of the stream turns (from the perspective of the main view stop, a continuing flow of water is shown), and in order to save space, the pavilion and the windowed veranda are connected but they seem to scatter back and forth on the main front view, which is a successful design. The arrangement of shadows is also worth noting. The main hall faces south, and the above-mentioned main front view faces north. Therefore, the stream, lying far away from the lake, is in a shadow and the backlight adds a subtle deepening effect.

Contrary to the previous example that has mountains but no water, the ground surface of this garden is shaped with water but no mountains. However, there are several stone peaks scattered around the lake, which are shaded by bamboo and trees, hinting at the distant surrounding mountains. On the east side of the garden, there are a half corridor and a half pavilion along the wall. On the west side, there are mountain features including planting beds made of lake stones along the high wall, stone peaks, and plants to serve as a background. In addition, there are ornamental perforated windows on the upper part of the high walls, which not only reduce the weight of the wall but also serve as a decoration. Vines such as Ficus pumila are also planted to increase the natural component of the scene.

Small as the waterscape garden is, it doesn't seem crowded with its lake features highlighted by scenes including the towpath-style bridge, the promontory, the inlet, and the stone footbridge. The static view stop is quite simple, consisting only of two opposite buildings in the north-south direction. However, the scenery contains scenes including the hall, the pavilion, and the windowed veranda. It was indeed a rare masterpiece but was unfortunately destroyed in the late 1980s.

Fig. 4.7 The flower hall (main hall) of Hu Yuan in Suzhou from the perspective of the southern boat hall

4.1.4 Garden of Natural Delight (Chang Yuan) in Suzhou, Jiangsu Province (Small Garden of Water Scape)

Garden of Natural Delight (Chang Yuan) is located in the eastern part of house No. 22, Miaotang Lane, Suzhou, covering an area of about 900 m^2 (Fig. 4.8). Though as a small garden, it is three times the size of the aforementioned Garden of Miniature Landscape as if contained in a Pot (Hu Yuan). In addition to small doors connecting with the back house and the front hall, the small garden also has an independent street door (Fig. 4.9).

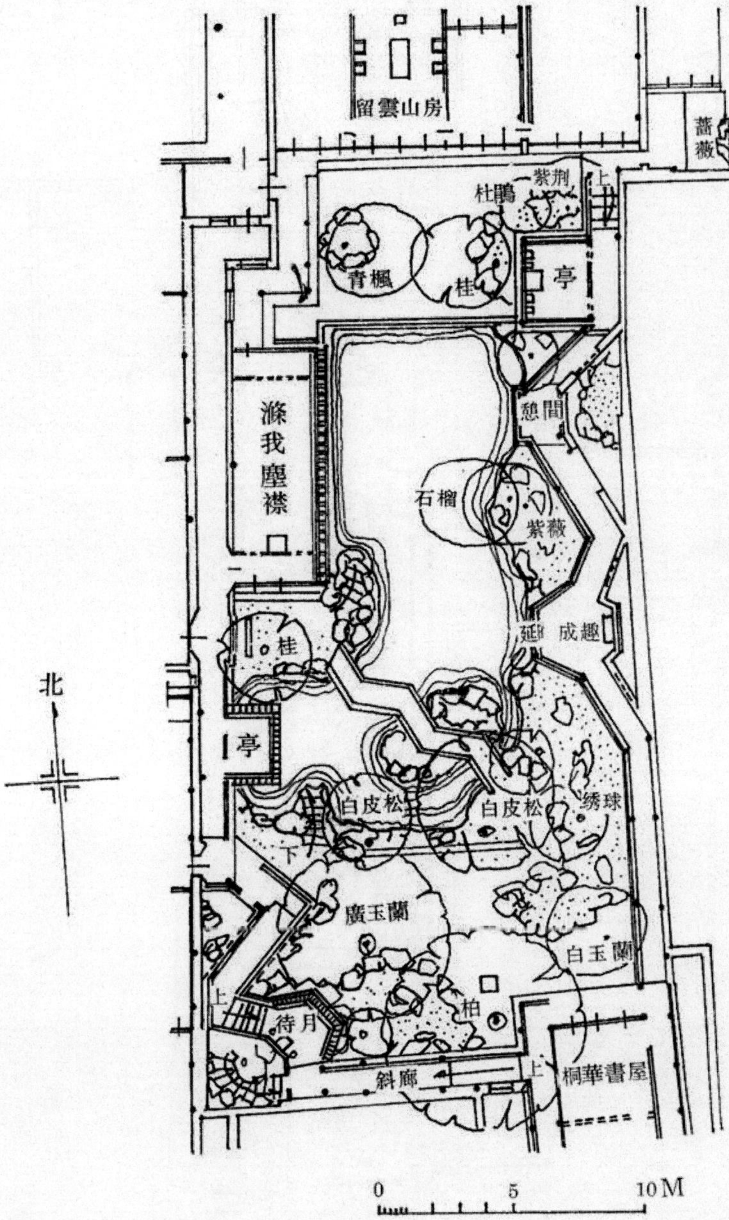

Fig. 4.8 Plan of Chang Yuan in Suzhou (From Suzhou Classical Gardens)

The entrance hall is immediately next to the garden door and there is a patio with skylight to the north of the entrance hall, and its main room is titled Study of Tung Blossom (Tonghua Shuwu). The surrounding corridor outside the study is directly

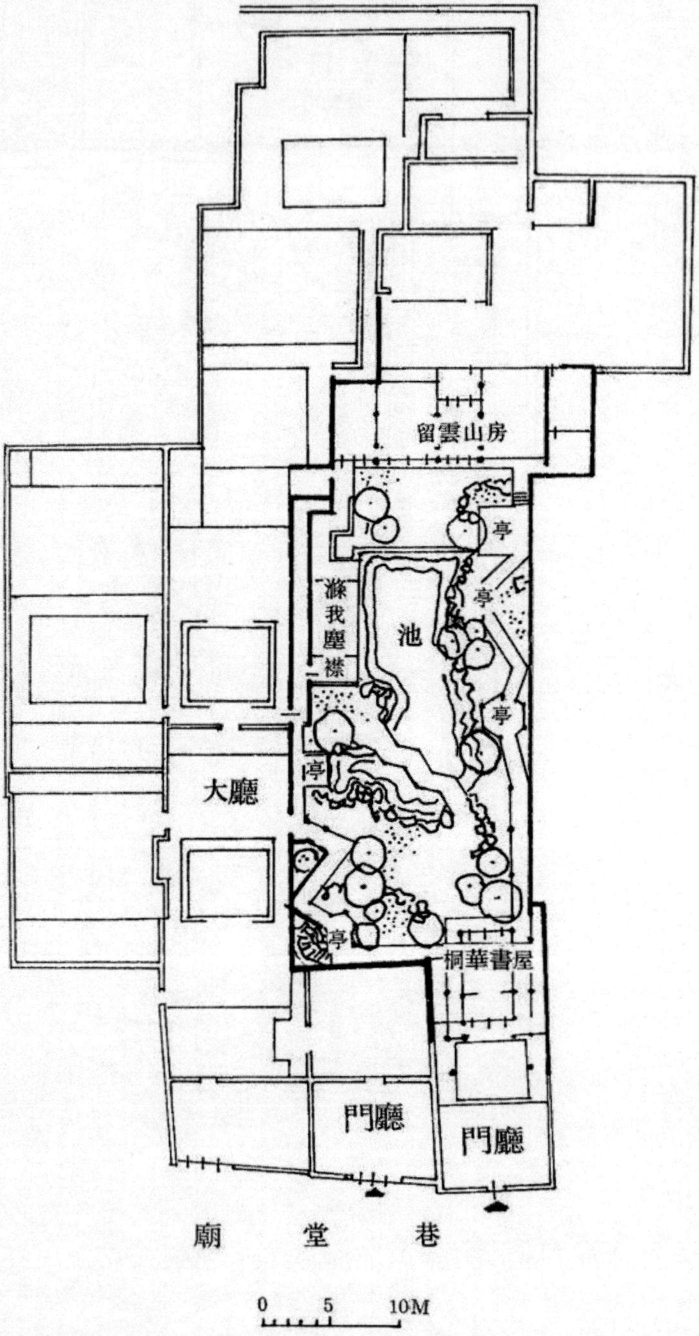

Fig. 4.9 The overall plan of the relationship between Chang Yuan in Suzhou and the residence (From Suzhou Classical Gardens)

Fig. 4.10 The main front view of the principal scenery of Chang Yuan in Suzhou (From Suzhou Classical Gardens)

connected to the covered walkway in the garden and there are long windows on the north side of the study, which has become a main view stop for the south view aspect of the principal scenery. The garden is relatively long in the north-south direction and the water surface covers 2/3 of the area from the northern border. The type of water surface is a lake, with the island, promontory, dock, and towpath-style bridge (fivefolds) serving as a foil. At the northern end, there is a moon-viewing platform near the water, as wide as a courtyard. On the east side of the platform, there is a warm pavilion (with half walls and sash windows on three sides) and on the north side, there are three halls facing south as the main view stop with a Bian entitled "Mountain House of Lingering Clouds (Liuyun Shanfang)". So the garden is still based on the basic pattern of small gardens, where the main hall faces south and the main front view of the principal scenery is northward (Fig. 4.10).

The main body of the scenery is water, and a curved corridor is arranged along the wall on the east side. The southern part of the corridor is a half corridor relying on the garden wall and the northern part is a zigzag half corridor separated from the garden wall. The northern half corridor's wall beside the garden wall is shaped with a moon gate and ornamental perforated windows, and spot scenes with banana trees, bamboos, and stones are created in the small space between the two walls. This extends the depth of field and creates an illusion that the corridor crosses the principal scenery. Two pavilions are arranged in the way of the corridor. The one in the south is hexagonal and takes the title "The Fun of Extending the Light (Yan Hui Cheng Qu)", which points out that it is a place for enjoying the afternoon and dusk scenery. The one in the north is square and entitled "Rest Room (Qijian)" and is not with a tented roof but a two-slope roof, modeling the roof of a simple folk hut.

On the west side of the principal scenery, a boat hall called "Cleansing my earthly filth (Di wo chenjin)" is positioned near the moon-viewing platform at the north end.

The hall is installed with many sash windows by the lakeside symbolizing the porthole of a boat and a long gooseneck chair at the lower part in the same fashion as a waterside pavilion. In order to avoid the imbalance with the cramped lake surface, the boat hall adopts the method of *xie yi* (artistic rendition) to avoid concretization. Instead of combining the building with the boat, it adopts a single roof and xianshan (revealed pediment) roof on the longitudinal east side. This reflects that there is some theoretical guidance and, therefore, depth in its creation. The boat hall is connected with a corridor, which leads north to the main view stop, Mountain House of Lingering Clouds (Liuyun Shanfang), and directs south to a square pavilion with revealed pediment roof. The pavilion, in contrast with the hexagonal pavilion "The Fun of Extending the Light (Yan Hui Cheng Qu)" on the opposite bank, is designed to be overhang the lake in the likeness of a building on stilts. The half corridor to the south of the pavilion is shaped into a zigzag up the mountain and the small space between the half corridor and the garden wall is also decorated with some spot scenes that can be seen through the leaf-shaped moon gate and ornamental perforated windows opened on the corridor wall. The half corridor is also connected with the hexagonal pavilion—Waiting For the Moon Pavilion (Daiyue Ting)—on the mountain at the southwest corner of the principal scenery. It turns into a half corridor leaning against the garden wall after going through the pavilion, and runs down the slope to the Study of Tung Blossom (Tonghua Shuwu).

The mountain in the southwest corner was created by stacking natural stones outside the base of the climbing corridor, shaping the base of the Waiting For the Moon Pavilion (Daiyue Ting) into a hollow mountain with piled stones and decorating it with planting beds, which is a very economical way. In addition, the small space between the pavilion and the southwest corner of the wall is used to arrange a winding path down the mountain, which goes through the cave under the pavilion and then reaches the plain to the south of the lake. The plain is planted with cypress, magnolia, white orchids, and hydrangea, among others, and there are also pinus bungeanas plants on the lakeside. At the south inlet of the lake, there is a dock with stone steps under the pinus bungeanas, which highlights the scenic character of the lake. Beside the Fun of Extending the Light Pavilion (Yan Hui Cheng Qu Ting), reeds, cattails and the like are planted by the lake, bringing out a flavor of wildness. Matched with the water lilies on the small lake, it not only conforms to the theme of the scene but also enriches decoration.

4.1.5 *Garden of the Pavilion of Unity with Heaven (Tianyi Ge) in Ningbo, Zhejiang Province (Small Garden of Mountain-and-Water)*

The garden is set in the front courtyard of the Pavilion of Unity with Heaven (Tianyi Ge) (Fig. 4.11). As a mountain-and-water landscape, its structure is relatively simple with the main front view facing the pavilion. There is a waterside moon-viewing

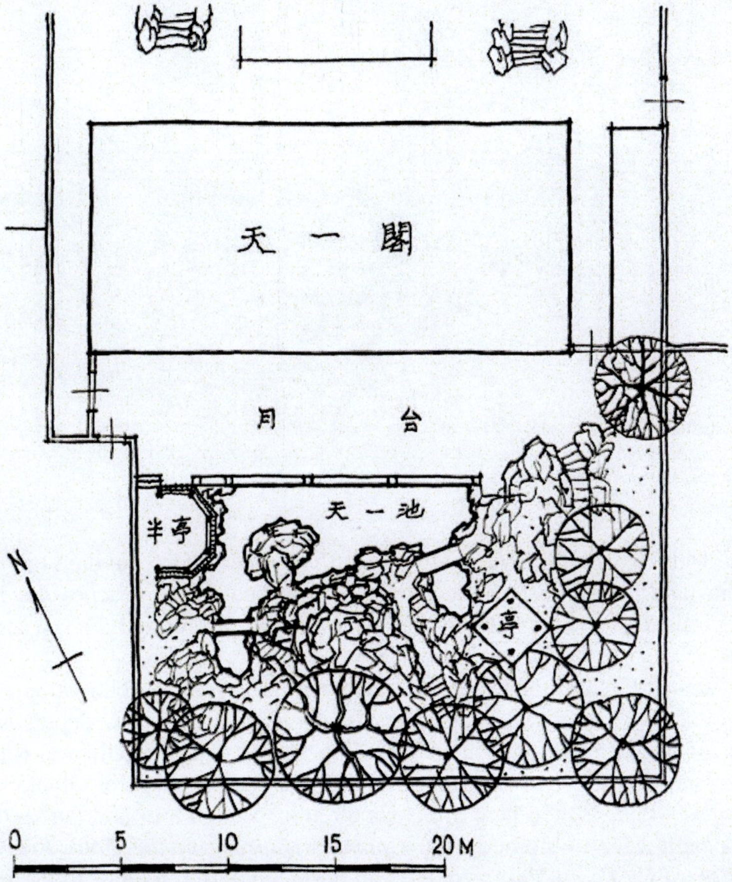

Fig. 4.11 Plan of Tianyi Ge in Ningbo

platform in front of the pavilion, from the perspective of which the water is nearby and the mountain is in the distance (Fig. 4.12). The water is called the Pond of Unity with Heaven (Tianyi Chi) and is shaped to look like a corner of a lake. On its bank, stones are piled to create a peninsular promontory and there are also stepping staircases near the water, serving for people to play at the waterside. A multi-angle half pavilion is set up against the west wall near the water with piled stones, providing a base point to view the lake and mountain in the east. The mountain is stacked with lake stones and a square pavilion is set up in the northern corner among the peaks. The pavilion is arranged obliquely so that the main view stop becomes the vantage point in perspective sightseeing (avoiding rigid front elevation). Though small, the garden masterfully draws visitors into its environment. The route into the mountain-and-forest composition is as follows: A natural mountain path leads from the east side of the moon-viewing platform in front of the pavilion up to the mountain, where located the square pavilion overlooking the pond and the Pavilion

Fig. 4.12 Scenery of Tianyi Ge in Ningbo

of Unity with Heaven (Tianyi Ge) in the north, and beside the pavilion the winding mountain path heads west to the mountain foot and finally reaches the lakeside stepping staircases and the half pavilion near the lake. In summary, it is a successful work.

The famous Pavilion of Unity with Heaven (Tianyi Ge) in Ningbo is a private library built more than 400 years ago. Its owner, Fan Qin, was deputy assistant minister of the Ministry of War in the Ming Dynasty. The pavilion is a low-rise building, showing an obvious style of the Ming Dynasty. Emperor Qianlong in the Qing dynasty once visited here in his southern inspection tour and the seven royal libraries built later to store up the *Complete Library in the Four Branches of Literature* (*Si Ku Quan Shu*) were thus all designed with reference to the pavilion, including The Belvedere of Literary Profundity (Wenyuan Ge) in the Forbidden City, Pavilion of Literary Source (Wenyuan Ge) in the Old Summer Palace, Wenjin Ge in the Imperial Mountain Summer Resort (the above mentioned three are in Beijing), Wensu Ge in the Imperial Palace in Shenyang, Wenhui Ge in Yangzhou, Wenzong Ge in Zhenjiang, and Wenlan Ge in Hangzhou.

4.1.6 Garden of Particles (Canli Yuan) in Suzhou, Jiangsu Province (Small Garden of Mountain-and-Water)

Garden of Particles (Canli Yuan) is located on the right side of house No. 34, Zhuangjia Qiao Lane in Suzhou (the whole group of residence faces southwest, so the garden is on the southeast side of the house) (Fig. 4.13). According to a survey by Professor Liu Dunzhen, the garden was built during the late Qing

Fig. 4.13 The overall plan of the relationship between Canli Yuan of Suzhou and the residence (From Suzhou Classical Gardens)

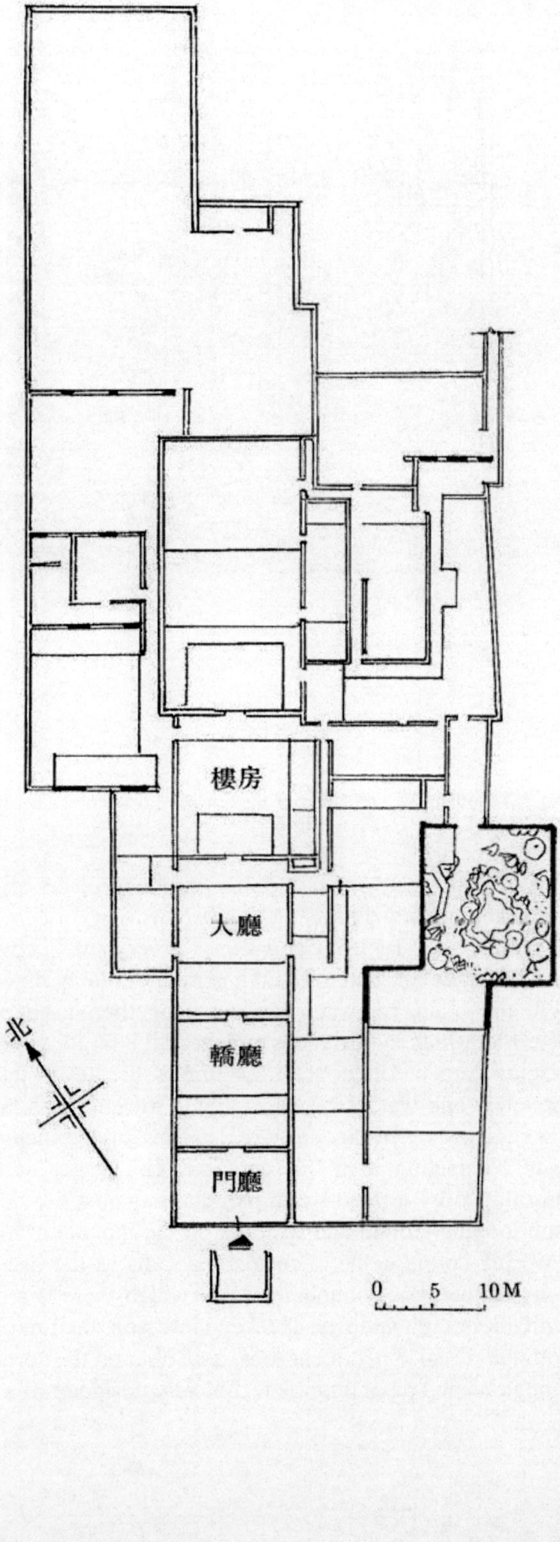

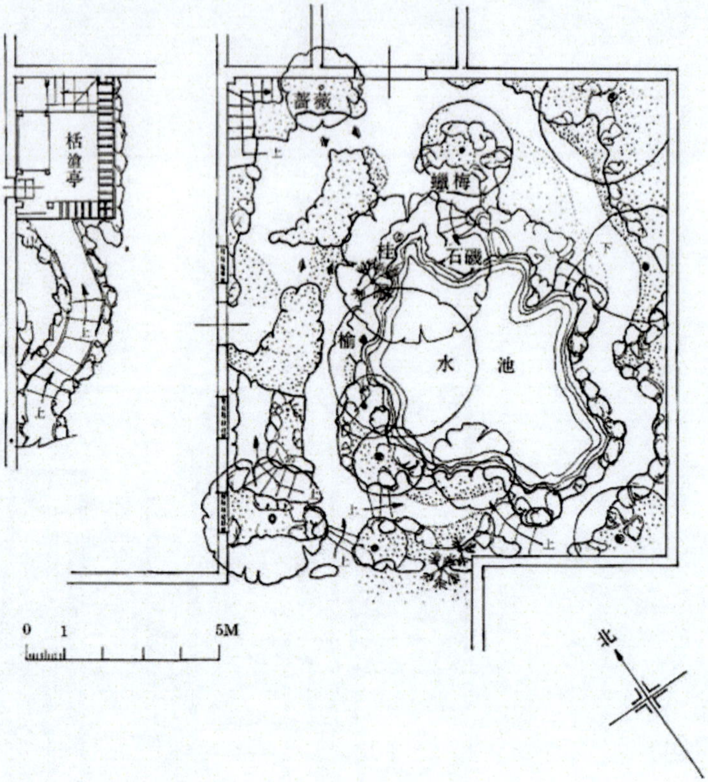

Fig. 4.14 Plan of Canli Yuan of Suzhou (From Suzhou Classical Gardens)

Dynasty and it, together with the residence, was originally the property of a salt merchant in Yangzhou.

Garden of Particles (Canli Yuan) is very small, covering only 140 m^2 (Fig. 4.14) and it is quite difficult to build a garden of mountain-and-water in such a small area. With mountains and water congested together, it is necessary to downsize as much as possible, which in turn tends to exceed the limits and leave the impression of a large penjing garden. Under such conditions, the key is the conception of the subject of mountain-and-water landscape, that is, the choice of landscape type. The designer of the garden showed ingenuity in determining the theme of sunken pool and successfully completing it in this creation. The center of the garden is a sunken pool surrounded by a piled-stone mountain against the boundary of the garden, with a path looping around and winding up the mountain. In the northern corner, with the towering corbie gable of residential halls as the background, positioned the main body of the stone mountain, inside which there is a cave, a natural stone chamber with doors and windows. The key view stop, the Luxuriant Pavilion (Guacang Ting) with the revealed pediment roof, is also set on the northern mountain (Fig. 4.15), and though named pavilion, it is technically designed as a greatly downsized windowed

Fig. 4.15 (**a**) Architectural drawings of Guacang Ting in Canli Yuan of Suzhou (From Suzhou Classical Gardens). (**b**) Guacang Ting in Canli Yuan of Suzhou

veranda without doors on the mountain. It replaces the general hall and serves as the main building, which is a fine example of creativity. The combination of the Luxuriant Pavilion (Guacang Ting) and the stone mountain is just enough to hide the high wall, while the door inside the pavilion leads straight to the inner residence. Drawing on the design of the residential mountain houses in the Jiangnan region, the

Luxuriant Pavilion (Guacang Ting) makes full use of the space: interiorly, a stairway leads down to the stone chamber and a bed is skillfully arranged on the roof of the stairway; externally, a stone bridge is set high on the gully to serve as a passageway connecting to the stone path winding down the mountain.

The plant layout of the garden is very luxuriant, perfectly setting off the shape of the sunken pool. The rise of the rocky shore is exactly what is required for a sunken pool to show the depth of water. This treatment is in stark contrast to the so-called "lacking" treatment given by some professionals. In order to break the monotony of the surrounding banks, a promontory overhanging the edge of the water is arranged, which not only depicts the eroded landscape of the limestone banks but also plays a role in creating remote and endless effects and at the same time serves as an intriguing guide to the water's edge. Since it is shaped into a sunken pool, there is no bridge on the small water surface.

The entrance to the Garden of Particles (Canli Yuan) is a moon gate, as the same with that of Garden of Miniature Landscape as if Contained in a Pot (Hu Yuan), and the plaque on it is titled "Beautiful Rendezvous (Jin ke)", which points out the artistic taste of the small garden. The difference between the two entrances is that the scene in contraposition of the moon gate is quite clear in Garden of Miniature Landscape as if Contained in a Pot (Hu Yuan), which means that upon entering the garden, a corner of the principal scenery (the waterscape) comes into sight-- a stream shaded by plants such as wintersweet and pomegranate, with a pavilion on the stream as the background. While the entrance to the Garden of Particles (Canli Yuan) adopts the skill of concealing. As such, looking inside from the moon gate of the garden, a stone peak representing the mountainscape shows, set off by trees and flowers. Scenes that conceal refers to concealing the principal scenery. While from the perspective of the garden gate, the plants and the stone peak can be viewed as the scene in contraposition of the moon gate. In summary, this small garden is a successful creation.

4.1.7 The Villa of Beautiful Surrounds (Huanxiu Shanzhuang) in Suzhou, Jiangsu Province (Medium Size Garden of Mountain Scape)

The Villa of beautiful surrounds (Huanxiu Shanzhuang) is a national key cultural heritage site and is listed on the world heritage list by UNESCO. Located at No. 280, Jingde Road of Suzhou, it is a famous garden worthy of emphatic introduction.

According to the research of Professor Tong Jun (*Records of Jiangnan Gardens*), this garden was originally the site of the Garden of Gold Grain (Jingu Yuan) owned by Mr. Qian, the local king of Guangling County in the Five Dynasties, and its ownership changed hands repeatedly through the Song, Yuan, and Ming dynasties until its reassignment to Jiangji, an official of the criminal department during the reign of Qianlong in the Qing dynasty. After that, it was transferred to minister Bi Yuan and then to Sun Shiyi (styled Bushan, posthumously titled Wenjing Gong). At the end of the reign of Daoguang in the Qing dynasty, it became the joint property of the Wang family. The Wang clan ancestral hall was set up there, so the garden became affiliated with the ancestral hall and was named "The Farming Land to Shelter Descensdants (Gengyin Yizhuang)", popularly called "The Land of Wang Family (Wang Yizhuang)". According to the book *Miscellaneous Notes in the Garden of Lv*, the lake stone mountain in the garden is the work of Ge Yuliang, an outstanding artificial mountain artist during the reign of Qianlong, and it is, therefore, a precious cultural relic. In fact, the creation of the whole garden is of extremely high caliber. It is obvious that the mountains and water, as well as buildings and plants, are an organic whole, so they should all be designed by Mr. Ge alone. It is a pity that most of the halls, pavilions, and corridors have been destroyed in modern times. Though they have been rebuilt now, the restoration is unfortunately not a full one. Due to poor protection, several key ancient types of wood such as crape myrtle and green maple on the mountain have died, thus destroying the integrity of the mountain-and-forest composition. Here, we will try to discuss the restored one.

The Villa of beautiful surrounds (Huanxiu Shanzhuang) originally occupied an area of 1600 m^2 (Fig. 4.16). As can be seen from the previous example, in most of the small gardens of less than 300 m^2, mountainscapes are built along the walls. Here, the principal scenery covers an area of about 800 m^2, making it feasible to arrange the stone mountain in the middle. The principal scenery indicates the theme of the Villa, so it is a rather bold idea to position a stone mountain in the middle to highlight the theme. As the saying goes, "Boldness of execution stems from superb skill" and so it is impossible for the designer to make such an assumption without the confidence of "working out the plot before putting pen to paper" (Fig. 4.17).

After restoration, the overall layout is divided into two spaces. The southern one is the courtyard of the main hall titled "Villa of beautiful surrounds (Huanxiu Shanzhuang)" (Fig. 3.284) and the other nothern one is the theme scenic space with a stone mountain as the main body. Although the main hall is set in the south and the principal scenery in the north, the relationship between the two is different. The main hall is self-contained and not directly organized in the principal scenery. There is a courtyard surrounded by corridors to the north of the hall. That is to say, the hall is far away from the principal scenery and concealed by an empty corridor, more or less implying that the mountain scenery is to be viewed from the hall. It can be said that the main view stop of the principal scenery is not the hall but the empty corridor near the water in its northern court. Therefore, the main front view of the stone mountain still faces south, taking into account the corridor near the garden gate on the west side. That is, its main front view faces both the west and the south. Strictly speaking, the principal scenery with the stone mountain as the main body forms a relatively independent and complete unit and the Mountain House Complementing Autumn (Buqiu Shanfang) integrated into the scene is the main view stop of immersion into the garden environment.

The principal scenery is shaped mainly by mountains and supplemented by water. The main body of the mountain is stacked with lake stones and is located in the

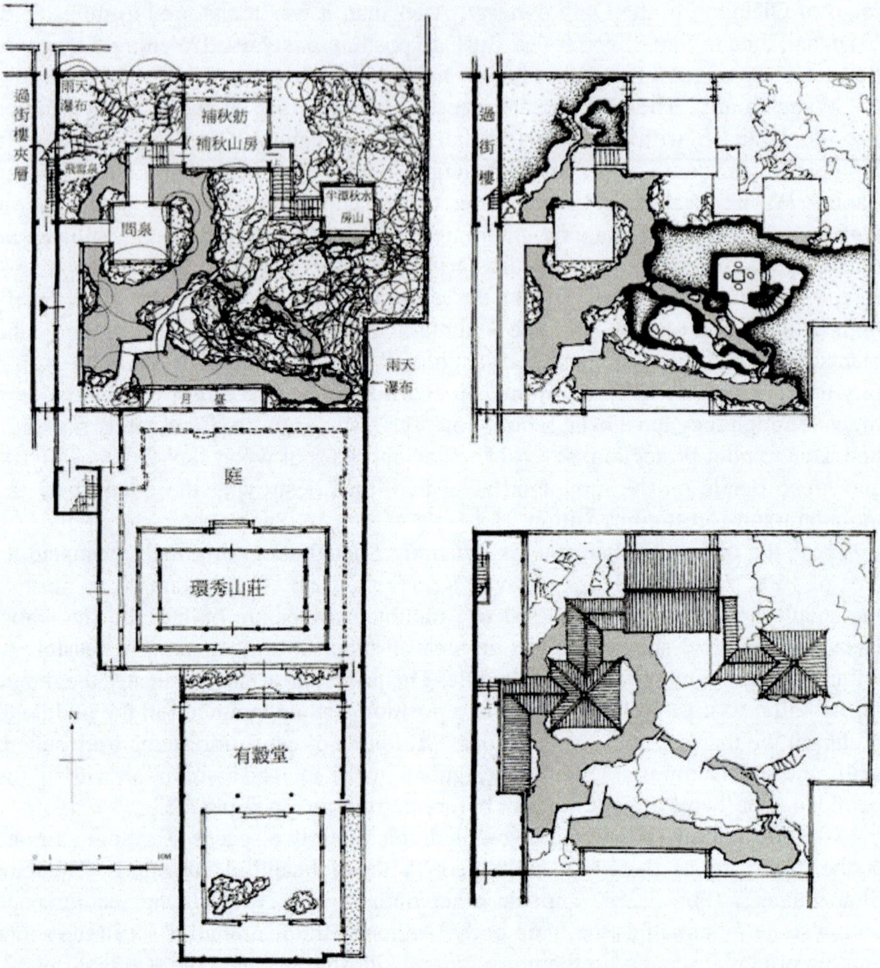

Fig. 4.16 The recuperative plan of Huanxiu Shanzhuang in Suzhou (Drawn according to the plan of 1930s in The Catalogue of Jiangnan Gardens and the garden relics of 1950s)

southeast near the empty waterside corridor. The rest of the mountain range, with its topsoil mostly exposed and interspersed with stones, extends to the northeast and directly reaches the garden boundary. The subordinate body of the stone mountain, on which an island with the Putting a Question to the Spring Pavilion (Wenquan Ting) is located (Fig. 4.18), is arranged in the northwest corner and is connected to the main body by a water surface formed by Flying-Snow Spring (Feixue Quan). The island is surrounded by a stream, on which a stone bridge straddles respectively at the west and north side of the island. The southeast water surface is slightly wider,

Fig. 4.17 Section of Huanxiu Shan zhuang in Suzhou (From Suzhou Classical Gardens)

Fig. 4.18 Wenquan Ting of Huanxiu Shanzhuang in Suzhou

portraying a corner of a lake, which is set off by a three-fold slab stone towpath-style bridge.

The main peak of the main stone mountain is 7.2 m above the water surface and 5.5 m higher than the current moon-viewing platform (Fig. 4.19). And inside the mountain, a gully, 1.5-m wide and 12 meters long, runs northwest-southeast. There is also a cave on each side of the gully. In order to avoid repetitiveness, one cave is natural and the other structured (Fig. 3.16a, b). The southwest one is a natural cave with natural stone tables and stools inside. There are also several stone gaps and holes in the cave wall for ventilation and lighting. Among the gaps and holes, the ventilation holes under the stone table and the space between the stools are the most

Fig. 4.19 The main peak of the stone mountain in Huanxiu Shan zhuang, Suzhou

unique and interesting ones. They not only provide visitors drinking or playing chess at the stone table with good ventilation but also direct down to a corner of the water surface with shining light outside the cave, which, as a spot scene, accurately reproduces landscapes similar to the luminous rocks in Guilin (Fig. 3.20). The northeast side of the gully is carved into a square stone chamber, showing the artistic taste of artificial digging. Towering trees such as Chinese hackberry, cypress, and green maple are planted on the mountain and their shade covers the gully, creating a deep and serene atmosphere in the cave. As for the guidance of the stone mountain, in addition to two inner view stops (the cave and the chamber), a circuitous mountain path is also organized outside as the dynamic view route (Fig. 3.39).

The route from the main hall to the mountain is: First, a slabstone folding bridge leads to a winding waterfront path below the precipice (Fig. 4.20), which crosses a stone hillock with natural stone stairs, turns to a valley (A stone bridge is on the cliff at entrance of the valley) (Fig. 3.21a, b) and then goes through a winding tunnel leading to a cave with natural stone tables and stools. At the exit of the cave, which faces a deep valley, there are stepping stones across the valley directing to a winding

Fig. 4.20 The precipice in the west of Huanxiu Shanzhuang, Suzhou

uphill path among cliffs, which leads upward across a short tunnel, and then forks: one way leads to the mountain shade and bents up to the top of the mountain, reaching the top platform below the main peak through the mountain ridge hanging on the northwest entrance of the valley; the other one turns south to climb the mountain and reaches the top platform below the main peak via the stone bridge on the valley. The top platform below the main peak can be reached by either way, and the other one is thus the downhill road. That is to say, the general mountain route is circular, which allows visitors to avoid backtracking. The top terrace below the main peak can be used for a short rest while providing a bird's eye view of the scenery and it can also serve as a base point in the circular mountain path. The stone mountain covers an area of only 200 m^2 but the main winding mountain path is more than 60 m long, which is a model for expanding touring routes and space by creating mountains with stacked stones.

As for the size or magnitude of the main stone mountain, the main peak is 7.2 m above the water surface, which cannot be considered high. It is the tall trees on it that more than double the mountain height, while the path guidance that keeps the mountain "close up" adds to the height. Among all the view stops, including the western corridor, the northwestern Putting a Question to the Spring Pavilion (Wenquan Ting), the northern Mountain House Complementing Autumn (Buqiu Shanfang), and the square pavilion in the northeast mountain shade, the farthest corridor is just less than 9 m away from the main mountain, thus leaving quite a large viewing elevation angle. The main view stop in the south is the waterfront corridor, which is even closer to the mountain. In addition, it is shaded by eaves, architrave, and hanging corridor so that the tops of the high trees on the mountain are hidden out of sight from the perspective of the corridor, which helps preserve the rich appearance of the mountain-and-forest scenery. The scenic aspect as taken from the

Fig. 4.21 The view taken at the western entrance of Huan xiu Shanzhuang, Suzhou (1956)

southern empty corridor (it can also be taken from the hall and the courtyard through the empty corridor) is not only perfect in terms of the mountain landscape composition (for example, the reason why the crape myrtle on the cliff leans towards the west, lies in that it takes into account the composition of the main front view, as well as the composition of the scenic aspect at the western entrance) but is also rich in terms of the overall composition including the water surface, pavilion, bridge, corridor, etc. (Fig. 4.21).

The main stone mountain is not only unique in its spatial organization but also complete in figuration. From a distant perspective, the cliff, peaks and ridges, valleys, and plant configurations are arranged with distinct gradation, thus creating an overall lofty artistic conception. This is a complete stone mountain in a greatly compressed scale but it has a real feel to it. Minimized as it is, it does not seem fake like "the clay shaped to fool children" as satirized by Zhang Nanyuan. As a famous garden creator in ancient China, Zhang Nanyuan advocates portraying a corner of the landscape, that is, the so-called "standing at the foot of a mountain and cutting off the stream and the valley so as to get a cut-off section of it". There are such scenes with a great degree of similarity with the actual scale of landscape sections in the Villa of beautiful surrounds (Huanxiu Shanzhuang), but they are not in the main

Fig. 4.22 The recuperative corridor in the west of Huanxiu Shan zhuang, Suzhou (Taken in 1989)

body of the stone mountain. In terms of waterscape, the southwest lake is an example. In terms of stone-stacked mountains, one example is the segmented landscape interspersed with stone in the northeast corner, which is just like a corner cut from the foot of the mountain outside the wall (Figs. 3.1, 3.2, 3.3, and 3.4). Another example is part of the stone mountain in the northwest corner. Although cliffs and caves echo the main mountain, the northwest mountain, as a near mountain (jinshan) creation combined with the mountain pavilion (actually a crossing building), is much larger in scale than the main stone mountain. When entering the mountain pavilion through the stairway, the scene feels like the corner of a mountain in actual scale.

The stone mountain is built by piled lake stones and the connections of the stone texture are works of ingenuity. It can be said that the creation is an integrated whole. The large concave-convex area mimics the dissolution of limestone, and the portraying of karst caves inside the mountain body is also very realistic (Figs. 3.25a, 3.26a, and 3.41a). The top of the natural cave is made of lake stone with an arch structure using the so-called "hook bel" construction method, giving an artistic taste like that of real caves. This is exactly what Ge Yuliang advocated: "Only by connecting large and small stone hook belts like building a ring bridge can the cave be like a real one and last forever and then it can be called a successful creation" (volume 12 of *Miscellaneous Notes in the Garden of Lv*) (Fig. 3.17a).

The main entrance of the Villa of beautiful surrounds (Huanxiu Shanzhuang) opens in the space containing the principal scenery, which is a moon gate that opens on the whitewashed wall in the lower corridor of the western pavilion (Fig. 4.22). The guide to the garden is straightforward as the moon gate faces the main stonemountain. Therefore, the west scenic aspect of the principal scenery is also well arranged, and the composition is perfect. Originally, there was a vestibule to position views suitable for viewing in winter to the south of the main hall. To the south of the vestibule, there was a hall called The Hall with Grains (Yougu Tang) and a courtyard was positioned to the south of the hall.

Finally, the creation of springs, waterfalls, and ponds in the Villa of beautiful surrounds (Huanxiu Shanzhuang) as cited in the previous chapter on creation principles is also worth mentioning again. The Flying-Snow Spring (Feixue Quan) was originally located in the northwest of the principal scenery under the lake stone cliff. It was a comprehensive picture formed by a small pool stacked with lake stones and the lakeside inscription “Fei Xue”[1] (Fig. 3.64). There used to be two waterfalls in the area of the principal scenery. One was on the stone mountain in the northwest corner, where water slides down the eaves of the mountain pavilion, forming a enjoyable scene of the spring in rain, while the other, also taking advantage of the rain, was located on the east side of the southeastern main stone mountain and was organized in the western and southern scenic aspects. All that remains of the two waterfalls in the present are thresholds and piled stones (such as water-bearing stones) aligned with the lie of the mountain. The skill of selecting and stacking stones is so superb that the stones from top to bottom look exactly like a natural landscape eroded by the falling water over many years so that even though there is no water, it is reminiscent of the falling waterfall formed by the mountain spring (Fig. 3.25). In addition, there is a small pool on the shady side of the main mountain and beside the pavilion named “Half pool filled with autumn water by the house on the mountain (Bantan Qiushui Yifang Shan)” in the mountain-and-forest composition. The name of the pavilion indicates that the pavilion was set up for this small pool. In addition, the pool is a half pool, with one side stacked with lake stones and the other side positioned at the stone base of the pavilion. It looks like half the pool is under the pavilion or to be specific, the pavilion is built by the pool (Fig. 3.60c).

4.1.8 The Master-of-Nets Garden (Wangshi Yuan) in Suzhou, Jiangsu Province (Medium Size Garden of Water Scape)

The Master-of-Nets Garden (Wangshi Yuan) is a national key cultural heritage site and is listed on the world heritage list by UNESCO. Located in Kuojia Tou Lane in Suzhou, the garden site was originally part of the Hall with Thousands of Books (Wanjuan Tang) in the mansion of Shi Zhengzhi, a Yangzhou native and an official of the Southern Song Dynasty (around AD 1140). After retiring from his post as an assistant minister, Shi lived at the site and the title of the garden was Fishing Hermit (Yu Yin). After the death of Shi, the garden was purchased by Mr. Ding and divided into four parts and the garden was abandoned henceforth. In the middle of the reign of Emperor Qianlong in the Qing dynasty (1770), Song Zongyuan, Vice Qing (Qing

[1] Flying-Snow.

is a official title of in ancient China) of the Guanglu Temple purchased part of it and built the Master-of-Nets Garden (Wangshi Yuan) (still meaning Fishing Hermit (Yu Yin). It was deserted afterwards. Qu Yuancun obtained it during the reign of Emperor Jiaqing and renovated it as well as enlarged it in the reign of Emperor Daoguang, laying the basic pattern seen today. At that time it was commonly known as the Garden of Qu Family (Qu Yuan). Since then, after several periods of prosperity and disuse, it was transferred to Li Hongyi and has since changed the owner again. During the reign of Emperor Guangxu, the Beauty Within Reach Tower (Xiexiu Lou) was added to it. After the founding of New China, it was expanded again with the Cloud Stairway Room (Tiyun Shi), Cold Spring Pavilion (Lengquan Ting) and Azure-Containing Spring (Han Bi Quan) added.

The garden is located in the west and north of the residence, with a current area of approximately 5500 m^2 (Fig. 4.23). Previously, the entrances from the residence to the garden included two. The first entrance, titled "The Fisherman's Small Abode" (Wang Shi Xiao Zhu), is the western garden gate of the sedan chair hall in the front of the residence and the gateway for relatives and friends to enter the garden. The other gate is in the Beauty Within Reach Tower (Xiexiu Lou) at the back of the residence, which is the passageway for the owner and his family to enter the garden. It is now open for public sightseeing and the back door leading to Shiquan Street has been converted into the garden gate.

According to the relationship between the original residence and the garden, the main hall for gathering activities is named the Small Hill and Osmanthus Fragrans Pavilion (Xiaoshan Conggui Xuan), the meaning of which alludes to the line "Osmanthus grove on a hill" from *Call For the Hermit At the Hill* in *Chu Ci*. The pavilion is adjacent and connected by a short corridor to the gate of the garden—The Fisherman's Small Abode (Wang Shi Xiao Zhu). It is actually a four-sided hall surrounded by piled stones. In the north, the main part of the piled stones, the yellow-stone mountain Cloudy Ridge (Yungang), is arranged with osmanthus-centered plant layout echoing the title of the pavilion. Additionally, another gully called The Stream of Retreat (Panjian) is beside the east side of the pavilion. The couplet on pillars in the front of the pavilion points out the landscape here: "The winding mountain really forms a grand painting; The tortuous stream is truly like calligraphy", further extending the artistic conception. The Cloudy Ridge (Yungang) in the north of the pavilion blocks the principal scenery, the shining lake and rippling waves, through real concealment so that the main hall here is not the main view stop for the principal scenery. From the perspective of the main hall, the scenery is arranged in the way of "near mountains and distant water", which is the opposite of the arrangement of "near water and distant mountains". The reasons for such an arrangement are twofold. Firstly, it functionally separated the hall for gathering activities to the south of the lake and the Watching Pines and Appreciating Paintings Studio (Kansong Duhua Xuan) for reading and painting quietly to the north of the lake. Secondly, it draws all relatively large buildings away from the water in terms of composition, thus maintaining the effect of a vast expanse of lake surface (Fig. 3.283d).

a

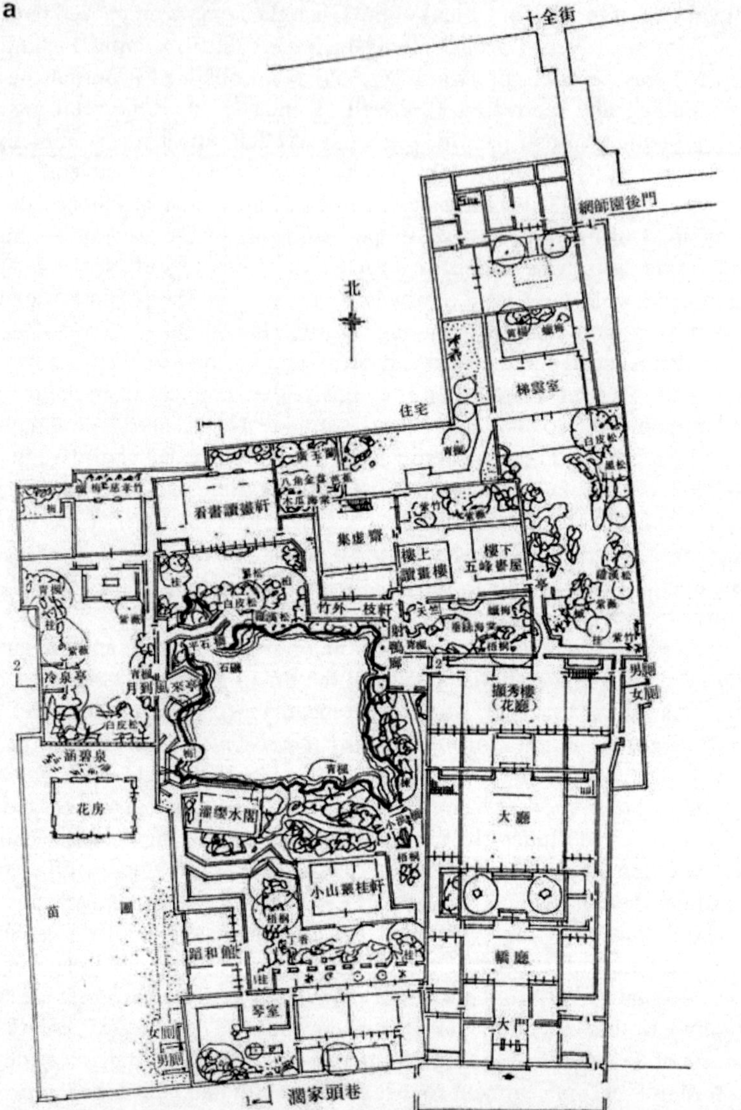

Fig. 4.23 (**a**) Plan of Wangshi Yuan, Suzhou (From Suzhou Classical Gardens). (**b**) Aeroview of the middle of Wangshi Yuan, Suzhou (From Suzhou Classical Gardens)

b

Fig. 4.23 (continued)

The main body of the principal scenery is the water surface: A concentrated lake, which highlights the theme of "master-of-nets" or "fishing hermit". In order to depict the characteristics of the lake, promontories, fishing platforms, and zigzag slab stone towpath-style bridges are arranged in the piled stones along the banks of the inlet extending northwest from the lake. In this way, not only can a picturesque view be seen from the Cloudy Ridge (Yungang) in the southeast but also a far-reaching effect can be achieved due to the increase in the absolute length of diagonal lines and gradation formed by the promontories and folding bridges. On the west side of the Cloudy Ridge (Yungang), a relatively small lakeside pavilion overhead the lake called the Waterside Pavilion of Washing Hat-Ribbons (Zhuoying Shuige), which, as an extension of the separated main hall, serves as a view stop for the main waterscape. On the west bank of the lake, a hexagonal pavilion (The Moon Comes with the Breeze Pavilion (Yuedao Fenglai Ting) is set up on the piled-stone ravine as the middle view.

The scenic structure of the north shore is different from that of the south bank. It adopts a formal barrier with sparse mountain forests. Black pines, Rohan pines, pinus bungeanas, junipers, and other trees are interspersed among stacked stones, slightly covering up the back of the Watching Pines and Appreciating Paintings Studio (Kansong Duhua Xuan). As a matter of fact, garden living mainly uses the Watching Pines and Appreciating Paintings Studio (Kansong Duhua Xuan), the

Meditation Study (Jixu Zhai), and Prunus Mume Pavilion (Zhuwai Yizhi Xuan), which are the main view stops for the principal scenery. The view to the south of the Watching Pines and Appreciating Paintings Studio (Kansong Duhua Xuan) follows a diagonal line with the largest possible length. With a harbor at the southeast corner and a small-scale stone arch bridge increasing perspective, it creates a pronounced far-reaching effect. The Meditation Study (Jixu Zhai) to the east of the Watching Pines and Appreciating Paintings Studio (Kansong Duhua Xuan) is the commanding elevation of the whole garden, in front of which a low-rise wide corridor, the Prunus Mume Pavilion (Zhuwai Yizhi Xuan), is added to serve as a partition between the lake, thus keeping the size of the lake unaffected. This shows a successful creation method. That is, in order to create a vast and far-reaching view of the main lake, all relatively large buildings will be kept a step back. As for the close view of the lake, the relatively small pavilion, corridor, and verandas, as well as the piled stones and plants representing mountain features are all used to provide static view places and block large-scale buildings, thus keeping them far away from the waterscape.

With the lake as its main body, the principal scenery of the garden is an open space. The Waterside Pavilion of Washing Hat-Ribbons (Zhuoying Shuige), The Moon Comes with the Breeze Pavilion (Yuedao Fenglai Ting), the Prunus Mume Pavilion (Zhuwai Yizhi Xuan), and the Duck-Shooting Veranda (Sheya Lang) are all directly organized in this open scene. The Watching Pines and Appreciating Paintings Studio (Kansong Duhua Xuan) and the Meditation Study (Jixu Zhai) are implicitly and indirectly connected with the theme scene and can all be described as semi-open spaces. In order to create a deep and colorful effect in the garden, there are also several relatively independent enclosed scene spaces. These include the Small Hill and Osmanthus Fragrans Pavilion (Xiaoshan Conggui Xuan), the Cloud Stairway Room (Tiyun Shi), the Five Peaks Library (Wufeng Shuwu), the Music Room (Qin Shi), the Peace and Tranquility Guest House (Daohe Guan), and the Peony Study (Dian Chun Yi). The aforementioned sceneries, separated from the principal scenery by real concealing views, become different types of "garden in a garden".

Such structures have complete art of composition due to their relative independence. Take the Peony Study (Dian Chun Yi) as an example, its main view stop, named after the poetic verse "Peony is still following the late spring wind (peony opens at the end of spring)" is in the form of a flower hall (Fig. 4.24). Peonys, matched with plants such as plum, bamboo, Panicum maximum and plantain, are planted in the lake stone flower terrace (the scene in contraposition of the study) to echo the title, and plantains are also planted under the window. Therefore, the couplet on pillars of the study reads "The birds prepare their nesting when air of spring fills; The plantains shelter windows as rain of night chills" (Fig. 3.259b). Small as the court garden is, it has both mountains and water. The stone-piled planting beds and stone peaks against the east, south, and west walls represent mountains. While in the southwest corner, a shadowy and unfathomable deep pool (Azure-Containing Spring<Han Bi Quan>) is positioned as the waterscape (Fig. 4.25 and Fig. 3.60e), beside which Cold Spring Pavilion (Lengquan Ting) is set up as the view stop for close viewing. The pavilion contains an elegant cultural

Fig. 4.24 Dian Chun Yi in Wangshi Yuan, Suzhou

Fig. 4.25 The scene in contraposition of Dian Chun Yi in Suzhou's Wangshi Yuan – the flower terrace of peony, stone peak, Lengquan Ting and Lengquan sunken pool

relic (the Lingbi Stone), reportedly a relic of the residence of Tang Bohu, a great painter of the Ming Dynasty (Fig. 3.244f). Modern painter Zhang Daqian once lived here and buried the remains of his beloved tiger under the sidewall of the study, adding a spot scene with a natural taste and sentimental vibe.

4.1.9 *Villa of Reposing Roars (Jixiao Shanzhuang) in Yangzhou, Jiangsu Province (Medium Size Garden of Mountain-and-Water)*

Villa of Reposing Roars (Jixiao Shanzhuang) is located in Garden Lane, Xu Ningmen Street, New City Region of Yangzhou. It is a court garden built in the ninth year of the reign of Emperor Guangxu (AD 1883) by its owner He Zhifang, the Daotai (an official title in ancient China) at the time. The garden, taking the meaning of Tao Yuanming's poetic verse "Leaning against the southern window to repose pride; Climbing the eastern hillside to roar aloud", is titled "Villa of Reposing Roars" (Jixiao Shanzhuang), commonly known as the Garden of He Family (He Yuan). The residence of He Family was originally another estate of Wu Jialong, where the famous Schistose Mountain House (Pianshi Shanfang)—Garden with two pagoda trees (Shuanghuai Yuan) once stood. There are still some remaining stone mountains, which are reportedly Shi Tao's works, and a hall made of nanmu built during the Ming dynasty in the residence.

The villa is located in the rear (north) part of the residence, with which it is connected. The north wall of the garden is adjacent to Diaojia Lane, where an independent garden gate is set up as access for guests (Fig. 4.26). The residence and the garden are separated by a two-way corridor with mixed types of ornamental

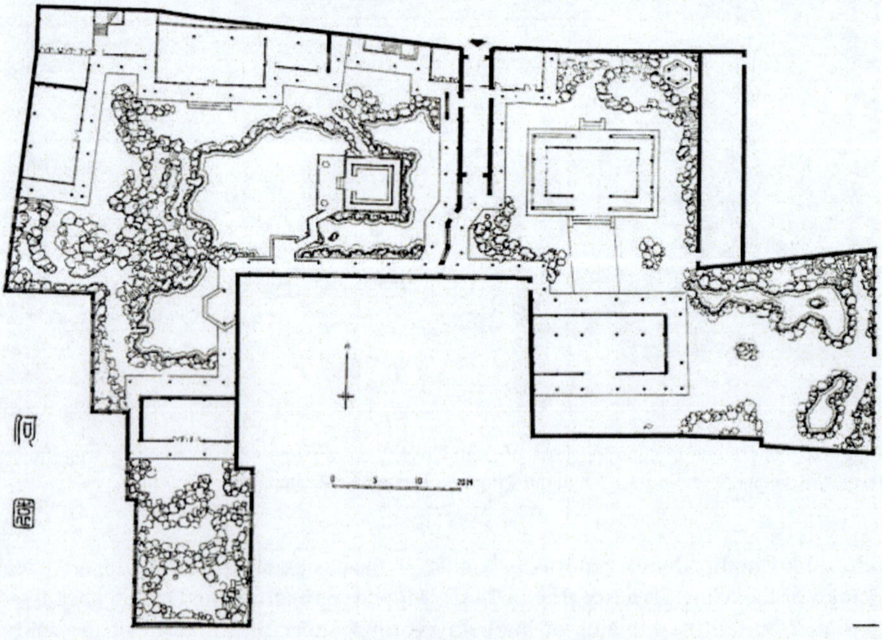

Fig. 4.26 Plan of Jixiao Shanzhuang in Yangzhou (He Yuan) (From The Catalogue of Jiangnan Gardens)

perforated windows and part of the garden can be faintly seen from the rear court of the residence. The principal scenery of the garden is in the west, with waterscape as the main body. The type of water surface is a lake, with the main water surface flowing southward forming a stream. An island is positioned in the east of the main water surface, on which a pavilion called "Little Square Pot (Xiao Fanghu)" is built for enjoying the cool air and to serve as a stage for performances (Fig. 4.27) since the owner likes traditional opera. The main view stop (a row of pavilions) is set to the north of the lake. It is rare in Suzhou gardens to set pavilions instead of halls as the main view stop as well as positioning the main view stop to the north of the main front view. The building has two floors, with three rooms as the main body and two subordinate rooms on the left and right wings, so it is commonly known as the Butterfly Hall (Hudie Ting) (Fig. 4.28). The corner between the main building and the east wing building is planted with tall palm trees and the suitable position helps it stand out (Fig. 3.81b). The unique characteristic of the scenery lies in the configuration of the two-way corridor. The well-proportioned zigzag corridor surrounding the principal scenery, connected by buildings, provides a way for a high-angle, continuous view overlooking the garden, thus obtaining a different sequence of scenic aspects, which is unprecedented in general gardens.

In the southwest of the water, there is a stone mountain with a tunnel and a mountain stairway, forming a self-contained area (Fig. 3.15e). A building is positioned to the south of the mountain that can be reached through the natural stone-piled stairway. The building connects to the winding surrounding two-way corridor and is also the entrance and exit of the residence connected to the two-way corridor inside the rear residence. The two-way corridor along the garden inside the rear residence was originally a place to watch operas and this is exactly the

Fig. 4.27 Xiao Fanghu of Jixiao Shan zhuang in Yangzhou (Taken in 1989)

Fig. 4.28 Main Building of Jixiao Shanzhuang in Yangzhou–Hudie Ting (Taken in 1989)

aforementioned place where mixed types of ornamental perforated windows are opened on the corridor.

To the east of the two-way corridor along the east side of the water, there is another scenic area (Fig. 3.217e and Fig. 3.111a). Large ornamental perforated windows of special styles are opened on the upper and lower floors of the two-way corridor, thus forming a penetrating partition between the eastern scene and the western principal scenery. The ornamental perforated windows in this garden are made with exquisite craft and unique pattern and their style is different from that found in Suzhou and other places (Fig. 3.216g, and Fig. 3.131a, b). The center of the eastern district is a four-sided hall and mountains are shaped along the east and north boundary walls. In the northwest corner, a natural stone step leads to the building and a spot for admiring the moon, The Platform Accompanying the Moon (Banyue Tai) in the corridor, which can also lead to the western scenery.

4.1.10 Garden that Is near Enough to Be Called a Garden (Jin Yuan) in Changzhou, Jiangsu Province (Medium Size Garden of Mountain-and-Water)

The garden was originally a private garden owned by Yang Qingyan, a Piling (the ancient name of Changzhou) native in the early Qing Dynasty. According to the postscript to Wang Shigu's *The Records of garden that is near enough to be called a garden (Jin Yuan)* written by Da Chongguang, a Runzhou (the ancient name of Zhenjiang) native of the same age with Yang Qingyan, the garden was designed by

Yang Qingyan, the owner of the garden. Since modern times, it has been owned by the Yun family, commonly known as the Garden of Yun Family (Yunjia Huayuan) (Fig. 4.29).

Its principal scenery is a mountain surrounded by water. The so-called mountain is actually a big island in the lake, where there is a mountain serving as the main body (Fig. 4.30). The main view stop facing the mountain is positioned to the north of the lake—a hall named after the residence of Shao Yong of the Song Dynasty as The Cozy Nest (Anle Wo) (Fig. 4.31). A waterside platform is positioned in front of the hall, while the water between the platform and the mountain on the island is not wide. In this way, the mountain is drawn closer to exaggerate its size and produces the effect of approaching a corner of the mountain. A pavilion is built behind the mountain as a foil, and there are mountains along the garden wall positioned on the opposite bank of the mountain on the island as its distant background from the perspective of the Cozy Nest (Anle Wo), which are also used to conceal the garden wall and expand the depth of field. At a corner of the lake, Getting the moonlight Veranda (Deyue Xuan), and Peony Pavilion (Tianxiang Ge) are arranged (Fig. 4.32), with the name of the veranda echoing the poetic verse "A waterfront pavilion gets the moonlight first" and the name of the pavilion indicating that there might be tree peony nearby the pavilion (In the minds of all Chinese, peony equals national beauty and heavenly fragrance, which reads Guose Tianxiang in Chinese, hence the name

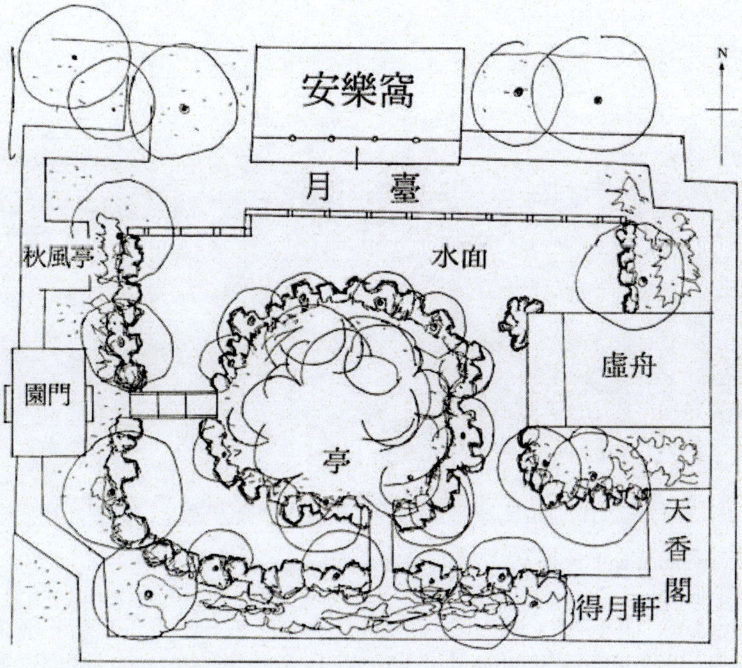

Fig. 4.29 Plan of Jin Yuan in Changzhou

Fig. 4.30 The main front view of the mountain on the island of Jin Yuan in Changzhou-- facing the main hall "Anle Wo"

Fig. 4.31 The main hall "Anle Wo" of Jin Yuan in Changzhou

Tianxiang Ge). There is a winding corridor leading to the garden gate beside the Getting the moonlight Veranda (Deyue Xuan), and the entrance directly faces a small bridge leading to the mountain on the island, where another bridge directs to the distant mountain along the garden wall.

Fig. 4.32 Deyue Xuan and Tianxiang Ge in Jin Yuan of Changzhou

Take the western garden gate as the base point, with the mountain on the island in the middle, the Virtual Boat Pavilion (Xuzhou Shuixie) is set up on the eastern bank, which serves as a view stop for side views of the mountain on the island, thus forming an axis from the garden gate to the Virtual Boat Pavilion (Xuzhou Shuixie) (Fig. 3.164b). The short folding corridor on the right side of the garden gate leads to the Getting the moonlight Veranda (Deyue Xuan) and the long corridor on the left which may not be the same as the original one as it is not connected with the Cozy Nest (Anle Wo), leads to the platform of the Cozy Nest (Anle Wo) in the main hall via Autumn Breeze Pavilion (Qiufeng Ting). The left and right corridors of the Cozy Nest (Anle Wo) are asymmetrical but a balanced layout is nevertheless created. The right corridor is one-folded and is complemented by a whitewashed wall with a moon gate (titled "A Song of Jian Lake <Jianhu Yiqu>") and a small courtyard. The left corridor is twofolded (with the decoration of stone carvings on the corridor wall) and leads to the Virtual Boat Pavilion (Xuzhou Shuixie). As far as the guidance of the whole garden is concerned, the Cozy Nest (Anle Wo), the Virtual Boat Pavilion (Xuzhou Shuixie), and the Getting the moonlight Veranda (Deyue Xuan) together with the Peony Pavilion (Tianxiang Ge) are the three main view stops positioned in a tripod layout, while the routes include the circle along the water and the small circle on the island.

The densely planted mountain on the island in the garden is indeed a work of ingenuity.

4.1.11 Garden of Art (Yi Pu) in Suzhou Jiangsu Province (Medium Size Garden of Mountain and Water Scape)

Garden of Art (Yi Pu) is a Key Cultural Relics Site Under the State Protection and is listed on the World Heritage List by UNESCO. Located at No.5, Wenya Lane, Suzhou City, it is an ancient and famous garden, owned by Yuan Zukang in Ming Dynasty, later purchased by Wen Zhenmeng (Wen Zhengming's great grandson), named "Herbal Garden" (Yao pu) at that time. In the transitional period between Ming Dynasty and Qing Dynasty, Jiang Runong, who is from Laiyang, Shandong province, lived here and changed its name to "Garden of Art" (Yi Pu) or "Jingting Mountain House" (Jingting Shanfang). The garden is located in the southwest of the house and covers an area of about 5 μ (more than 2000 m^2) (Fig. 4.33). In the early

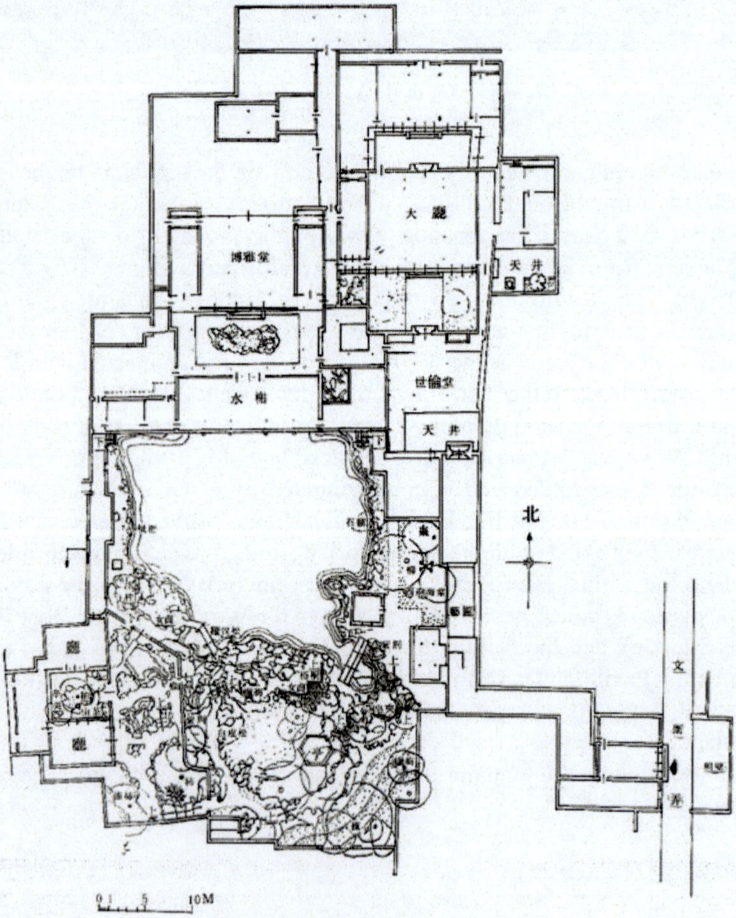

Fig. 4.33 Plan of Yi Pu, Suzhou (From Suzhou Classical Gardens)

1960s, according to Mr. Liu Dunzhen,[2] the garden was relatively intact but different from the drawing by Wang Yun (a.k.a. Shi Gu) during the reign of Emperor Kang Xi of the Qing Dynasty. The existing main building by the lake, Shui Xie (waterside house), replaced Yue Tai (viewing platform) in the drawing, on the west of which used to be the "Jingting Mountain House" (Jingting Shanfang). According to Mr. Liu's book *Suzhou Classic Gardens*, there were also changes in waters and folding bridges.

About three fifths of the Garden of Art (Yi Pu)'s main landscape area is one water, which creates a wide visual effect. This water area belongs to the type of lake-swamp, and a waterside house is set on the north side of the lake as a main viewing point. The cantilever pavilion facing lake reminds people of the Ganlan (pile-supported) dwellings in the water village. It is ordinary, which is reflected in the shape of a row of 31-m long straight facades formed of five rooms together with two side chambers, the smooth roof and the straight eaves. Even compared with around buildings, the different heights of ridges, which are naturally required by structure and function without any organized modification, further strengthen its simplicity, quality and rugged local flavor (Fig. 3.159). The main building is in harmony with water and highlights the theme of marsh. The arrangement of aquatic animals and plants is also extremely successful—lotus, algae and wild swimming fish (other than cultivated goldfish), which all focus on the same theme. It is rare in the existing examples that the main building is not a hall, and there is no atmosphere of a rich mansion. It is the detail treatment that established the natural and idyllic wild interest of the main scene, which was once common in the Ming Dynasty. So it is rare to find the treatment in the reconstruction after the reign of Emperor Kang Xi of the Qing Dynasty.

The southeast peninsula, the rocky ledge, the ladder-shaped stone bridge at the bay, and the zigzag bridge lowly above the water surface at southwest bay, they all emphasized the marsh features (New works with neat decoration have now replaced some precious ancient stone bridges and stone beams) (Fig. 4.34, Figs. 3.170b and 3.174). To the south of the lake, where is far away from the main scene, an earth slope lies down the south wall looking like a corner of a mountain, with scattered stone bones buried. This is exactly the embodiment of Zhang Nanyuan's[3] theory (Fig. 4.35). Beside the hillside, another earth hill is built to serve as the background of water landscape, set with the same scattered stone bones, lush trees, and small pavilions on the hill. It is indeed very harmonious to set such plain hills here, as the stones and lush trees of the hill makes it look like a corner of the the natural foothills. Facing to the north, the main scene is on the shady side of the backlighting, so that it looks even more profound. This, as Zhang Nan Yuan said, created the illusion that the outside of the garden was the mountains, which produced a masterpiece with *yi jing* (art conception). Along the south bank of the lake, the stacked stone hills may be increased in the Qing Dynasty. This superfluous work is even more deplorable with

[2] Liu Dunzhen 1897.9.19–1968.5.10, Chinese architect.

[3] Zhang Nanyuan, garden artist in the transitional period between Ming Dynasty and Qing Dynasty.

Fig. 4.34 The zigzag bridge with one bend in Yi Pu, Suzhou

Fig. 4.35 The remaining corner of the earth mountain in the southeast corner of Yi Pu, Suzhou (it has been changed from the scene that stones were exposed out of the soil cover at the foot of the mountain to the recent Shilejiao practice) (Taken in 1989)

trivial stones piled up without harmonious texture (however it is much better later with ornamental plants after the renovation) (Fig. 4.36). We have to say that the stone hills along the lake destroyed the natural interest of the earth hills, and even destroyed the artistic conception of the whole scene. There is a great disparity between earth hills and stone hills, obviously the last one was built by later generations who did not understand the original theme of the scene, making an unworthy continuation of a great work. It was not the original work at the beginning of the garden making in Ming Dynasty.

In the southwest corner of the main scene, there is a detached scene separated by a whitewashed wall. A rimless moon gate is opened on the wall facing the lake in the northeast, connecting the detached view in the enclosed space with the open main scene of the lake. To lend a sense of villa, there are streams and stone beams among the rocks in the courtyard (Fig. 4.37). To deepen the scene, there is a small courtyard when entering through another moon gate (Fig. 3.293b) in the courtyard. The inner courtyard has two halls face to face, and there is a flower terrace made of lake stone to show the mountain environment. In the 1950s, rabbits were naturally raised here, which added to the wild interest. The north hall of this deep villa courtyard has a gate leading north to a half corridor, which is the western background of the main scene. You can watch the water village scene along the corridor. This corridor leads to the main building, the waterside house, where a flower terrace made of stone is set in the middle of the backcourt, facing the former main hall, the Boya Hall (Boya Tang) to the north. In the past when there used to be no waterside house in front of the hall, the main scene could be seen directly from the hall, a marsh in the background of distant flat earth hills (added with the stone hills in the front).

On the side of Boya Hall (Boya Tang), there is a door leading to the inner residence, which is the main passage for the owner to enter the garden. In the northeast corner of the main scene of the garden, there is a gate leading to the front hall of the residence, Shilun hall (Shilun Tang), which is the way for guests to enter the garden. The winding corridor and the hallway leading to the garden provide a tertium comparationis for viewing the main scene. When entering the main scene

Fig. 4.36 The main front view of the stone mountain in Yi Pu, Suzhou (From Suzhou Classical Gardens)

Fig. 4.37 The southwestern courtyard of Yi Pu, Suzhou

from the hallway, you can see the lake and sky at first sight and feel relaxed and happy.

4.1.12 A Small Heaven and Earth Created (Xiao Pangu) in Yangzhou Jiangsu Province (Medium Size Garden of Mountain and Water Scape)

Some gardens in Jiangnan area are named after the main buildings, such as Little Lotus Manor (Xiaolian Zhuang) in the town of Nanxun and Garden of Mist and Rain Pavilion (Yan Yu Lou) at Nan hu in Jiaxing. However, this garden is named after the mountain. In Yangzhou, there were originally two places with the mountain title of "A Small Heaven and Earth Created" (Xiao Pangu), one was the Yellow Stone Mountain of Qin's Yiyuan Garden in Tangzi Lane at the South Gate of the Old City. The garden was built by Ge Yuliang,[4] but unfortunately ruined later, only remaining relics of the A Small Heaven and Earth Created (Xiao Pangu).

The existing garden named "A Small Heaven and Earth Created" (Xiao Pangu) is located in Dashu (big tree) Lane, which was the residence of Zhou Fu, an official who served as Governor during the reign of Guangxu in the late Qing Dynasty. Zhou bought the garden from the He family and rebuilt it, so it can be deduced that the

[4]Ge Yuliang, a garden master, born in Changzhou in late period under the reign of the Emperor Qianlong in the Qing Dynasty.

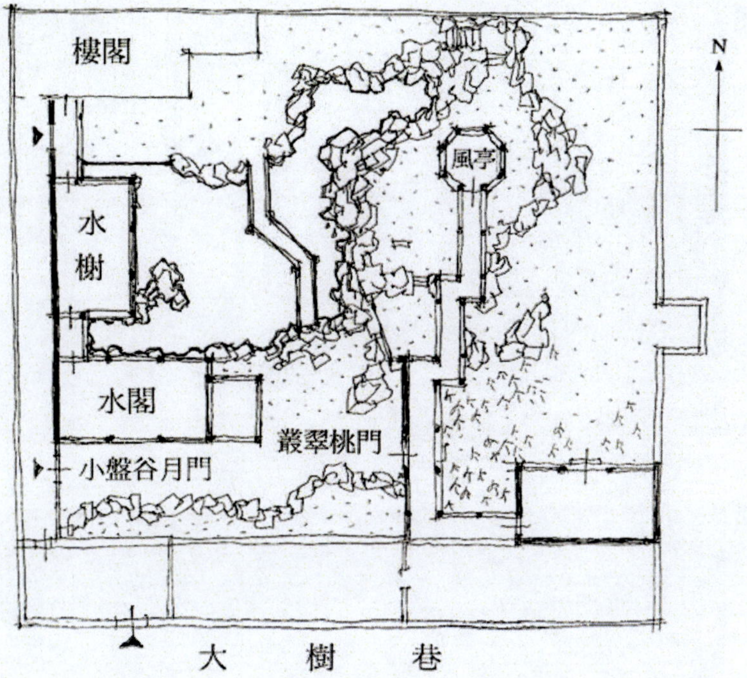

Fig. 4.38 Plan of Xiao Pangu in Yangzhou (before reconstruction, drawn according to the illustration of Zhu Jiang' article in Cultural Relics: A Brief Talk on Yangzhou Classical Garden)

garden was built even earlier. In recent years, common faults of arbitrary changes in the ancient gardens has also happened in the reconstruction of this garden, such as replacing Xie (house on a terrace) with pavilions, changing octagonal pavilions on the mountain to quadrangular pavilions, etc. The basic pattern is still the same as that of the late Qing Dynasty.

The garden covers an area of less than 2000 m^2 and is located in the east of the residence. You can enter the garden through the rimless moon gate near the hall inside the residence (Fig. 4.38). The moon gate is inscribed with the inscription "Xiao Pangu",[5] which is said to be written by Chen Hongshou (a.k.a Mansheng), one of the "Eight seal carvers of Xiling Seal-engravers' Society (Xiling Yinshe)" during the reign of the Emperor Qianlong in Qing Dynasty. The design of the garden especially emphasizes the convenient connection between the garden main scene and the residence, with the main scene arranged on the north side of the garden gate. After reconstruction, main buildings are a hall facing the water in the north and a square warm pavilion facing the water in the east (the pavilion is reconstructed from Xie), which are connected by a short corridor (Fig. 4.39). The water is concentrated near the main scene, winding its way to the northeast and forming the a style of

[5] A Small Heaven and Earth Created.

a

b

Fig. 4.39 (**a**) A corner of the main hall of Xiao Pangu in Yangzhou; (**b**) The warm pavilion in Xiao Pangu of Yangzhou

Haopu[6] combined with the stone hills. The main viewing direction of the north-facing hall is in the northeast, extending the depth of the scene obliquely, and showing a scene of rolling hills and endless streams. At the northern end of Haopu, an L-shaped building is set up as a northeast corner (Fig. 4.40) of this scene. The main viewing direction of the east-facing pavilion is due east, with a view of the cliff from water. A stone-slab zigzag bridge was built as the middle scene on Haopu (Fig. 4.41 and Fig. 3.3c). Along the water is a stone mountain on one side and

[6] A narrow water between mountains.

Fig. 4.40 The curved-shaped building in Xiao Pangu of Yangzhou

its buildings are low near the water surface, which also contributes to the feeling of facing the cliff. And the bridge on the water surface is not abrupt, which is different from the that of Couple's Retreat Garden (Ou Yuan) in Suzhou.

The overall structure of the garden is divided into two major scenes in the east and west, with the west being the main scene. The north-south line partition is formed by corridors, *yun qiang*[7] and stone hills. The southern section of the veranda is a *fu lang,*[8] while the northern section is a *pashan lang*[9] (Figs. 3.144f and 3.145). The stone hills are piled up of lake stones, with a peak height of more than 9 m, and there is a cave with furniture in it. Lighted by rock cavity, the cave has three entrances (which are also used for lighting), so it is a passage of the tunnel style that you can pass through it. As the stones to build mountains mostly comes from other places to Yangzhou, for the convenience of transportation, they mostly in small pieces. Therefore, building a large concave-convex stone mountain with small stones is typically Yangzhou's own style. The mountain body is composed of nine combination individuals of small stones. To prevent the plain appearance, each combination has its own changes of shape, like a lion's head, body and tail, so this mountain is also called a "nine-lion mountain". It is a popular meaning of building a stone mountain in Yangzhou. Volume 2 of *Yangzhou Painted Boat Record* reads: "Taoist Dong from Huai'an was known for building Nine-lion mountain in the reign of the Emperor Qianlong in Qing Dynasty."Therefore, he was said to be the producer of this "Nine-Lion Mountain" in the garden. It is a representative work of the high-level

[7] Cloud wall, with curved decorations like cloud on the top of the wall.

[8] Double corridor.

[9] Hill climbing corridor.

Fig. 4.41 The main peak of "Jiushi Shan" (probably by Dong Daoshi during the reign of Emperor Qianlong) and the zigzag bridge on the Haopu in Xiao Pangu of Yangzhou

stone mountains in the existing Yangzhou gardens. Its peaks, cliffs, caves, valleys, cliff rocky, stepping-stones and terraces are all impressing, and the organization is quite natural, compact and artistic.

Nine-lion Mountain is the main body of the composition of the main scene, and other parts are arranged on the right hand side of the garden in the south, with a double corridor connecting the two. A cloud wall is arranged between the foothills at the northern end of the double corridor, which can set off the mountain scenery and add interest of a small space. There are assorted windows on the cloud wall, which weakens the wild interest of forest landscape. This is the situation now after reconstruction, whatever the unknown past decorations were, it is a manifestation of the traditional local style.

Overlooking the whole park, an Octagonal pavilion is set on the flat ground at the top of the mountain between the peaks of Nine-lion Mountain. Originally known as the "Wind Pavilion" (Feng Ting), the pavilion is now rebuilt (Fig. 4.42).

Fig. 4.42 Feng Ting on the mountain of Xiao Pangu in Yangzhou

The eastern scene has been completely destroyed and is still being rebuilt. The connection between the eastern and western scenes is mainly through "Congcui", the peach-shaped rimless gate (Fig. 3.117). The gate is on the middle wall of the double corridor, which is opposite to the entrance to the moon gate. In the past, through Congcui, you could enter the eastern scene of another interest with plants from the main scene of stone hills and waters. The peach-shaped gate inscribed "Congcui", which means green bush, indicates the theme of the eastern scene. It can be restored and renovated into a bamboo forest in the future, so that the small hall in the southeast corner can be used as a good place to enjoy the bamboo, the wind and the piano, or the rain at night.

4.1.13 Couple's Retreat Garden (Ou Yuan), Eastern Garden in Suzhou, Jiangsu Province (Medium Size Garden of Mountain and Water Scape)

Couple's Retreat Garden (Ou Yuan) is a Key Cultural Relics Site Under the State Protection and is listed on the World Heritage List by UNESCO. The garden is located in the No.6 residence of Xiao xin qiao (small new bridge) lane in Suzhou. The residence has two gardens, east and west, so they are called "Couple's Retreat Garden" (Ou Yuan) (Fig. 4.43). The western garden is called "Western Garden", consisting of two small courtyards before and after, one of which is the library and the other is "Old House with Woven Curtains" (Zhi Lian Lao Wu). Both courtyards are small gardens of mountain landscape (Fig. 4.44) with stone flower terraces and

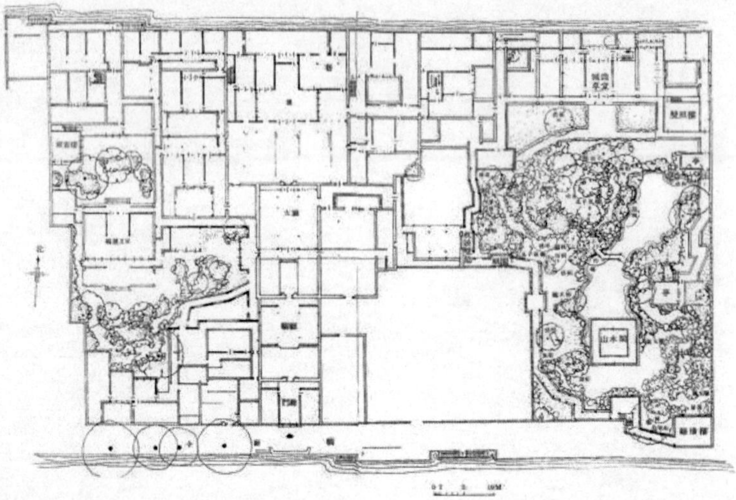

Fig. 4.43 The General plan of Ou yuan in Suzhou–with residence in the middle, and the eastern and western gardens on the left and right (From Suzhou Classical Gardens)

Fig. 4.44 The hall of learning in the Western garden of Ou yuan in Suzhou

peaks. What is discussed here is the large garden in the east of mountain and water landscape, commonly known as the "Eastern Garden".

The East Garden covers an area of about 2400 m^2. According to Professor Liu Dunzhen's research, the garden was built in the early Qing Dynasty and was the residence of Lu Jin, the governor of Baoning prefect. It used to be called "She Yuan", also known as "Xiao yu lin". The garden was once abandoned, but after many times of changing the Lord, in the late Qing Dynasty, Shen Bingcheng, an official, owned the garden and rebuilt it. At the same time, the western garden was added, thus forming a pattern of east and west in the house. Therefore, it was called "Couple's Retreat Garden" (Ou Yuan) (*Suzhou Classical Gardens*). The garden has been in disrepair for a long time since modern times, and has been restored after the founding of the People's Republic.

The creation of the Eastern Garden is unique in conception and originality. The main building is designed as a tower instead of a common hall style. This building faces south, and there are three main halls downstairs, which are entitled "Thatched Cottage at the City Corner" (Chengqu Caotang). The east compartment of the main building is equipped with a side building, called "House of Double Reflection" (Shuangzhao Lou) (Fig. 3.81e). The outer corridor of the mansion is connected with the front corridor of the main building, forming an L-shape on the ground floor, but the top is independent of each other. The two buildings connected inside through double corridors, with three patios for ventilation and lighting. The building opens to the inner residence to the west. In the front of the building, a viewing terrace is set against the main hall. Beside the building, there is no flower terrace but a lawn, which is also a way to break the routine (It has been renovated in recent years and floors have been paved).

The view of the main building is not usually arranged according to the law that water is near and mountain is far away, but is instead treated near the mountains and far from water. The building is slightly facing the west, about six or seven meters apart, there is a yellow stone mountain covering an area of about 500 m^2. Looking from Thatched Cottage at the City Corner (Chengqu Caotang), you can see the mountain straightly, and you are in the shadow of the mountain. This is a cool but strong-color backlit scene suitable for summer. This stone mountain is vigorous and majestic in shape, with high skill in stacking stone textures and simple style. It seems to be an early relic. The mountain does not set up pavilions as general, but imitates the wild scenery of the empty mountain forest. The main peak of the stone mountain is slightly to the west, on the east side of which lies a platform on the top of the mountain. Under the main peak there is a cave opening to the southeast. This is a natural place of the wild interest to take a rest instead of ordinary pavilions. It shows a high image through the sloping peaks (Fig. 3.6). To the west of the main peak is a steep wall under the winding road. Along the road to the west is a valley running southeast and northwest, which has the cliff inscription "Sui gu" (deep valley) (Fig. 3.23). The valley is about one or two meters wide and 13 m long. On the west of the valley, the mountain range gradually slopes gently, and its scope is also small to imitate the stretching branch.

The southeastern part of this yellow stone mountain has been treated as a steep cliff, which is organically connected with the Haopu water (Fig. 3.2a). The whole mountain is filled with soil to create a rich vegetation landscape. As for plants, except vines and the like used at the cliff, the trees all over the mountain are appropriate deciduous and evergreen matching considering the seasonal phase. Common groundnut trees include conifers, arborvitae, melon seeds, yellow poplar, camellia, privet, etc. Fallen leaves include plum, crape myrtle, junqiu, sophora japonica, celtis sinensis, tung tree, etc. Trees of inclined exploration postures are set with purpose on cliffs and plateaus. However, there are not enough trees to provide shades on the top of the valley. The mountain entrance road from the main scene of the Thatched Cottage at the City Corner (Chengqu Caotang) building is located in the northwest of the mountain. There are three mountain paths in total: the westernmost one is the gentle hill to the west of the Sui gu valley, which is located on the top of the mountain by the ridge road under the west porch of the garden boundary. Another path is the Sui gu (deep valley) standing between both sides of the rock wall, which is the passage from the stone mounatin to the south. The other is the whole mountain path entering the main part of stone mountain. This path starts at the north entrance of Sui gu (deep valley), with spiral stone-grades to climb the mountain leading to a small platform on the top of the mountain, to the northwest of which is a large piece of yellow stone overlapping peaks.

As mentioned earlier, this main peak is hollow with caves underneath. Continue climbing from the left hand of the cave entrance, you can arrive at another larger terrace on the top of the mountain. Looking southeast from here, you can clearly see all of Haopu water. The summit can be reached from this terrace. Descending from the northeast of this terrace, there is a mountain road on the side facing Haopu leading back to the southwest, where are two roads. One road leads upstairs to the small platform, and then follows the original road to descend to Haopu Stone Mine at the bottom of the cliff. There are three more routes at the rocky ledge, one northeast is stepping down to the water side, another facing west leads to the flat land to west of Haopu in the south of the mountain, and the other towards southwest crossing the stream by the stone bridge leads to the other side of the mountain forest pavilion corridor. The stone slab bridge is elevated above the rocky mountain to set off the high mountain and the deep water, which is in line with the theme of the scene, but the zigzag bridge won't work (Fig. 3.59). If it is changed into a combination of Ji (rock), Ao (bay strip), stone beams and stone stepping, which are located on the lower surface of the water to cross Haopu, the unity and integrity design of the scene will be ensured.

As this garden site has a high terrain, the water level is relatively low. The creation of water is adapted to local conditions, not limited to the lake-based method of creating bodies of water in large and medium-sized gardens. For example, haopu is created due to the low water level. From the front of the main building "House of Double Reflection" (Shuangzhao Lou) in the north of haopu, the water bends southwesterly along the cliff until it reaches the "Four Sides Hall" type pillar-supported water House, Hall Among the Mountains and Water (Shan Shui Jian), which are erected on haopu in the south. The water extends from the house, but does not reach the end (Fig. 3.157a, b). This narrow and long water surface with lower water level is divided into two parts by the stone bridge in the middle, with clear priorities in perspective. The Hall Among the Mountains and Water (Shan Shui Jian) and the House of Double Reflection (Shuangzhao Lou) at the north and south ends of the water body are the two main stops for the enjoyment of this scene. The shadow of the forest covers the water surface after noon, showing the depth of haopu, which is the typical artistic conception of haopu scene at its most vivid moment. The scenery seen from the two stops in the north and south is far-reaching and interesting.

To the east of the water body, from House of Double Reflection (Shuangzhao Lou) to the south, to the vicinity of Hall Among the Mountains and Water (Shan Shui Jian), pavilions and corridors are set up along the wall to guide people to look at the mountain near the water. On the west and south sides of the Hall Among the Mountains and Water (Shan Shui Jian), there are half corridors along the wall from the main building in the north. The corridor connects the small hall in the southeast corner of the garden with the Ting Lu Tower (Ting Lu Lou) (Fig. 3.304). Between the small hall and the Ting Lu Tower (Ting Lu Lou) is a hill climbing double road, with flower terraces arranged downstairs in the shape of a small stone mountain to echo the yellow stone Mountain in the northwest. It is very important to the scene of haopu, showing that water is between the two mountains. Therefore, considering the stone mountain in the garden, its structure is that the main body is located in the northwest, while the subordinate part is located in the southeast, echoing each other from afar. Ting Lu Tower is built with the background scene of the neighboring river outside the garden. Looking at the garden upstairs, you can see the subtle scenery between the mountains and rivers.

The west corridor of the garden is connected with the inner residence in many places. The main entrance to the east of the pavilion Living in Seclusion in Pairs (Zhen Bo Shuang Yin) in the southwest corner of the mountain is the garden gate (Fig. 4.45) inscribed with the inscription "Ou Yuan" (Couple's Retreat Garden). Stone hills are set near the gate to form a scene that conceal. Walking left or right, you can have a garden tour along the half corridor.

Fig. 4.45 The rimless moon gate at the entrance of the Eastern garden of Ou yuan in Suzhou

4.1.14 Garden of Surging Wave Pavilion (Canglang Ting as in the Writing of Mengzi) in Suzhou Jiangsu Province (Large Garden of Mountain Scape)

This garden is a national key cultural relics protection unit, and has been included in the world heritage list by UNESCO. Located in southern Suzhou, it was the garden of Yuanli, the king of Guangling in Wuyue of the Five dynasties, and later the villa of Sun Chengyou, the military governor of the Wu Army. In the Qingli period of the Northern Song Dynasty (AD 1044), poet Su Shunqin (Zimei) first built the Garden of Surging Wave Pavilion (Canglang Ting) near the water.

At the early Southern Song Dynasty, it was the mansion of Han Shizhong, a famous anti-Jin general, which got a great expansion. Later, it was changed into a Buddhist temple and abandoned gradually. In the early Qing Dynasty, in the 35th year of the Emperor Kangxi's reign (1696), it was rebuilt on the hill which is the position we see now, surrounded by bamboos and stones. In addition, corridors, pavilions and stone bridges at the entrance have been built, which formed the basic pattern of the remains (Fig. 4.46). In the late Qing Dynasty, it was destroyed and restored.

Garden of Surging Wave Pavilion (Canglang Ting) now covers an area of 14 μ, about 9400 m^2 (Fig. 4.47). A stone memorial archway inscribed "Cang Lang Sheng Ji" is stand at the bridgehead in front of the gate. In the hall after entering the gate, inscriptions of *Cang Lang Ting Tu*, drawn by monk Canglang during the reign of Emperor Guangxu in the Qing Dynasty, *Cang Lang Ting Ji*, written by Su Zimei and Gui Youguang, and *Xu Xiu Cang Lang Ting Ji*, written by Song Luo and Liang Zhangju, are inserted in the east and south wall respectively (Fig. 4.48). The garden is dominated by mountains, which are the original highland. The eastern part of Garden of Surging Wave Pavilion (Canglang Ting) is the main body of the garden, which is decorated with yellow stones, indocalamus and arbores, forming a mountain-and-forest composition that is full of wild appealing. Square kiosks here are composed of stone pillars and tile roof, also full of wild appealing. Its name still

Fig. 4.46 Scene of Canglang Ting during the reign of Emperor Kangxi in the Qing Dynasty (Looking from the north across the water to the south)–Restored according to The Picture of Canglang Ting painted by Wang Hui in the 39th year of the reign of Emperor Kangxi (1700), the illustration in Annals of Cang Lang Ting of the 35th year of the reign of Emperor Kangxi (1696) and its present scene

a

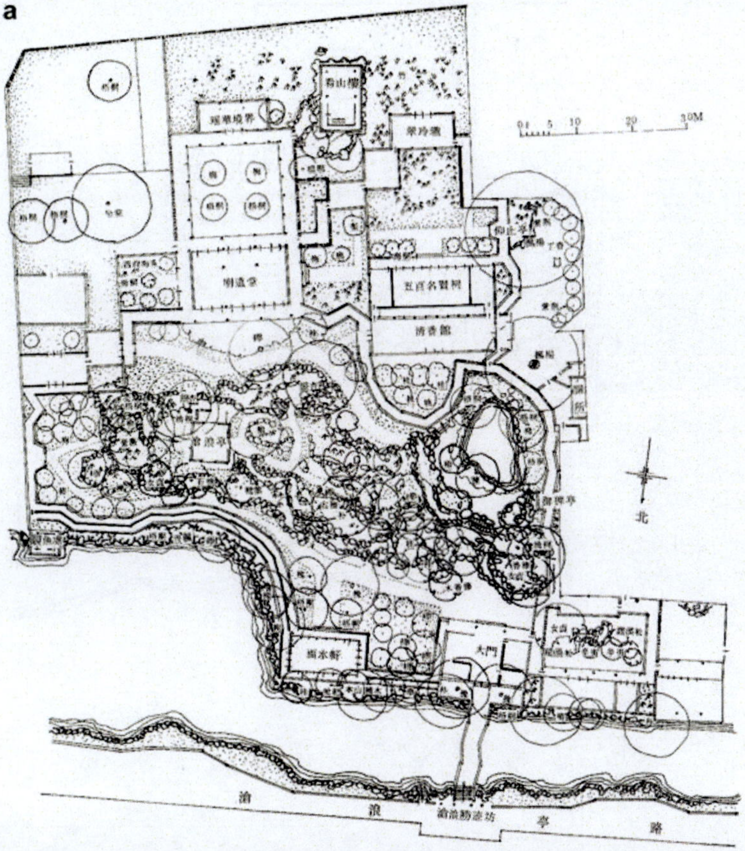

Fig. 4.47 (**a**) Plan of Canglang Ting in Suzhou (From Suzhou Classical Gardens); (**b**) The general section line drawing of Canglang Ting in Suzhou (From Suzhou Classical Gardens); (**c**) Section 1–1 of the overall scenery of Canglang Ting in Suzhou (From Suzhou Classical Gardens); (**d**) Section 2–2 of the overall scenery of Canglang Ting in Suzhou (From Suzhou Classical Gardens)

follows the name when it was firstly built near the river in the Northern Song Dynasty—"Garden of Surging Wave Pavilion" (Canglang Ting). The tablet was written by Yu, a famous scholar in the Qing Dynasty. A couplet carved on stone pillars read, "Qing Feng Ming Yue Ben Wu Jia, Jin Shui Yuan Shan Jie You Qing",[10] which tells the close connection between the pavilion and water before the waterside pavilion was built.

The plant layout here is outstanding. Tall arbors formed the shade (Fig. 3.309c), with the emphasis on the hanging vines. In particular, between the earth and rocks on both sides of the mountain path, clusters of indocalamus with low poles and broad

[10]The moon and breeze are priceless, near rivers and distant mountains are sentient.

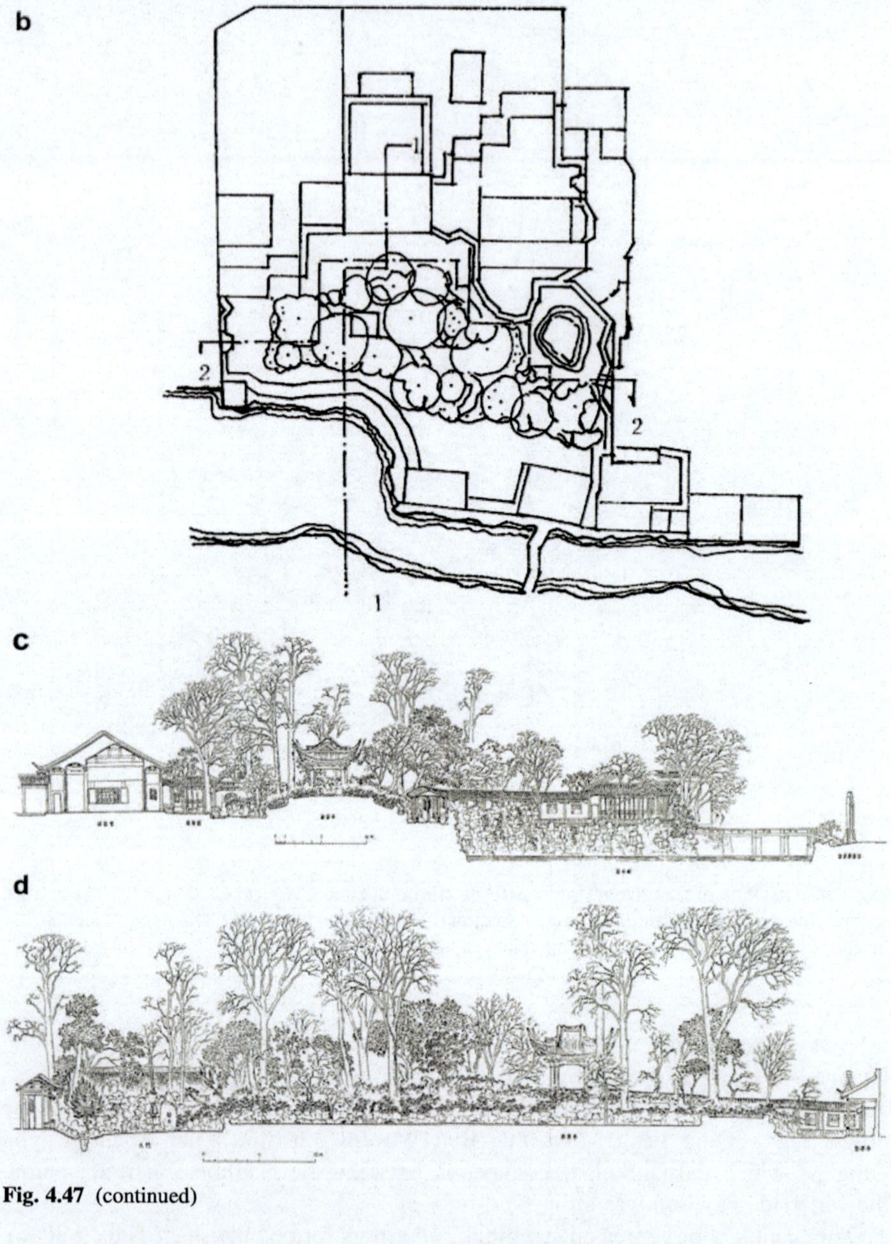

Fig. 4.47 (continued)

leaves were planted, reproducing the scene of deep mountains with dense vegetation (Fig. 3.260). Moreover, it has the management foundation of the Five Dynasties, especially the Song Dynasty. Stack stones looks natural and antique, making the whole mountain seems deep and secluded like true mountains (Fig. 4.49). The west part of the garden is mainly built with lake stones, forming a scene of half mountain

Fig. 4.48 Rubbing of The Picture of Cangang Ting by a monk during the reign of Emperor Guangxu in the Qing dynasty

surrounding deep pool according to the terrain. In general, concentrated and low water, with high-raised stone banks, forms an effect of closing to the abyss (Fig. 3.60d). However, the details are relatively poor. Stack stones here were mended in the late Qing Dynasty, which were quite poor in the artistic and technical

Fig. 4.49 The double corridor and a corner of Canglang Ting

standards. And, this foil waterscape is compared to "Liu Yu"[11] as the inscriptions on cliff, which was unqualified because it didn't catch the theme of abyss. If the integrity of stack stones strengthened, and the inscription on tablet closely linked the theme, this scenery will be a comparable masterpiece as the mountain-and-forest composition in the eastern part.

Since the Southern Song Dynasty, the garden has been abandoned and used as a Buddhist temple. In the late Qing Dynasty, the "Five-Hundred-Sage Temple" (Wu Bai Ming Xian Ci) was built. So it is a public garden, rather than a general private garden. It is relatively simple and focused in terms of the structure of gardens. A mountain-and-forest composition with large space capacity can be seen after entering the garden gate, surrounded by covered walkway. The garden mainly features mountainscape that approaches the waterscape in front of the gate in the north by architectural method. It starts from the "Fish Viewing Spot" (Guan Yu Chu) (commonly known as "Fishing Terrace" (Diao Yu Tai)) in the east, and ends at the garden gate in the west. In the middle, double corridor and "Facing Water Pavilion" (Mian Shui Xuan) ("Luzhou Shuiwu") were built for waterscape view. To the west of the garden gate, waterside pavilions were built with the layout of arbores like Chinese parasols and huge ancient trees as foil. This is a prominent feature, that is, the landscape is not closed, but open to the city and beautify the city, which is a great inspiration for the design of modern parks.

To the south of the main mountain-and-forest composition of the "Garden of Surging Wave Pavilion" (Canglang Ting) is dense architecture area with large variations in history. Hall of Illuminating the Way (Mingdao Tang), a huge-magnitude architecture located on the south of the Garden of Surging Wave Pavilion

[11] Flowing Jade.

Fig. 4.50 Mingdao Tang of Canglang Ting, Suzhou

(Canglang Ting), is the main viewing point of the principal scenery of Garden of Surging Wave Pavillion (Canglang Ting)[12] (Fig. 4.50). Pavilion of Fragrance (Qingxiang Guan) is disposed accordingly to the south of "Flowing Jade" (Liu Yu) that is in the west of the mountain-and-forest composition of Garden of Surging Wave Pavillion (Canglang Ting) (Fig. 4.51). In front of which, a zigzag half-corridor with traceries was set as a partition, thus forming an implicit and pervasive transition space. Pavilion of Fragrance (Qingxiang Guan) is connected to the Five-hundred-Sages Temple (Wubai Mingxian Ci) in the south. In the south end of the principal axis is the "Exquisite Emerald Bamboo House" (Cui Ling Long),[13] three pavilions among bamboo grove. Compared with the scenery group of Hall of Illuminating the Way (Mingdao Tang) to the east, the scenery here is free in form, with a combination of half-corridors, half-pavilions, anterooms and whitewashed walls which are dis-symmetry between east and west (Fig. 4.47).

At the southern end, "Mountain Outlook Tower (Kanshan Lou)" was built between the "Yao Hua Jing Jie" and "Exquisite Emerald Bamboo House" (Cui Ling Long) to view the rural scenery in the south of the garden.[14] Below the tower is the grotto scenery built with natural stones. There are two stone chambers in grotto.

[12]The hall was set here as the basis of the ancient site.

[13]Named after the poem of Su Zimei "Ri Guang Chuan Zhu Cui Ling Long".

[14]Now replaced by scattered modern buildings.

Fig. 4.51 Qingxiang Guan of Cang lang Ting, Suzhou–with Xieshan top and an entrance at the side of the top

And "Yin Xin Shi Wu" written by Emperor Daoguang in the Qing Dynasty was inscribed on the forehead of the door (Fig. 3.142d). In front of the stone chambers was a small courtyard surrounded by stack stones. The entrance was a deep cave with four characters inscribed on it "Yuan Ling Zheng Jian" written by Lin Zexu.

The essence of this garden lies in the proper disposition of mountain-and-forest composition and the waterscape outside the garden. It is a typical example of the "near borrowed" in borrowed scenery.

4.1.15 West Garden of Jiechuang Si Temple (Monastery of Ordination Pennant) in Suzhou Jiangsu Province (Large Garden of Water Scape)

West Garden is an example of temple garden. It stands in Jiechuang Si Temple (Monastery of Ordination Pennant) on the Liuyuan Road out of the Chang Men of

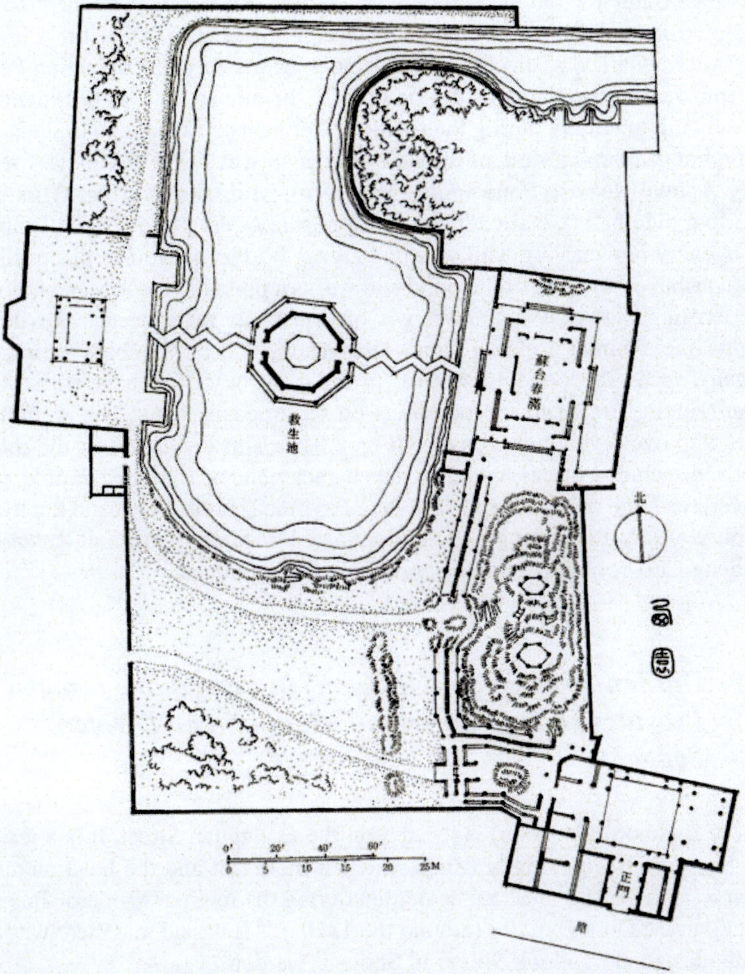

Fig. 4.52 Plan of the west garden of Jiechuang Si Temple in Suzhou (From The Catalogue of Jiangnan Gardens)

Suzhou City, which is the west *hua yuan*[15] of the monastery (Fig. 4.52). Jiechuang Si Temple (Monastery of Ordination Pennant) was originally known as Guiyuan Temple, built between the foundation of the Yuan Dynasty to the time of Yuannian (1271–1294), and destroyed in the Ming Dynasty. During the Jiajing period (1521–1565) of the Ming Dynasty, this place was the private garden of Taipusi Xu Taishi (Taipusi high officer), which was called the "West Garden". Xu also

[15]Flower garden.

owned "East Garden"[16] on the east of West Garden. Xu's son abandoned the garden and restored it to Guiyuan Temple. In the eighth year (1635) of the Chongzhen period, it was renamed as Jiechuang Si Temple (Monastery of Ordination Pennant). In the tenth year of the Xianfeng period (1860), the monastery was destroyed in war and was rebuilt gradually during the reigns of Tongzhi and Guangxu in the late Qing Dynasty, but the former view of the west *hua yuan* was never completely restored.

Only Animal Release Pond (Fangsheng Chi), Mid-lake Pavilion (Huxin Ting) and the four-side hall "Sutai Chunman"on the side of the pond (Fig. 3.103b) exist. Part of scenery like curving walkway and zigzag bridge, as well as places like both sides of Qushui on the east of the pond and western pond have not been repaired yet.

The existing part of West Garden is obviously its main scene, and its theme highlights the Animal Release Pond (Fangsheng Chi). The base point, "Sutai Chunman", faces the pond westward, forming an east-west axis with Mid-lake Pavilion (Huxin Ting) and the buildings on the opposite bank, and strengthening this axis with two seven-zigzag bridge (Fig. 3.91a). This layout shows the solemnity in quiet and secluded places, which is the characteristic of Buddhist temple gardens. Moreover, the huge magnitude of Mid-lake Pavilion (Huxin Ting) and the treatment of double eaves further strengthen the dominated position of Animal Release Pond (Fangsheng Chi), which is representative in temple gardens.

4.1.16 Garden of Bamboo (Ge Yuan) in Yangzhou, Jiangsu Province (Large Garden of Mountain and Water Scape)

Garden of Bamboo (Ge Yuan) is located on the Dongguan Street. It is a residential garden built by Huang Yingtai (Zhijun), a salt merchant and the head merchant of *Lianghuai,*[17] on the site of Shouzhi Garden during the reign of Emperor Jiaqing and Daoguang in the Qing Dynasty (around the 1820s).[18] It is said that there was a stone mountain designed by monk Shitao in Shouzhi Garden.

Garden of Bamboo (Ge Yuan) is located at the back (north) of the residence. Go north along the *Huoxiang*[19] that is on the east of the residence and then turning west, people can enter the garden through the gallery (not preserved) under the double road. The existing garden area is 4500 m^2, the original area was even larger (Fig. 4.53). A large number Yazhu bamboos were planted in the garden in those years, also known as "Wangan". So the garden was named after Garden of Bamboo (Ge Yuan) (Huang's alternative name is also "Geyuan"). Garden of Bamboo (Ge Yuan) is known for "Seasonal Artificial Mountains" (Si Ji Jia Shan). The

[16]Garden that was preserved miraculously (Liu Yuan) in the late Qing Dynasty.

[17]Huainan and Huaibei.

[18]*Record of Ge Yuan*, Liu Fenghao, Qing Dynasty.

[19]The lane that block the fire.

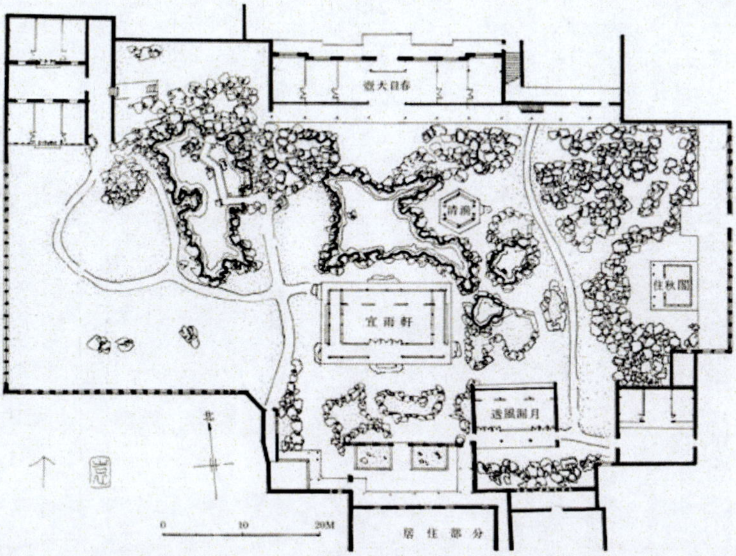

Fig. 4.53 Plan of Ge Yuan in Yangzhou (From The Catalogue of Jiangnan Gardens)

so-called Seasonal Artificial Mountains (Si Ji Jia Shan) refer to four mountain-scape of spring, summer, autumn and winter that were made of stalagmites, lake stones, yellow stones and Xuan shi (Xue shi) respectively. The guidance program is also arranged in the order of four seasons. So people see the Spring Hill first before entering the garden from the north of the residence. The scenic structure is simple. That is, in the courtyard that is composed of three double roads on the east, north, west and the white wall on the north, stalagmites are scattered on both sides of moon gate of the white wall, accompanied by sparse bamboos (Fig. 4.54 and Fig. 3.8c). It is a symbolic design. Stalagmites refer to vigorous Spring Hill. And the word *Sun*[20] in *Shisun*[21] matches with bamboo forest leading to the association of "bamboo shoots spring up after a spring rain".

"Ge Yuan" (Garden of Bamboo) was inscribed on the moon gate, which is the entrance of the garden. There is a natural planting beds vestibule stacked with lake stones. Veranda of Rain Delight (Yiyu Xuan), the base point, commonly known as "Guihua Ting" (Fig. 3.230b), is designed in the north of the vestibule. Veranda of Rain Delight (Yiyu Xuan) is in the form of a four-side hall, with outer corridors on the east, south and west sides, and gooseneck chairs on the east and west corridors, which back are similar to palace chairs, and have not been seen elsewhere (Fig. 3.220e, f). There is a small lake-shaped water body on the north of Veranda of Rain Delight (Yiyu Xuan), which stretches out bays, branching streams and

[20]Bamboo shoots.

[21]Stalagmite.

Fig. 4.54 The rimless moon gate of Ge Yuan in Yangzhou (Flanked by "Spring Hill")

creeks. Stones in the water symbolize islands. Across the water, there is a seven-room building—"Hu Tian Zi Chun" a base point overlooking the garden is provided upstairs (Fig. 3.81a). On the west side of the lake, there were originally two connected boats named "Mandarin Duck"[22] (not preserved), facing the "Qingyi" Pavilion across the water. At the end of the stream on the south of the pavilion, there are two small pool-shaped water bodies. The treatment of water here, turning one water into groups of water, enriches the scenery.

The huge magnitude lake-stone mountains on the south-west of the building adopt the method of *shuijiashan.*[23] On the south of lake-stone mountains, there is ravine type water. With the combination of water and mountain, water flows into open cave, and zigzag stone footbridge near the water surface inserts into the cave. Stalactites hanging upside down in the cave, tunnels are connected to each other, which have been called "twelve caves" formerly. It is a kind of dark, moist and cold

[22]Yuanyang.

[23]Water rockery.

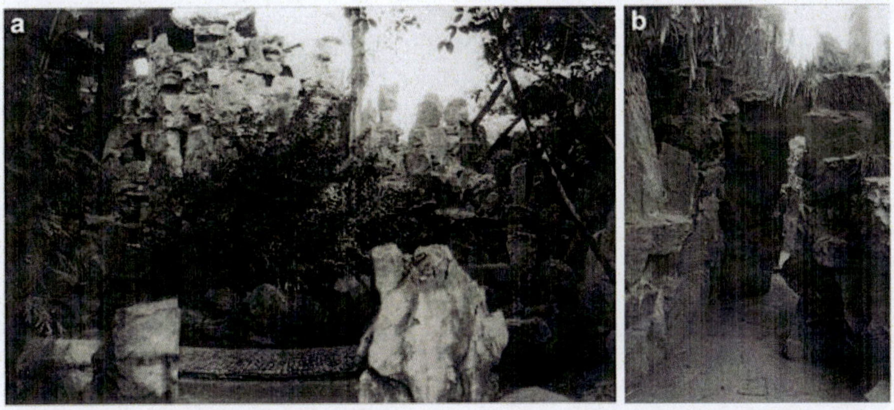

Fig. 4.55 (**a**) A Glance at Autumn Hill of Ge Yuan in Yangzhou. (**b**) Entrance to the valley of Autumn Hill in Ge Yuan of Yangzhou

Karst cave scenery with undercurrent inside. Pines, cypresses and deciduous trees are laid out among mountains, which grow verdant in summer, making people unwilling to leave. This is the so-called "Summer Hill" (Fig. 3.15d). The landscape design of Summer Hill is advisable. The southward arrangement has achieved a distinctive contrast between the mountain rocks under the sun and the shady caves, which strengthen the charm of the water caves in summer. But the stacked stones are lack of vitality. This work is a reflection of appreciation style of the so-called "the man in charge" in *The Craft of Gardens (Yuan Ye)*, that is, the salt merchant. This hill has a walkway to climb. Move eastward from the top of the hill, one can approach to the front of the floor.

The west section of the Summer Hill is cut off in front of the building, east section resumed on the other side, which then transitioned to the Autumn Hill in the east part of the garden (Fig. 4.55). The "Autumn Hill" is purely a stack of yellow stones, which can access to the building through the double road on the north. The "Autumn Hill" was built near the garden wall. Different from the Summer Hill, it has a large roof flat, which can be used for climbing in autumn."Zhuqiu Ge" is built among the forest on the roof flat as a base pint for climbing. One can not only overlook the garden here, but view the sceneries afar like Slender West Lake, Pingshan Tang and Guanyin Shan, etc. The main scenery of this hill is facing west, so the ideal effect can be achieved by setting sun. The yellow-stone mountains were casted with orange sunlight, creating a poetic artistic conception, which is a charming scene in contraposition of the west part of the "Veranda of Rain Delight" (Yiyu Xuan), the main hall.

The "Autumn Hill" has a huge magnitude, with dignified and magnificent shape. Actually, the stones used is far less than expected as there are many tunnels and valleys inside, which is quite economical (Fig. 3.16d, e). The creation of this yellow-stone mountain--Autumn Hill is very successful. The "Winter Hill" is to the south-west of the "Autumn Hill", and it is the scene in contraposition of the south part of a small three-room hall (Fig. 3.13a). It is stacked by snow-white Xuan Shi (also known as Xue Shi). It's main scenery is northward nightside, which leading to the association of a snow-covered mountain. The "Winter Hill" is located at a east-west narrow courtyard, and its overall structure is traditional "nine-lion" shape in Yangzhou, which is called *Jiu Shi Tu*. There are several round holes on the white wall behind (south) the hill. These windows are not used for framing, so their positions are not in the general field of vision. The plaque hanging in the base point hall—"Tou Yue Lou Feng" brings out its theme.

The small gate on the west of the nathex of the "Tou Yue Lou Feng" is just connected with the side door on the white wall of the entrance to the garden. One can return to the small courtyard before entering the garden through this gate.

Garden of Bamboo (Ge Yuan) can be said to be the representative of gardens of Yangzhou.

4.1.17 Garden of Look Forward (Zhan Yuan) in Nanjing Jiangsu Province (Large Garden of Mountain and Water Scape)

Nanjing originally had two Gardens of Look Forward (Zhan Yuan), one was built by Qin Jianquan, a grand secretary during the reign of Emperor Jiaqing in the Qing Dynasty, on the east of Wuding Bridge in the south of Nanjing, has not existed. Currently, Garden of Look Forward (Zhan Yuan) is in the north side of Zhanyuan Road that is on the east of Zhonghua Road, which is adjacent to the History Museum of Taiping Rebellion on the east. This place was originally the mansion of Xu Da, Guogong of Wei in the early Ming Dynasty. It belonged to the Office the Rectifying Armed Force in the early Qing Dynasty, and later be used as the Office of Chief Secretary. During the Taiping Heavenly Kingdom (1850–1864), it was once used as the mansion of Yang Xiuqing, the East King, and Lai Hanying, the Summer Officer (prime minister). From the end of the Qing Dynasty to the Republic of China, the residences on the east were either offices or private houses. Although the garden on the west is deserted, it has been preserved. After the founding of the People's Republic of China, the "History Museum of Taiping Rebellion" was established in the east, Garden of Look Forward (Zhan Yuan) in the west was renovated and opened.

Garden of Look Forward (Zhan Yuan) is a famous garden with a long history in Jiangnan. Its name took from the poem of Su Dongpo, "Looking forward the Yutang as if in heaven". Emperor Qianlong of the Qing Dynasty had visited Garden of Look Forward (Zhan Yuan) during his two inspection tours in the south and inscribed "Zhan Yuan" (Garden of Look Forward). Later, craftsmen were ordered to built "Ru Yuan" in Changchun Yuan of the Old Summer Palace, Beijing, by imitating the shape and structure of Garden of Look Forward (Zhan Yuan). Garden of Look Forward (Zhan Yuan) was built in the early or middle period of the reign of Emperor

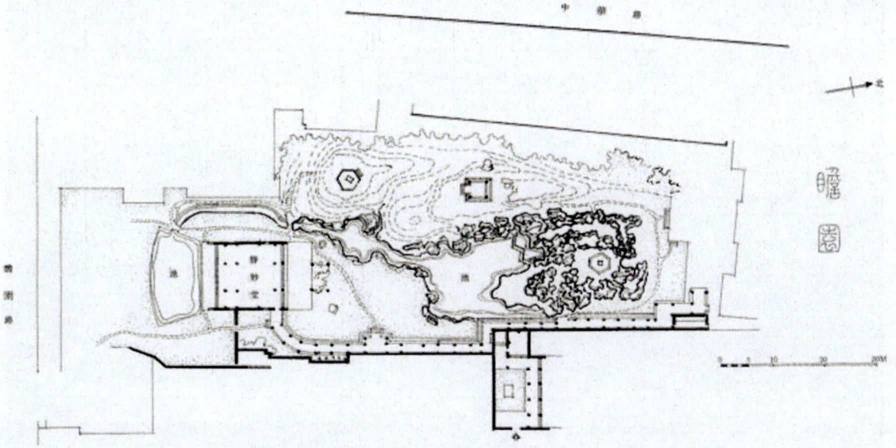

Fig. 4.56 Plan of Zhan Yuan in Nanjing (From The Catalogue of Jiangnan Gardens)

Jiajing of the Ming Dynasty by Xu Pengju, the seventh generation of Xu Da, which was referred in "West garden in the masion of Wei" in *Tour of Gardens in Nanjing* written by Wang Shizhen (*Study of Zhan Yuan in Nanjing*, Liu Xujie). *Zhan Yuan Tu*, drawn by Yuan Jiang during the reign of Emperor Yongzheng in the early Qing Dynasty, has a more realistic record of the early scenes of Garden of Look Forward (Zhan Yuan). Here, we only briefly review the existing scenes.

With an area of 5500 m^2, Garden of Look Forward (Zhan Yuan) is relatively simple in structure (Fig. 4.56). Hall of Hiding Beautiful Scenery (Jingmiao Tang), the main base point, was designed in the south according to its south-north narrow shape. A lake-stone mountain[24] was stacked in the north, the front of which combines with the lake-shaped water body. Cliff winding path and rock patch were designed near the mountain, as well as the zigzag bridge (Fig. 3.263b). At the far end of the lake where near the hall, it is transformed into a crooked stream which interspersed with stepping stones (Fig. 3.197). The stream flows around the west side of the hall and pours into the flabellate lake on the south of the hall. The relation between hall and lake is different from the Suzhou style. There is an open lawn with no flowers and trees between the hall and the lake, instead of setting the platform in the north of the hall by the water (Fig. 4.57).

The design of landscape of this garden is excellent, especially the connection of mountain and water. The structure of this mountain is similar to that of yellow-stone mountain in Garden to delight one's parents (Yu Yuan), Shanghai. There are seven grottoes and tunnels in the lake-stone mountain (Fig. 4.58). The stack of winding climbing path and the layout of plants are advisable (Fig. 3.38b). There is a pavilion on the mountain, and the composition location is hidden, which is different from the River Outlook Pavilion (Wangjiang Ting) in Garden to delight one's parents

[24] It is said that lake stones were remains of Emperor Xuanhe.

Fig. 4.57 The lakeside lawn to the north of Jingmiao Tang of Zhan Yuan in Nanjing

Fig. 4.58 The entrance of the western tunnel of the lake in Zhan Yuan of Nanjing

Fig. 4.59 The lakeside pavilion is right behind the garden gate of Zhan Yuan in Nanjing, so the principal scenery will leap to the eye when entering the garden

(Yu Yuan), Shanghai. "*Qiaobi shan*" in the northwest corner of the garden echoing the main mountain is a masterpiece with touching artistic conception (Fig. 3.3a).

The entrance of Garden of Look Forward (Zhan Yuan) is on the east side near the mountain-and-water scenic imagery. So, the main hall is first to be viewed as the south-facing gate. Once in the garden, visitors are placed in the half corridor of the east wall, and they will reach the key viewpoint, Hall of Hiding Beautiful Scenery (Jingmiao Tang), along the guidance of the corridor. After entering the garden, ones can go north through the corridor and approach to the lakeside and the cliff. Here, the garden gate is exactly at the S-shaped twist of the half corridor. So, the guide in the garden is clear in sequence, which gives us a great inspiration in the organization of guide program. Unfortunately, this smart guide in the garden was destroyed during the renovation. Instead, visitors approach to the lakeside pavilion after entering the garden gate directly, which displays the main mountain-and-water scenic imagery first (Fig. 4.59).

4.1.18 Garden to Delight One's Parents (Yu Yuan) in the City of Shanghai (Large Garden of Mountain and Water Scape)

Garden to delight one's parents (Yu Yuan) is a private garden owned by Pan Yunduan, governor of Sichuan in the Ming Dynasty. According to "*Yu Yuan ji*" (Notes on the Yu Yuan) written by Pan, the garden was built by his father Pan En (Minister in the reign of Emperor Jiajing in the Ming Dynasty), so the name of the

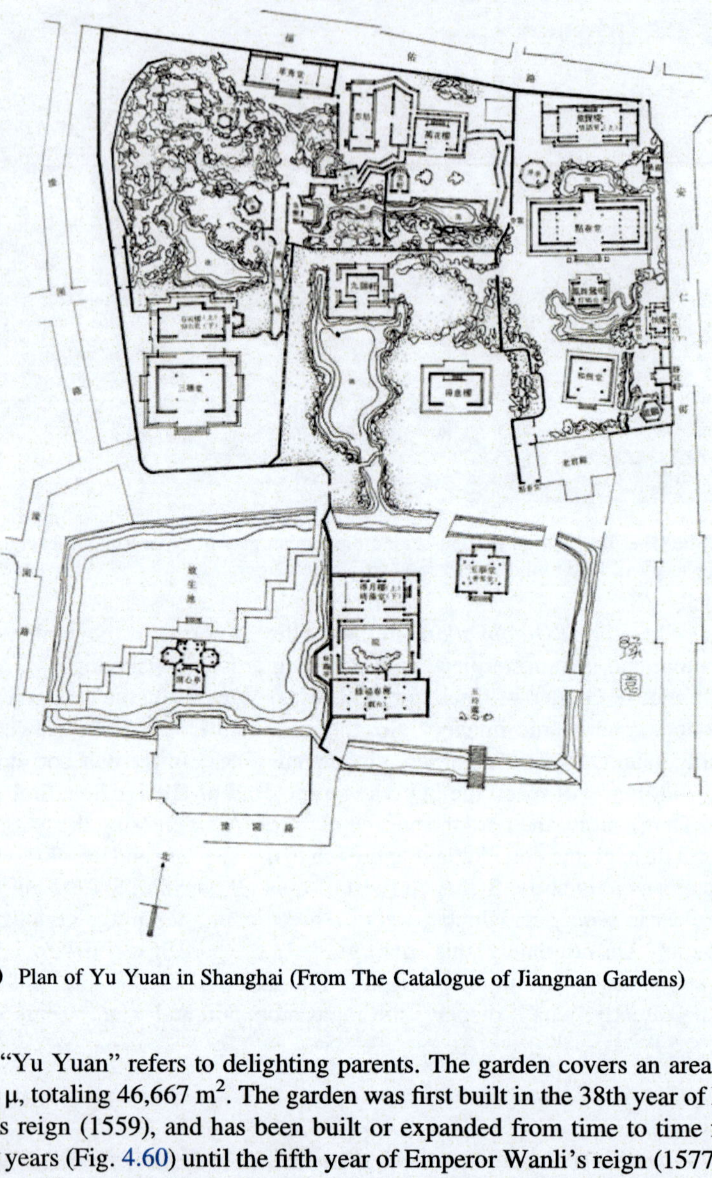

Fig. 4.60 Plan of Yu Yuan in Shanghai (From The Catalogue of Jiangnan Gardens)

garden "Yu Yuan" refers to delighting parents. The garden covers an area of more than 70 μ, totaling 46,667 m^2. The garden was first built in the 38th year of Emperor Jiajing's reign (1559), and has been built or expanded from time to time for more than 18 years (Fig. 4.60) until the fifth year of Emperor Wanli's reign (1577). Zhang Nanyang, a famous gardener at that time and also known as "Zhang Shanren",[25] was hired to create mountains inside the garden. And after the garden was completed, it became quite famous all over Jiangnan region (Fig. 3.263a). At the end of the Ming Dynasty, the garden fell with the owners. In the 48th year of the reign of Emperor Kangxi of Qing Dynasty (1709), about 2 μ of land in the eastern part of the garden

[25]Meaning mountain creator.

was handed to other person and built into "inner garden". In the 25th year of the reign of Emperor Qianlong of Qing Dynasty (1760), the western part was handed to another person and renovated, thus forming the pattern of two gardens respectively in the east and west. The renovated Garden to delight one's parents (Yu Yuan) is called "West Garden" and the Inner Garden (Nei Yuan) is called "East Garden".

In the Qing Dynasty, there are often gardens set beside city temples in Jiangnan region. For example, town god's temple of Changshu is attached with a back garden, town god's temple of Jiading is attached with Garden of Autumnal Glow (Qiuxia Pu), and town god's temple of Qingpu is attached with a Qushui garden. Around the middle of the Qing Dynasty, the Garden to delight one's parents (Yu Yuan) and the Inner Garden (Nei Yuan) also became the back gardens of the town god's temple of Shanghai. During the reign of Emperor Daoguang, the Garden to delight one's parents (Yu Yuan) worn down by the years without repair. The government ordered industrial enterprises to raise funds and manage it. At that time, enterprises from 21 industries renovated different part of the garden. After completed, it served as the place for enterprises to hold gatherings, banquets and discussions. In 1842, it was destroyed by the British invaders. In 1853, Shanghai people's uprising organization, the Small Knife Association, once set up its headquarters at Dianchun Hall (Dianchun Tang) of the garden. In 1860, it was further ruined by Anglo-French Allied Force. At the end of the Qing Dynasty, the ruined Ning Hui Pavilion (Ning Hui Ge), the Boat Hall (Chuan Fang Ting) and the Jiushi Pavilion (Jiushi Ting) were opened to the public.

The gardens that remain today are only the Mid-lake Pavilion (Huxin Ting) and the part to the north of it. The Mid-lake Pavilion (Huxin Ting) is separated outside the garden and used as a teahouse in the city. The existing Garden to delight one's parents (Yu Yuan), including the Inner Garden (Nei Yuan), has been renovated and opened. In terms of pattern, the garden still reveals the situation that it had been separately used and redundantly decorated by industrial enterprises. For example, the top of *yunqiang*[26] is shaped like a dragon, which reflects the common customs of businessmen (Fig. 3.149). Fortunately, a group of scenes of Zhang Nanyang's masterpiece "big jia shan" (big artificial mountains) are still well preserved, and some of the rest scenes are also desirable.

The western scenery in Garden to delight one's parents (Yu Yuan) is noteworthy: the main building of its principal scenery is the Rain-rolling Building (Juanyu Lou), which is located in the south, and the main viewing objects-a group of yellow-stone mountains, which is located in the north, adopt the usual layout of gardens in Jiangnan region. The downstairs of the Rain-rolling Building (Juanyu Lou) is called Yang Shan Hall (Yangshan Tang), which directly points out that the place is where the mountainscape is observed (Fig. 4.61). The principal scenery is dominated by mountains and supplemented by water, and the arrangement is still near water and distant mountains. The water surface right beside the Yang Shan Hall (Yangshan Tang) is generally shaped into a lake set off by a three-fold slabstone towpath-style

---

[26] Cloud hill-climbing wall.

Fig. 4.61 Juanyu Lou Yangshan Tang of Yu Yuan in Shanghai

bridge. And the water bends far along the foothills and forms a Haopu set off by a small stone footbridge. The mountain road turns and twists and the mountain is stacked on both sides of the winding road. At the bend of the road, stones are folded high like a stone peak, which shows the interest of "*feng hui lu zhuan*"[27] (Fig. 3.38a). On the back side of the mountain, decorated with the dim light through the forest shade, the mountain seems even more selusive. The River Outlook Pavilion (Wangjiang Ting) is set up at the top of the mountain to overlook the Huangpu River, which is a borrowing from afar. At the foot of the mountain there is a pavilion called "Yi Xiu", which fronts water and with the mountain on the back, creating a feeling of being immersed in real mountains. On the east side of the pavilion, there is a winding empty corridor titled "go into a more beautiful place" that leads directly to the scene with the title. According to the guide of the garden, after entering the southeast gate, the corridor will leads to the scene. So looking to the northwest from the corridor, the scenery is absolutely the largest in depth. Furthermore, the water beside it keeps narrowing as it reaches out, which strengthened the perspective effect, thus creating the effect that the lake water goes far into the valley (Fig. 4.62).

Against the north wall of the garden stands the Cuixiu Hall (Cuixiu Tang), which not only conceals the boundary, but also provides a base point for viewing the scenery at the mountain shade.

The Rain-rolling Building (Juanyu Lou) and Yang Shan Hall (Yangshan Tang) are small in size to maintain harmony with the size of the landscape. So in order to meets the needs of public gathering activities, a large Sansui Hall (Sansui Tang) is arranged next to the south of Yangshan Hall (Yangshan Tang). In this way, people

[27] The path winds through high peaks.

Fig. 4.62 The mountain scene from the perspective of "Jianru Jiajing" corridor in Yu Yuan of Shanghai (The close view is Yixiu Ting)

can get together to play in the south hall and enjoy the view in the north hall. Therefore, the main activities in the garden and the viewing point are connected by the combination of halls and pavilions.

4.1.19 The Humble Administrator's Garden (Zhuozheng Yuan) in Suzhou, Jiangsu Province (Large Garden of Mountain and Water Scape)

Located in the northeast street inside Loumen in Suzhou city, the garden is a national key cultural heritage site and has been added to the world heritage list by UNESCO in recent years. Originally, Dahong Temple is on the site. During the reign of Emperor Zhengde in the Ming dynasty (1506512), Sensor Wang xianchen (Jing zhi) was frustrated in politics, and he then imitated Pan Yue to live in seclusion. Therefore, he built the villa garden here and, taking "watering the garden to grow vegetables is the politics of the humble administrator" in *Ode to Living in Seclusion* written by Pan Yue, titled it "Zhuozheng Yuan".[28] Wang Xianchen owned the Humble Administrator's Garden (Zhuozheng Yuan) for about 30 years before the garden frequently changed its owner and was divided into private gardens, government bureaus and residential buildings. In the eighth year of Emperor Chongzhen in the late Ming dynasty, the Wang family returned to the countryside and lived in the eastern part of the garden, which was deserted in modern times and not restored as original after the founding of the people's Republic of China. The western part was named Leaf Garden (Ye Yuan) in the Qing Dynasty. During the Taiping Rebellion, its central and western part became the garden of the loyalty king Li Xiucheng's residence. After the rebellion failed, the middle part became the guild hall of officials of the "Eight Banners",[29] while the western part was transferred to Zhang's family at the beginning of the reign of Emperor Guangxu and renamed as Subsidiary Garden (Bu Yuan). At present, the Humble Administrator's Garden (Zhuozheng Yuan) includes all the middle, eastern and western part, covering 40,000 m^2 (Fig. 4.63).

The principal scenery of the Humble Administrator's Garden (Zhuozheng Yuan) is in the central part. When the garden was first built, the lake-and-mountains principal scenery was created by taking advantage of stagnant water and craggy land. According to *Records of Wang's Zhuozheng Yuan* and *Zhuozheng Yuan tu*[30] produced by Wen Zhiming in the Ming Dynasty, there were few buildings in the garden at that time, reflecting much wild appeal (Fig. 4.64). Judging from Wang Luan's *Zhuozheng Yuan Map*, Wu Qiao's *Zhuozheng Yuan Map* and *The Guild Hall of Officials of the "Eight Banners" Map* in the Qing Dynasty, the number of buildings increased and the style was thus changed (Fig. 4.65). But up to now, the basic pattern of mountains and rivers in central part of the garden has not changed much since the Ming Dynasty. While the western part has changed a lot, and the existing landscape is basically restored according to the pattern of Subsidiary Garden (Bu Yuan) in late Qing Dynasty.

[28] Also known as the Humble Administrator's Garden.

[29] Military-administrative organizations of the Man nationality in the Qing Dynasty.

[30] Landscape of Zhuozheng Yuan.

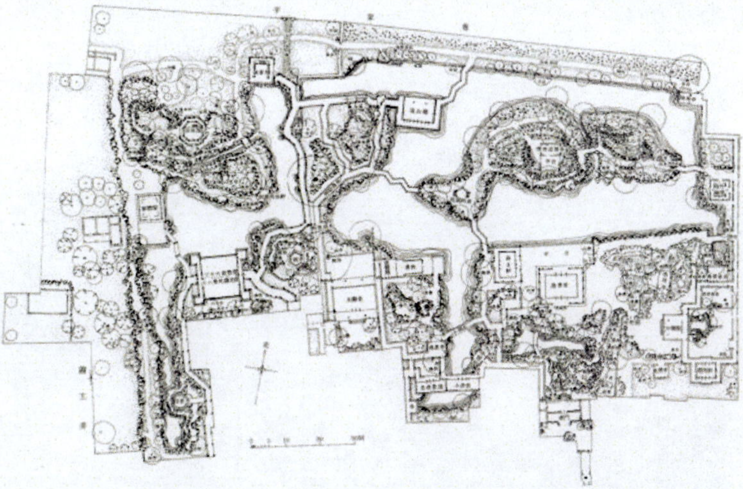

Fig. 4.63 Plan of Zhuozheng Yuan and the western Bu Yuan (From Suzhou Classical Gardens)

Fig. 4.64 Four Scenes in "The picture of Zhuozheng Yuan" painted by Wen Zhengming in the Ming Dynasty–the sparse forests interspersed with wooden bridges, bamboo hedges, thatched pavilions, humble cottages and some halls and buildings, revealing a plain pastoral style

a

b

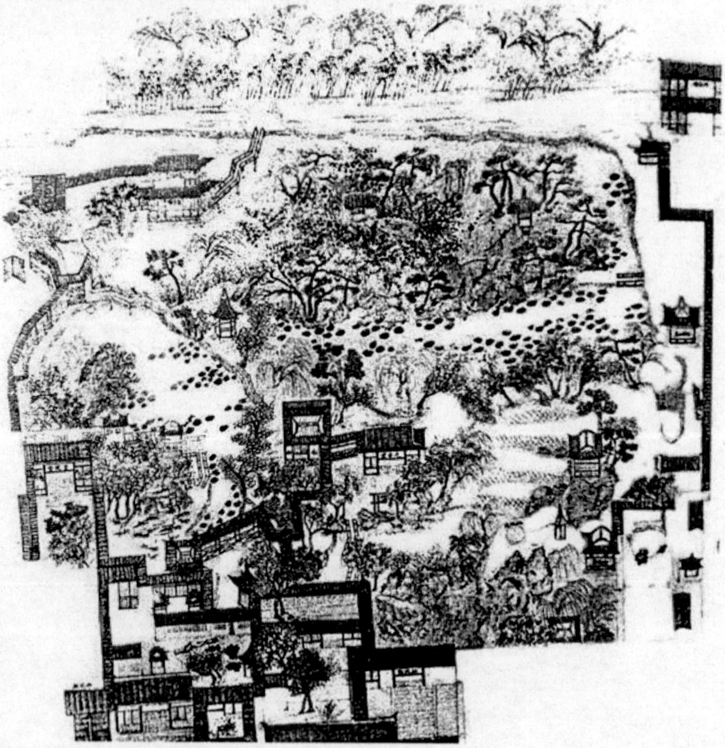

Fig. 4.65 (**a**) The Picture of Zhuozheng Yuan painted by Wangjun in the Qing dynasty (From Suzhou Classical Gardens). (**b**) The Picture of the Guild Hall of Officials in the "Eight Banners" during the reign of Emperor Guangxu in the Qing Dynasty (From Suzhou Classical Gardens)

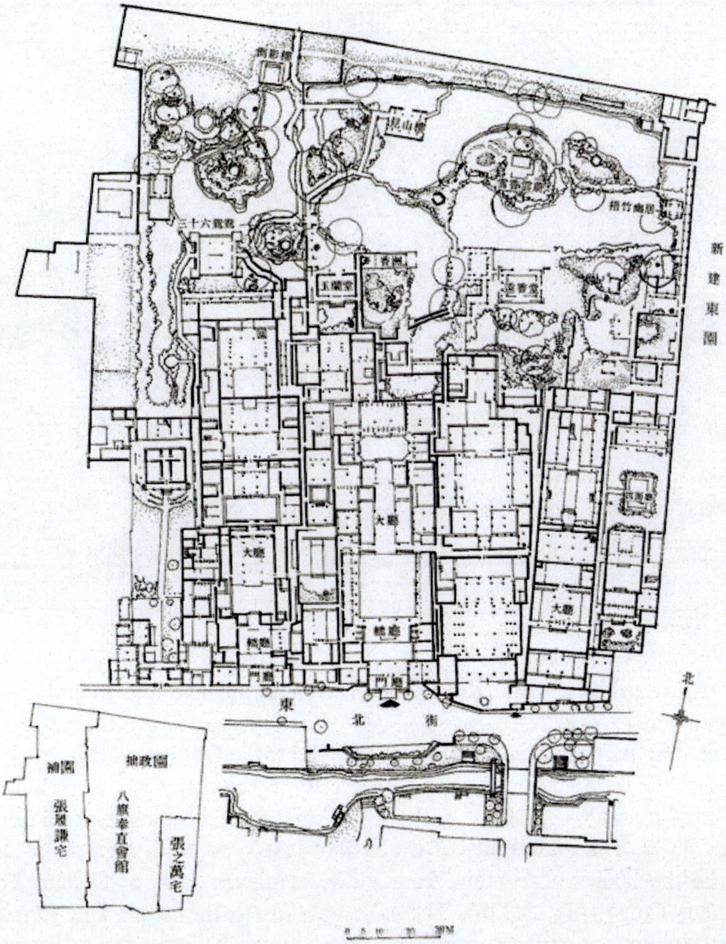

Fig. 4.66 The general plan of the relationship between Zhuozheng Yuan, Bu Yuan and the residence (From Suzhou Classical Gardens)

As the back garden of the residence located in the rear (north) part of it, the Humble Administrator's Garden (Zhuozheng Yuan) connects with the residence by doors at different places. At that time, private gardens in Jiangnan region were sometimes open for people to visit, so there was another independent garden gate leading directly to the street in front of the house (Fig. 4.66). Entering the garden from the garden gate on the east side of the house gate facing the street, there is a long lane leading to the waist gate. And inside the waist gate, a yellow-stone mountain is folded as the scene that conceal, and the main base point of the central part of the garden, Hall of Distant Fragrance (Yuanxiang Tang), stands to the north of the mountain with a small pond separating them (Fig. 3.76). The hall is not only close to the entrance, but also close to the back house, which is extremely convenient

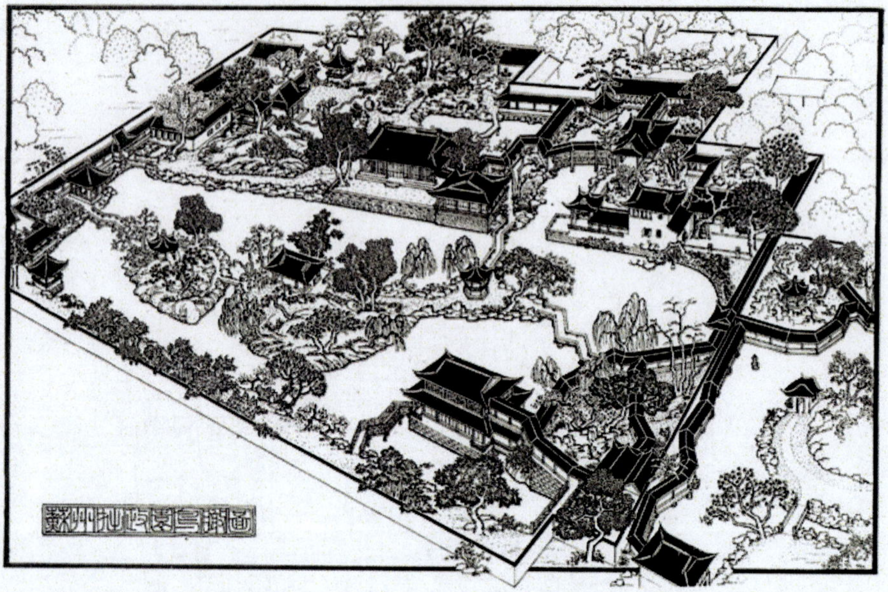

Fig. 4.67 Aeroview of Zhuozheng Yuan

to use. To the north of the hall, the principal scenery, the lake-and-mountains scenery is arranged here.

The central part of the Humble Administrator's Garden (Zhuozheng Yuan), which keeps the foundation of the Ming Dynasty, is the essence of the whole garden, with an area of about 13,000 m^2 and a water surface accounting for one third of the area (Fig. 4.67). The main body of water is of lake type, with a group of mountain features in the lake whose main front view facing the Hall of Distant Fragrance (Yuanxiang Tang) (Fig. 3.32b). The lake extends to the south and forms a belt-shaped water surface as the subordinate body, which is similar to a Haopu. And it ends under a half pavilion to imply the source and flow of water. A wind-rain bridge (or shelter bridge)—the Small Flying Rainbow (Xiao Feihong) and a variant of the covered bridge—Minor Surging Waves Pavilion (Xiao Canglang) are built across the Haopu, which strengthened the appeal of the watery region (Fig. 4.68). The southeast corner of the main water surface bends southward and forms an inlet with a stone bridge[31] built on it to set off its far-reaching effect. And it also ends under a pavilion (Malus spring castle) to imply the source and flow of water (Fig. 3.168).

The central mountain of the lake is mainly earthen mountain, which is divided into the east one and the west one by a meandering stream in the middle, and the stream is arranged meandering so that it is more unpredictable and profound. A small stone footbridge is built across the stream to set off the high mountain and the deep

[31] It is quite ancient and probably a relic of the Ming Dynasty.

Fig. 4.68 Xiao Feihong from the perspective of Xiao Canglang in Zhuozheng Yuan

water as well as to serve as an guiding route (Fig. 3.193b). The earthen mountain provides conditions for extensive planting of trees. Tall arbores are planted to create shade and to strengthen the lie of the mountain so as to form the mountain skyline, shrubs are planted to cut the horizontal line of sight between mountain roads, and indocalamus and climbing plants are planted to decorate rocks and cliffs, which together forms a deep mountain-and-forest atmosphere with abundant vegetation, creating wonderful shade effect (Fig. 3.253b). With a "Snow-Like Fragrant Prunus Mume Pavilion" (Xuexiang Yunwei) highlighting the peak to set off its wildness, the western mountain is the main one of the two earthen mountains. As for the subordinate eastern mountain, a six-corner pavilion called Daishuang Ting or Waiting for the Frost[32] is set up on its peak.

The western mountain beside the Mountain-in-View Tower (Jianshan Lou) is the subordinate body of the lake mountain, and it forms the distant mountain background overlooking from Hall of Distant Fragrance (Yuanxiang Tang), the main base point. The scene in southern contraposition of Hall of Distant Fragrance (Yuanxiang Tang) is also a stone mountain, which winds east to connect to the earthen mountain around Embroidered Silk Pavilion (Xiuqi Ting) on the east side of the Hall, and also corresponds to the lake-and-mountains scenery (Fig. 3.36a). The loquat forest in the southeast corner is separated from the principal scenery by *yunqiang*.[33] On the one hand, the low and intermittent stacked stone can add to the majesty of the mountain, and on the other hand, the curved contour line of the

[32] Once titled "Diligent in Farming" in modern times, commonly known as Pavilion on the North of the Mountain (Beishan Ting).

[33] Cloud hill-climbing wall.

wall emphasizes the mountain environment. The treatment is quite appropriate and can be viewed as one of the representative examples of *yunqiang* creation in the existing Jiangnan gardens (Fig. 3.148a).

The rebuilt buildings in the garden features the pattern and style since the Qing Dynasty, so generally speaking, they maintain the style of the Qing Dynasty. The "Xiangzhou" is a mature and typical boat-like work in the picturesque waterside buildings in Jiangnan gardens created with reference to the composition of original boats. The covered walkway overhanging water surface in the western Subsidiary Garden (Bu Yuan) not only enables the narrow section along the waterside wall to cover the boundary and expand the space, but also forms a scene with the fun of seemingly wading. This zigzag waterside corridor is quite successful in terms of the combination of architecture and water (Fig. 3.199).

4.1.20 Garden that Was Preserved Miraculously (Liu Yuan) in Suzhou, Jiangsu Province (Large Garden of Mountain and Water Scape)

Located in Xiatang outside Changmen of Suzhou City and beside the Liu Yuan road, the garden is a national key cultural heritage site and is listed on the World Heritage List by the UNESCO.

During the reign of Emperor Jiajing in the Ming dynasty (15211565), Xu Taishi, the official of the taipu temple, first founded the "East Garden".[34] There is a "Shiping"[35] mountain created by gardener Zhou Bingzhong (Shi Chen) in the garden. According to the description of Yuan Hongdao (Zhong Lang) in the Ming dynasty, it is as "three zhangs high, 20 zhangs wide, exquisite and steep, like a landscape painting, without intermittent traces, it is really wonderful" (*Complete Works of Mr. Yuan Zhonglang*, Volume 14). At the turn of the reign of Emperor Qianlong and Emperor Jiajing in the Qing Dynasty, the garden was turned over to Liu Shu (Rong Feng), an official, who rebuilt it and added 12 Taihu lake stone peaks (Fig. 4.69). With its "cold and clear bamboo and green waves" (*Preface to Banquet of Hanbi Villa* by Qian Daxin), the garden is also called "Cold Green Villa" (Hanbi Shanzhuang) (Fig. 4.70). And since it is located in Huabu neighbourhood, it is also called "Hua Bu Xiao Zhu", commonly known as "Liu Yuan". At the beginning of the reign of Emperor Guangxu, official Sheng kang (xu ren) became the owner of the garden and further expanded it. After the Taiping Rebellion, of all the gardens outside the Changmen, only this garden was preserved. Hence, the name of "Liu

[34]Now the temple is the West Garden.

[35]Stone Screen.

Fig. 4.69 (**a**) The Picture of Liushu Yuan drawn by Wang Xuehao in the 54th Year during the reign of Emperor Qianlong (From Suzhou Classical Gardens). (**b**) The Picture of Liushu Yuan drawn by Zhai Dakun in the 59th Year during the reign of Emperor Qianlong (From Suzhou Classical Gardens)

Fig. 4.70 The Picture of Hanbi Shanzhuang drawn by Liu Maogong in the middle of the Qing dynasty (From Suzhou Classical Gardens)

Yua" was changed to "Liu[36] Yuan" (a plaque inlaid at the entrance was inscribed with "long preserved in the world").

The existing garden was basically restored according to its appearance in the late Qing dynasty, with an area of about 30 μ, or 20,000 m^2 (Fig. 4.71). Based on the foundation of Cold Green Villa (Hanbi Shanzhuang) in Ming Dynasty, the central district of the garden is planted with many tall and ancient trees which creates seclusive *yi jing*, so it is the essence of the whole garden. The central district includes the western part of the lake -and-mountains scenery and the eastern part of the combination of small courtyards. The most distinctive part in Garden that was preserved miraculously (Liu Yuan) is the eastern district, which provides an example of highlighting activity space with architectural interest (Fig. 4.72).

Garden that was preserved miraculously (Liu Yuan) is also located in the rear (north) of the mansion, and can be reached through the gate beside the "the House of Cranes" (He Suo) connecting the back of the residence (Fig. 3.106).

Outsiders can enter the garden through an independent garden gate facing the street. Entering the garden gate, walking through the sedan chair hall, passing the winding lane and the patio with skylight, then the "Interwined Old Trees" (Gumu Jiaoke) is right before the eyes. With a roof and two small patios with skylight at the opposite corners, the sluggish lane becomes an S-shaped corridor, which is an outstanding creation. "Interwined Old Trees" (Gumu Jiaoke) is a scene with the theme of cuddling Cooper and Privet, and the building there is a corridor as well as a *xuan* (Figs. 3.255 and 3.269). Unfortunately, the ancient trees existing for hundreds of years there have been cut down. Scenes that conceal like mountain partition are not arranged as usual between the corridor and the principal scenery, the lake-and-mountains scenery. Instead the architectural method is adopted-the faintly transparent leaking wall is built to create infiltration effect as well as guidance.

The main base point for the principal scenery—Mingse Lou of Verdure Mountain House (Hanbi Shanfang), is on the west side of Interwined Old Trees (Gumu Jiaoke) (Figs. 3.71 and 3.84), and can be reached by crossing the transitional space, "green

[36] Another Chinese Zi with the same pronunciation but different form, namely, the homophonic Liu. It means preservation.

a

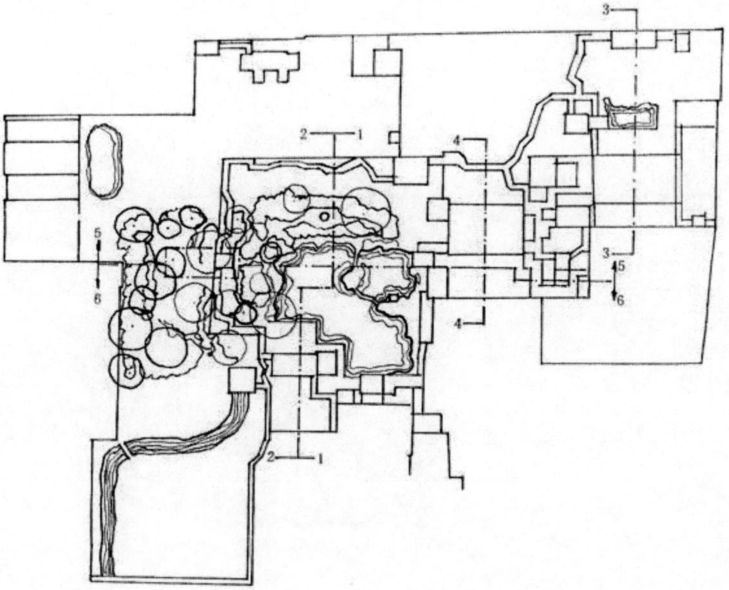

b

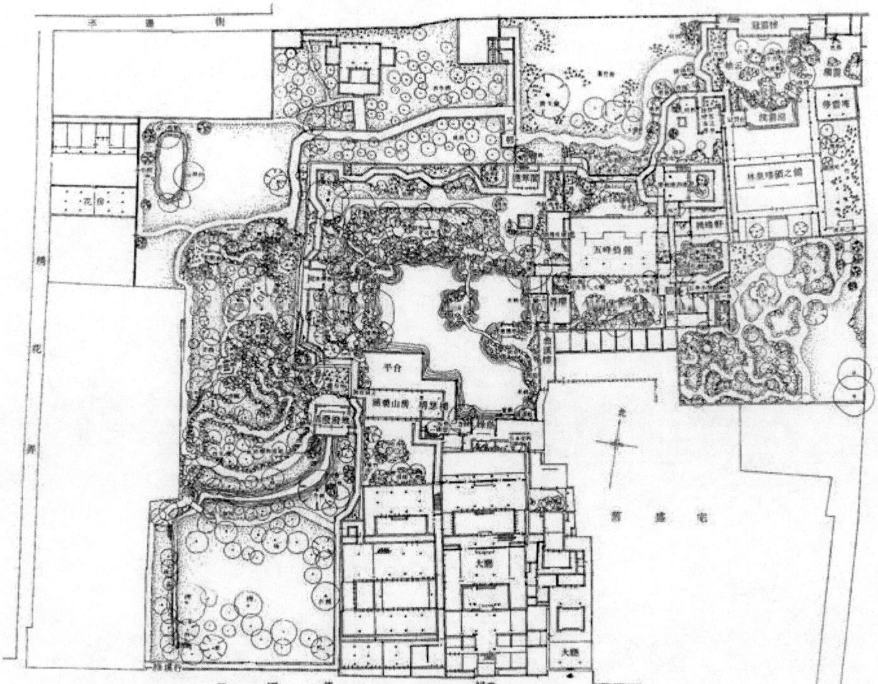

Fig. 4.71 (**a**) The general section line drawing of Liu Yuan (From Suzhou Classical Gardens). (**b**) Plan of Liu Yuan in Suzhou (from Suzhou Classical Gardens). (**c**) General sect. 1–1 of Liu Yuan. (**d**) General sect. 2–2 of Liu Yuan. (**e**) Section 3–3/382. (**f**) Section 4–4/383. (**g**) Section 5–5/382. (**h**) Section 6–6/382

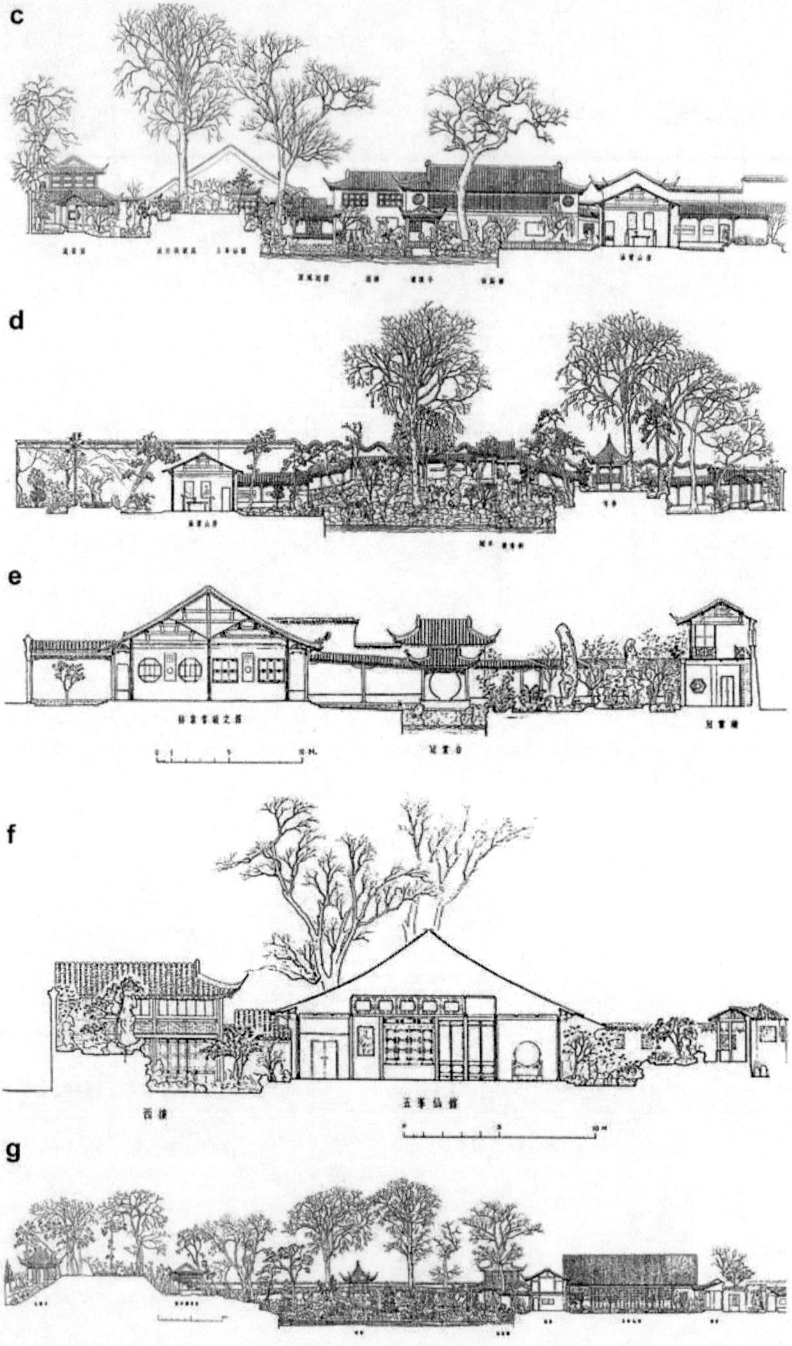

Fig. 4.71 (continued)

h

Fig. 4.71 (continued)

shade", which takes the allusion of "insight" in Buddhism (Fig. 3.106a). The "green shade" is a waterside pavilion also with the theme of ancient green maples, which form a good composition together with the pavilion. The guiding way of the principal scenery is the same as that of other Jiangnan gardens, that is, the main base point is Verdure Mountain House (Hanbi Shanfang), and there is a moon-viewing platform to the north of the house facing the lake, while the mountain is on the other side of the lake. The water-front part of the mountain is piled up by yellow stones, but with some lake stones added in the later period, which was slightly inconsistent. The mountain is an earthen mountain with dense vegetation, and a hexagonal pavilion "Passable Pavilion" (Ke Ting) is set up among the trees. The pavilion is built small in size to set off the larger size of the mountain-and-forest composition, which is a good example of scale selection of buildings (Fig. 3.254b). Osmanthus trees are planted in the western part of the earthen mountain, and a pavilion titled "Scenting the Sweet Osmanthus Fragrance" (Wen Muxixiang) is set up among them.

There is an island in the lake called Little Penglai (Fig. 3.105) and a peninsula connected the east bank with Haopu Pavilion set up on the waterfront of it. In the northwest corner of the lake, a belt-shaped water surface forms a ravine, and there is a stone island at the mouth of the ravine, which is connected to the northern and western mountain respectively by a stone footbridge. As a background, there are also successively raised stone footbridges on the ravine matching the lie of the mountain perfectly (Fig. 3.192a). The scene at the end of the ravine is shaped as the mountain spring and waterfall. On the waterfront sloping mountain at the west bank, the ancient gingko, camphor and cypress trees set off the Osmanthus forests, creating *yi jing* of mountain features, which is an excellent creation (Fig. 3.1b).

The Winding Stream Tower (Quxi Lou) on the east bank, facing Verdure Mountain House (Hanbi Shanfang) and the moon-viewing platform, is a building mainly for shape appreciation, for its whitewashed wall and wooden structure form a bright composition, and it serves as a beautiful background together with the west building and the Qingfengchi Hall on the north. In particular, with the old maple trees in front of the building (unfortunately dead) and the stone buildings and stone footbridge on the waterfront, the whole scene becomes full and integrated (Fig. 3.273).

The eastern district is made up of the combination court gardens with Hall of the Peak of Five Immortals (Wuxianfeng Guan) as the main body, and its essence lies in the area from the House of Cranes (He Suo) and the Hut of Stone Forest (Shilin Xiaowu) to the "Give Me Back the Place to Read" (Huan Wo Du Shu Chu)

a

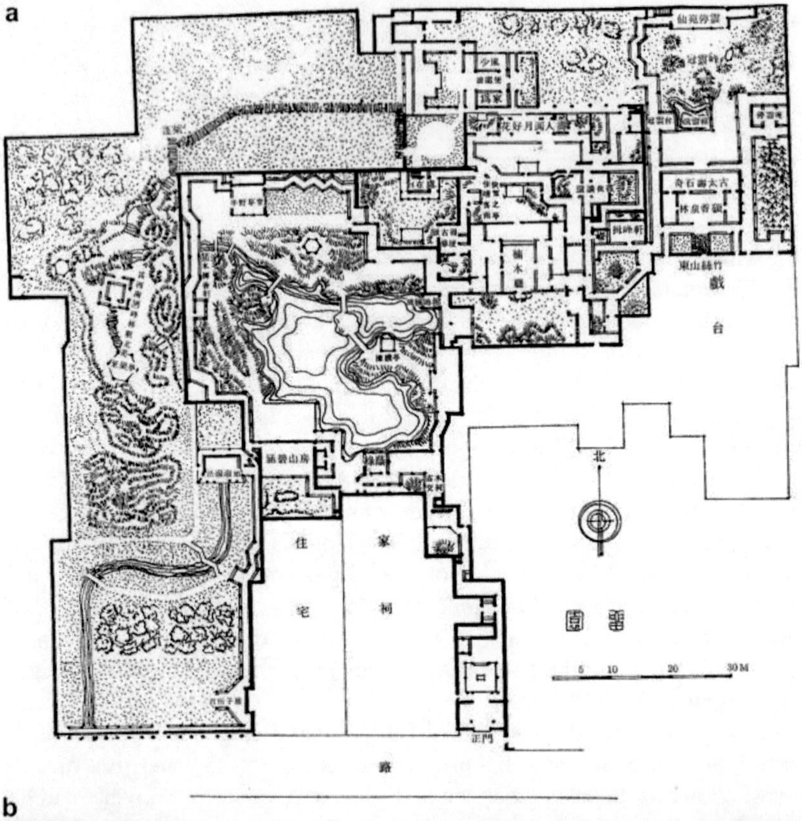

b

Fig. 4.72 (**a**) Plan of Liu Yuan in 1930s (From The Catalogue of Jiangnan Gardens). (**b**) The principal scenery of Liu Yuan in Suzhou–Hanbi Shanfang and its northern lake

courtyard in the north. It has reached the state of blending indoor and outdoor space, reflecting the philosophy of the relationship between artificial and nature, and is worthy of serious exploration (Fig. 3.11b).

In addition, it is worth mentioning that the mountain-and-forest composition where Free Roaring Pavilion (Shuxiao Ting) and Pavilion of Supreme Delight (Zhile Ting) are located in the west is a very prominent example of earthen mountains among the existing Jiangnan gardens (Fig. 3.261a), where stack stones like rugged denuded rocks, the intertwined maple forest projection and the ancient flaky stone on the floor are unified to create a tranquil mountain-and-forest *yi jing* (Fig. 3.251). The creation of the *yunqiang*[37] here is quite successful. Not only the curved wall strengthens the mountain atmosphere, but also the low wall, slightly higher than the mountain elevation, effectively foil the mountain height. It is the best example of *yunqiang* adoption in the existing Jiangnan gardens (Fig. 3.106a, b).

4.1.21 Garden of Pleasure (Yi Yuan) in Suzhou, Jiangsu Province (Large Garden of Mountain and Water Scape)

Garden of Pleasure (Yi Yuan) is located the west of Renmin Road and the north of Shangshu Lane in Suzhou. The owner of Garden of Pleasure (Yi Yuan), Gu Wen Bin (Zishan), was an official in the late Qing Dynasty and had once suppressed Taiping Rebellion with officials including Li Hongzhang. During the reign of Emperor Tongzhi and Emperor Guangxu, Gu Wenbin built Chunyin Yizhuang in the north of the residence across Shangshu Alley, and open up a garden in the east of the ancestral hall, "to Yixing Yangshou"[38] (quoted from *Record of Yi Yuan* by Yu Yue in the Qing Dynasty), so it was entitled "Garden of Pleasure" (Yi Yuan) (Fig. 4.73).

The total area of the garden is about 9 μ, or 6000 m^2 (Fig. 4.74). The new garden gate, rebuilt in 1968, faces Renmin Road in the east, while the original garden gate is adjacent to Shangshu Lane and diagonally opposite to the rear house gate (Fig. 4.75). Entering from the original gate, walking through the sedan chair hall and the main hall successively, the garden is just behind the corner gate on the west side of the main hall. After finally arriving at the garden, Poxian Music Studio (Poxian Qinguan) and Veranda of Stone Worship (Baishi Xuan) will first leap to the eyes. Poxian Music Studio (Poxian Qinguan), also titled "Shitingqin Room" (Fig. 3.77b), stores ancient Qin of the famous scholar Su Dongpo of the Song Dynasty. And there are two vigorous Taihu stone peaks under the window which look like two old men standing to listen to music and are quite vivid (Fig. 3.11a). In the northern courtyard of Veranda of Stone Worship (Baishi Xuan) positioned a number of strange stones, so the name takes the meaning of "Mi Dian (Nangong) worship stones". Evergreen

[37] Cloud hill-climbing wall.

[38] Take care of the spirit and keep fit.

1

2

3

Fig. 4.73 The picture of Yi Yuan in the Qing dynasty (**1**) Ouxiang Xie. (**2**) Miansu Ting. (**3**) Jinbi Ting

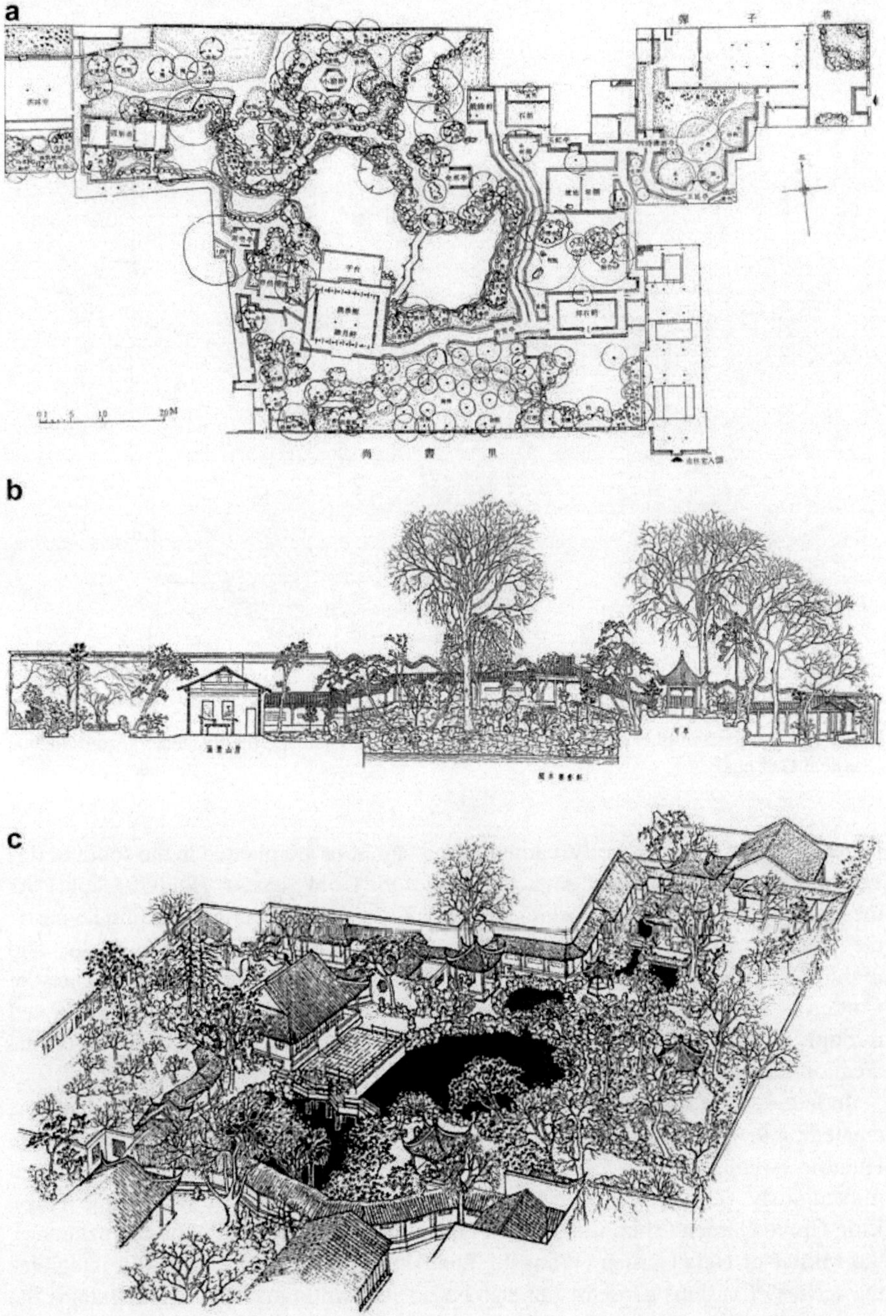

Fig. 4.74 (**a**) Plan of Yi Yuan (From Suzhou Classical Gardens). (**b**) North-south section of Yi Yuan (From Suzhou Classical Gardens). (**c**) Aeroview of Yi Yuan (From Suzhou Classical Gardens)

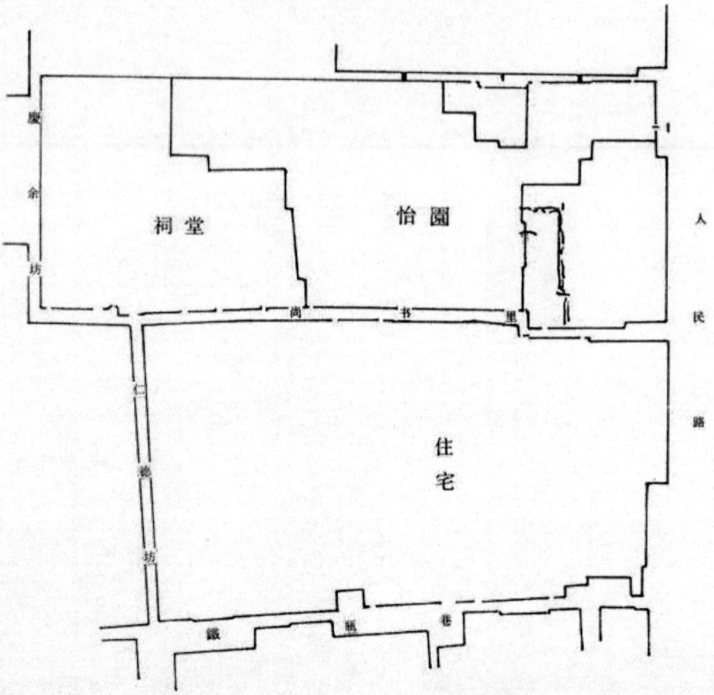

Fig. 4.75 The general plan of the relationship between Yi Yuan and the residence (From Suzhou Classical Gardens)

plants like pine, cypress, holly, camellia, and bamboos are planted in the south of this veranda, so it is also named "Thatched Hut of the Cold Season" (Suihan Caolu). As the reconstruction of the old residence of Wu Kuan, an minister in the Ming Dynasty, the place is arranged with simple scenes like old yet lovely stone peaks, and highlights "listening to music" and "worshipping stones". The principal scenery of Garden of Pleasure (Yi Yuan) is in the west of the area, which is slightly infiltrated through a two-way corridor with ornamental perforated windows, drawing on the creation method of Garden of Surging Wave Pavilion (Canglang Ting).

In fact, since it is built in relatively late period, Garden of Pleasure (Yi Yuan) was created on the basis of drawing extensively on scenes in famous gardens such as the Humble Administrator's Garden (Zhuozheng Yuan), Garden that was preserved miraculously (Liu Yuan), Garden of Surging Wave Pavilion (Canglang Ting), Lion Grove Garden (Shizi Lin), Villa of beautiful surrounds (Huanxiu Shanzhuang), the Master-of-Nets Garden (Wangshi Yuan). So it not only successfully integrates the scenes of various gardens, but also boasts its own innovation and maintains the integrity and unity of the scenes. It is indeed an excellent work.

The principal scenery of Garden of Pleasure (Yi Yuan) is also a lake-and-mountains scenery facing the southern main hall as that of the Humble

Administrator's Garden (Zhuozheng Yuan) and the Garden that was preserved miraculously (Liu Yuan). The main hall is *yuanyang ting* (tandem hall), with the Bian of its south door titled "Chuyue Xuan" (Hoeing-the-Moon Windowed Veranda) and the Bian of its north door titled "Ouxiang Xie" (Pavilion of the Fragrance of Lotus Roots) (Fig. 3.72). A six-corner pavilion with stone pillars—"Minor Surging Waves Pavilion" (Xiao Canglang) (Fig. 4.76) is built on the mountain as an variant of Garden of Surging Wave Pavilion (Canglang Ting), and though not as old as Garden of Surging Wave Pavilion (Canglang Ting), it boasts special *yi jing* of fresh. What is innovative about the mountain here is that the main peak is not in the main position. Instead, a lake-stone mountain on the west side of Minor Surging Waves Pavilion (Xiao Canglang), which refers to the lake-stone mountain of Villa of beautiful surrounds (Huanxiu Shanzhuang), is stacked high and most of the stones are well integrated with properly arranged climbing plants. At the foot of the mountain, there is a tunnel with a Spiral Hairdo Pavilion (Luoji Ting) at its top, and the pavilion is small in size just like the Passable Pavilion (Ke Ting) in Garden that was preserved miraculously (Liu Yuan), which is very appropriate (Fig. 3.291b). The overall mountain chain is consciously designed in its creation. The branch of the main mountain (or the principal scenery) crosses the lake (in the form of a stone cave) south, goes by the "Biwu Qifeng" hut and then directs straight to the south of Hoeing-the-Moon Windowed Veranda (Chuyue Xuan) with the connection of *yunqiang*[39] (Fig. 3.148b). The piled stones to the south of the Hoeing-the-Moon Windowed Veranda (Chuyue Xuan) is further connected to the stones stacked along the lake on the east side of the veranda, and the stones are stacked all the way along the east side of the lake, so they finally return to the main mountain in the north. The ring range sets off the majesty of the mountain, which makes it outstanding.

The main water surface of the lake extends to the west, forming a stream (named "Baolv Bay"), and it slightly expands after turning back, forming a harbor. The Shufang Zhai (with a plaque titled "Songlai Ge" (Songlai Pavilion) on it, originally the plaque with Yu Yue's seal script "The Song of Green Stream and the Shadow of Old Pine" hang here) is slightly similar to the Xiangzhou of the Humble Administrator's Garden (Zhuozheng Yuan) but is smaller in size (Fig. 3.291a). At the turning area where the stream is narrowest stacked the stone mountain where the Spiral Hairdo Pavilion (Luoji Ting) is located, with overhanging cliff beside the stream, and they are an ingenious combination of mountains and waters. This is a scene worth staying for fun, so a "Facing the Wall Pavilion" (Mianbi Ting) is set up to guide the viewing. Where the lake type is transformed into a stream, a rill-and-moon grotto is formed naturally by stones stacked across the water, which is an imitaion of scenes in Lion Grove Garden (Shizi Lin). Decorated with climbing plants, the grotto reproduces the natural landscape of "Immortals' Bridge" (Xianren Qiao) in Guangxi (Fig. 3.70m).

[39] Cloud hill-climbing wall.

Fig. 4.76 "Xiao Canglang" pavilion in Yi Yuan of Suzhou

It is worth noting here that the Locking Green Pavilion (Suolü Xuan) at the north end of the two-way corridor is set facing west, and bamboo groves are arranged on the west side of it, which makes the interior of the pavilion glow with green light in the afternoon. Unfortunately, renovation in recent years have destroyed such a poetic scene.

4.1.22 *Lion Grove Garden (Shizi Lin) in Suzhou, Jiangsu Province (Large Garden of Mountain and Water Scape)*

This garden is a national key cultural relic protection unit and is listed on the World Heritage List by UNESCO. Located in Suzhou City, the garden is now on the garden road. In 1342, disciples of a Zen master Tianru built a temple as his residence. The terrain here undulates, and there are trees and large bamboo forests. Many grotesque stones remain under the bamboos. Some are like lions, so it is called "Lion Grove Garden". At the same time, this name is also to commemorate the fact that Weize learned Buddhism from Zen master Zhongfengben. Because Zhongfengben did Buddhism practices in Shizi Yan of Tianmu mountain, the name Lion Grove Garden (Shizi Lin) is a wish of inheriting the master's Buddhism. When it was first built, the temple was named Shilin Temple (Shilin Si), later renamed Bodhi Orthodox Temple (Puti Zhengzong Si). It is also called Lion Grove Garden (Shizi Lin) because of its garden surroundings.

At the beginning, "bamboo and stone occupied more than half of the land around the temple, so there were not many houses", wrote Ouyang Xuan (1283–1358) in *Book of Shi zi lin*, which was depicted in *The Painting of Shi zi lin* (Fig. 4.77) by Nizan (1301–1374) in 1368, the first year of the Ming dynasty (1368–1644). According to *The Painting of Shi zi lin*, it is speculated that Nizan once stacked stones to built mountains here, or it was created by about ten people: Nizan, Zhu Runde, Zhao Shanchang, Xu Youwen, etc. It became a private garden during the reign of Jiajing (1522–1566), the 11th emperor of the Ming dynasty, and was restored to an temple, named Shengen, during the reign of Wanli (1573–1620), the 13th emperor of the Ming dynasty. In 1758, the name was changed to Hua Chan Temple and walls were built to separate the temple from the northern garden. Since then, Lion Grove Garden (Shizi Lin) only refers to the gardens. In 1918, Lion Grove Garden (Shizi Lin) was purchased by the Bei Runsheng. In the east of the garden, he built family ancestral temples and family schools. The west part was expanded, the mountain-and-forest composition being added. At the same time, the whole garden was rebuilt on a large scale. In stone mountain project, iron parts were often used to

Fig. 4.77 The Picture of Shizi Lin painted by Ni Yunlin (From Suzhou Classical Gardens)

hook and hang, and cement was used to caulk. In building decoration, the prevailing European style was also used. The marble boat completely adopts the European style of the Qingyan boat in the Summer Palace of Beijing. This renovation of the garden completely destroyed the integrity of the tradition. This is what we have seen so far.

The main entrance to the garden is on the west side of the main hall of the ancestral shrine (Fig. 4.78). Behind the entrance is the vestibule of Yanyu Hall (Yanyu Tang). Entering from the west of the vestibule, one can seethe scenic space of Standing-in-the-Snow Hall (Lixue Tang). "Lixue" is the old name of Lion Grove Garden (Shizi Lin) in the early days. It is taken from the story that Master Huike devoutly stood in the snow and wait for Dharma until next morning. This shows that this garden was born out of a Buddhist temple garden. The north of Standing-in-the-Snow Hall (Lixue Tang) is the mountain area, and the main base point in the mountain is the Chamber of Reposing Clouds (Woyun Shi) (Fig. 4.79). In the north of the mountain, the main base point for viewing the mountain-and-forest composition is called Veranda of Pointing to the Cypress (Zhibai Xuan). On the west side of the mountain is the lake type water surface, and its main viewing point-the lotus hall, with a platform adjacent to the water, is also located in the north. The structure of this garden is different from others. When one enters the garden, the mountain scenery is seen first and then the water scenery. From the perspective of the relationship between principal scenery, the main viewing base point is in the north. In other words the main viewing scenic plane faces north—achieving a backlighting effect (Fig. 3.10b).

Lion Grove Garden (Shizi Lin) is famous for its jia shan (artificial mountains). In its striking main mountain area, lake stones/hu shi[40] are stacked. Its features are clear and empty, with attractive guidance. On the top of it, there are many stone peaks such as "Hanhui", "Tuyue", "Xuanyu" and "Angxiao". The "Lion's Peak" (Shizi Feng) is the head of the group of peaks. Grottoes and tunnels, gullies, and stepstone mountain paths, formed by stacked lake stones, are interspersed and circuitous. The guidance of mountain tours varies greatly (Fig. 4.80). Among them, there are several scenes that will make visitors lose their way, creating the fun of walking in a maze. There are many caves and valleys in this mountain, with the reputation of "18 scenes of peach garden" in the past. It also has various plant layout, especially the ancient tree with a history of 100 years rooted in the stone crevice. The whole scenery should have the poetic effect, but the scenes do not have the charm it should have. This is due to the excessive pursuit of the labyrinth interest of guidance, which damages its poetry and painting artistic conception. It is also because of the improper coordination between the architecture and the mountain and forest in the garden and the low style.

European architecture in colonial style, stacked stones like "lions", "frogs", "cattle", "crabs", etc. which aim at *xing si*[41] in front of Standing-in-the-Snow Hall

[40]Lake stone, the water-eroded rocks.

[41]The formal likeness to nature, or the "realistic truth".

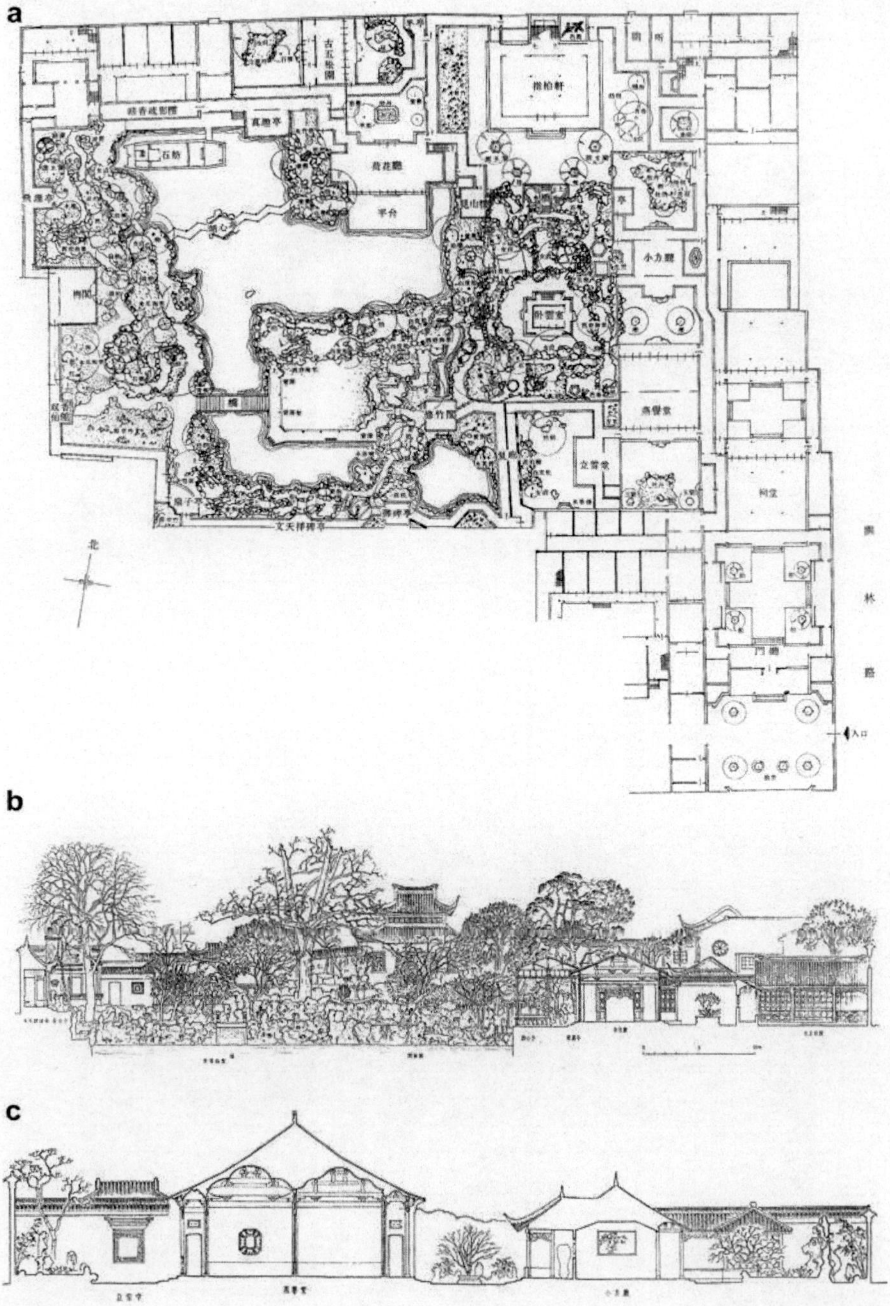

Fig. 4.78 (**a**) Plan of Shizi Lin in Suzhou (From Suzhou Classical Gardens). (**b**) Section 1–1 of Shizi Lin. (**c**) Section 2–2 of Shizi Lin

Fig. 4.79 Woyun Shi of Shizi Lin

Fig. 4.80 Numerous peaks–Among peaks including Hanhui, Tuyue, Xuanyu, Angxiao, Shizi peak is the highest (Shizi Lin of Suzhou)

(Lixue Tang) and the small palace hall of "real interest"[42] all reflect the characteristics of upstarts with comprador nature in the past century. The above is so destructive to the whole that some excellent parts have been neglected. In fact, there are many outstanding scenic imageries in Lion Grove Garden (Shizi Lin). For example, the mountain and forest in the west, Xiuzhu Ge and Little Red Cliff (Xiao Chibi) (Fig. 3.70a) and Shanzi Ting (Fig. 3.93a) on the water grotto in the south, the Ancient Five Pines Garden (Gu Wu Song Yuan) and Building of Anxiang Shuying (Anxiang Shuying Lou) in the north (Fig. 3.249z), Yanyu Hall (Yangyu Tang) and its surrounding area in the east, and the begonia cave gate in front of Veranda of Pointing to the Cypress (Zhibai Xuan) (Fig. 3.116a) are all of attractive *yi jing*. Therefore we get a useful conclusion, that is, how important is the integrity of a garden work.

4.2 Examples of Garden Mountain and Wooded Area

The Craft of Gardens (Yuanye) proposed the site selection of "mountain forest land" to illustrate that garden making in woods-and-mountain environment can take advantage of favorable natural conditions. Compared with the gardens in urban areas, such gardens have their own characteristics in scenic structure. Here are just three examples:

4.2.1 Villa that Embraces Greenery (Yongcui Shanzhuang) at Tiger Hill (Hu Qiu) in Suzhou Jiangsu Province

Villa that embraces greenery (Yongcui Shanzhuang) is located on the west side of the Er shanmen of Tiger Hill (Hu Qiu), and is adjacent to the ancient Spring of Simplicity and Honesty (Hanhan Quan) in the east. According to Yang Xian's *Records of Yongcui Shanzhuang*, the villa was built in 1884 and was renovated after the founding of the People's Republic of China.

The villa is located on the south slope of Tiger Hill (Hu Qiu), next to the stepstone mountain path (Fig. 4.81). The whole garden is a rectangle with a depth of north and south along the direction of stepstone mountain path, and the main axis of the garden gate is north by east. The villa covers an area of about 700 m^2, rising gradually from the entrance to the rear with a height difference of about 8 m. Accordingly, five terraces were built in the garden (Fig. 4.82).

[42] Emperor Qianlong visited this garden during his southern inspection tours. "Real interest" was his calligraphy. Later, the owner of the garden Bei Runsheng continued to use this name and built this magnificent little palace.

Fig. 4.81 The "mountain and wooded area" in Hu Qiu, where Yongcui Shanzhuang of Suzhou is located

The entrance of the garden is similar to the entrance of a manor or a nunnery. The vestibule is built with a high platform base made of strips of stone and outside the garden are long steps to the entrance (Fig. 4.83). The layout of this garden is basically that of a court garden mountain villa. Facing the entrance is Baoweng Xuan, which is composed of three square front halls. On the east side of the fence behind the veranda, there is a side gate leading directly to the outside of the garden. And outside the gate is the stepstone mountain path stacked by natural stones (Fig. 4.84). In the northeast corner behind Baoweng Xuan, there is an alley leading to Yongcui Ge, which was built to meet the needs of tourists for sightseeing. In order to facilitate business, on the side of Yongcui Ge a door was opened, which is near Tiger Hill (Hu Qiu) Mountain road. (Fig. 4.85). The second platform containing a retaining wall can be reached by the small steps in the east. From the second platform to the fourth platform is a court garden with free layout. The second platform is a transitional site, which is paved with ice cracked stone without special treatment. The third platform is the principal scenery, where Putting a Question to the Spring Pavilion (Wenquan Ting) is built to extend *yi jing*/artistic conception by overlooking Spring of Simplicity and Honesty (Hanhan Quan). On the west side of the pavilion is stepped mountain path of stack stones, with flowers and trees such as crape myrtle, pomegranate, oleander, tung tree, white bark pine, populus et folium bovis seu bubali, privet, etc.

On the mountain, Yuejia Xuan is built along the garden wall, which forms the main scenic plane to the west of Putting a Question to the Spring Pavilion (Wenquan Ting). The main architecture of the whole garden, Linglan Jingshe, is built at the fourth platform that enters from the north of the stone mountain (Fig. 4.86). This is the main viewing point which can not only overlook the vestibule and the group of sights: Putting a Question to the Spring Pavilion (Wenquan Ting), Yuejia Xuan and

a

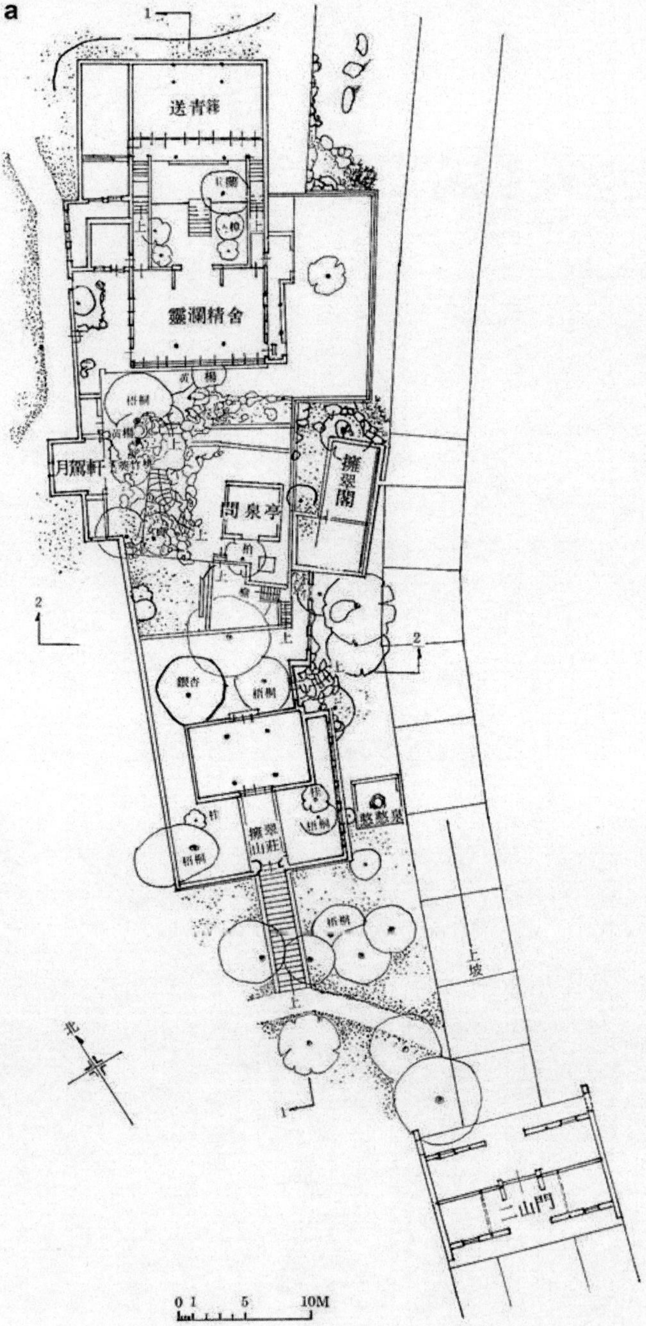

Fig. 4.82 (**a**) Plan of Yongcui Shanzhuang in Suzhou (From Suzhou Classical Gardens). (**b**) Section 1–1 of Yongcui Shanzhuang (From Suzhou Classical Gardens)

b

Fig. 4.82 (continued)

Fig. 4.83 Aeroview of Yongcui Shanzhuang

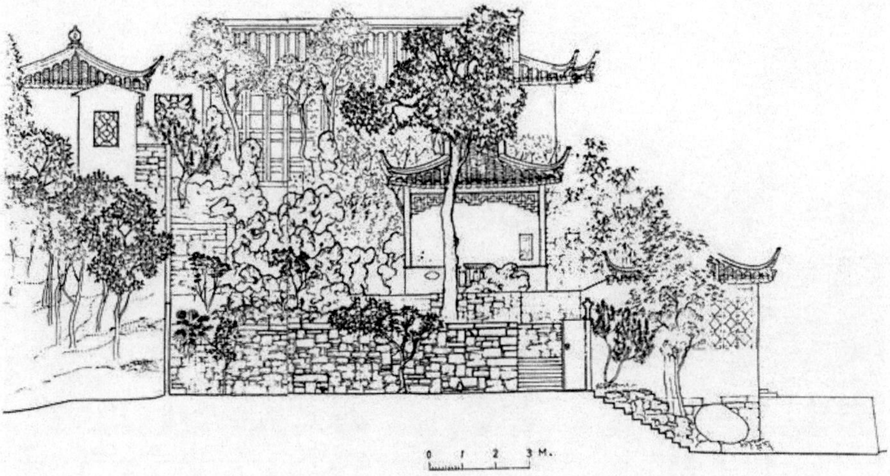

Fig. 4.84 Section 2–2 of Yongcui Shanzhuang (From Suzhou Classical Gardens)

Fig. 4.85 Yongcui Ge of Yongcui Shanzhuang

Fig. 4.86 Linglan Jingshe of Yongcui Shanzhuang

stone mountain, but also borrow from afar the scenery of Lion Mountain and the foothills of Tiger Hill (Hu Qiu) Mountain (Fig. 4.87). The terrain here is very high. Standing here is like standing on a balcony. It is only by virtue of the mountain that one can get such an viewing effect. There is a large balcony on the east side of Linglan Jingshe, which has crossed the boundary of the garden, just like a belvedere in the Tiger Hill (Hu Qiu) mountain. There is a very wide view here. One can not only overlook Lion Mountain in the south, but also look up to the Tiger Hill (Hu Qiu) Tower. This example clearly reflects the advantages of gardening making in mountain and wooded area.

In the center of the backyard of Linglan Jingshe, there is a stepping road that goes straight up to the fifth platform. Opposite Linglan Jingshe on the platform is the back hall, Songqing Yi, which is the back bedroom for the mountain residence. Here is the northern end of the villa.

Fig. 4.87 The eastern moon-viewing platform of Linglan Jingshe borrowing scenery from Lion Mountain

4.2.2 Xiling Seal-Engravers' Society (Xiling Yinshe) at Gushan in Hangzhou, Zhejiang Province

Xiling Seal-engravers' Society (Xiling Yinshe) is a seal cutting society founded in 1903, located in the west of Gushan Mountain, West Lake, Hangzhou, Zhejiang province. It is completely operated as a garden. The design of this garden has a strong style of seal cutting, and it was created under the guidance of Deng Shiru, Ding Jing and Wu Changshuo, etc.[43] This is a rare garden masterpiece in modern

[43] The three are masters of seal cutting in ancient China.

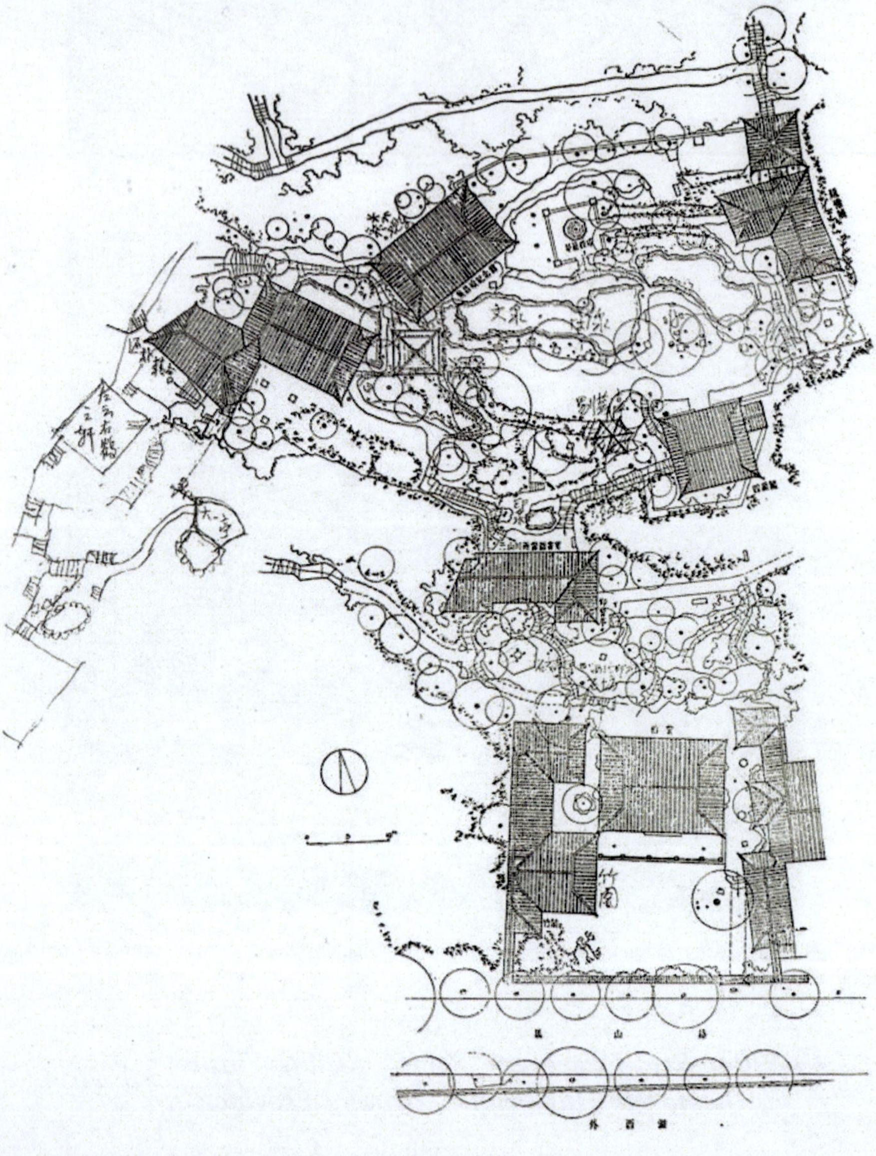

Fig. 4.88 The general plan of Xiling Yinshe (From A Pictorial Catalogue of Jiangnan Gardens)

times. "Xiling Seal-engravers' Society (Xiling Yinshe) boasts both natural beauty and cultural treasures." This is a proper comment.

Xiling Seal-engravers' Society (Xiling Yinshe) is located from the southern foot of Gushan Mountain to the top of the mountain. There is no garden wall, which shows it is open to the public (Fig. 4.88). A group of buildings called "Bai Tang" are laid out on Gushan Road near the West Lake as the working place of Xiling Seal-

Fig. 4.89 A small dolmen of Xiling Yinshe–The entrance to the north foot of Gushan

engravers' Society (Xiling Yinshe). Behind the Bai Tang courtyard, the garden area starts from the foothills. The first is stepstone mountain path marked by the entrance of small Shifang (Fig. 4.89). With a simple structure, on the top of the Shifang are "Xi Ling Yin She" (four Chinese characters). Going up by the steps, the first scenic spot is Shijiao Ting (or Yangxian Ting) and the connected Library of Shan Chuan Yu Lu (Fig. 4.90). There is a hall door. Going north from here, under the shade of buildings and bamboo bushes, there is a deep pool, Seal Spring (Yin Quan), which is remote and quiet and vivid (Fig. 3.61b). There are stepstone mountain paths on the left and right sides of the pool. The right stepstone mountain path is called Hongxue Jing (Fig. 3.278b), leading to Sizhao Ge, the main hall, which is arranged on the top of the mountain and near the steep wall. This pavilion is a four-sided hall with an excellent view. When you are in it, you don't feel that you are in the garden, because it is integrated with the lake and mountain scenery of the West Lake. It is an excellent example of overlooking and borrowing scenery (Fig. 3.307).

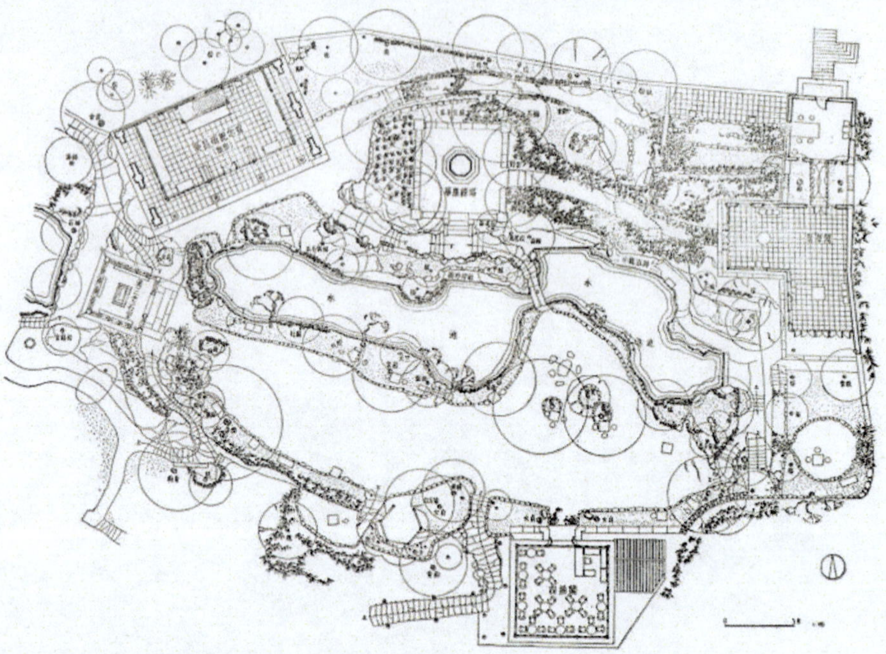

Fig. 4.90 Plan of the garden at the summit of the mountain, Xi ling Yinshe (From A Pictorial Catalogue of Jiangnan Gardens)

The pavilion overlooks the West Lake to the south and the principal scenery of the garden to the north. The Tixian Ting beside the pavilion and facing the open area of the West Lake is also built for viewing the scenery of the West Lake. The garden is decorated with sculptures of figures, such as the bronze statue of Wu Changshuo and the stone statues of Deng Shiru and Ding Jing, which are rare among the existing old gardens in Jiangnan region. In addition, the design of small buildings includes the Huayan Sutra Tower with beautiful sculpting (Fig. 3.247). More special, there are antique four-door pagoda-style "Han San Lao Chamber" (Han San Lao Shi Shi). These buildings are simple and dignified, so they have appreciation value (Fig. 4.91).

The scenic spot near the north of Sizhao Ge is a fish pond chiseled out of rock terrace under the big rock near the top of the mountain. The fish pond is of the deep pool type to show the "Wen Quan" and "Xian Quan" here. On the big rock beside the fish pond, as a garden theme, the four characters "Xi Ling Yin She" are engraved (Fig. 3.61a and Fig. 3.315d). Beside the pond, there is a low stone fence, also of primitive simplicity. And the viewing route of *xieyi* rocky stone and cliff road is chiseled. The stepstone mountain path is also made of mountain rock. On the east wall of the pond is inscribed "Guiyin ya", and a cave niche is chiseled beside it, in which is a bronze statue of Wu Changshuo donated by Japan. The rock on the top of

a

b

Fig. 4.91 (**a**) The north cross-sectional view of the landscape of Xi ling Yinshe (From A Pictorial Catalogue of Jiangnan Gardens). (**b**) The west cross-sectional view of the landscape of Xi ling Yinshe (From A Pictorial Catalogue of Jiangnan Gardens)

the mountain on the north side of the pond is chiseled into a grotto, and the internal wall is inscribed with "Xiao Long Hong Dong" (four Chinese characters) and Buddha statues, which are somewhat like grotto temples (Fig. 3.15f). On the west side of the entrance of the grotto is a standing statue of Deng Shiru. From the north of Xiaolonghong Dong, there is a pavilion as the back door of the garden. On the way out, there is a stepstone footpath leading to the foot of the mountain and reaching to the inner West Lake.

The pools, niches, grottoes, caves, stone steps and inscriptions, etc. in the garden are all carved by means of seal cutting, which has the characteristics of a garden owned by a master of seal cutting and calligraphy.

Fish pond, Xiaolonghong Dong, Deng Shiru statue, inscription of "Xi Ling Yin She" and the stone tower in tall trees directly above constitute a perfectly composed scenic plane when one looks from Sizhao Ge to the north. On the east side and above Guiyin ya is the Tijin Guan, which is also an important base point for borrowing scenery from nearby Gu Shan and afar from the West Lake. There are pavilions on the west side (which have been designated as Wu Changshuo Memorial Hall) overlooking the fish playing in the pool, hence the title “Guan Le” (Watching Playing). In the south of the building is the Han San Lao Chamber (Han San Lao Shi Shi), which houses carved stone drums and other cultural relics. In the south of the chamber is the statue of Ding Jing. The principal scenery space is closely laid out in the east, north and west, forming an open space towards the West Lake to the south.

Gushan is densely covered with vegetation. Stepstone mountain path and slope protection walls are hidden among them, and mosses grow on the path and walls, which is more simple and vigorous and full of wild interest.

4.2.3 Garden for Ease of Mind (Jichang Yuan) at Huishan in Wuxi, Jiangsu Province

This garden is different from the former two. Although it is located in the mountain forest, it is not at the mountainside or the top of the mountain, but at the flat area at the foot of Huishan mountain. In the Yuan Dynasty (1271–1368), the garden site was two monasteries, formerly known as “Nanyin” and “Ouyu”. In the early years of the Jiajing reign of the Ming dynasty (around 1527), Qin Jin, minister of the Nanjing Ministry of war, built a villa here and named it “Feng gu xing wo”. When Qin Jin died, the garden belonged to his nephew Qin Han and his son Qin Liang, governor of Jiangxi Province. In the 39th year of Jiajing’s reign (1560), Qin Han had the garden house repaired, got the pool chiseled and had the rockery built, which was called Fenggu Villa (Fenggu Shanzhuang). After the death of Qin Liang, the garden belonged to Qin Yao, the imperial censor of Department of Supervision and the governor of Hunan. In the 19th year of Wanli’s reign (1591), Qin Yao was dismissed from his post because of the case of his teacher Zhang Juzheng. He returned to his hometown, Wuxi, to relieve his depression through garden living and changed the villa into a garden with a total of 20 scenes. They are Hall of Tea (Jiashu Tang), Qingxiang Zhai, Embracing Beauty Lake (Jin Hui Yi), Qingzhuan, Knowing-Fish Waterside Pavilion (Zhiyu Jian), Qingchuan Huabo, Hanbi Pavilion (Hanbi Ting), Xuancong Jian, Woyun Hall (Woyun Tang), Linfan Pavilion (Linfan Ge), Large Shishang House (Da Shishan Fang), Danqiu Xiaoyin, Huancui Lou, Xianyue Xie, Crane Steps Shoal (Hebu Tan), Hanzhen Zhai, Shuang Tai, Flying Spring (Fei Quan), Lingxu Ge, Qixuan Hall (Qixuan Tang), which were completed in the 27th year of Wanli’s regin (1591). “Ji Chang was derived from the verse of Wang Yizhi’s poem Answer Xu Chuan ‘Qu Huan Ren Zhi Le, Ji Chang Shan Shui Yin’. That’s the origin of the name of the garden”. (*Qin Family’s Genealogy*).

Thirty-two years of Wanli (1604) Qin Yao died, and the garden was once separated. In the early Qing Dynasty, Qin dezao, his great grandson, merged the garden and asked Zhang Lian, a famous craftsman of hill-stacking of the time, and his nephew, Zhang Shi, to pile stones and lead springs for reconstruction. This

Fig. 4.92 The picture of Jichang Yuan in Nanxun Shengdian compiled during the Qianlong reign of the Qing Dynasty

garden has been owned by Qin family for hundreds of years, so it is commonly known as "Qin garden" in history. Emperor Kangxi and Emperor Qianlong of the Qing Dynasty visited the South of the Yangtze River 12 times in total, and each time they visited the Qin Garden (Fig. 4.92). Qianlong believed that "of all the scenic spots in the south of the Yangtze River, only the Huishan Qin Garden is the oldest". Because he "loves its seclusion", he ordered to build Huishan Garden (Huishan Yuan)[44] at the eastern foot of Wanshou Mountain in Beijing's Qingyi Garden.[45] Garden for Ease of Mind (Jichang Yuan) was divided and united in history, and was damaged because of the wars between the Tongzhi and Guangxu periods in the late Qing Dynasty. Fortunately, the landscape structure remained basically unchanged, but there were few ancient trees and buildings left.

Garden for Ease of Mind (Jichang Yuan) covers an area of about 10,000 m^2 at the eastern foot of Hui Mountain and the western foot of Xi Mountain (Fig. 4.93). There are two areas of scenic imagery, with waterscape as the main part. The waterscape is of lake type. In those days, there were ornate pleasure boats in the villa garden, and the reflection of the scenery was like colorful brocade, so the water was called "Embracing Beauty Lake" (Jin Hui Yi).[46] The bank of Embracing Beauty Lake (Jin Hui Yi) was originally disposition of dispersive stones (similar to the stack stones

[44]Later renamed as Fun Garden (Xiequ Yuan).

[45]Later renamed as Summer Palace in the late Qing Dynasty.

[46]*The record of Jichang Yuan* by Wang Kaideng at the end of Ming Dynasty.

Fig. 4.93 The recuperative plan of the principle scenery of Jichang Yuan

scattered along the bank beside the back lake of the Summer Palace in Beijing during the Qianlong period), which are quite natural and wild. It should be a legacy of the Ming Dynasty. Especially the stacked stones and stone beam in the bay which was connected with the mountain, and the stacked stones of the low and flat Crane Steps Shoal (Hebu Tan) with two sloping maple trees, which are rare masterpieces. Unfortunately, in the maintenance of the 1950s to the 1960s, a large number of stones were added, and the scenery was totally different. It is hoped that relevant departments will pay attention to this serious problem and restore the scene in a reasonable way. It is also hoped that the state will list the garden as a national key cultural relic protection unit as soon as possible in order to avoid "constructive destruction". The big maple trees have also died and in recent years been replanted. It will take several years before the original scenery effect can be restored.

The entrance to the garden was originally adjacent to Qin Yuan Street (now called Huishan Hengjie) and at the south of the present brick gate. When the Hengjie was widened in 1954, the east garden wall was shortened by 7 m and the brick gate was moved northward. Behind the gate of the garden is the Knowing-Fish Waterside Pavilion (Zhiyu Jian) that faces the lake. Once you enter the garden, you can see the

Fig. 4.94 The renovated Jiashu Tang in Jichang Yuan

scenery. It is not clear whether there are garden gates in the living buildings of Hunan villas, but now there is no garden gate in this area. In the 1950s, a new gate was opened on the left side of the entrance road of Huishan Temple in the south, and a group of small gardens such as Bingli Tang Hall were newly built.

The water surface of Embracing Beauty Lake (Jin Hui Yi) is longer in the north-south direction, and the main base point, Hall of Tea (Jiashu Tang), is located on the north bank of the lake. When one views Embracing Beauty Lake (Jin Hui Yi) from here, the nearby scenery is the Seven Star Bridge (Qixing Qiao) at the corner of the lake (see Fig. 3.172). The scenery in the middle is the big maple and Crane Steps Shoal (Hebu Tan) (Fig. 3.29b). In the far distance, there is the Xishan Longguang Tower as the background, with the largest and diverse depth of field (see Fig. 3.305). Hall of Tea (Jiashu Tang) has not existed for a long time. Now it is rebuilt as a teahouse (Fig. 4.94). There are pavilions, verandas and huge ancient trees on the east bank. To the south of the waterside pavilion Knowing-Fish Waterside Pavilion (Zhiyu Jian) (Fig. 4.95), there is Yupan Pavilion (Yupan Ting), and to the north, there is Hanbi Pavilion (Hanbi Ting). According to *Records of Southern Inspection Tour* and *Records of Hongxue Yinyuan*, the living buildings in the southwest corner of the garden in Qianlong's regin include Chenhan Hall (Chenhan Tang), Peony Pavilion (Tianxiang Ge) and Woyun Tang (Woyun Hall). The southeast corner is near the garden wall, and there is Lingxu Ge. Mei Pavilion (Mei Ting), which is

Fig. 4.95 Zhiyu Jian in Jichang Yuan

located on the northwest corner of the garden and can show the peak of the mountain, no longer exists.

On the west side of the lake, the earthen mountain were created on the base of the original foothill. This was created by Zhang Lian's nephew Zhang Shi, whose uncle was a famous gardener in Jiangnan region at that time, when the gardens were renovated in the early years of Kangxi of the Qing Dynasty. This mountain reflects the idea of Nanyuan, which is mainly made of soil and properly matched with stack stones. In the middle of this mountain there is a winding ravine and gully. The two walls and the bottom of the stream are stacked with *shan shi.*[47] In the meantime, streams and stepping stones are laid out (Fig. 3.22b) and water of Er Quan is channeled into it to form streams that flow along the valley stream. Although it is a trickle, it falls on stones of different heights and gurgles soundly (Fig. 3.63). On the rock at the end of the stream and near one side of Hall of Tea (Jiashu Tang), "Ba Yin Jian"(three Chinese characters) (Gully of Eight Sounds) is carved to show the theme of the stream (formerly called "Xuanzong Jian") (Fig. 3.315e). This is a rather successful *xieyi* work of a stream.[48] Making use of the fall of mountain springs to form a number of gurgling sounds of nature is the most wonderful design in the creation of Garden for Ease of Mind (Jichang Yuan). Wang Zhideng's description of the garden in the late Ming Dynasty was: "The most wonderful is stream. The

[47] Mountain stone.

[48] In the variant creation of Huishan Garden in Qingyi Garden of Beijing in Qianlong period, the stream was changed to a realistic style.

streams are not only rich in number, but are skillfully designed". There are dense forests on the mountain. Among them it is said that there is also a thousand-year-old camphor tree. Unfortunately, they were destroyed by those disasters since the late Qing Dynasty. At present, there are few large arbores left on the mountain and they have not been replanted. This makes the mountain not only lack the remote atmosphere of the mountain forest, but also makes the Gully of Eight Sounds (Bayin Jian) less interesting when exposed to the sun.

It is often criticized by critics to build jia shan (artificial mountains) among natural mountain forests. Garden for Ease of Mind (Jichang Yuan) is located in a mountain forest. The creation of the earth mountain in the garden has played a role in blocking the wall.[49] Therefore, the scenery in the park is linked with Huishan, thus the scene space is expanded (Fig. 3.301). This shows that the criticism is not correct. In 1959, I created the Crescent Building (Yueya Lou) in Qixingyan, Guilin. At that time, the main building had to be placed on the flat ground under the rising stone mountain (crescent rock), but it was required to combine with the stone mountain. The reason why my work was selected from many schemes and put into practice, and was praised by Professor Liang Shisicheng, is that jia shan (artificial mountains) was built beside the real mountain. That is to say, some rocks were taken to build a natural staircase in front of the building, so that the building was "embedded" between the steep walls of the stone mountain. My inspiration is from the example of Garden for Ease of Mind (Jichang Yuan).

4.3 Examples of Garden Besides River and Lake

Garden construction besides river and lake is similar with that on mountains and hills. Both can be made by borrowing of scenes, with difference lying on means. Two examples will be discussed here.

4.3.1 Three Ponds Mirroring the Moon (Santan Yinyue) at West Lake in Hangzhou, Zhejiang Province

Located at the east of the West Lake Dyke, San tan yin yue has been a synonym of Xiao yingzhou. In other words, what we talk about is Xiao yingzhou, which is a public garden, a scenic spot in the West Lake (Fig. 4.96).

Formed by small islands and causeways, Three Ponds Mirroring the Moon (Santan Yinyue) is "lake in the lake, island in the island". When you look far into its distance, it is an oasis with dense forests being interlaced with pavilions, which is also known as Xiao yingzhou; when you set foot in it, fishpond, lotus pond and

[49] From the main scene—the lake area, looking towards Huishan.

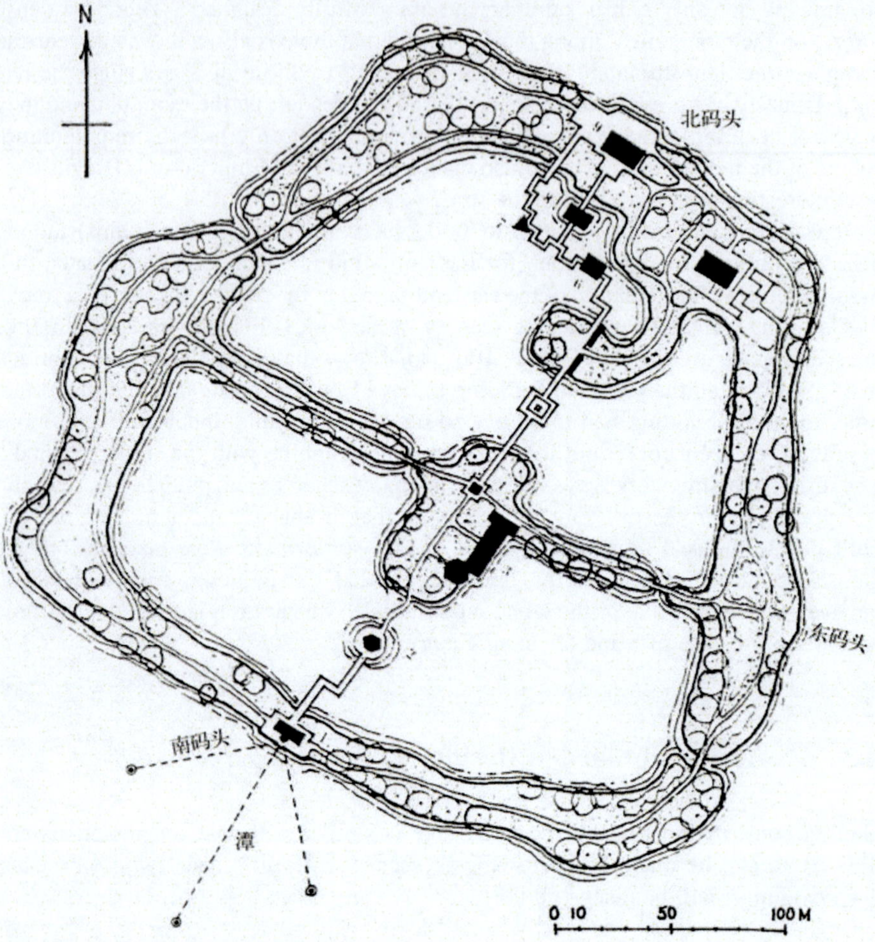

Fig. 4.96 Plan of Santan Yinyue (Xiao Yingzhou) in West Lake, Hangzhou (From Chinese Landscape Architecture)

islands in these ponds come into view. There is no need for fences or corridors as it serves as a continental island as well as public scenic spot. Compared with ancient gardens in Qing Dynasty, construction marks distinctions with that at present (Fig. 4.97). An interesting triangle pavilion-Kai Wang Ting-with a crane on its top is built. Corridors, in the form of wan (卍) and zigzag bridge, which is convenient for watching fishes, decorate the construction.[50]

[50] Square angle zigzag bridge, opposed by Wen Zhenheng, a noted painter and garden designer in Ming Dynasty, is quite suitable here.

Fig. 4.97 Kaiwang Ting of Santan Yinyue in West Lake, Hangzhou

Three Ponds Mirroring the Moon (Santan Yinyue) refers to three stone pagodas rising from the south side of the water, in the style of Stupa. These pagodas are said to be built when Su Dongpo was a local official in the Song Dynasty, while what we see now is set in the Ming Dynasty. In terms of gardening art, the three stone pagodas play a role of expanding space, thus extending scenery to the outside of the West Lake.

As for the creation of Three Ponds Mirroring the Moon (Santan Yinyue), what attracts viewers lies in scenery of the West Lake surrounded by green hills rather than its interior scenes, according to the treatment without any barriers. Li Yuan and Yu Zhuang, which were built in modern times on the banks of Taihu Lake in Wuxi, also combine the garden with the lake scenery by enclosing the water surface with dikes. The key of making garden lies in borrowing of scenes.

4.3.2 Garden of Mist and Rain Pavilion (Yan Yu Lou) at Nan Hu in Jiaxing, Zhejiang Province

Surrounded by water, Garden of Mist and Rain Pavilion (Yan Yu Lou) is built on Yuanyang Island—a highland at Nan hu in Jiaxing. First built in the Five dynasties by Qian Yuanliao, the king of Guangling, it was renovated by Wang Xizhao in the Southern Song Dynasty when Emperor Ningzong was in office. During the Jiajing and Wanli periods of the Ming Dynasty, the area of the island was enlarged, with pavilions, halls and “Diao’aoji” platforms and “Yu Leguo” free life pond in front of

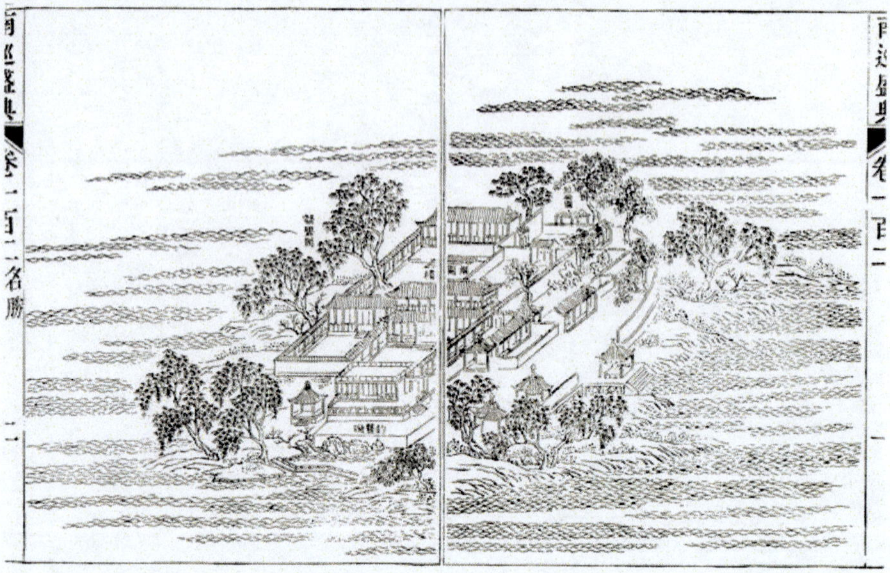

Fig. 4.98 The picture of Yan Yu Lou at Nan hu in Jiaxing published in Nanxun Shengdian

the Garden of Mist and Rain Pavilion (Yan Yu Lou) added and the pool of releasing built. The tower underwent several renovations during the inspection in the southern China of the emperor of Kangxi and Qianlong in the Qing Dynasty. After that, an imitative tower was ordered to be built on Qinglian Island in Chengde Summer Resort by Qianlong. During the reign of Xianfeng, it was destroyed and later rebuilt (Fig. 4.98).

Although the current situation of this tower is completely different from that in *Records of Southern Inspection Tour*, the style-construction is arranged in accordance with axes-is identical. Not too much artistic scenery can be recognized from the perspective of garden. However, the main reason why it has become a scenic spot in Jiangnan gardens lies in borrowing of scenes, which can be known from its title.

Rurality set off Nan hu, with limited water surface (Fig. 4.99). However, surrounding buildings, nowadays, are in disorder, leaving no natural interests. The main purpose of building Garden of Mist and Rain Pavilion (Yan Yu Lou) on the land lies in borrowing of scenes, instead of the management of the land itself. Apart from borrowing of scenes from the perspective of space—standing at a highland or surrounded by a lake, it emphasizes more on borrowing of scene in relation to time. The wonderful thing is that "when the mist covers and rain will be heavy, fishing boats are looming, while the sound of oars begins noisily". (*Elegant works at Chao Cai Guan*). That is to say, limited water surface in the mist can be reminiscent and suggestive of being in the vast sea and misty wave. Garden of Mist and Rain Pavilion (Yan Yu Lou) powerfully prove that garden is an art of time and space.

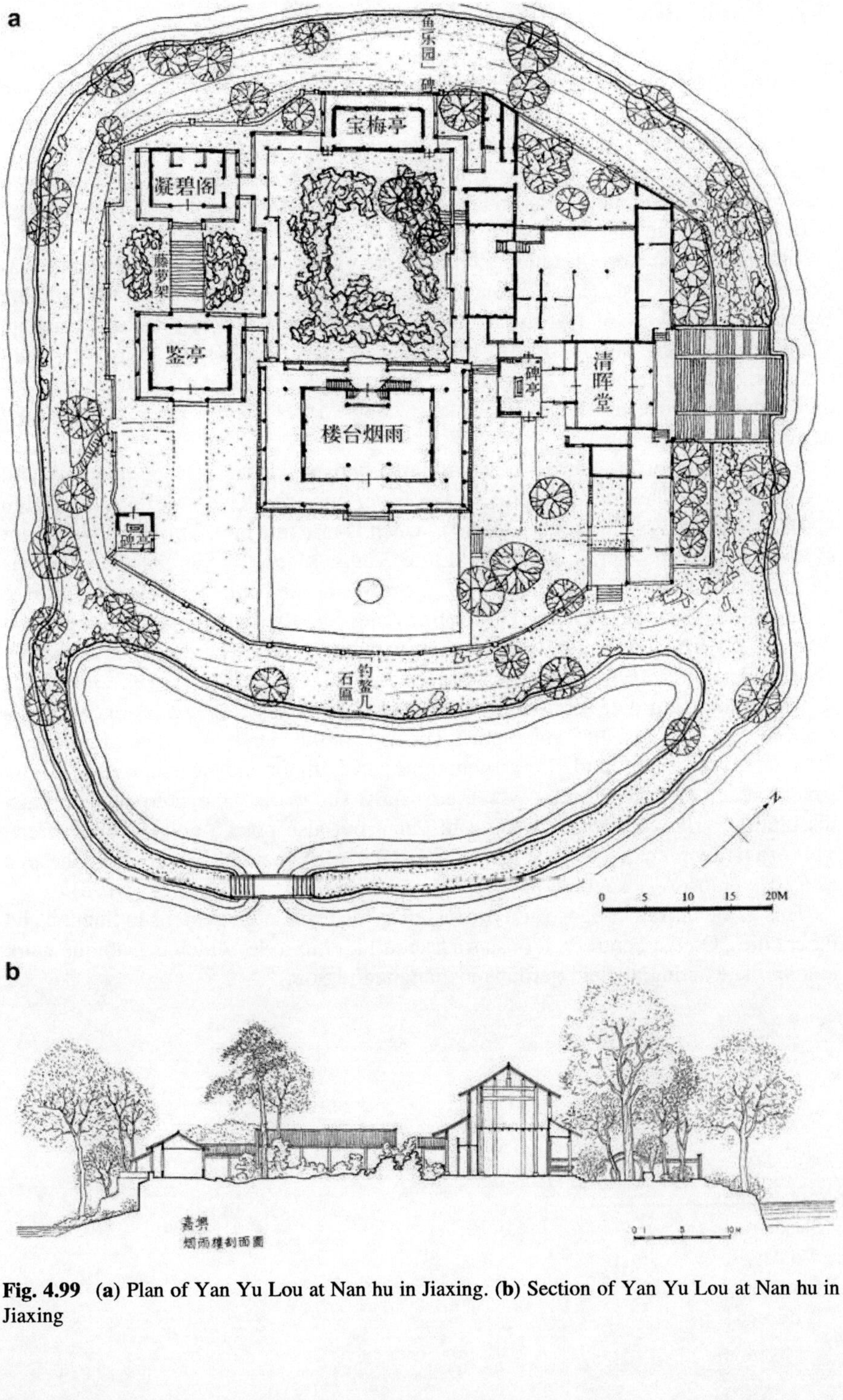

Fig. 4.99 (**a**) Plan of Yan Yu Lou at Nan hu in Jiaxing. (**b**) Section of Yan Yu Lou at Nan hu in Jiaxing

4.4 Examples of Garden in Villages

The site of garden in village is generally between the relatively flat crisscrosses and fields in Jiangnan region. With vast land, the gardens are always larger than that of the city, creating more sparse scenic structure. Burnished by idyllic scenery, these gardens always borrow the scenery of water villages outside the park and acquire a high view by raising mountains and pavilions. Little Lotus Manor (Xiaolian Zhuang) in the town of Nanxun is an example.

Little Lotus Manor (Xiaolian Zhuang), commonly known as Garden that was preserved miraculously (Liu Yuan), was built by Liu Guanjing at the end of the Qing Dynasty in the west of Nanzha Wan Gu Qiao in the town of Nanxun, Huzhou, Zhejiang Province (Fig. 4.100). In the 1920s, the Jiaye Hall (Jiaye Tang) was built next to the garden to collect books by his posterity (Fig. 4.101), and in the north was the Yizhuang Ancestral Hall (Yi Zhuang Zong Ci). However, the garden is now deserted.

The majority of the garden is waterscape, with vast river surface, marshes and idyllic scenery. The ancient name of the water surface, "Guapiao Pond (Guapiao Chi)", is known as 10 µ (about 700 m^2), which is rare in a city. The main viewpoint of this waterscape—principal hall "Little Lotus Manor" (Xiaolian Zhuang), is located on the south bank (Fig. 4.102); on the north bank, there is also a base point called "Jingxiang Poetry Hall" (Jingxiang Shi Ku) with a pavilion beside it (Fig. 4.103). The pavilion beside the lake, which used to be on the water, has been destroyed (Fig. 4.104). The lakeshore is built by overlapping lakes and rocks, depicting the natural erosion landscape and making artistic exaggeration, which is quite successful (Figs. 3.50 and 3.54a). The west of the marsh is a mountainous area, dominated by earth mound. Stone mountains and earth mound are usually stacked by small pieces of lake stones, which not only show the characteristics of karst dissolution—*shou,*[51] *zhou,*[52] *lou,*[53] and *tou,*[54] but also pose a good overall effect. This is a rarely precious work (Figs. 3.3b and 3.33c). Yanzui Xuan is built in the area as a base point. A hexagonal pavilion can be seen in the garden (Fig. 4.105).

This garden is not as dense and processed as the residential gardens in Suzhou and other cities. On the contrary, it boasts a tranquil idyllic style, which is a unique work among the existing ancient gardens in Jiangnan region.

[51] Slender.

[52] Wrinkly.

[53] Cavernous.

[54] Perforated.

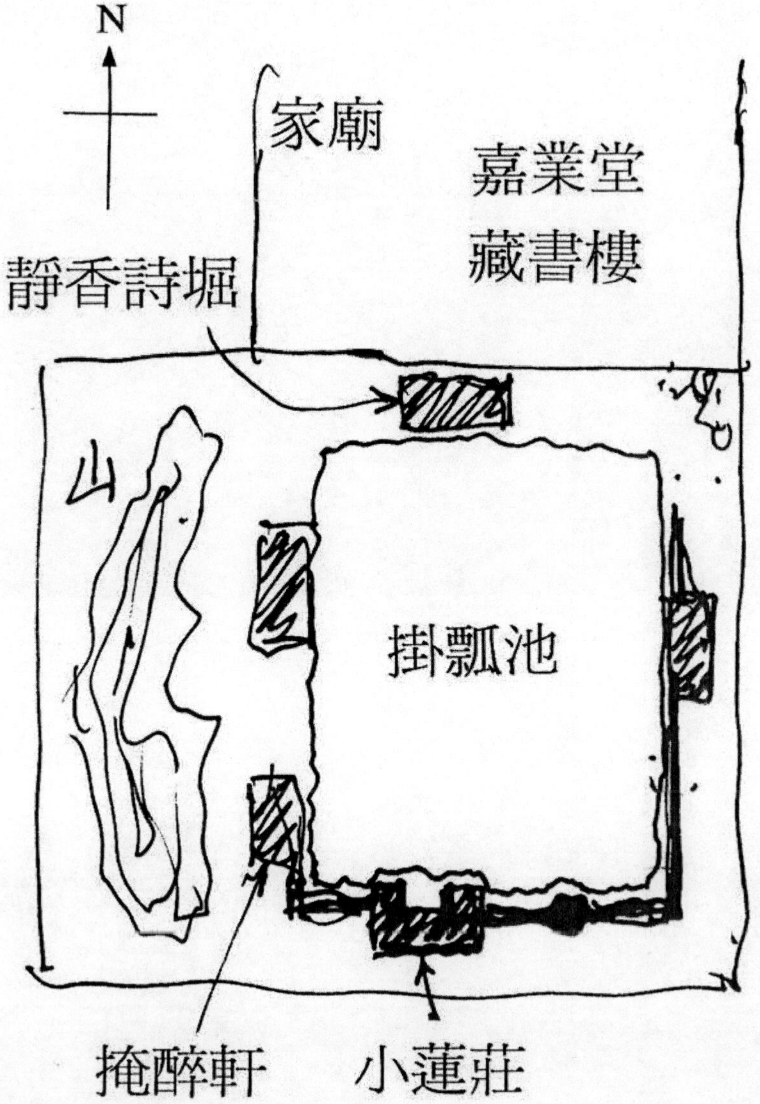

Fig. 4.100 Plan of Xiaolian Zhuang in Nanxun

Fig. 4.101 Jiaye Tang, the book-collection building beside Xiaolian Zhuang in Nanxun

Fig. 4.102 The main building–"Xiao lian Zhuang" (a sketch of photographs taken in the 1930s in The Catalogue of Jiangnan Gardens)

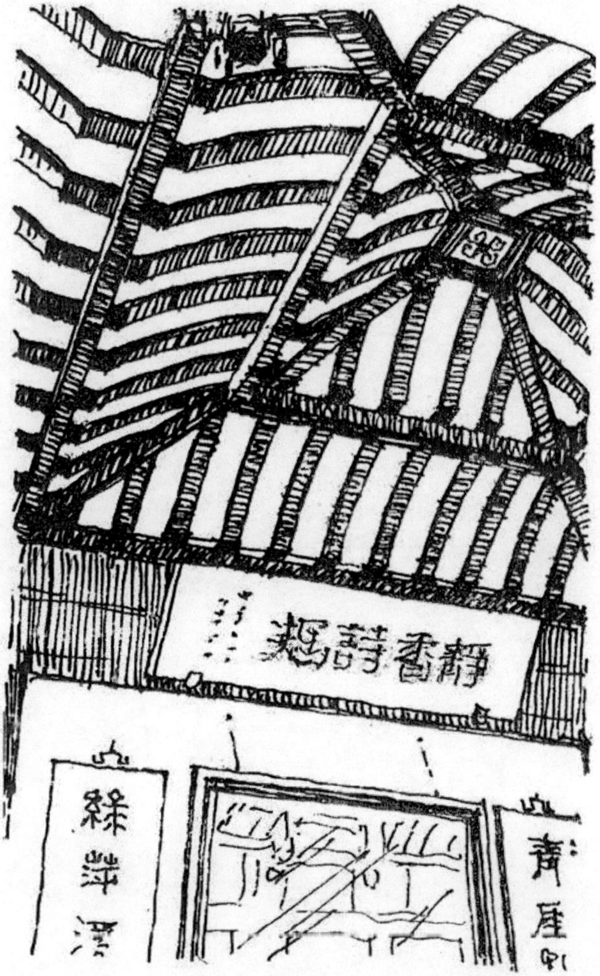

Fig. 4.103 One of the sunk panels in the yuanyang ting of "jingxiang shiku" in Xiao lian Zhuang in Nanxun (a sketch of photographs taken in the 1930s in The Catalogue of Jiangnan Gardens)

Fig. 4.104 The site of the pavilion overhanging the lake in Xiaolian Zhuang

Fig. 4.105 The hexagonal pavilion remained in Xiaolian Zhuang (Taken in 1978)

Chapter 5
Conclusion

Just as all scientific researches are abstract, we have spent the above four chapters to analyze the principles of the artistic composition of the classical gardens in the Jiangnan region, which can be summarized as follows.

Firstly, for the classical gardens in the Jiangnan region, profound principles of artistic composition can be found in the objective entities such as landscapes, buildings, plants and animal life that we can see intuitively. The garden scene, which embodies the theme of *shi qing hua yi,*[1] is the basic feature of garden art. Scenery consists of structural foundation and guidance, which are interdependent and mutually restricted, of which guidance is the most active leading aspect and can be regarded as the soul of the garden. From the perspective of a visitor to the garden, guidance is an arranger of scenic images. As for the scenic structure itself, guidance manifests not only spatial but also temporal relationship. Guidance mainly determines the transformation of space and time of classical gardens in Jiangnan region, as well as the integration of nature and artificial art.

Secondly, at that time, classical gardens in the Jiangnan region are touching, which serve as not only the object of natural aesthetic enjoyment, but also the environment for pleasuring and living functions. The reasons for its durable artistic charm are as follows:

1. The creation of scenery originates from nature and life. Those gardens employ themes that are compatible with the aesthetic tastes of the time and can be applied in poems and paintings. The landscape, flora and trees in the garden are the portrayal of the natural scenery that people are familiar with. Many of them are representation of garden art by means of other artistic forms, such as landscape painting depicting scenery and splendid fragments extracted from poems. Combined with buildings setting in landscape, these scenery are beautification of life. For example, halls, pavilions, towers, belvederes, corridors, and rooms imitate certain buildings such as mansions, temples, etc.; pavilions, palaces and corridor

[1] Poetic sentiment and picturesque overtone.

© Springer Nature Singapore Pte Ltd. 2022
H. Yang, *A Treatise on the Garden of Jiangnan*,
https://doi.org/10.1007/978-981-16-6924-8_5

bridge etc. imitatepavilions for resting on the roadside or ferry and shelter bridge etc.; zigzag bridge imitate slabstone bridge; *yunqiang* (cloud hill-climbing wall) imitatehills and walls; what's more, fences, canopies, *lou chuang* (openwork window), stone beams, stepping stones, parquet, etc. All stem from reality. These processed and typified scenery, taken from the reality that people are familiar with, can offer visitors aesthetic enjoyment full of life interest in a more concentrated and prominent way.

2. Jiangnan classical gardens have sufficient practical settings to adapt to specific garden tours and residence modes. Pleasure in gardens, including residual gardens, should be premised on comfort. A garden, even with charming objects of viewing, can't offer full aesthetic enjoyment if it lacks sun and rain shelters, supplies for the thirsty and hungry or proper ways of rest for the exhausted. As pleasure in gardens is a recreational activity featured with nature, functionality is a must for gardens in Jiangnan region. According to high officialdom's garden living style, some settings are a must, such as a small room for the young servant to bake tea, poetry club for reciting poems with friends, a room for plucking a qin zither, a study for stepping into the past and present, an abode for playing a game of weiqi (an ancient Chinese board game) and engaging a leisure conversation, a high pavilion for leaning on and dreaming, a hall for receiving guests, a waterside pavilion for viewing fishes or even rocks and boats for fishing. Classical gardens in the Jiangnan region, because of these practical settings, have satisfied the bookish, ideal and elegant enjoyment of high officialdom—have offered a full sense of affinity for them. Even for modern people, the park is still comfortable to visit because it has paths with sun and rain shelters and facilities for resting, such as seats on the verandah, stone benches in the mountains and caves, as well as beside ponds and roads, even goose neck chairs out of and furniture for sitting and lying inside pavilions.
3. The scenery of Jiangnan classical gardens is a unity of appreciation and functionality, with objects of viewing being settings with practical value. All kinds of bridges in the garden serve as embellishments of waterscape as well as paths for sightseeing; the pavilions and halls in the mountain or beside water are not only an integral part of pictorial structure of scenic imagery of the village or watery region, but also a shelter from the sun or rain, a place for resting or pleasuring etc.. This combination of appreciation and functionality involves pleasuring into the rhythm of the scenic art, leaving a comfortable, cordial, and touching impression.

Thirdly, the classical gardens in the Jiangnan region is a form of art of time and space with practical value. In terms of its spatial feature, the garden portrays a space in the space—to represent artfully the living space of nature and society by means of artificial space. In the art of the classical gardens in the Jiangnan region, natural and living space are integrated, which can be regarded as artful unification between natural and artificial environment which is the most fundamental achievement of Jiangnan classical gardens, or the essence of Chinese classical garden art.

As both the opposite and community of nature, human beings are a member of the objective nature that stays with the planet. In ancient times, unable to control the nature due to the shallow scientific knowledge, human beings were subject to the control of natural laws as a link in the natural ecological chain. Therefore, the ancients were awed by nature. Since China has its own written history, emperors with supreme power in the world have always humbled themselves in front of nature, offering sacrifices to heaven, earth, sun, moon, mountains and rivers, which has become China's national grand ceremony and imperial etiquette that have been implemented for thousands of years. The broad masses of the people also pay homage to mountain gods and the land. Therefore, the creation of the ancient man-made environment wouldn't dare to destroy the natural environment and "beard the lion in his den".

In ancient times, some philosophical Chinese did not respect nature out of superstition, which had a more profound impact on the planning of the artificial environment in ancient China. Whether it was architectural design or urban and rural planning, or whether it was the living residence or the tomb for the dead, this planning emphasized the harmony and unity with the natural environment, especially the planning of gardens and natural scenic spots. With the progress of human civilization, the development of science and the growth of the ability to transform the objective, human beings, as opposite of nature, won't fear and worship nature. China's respect for nature, based on philosophy, is no longer admired by people. On the contrary, people even arbitrarily seize and plunder life in nature.

Today, after human beings have suffered the consequences of their pollution of the environment in which they live and the destruction of the ecological balance, and have received merciless revenge and lessons from nature, they rediscover the authority of "mountain gods" and "land" as well as China's philosophy of emphasizing nature, what's more, they begin to re-establish the concept of respecting nature on the basis of science. Since the 1970s, environmental science has emerged worldwide. Some new topics need to be solved urgently, including how to eliminate the pollution of human environment, how to protect the balance of natural ecology, how to coordinate man-made environment with natural environment, and how to improve the quality of environmental culture—environmental art. Therefore, the significance of the achievement of Chinese classical gardening represented by Jiangnan gardens goes beyond gardening itself. In the modern times and future, it will boast a valuable and important reference for human environmental creation.

Fourthly, the existing ancient gardens in the Jiangnan region, whether they are renovated and opened or dilapidated, and whether they have been listed as Cultural Relics Unit or not, are important historical relics and outstanding cultural heritage which belong to both China and the world. What proves this is that the nine famous classical gardens in Suzhou have been listed in the World Heritage List by UNESCO. What we possess is responsibility to protect these gardens rather than right to destroy or tamper with them. There is no doubt that the department of cultural relics and gardens take direct responsibility, however, all the people have the obligation to supervise and protect the gardens. The author calls on all members of the academic community, apart from studying and utilizing these precious ancient gardens, to take active actions to do non-must, that is, to publicize their values and care about their protection, so as to pass these precious heritages on to future generations.

Appendix: A Treatise on the Garden of Jiangnan—A Synopsis

Introduction

A Time-Space Art of Practical Value

Scenic painting, that is the "landscape painting" of China, attempts to recreate the three-dimensional natural scenery on two-dimensional paper with skillful brush-works, a manipulation of spaces on flat surface. Poetry and prose attempt to describe spaces with words and scripts. Chinese garden that embodies the sentiment of poetry and painting, described in the West as landscape in "natural" or "scenic" style, is not only a real space, it is also a description of the art of space, an artistic space.

On the stage, an act of comedy makes one laugh, a tragedy makes one cry; in these acts, the man of real life-the actor, enacts the man of dramatic art-the role in accordance to the rules of drama in order to give the audience artistic sensation that is not far remove from the real life. Similarly, Chinese landscape art attempts to create an artistic imagery within a limited space. Utilizing the elements of nature, such as sand, stone, water, earch, plants and animals as material, the sentiment of poetry and painting and the infinite illusory sentient possess the visitors to the garden. The movement of the stage art recreates life while the imagery of landscape art recreates nature.

In the Jiangnan (south of Yangtze River) area of China, the natural scenery that is portrayed by even a small garden squeezed among a clutter of buildings would never miss out the intrigue of nature. Even if the site of Jiangnan garden is far from ideal, often a few plants and a piece of rock can be used to provide the buildings a 'natural' environment. That is why the garden of Jiangnan are often praised as the "garden suitable for inner-city". The garden of Jiangnan that recreates natural spaces with spaces of scenic imagery on a plot of and in inner-city epitomizes and typifies mother nature, just like a painting that reduces the scenery of thousand miles onto a scroll of

© Springer Nature Singapore Pte Ltd. 2022
H. Yang, *A Treatise on the Garden of Jiangnan*,
https://doi.org/10.1007/978-981-16-6924-8

a few feet. Thus, the "city-garden" must surly be also the 'garden that brought scenery afar into near distance'. Stage art is life in miniature while landscape art is nature in miniature. The garden of Jiangnan can also be compared to Beijing opera or a style of painting that concentrates on mood-capturing which utilizes artistic epitome as a necessary mean to its creation. Simple and abstract pattern on the mask of traditional Beijing opera gives the character of the role away, and at the same time intensifies the realism of the act. A dance in the opera helped by just a horse whip or an oar can delineate a particular scene or storyline graphically. Similarly, an "undulating cloud-patterned wall" of a garden in Jiangnan will remind one of a village in the mountain, and a crescent of clear water with a few rocks will give the impression of wild landscape.

The garden of Jiangnan may either be inspired by a landscape painting, or take its cue from pastoral or scenic poetry and prose, or even uses natural scenery as its blueprint. The creation of a garden is definitely based upon a particular theme, be it a scene of serene forest and limpid spring, or a scene of tranquil lake and mountain, or a scene of mountain village and winding path, or even a scene of a hall at the water edge with leaves of lotus swaying in the wind. One can thus conclude that the scenic imagery of a garden is subjectivised imagery. Although mother nature is vast in scale, more interesting and richer in scope, a garden is characterized by its concentration, symbolism, idealized form and is full of fascination. In the scenic imagery of a garden, it must contain a certain element of thought and feeling, which is what many commentators happily called the "sentiment of poetry and painting". This is precisely why a landscaped garden can surpass natural scenery in moving one's heart and leading one into detail appreciation.

When one enters a Jiangnan garden, one can immediately see pavilion, platform, tower and other buildings hidden among rocks, water, plants and trees; as one progresses to tour the garden, the scene changes with every step one takes. The rhythmic and melodic sensation created by the scenic imagery remind one of the saying that: "Architecture is a piece of frozen music". The garden of Jiangnan is more than a piece of frozen music,it is frozen poem. Within the boundary of a particular plot, one can create an ideal, natural and intriguing environment through carefully moulded earth surface, nicely arranged plants, skillfully constructed buildings and intersperse with birds, animals, reptiles and fishes, all in accordance with the thematic requirement of creating an environment full of poetic sentiment. Isn't this principle of landscape design much more profound than the design principle of European classical garden that preoccupies itself with geometrical disposition consisting of only "embroidered flower beds" and "sculpture in green?" This is also why William Chambers, the English architect and landscape designer, commented as early as in the middle of the 18th century that: "The landscape arrangement of Chinese gardens is superior to our own, the intrigues that are found in these gardens is what the English longed for in a long time but never quite achieved". (W Chambers, Designs of Chinese Buildings, Furnitures, Dresses, Machines, And Utensils, 1757).

The garden of Jiangnan can be likened to a form of painting in three-dimensional space with sand, rock, water, earth, plants and animals as materials, it is in fact a three-dimensional scroll of painting into which one may enter physically. The functional value of the Jiangnan garden is built upon its spatial characteristic. Therefore the art of landscaping is a form of pragmatic art. This functional value of the garden that supports the activities of living and appreciation is the very basis of its existence. Although miniature landscape contained in a clayware is also a three-dimensiona art form, it is only for visual appreciation and lacks practical value. Whereas a landscaped garden can induce one to make a tour as if into the painting by utilizing scenic imagery created in human scale. Even though the plot of land on which the garden is built is tiny, it would never lose its practicality as far as the activities of living, touring and appreciation are concerned. It is therefore different in essence from the large miniature landscape of Guangdong that is also complete with mountain and water imagery. In the classical commentary on painting, it was said that: "Whenever a landscape painting has attained the level of being able to entice one to walk through it, to tour in it, to dwell in it, it must be classified as the most excellent work of art". (Guo Xi, of Song dynasty: LIN-QUAN GAO-ZHI-JI) The so-called being able to walk through, to tour in and to dwell in the landscape painting is really an imaginary mental state of the appreciator of the painting based upon his experiences of life. Whereas the Jiangnan garden provides one with actual spaces to fulfill the desire to walk through, to tour in and to dwell in natural environment full of the sentiment of painting.

One can often hear praises that the Jiangnan garden affords one to "perceive large scenery through elements in small scale", and thereby concludes that this is a principle of garden design. Actually, "perceiving large scenery through elements of small scale" is similar to "surpassing a multitude of elements with just a few", they are simply end-effects. The basic principle is: "to describe the unlimited space with limited space", an unity of the dialectic relationship between the limited plot of land on which the garden is built and the unlimited natural scenery it attempts to recreate.

Again, many praise that the garden of Jiangnan possesses the quality of "leading one into serenity with winding path", and thereby conclude that this is the theory of garden design. Actually, this quality is just an appearance of the garden rather than its essence. It is through this appearance that one realizes that the garden contains "serenity"-the object of appreciation(taking natural beauty as its pre-occupation) and also the "path"-the mean of appreciation(the buildings and other elements by which one may make tour or rest). With the unity of the dialectic relationship between "serenity" and "path", one may perceive the paradoxical relationship between scenery(natural beauty) and architectural elements(man-made beauty) within the unified whole of the garden of Jiangnan. Precisely because of this, the gardens that possess the quality of "winding path leading to serenity" and developed scenic area are much easier for one to identify with than the wild mountain and primitive forest. What one sees as beautiful is actually related to its usefulness. In the quality of "winding path leading to serenity", its practical value lies in that one may dwell within the natural intrigues of sinuosity and variability. Thus in this quality, one can perceive the active dynamism of the means of appreciating the garden(referring to the path and place of stay) that possess practicality.

While mentioning the two principles of "perceiving large scenery through elements of small scale" and "winding path leading to serenity", some researchers of Suzhou gardens also refer to these two paradoxes as the inevitable weakness of Jiangnan gardens. What they do not realize is that these two qualities are precisely what made Jiangnan gardens so special, such paradoxes are the very essence of the Jiangnan gardens. If these contradictions were resolved (meaning total destruction of the quality) as suggested by the accusers, then Suzhou gardens (the cream of Jiangnan gardens) will be totally devoid of their intrigues.

The difference between landscaping art and the art of painting lies not only in that landscaping art attempts to describe space with space, from the point of view of appreciation, landscaping art is not static, the way to savor and enjoy such art form is to be in movement, which is what is commonly called "touring". Because of the continuous and discontinuous movement during the process of touring of the garden, landscape art can also be described as an art form organized around the factor of time, one can thus discern the inner structure of landscape art through its elements such as orientation, timing and the order of presentation of scenic picture. Just like in music, the organization of scale, timbre and volume with time produces melody and rhythm to represent certain moods that the composer wishes to bring across, the artistic effect of garden is achieved through the organization of a series of overlapped scenic impressions with the moving on of time during the tour of the garden. At the same time, present in the appreciation process of the scenic imageries of landscaped garden is the presentation of scenes under difference seasons and time, for example, the appearance of the garden is altered during each of the four seasons, morning and evening, raining and snowing time, bright moment and overcast time, and other climatic variations. This is also part of the effect of time. Just like a piece of sculpture which only shows its peculiar artistic effects under certain lighting condition, some garden scenic imageries can only show its appeal during a particular time. The classical Jiangnan garden has consciously captured this characteristic of time in its creation. Thus, landscaping art is not only an art form that manages space, it is also an art form that manages time. It is a practical art of space and time.

The garden of Jiangnan as a piece of artistic work out only performs the practical function for appreciation and living, it is at the same time a piece of combined engineering work. It involves soil, rock and water engineering in the techniques of stacking mountain and managing water; construction engineering in the building of pavilions, platforms, towers, halls, water-pavilions, verandahs, bridges and others; and also techniques relating the gardening works. These engineering techniques combine to create a garden, at the same time, with the improvement of these techniques, landscaping art obtains new inspiration.

On The Functional Content of the Garden of Jiangnan

The Form of Garden-living, the Conception of Landscape Design and the Philosophy of Its Creation

The functional content of a garden that will fully satisfy a certain form of garden-living has two aspects, the material and the spiritual. The material aspect refers to the practical content of the garden, a guarantee that the material condition of the garden is conducive to living. These are the practical facilities to ensure a smooth tour in appreciation, to satisfy the need to shelter in rain and under scorching heat and the need to rest or sleep and other activities of garden-living. These facilities that provide for rest, entertainment, sport and other activities can actually be obtained from either man-made or natural environment outside the boundary of its garden, thus as far as the garden itself is concerned, though these facilities are necessary, they are not the fundamental attribute of a garden. But since these facilities are organized within the garden, such practical content becomes an inseparable constituent of a garden through fusing with the philosophy and the spiritual content of the garden.

As for the spiritual aspect, it refers to the philosophical content of the garden, that is the thematic thought that unites the scenic imageries of the garden, which is also the sentiment of poetry and painting that moved the visitor of the garden greatly. The spiritual aspect is embodied in the scenic imageries of the garden. Due to the fact that the spiritual content of a garden gives the visitor an aesthetic enjoyment solely through the activity of appreciation, it is generally referred to as the appreciation content, which is the basic content of a garden. This spiritual content of the garden not only is in concord with the material content, it is also united with the scenic imagery and thus creates a work of landscape design.

Although the success or failure of a work of landscape design that embodies a certain appreciation and practical content depends to a large extent on the level of workmanship, it is, however, mainly decided by the level of design ingenuity. In the Ming dynasty book on landscape design, "Yuan-ye", written by Ji Cheng, it is said that there is a proverb relating to architectural design that goes: "Three parts out of ten depend on workman(i.e. workmanship), and seven parts those who mastered them(i.e. the designer and the insistent owner)". Which is to say design takes 70% of the credit while workmanship takes 30%. As for landscaping works, Ji said: "The master of the landscaping has to involve himself with nine parts out of ten of the works, while the workman only gets involved in one part". In the creation of the landscape garden, the landscape designer is responsible for 90% of the end result, and workmanship only accounts for 10%. In this instance, Ji is referring to the designing on site when one has to fit the elements into the limited plot of land. This observation is the result of Ji's wealth of experience. A garden reflects more of the taste and skill of the designer than a building would. To a greater extent, a garden represents more of the culture and sophistication and the style of garden-living of the owner of the garden. As the landscape designer is employed and controlled by the owner directly, the design philosophy of the designer inevitably has to be in accord with the garden-view of the owner. Furthermore, an old garden would have gone through many changes in ownership, many renovations and even major restructuring. Changes occur not only in the plant layout, which is the easiest to alter, but also those elements that influence the structure of the garden to a greater degree, such as the mountains, water and buildings. Particularly the small details that decorate scenic imageries are most prone to be changed by the subsequent owners in accordance to their likes, dislikes and the level of landscape appreciation.

The owners of the classical Jiangnan gardens in ancient time are mainly the court officials, that is the feudal gentry either in office or already retired, and also the powerful family and rich merchants; owners of small gardens are usually poets or writers that possessed certain wealth. The feudal officials that were cultured not only worked after wealth and status, they also desired to retreat in the fields, mountain or forests to show their integrity and that they were detached from the vulgar world. A house-garden in the city fits exactly to their requirement, the enjoyment of both material and spiritual realms. Those powerful family and rich merchant who wished to associate themselves closely with the gentry also follow suit. The poem of Su Dongpo- "I will rather live without eating meat than to live without bamboo surrounds, lack of meat will only cause one to slim, lack of bamboo makes one vulgar", became very popular among those "meat-eaters" of Jiangnan that attempted to follow the way of the cultured. In reality, they wish to eat meat and at the same time enjoy the presence of bamboo. They have sufficient wealth to fulfill their ideal of garden-living. What Wang Shizhen(Yuenmei), the owner of Yanshan garden in Taicang of Ming dynasty, said to Chen Meigong is most representative: "Living in the mountain is too lonely, living in the city is too noisy, only living in the garden is just right". They wish to enjoy the civilization afforded by the city and at the same time the tranquility of mountain and lake, and the "city-garden" resolves this dilemma for them perfectly.

In some of the private garden, while enjoying the garden by himself, the owner would usually not refuse other intruders that also love the garden. This reflects the spirit of open-mindedness of the owners at that time. Those who keep the garden to themselves are too selfish to be called lovers of nature one may even doubt their ability to appreciate landscape art. They only wish to show off their wealth by following the example of the cultured, garden to them is just the tool to earn fame.

Actually, the private gardens of Jiangnan, expressed in the form of "city-garden" or "miniature mountain and forests", in themselves, indicate the characteristics of the social class to which the owners belong. The "city-garden" is in fact a highly abstract form of art that attempts to recreate natural scenery that changes incessantly in the limited plot of land in the city. The scene of mountain and water in reduced scale, i.e. the "miniature mountain and forest", enables the visitor to the garden to enjoy a natural environment without going through the agony of long distance travel. Therefore the garden meets the requirement of the owner who live the life of a parasite perfectly, as they are too lazy to move their limbs about and only wish to indulge in the form of garden-living that allow them to "tour while lying down" and "appreciate in spirit". From the viewpoint of modern landscape design, it is not difficult to notice that for these reasons, the scenic imagery of classical garden of

Jiangnan emphasizes only the appreciation value of sight, and lacks the active function of rest through activities of sport.

The classical garden of Jiangnan can be divided into the following categories in accordance to their practical functions, they are the "courtyard garden" that is designed to fit in the courtyard of a dwelling, monastery, government office, temple, college or guildhouse; the "house-garden" that is constructed on a separate piece of land adjacent to the house; and "garden of the villa" (the scenic imagery of which is roughly similar to the house garden, but the content of its scenic imagery is usually combined with the requirement for living). Other than these, some monasteries, government offices, temples, colleges and guildhouses have independent gardens associated with the buildings. In general, the artistic structure of these gardens is similar to house-garden due to the fact that these gardens are not meant for living in, thus accordingly, they have less building.

The design of dwelling of upper class of the society in the Jiangnan cities in ancient time is similar to other regions in the country, in that it is a strictly symmetrical layout in the form of successive courtyards planned to the exact requirement of the feudal law. In these large houses, courtyard garden is usually placed in the courtyard of subsidiary axes in front of a small building, a study or inner living hall. Those few gardens that are part of large dwellings, monasteries, guild houses or colleges are usually designed with symmetrical layout. The decorative elements such as trees, plants, mountain, rock and miniature landscape in pots are placed in relation to the central axis. Even the shape of ponds follows the orthogonal architectural setting taking on geometrical shape, mainly rectangular, and generally placed on the central axis.

The courtyard garden in the secondary courtyard of large dwelling, or garden in the central courtyard of small dwelling is usually much free in arrangement. The structure of scenic imagery of house-gardens (and villa-gardens) is usually complete in itself as these gardens are normally relatively independent. It is therefore said that house-garden is the typical example of Jiangnan classical garden. Thus, our discussion will use the house-garden as the main object of analysis.

House-garden is, first and foremost, a place for relaxation and play for the family of the owner in times of festivity. For those ladies of the household who are kept in the rear of the house all year long, the "back garden" is an ideal place for them to pass time as well as a heaven for lover's rendezvous. For the master of the house, it serves as a place of rest from the daily rat race or during his retirement. Beside using the garden for stroll and appreciation, the typical manner of garden-living for the feudal gentry consists of reading, painting, chess-playing. playing of musical instruments, recitation of poems, chatting, drinking of tea and wine and activities such as meditation and day dreaming and etc. According to the social status and particular interest of the owner, the garden can also be used for the performance of opera, gambling and playing of other games. It is thus a place for social gathering as well.

For the typical way a feudal gentry spent his time in the garden in the late feudal era, we can rely on the records of Chen Fuyao (Huangzhi) of Hangzhou- "Daily lesson in the flower beds" contained in his book- "The mirror of flowers that is secretly kept", written in the years of Kangxi of Qing dynasty. In this work, he describes the various forms of garden-living in the four seasons.

In the last hundred years after the industrial revolution, particularly in the cities of industrialized nations, highrise buildings sprang up everywhere on the face of earth, which block off much sunlight and create a man-made valley of darkness: noisy factories emit smoke and dust in haste which has created artificial haziness. Gardens that are created in these cities would do much to restore health into the cities. As for classical garden of Jiangnan, its main function has always been to satisfy the spiritual needs of man. A garden can quench the thirst for nature-the forest and spring, in which one can seek spiritual fulfillment and also the cultivation of one's character. The way a garden fulfills spiritual requirements is through the conceptual content of the scenic imagery-its philosophy, mood and interest. For the garden of Jiangnan, particularly the representative house-garden, in accordance to the imagries it contains, one can distinguish the following categories: the garden of mountain-scape that has mountain as its main scene (e.g. the Villa of beautiful surrounds (Huanxiu Shanzhuang) in Suzhou, the Yan Yuan in Changshu and etc.); and the garden of water-scape that has water surface as its main attraction (e.g. the Master-of-Nets Garden (Wangshi Yuan) in Suzhou, the Yi Yuan in Nanxun and etc.); there is also garden of mountain and water-scape that has both scenes in equal proportion (e.g. the Humble Administrator's Garden (Zhuozheng Yuan) in Suzhou and Garden for Ease of Mind (Jichang Yuan) in Wuxi and etc.). As for the garden that depends mainly on flowering plants as centre of appreciation, we lack extant examples and thus it remains outside the scope of our discussion on the garden of natural scenery.

The artistic appearance of the garden depends on the concept of creation, which in turn depends on the landscape philosophy. During the time of Qin and Han dynasties, the rulers were deeply impressed by the techniques to achieve immortality. Under such influence, the art of land-scaping was used to create an environment that resembles one that the immortals dwell. Thus we are able to see the immortal mountains of the Eastern sea, the Tai-ye and Pong-laj, or the "three mountains in one lake" (Lake Tai-ye and the mountains of Pong-lai, Fang-zang and Yin-zhou) that developed from the prototypes in the Han dynasty as typical expression of such concept. This provides a sure foundation for the development of scenic garden in China in the ensuing thousands of year.

From the time of Wei and Jin dynasties, the love for nature and the concept of retreat in the mountain became in vogue among the class of gentry, which in turn became the guiding philosophy of the landscaping art. Through the inspiration of the natural, pastoral literature and landscape painting that was developed during the period, landscaping art had produced many realist's works that described the natural scenery with full vigour, a vast improvement and enrichment of the scenic imagery of mountain and water of the former era that was mainly based on the illusory concept of immortality. During the time of Song and Ming dynasties, with the infusion of the morality of Confucianism and Chan Buddhism (the integration of the ingenious Confucius and Taoist philosophy with the imported Buddhism), what the West called "natural" or "scenic" style of garden had matured both in terms of theory and techniques. The garden of Jiangnan of Qing dynasty inherited this tradition. The creation of garden utilized the sentiment of tranquility and retreat as its ideal and the nature that had been subjectivised by poem and painting as the blueprint for the creation of scenic imagery.

During the late feudal period, especially from the mid-Qing dynasty onward, the pre-occupation of the pursuit of wealth, status and enjoyment of the declining feudal class had prevailed over the pursue of simple and natural scenery. As a reflection of this transformation, the proportion of building element of scenic imagery that is mainly to serve garden-living and the pursue of enjoyment had increased in relation to the main element of the garden-plants. From some actual examples, we are able to see that the gardens of this period are much different from the gardens of Ming dynasty. The scenic imagery of Ming's gardens depended mainly on the elements of mountain, water surface, forest and plant, only in between these main elements, a small number of flower pavilion, bamboo screen and thatch huts that expressed simplistic natural wildness were placed among the natural landscape. The gardens of late Qing dynasty highlighted the building element in a garden, thus the natural elements have to play second fiddle to the twisting and winding architectural spaces, such as in the eastern part of the Garden that was preserved miraculously (Liu Yuan) in Suzhou.

When the practical function of the garden is being emphasized, architectural treatment became the main mean of organizing scenic imagery. The large increase in the number of building in the garden reduces its natural appearance: when the mountain, water and other natural elements of the garden were reduced relatively, the imagery of the garden was inevitably altered. The sentiment of nature had to be found from just a piece of rock or a pool of clear water, thus the creation of natural imagery became more symbolic. Examples of such symbolic design can be found in the placing of a solitary rock named "listening to the sound of lute" under the window sill of the "Hall of lute-the room of stone named listening to the sound of lute" and also the sole "rock of bowing" in the "hall of the stone of bowing" in the Garden to delight one's parents (Yu Yuan) in Suzhou, and also the stone "Yi-boat" on ground in the enclosed courtyard in Garden to delight one's parents (Yu Yuan) in Shanghai.

There are two main schools of thought in the creation of naturalistic or scenic garden, namely the schools of naturalism and realism. From the 18th century onwards, under the influence of Chinese and Japanese landscaping art, the trend of European landscape design had turned towards the naturalistic and scenic approach. If we leave aside the exotic works of "Chinoiserie", "Japanesque" or the "Anglo-chinese stule" that are full of baroque and rococo interest, we can discern that the mature, realist works of the later period, such as the large landscape garden of England that attempts to describe the vast Scottish pasture, were influenced by naturalism inspire by design in the East. On the other hand, the Chinese garden stresses on artistic realism rather than resemblance of form; that is the so called "transmission of spirit" rather than copying to the last detail. Thus, this can be termed "spirited" works.

Actually, there are also works influenced by realism in the classical landscaping art of China. In Jiangnan gardens, there are those which attempts to describe the vastness of natural scenery of lake and mountain in abstraction, such as the Humble Administrator's Garden (Zhuozheng Yuan) in Suzhou. There are also those gardens that describe part of the natural scenery realistically, such as the Garden of Art (Yi Pu) in Suzhou. In the case of the former attempt, the dimensions of the imagery are necessarily reduced, and the scenic imagery is more orderly arranged, thus it belongs to the class of spirited works. In the case of the latter attempt, it resembles actual natural scenery, thus it belongs to the class of realist works. Chang Nanhuan, a landscape theorist of early Qing dynasty from the region of Jiangnan, objected strongly to the use of limited space in the garden to express the large scale natural scenery. He wrote: "The mountain that soars high into the sky and the ravine allows no sunlight to penetrate its deep are both works of the creator-god, these can never be achieved through human effort. Furthermore, these natural landscape usually span a few hundred miles geographically, and yet, we attempt to imitate its form within the site of no more than ten square feet, or even in a ditch of mere five feet. This is just like the man in the market who kneads earth into small figures, both are done just to deceive kids". (From the biography of Chang)The extremist view of Chang expresses clearly the philosophy of the realist. Concerning the creation of mountain and woods in the garden, he suggested: "As for the works of creating ridges, plateau, slopes and mounds that sweep upward and downward, it can be completed in a matter of days by using the technique of stamped earth. Rocks can then be placed and set out in between the earthworks. It can be finished off with enclosed low wall or dense bamboo fences. It will in the end resemble the steep mountain and impassable mountain barrier, from outside of the enclosure,one may catch a glimpse of its soaring height.....It seems that part of the mountain slope and ravine has been cut off from the foot of the mountain and placed within in one's reach,one may then realize that this part of the mountain belongs to one solely." (From the biography of Chang) Again, this is another typical realist theory, to create mountain, one only needs to take a corner of the mountain and ravine; to create water,one only needs to imitate a corner of the lake. This way of describing a corner of the natural scenery possesses both form and spirit and can be very realistic at the same time. Yi-Du of Suzhou can be said to be a representative work of the school of realism. (Those rock mountain near the pool side were added at a later time which to me is totally unnecessary). These works remind one of the landscape paintings of Xia Gui and Ma Yuan of the Southern Song which also portrayed part of the natural landscape, and were being scorned at as "painting of broken mountain and left-over water."

Whether the creation of Jiangnan gardens is by the method of spirited delineation or the realist approach, the highest goal is always to achieve a particular milieu, one that will add colour to the natural scenery. Some recent European landscape schools attempt to recreate historical happenings, myths and fables, or foreign fashion by using elements like simulated castles, temples, tombs and ruins. This is termed the romantic school, or sentimentalism in the words of some Western scholars. Such design method also has its counterpart in China. From what is extant in Jiangnan gardens, we can still find scenes that take after myths and religious story as their theme. There are stepped rocks that reflect the "eight immortals crossing the sea," "the story of the monkey" and the likes which are mythical subjects. Of course,these may well be the interpreter's opinion rather than the designer's original intention. There are also those images which are based upon Buddhist themes as well, such as the Lion Grove Garden (Shizi Lin) of Suzhou which takes after the mountain of the Lion Grove described in the scripture,the stone scripture pennants of the Pu Yuan (western part of the Humble Administrator's Garden (Zhuozheng Yuan)) and in the lake of Garden that was preserved miraculously (Liu Yuan), and also the Xiling Seal-engravers' Society (Xiling Yinshe) in Hangzhou that is Rill of cliff images, inscriptions, pagoda, cave chambers, steles and etc. As in other form of art, the creation of a garden hinges on its philosophical content, which not only determines the theme of the garden, but also the structure of the scenic imagery.

On Scenic Imagery

The Basic Ingredient of Landscape Art and the Form of Garden-living

Within the bounds of art, one can say that its basic ingredient is the presentation of image, and for landscaping art, correspondingly, is the scenic imagery. The basis of the existence of a garden is the combination of philosophical content and practical content, a summation of all constituent parts of a garden. And scenic imagery is the form of the existence of a garden, the internal structure of the constituent parts of a garden. In another word, the philosophical and practical contents of landscaping art are expressed through scenic imagery.

Scenic imagery is a spatial concept, conveyed through the coexistence of the various structural elements of the scenic imagery, and also through the capability of being able to extend in the direction of up-down, left-right and front-rear of each element. Whether it is a house-garden or a courtyard-garden,the scenic imagery and its resulting spaces of a Jiangnan garden are always subsidiary to the formal living space, the spaces created by the imagery are additions or extensions to the house and serve as an environment for play, fun and appreciation of nature. Time is also an attribute of scenic imagery which is expressed in the interplay of scenes under various climatic conditions, such as the four seasons,dawn and dusk,and sunny day and overcast day; it is also expressed through the continuous presentation of scenes in the process of tour with the designed orderly introduction of the scenic imageries. Therefore,the creation of the Jiangnan garden not only involves the organization of the spaces of scenic imageries, it also involves the organization of the timing of introducing the scenes and also the organization of the aspect of time of scenic imageries in the interplay of scenes with the climatic condition. Thus, the landscaping art of Jiangnan garden is an art of time and space.

There are two aspects to the scenic imagery,namely the elements and the guiding component of the scenic imagery. The making of scenic imagery requires that the elements and the guiding components of the imagery to be in the relationship of unity and also in antithesis. In a way, the creation of Jiangnan garden is a process of unification of the elements and the guiding component of scenic imagery directed by practical requirements and the conceptual requirements of the theme.

The elements of the scenic imagery are the material basis of its structure. It can be divided into two categories, namely the natural elements and the man-made elements. The natural elements refer to terrain,plants and animals. The man-made elements refer to buildings and other architectural objects. The scenic expression of a garden, whether it is of mainly mountain-scape or water-scape, though dependent on its functional characteristic and the manner it is arranged, is largely determined by the elements of the scenic imagery.Scenic imagery is achieved by the organization of its elements directed by the principles of landscape design and also the guiding component of the imagery. Each of the groups of the scenic imagery possesses a certain thoughts and sentiment,the combination of them forms a garden. Generally speaking, therefore, the elements of scenic imagery are the material for and the means of making a garden.

The guiding component of the scenic imagery has to co-exist with the elements of scenic imagery. As far as scenic imagery is concerned it plays the role of an editor of the scenes, thus it is related to the structure of the scenic imagery. In relation to the garden and its visitors, the guiding component performs the function as the means and guide to the appreciation of the scenic imagery in both continuous movement and static observation. To the visitors it thus plays the role of a path. The guiding component is an active constituent of the scenic imagery. It is only through the guiding component of scenic imagery that its elements may be organized to form the scenic imagery, without which they are just a group of material formed purely by chance, and do not express the practical function and artistic conception of the scenic imagery. Indeed, through the guiding component possessed by the scenic imagery itself the elements of the imagery may become an interesting, practical artistic space for appreciating the scenes, thus the scenic imagery may be unified with the underlying concepts to create a milieu that wields artistic influence on its visitors.

The elements and the guiding component are mutually dependent on each other, and also each is the premise of the other. This relation is the real essence of the expression: "place a path where the scenes are and because of the path,the scenes are accessible", a traditional design principle. Just like all objects in the universe, the two constituents of scenic imagery-the elements and guiding component, are by nature relative in essence, rather than being absolute. The role of each interchanges along the process of appreciation of the garden. For example, when one makes a tour of the Humble Administrator's Garden (Zhuozheng Yuan) of Suzhou and stands in the "Hall of Distant Fragrance(Yuanxiang Tang)" looking north towards a group of scenes comprising of lake and mountains, the scene of "Fragrant snow and gathered cloud," the hall is the agent for touring the garden, a viewpoint in the act of appreciating the garden. With its particular characteristic, it qualifies as a guiding component. And the scene of "Fragrant snow and gathered cloud" is the object of appreciation, with its particular characteristic, it serves as the element of the scenic imagery. When the visitors stand in the vicinity of the scene of "Fragrant snow and gathered cloud" looking south towards the "Hall of faraway fragrance," then the

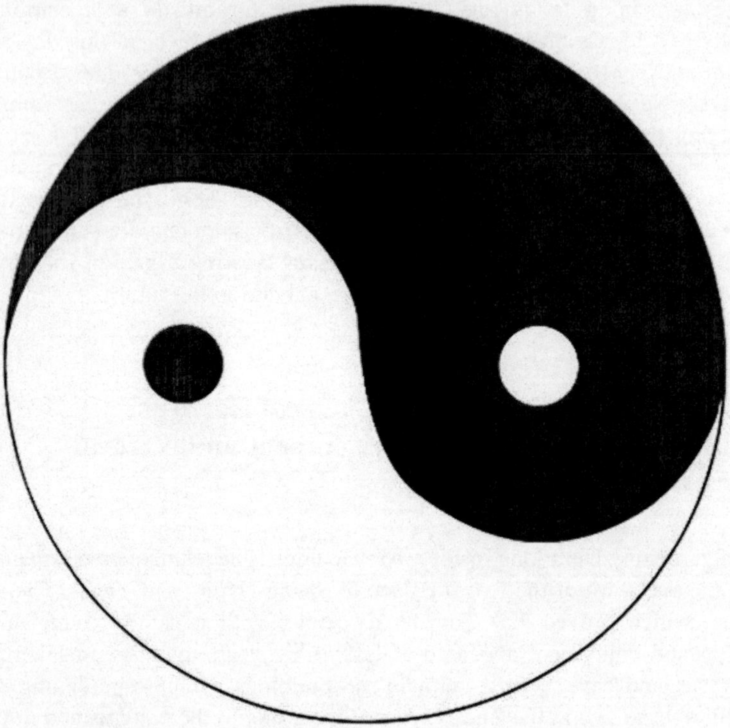

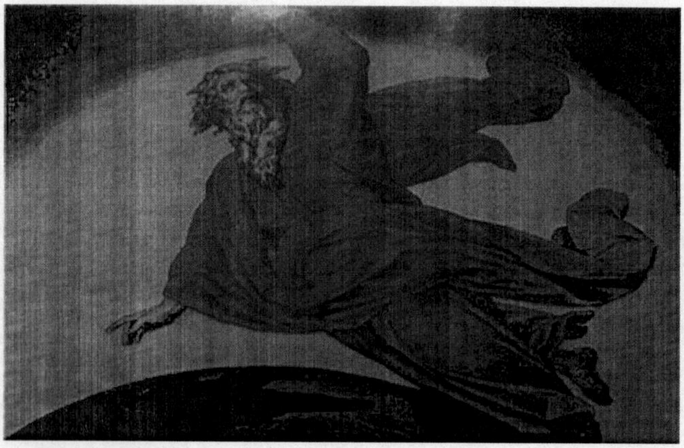

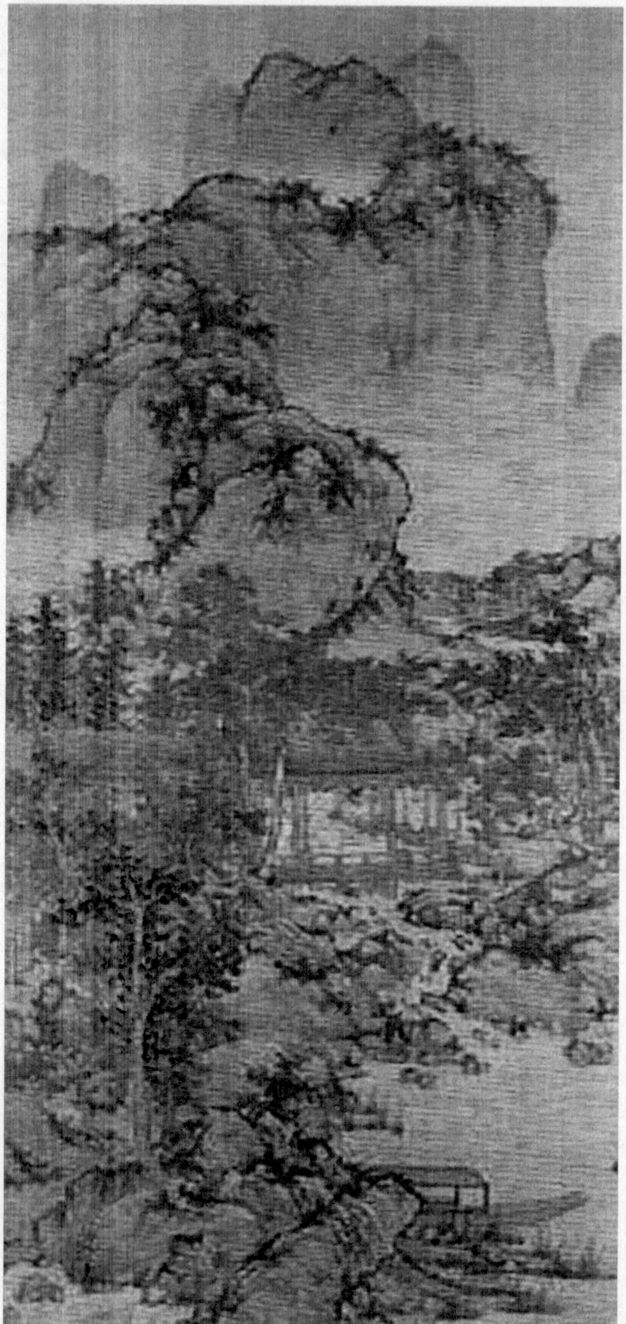

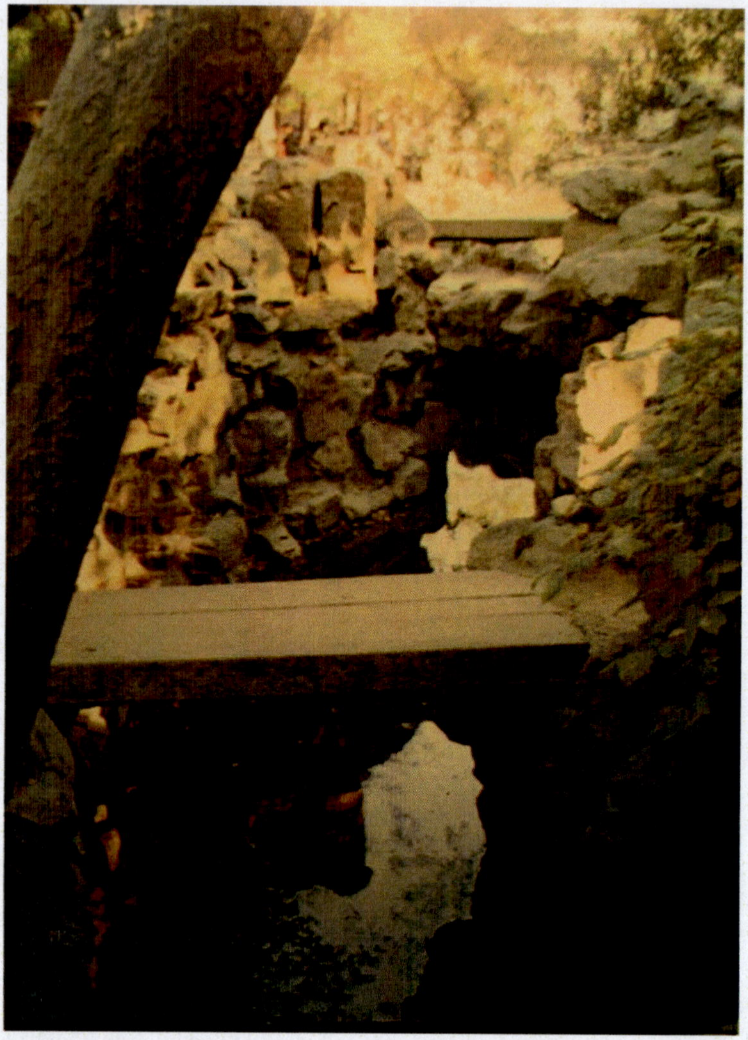

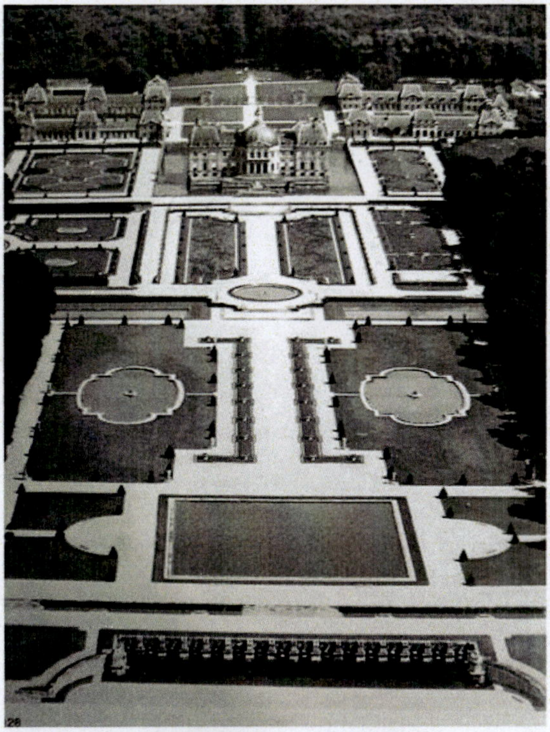

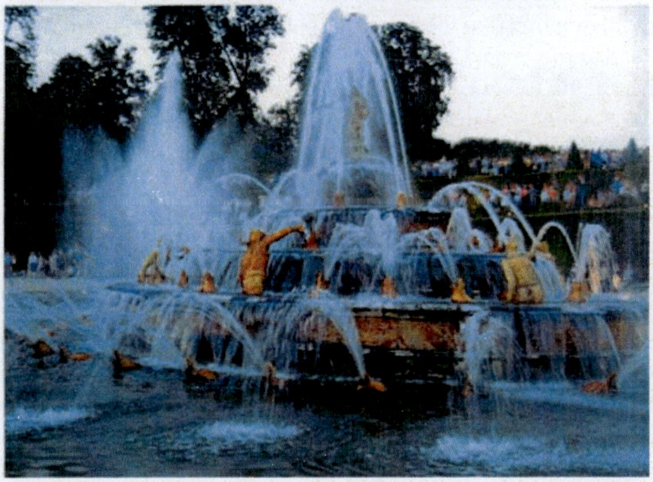

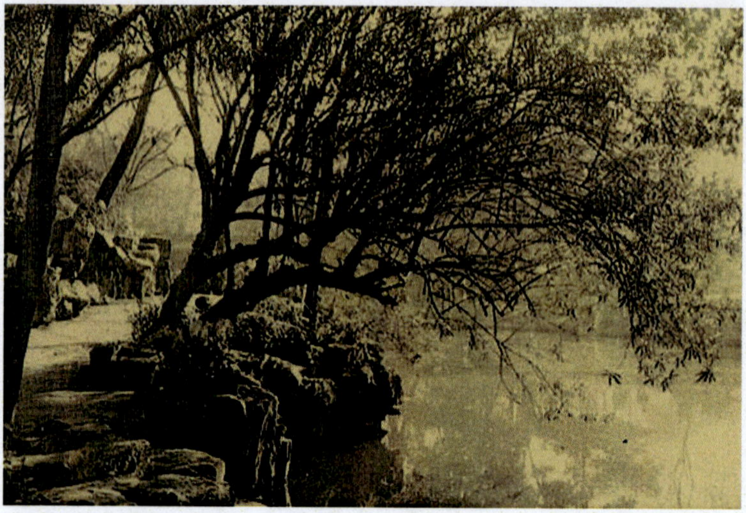

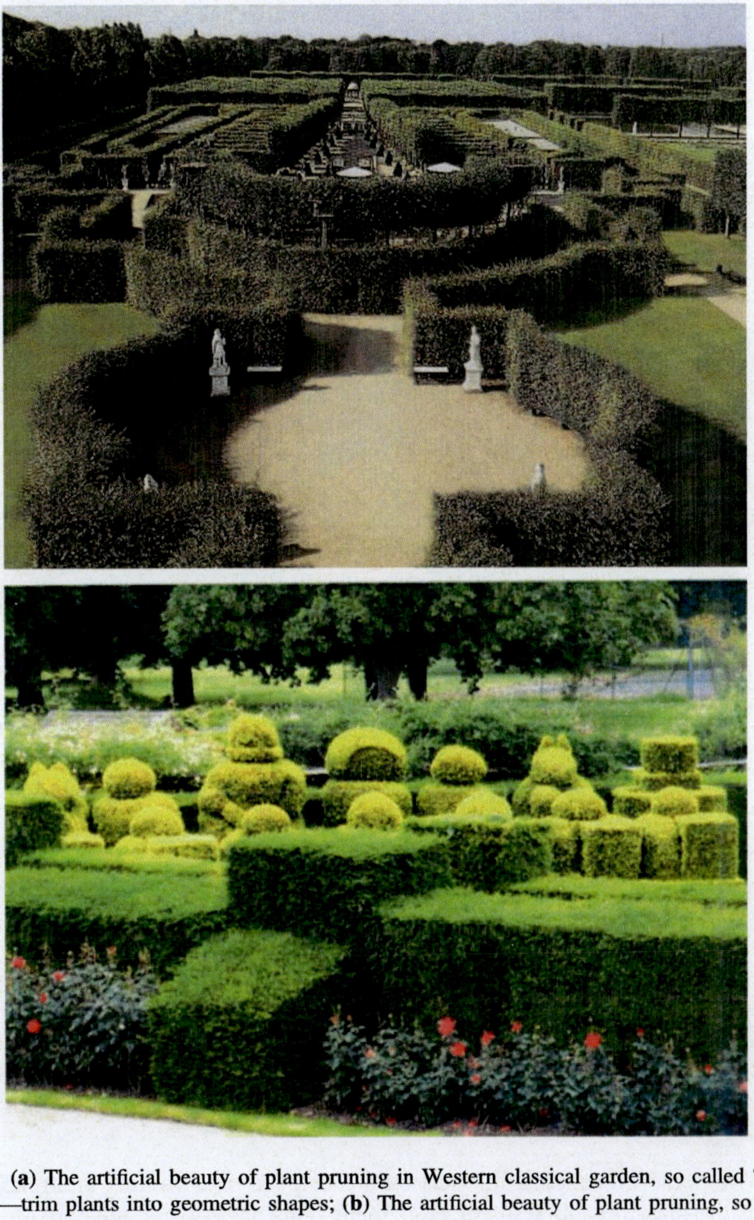

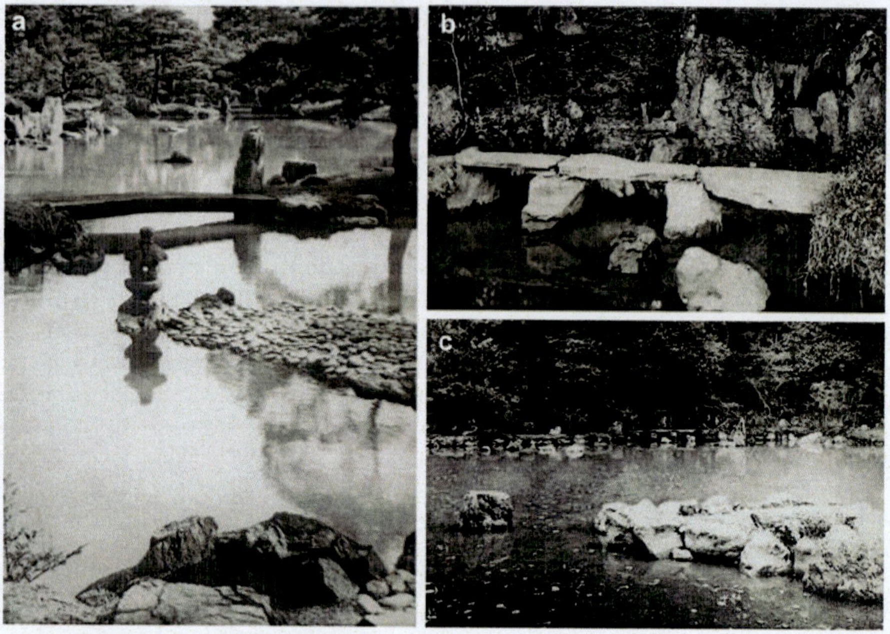

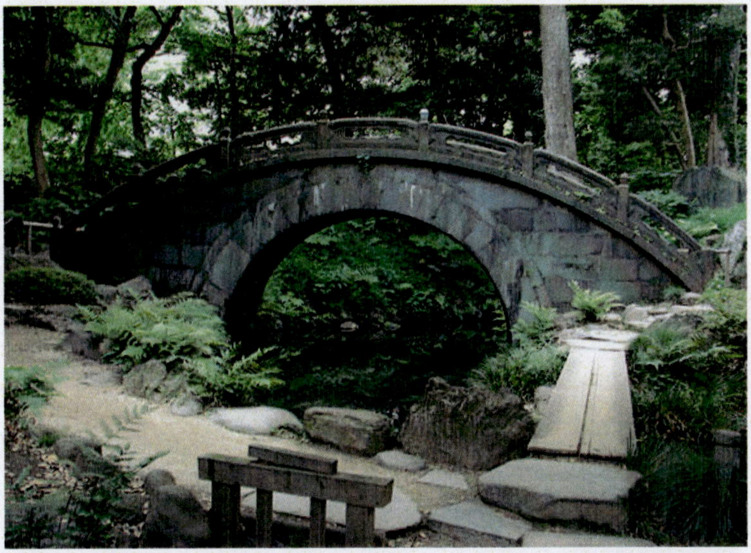

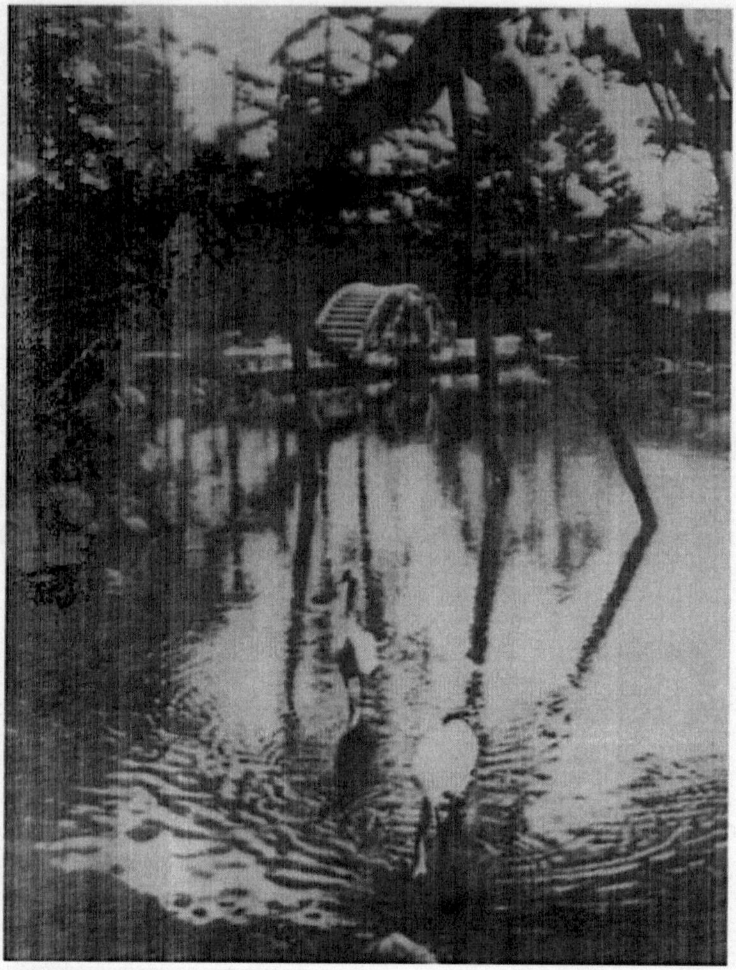

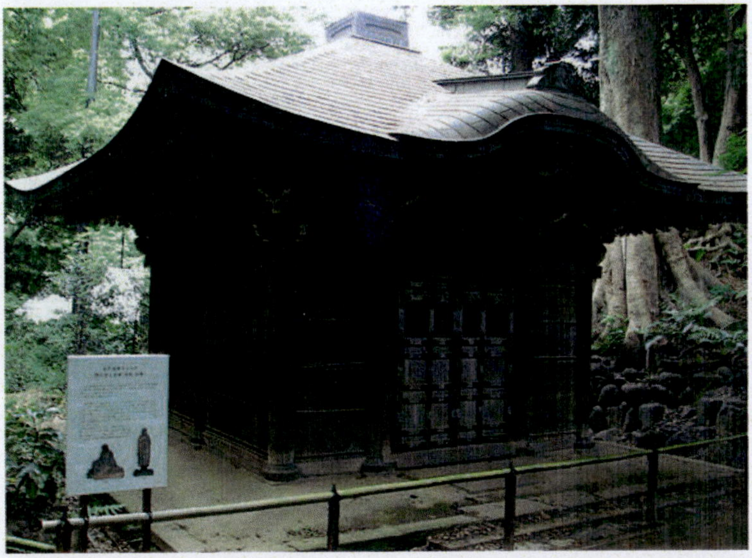

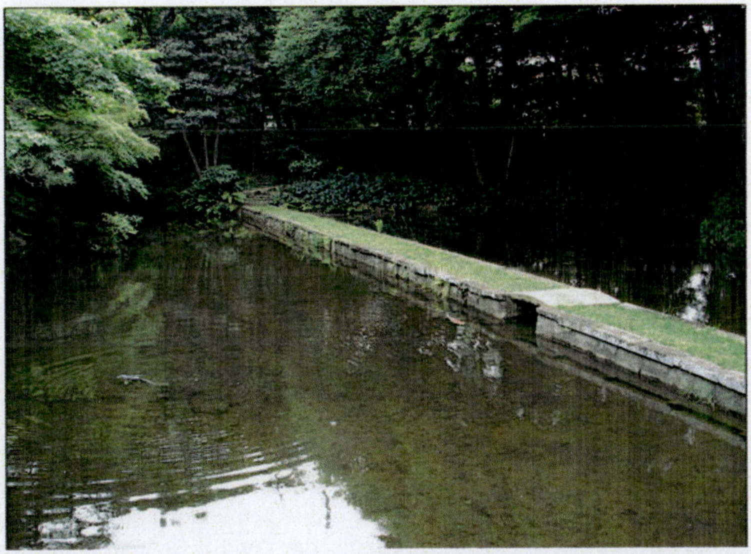

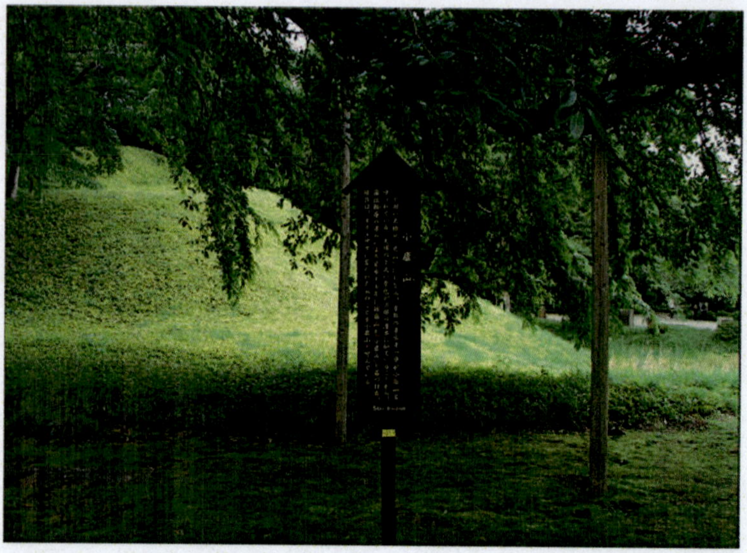

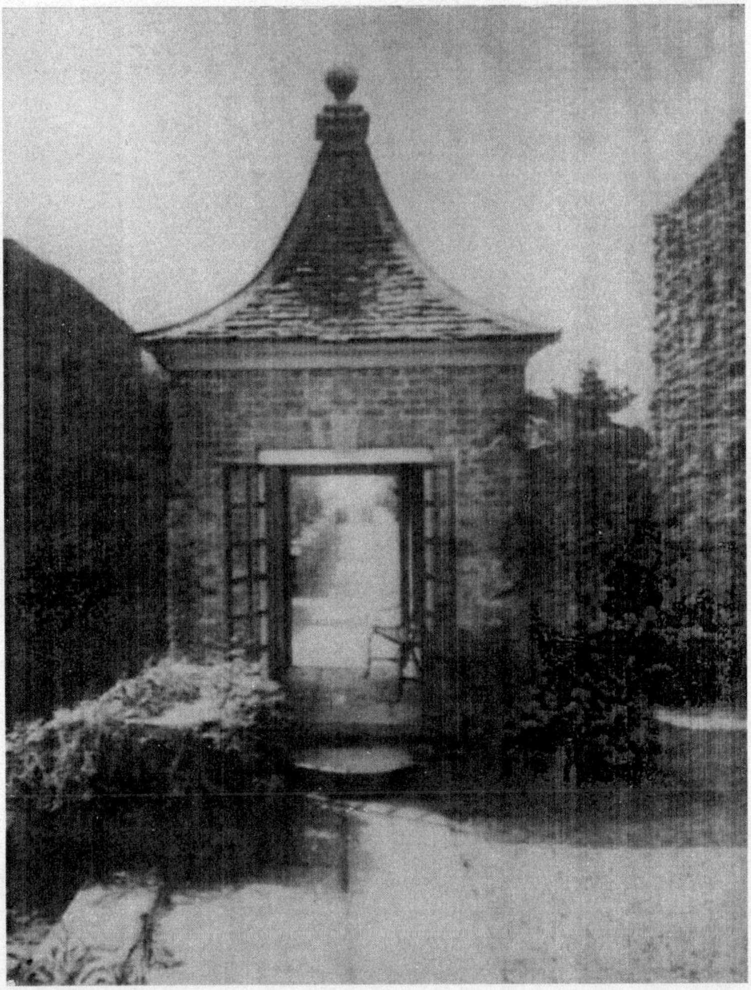

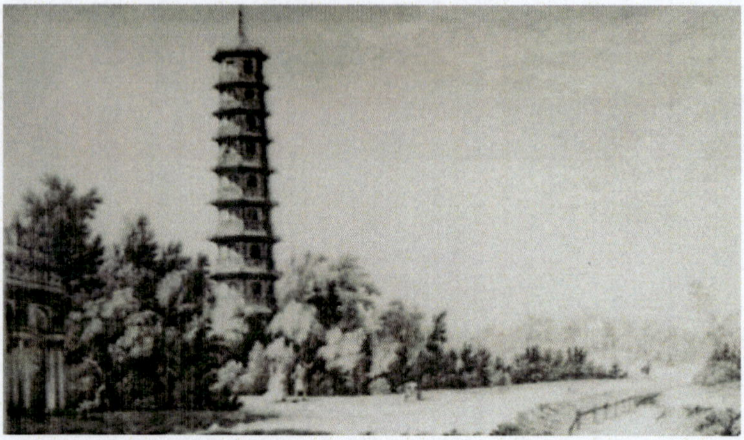

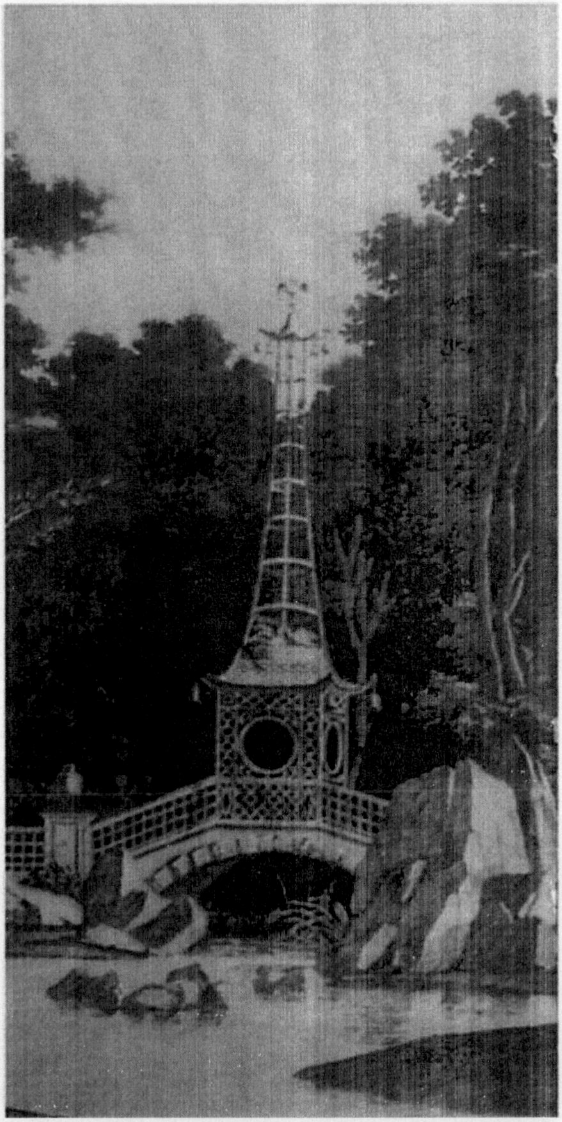

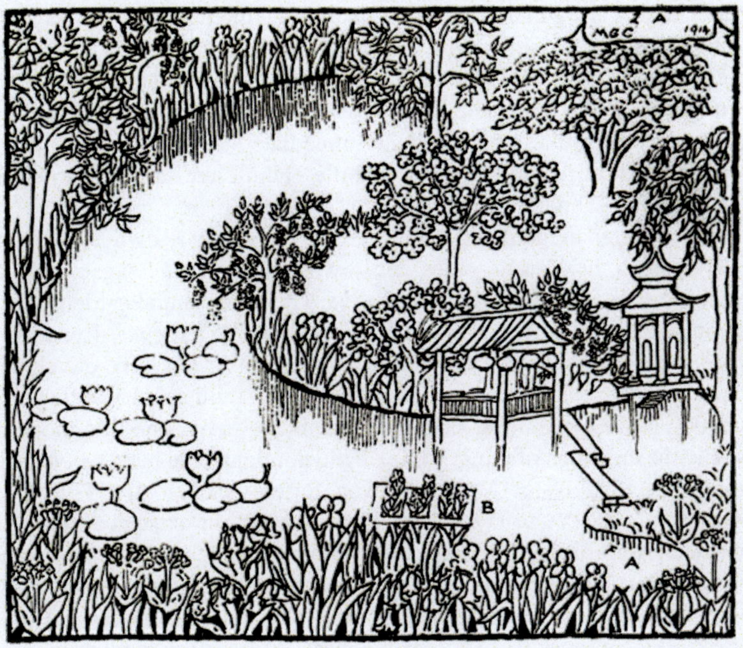

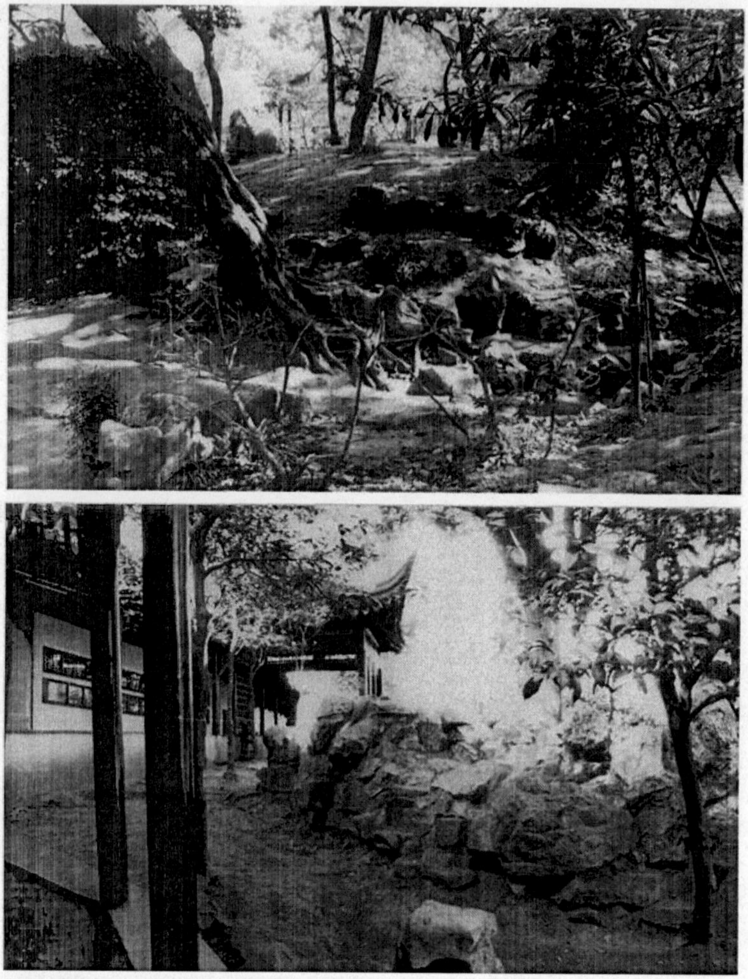

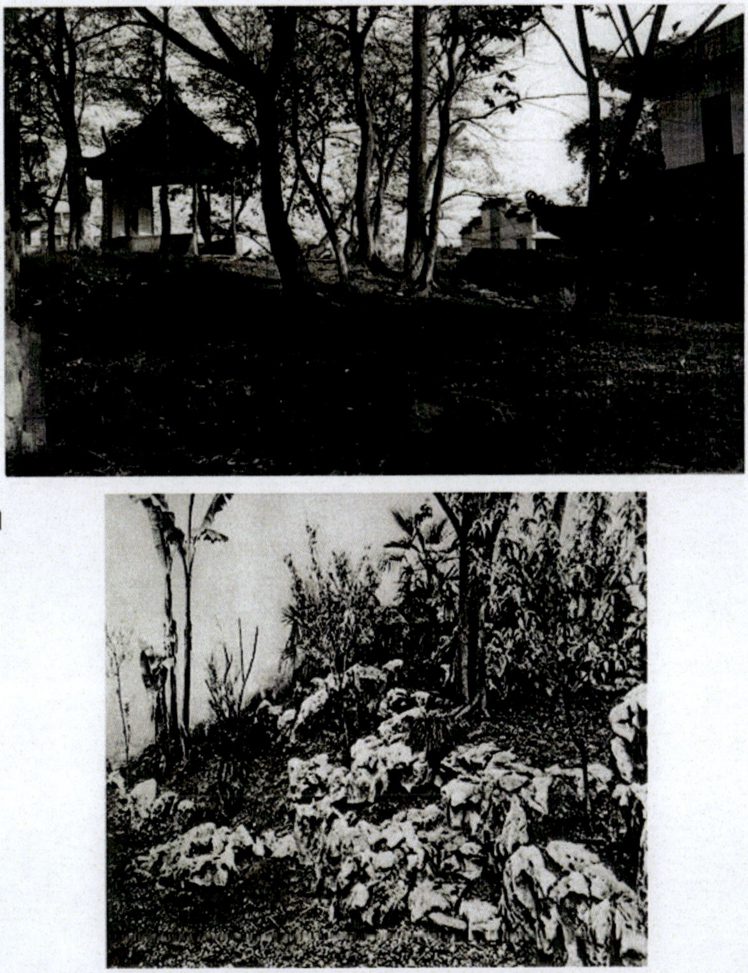

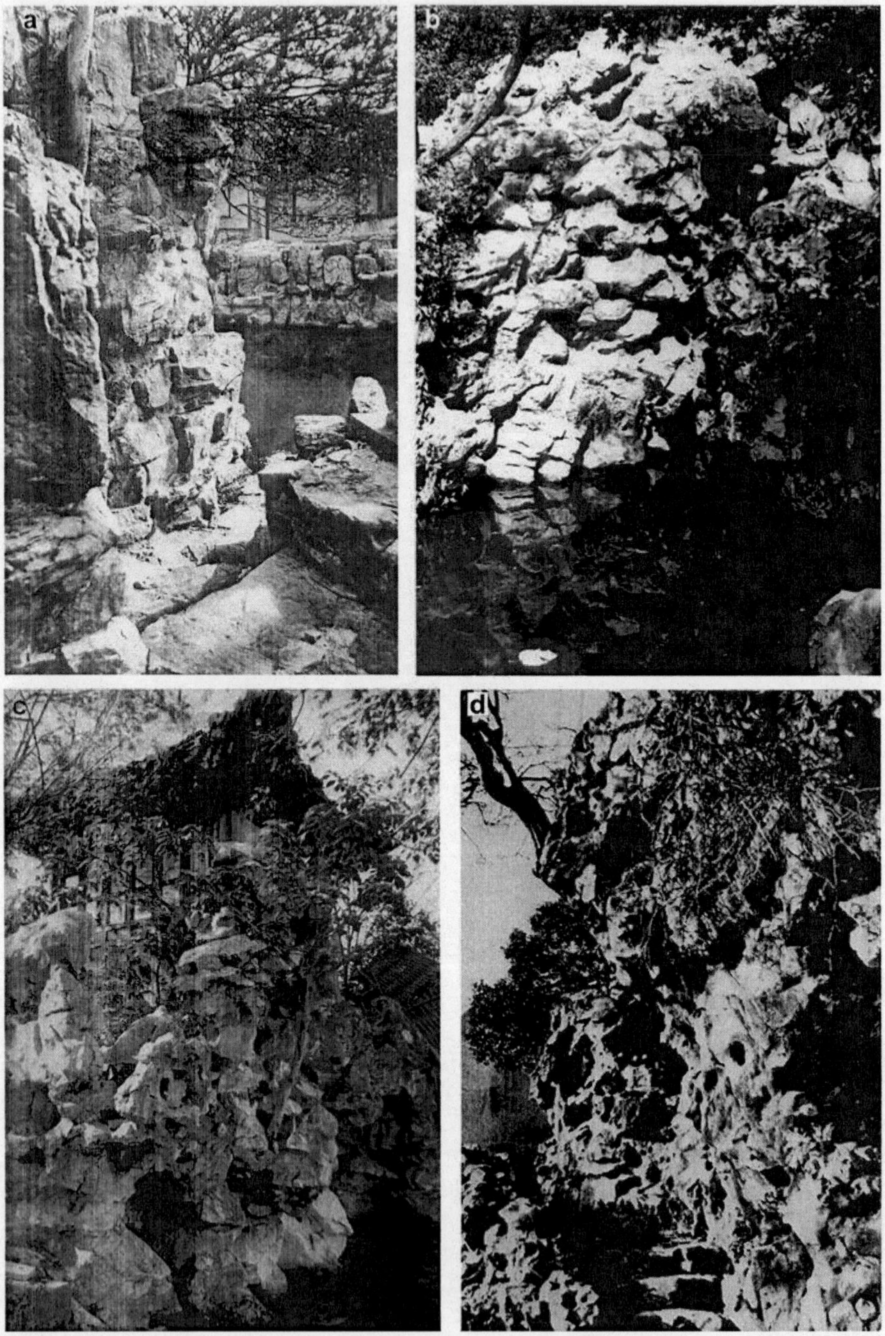

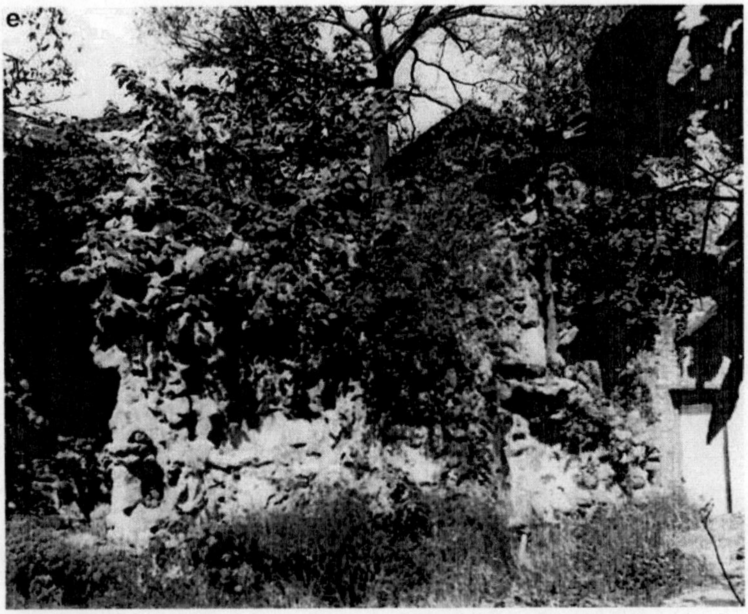

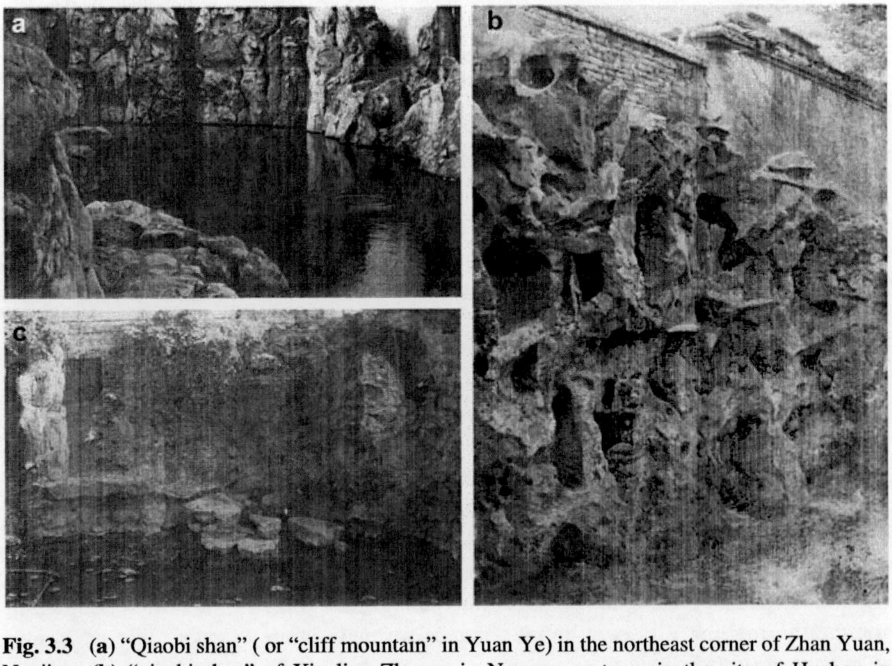

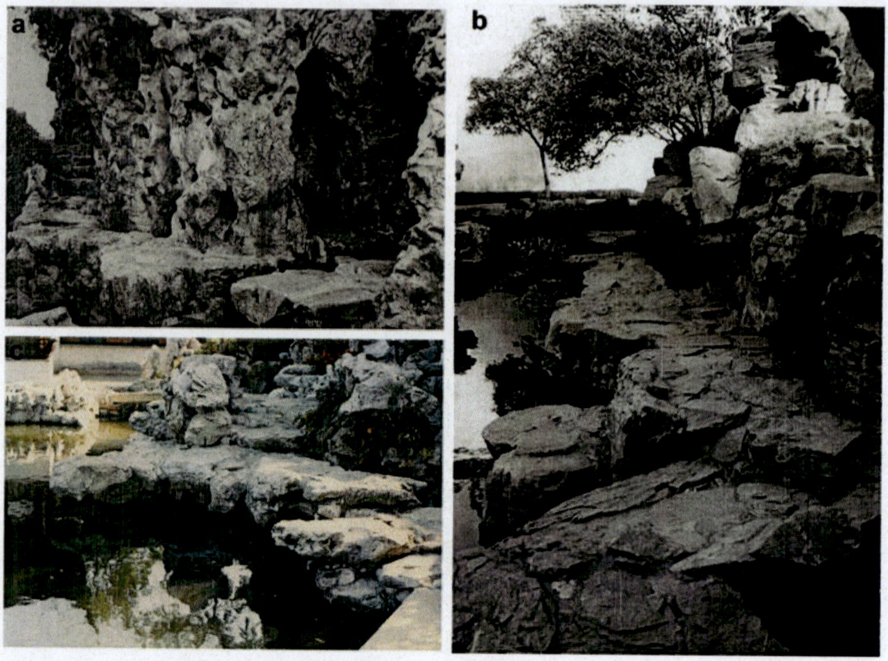

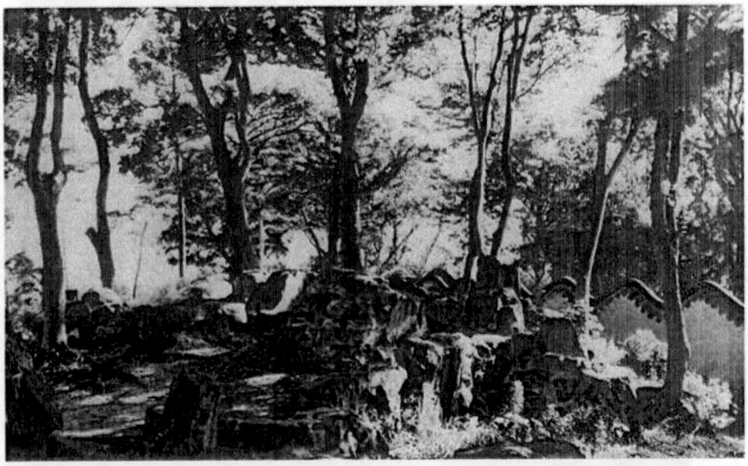

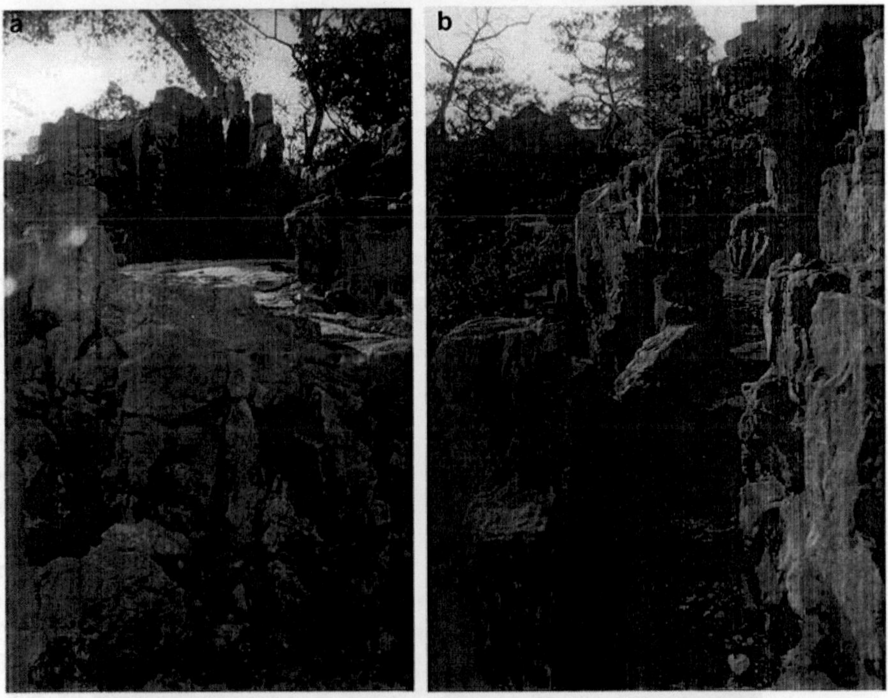

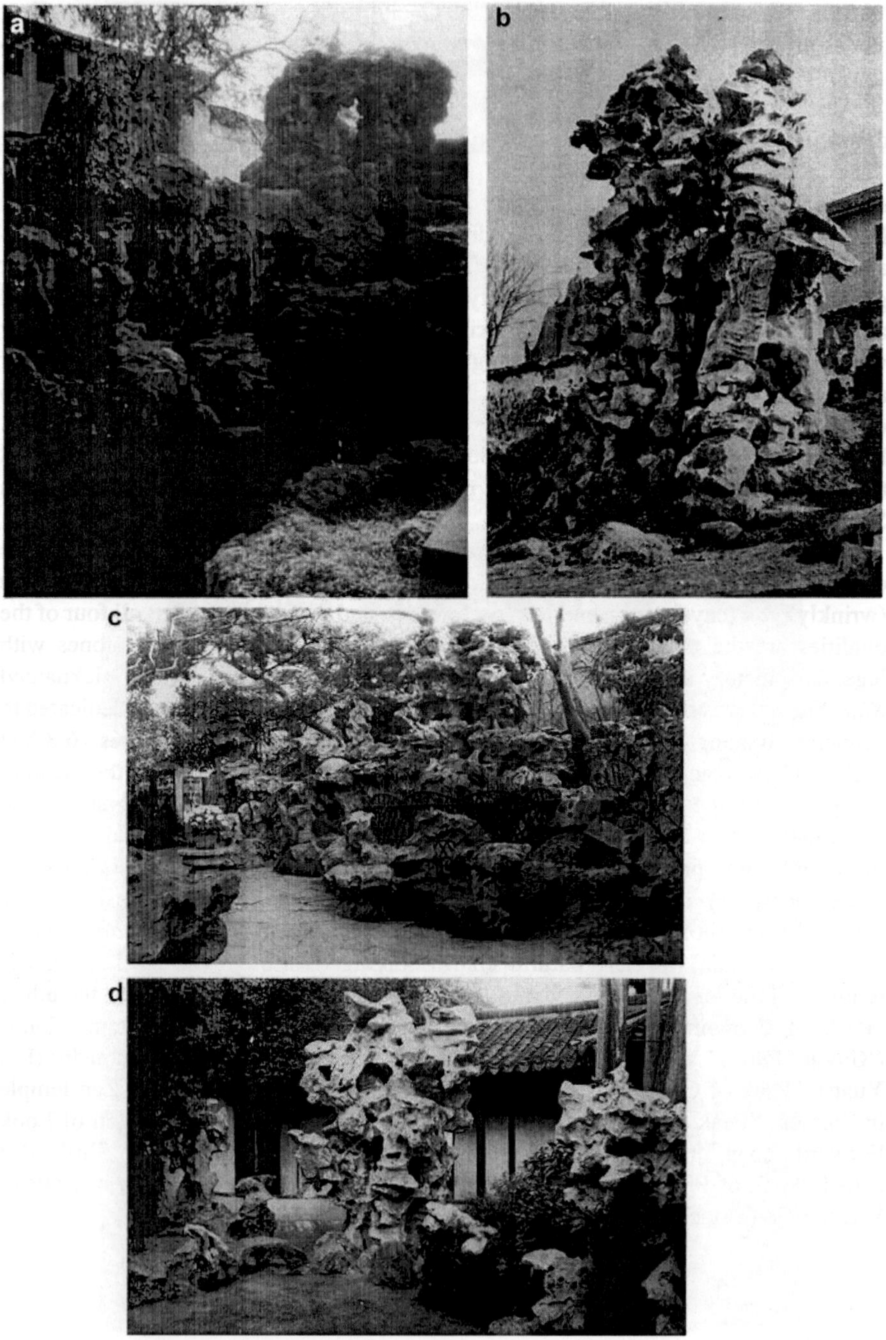

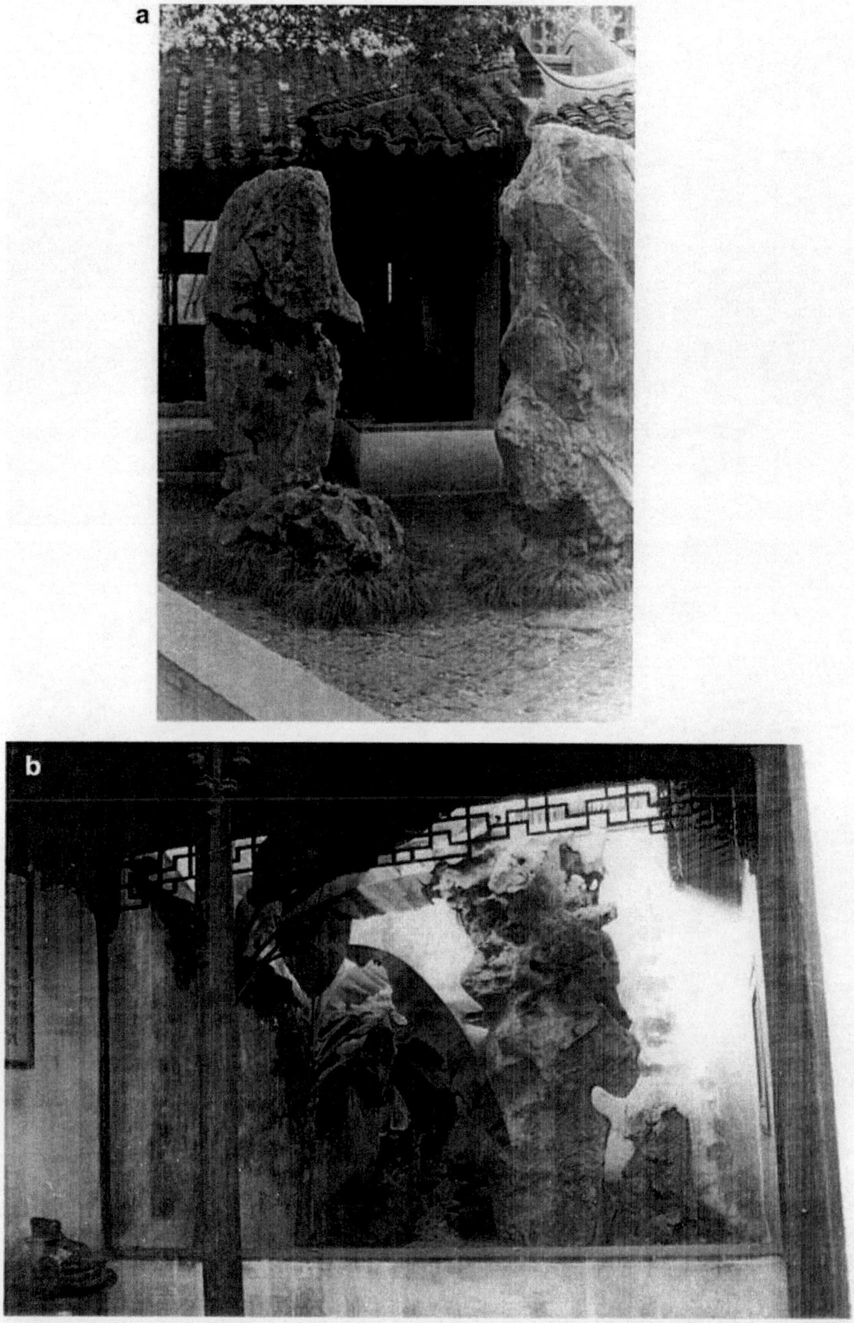

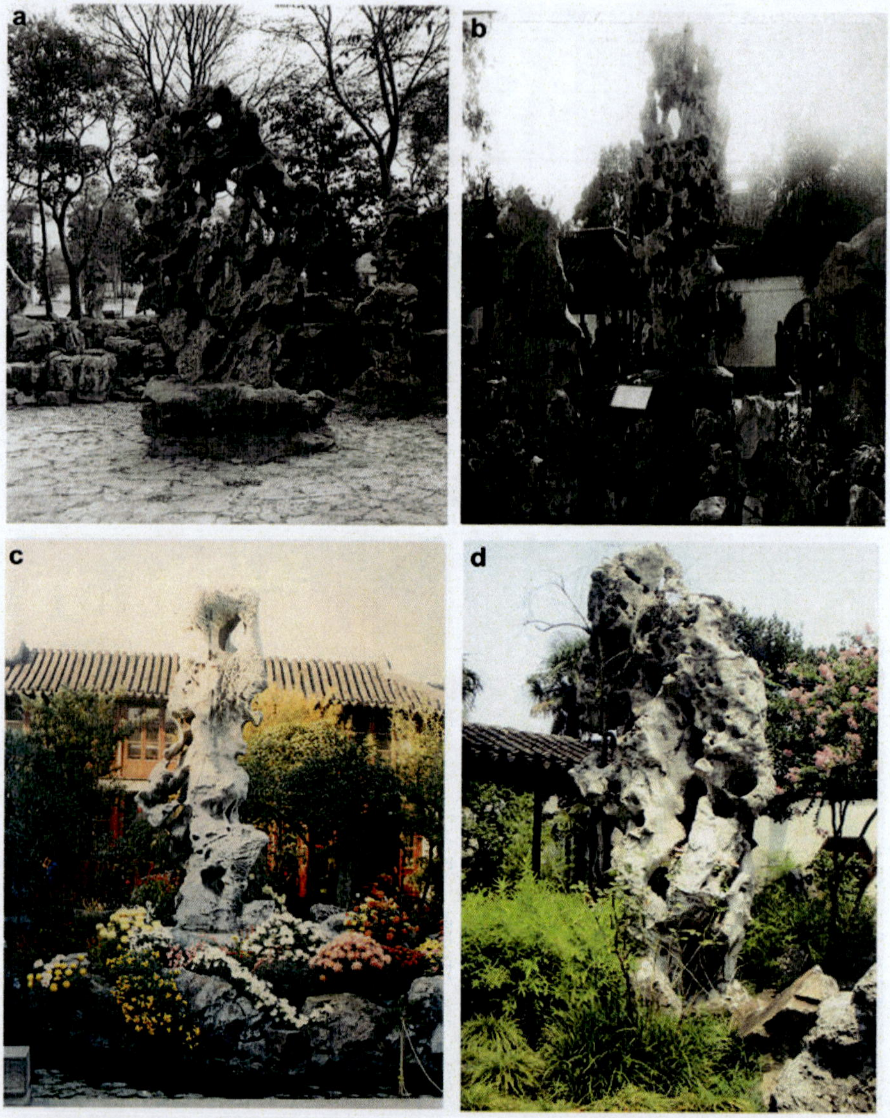

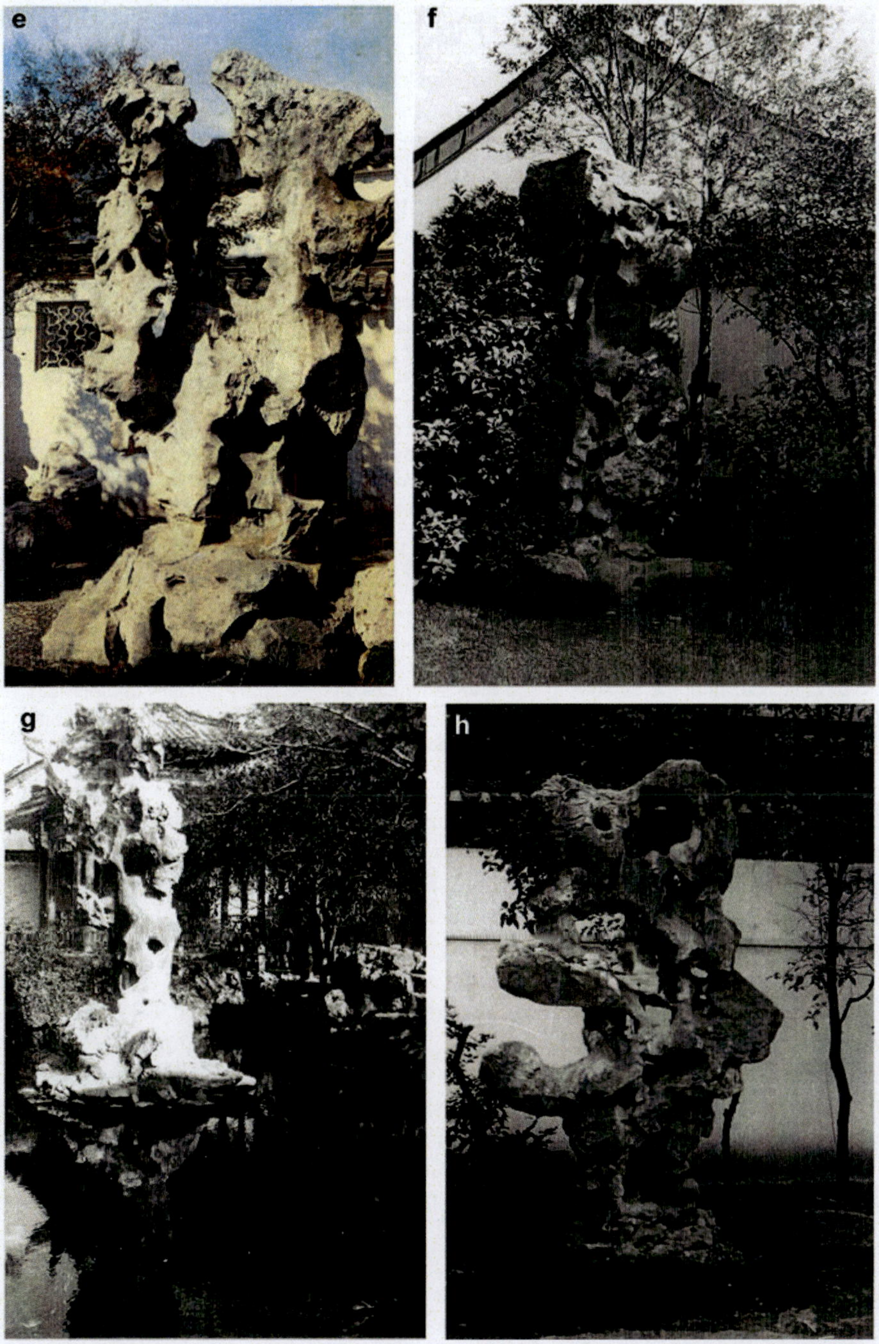

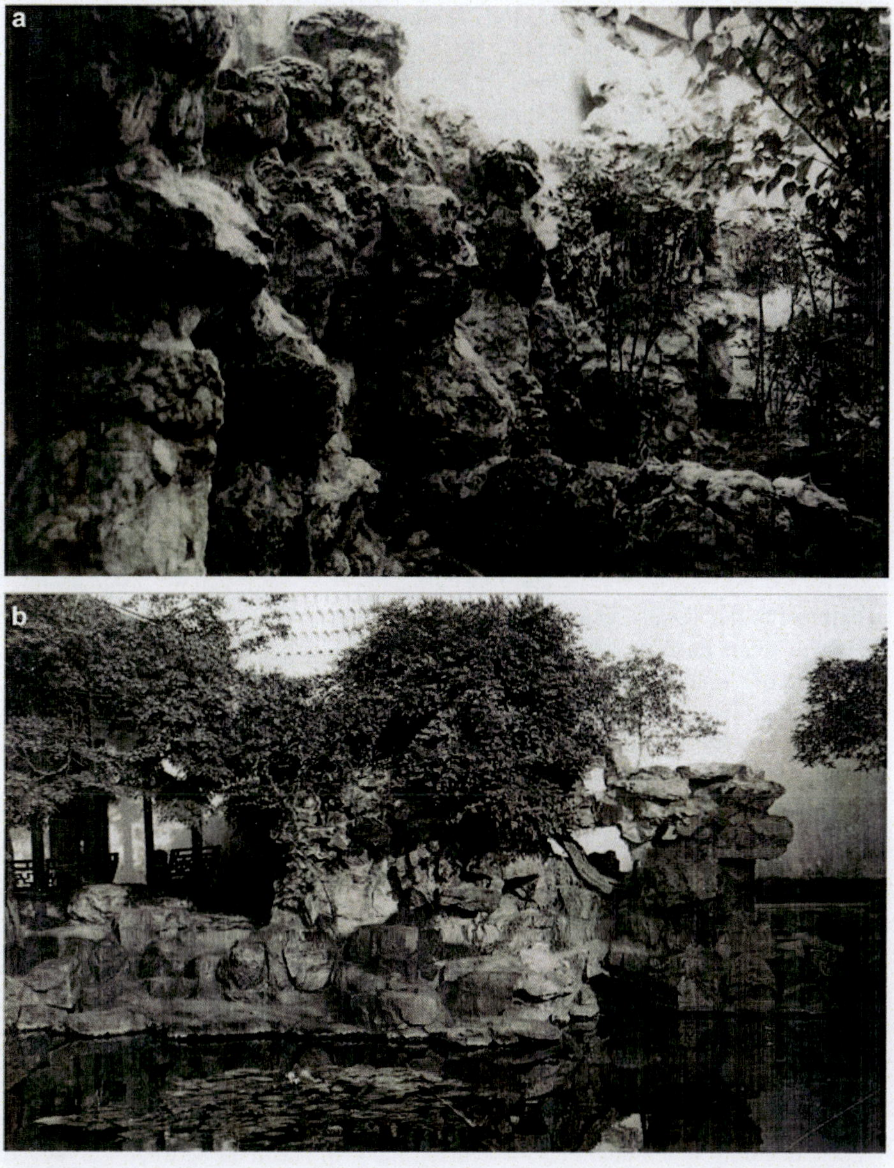

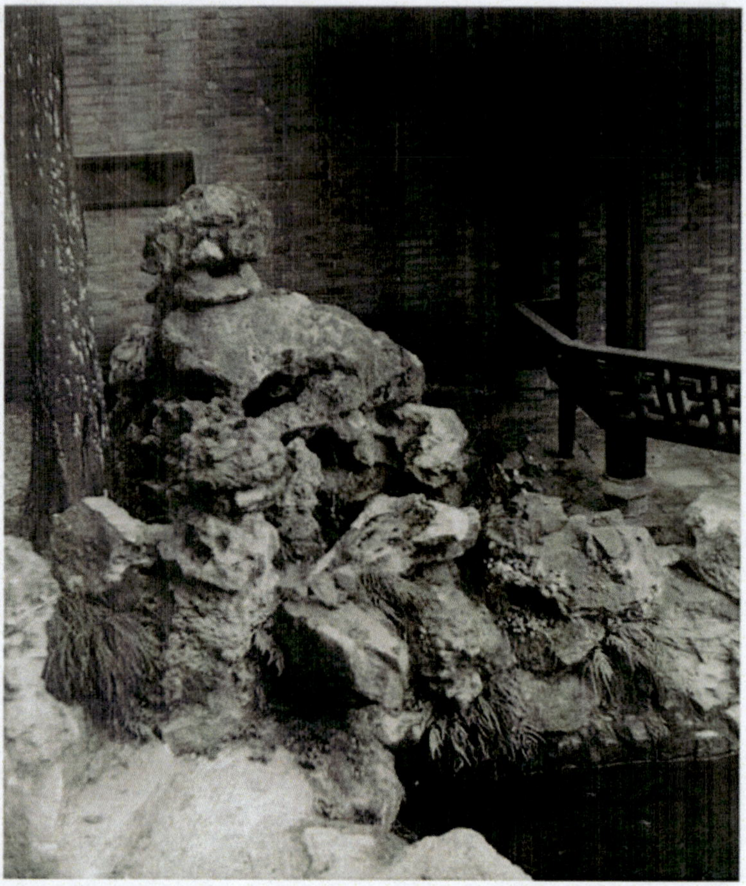

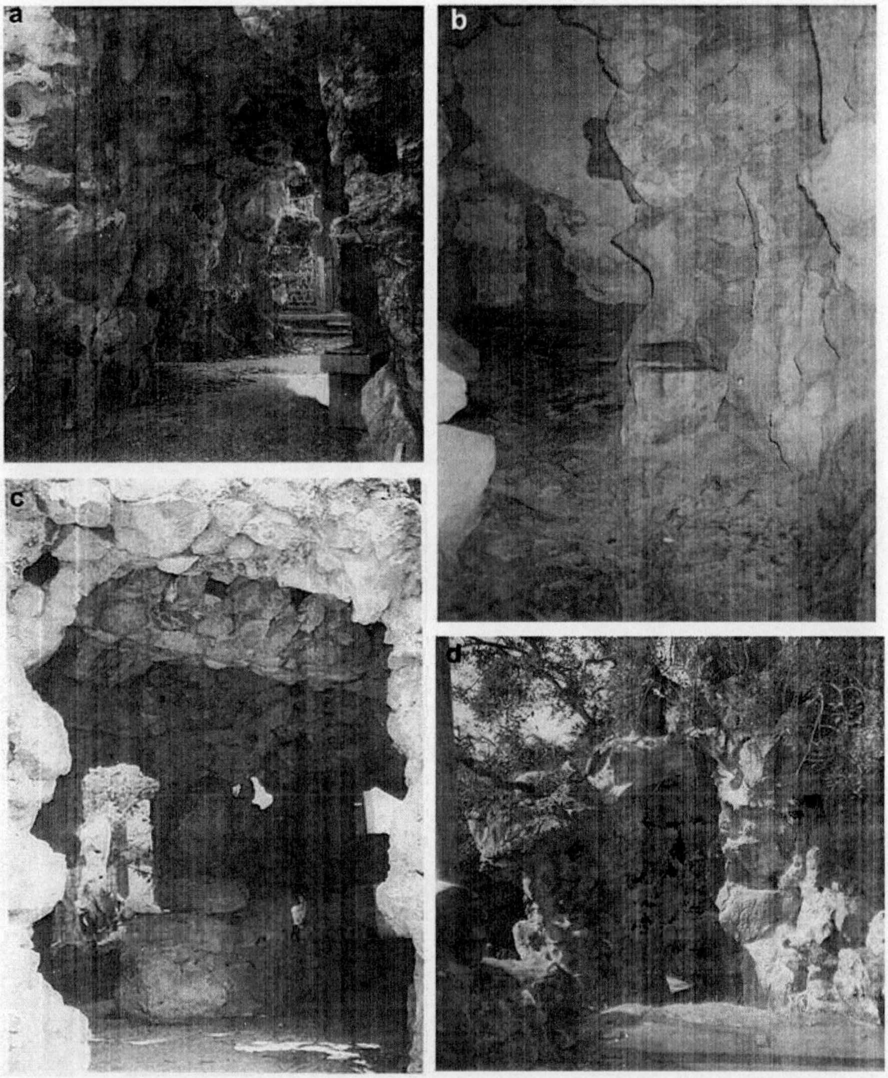

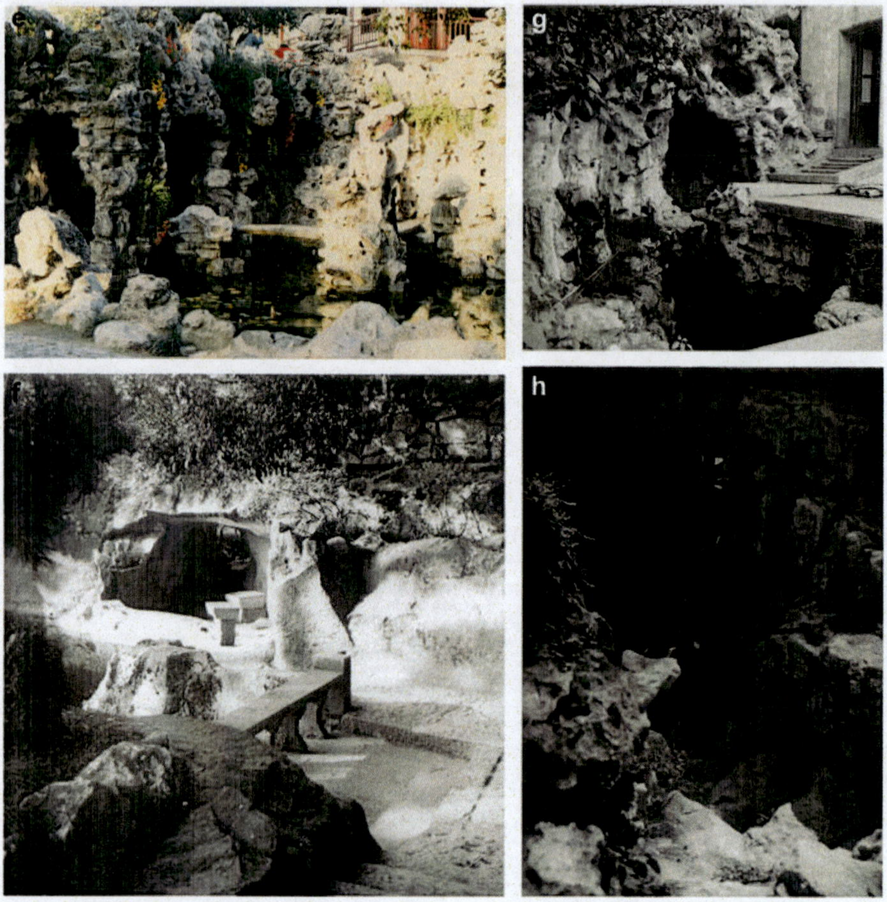

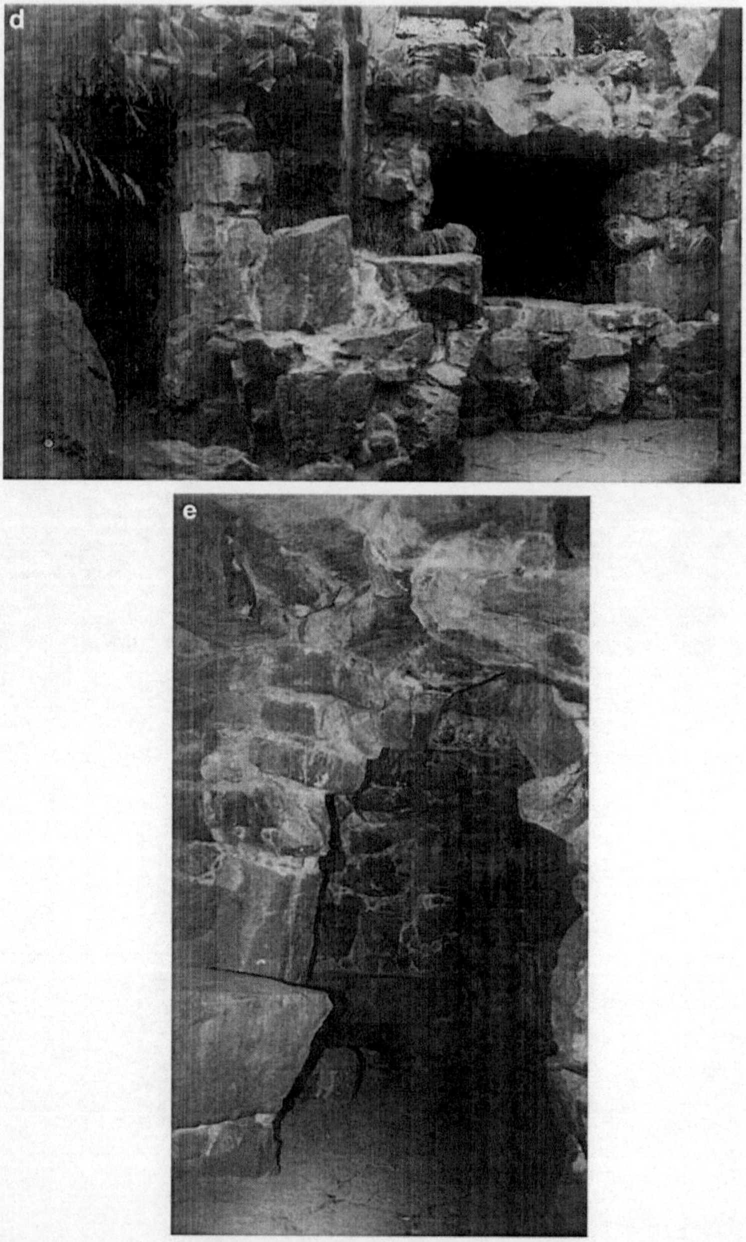

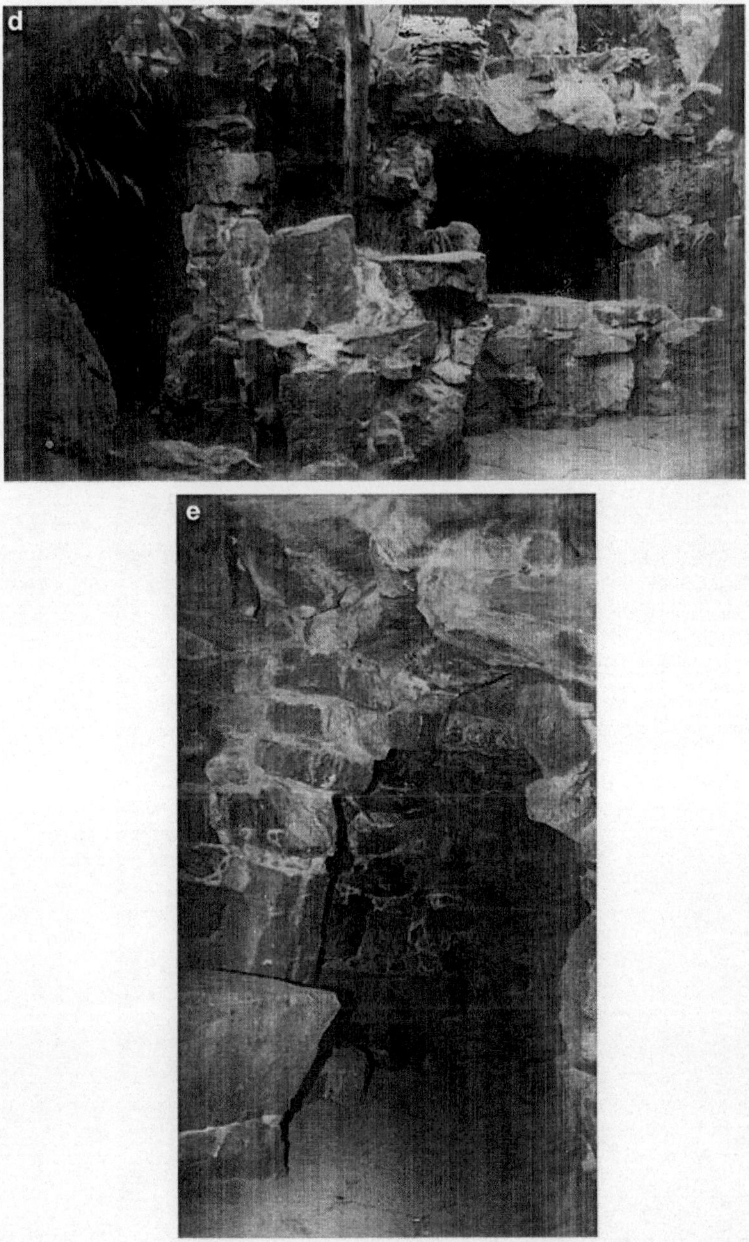

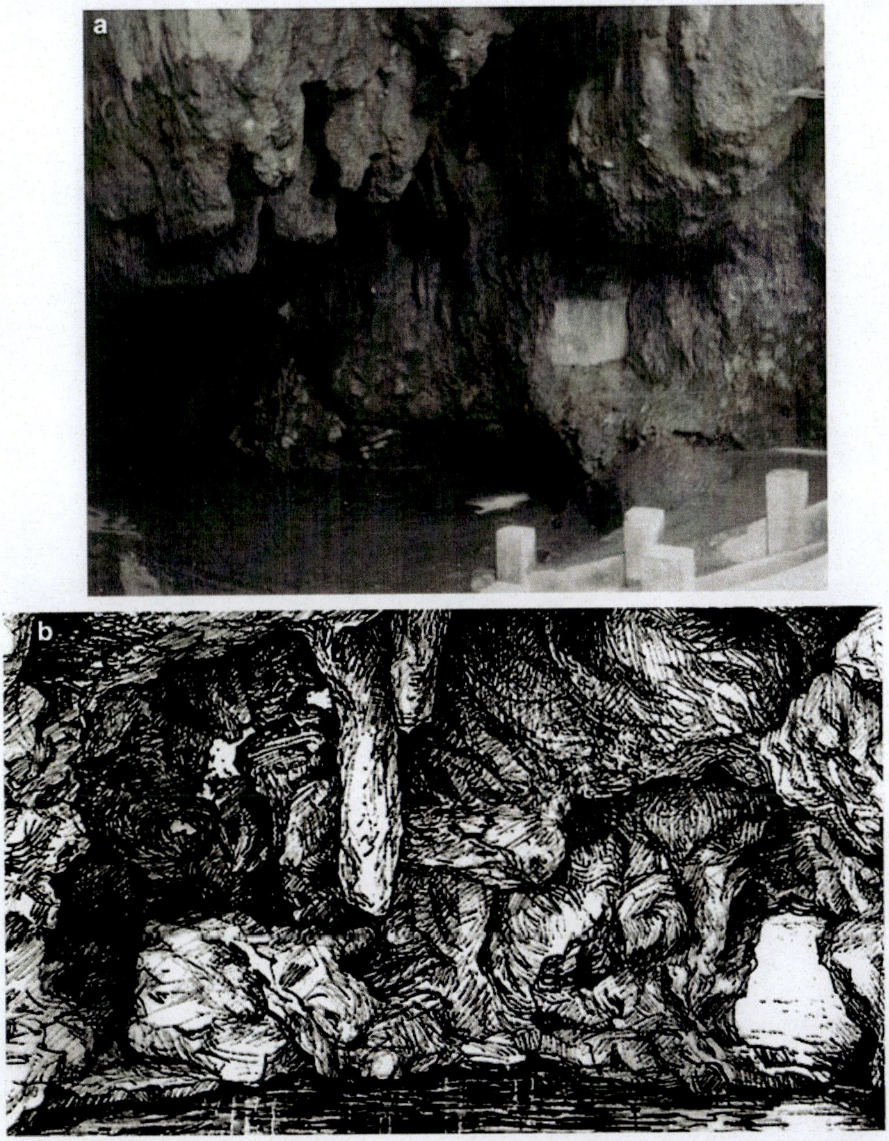

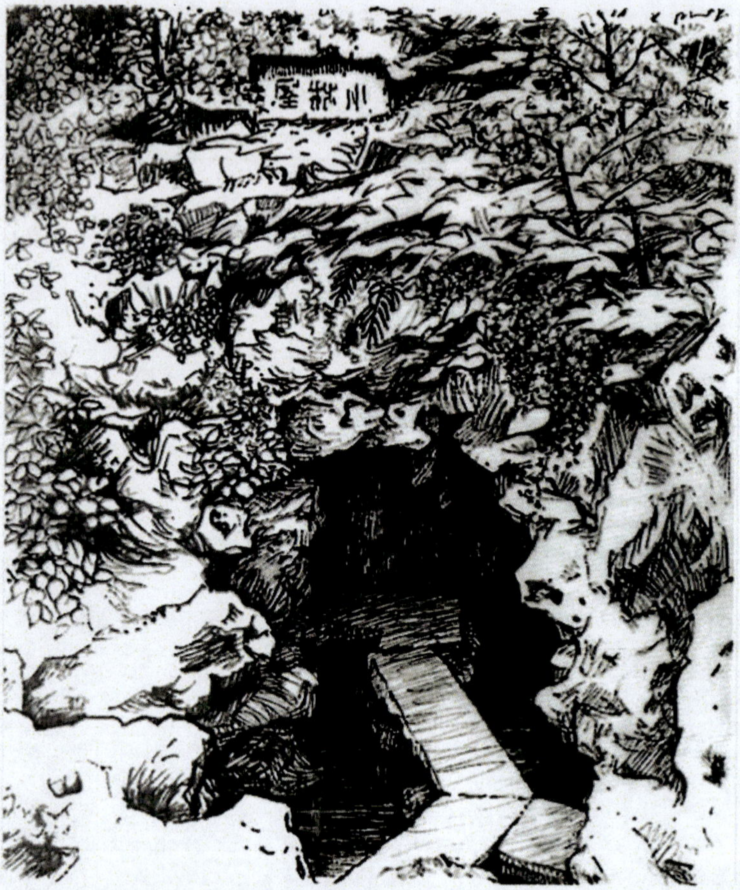

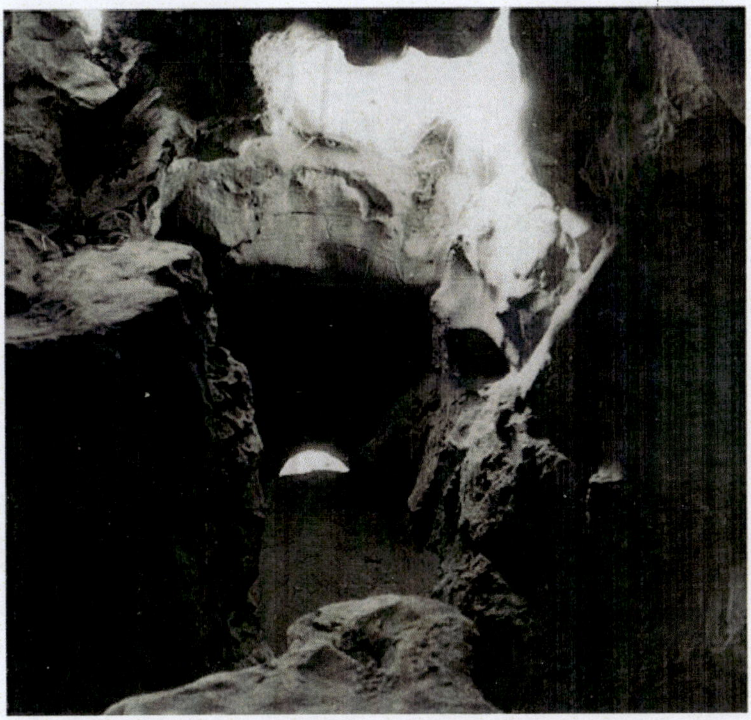

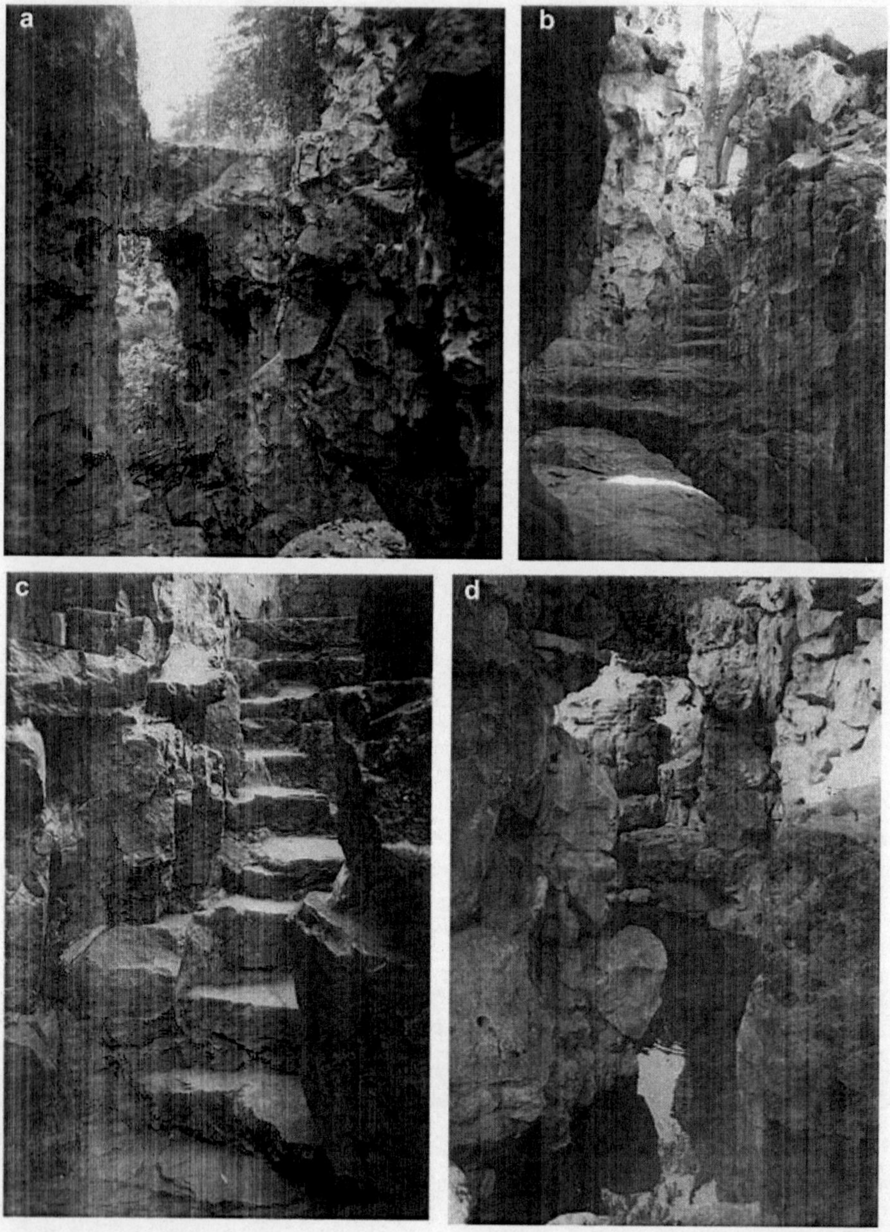

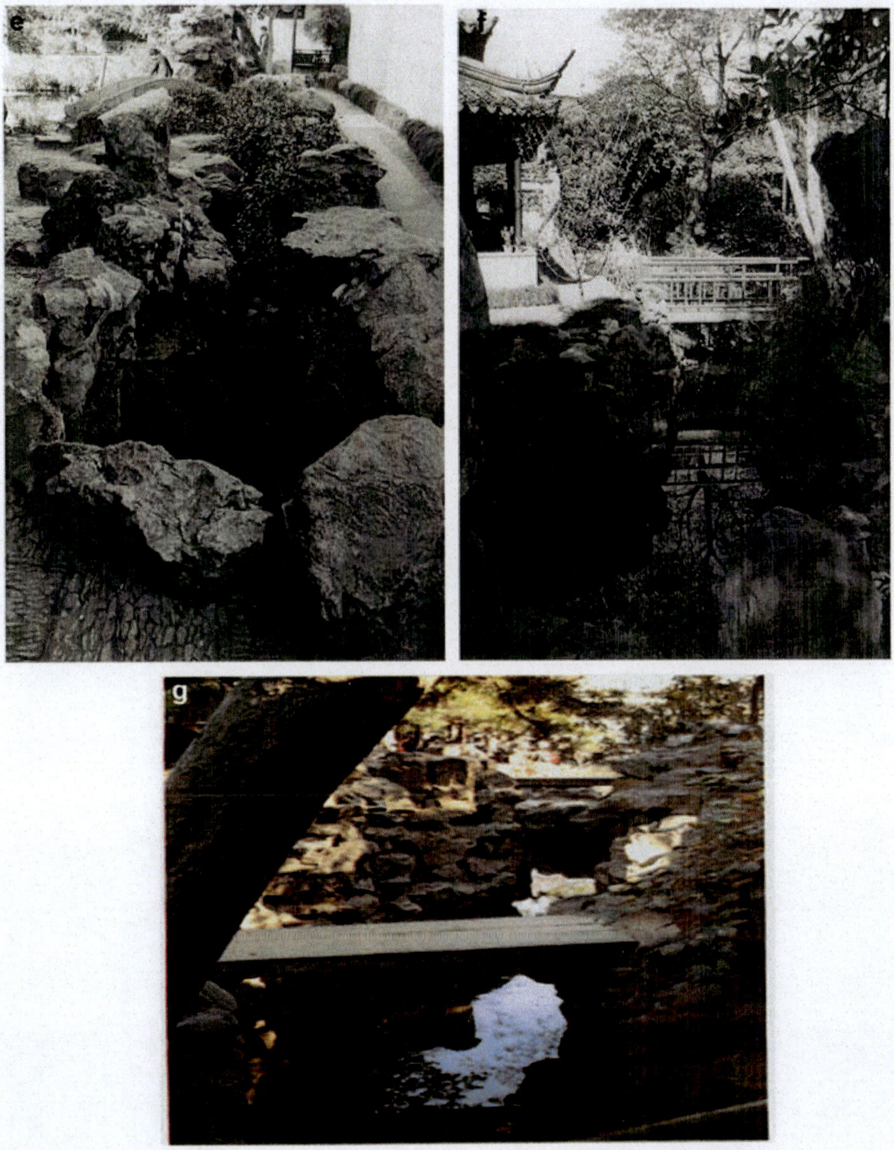

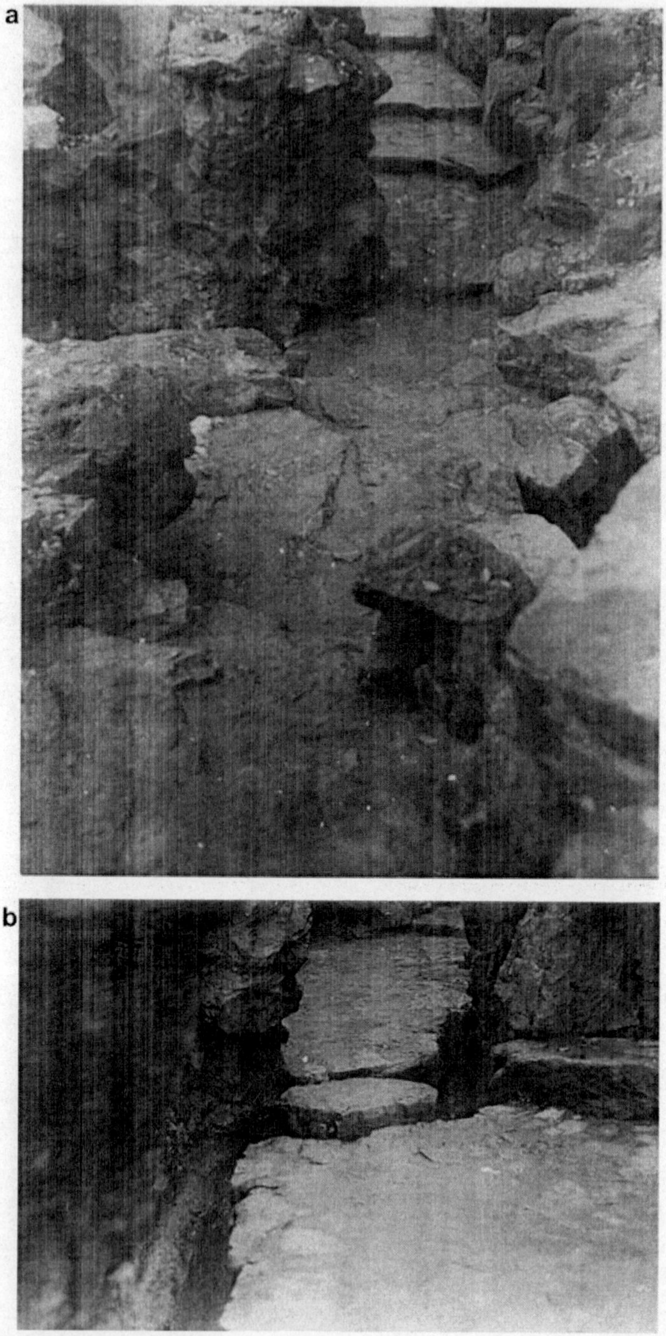

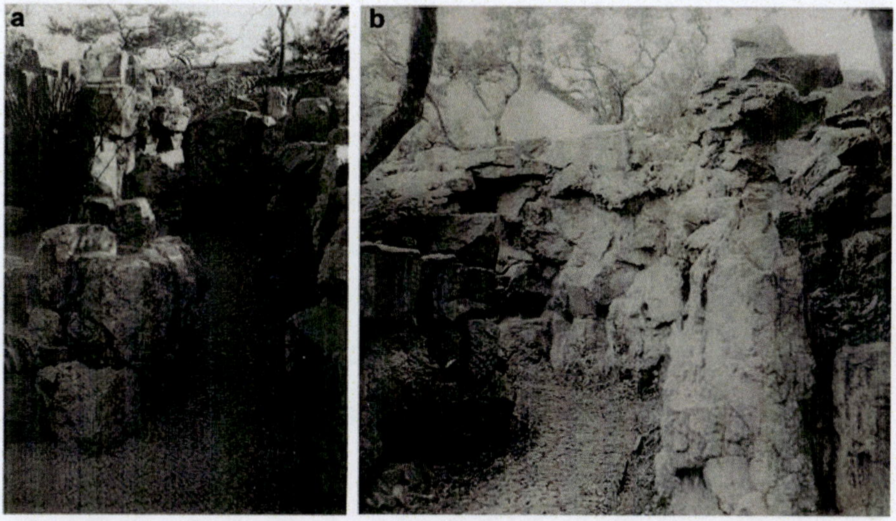

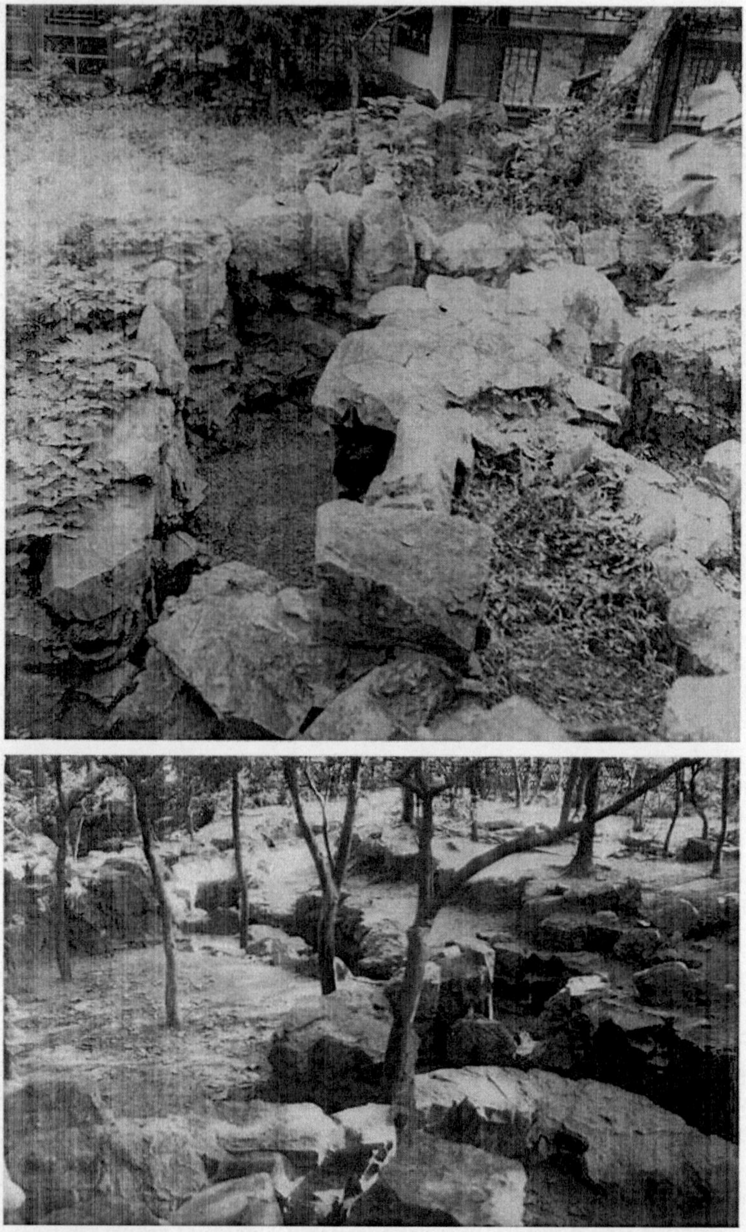

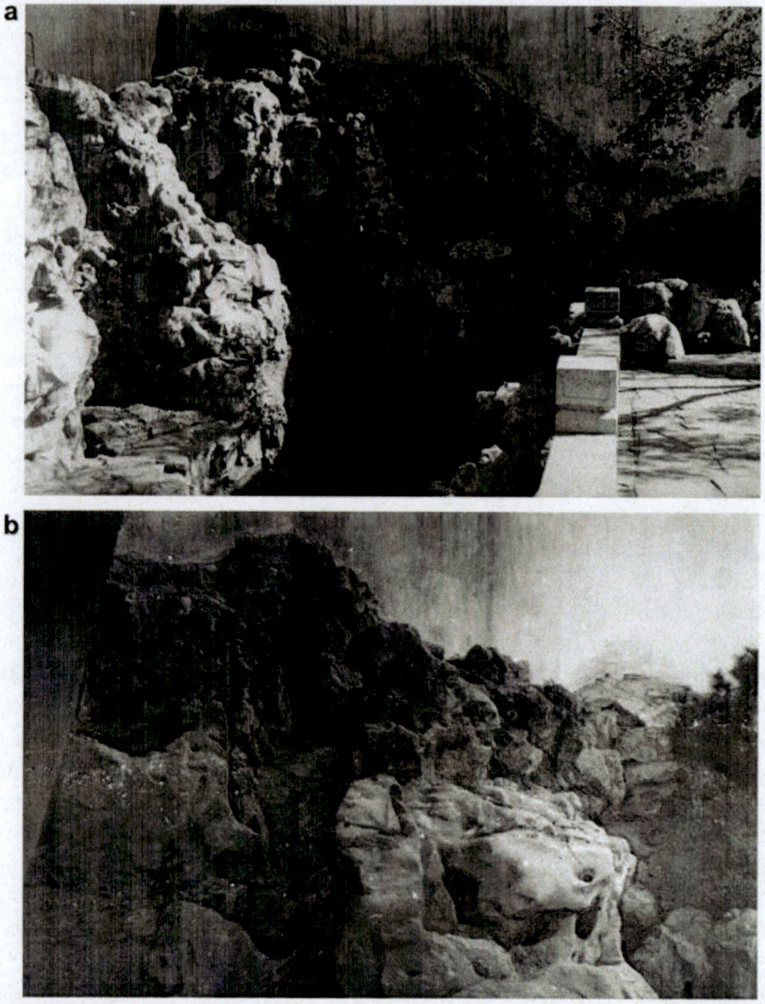

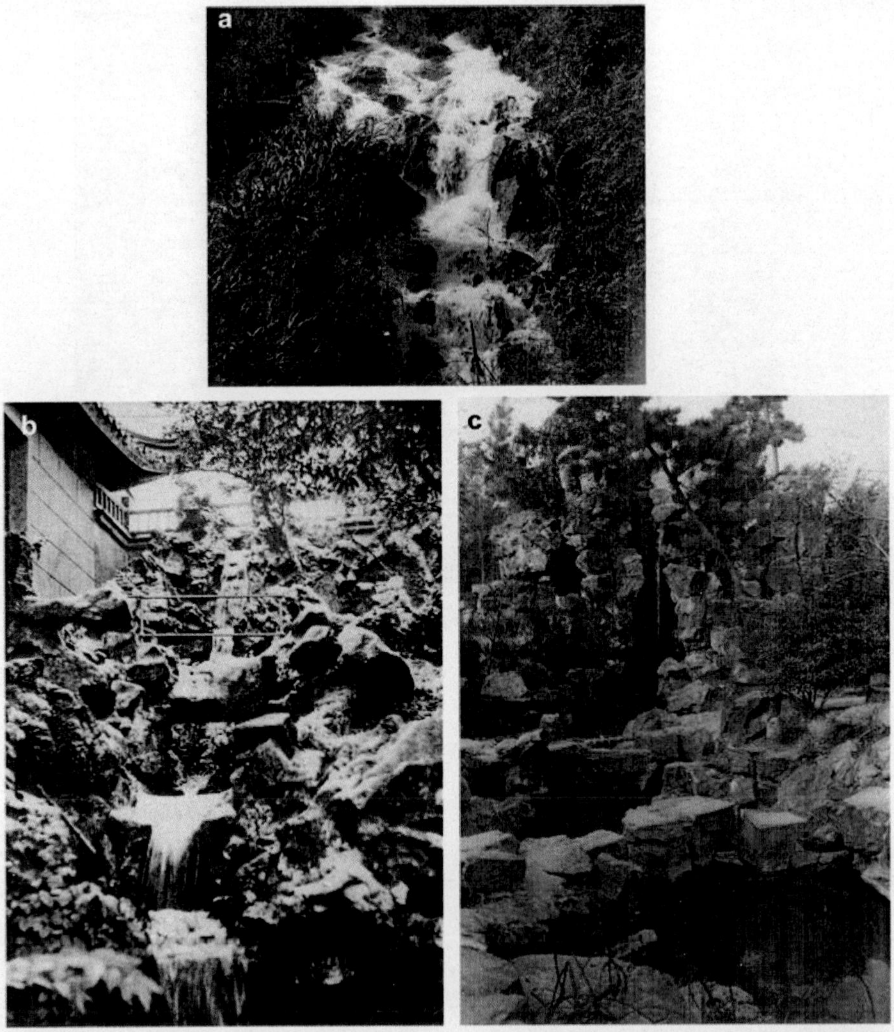

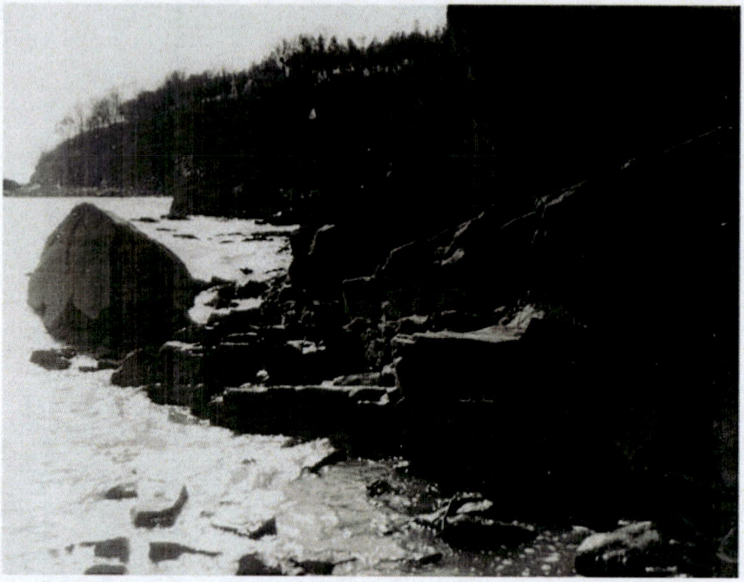

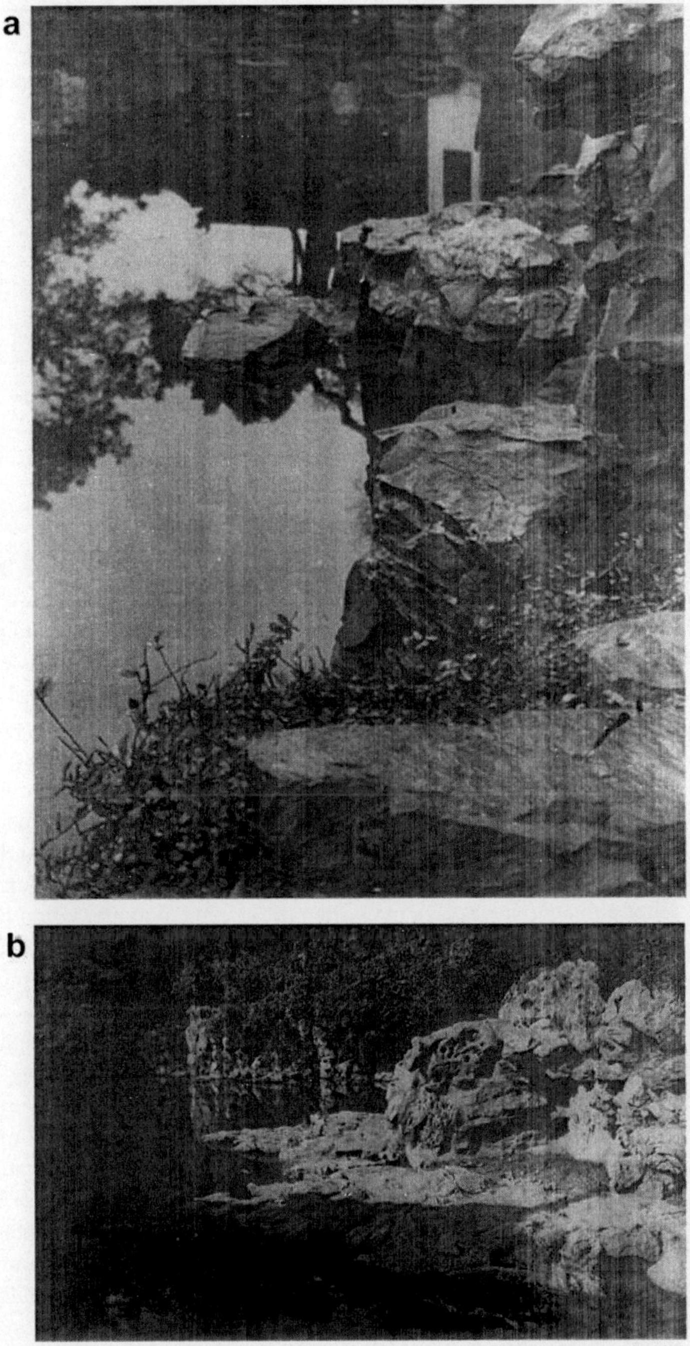

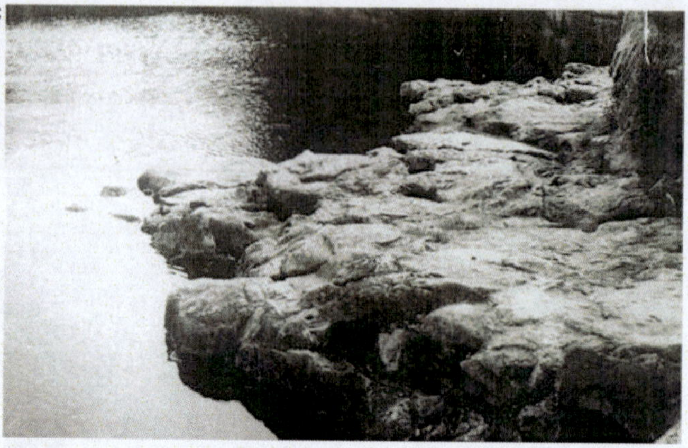

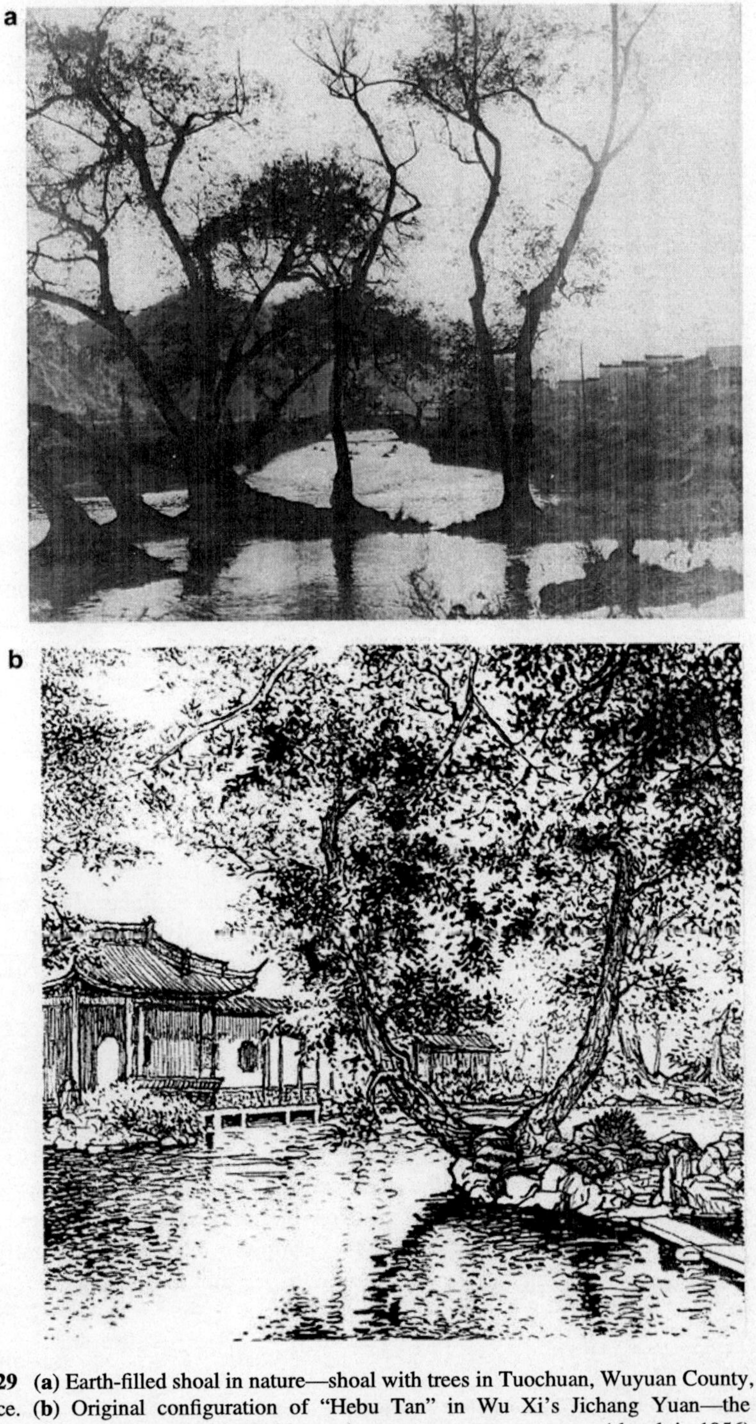

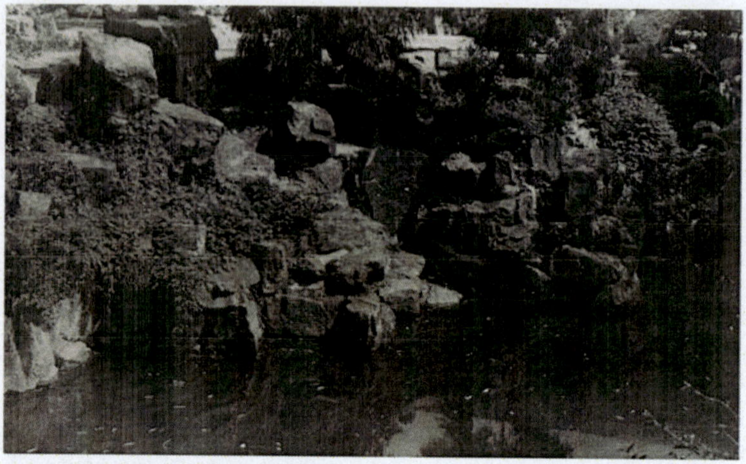

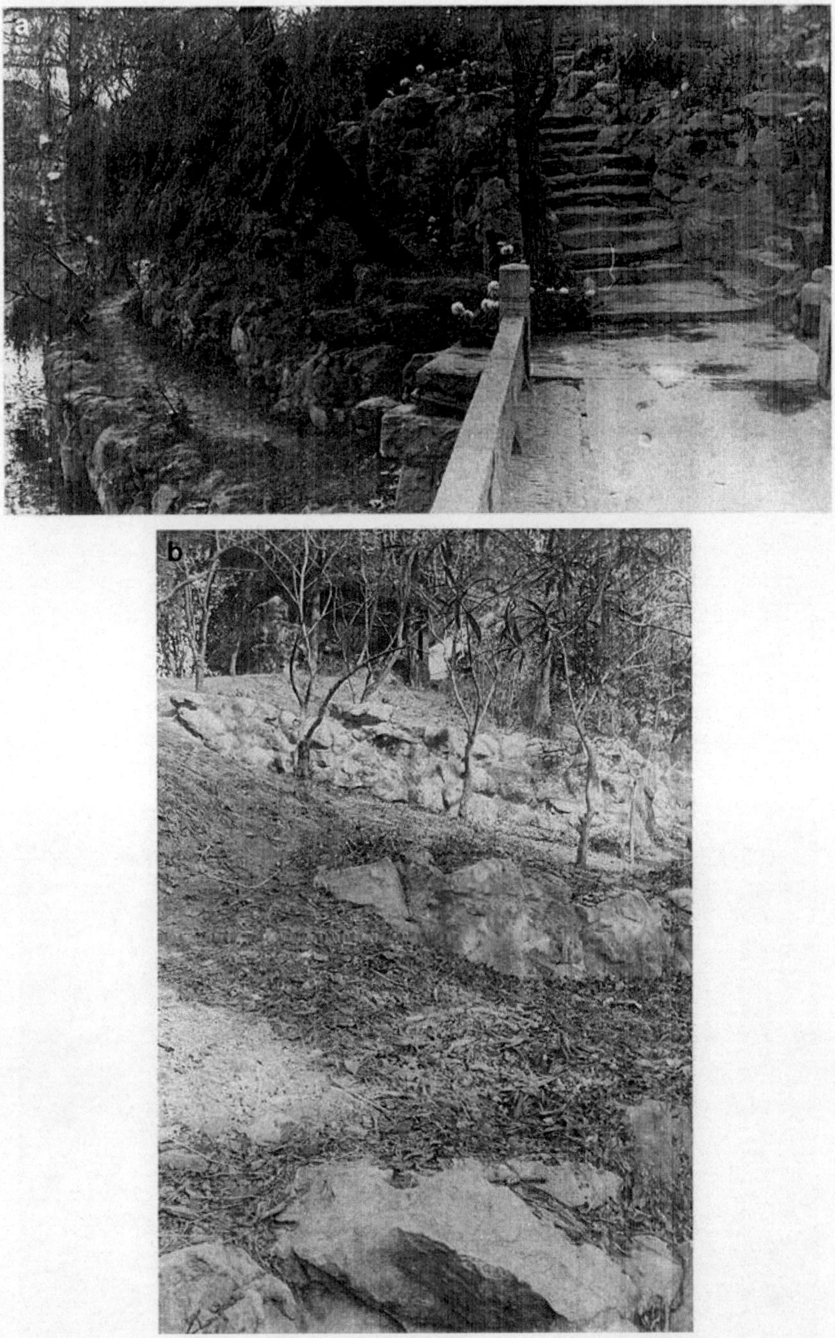

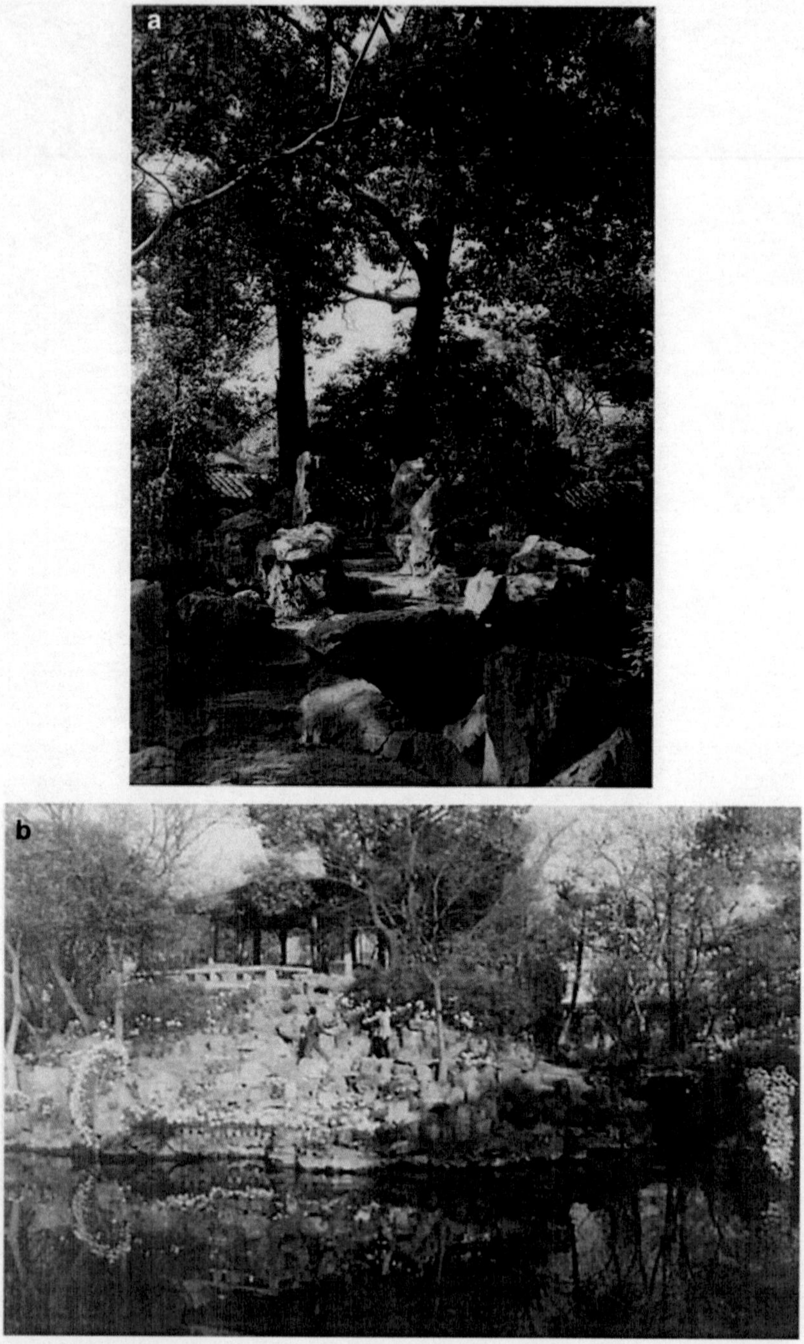

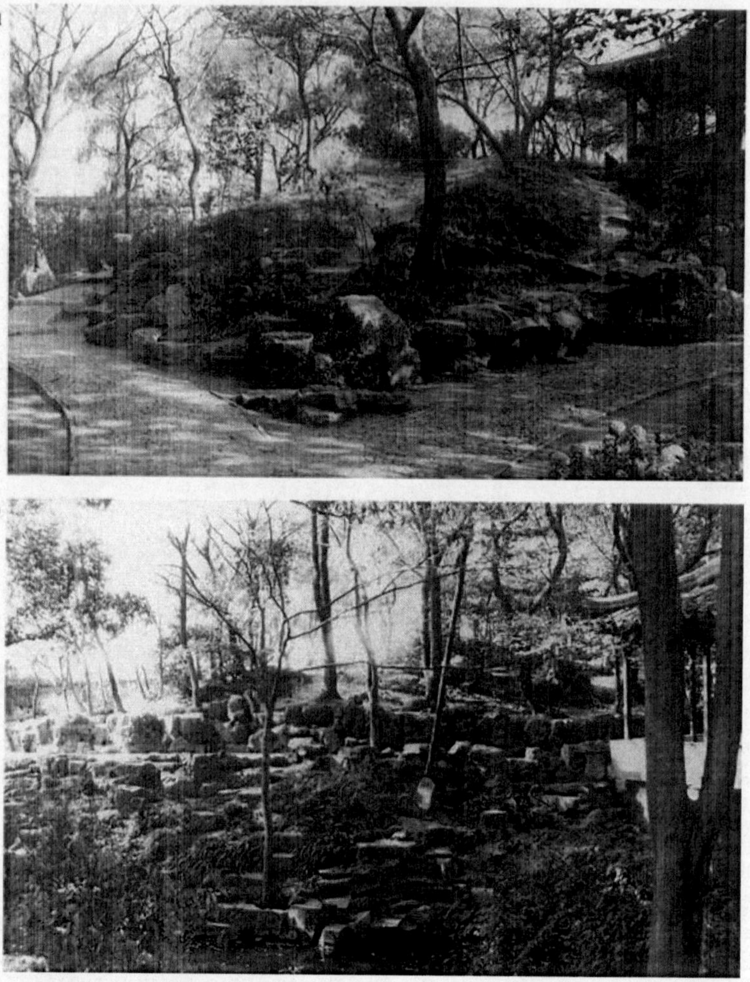

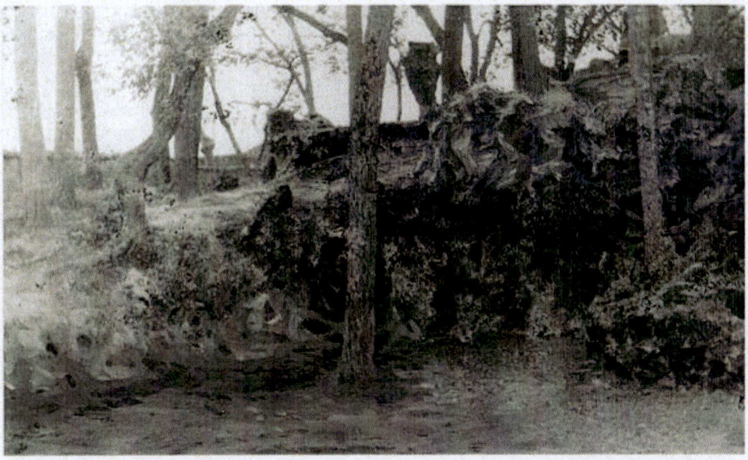

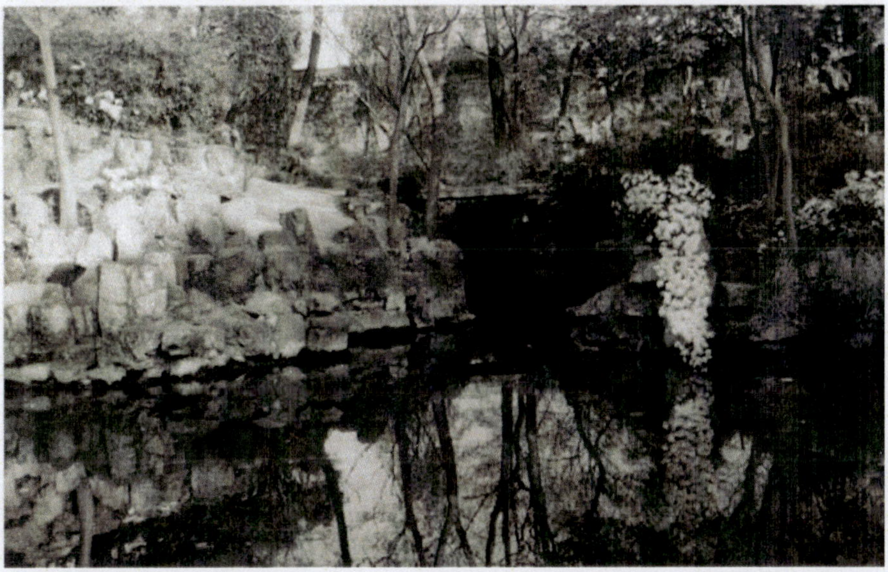

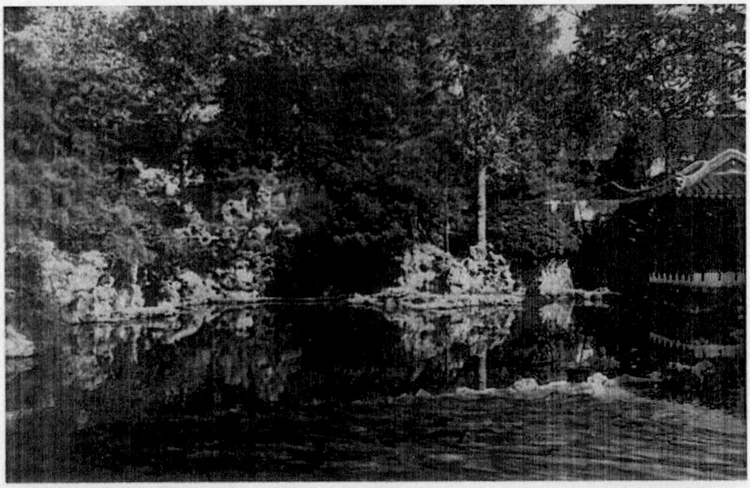
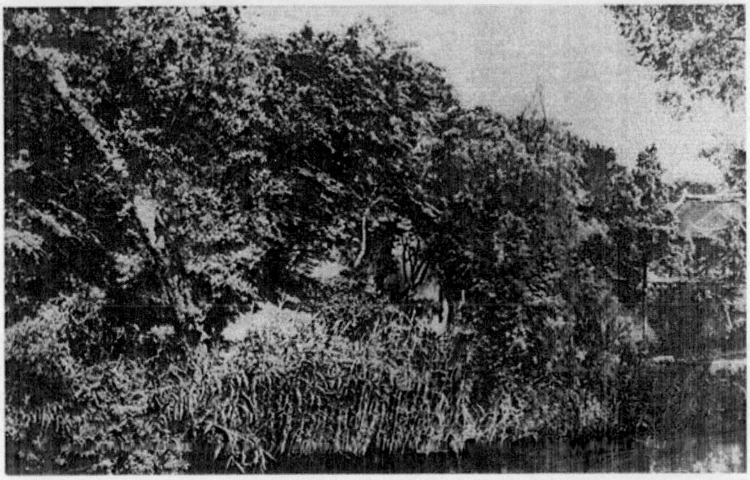

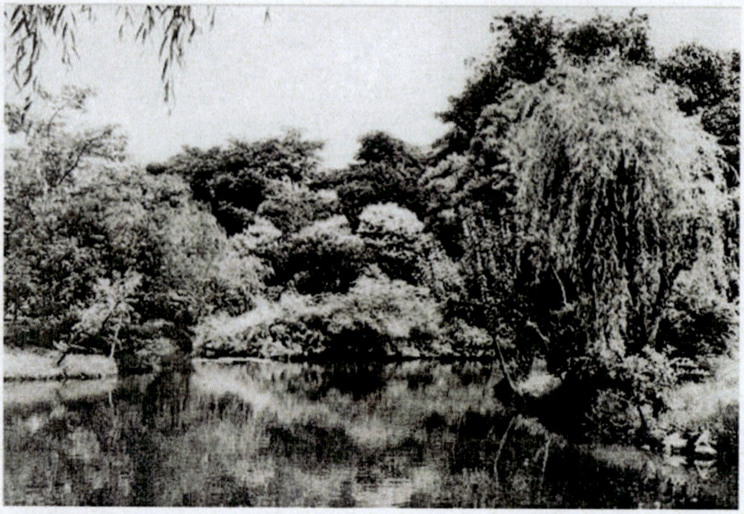

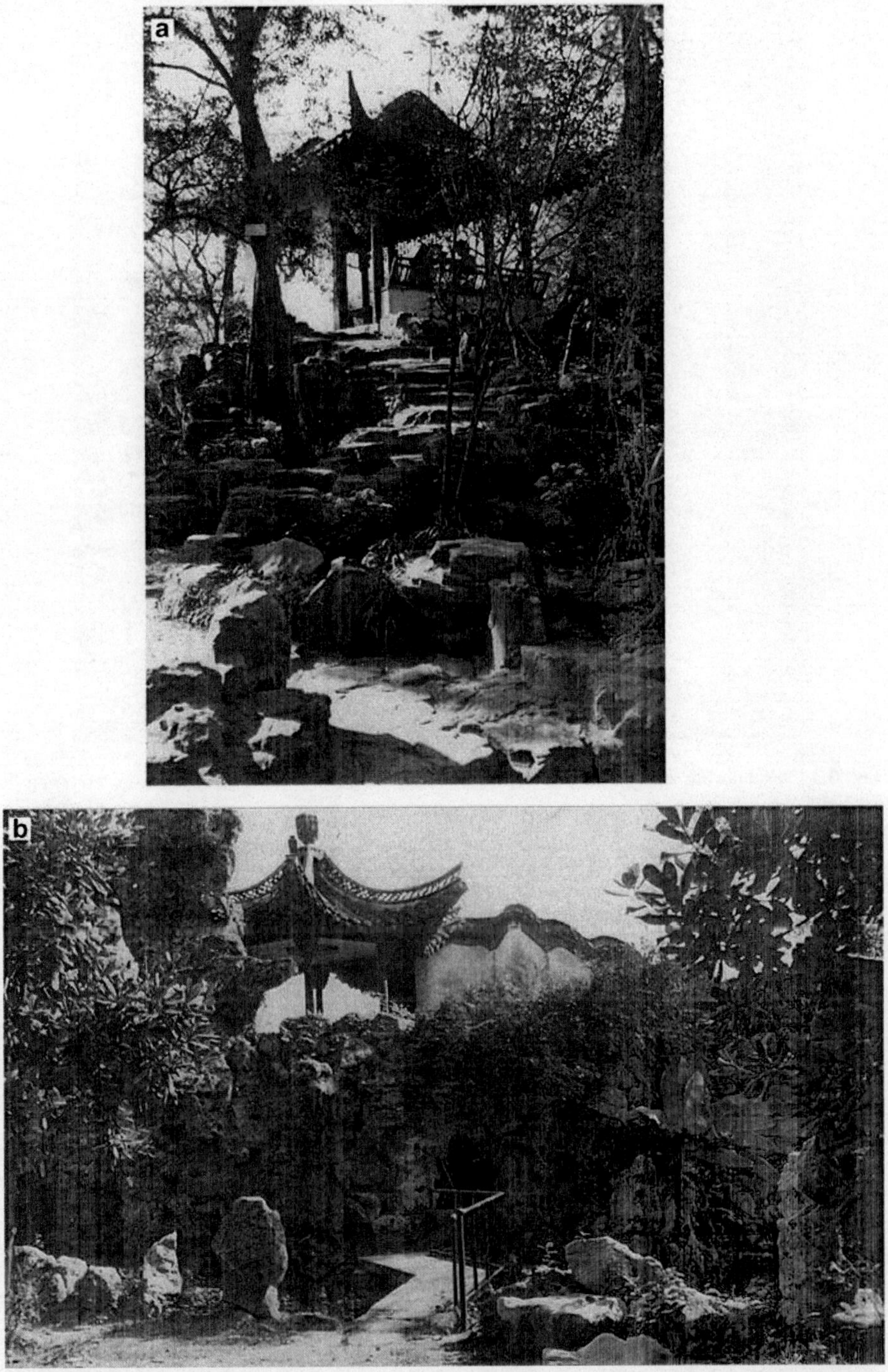

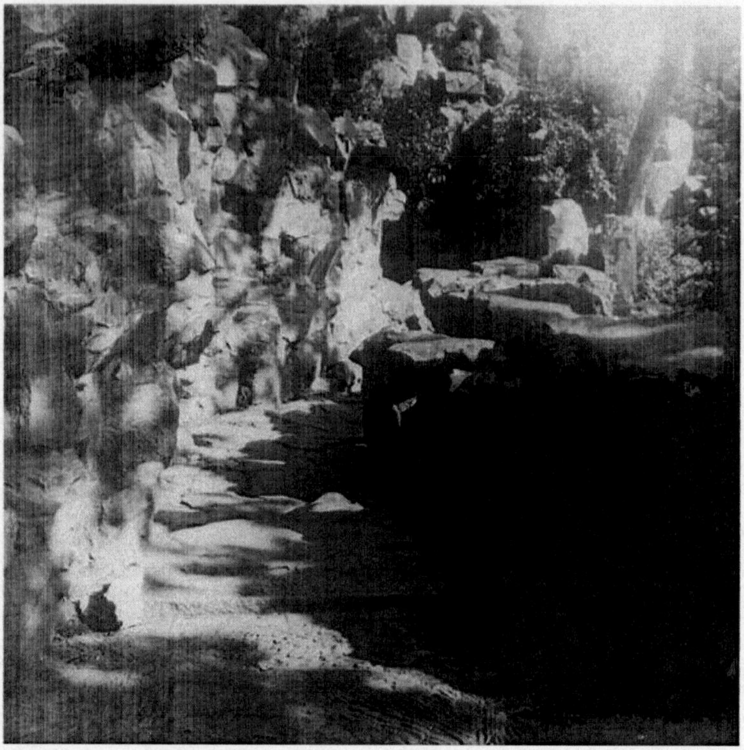

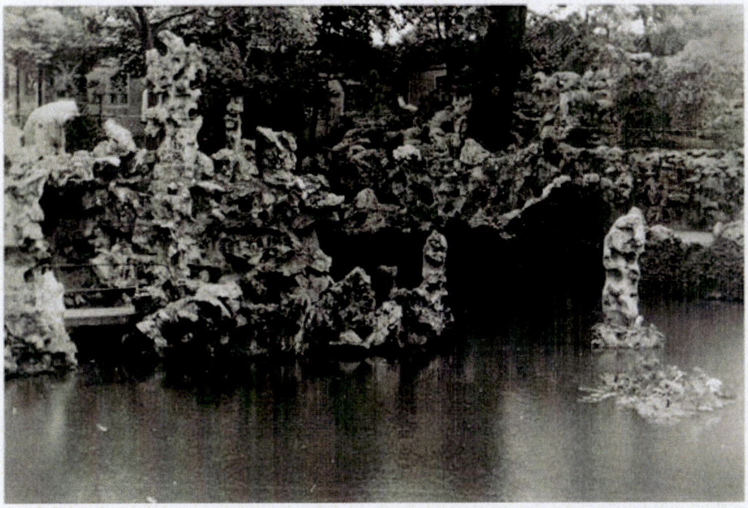

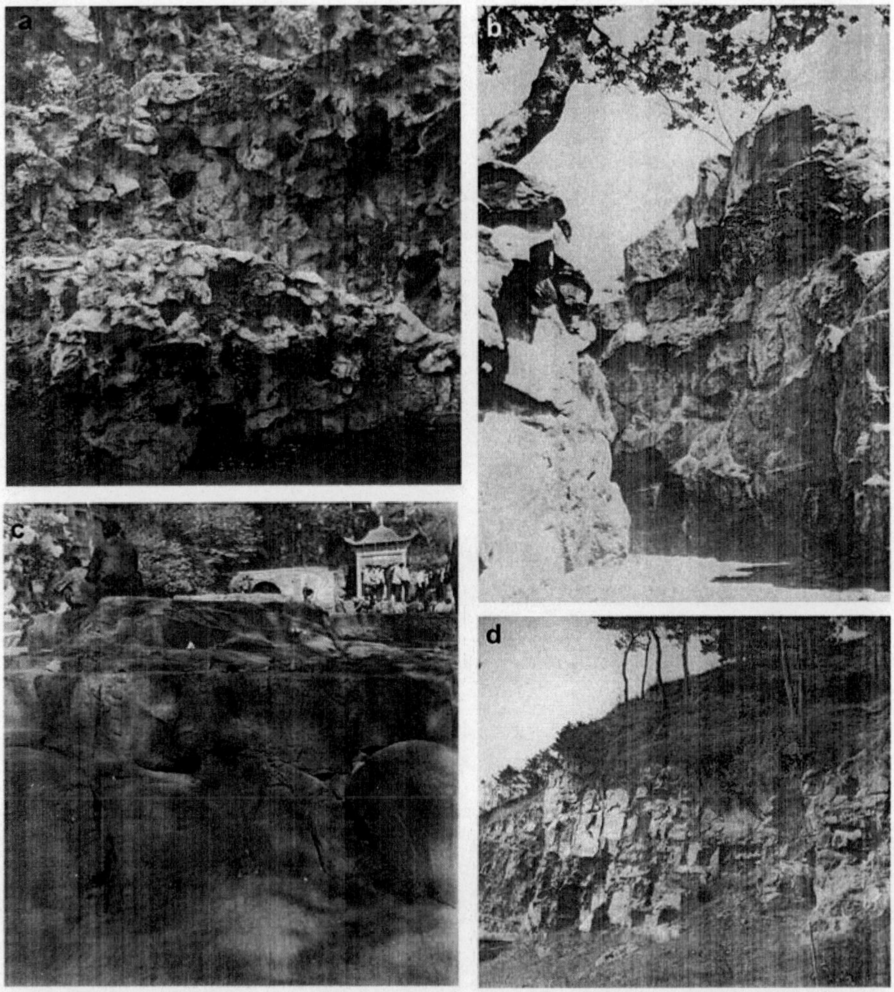

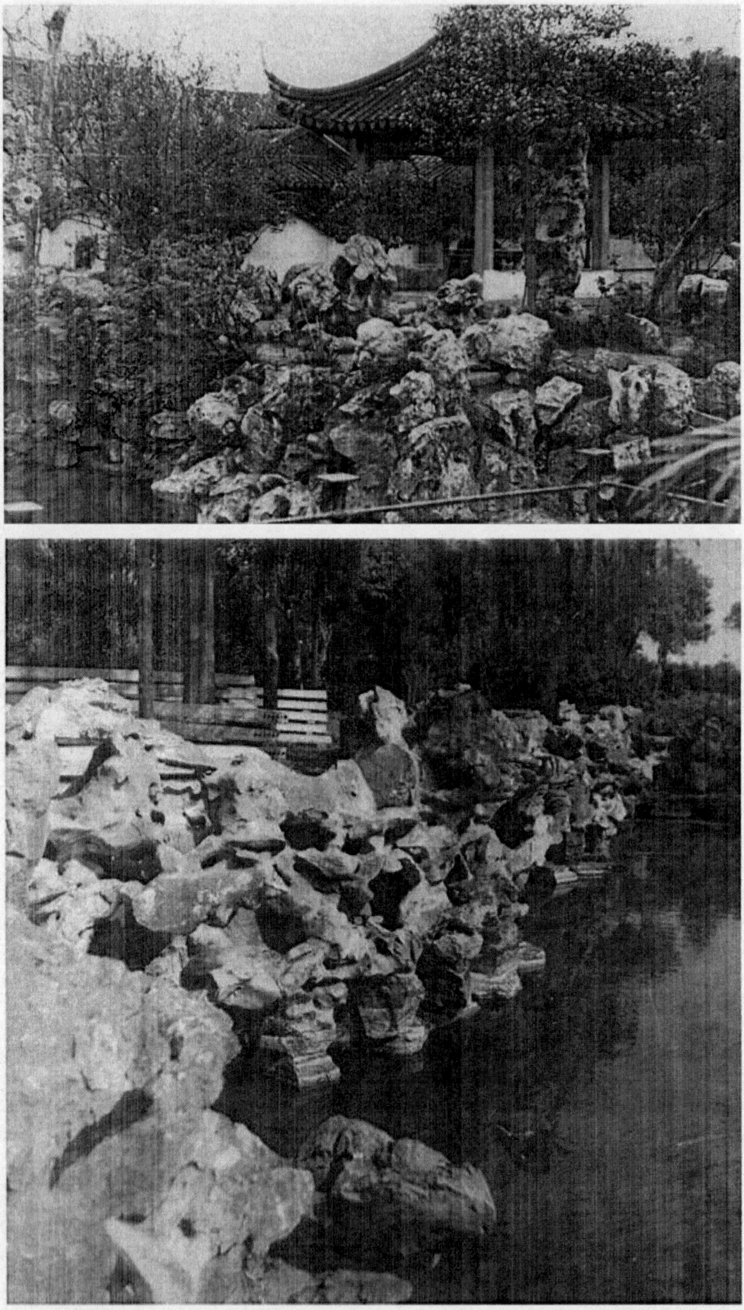

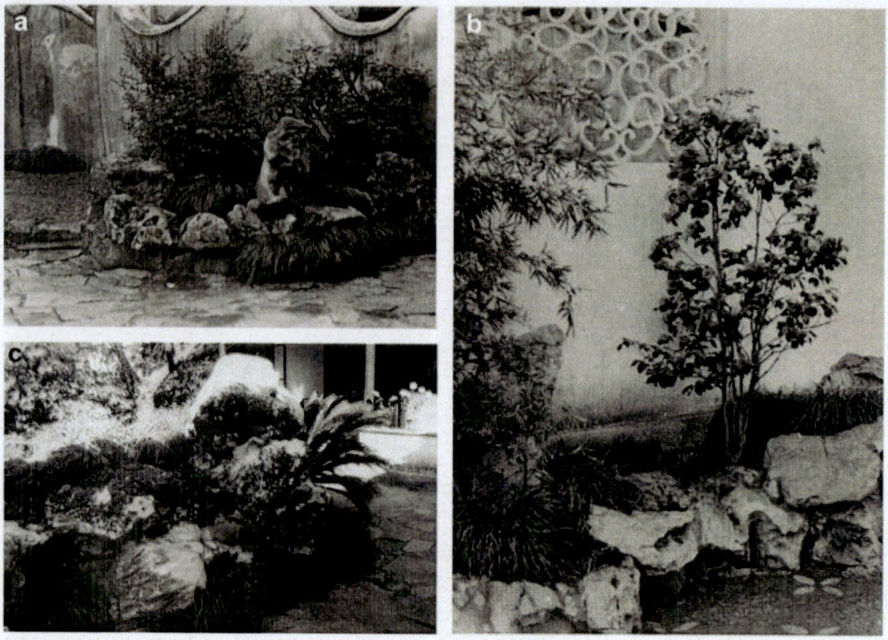

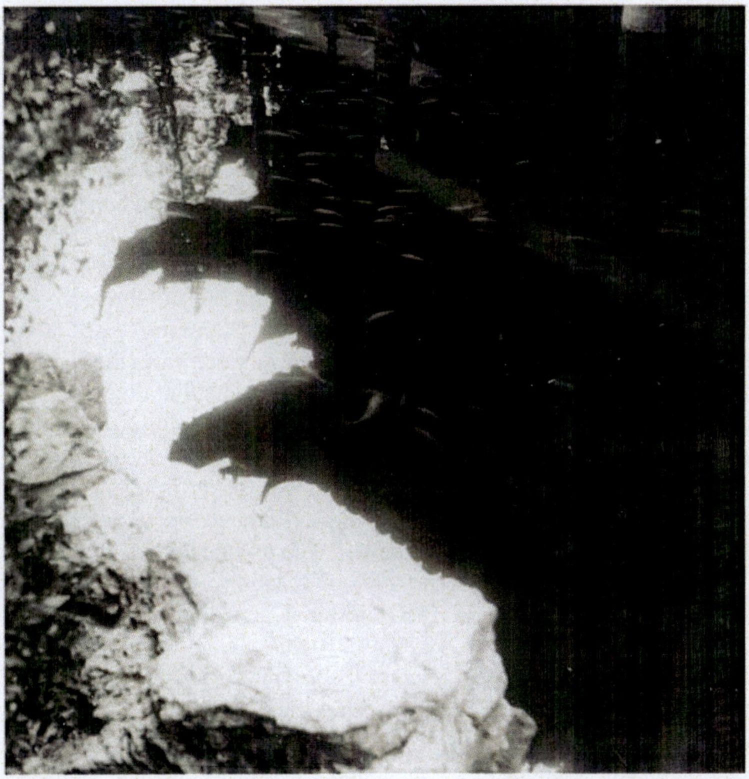

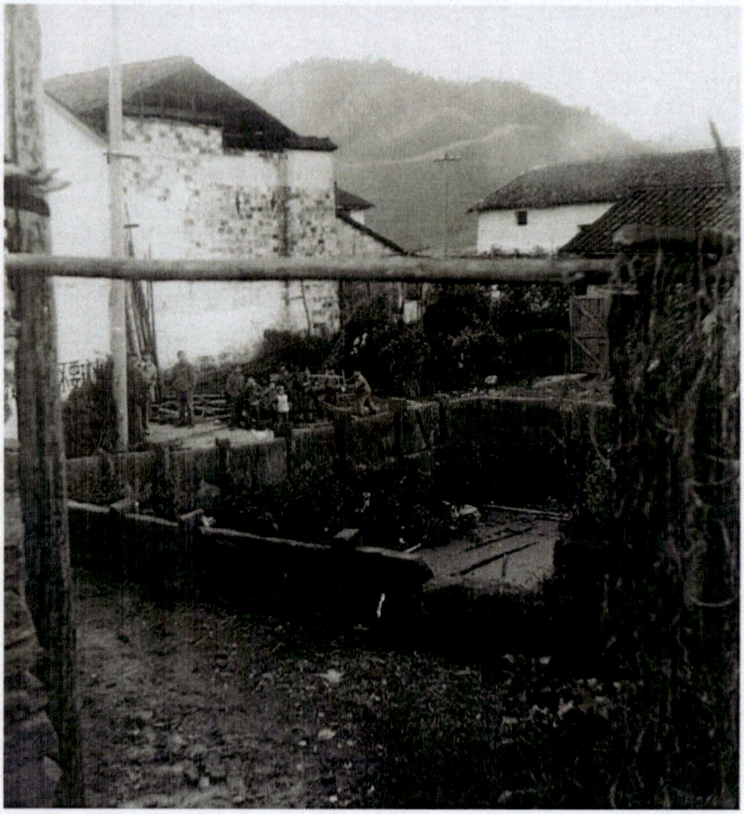

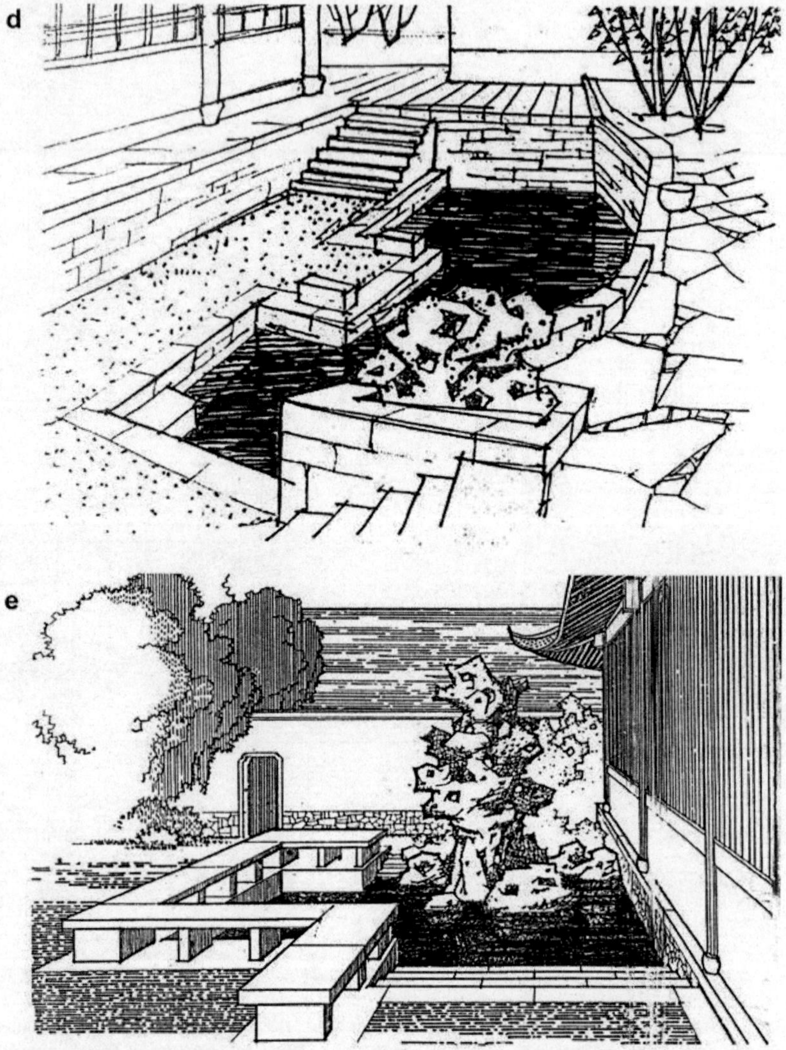

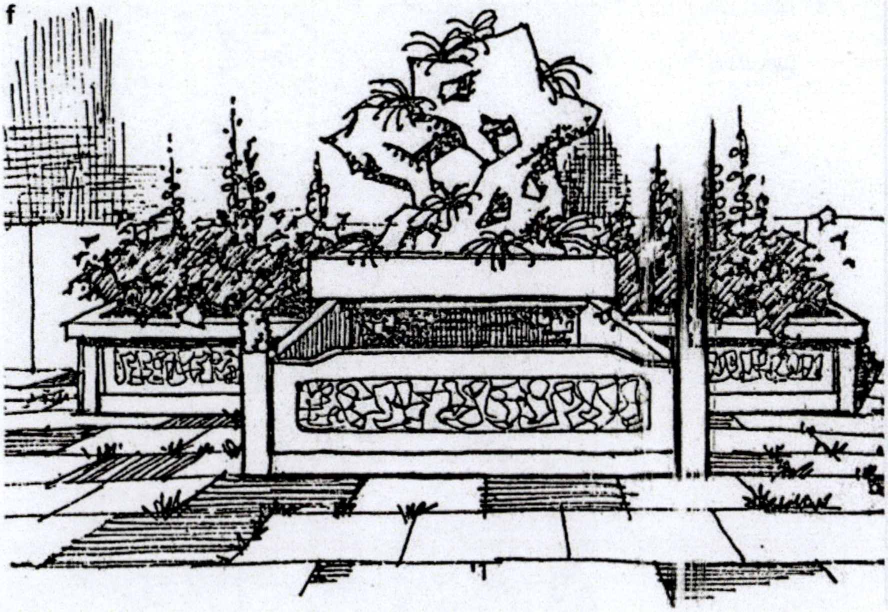

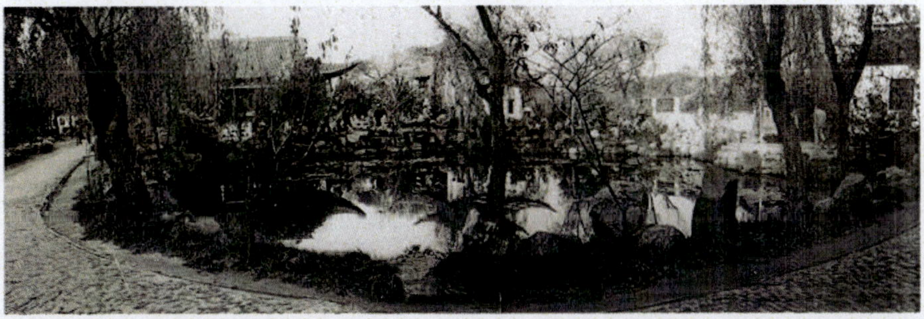

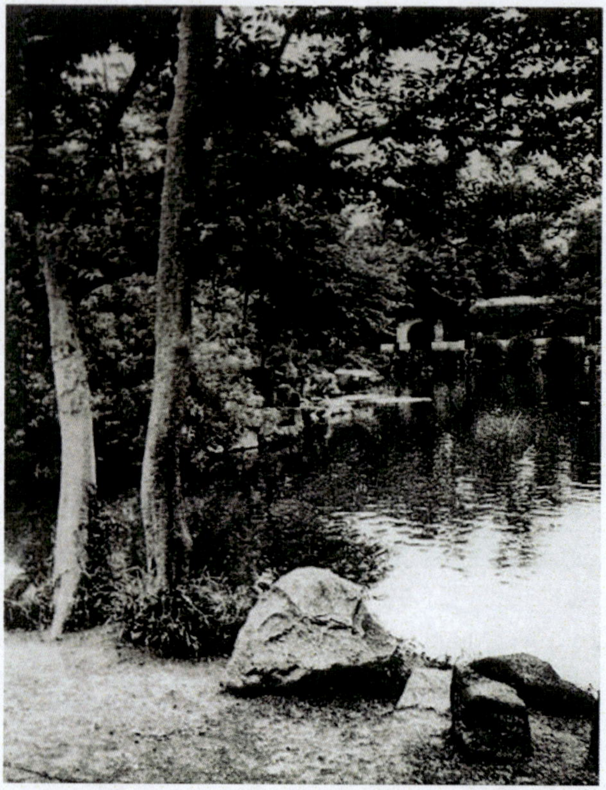

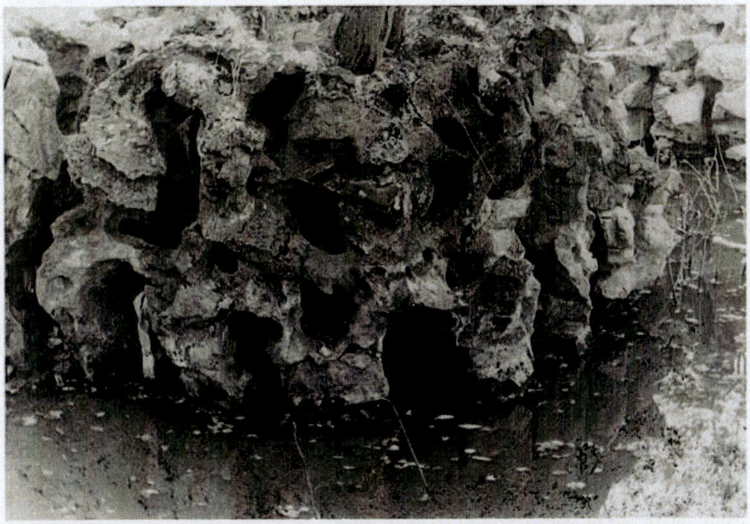

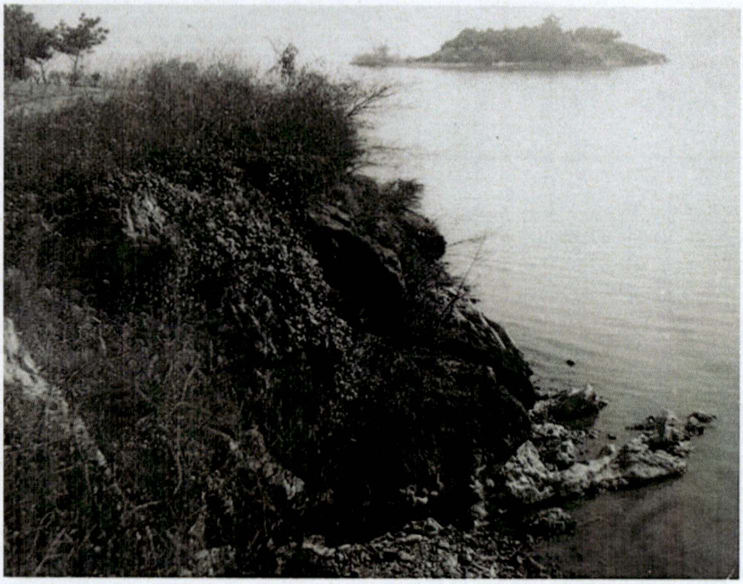

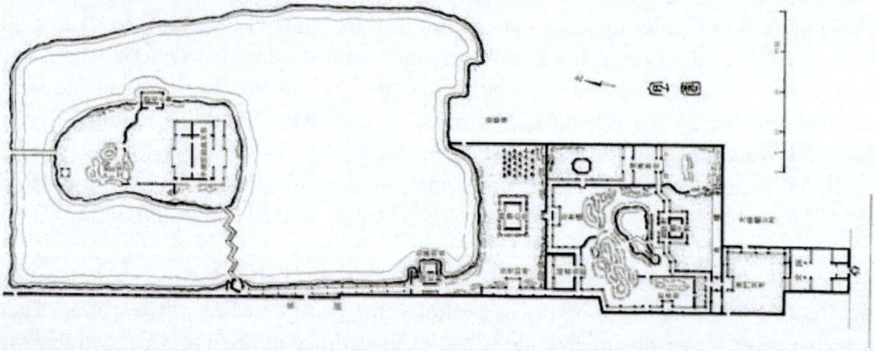

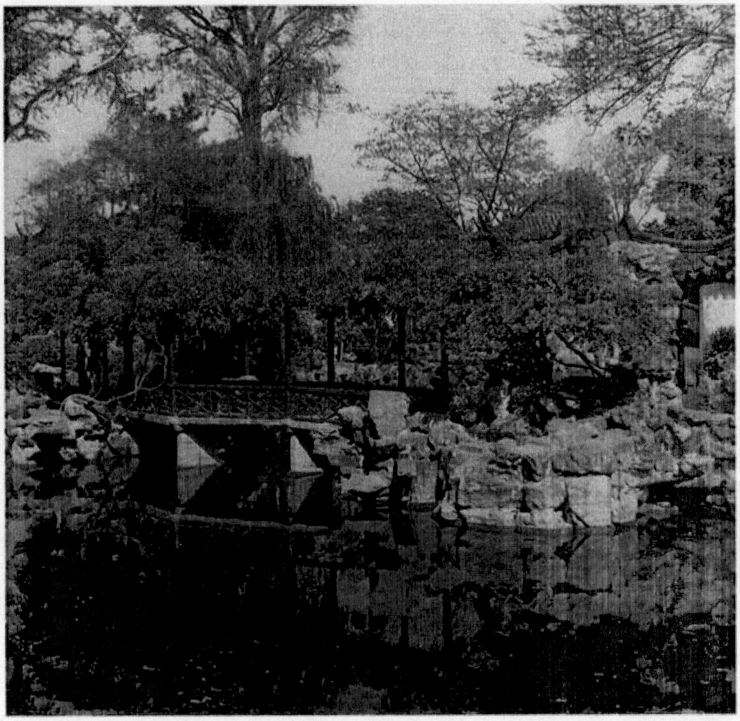

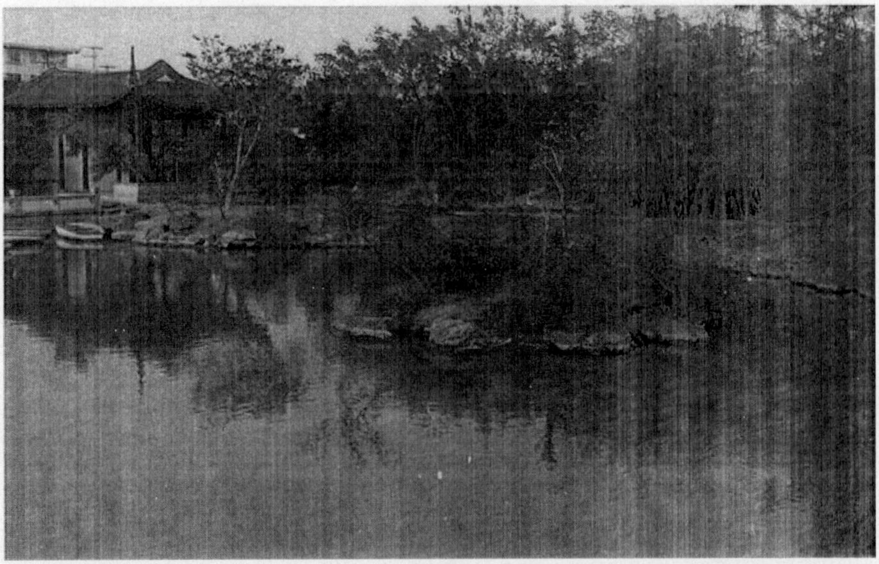

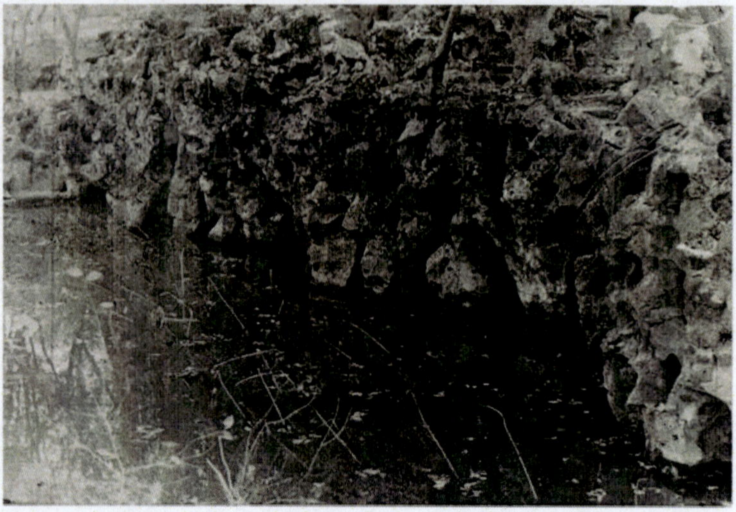

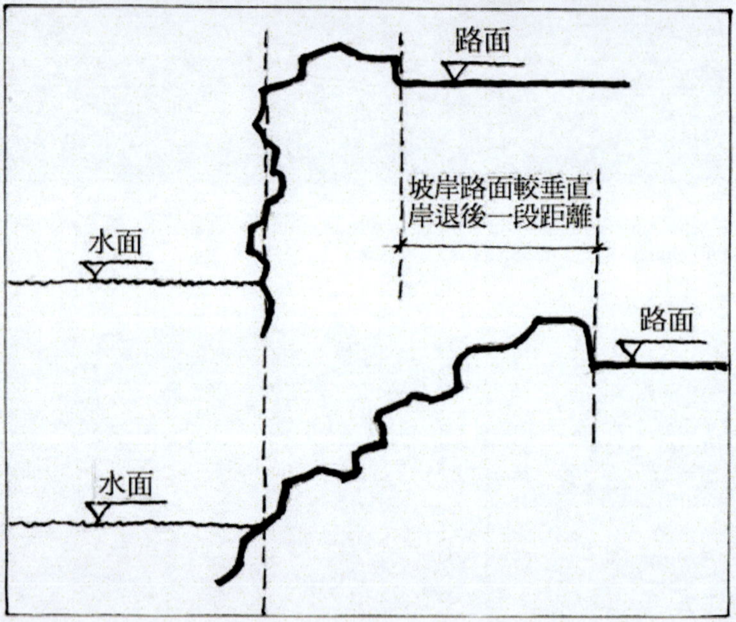

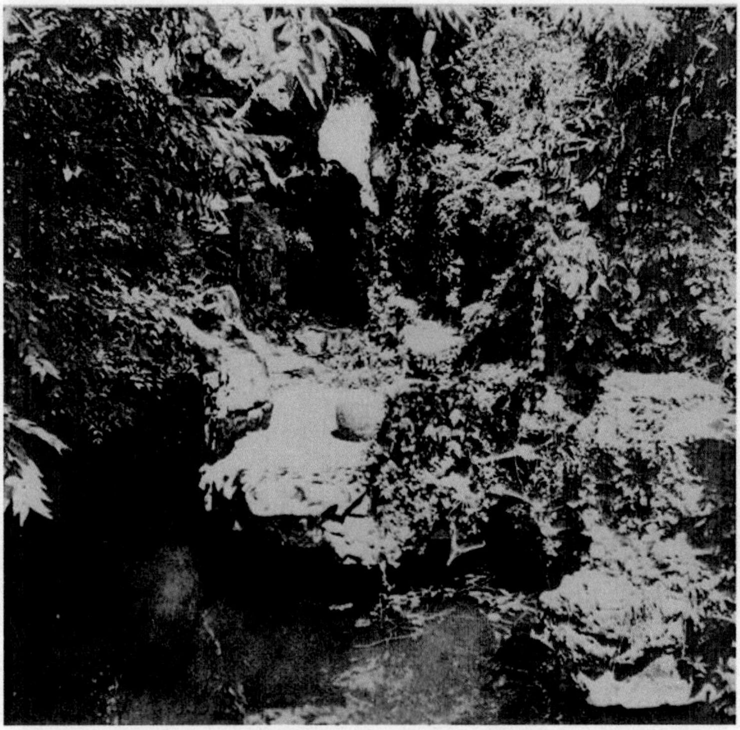

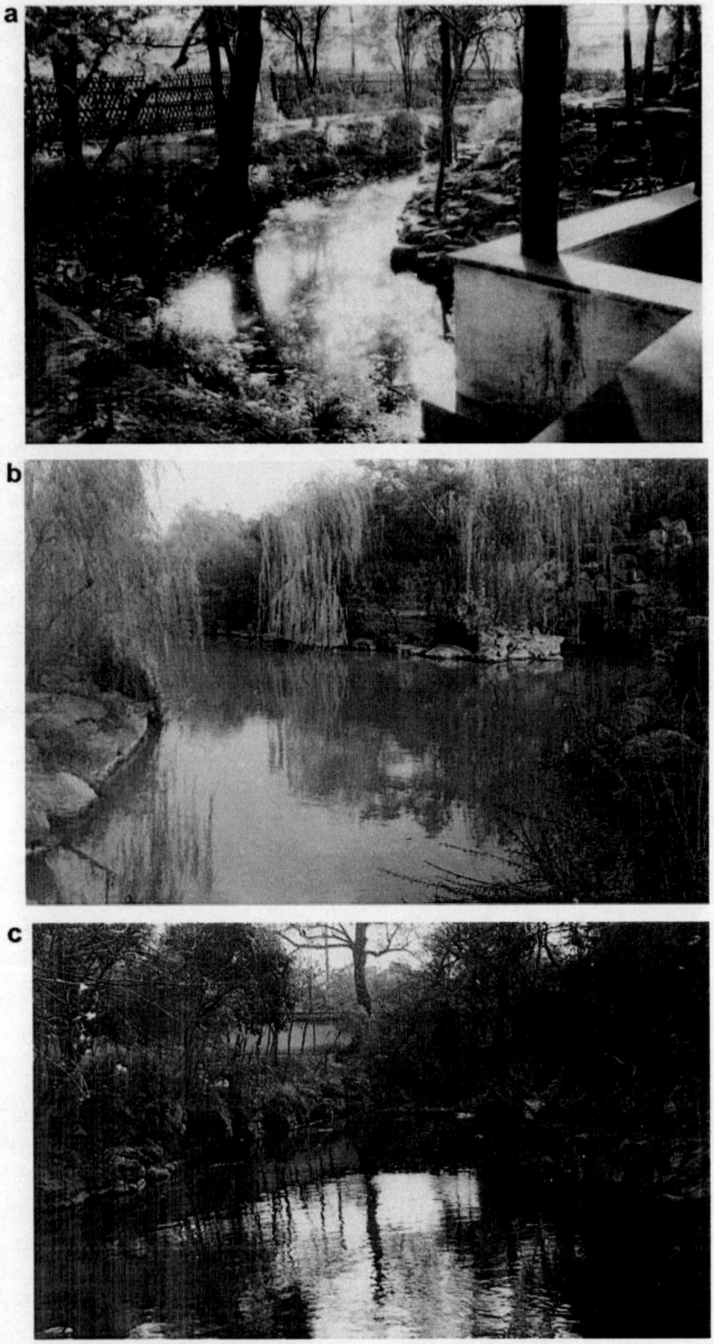

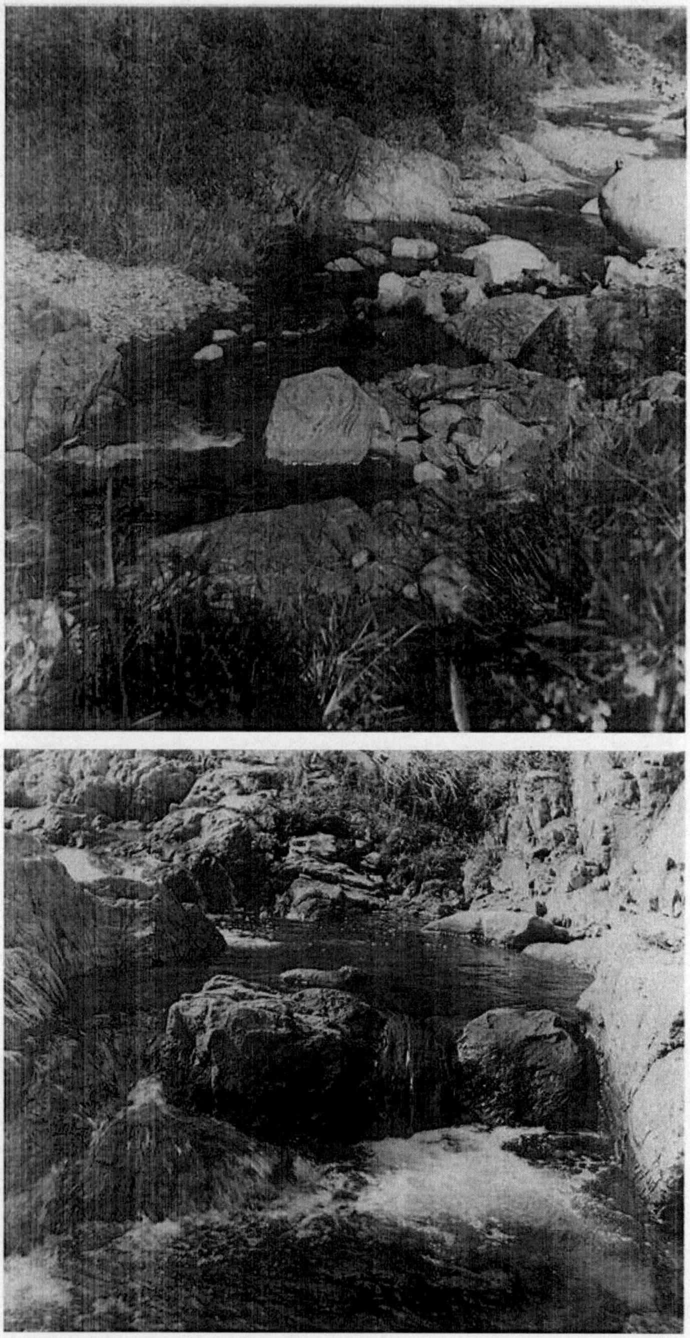

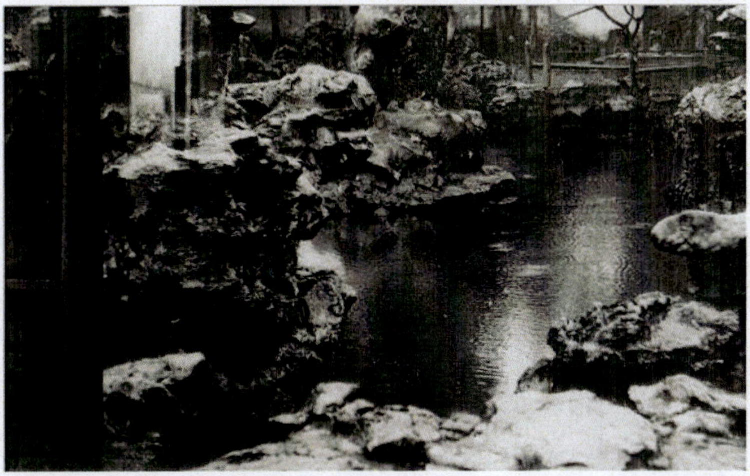

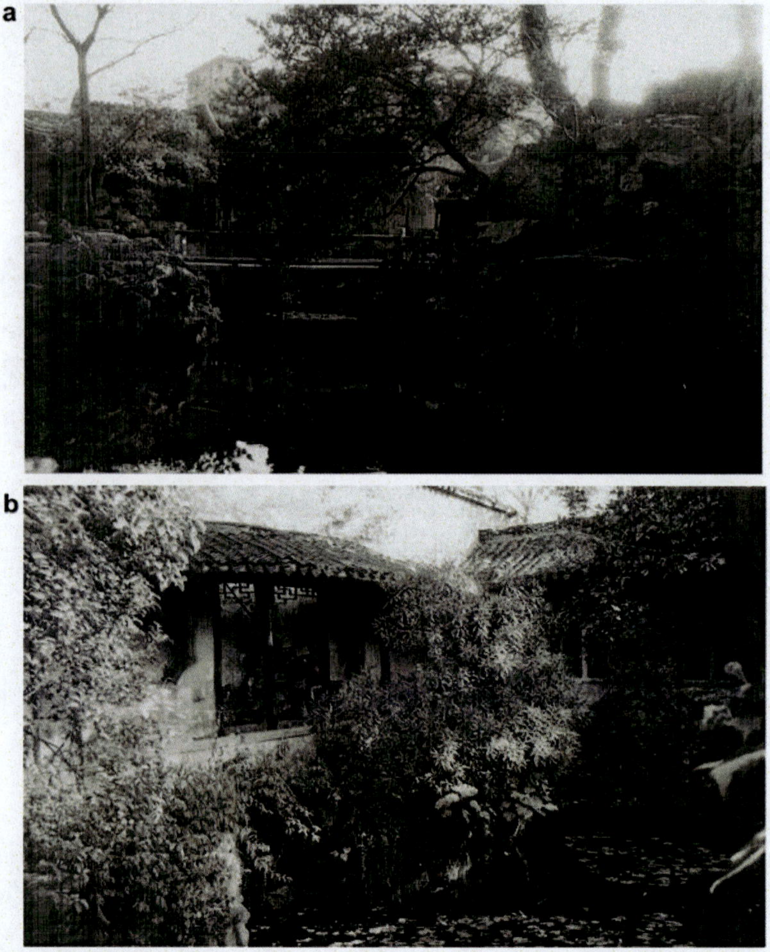

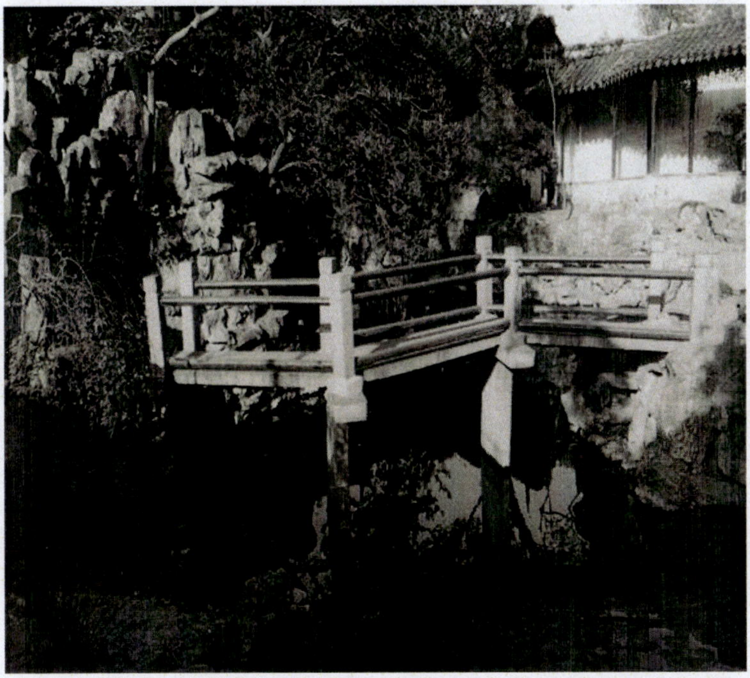

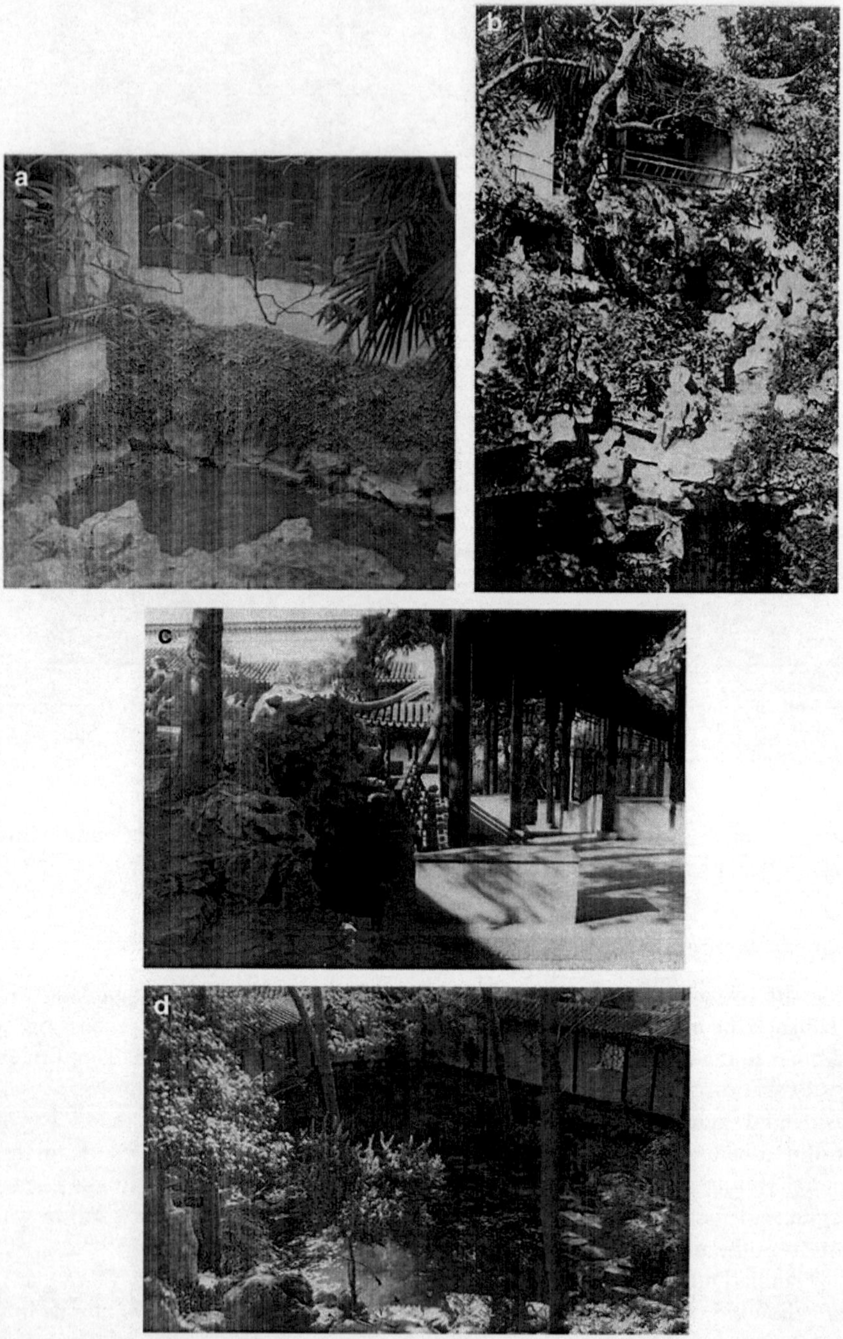

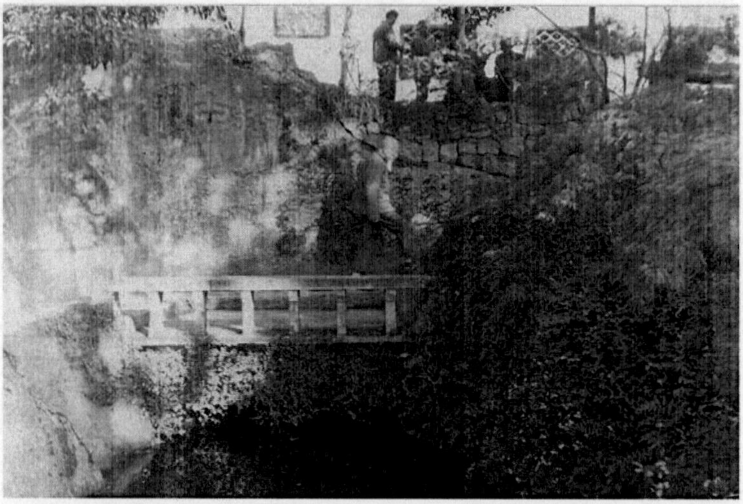

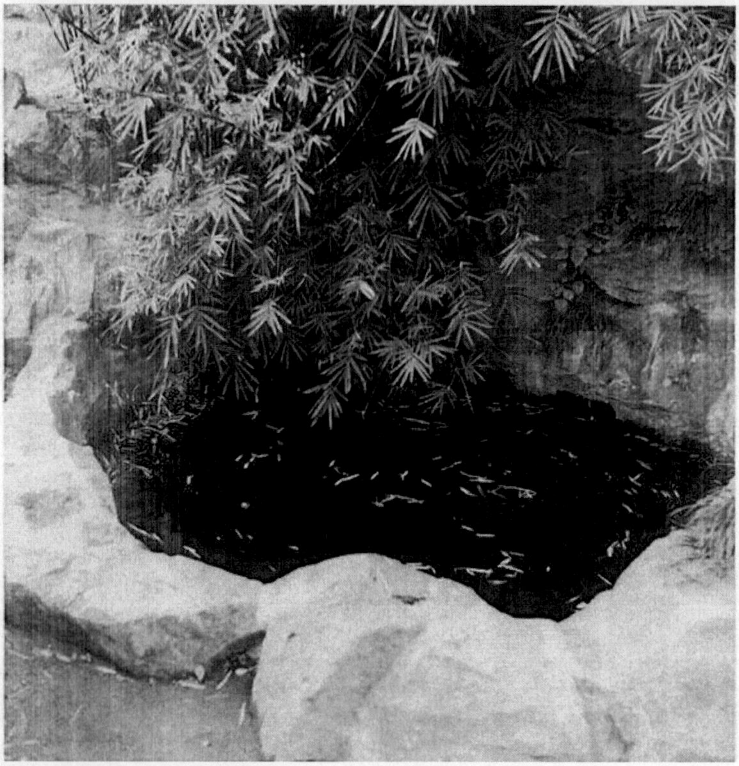

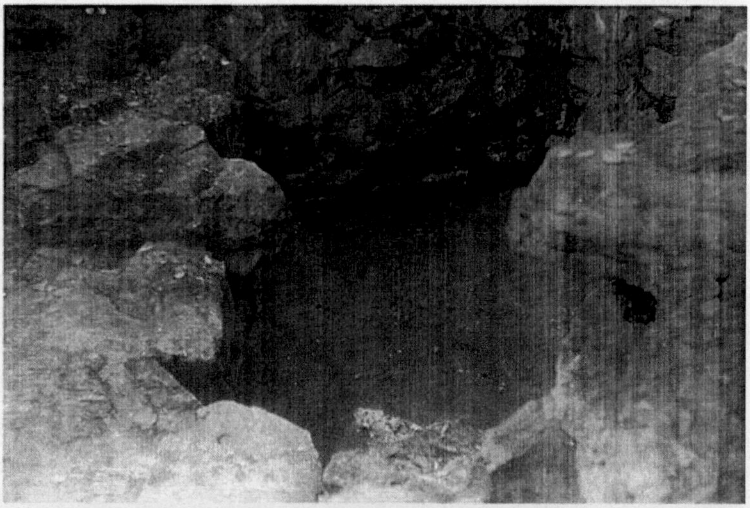

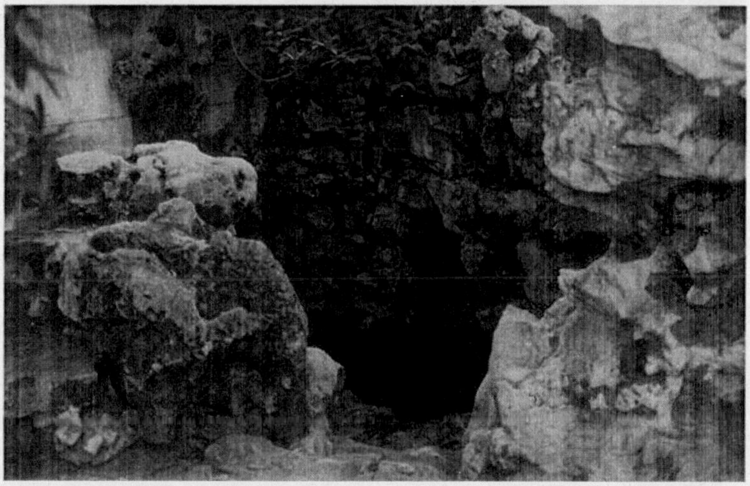

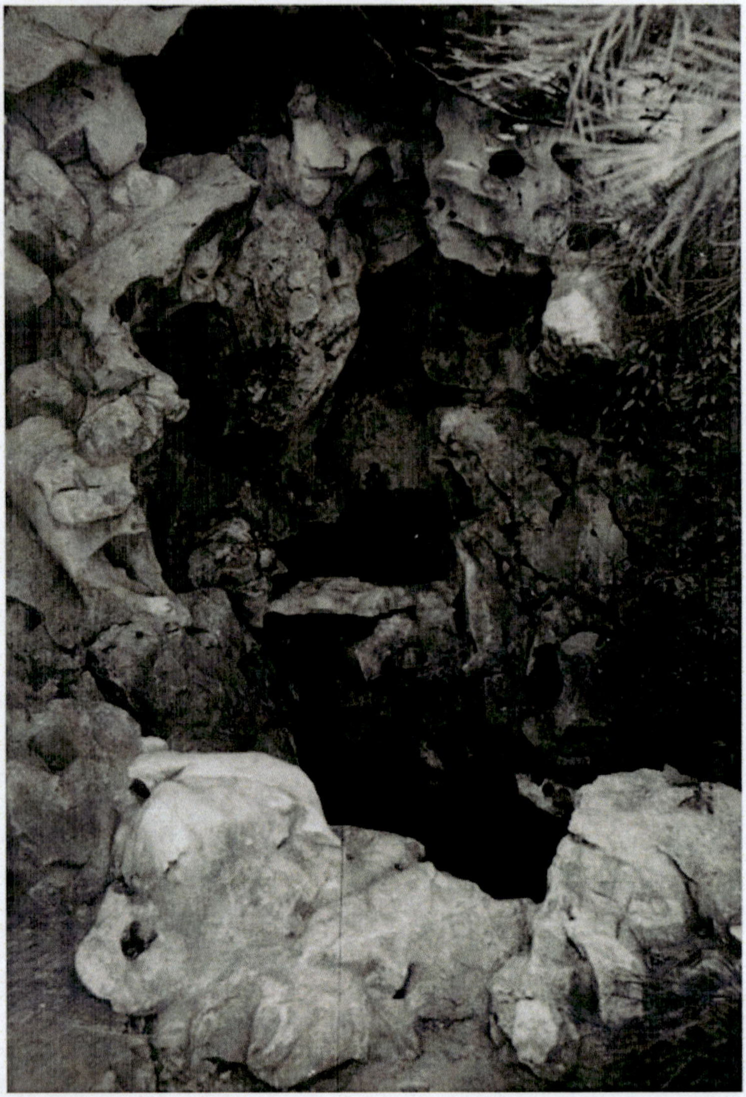

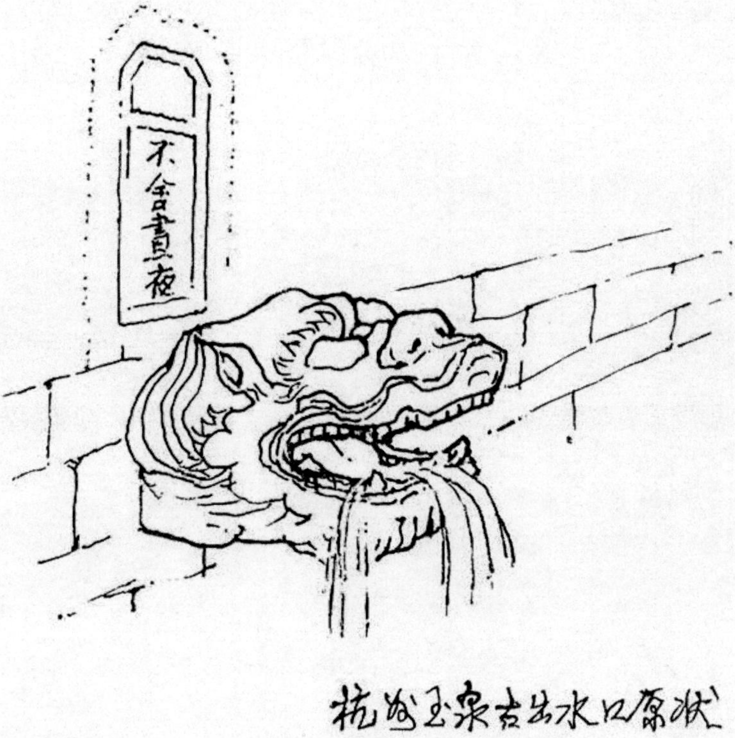

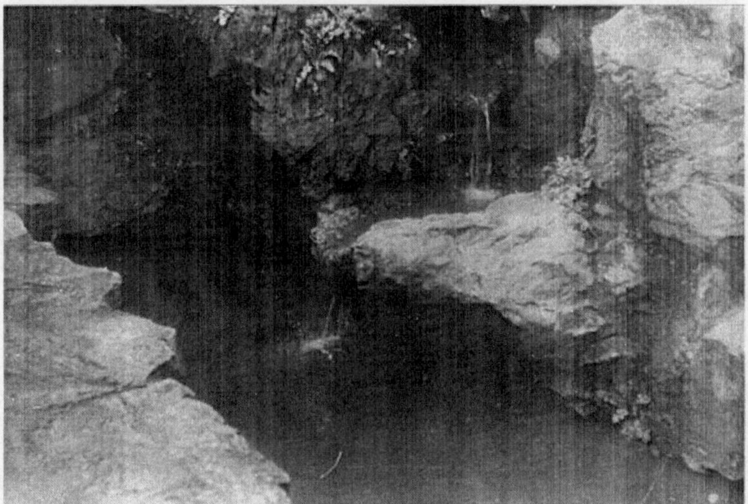

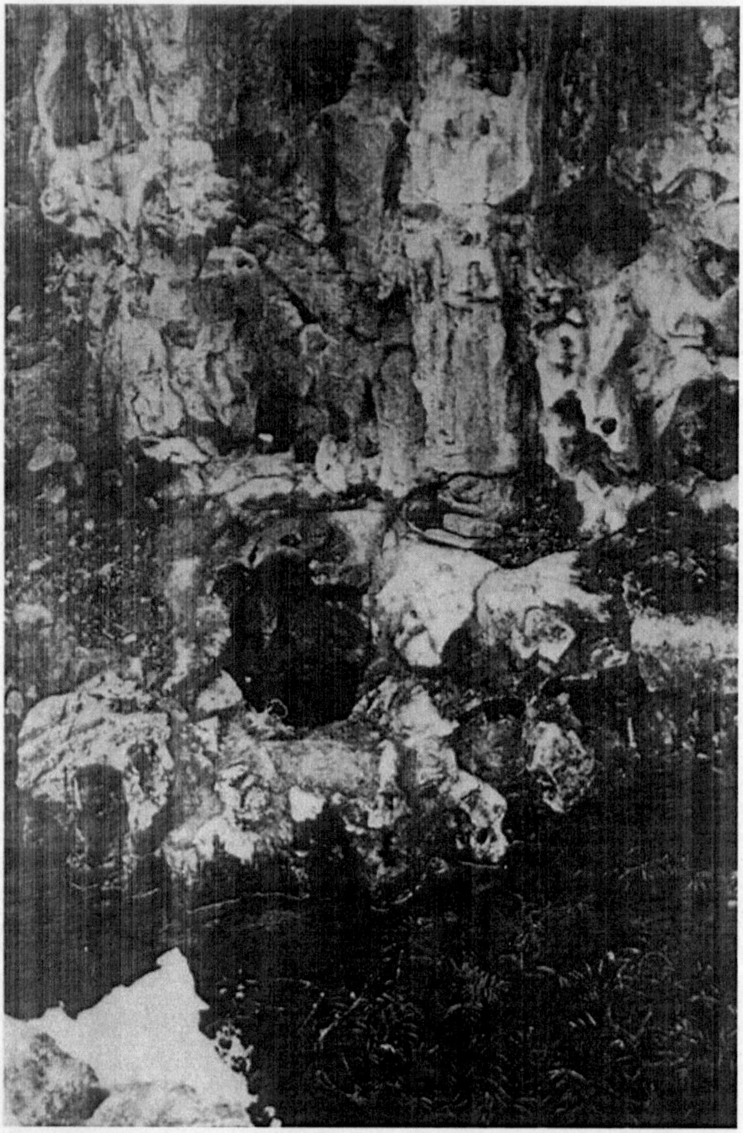

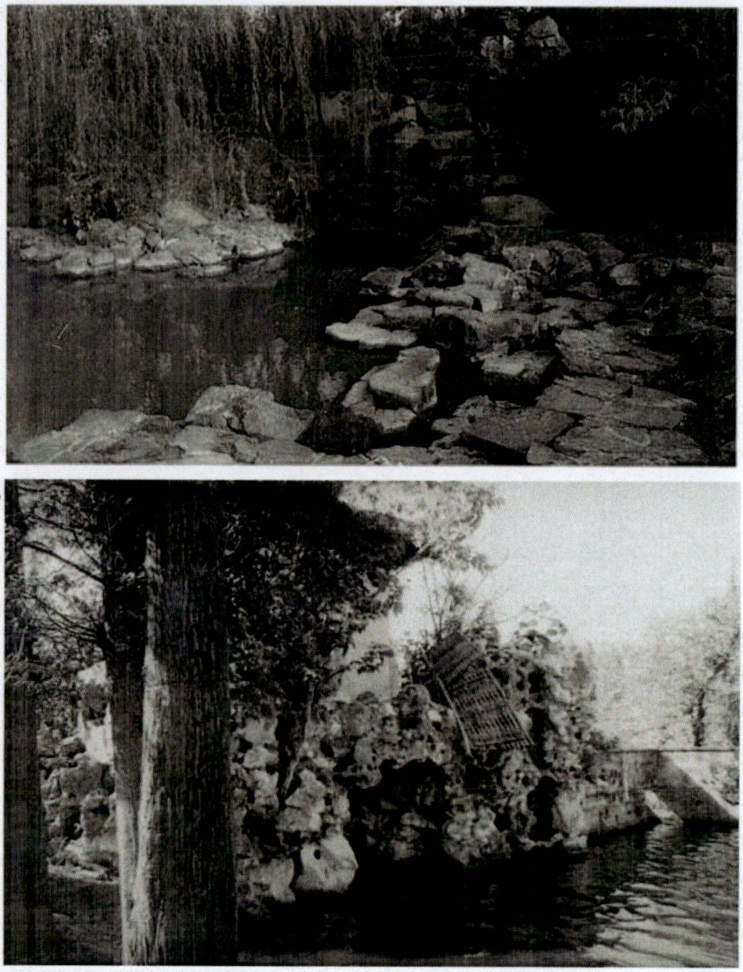

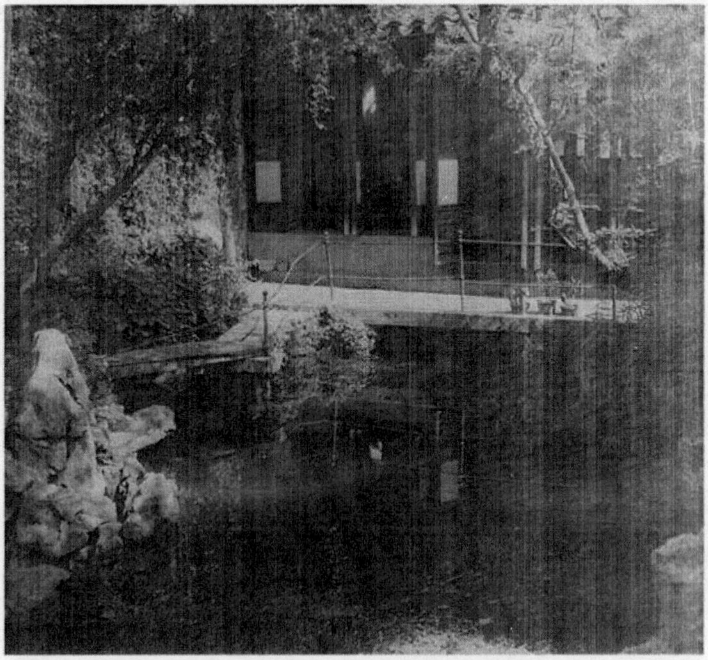

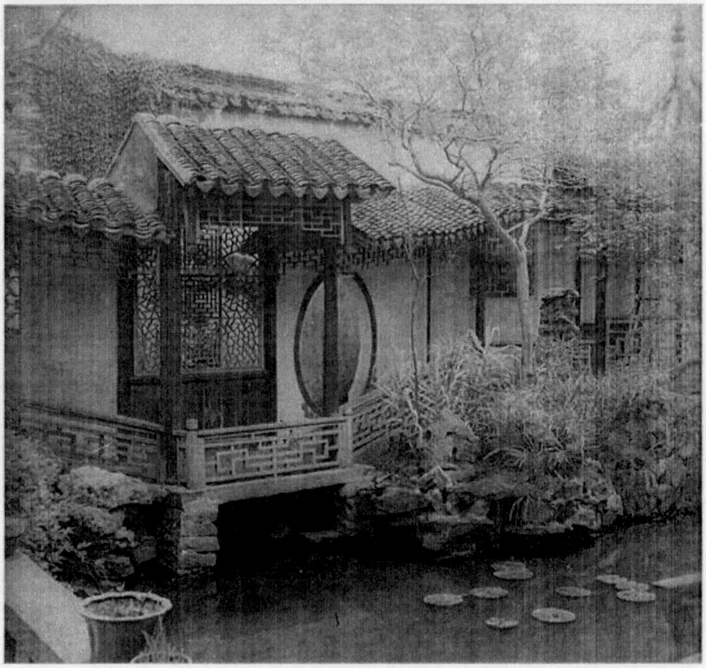

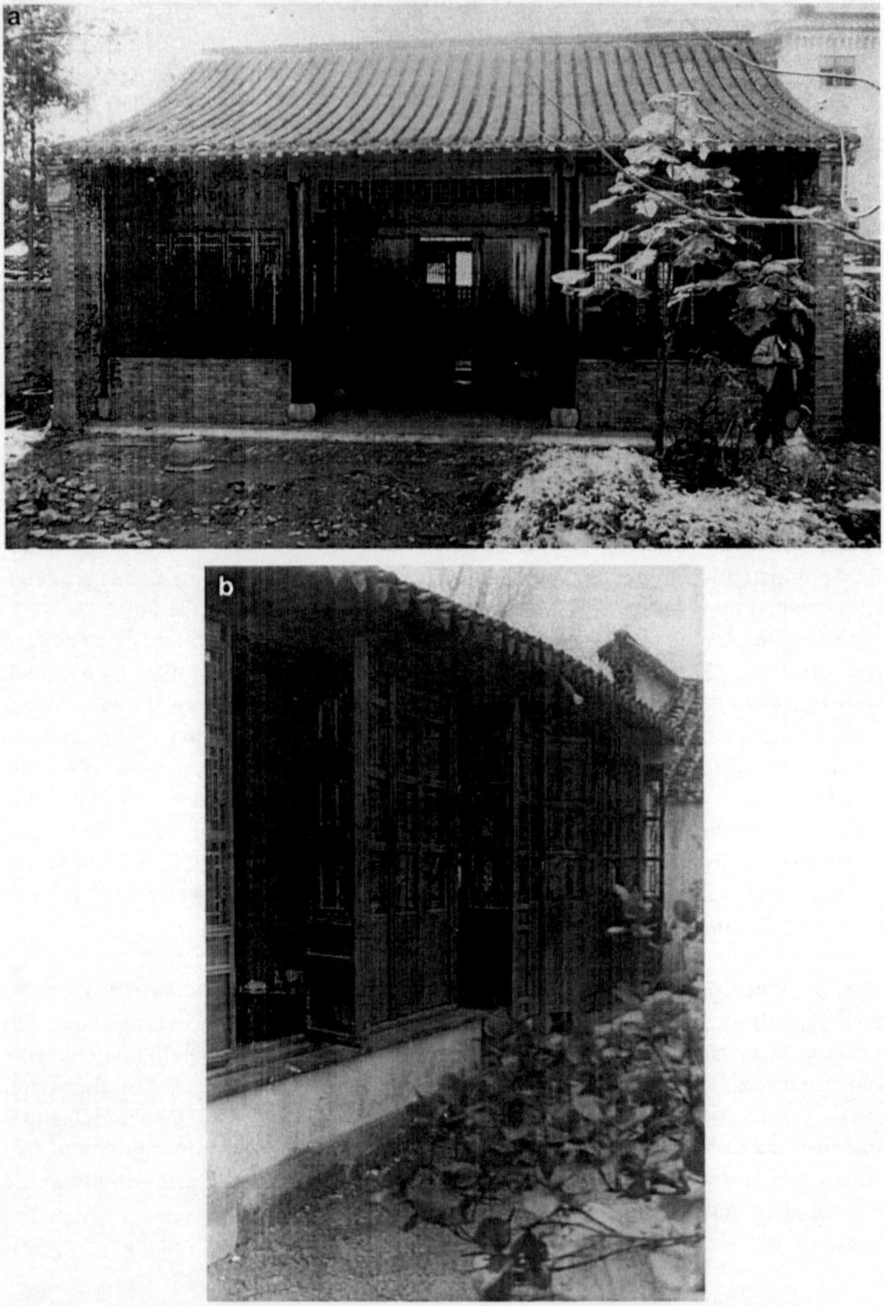

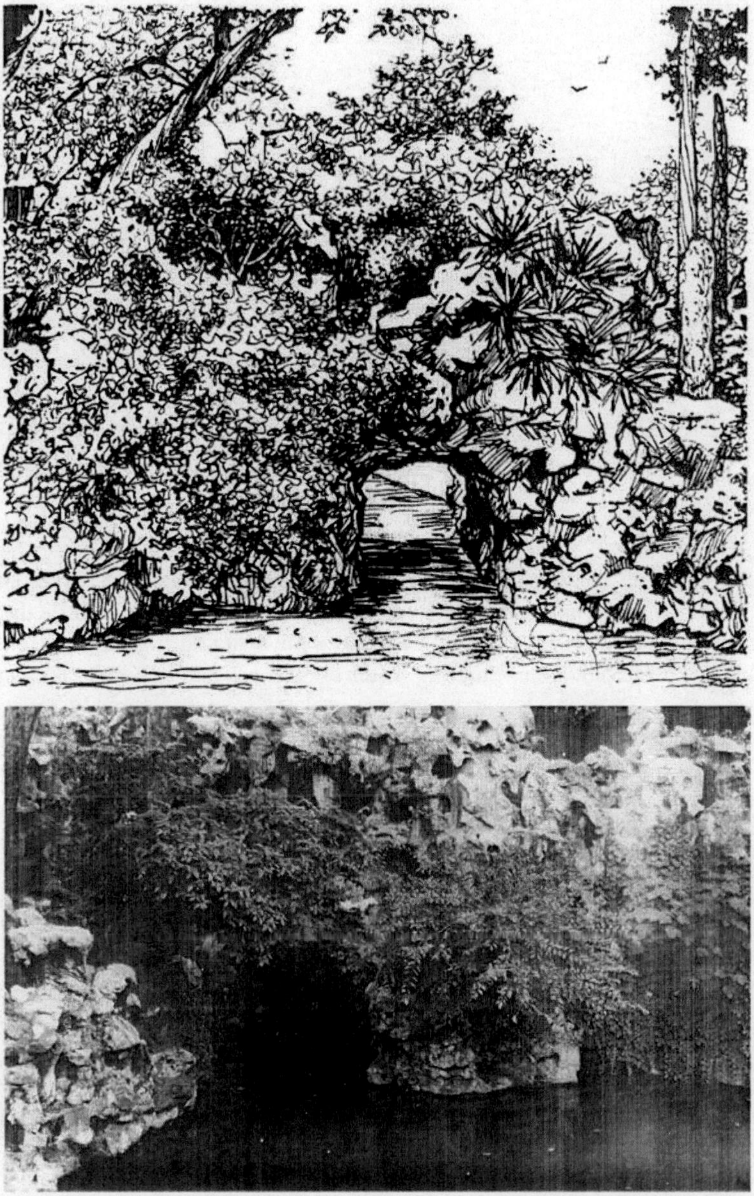

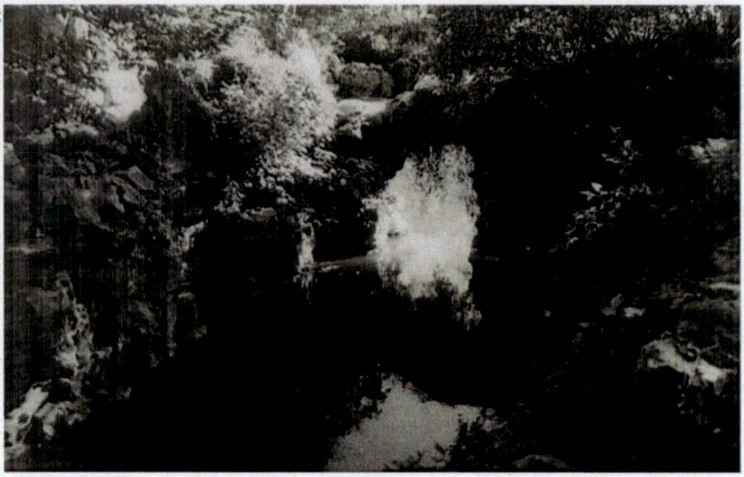

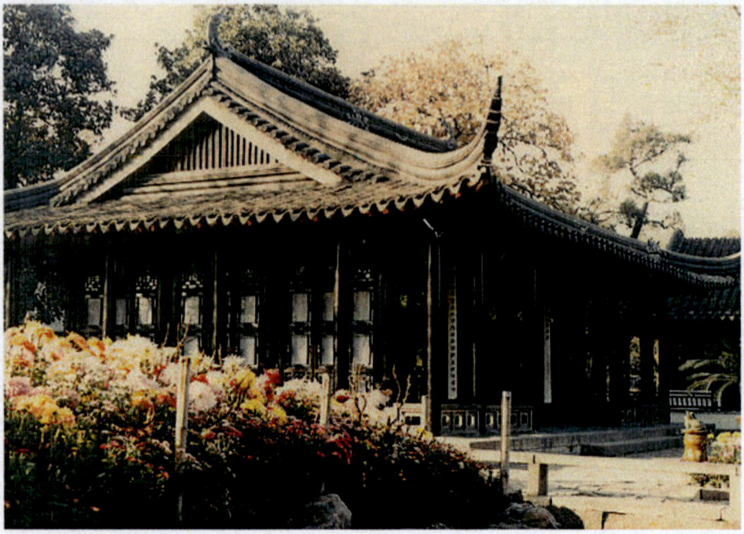

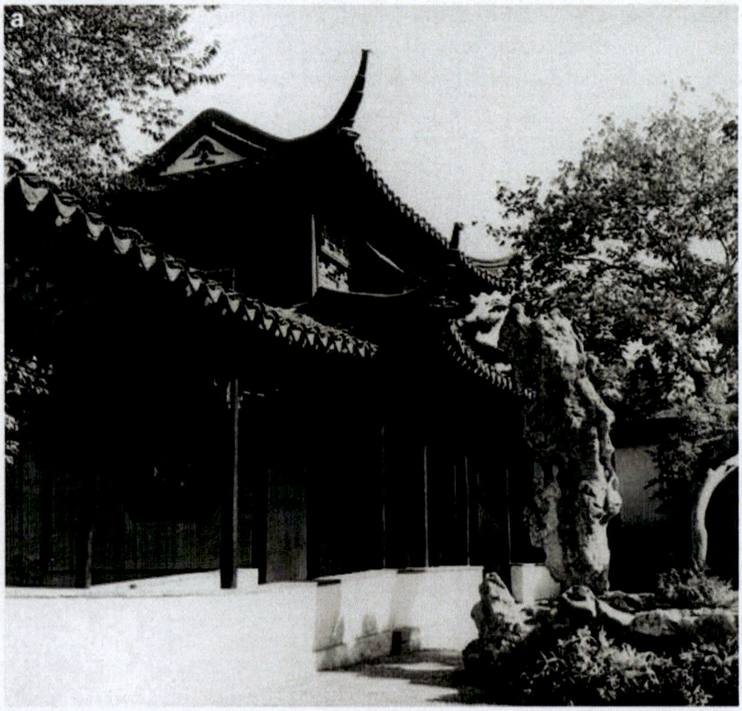

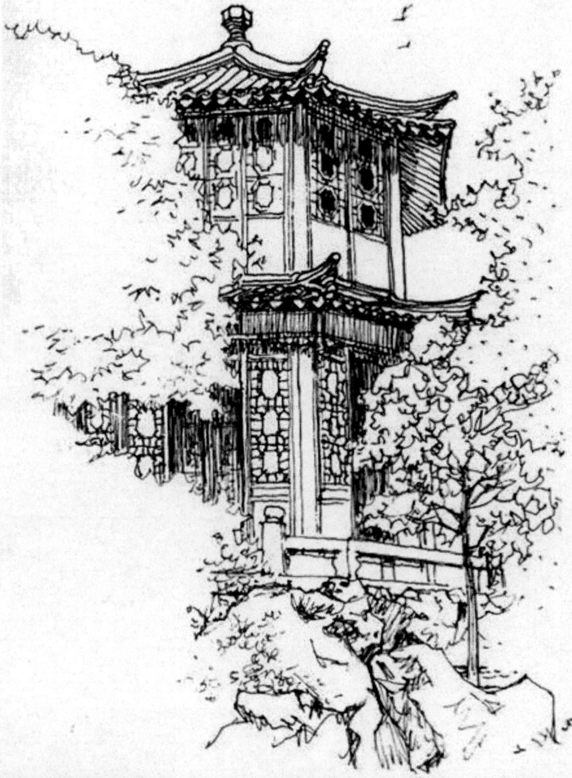

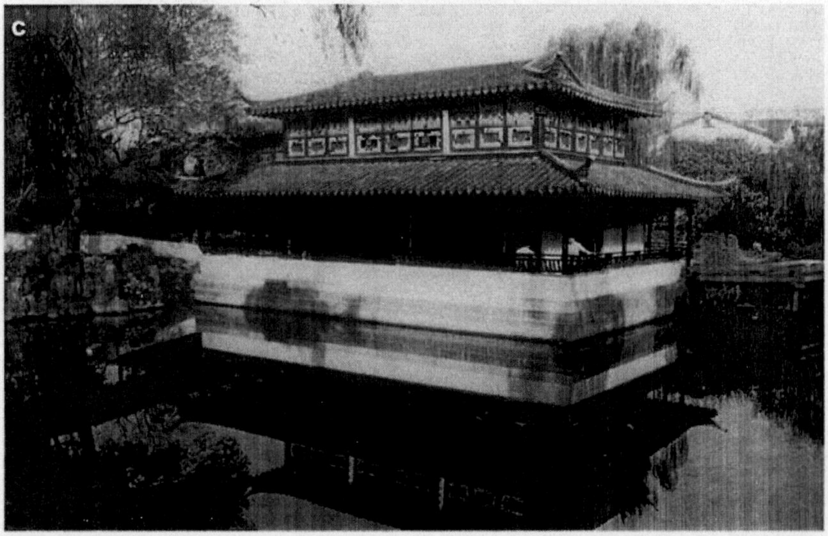

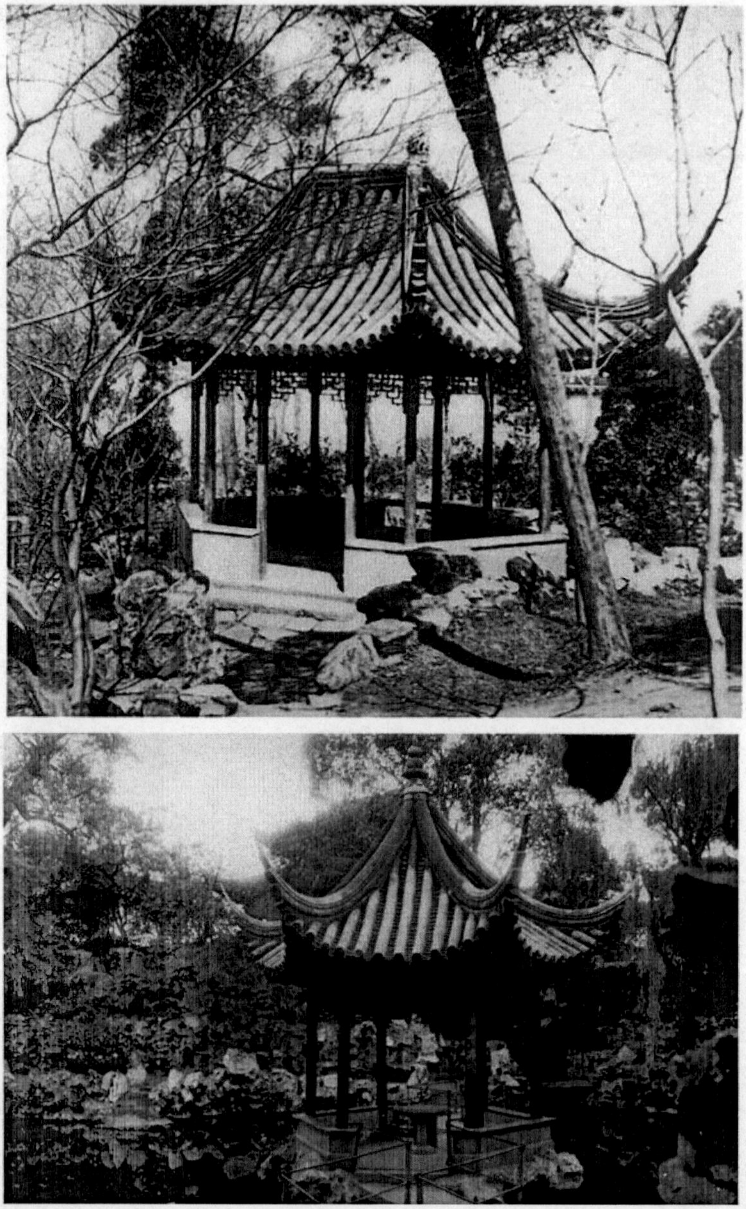

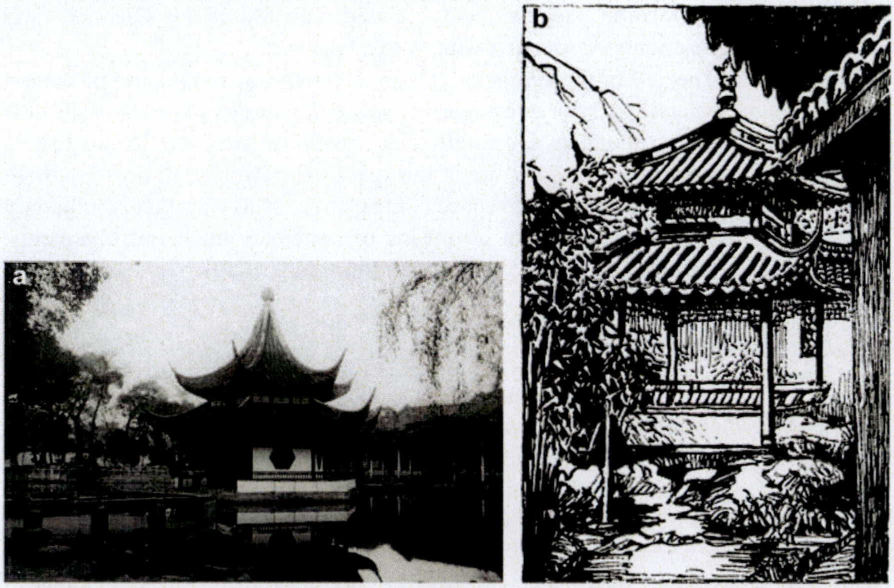

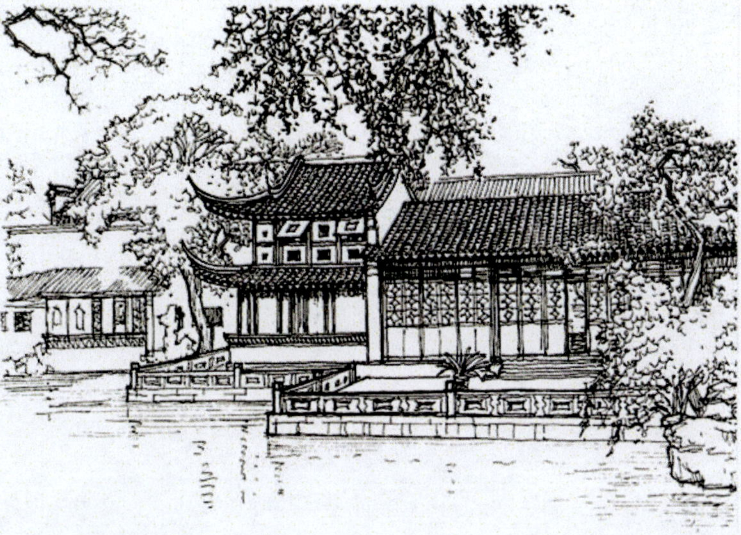

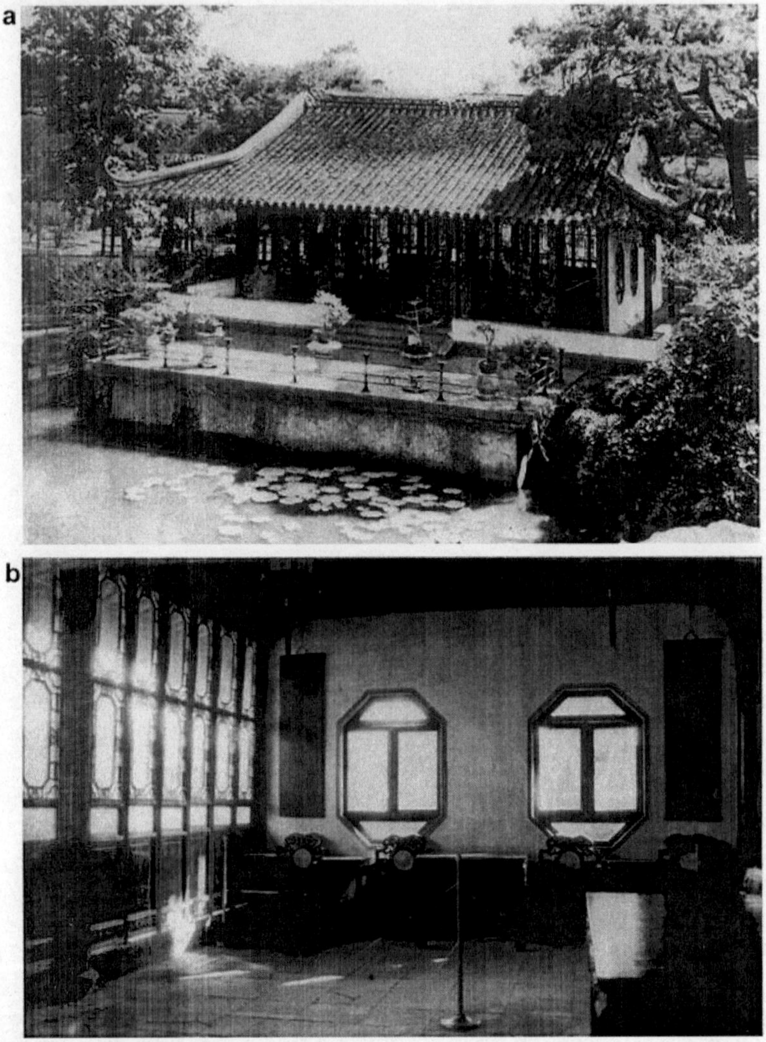

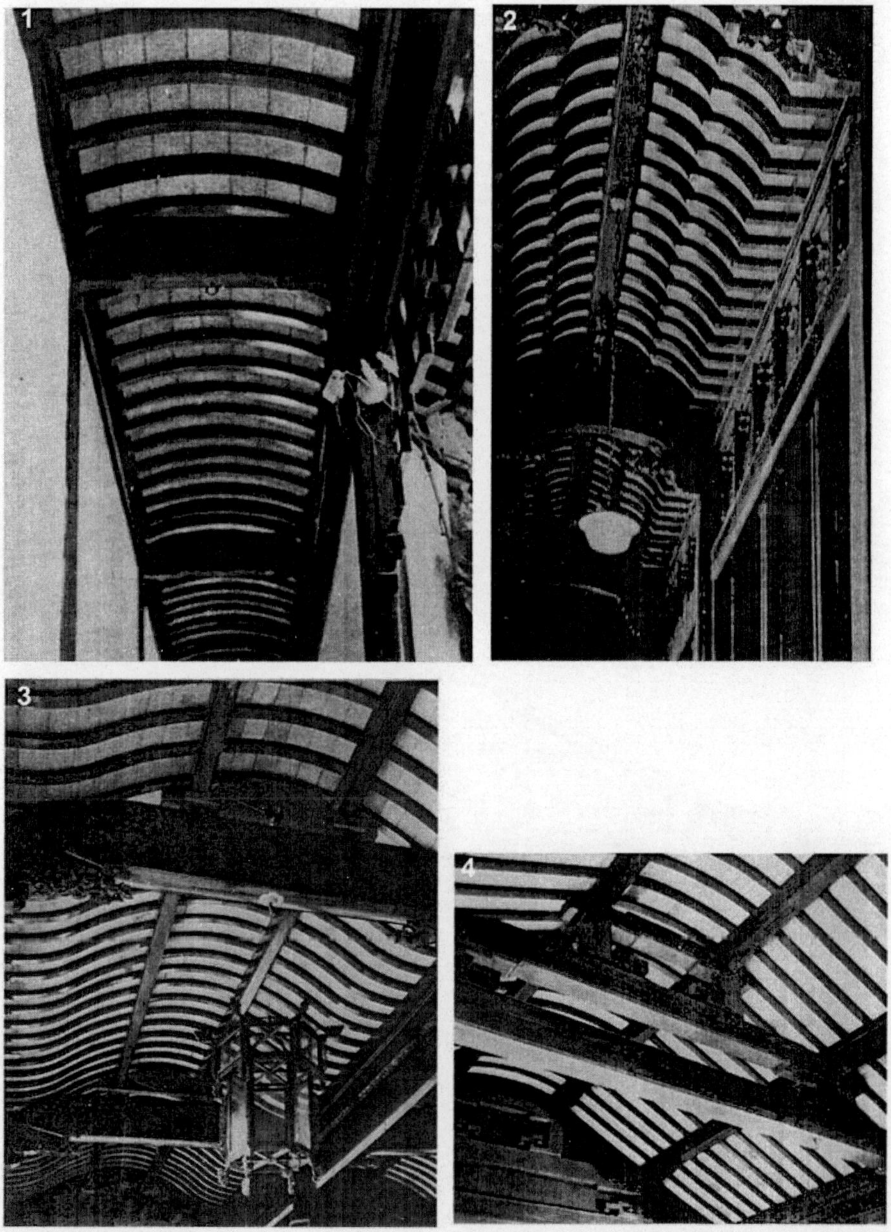

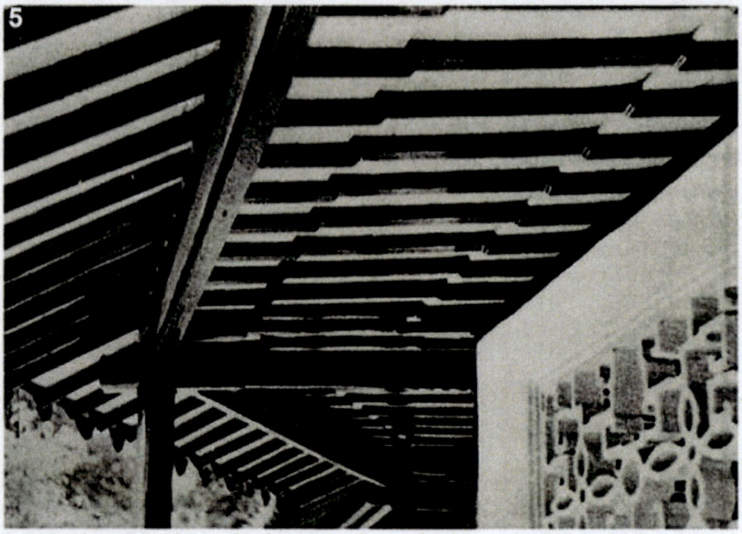

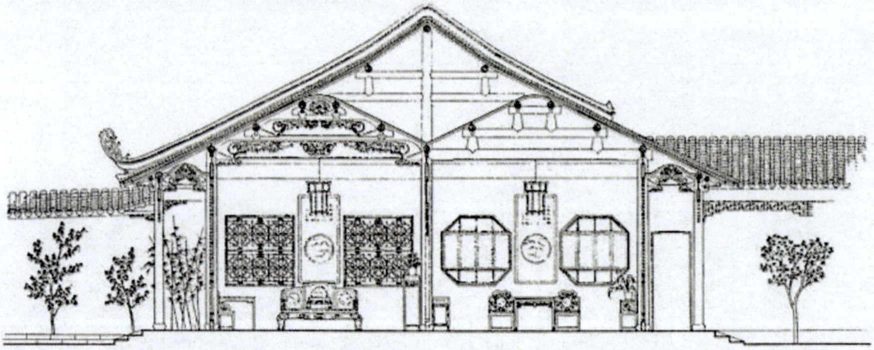

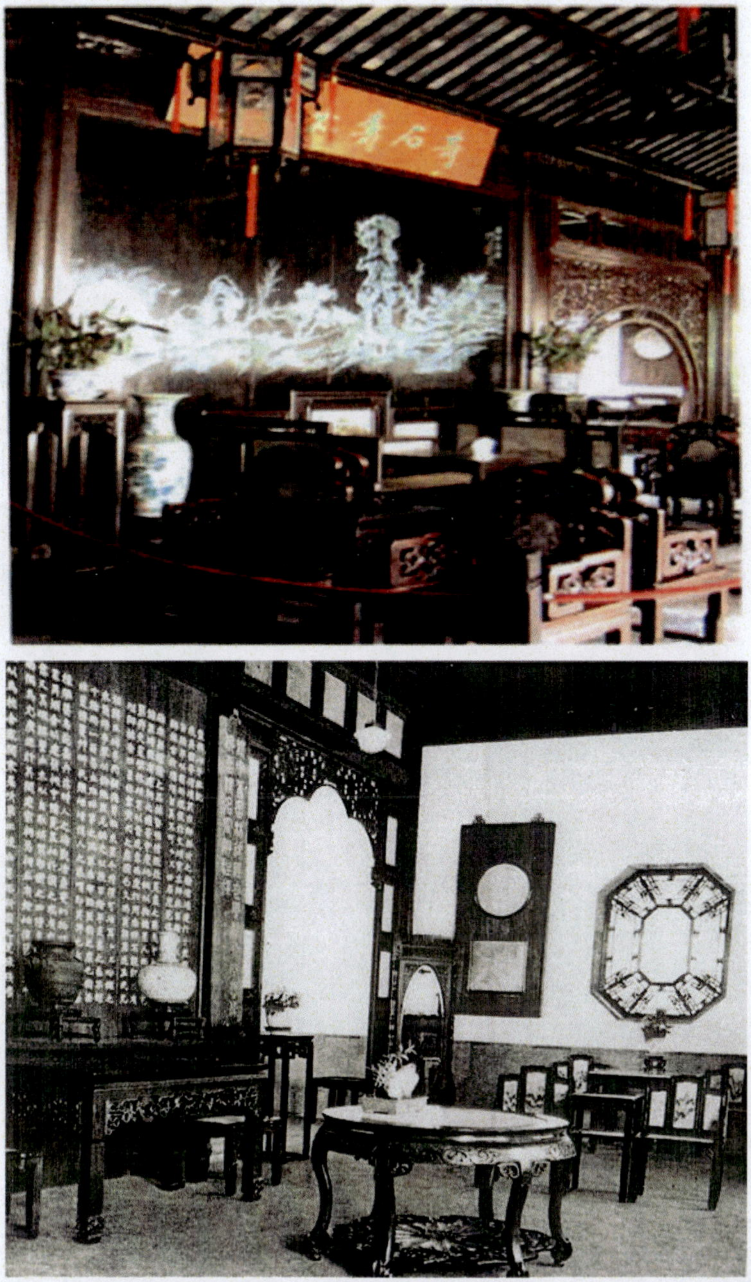

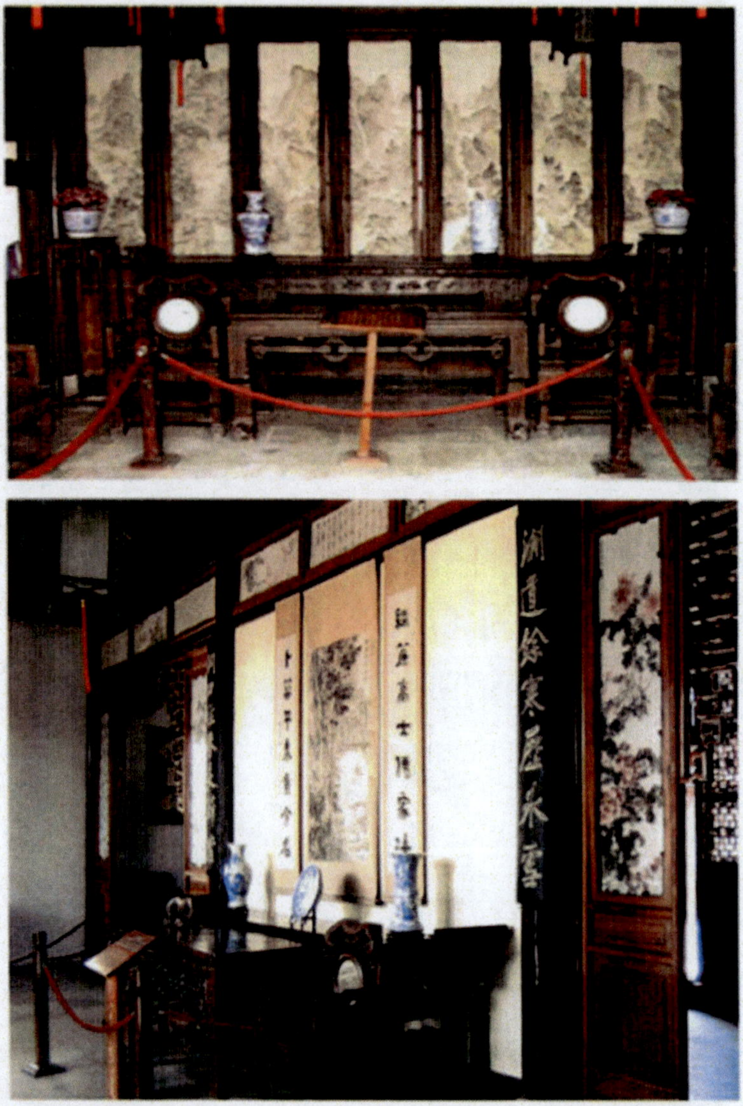

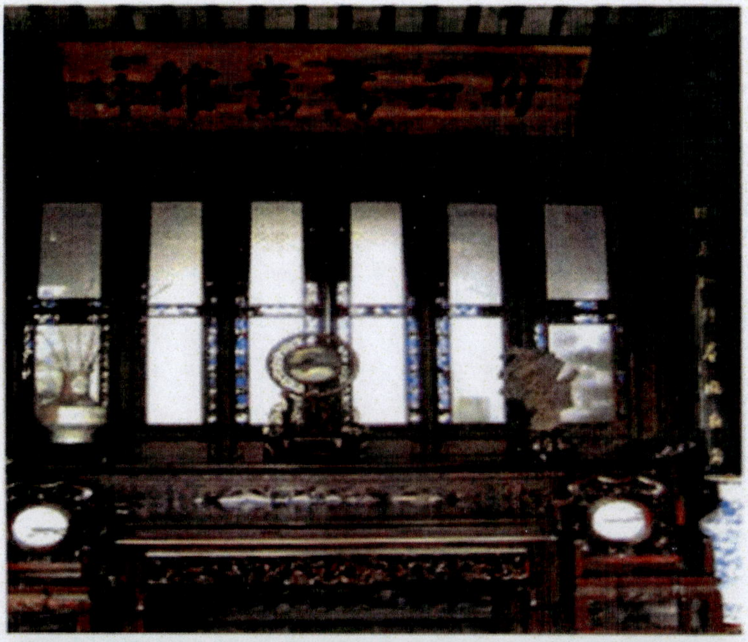

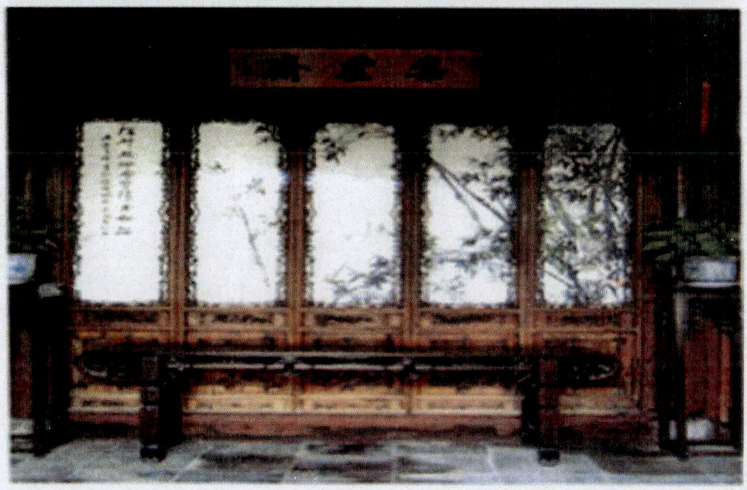

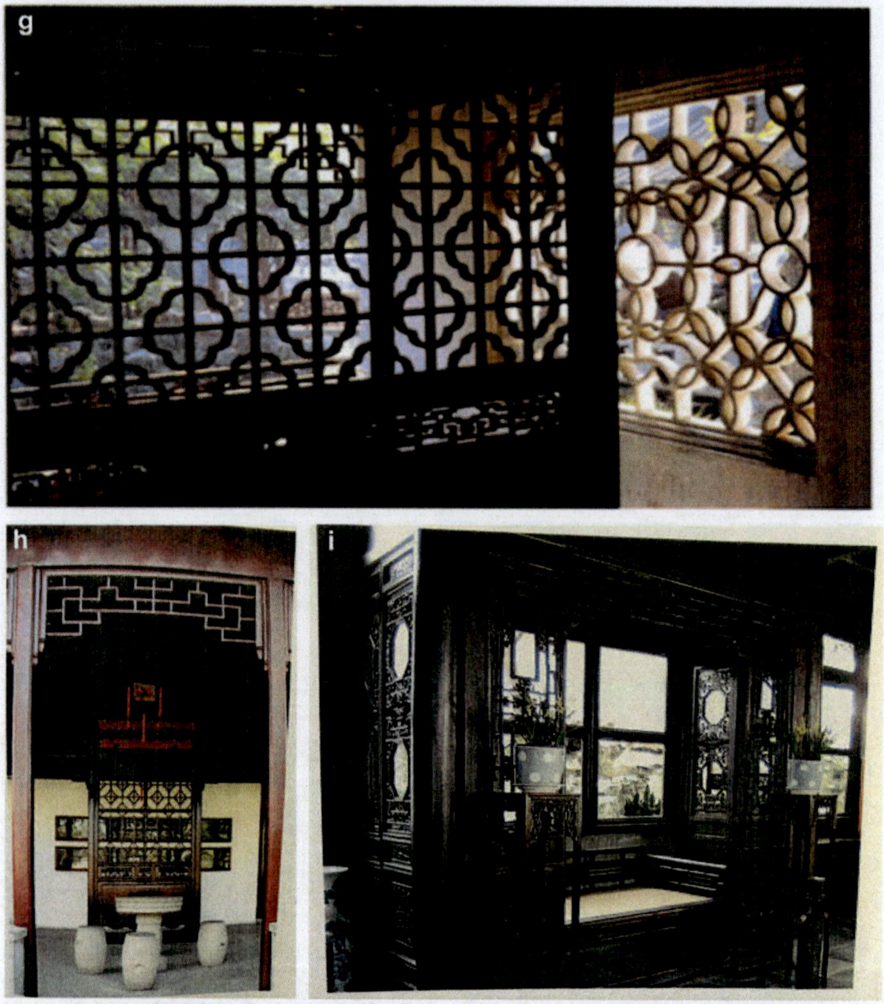

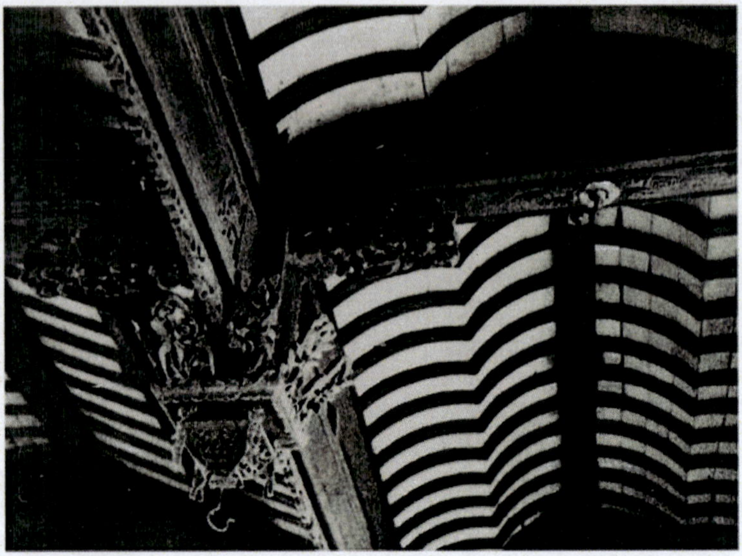

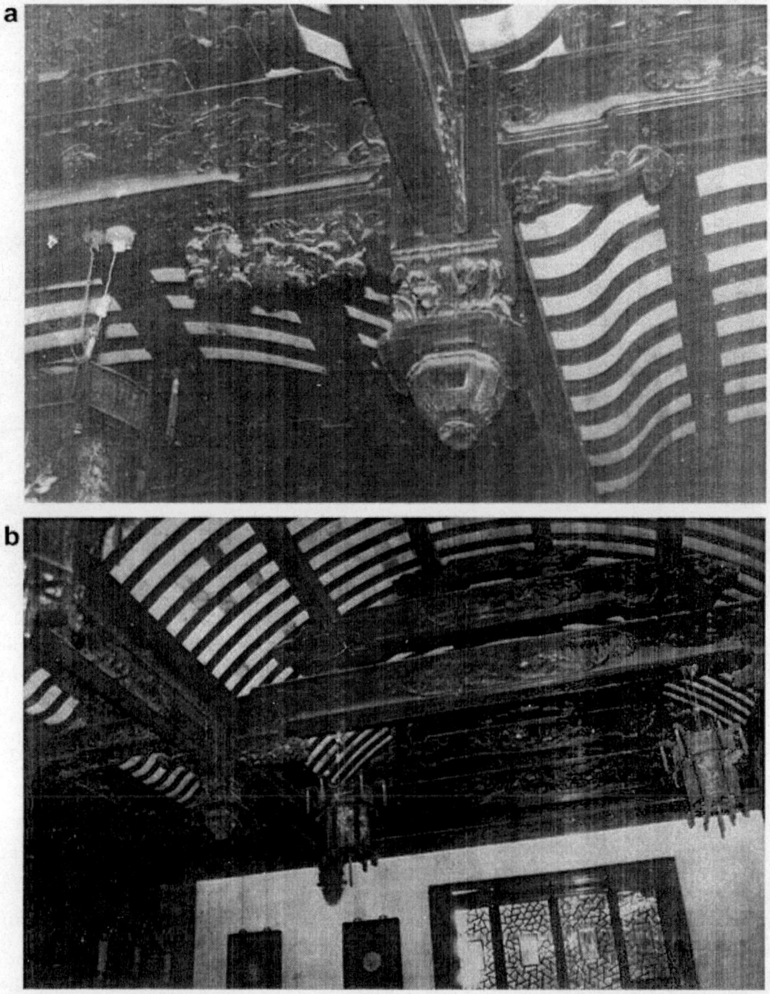

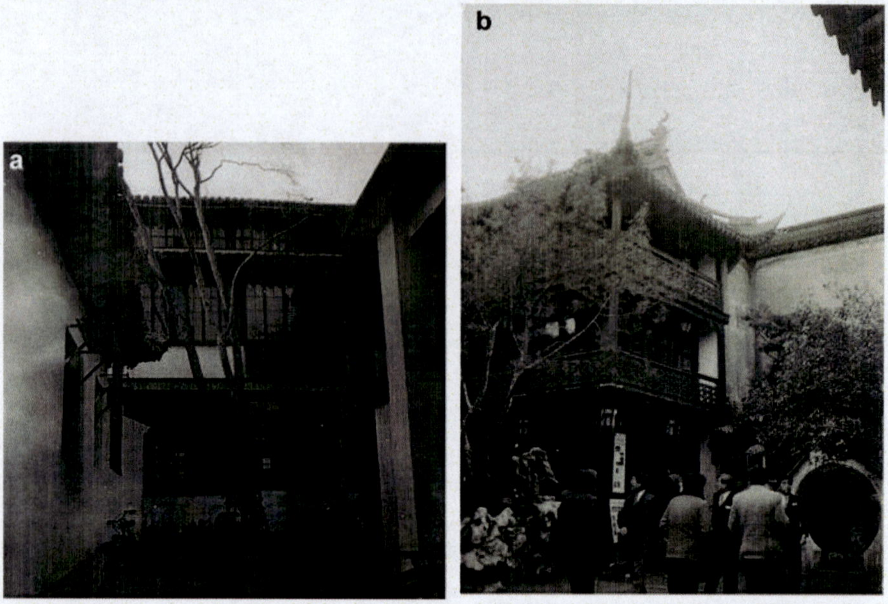

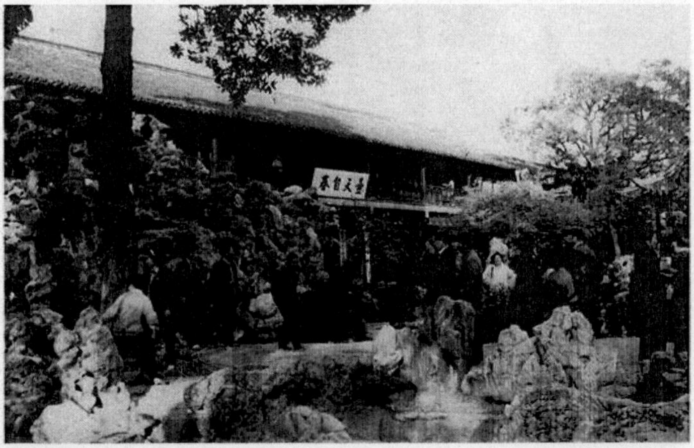

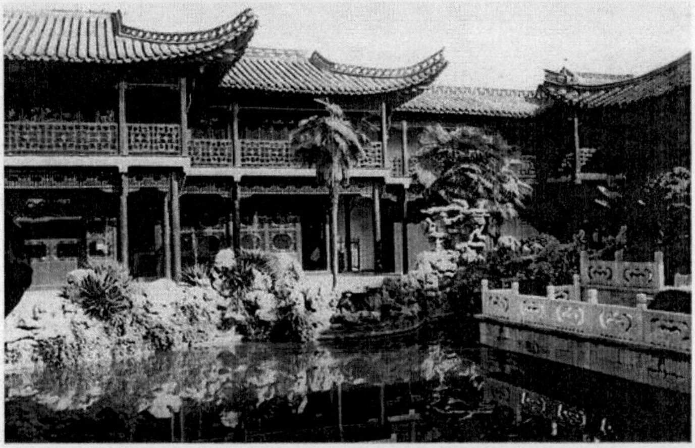

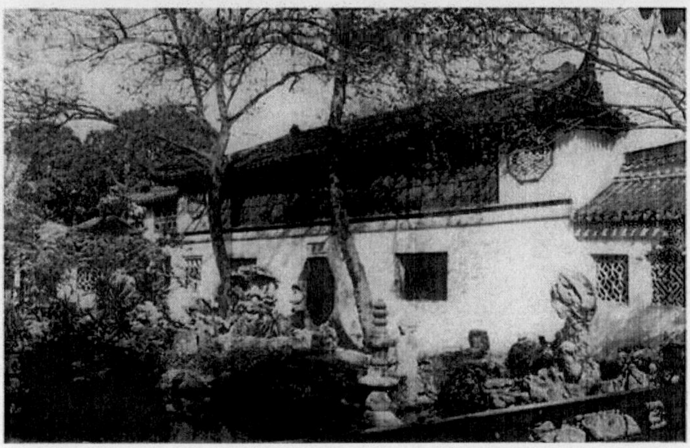

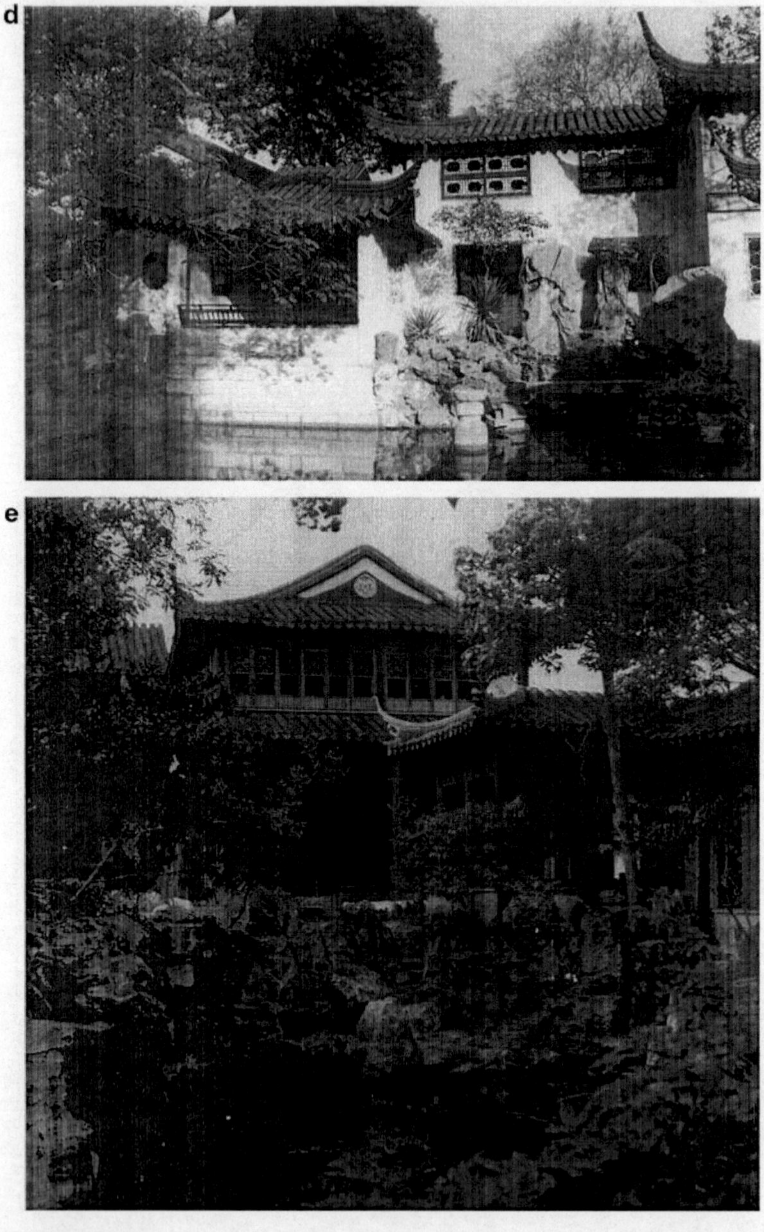

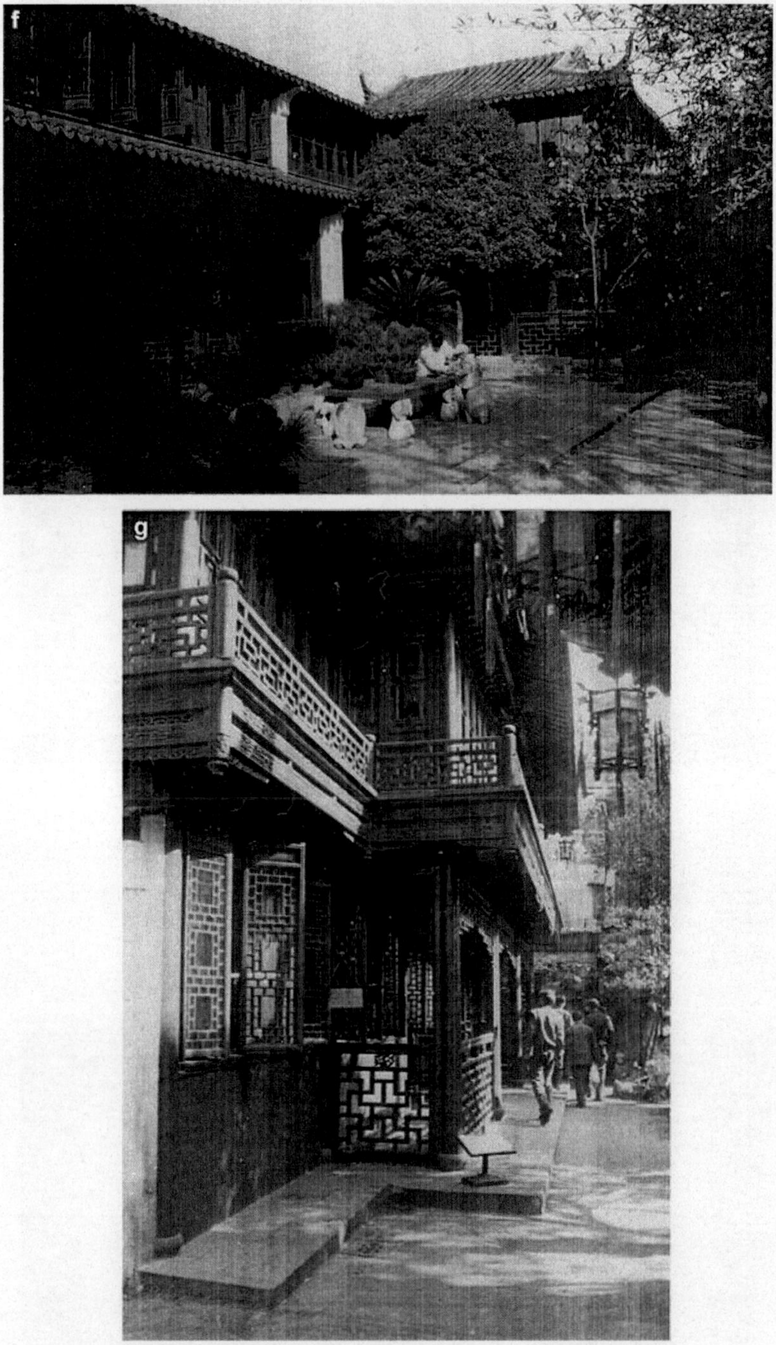

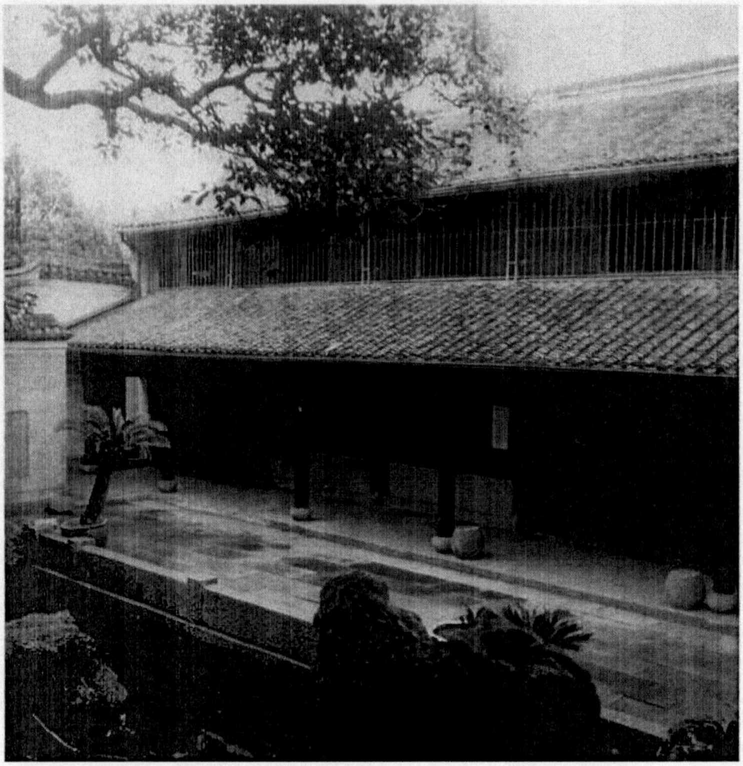

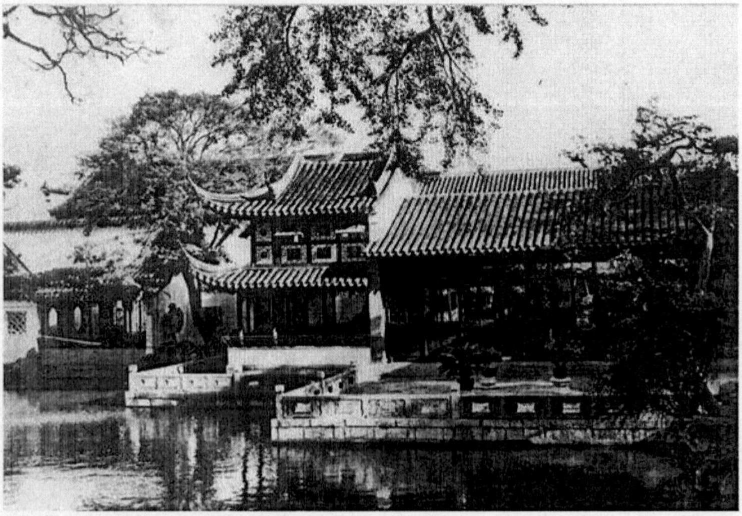

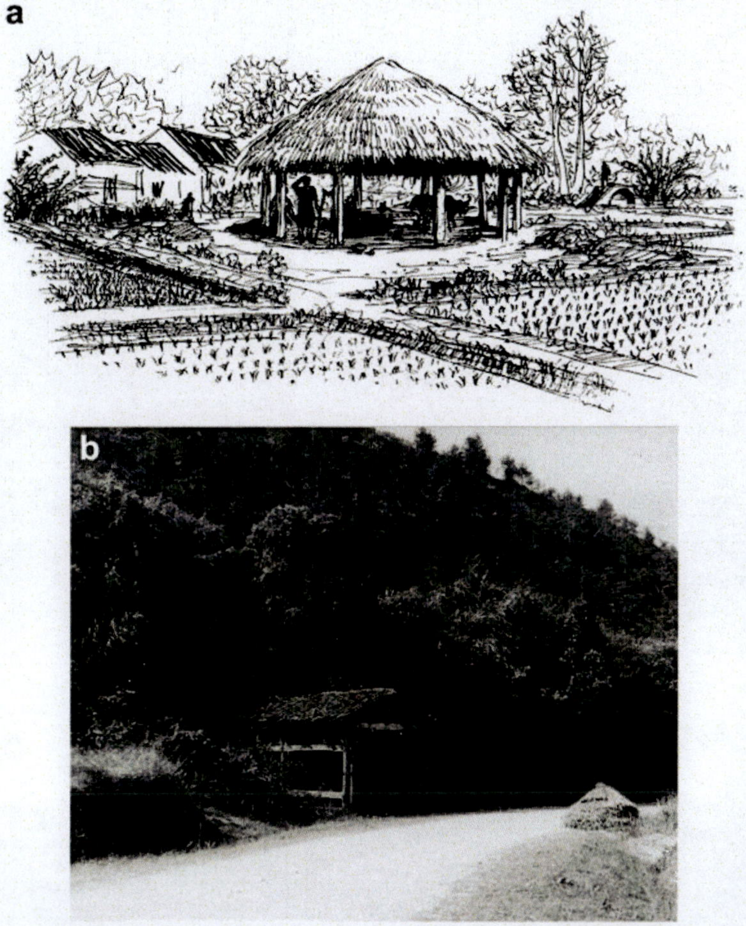

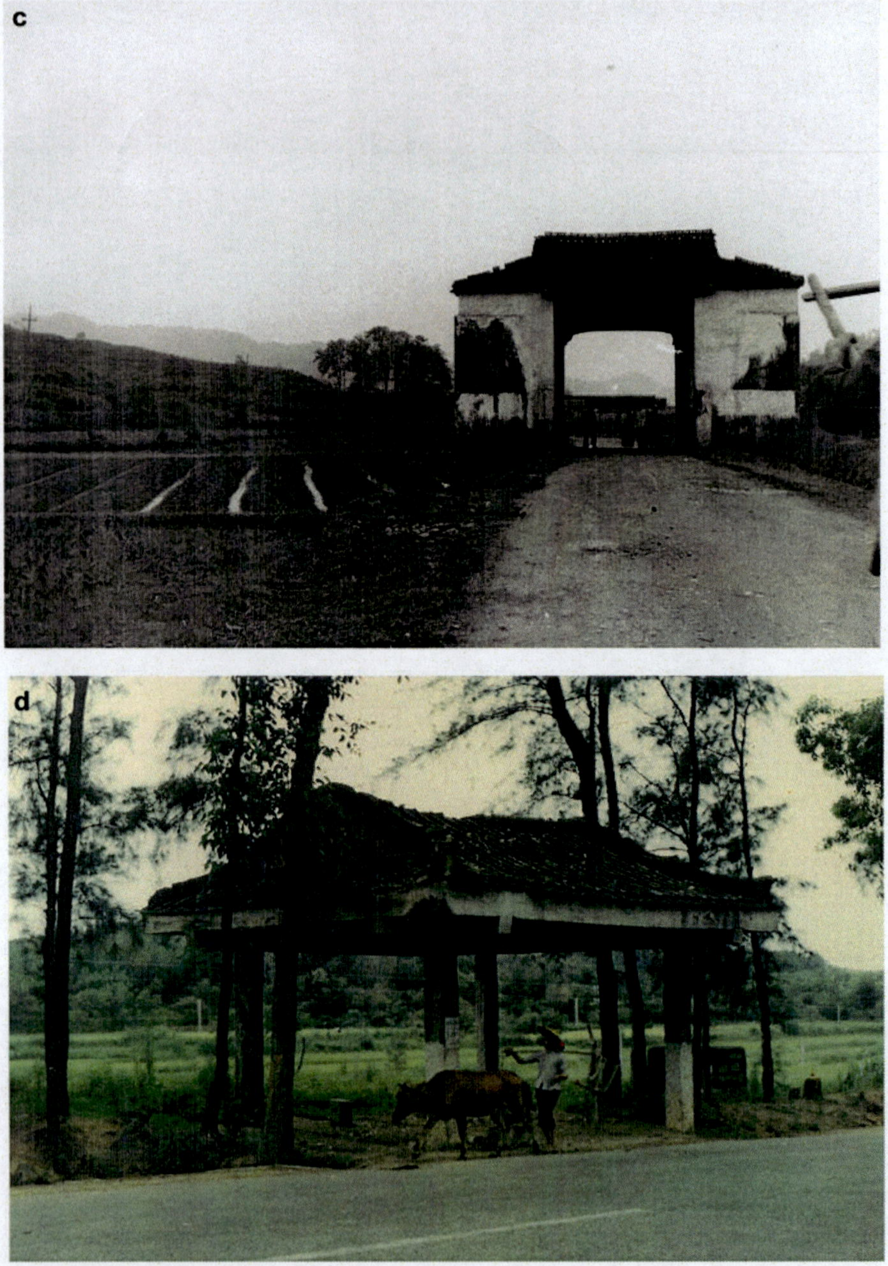

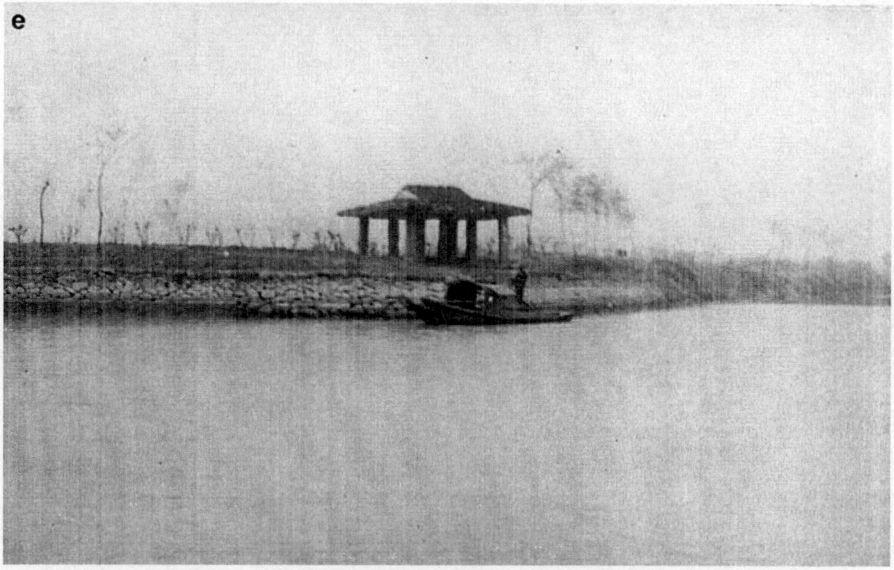

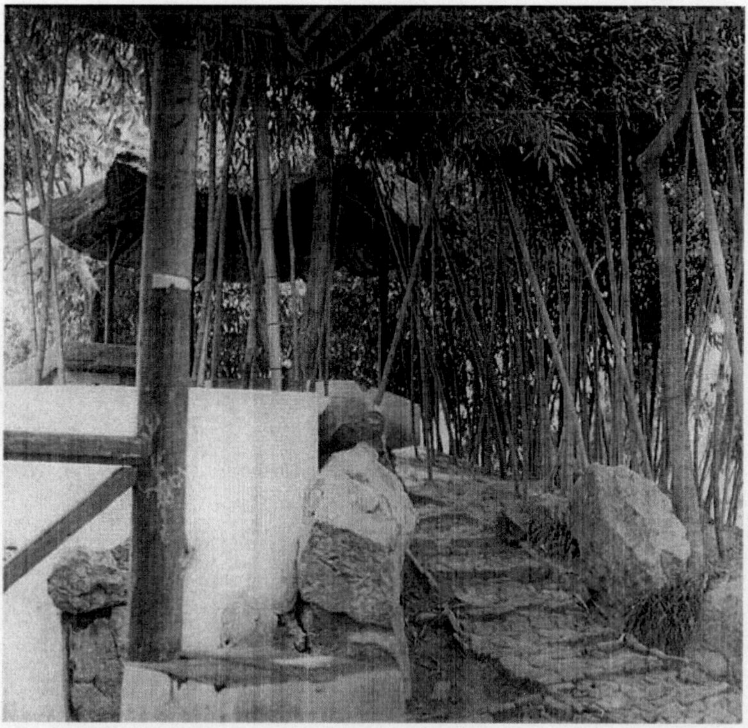

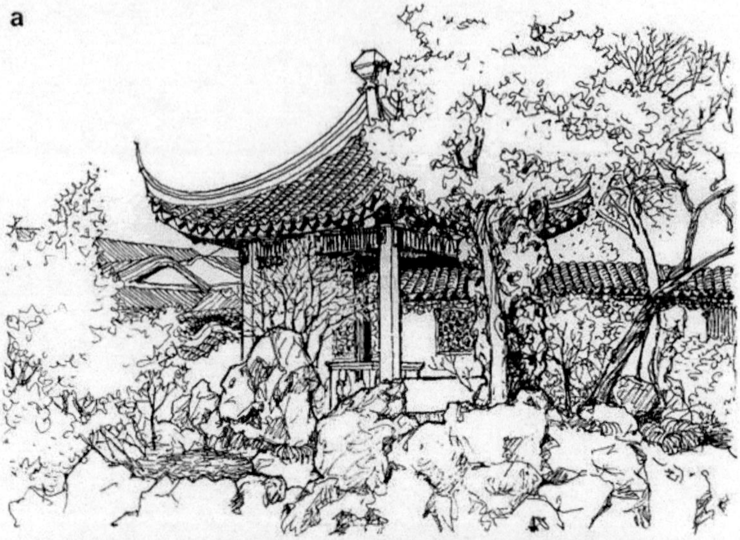

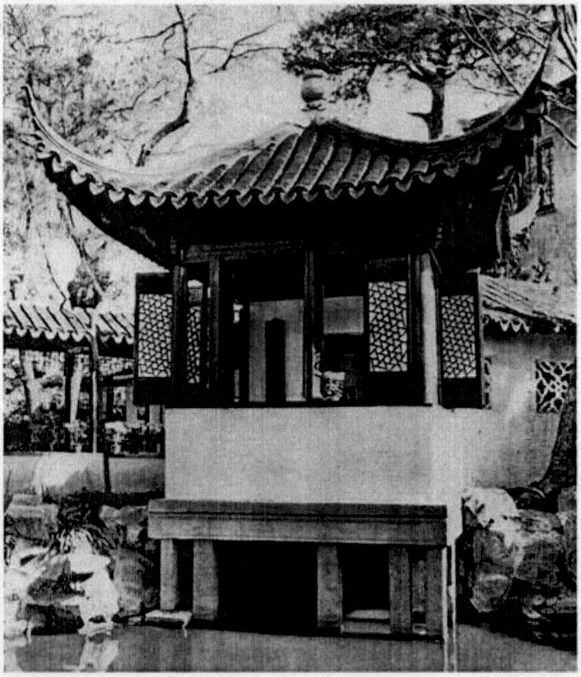

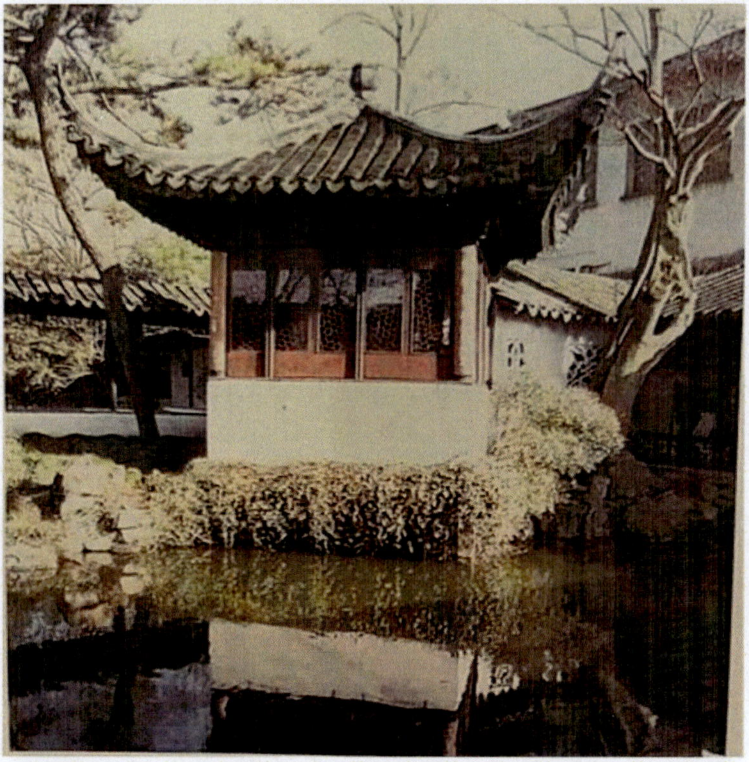

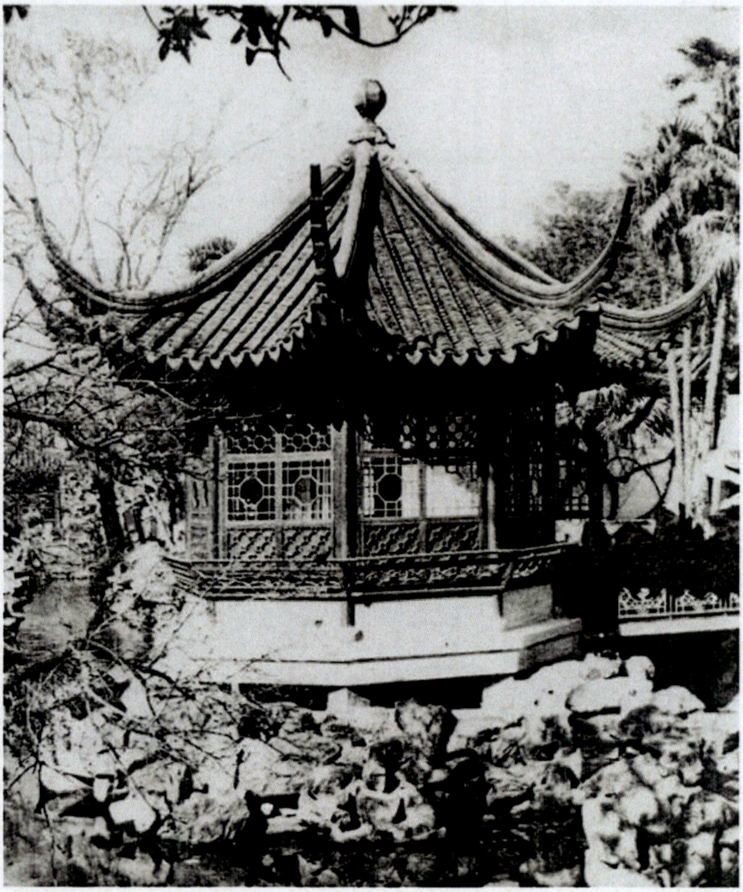

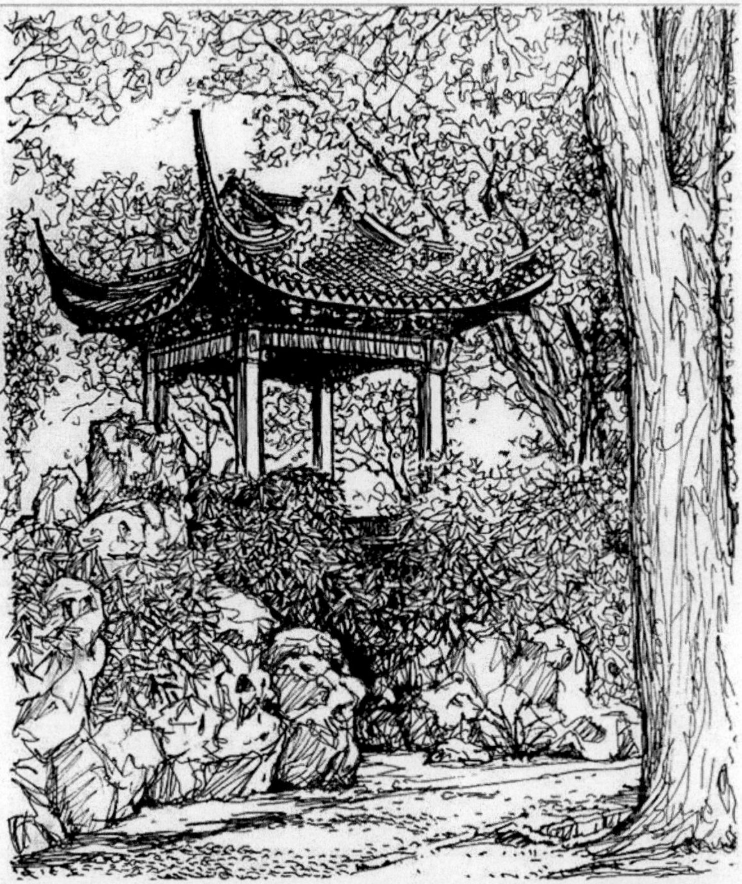

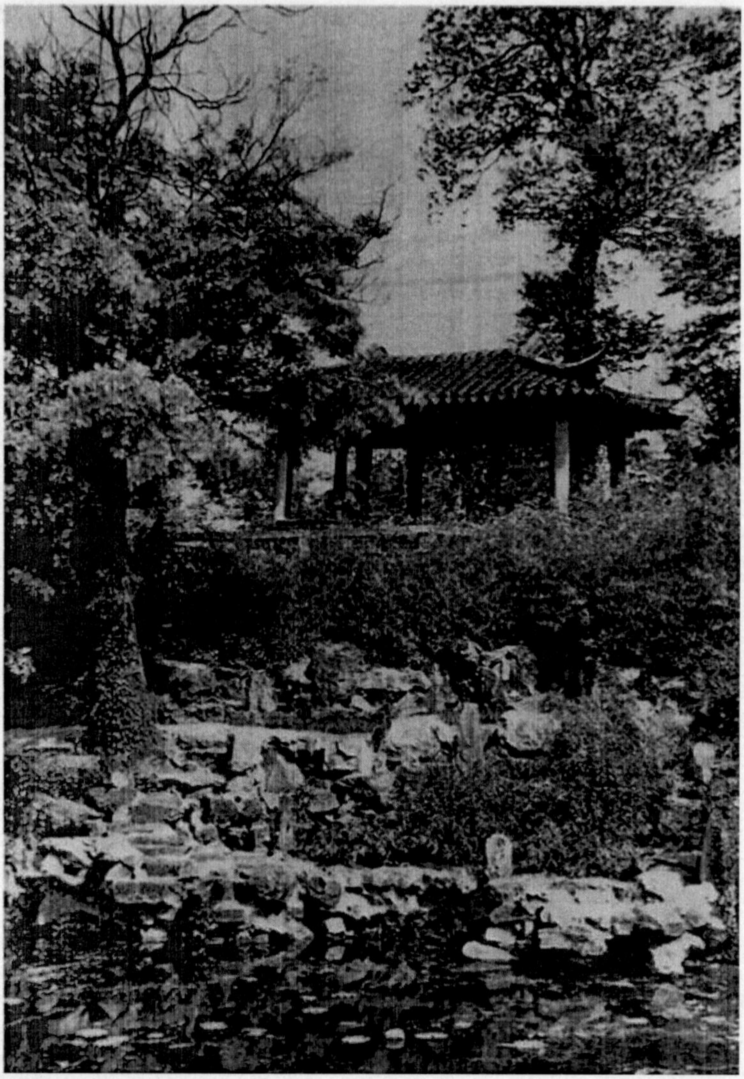

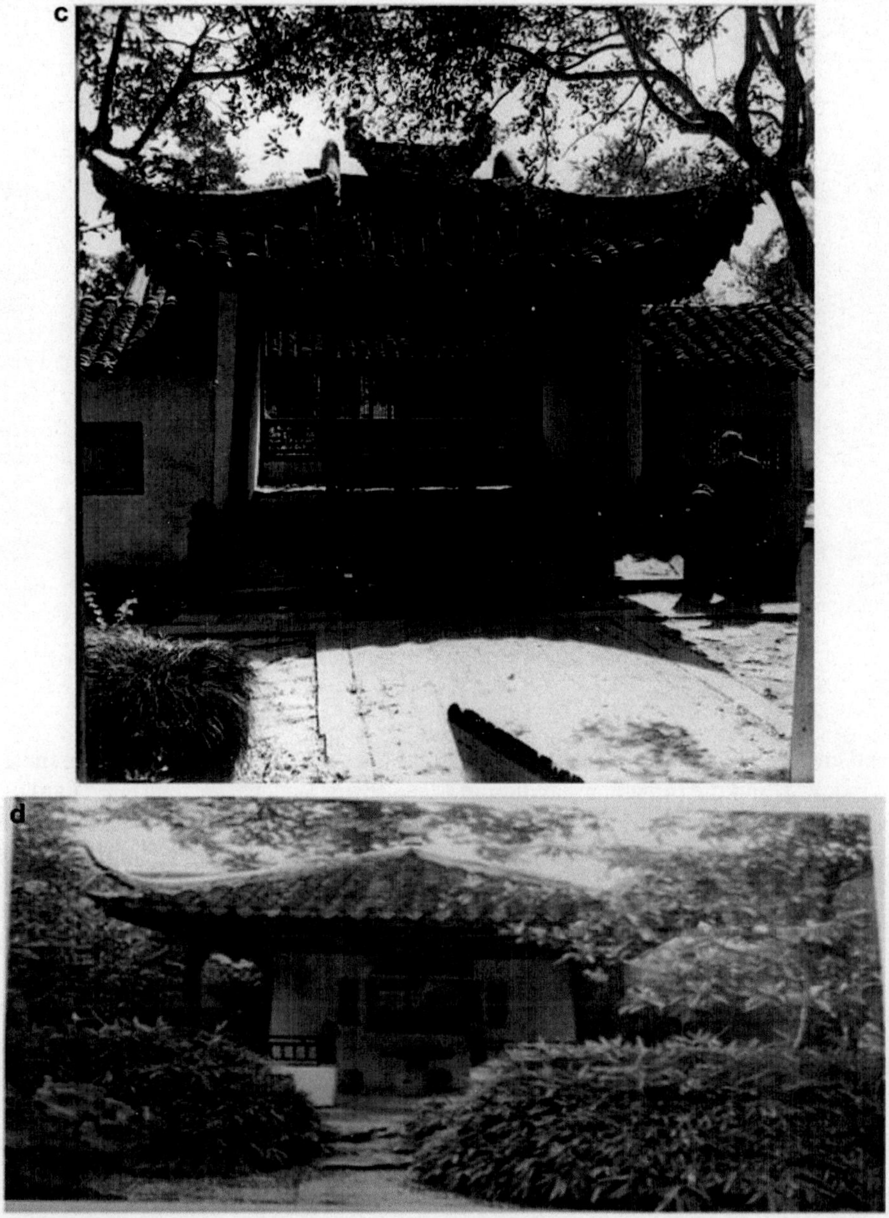

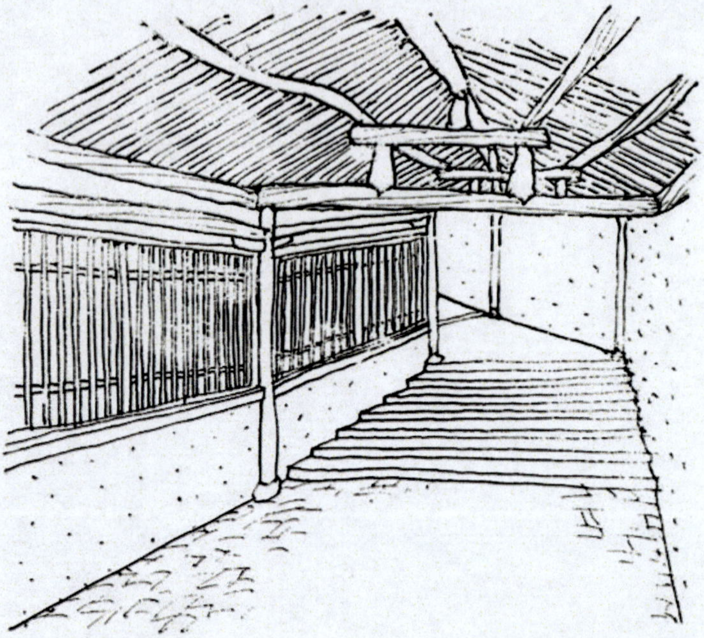

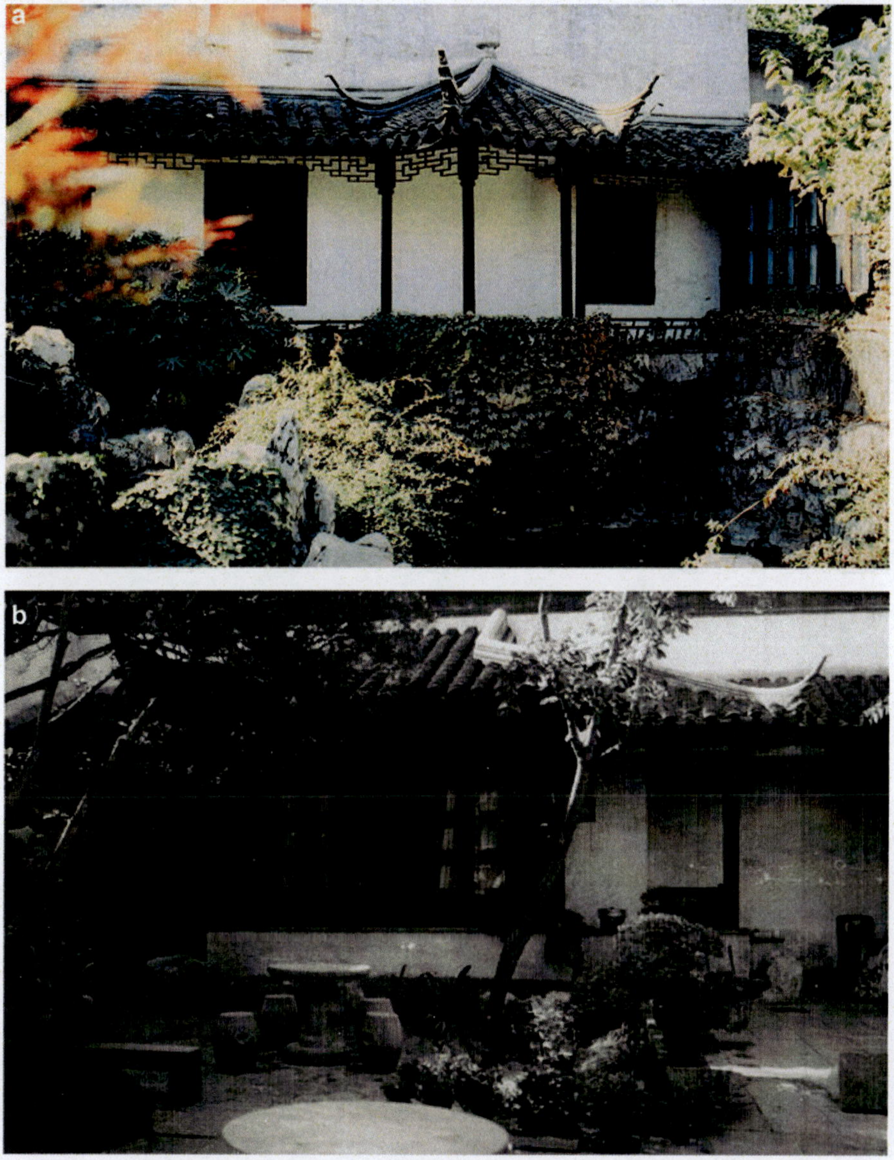

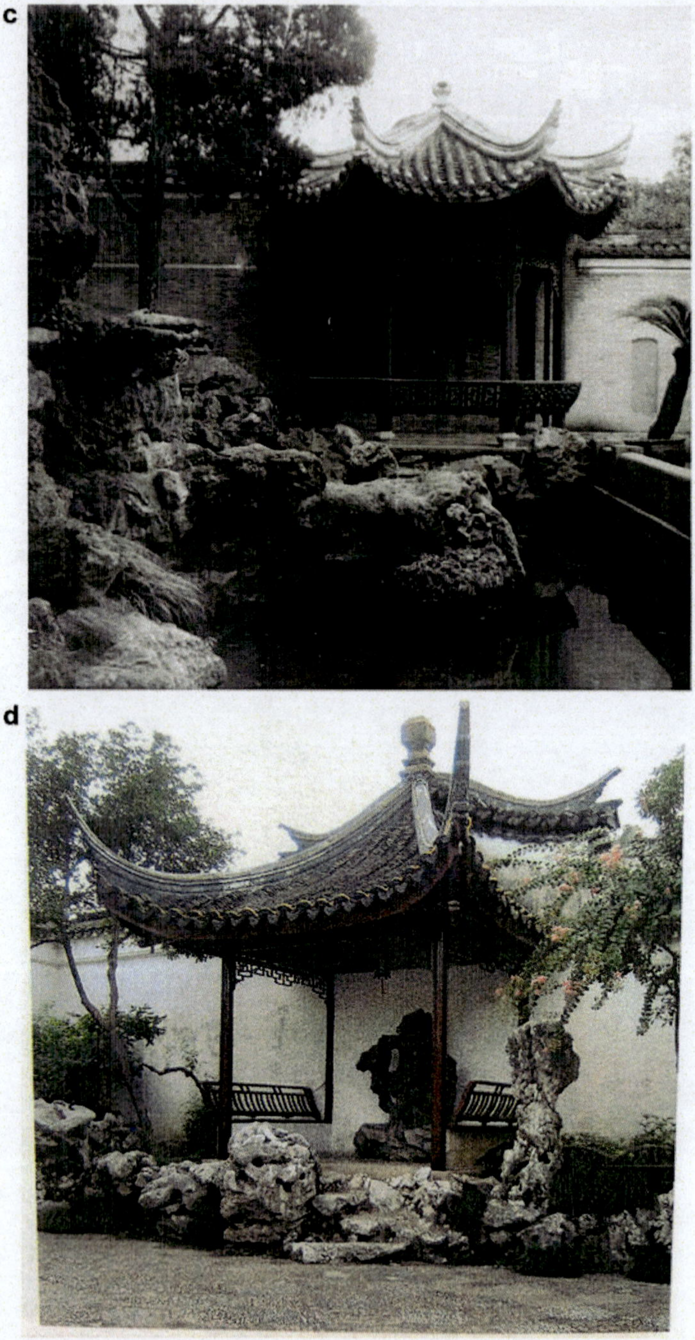

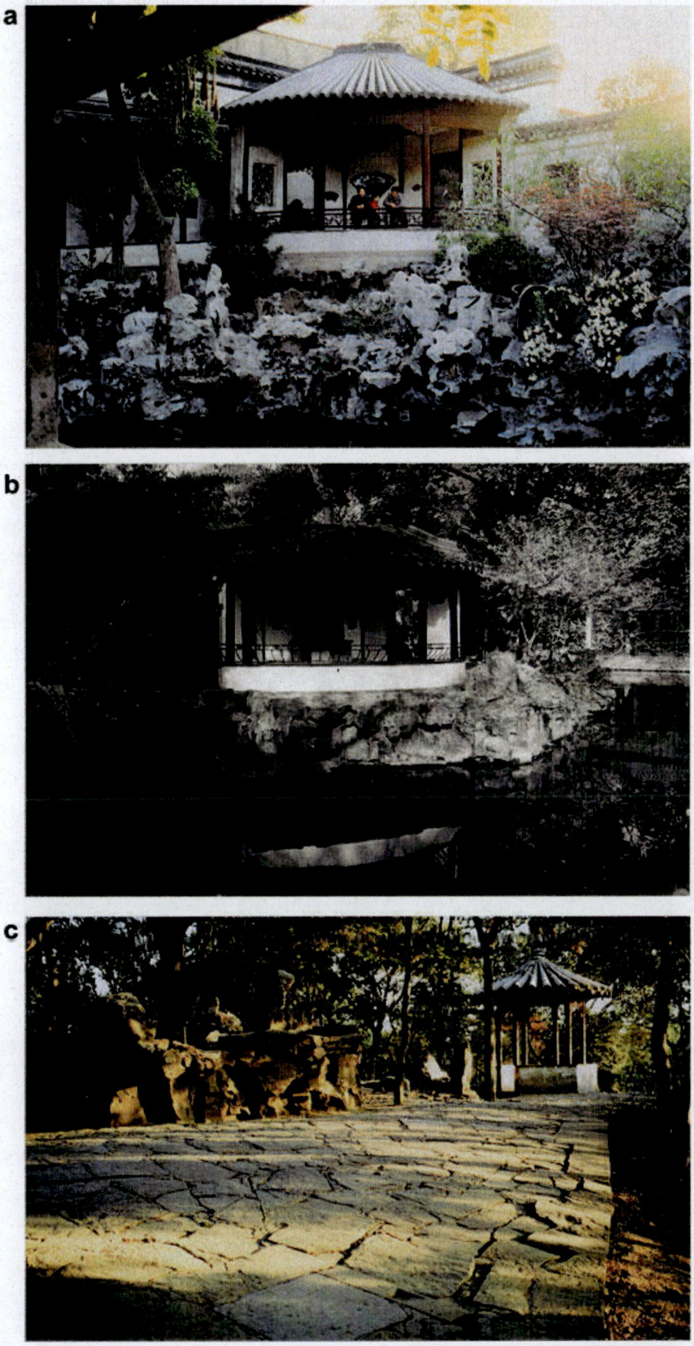

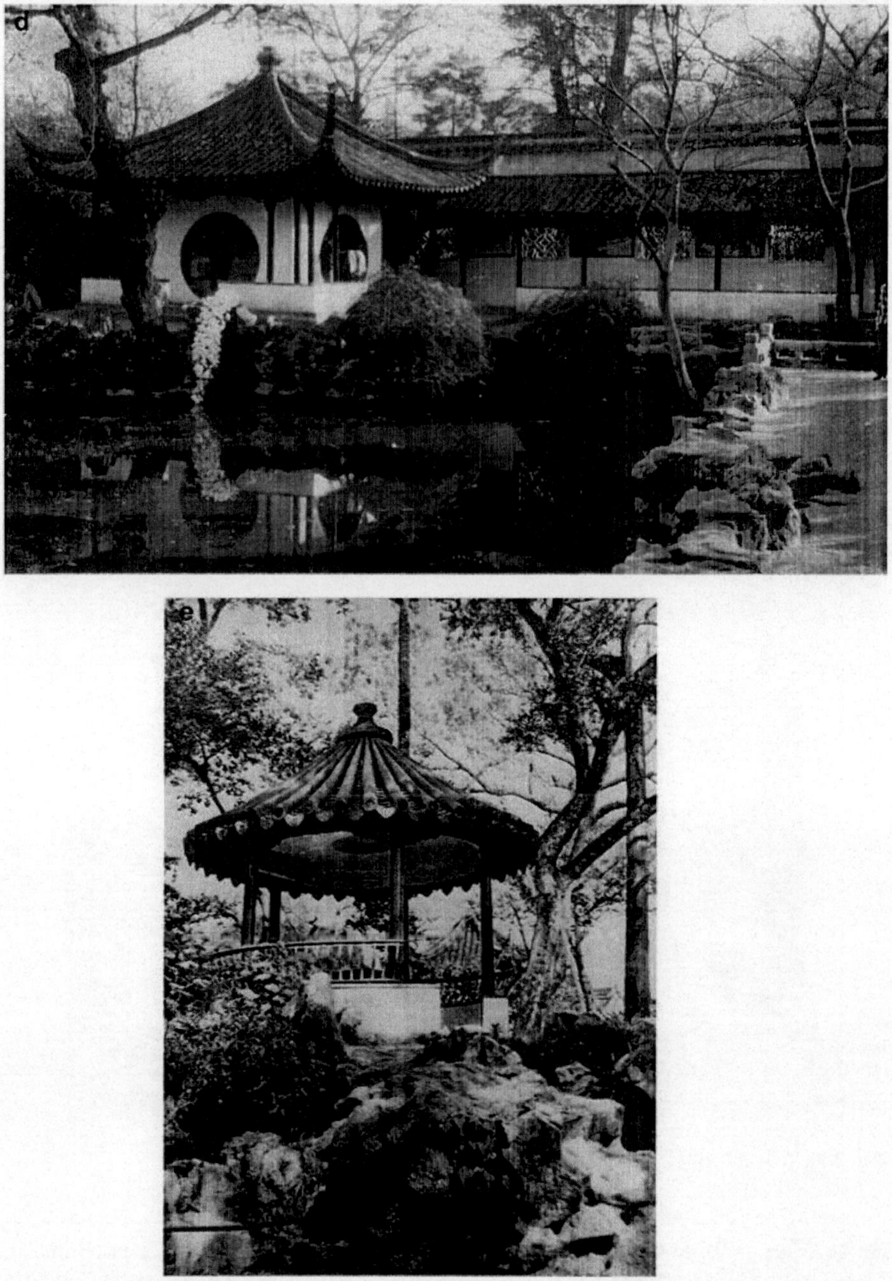

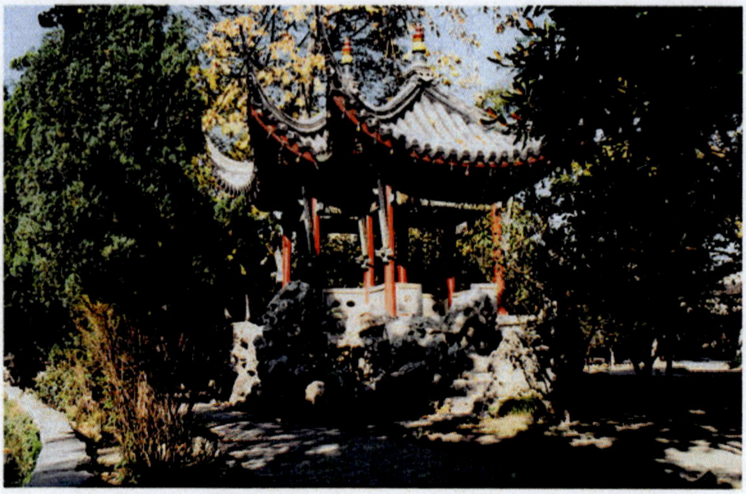

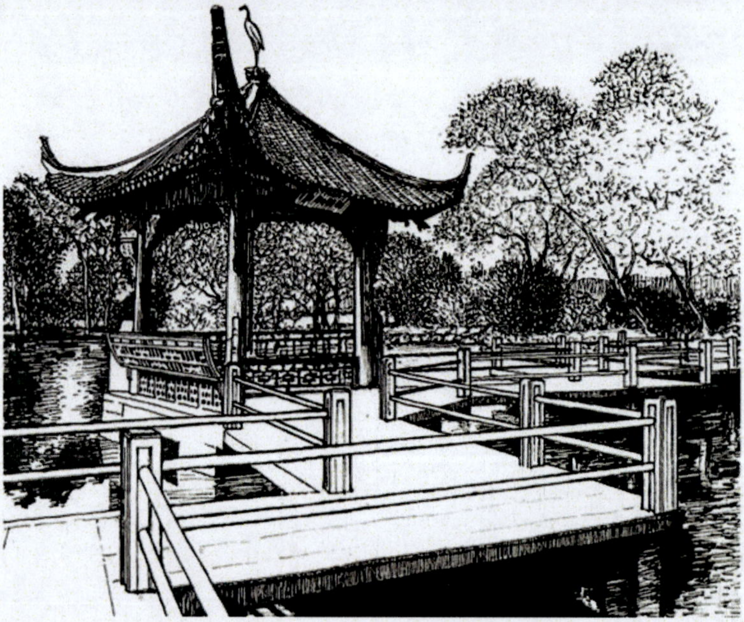

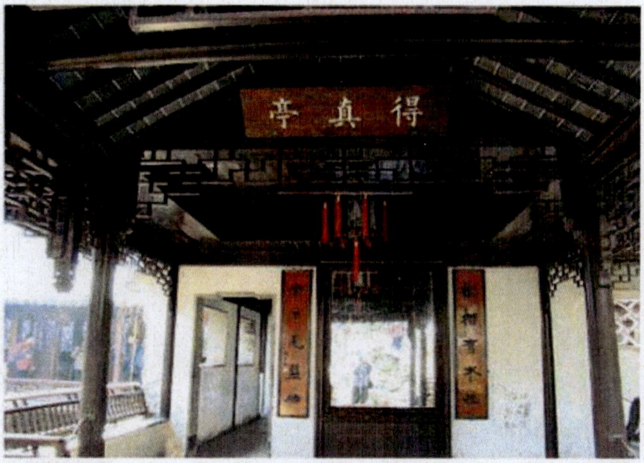

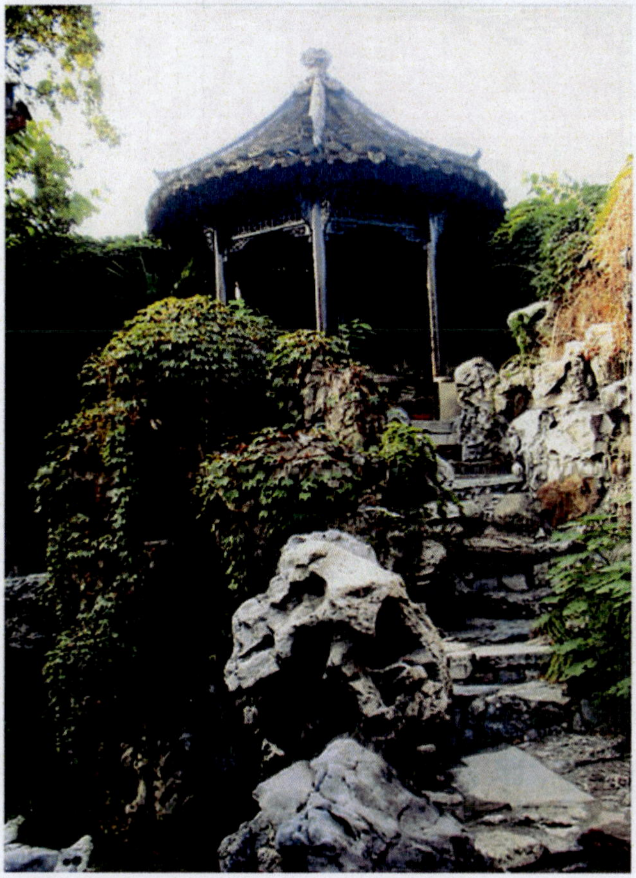

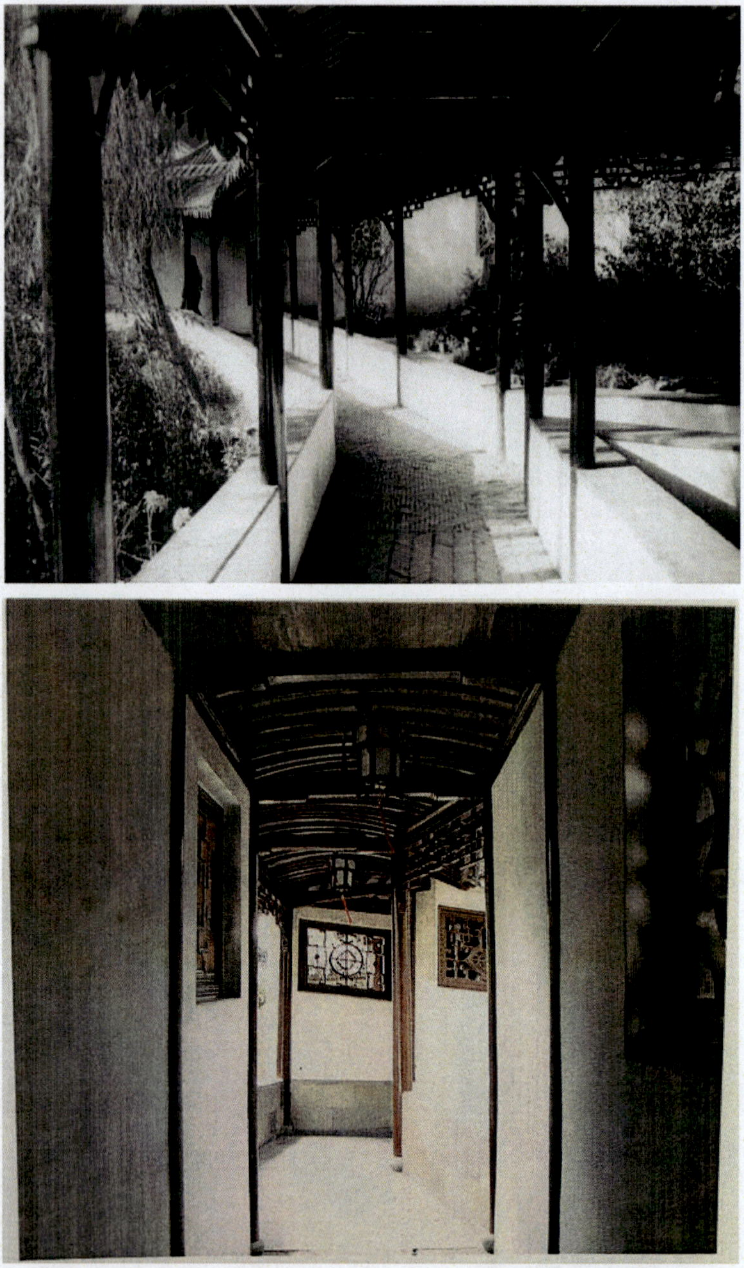

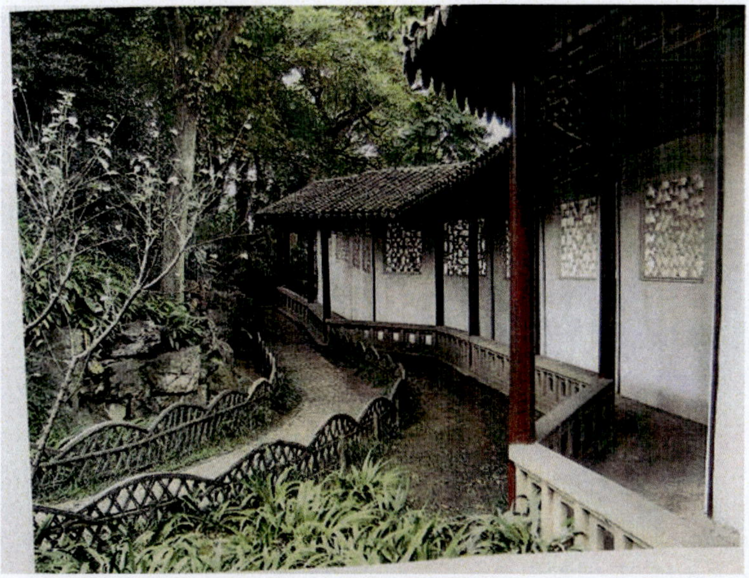

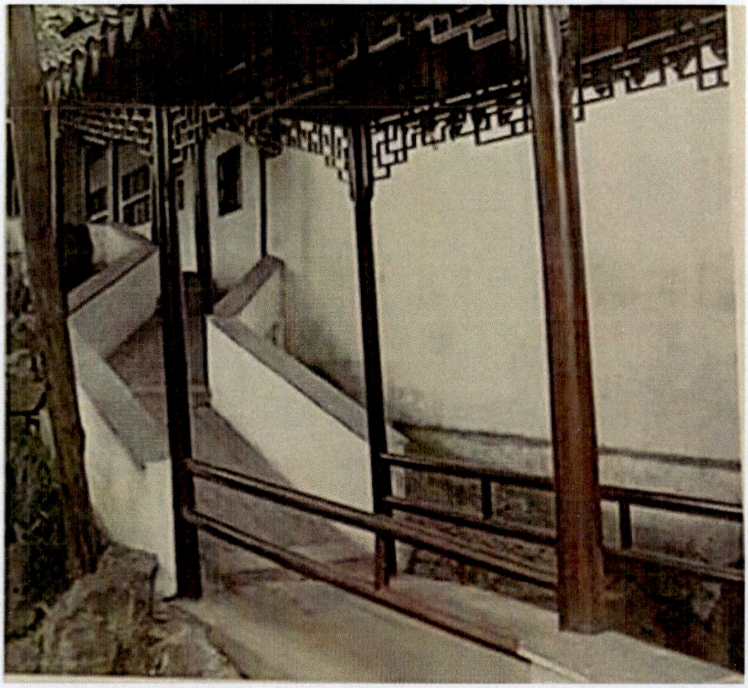

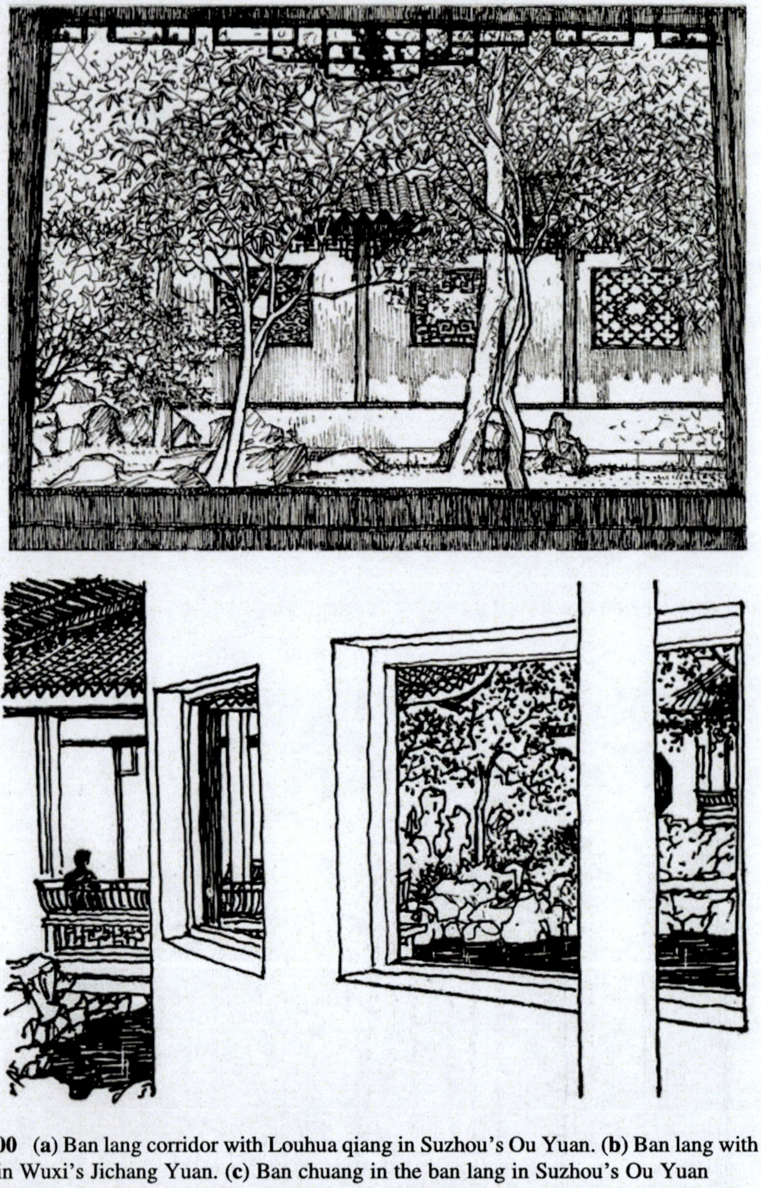

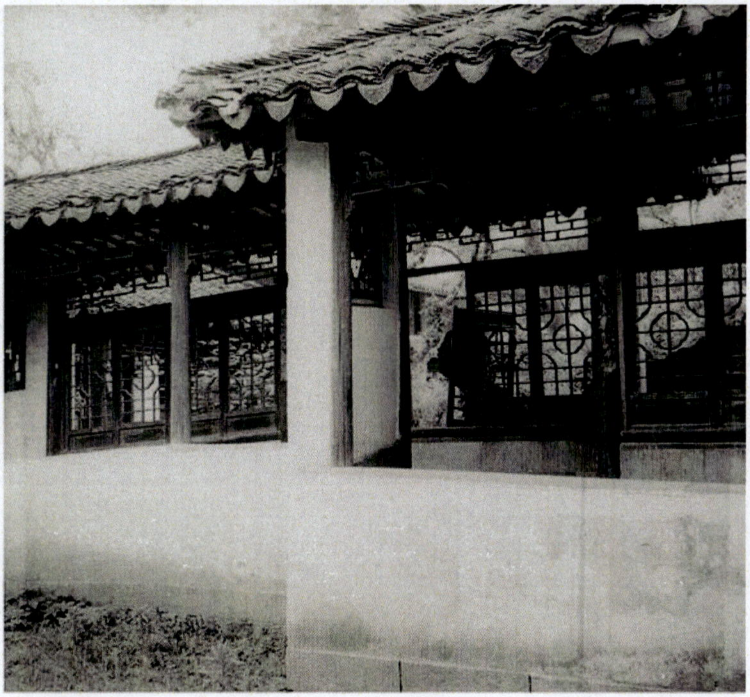

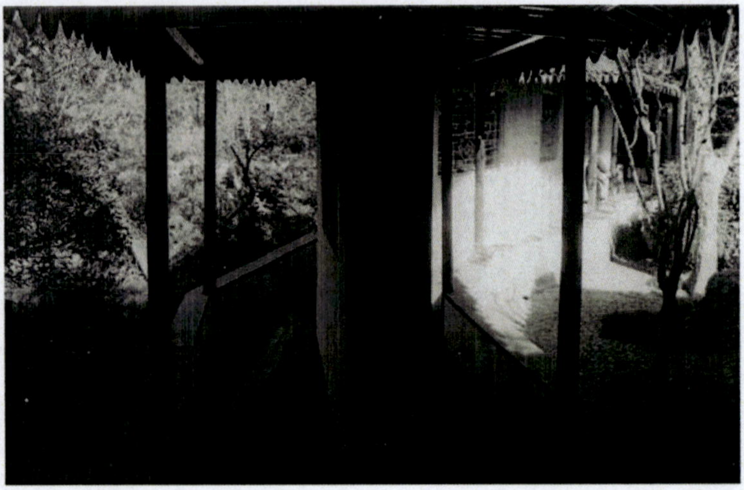

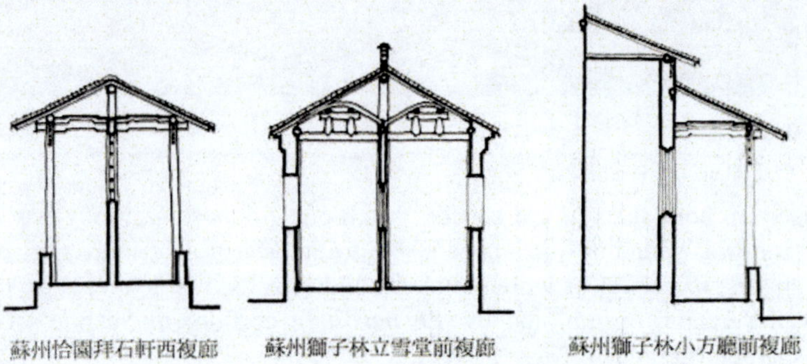

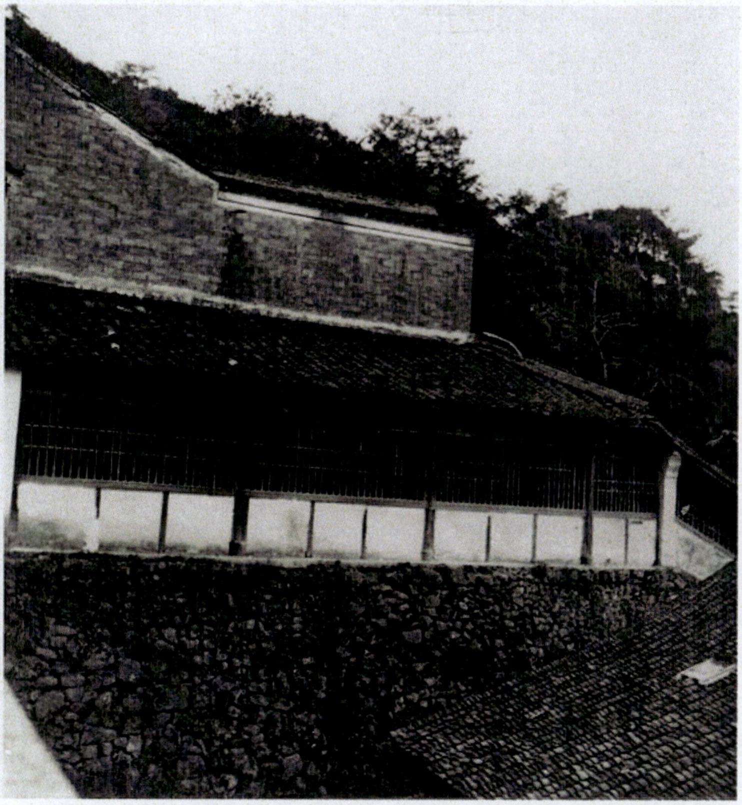

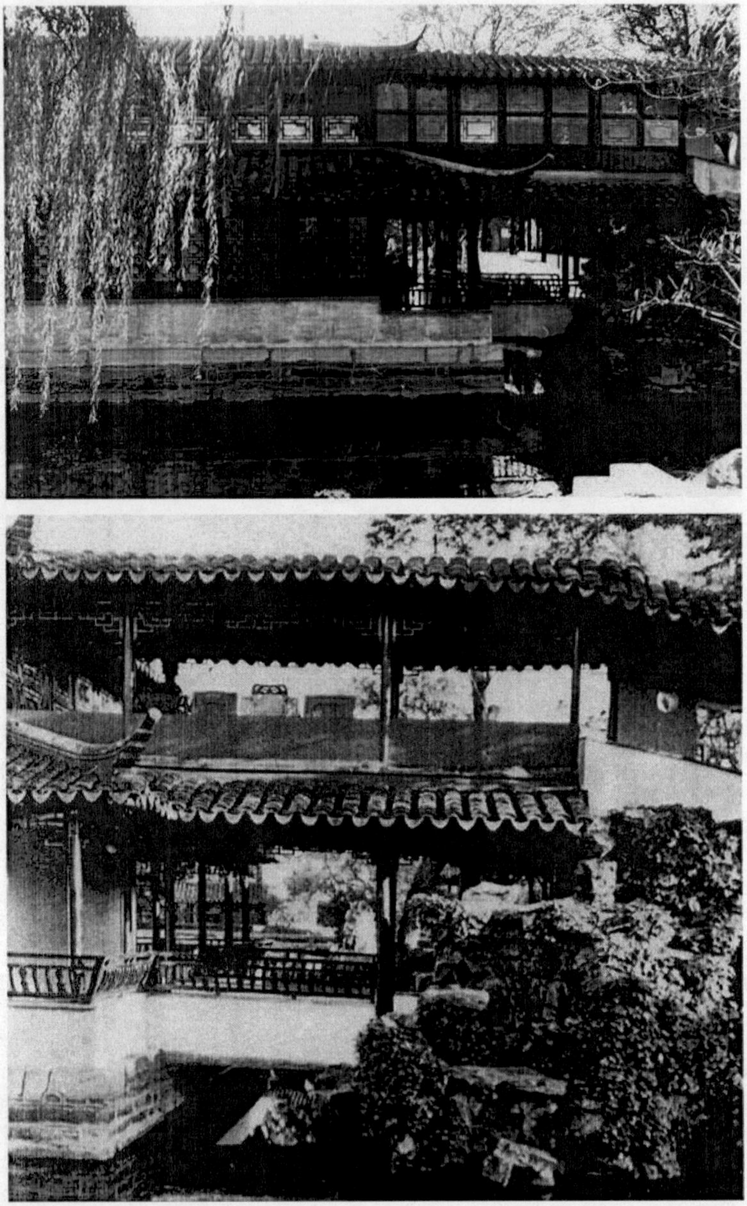

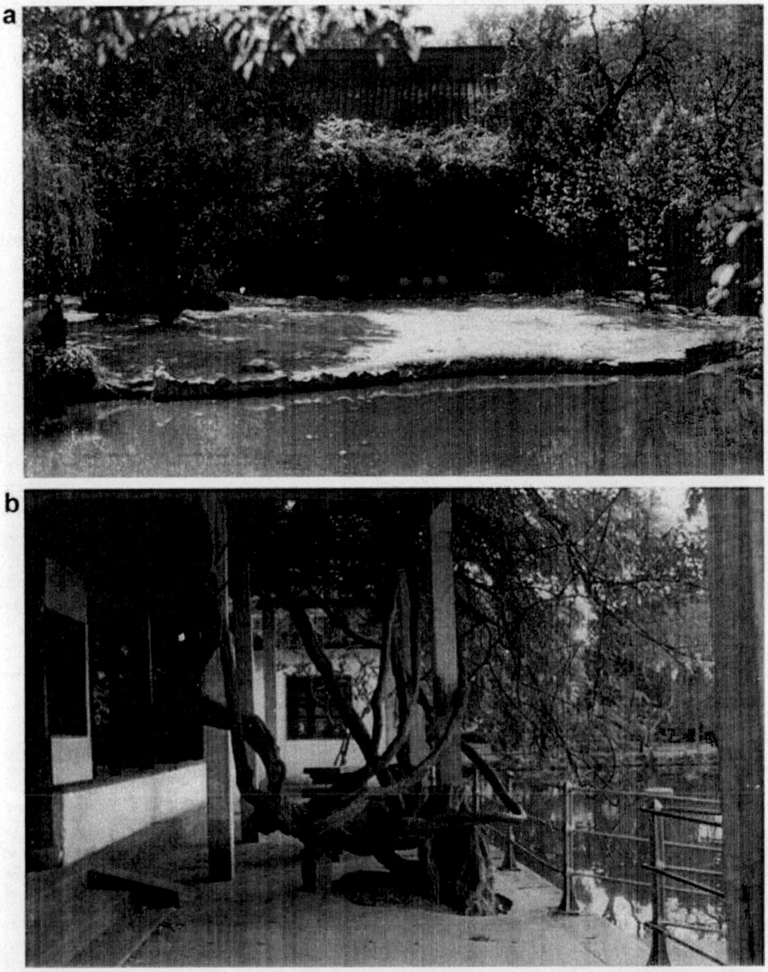

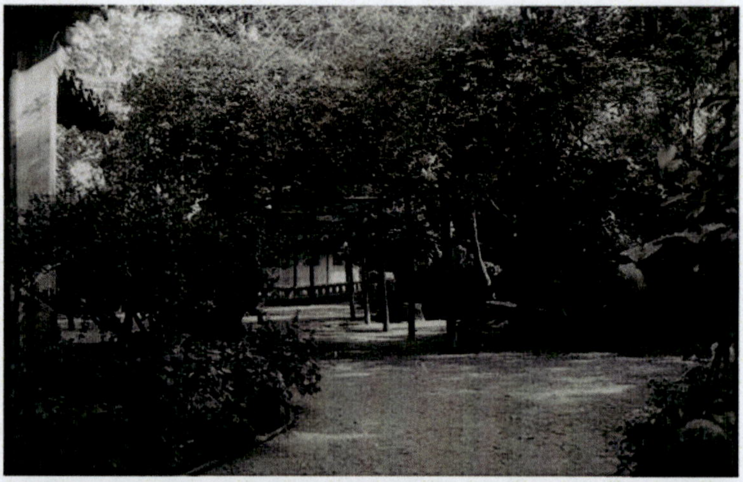

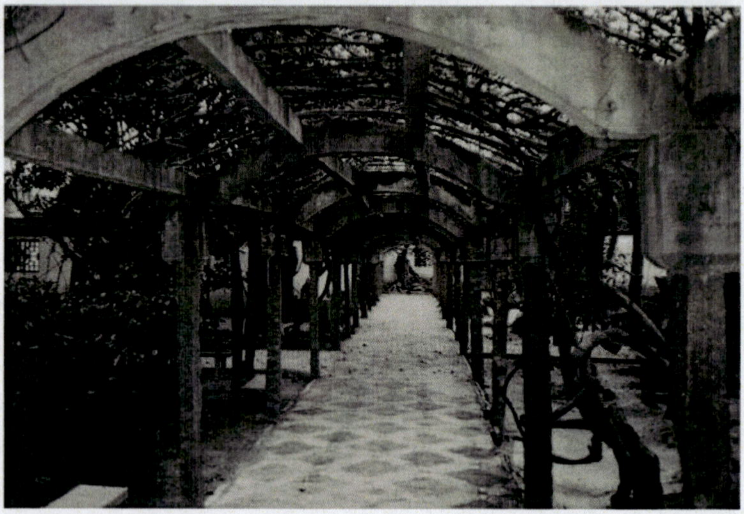

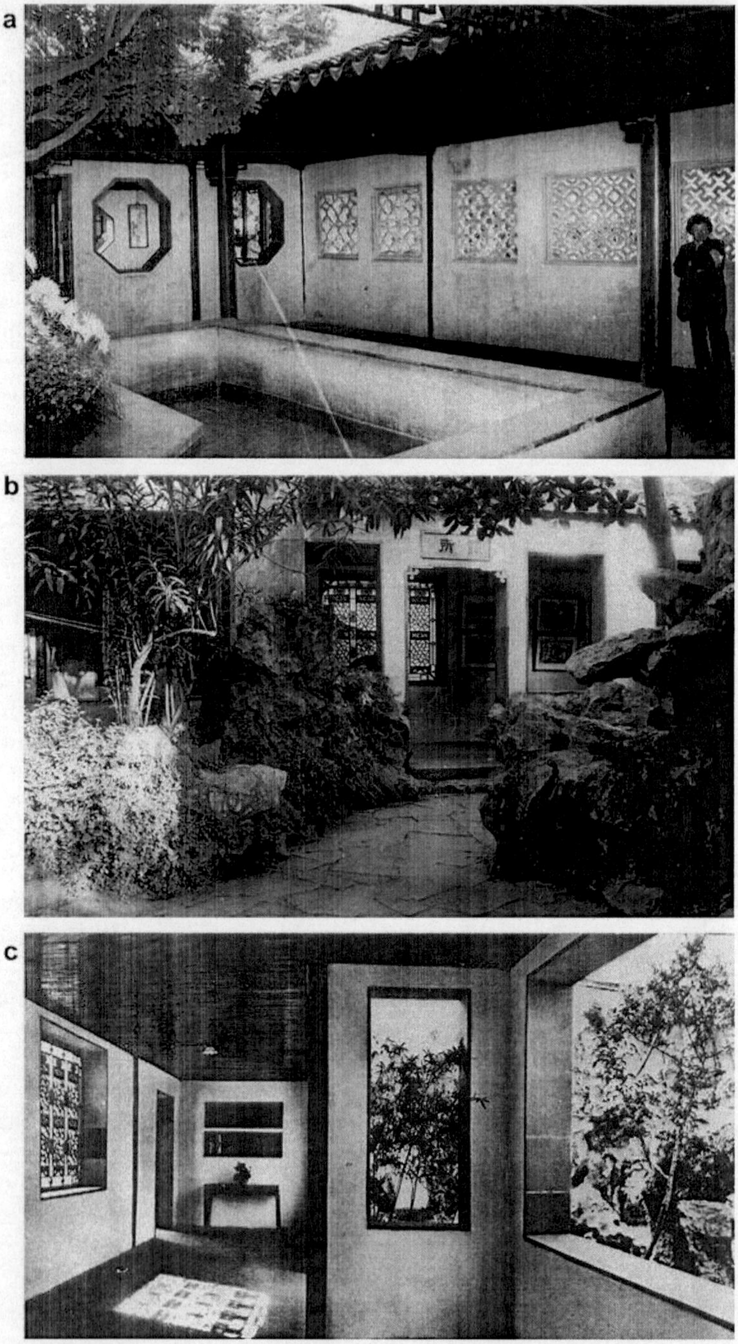

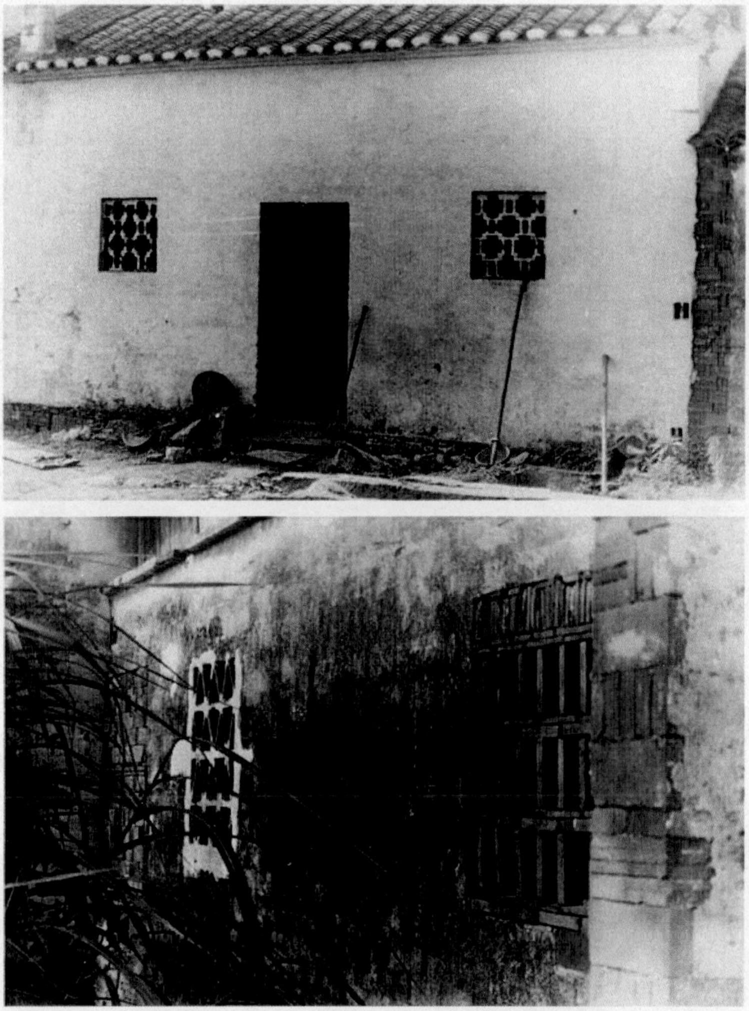

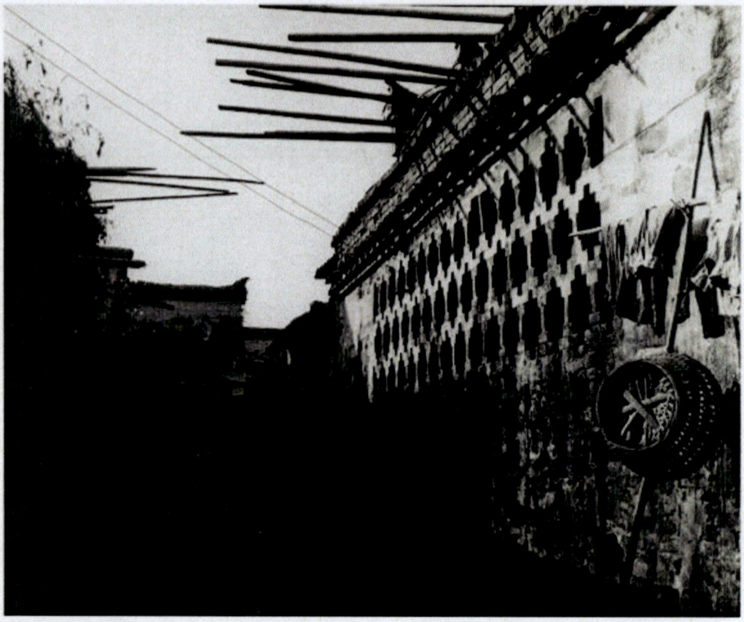

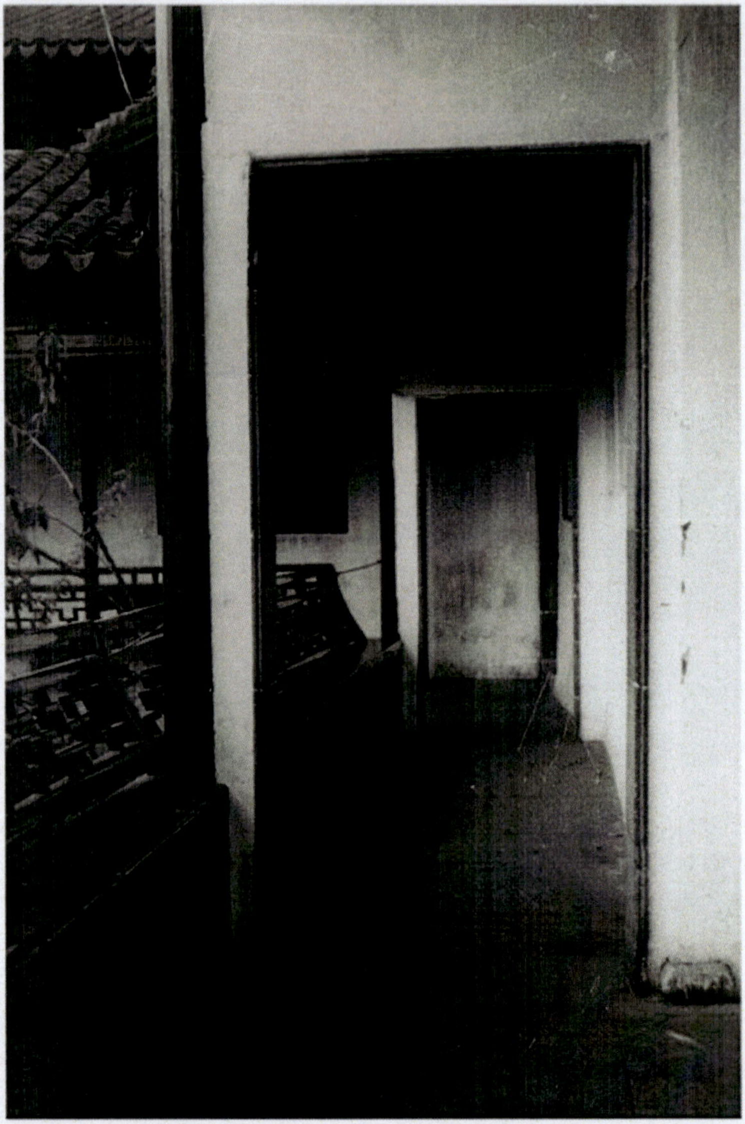

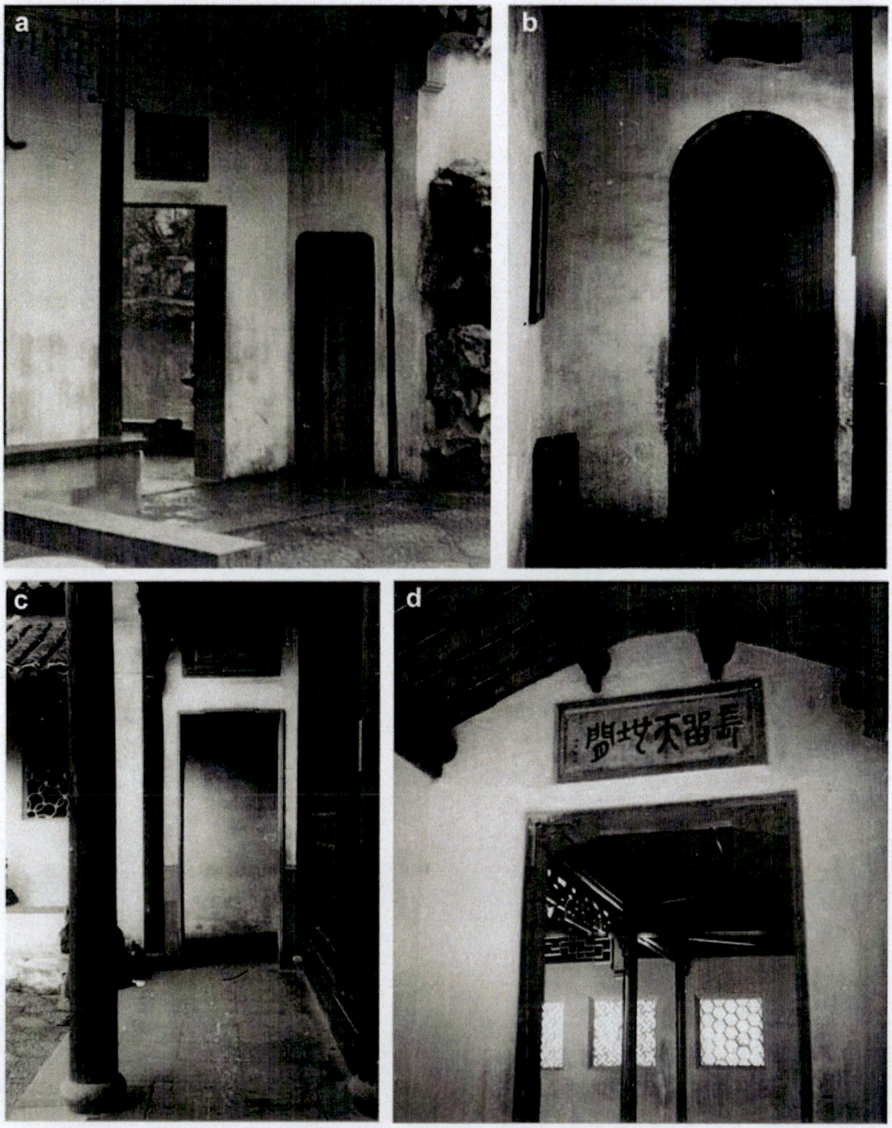

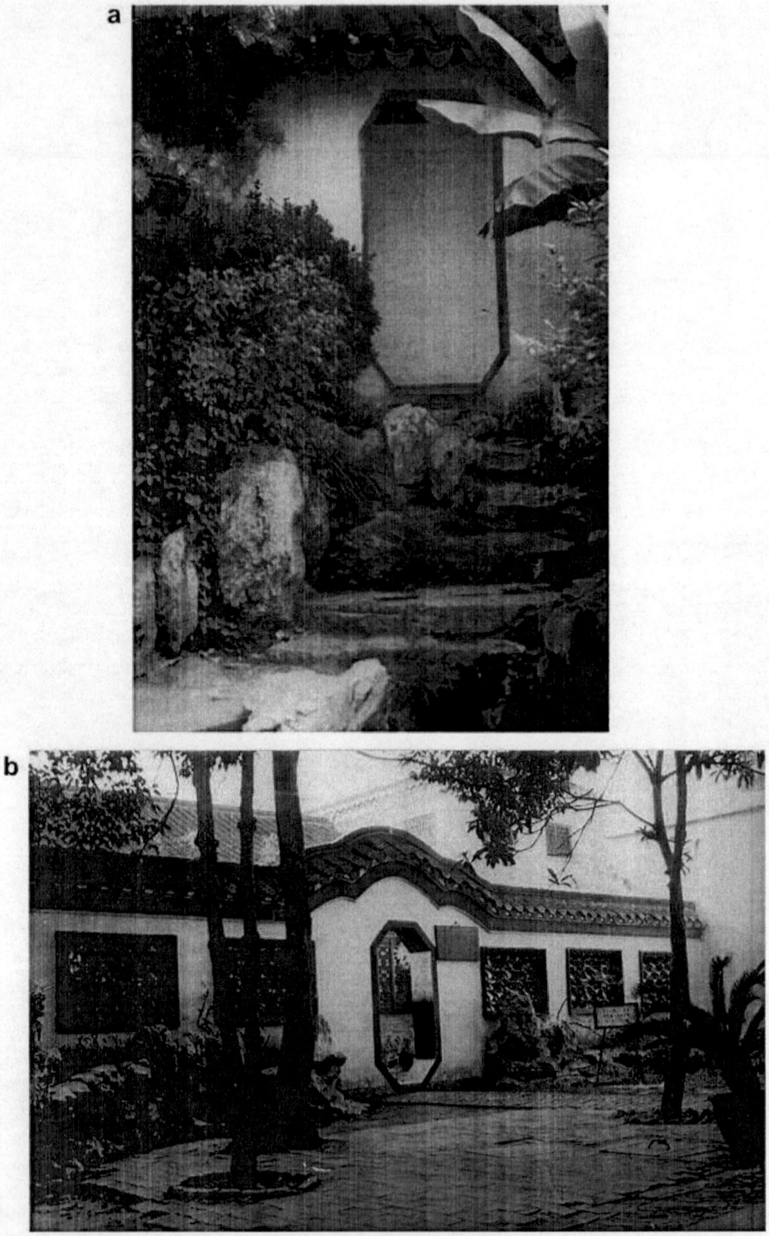

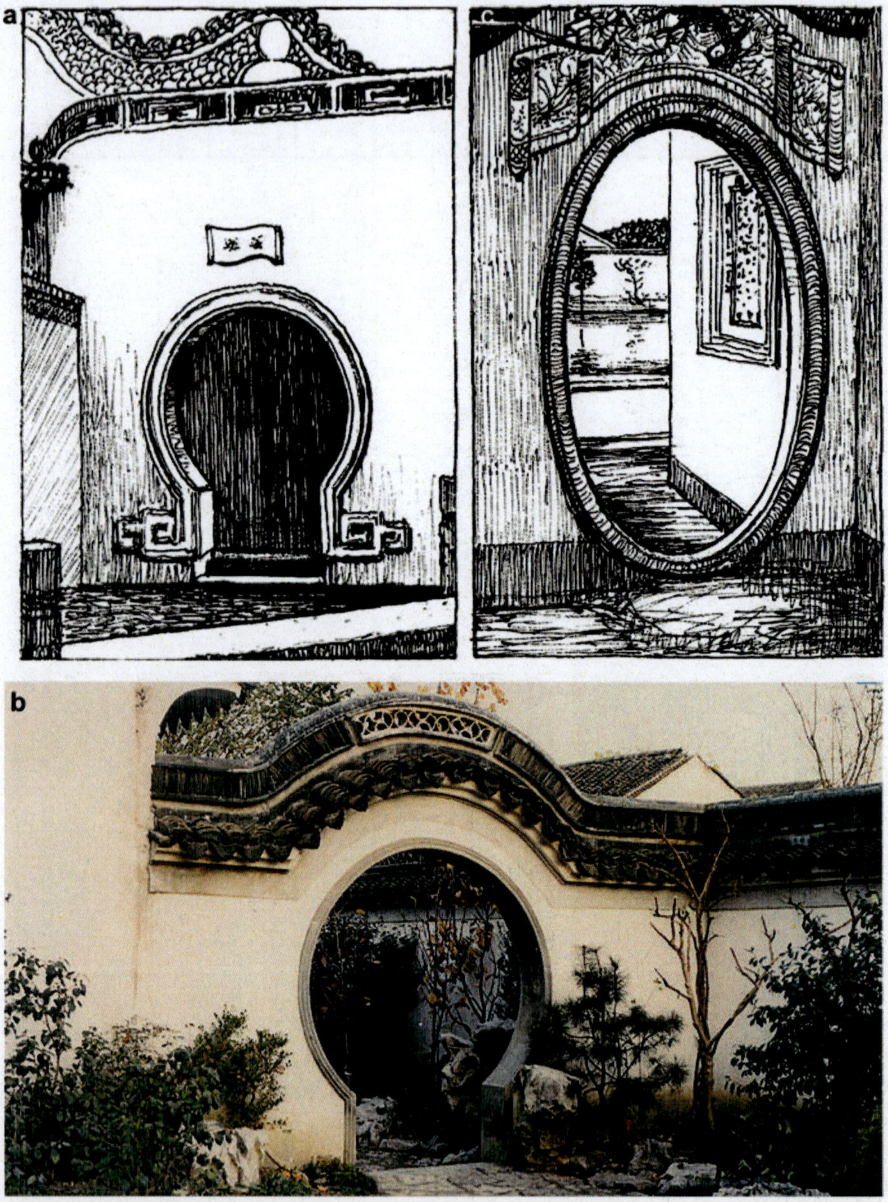

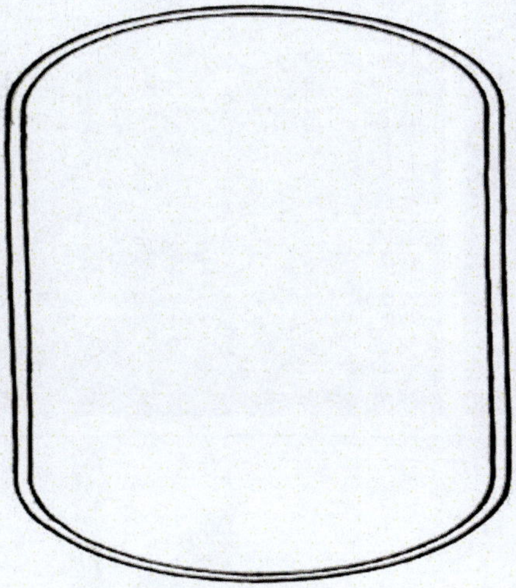

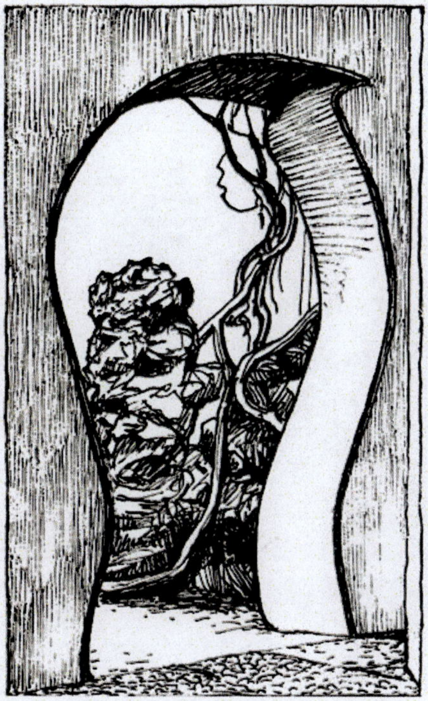

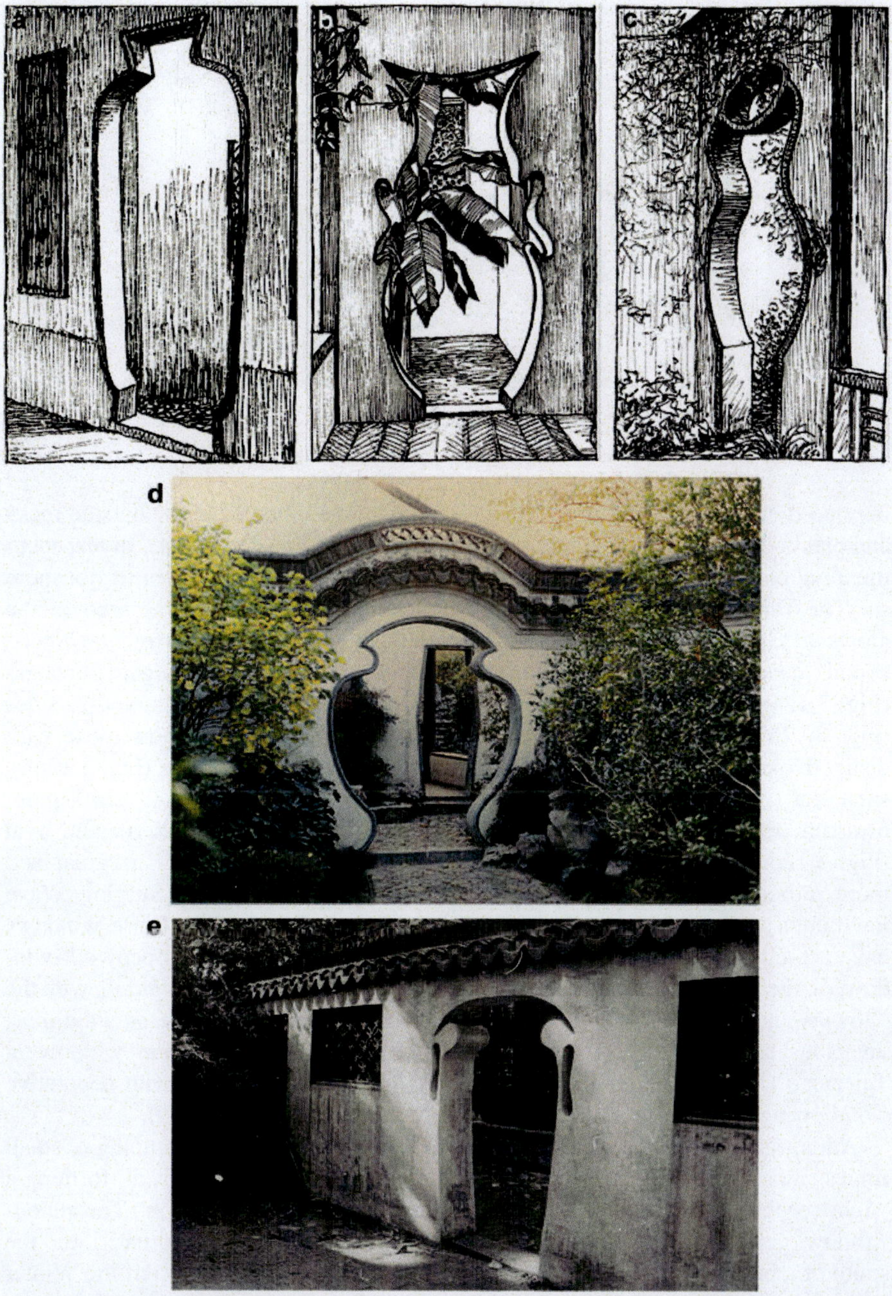

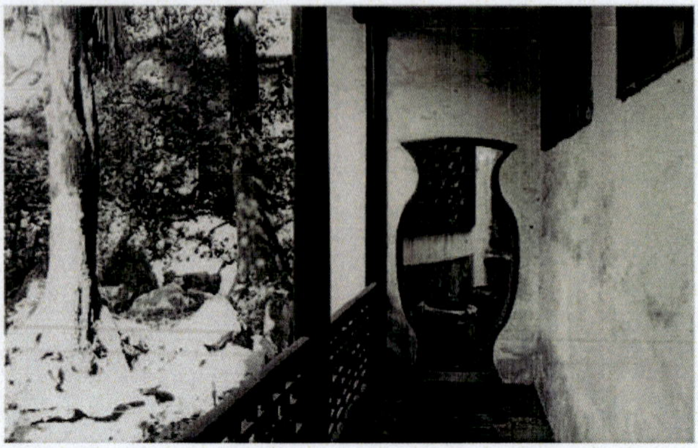

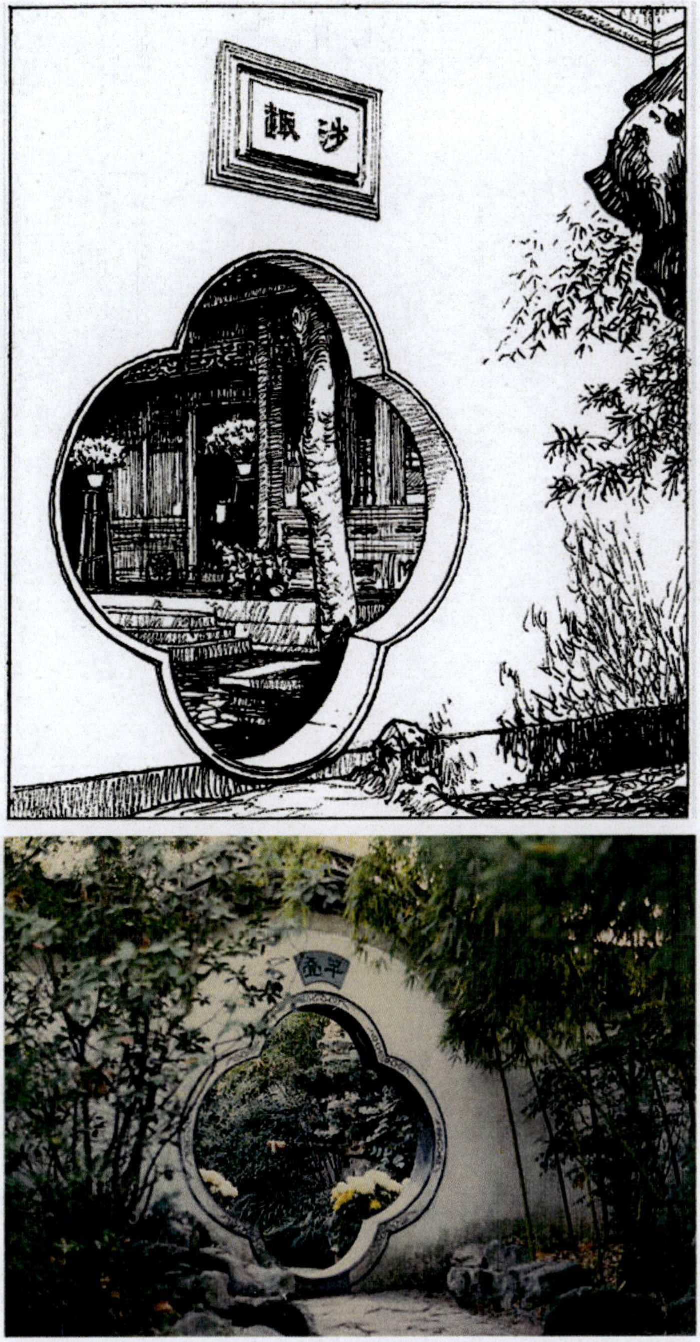

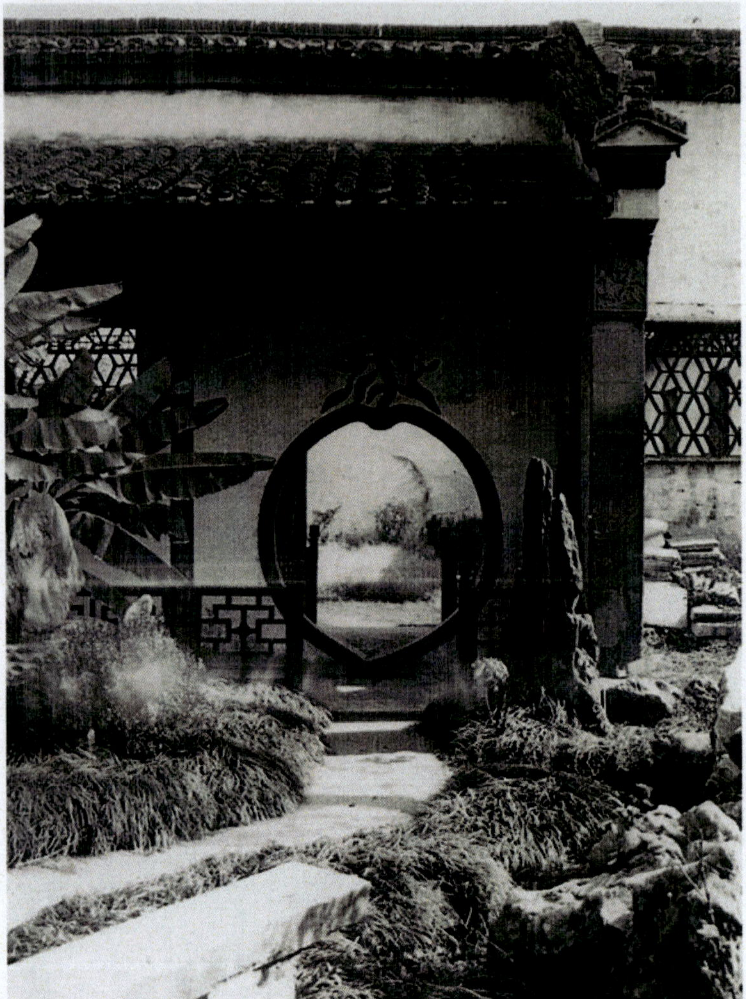

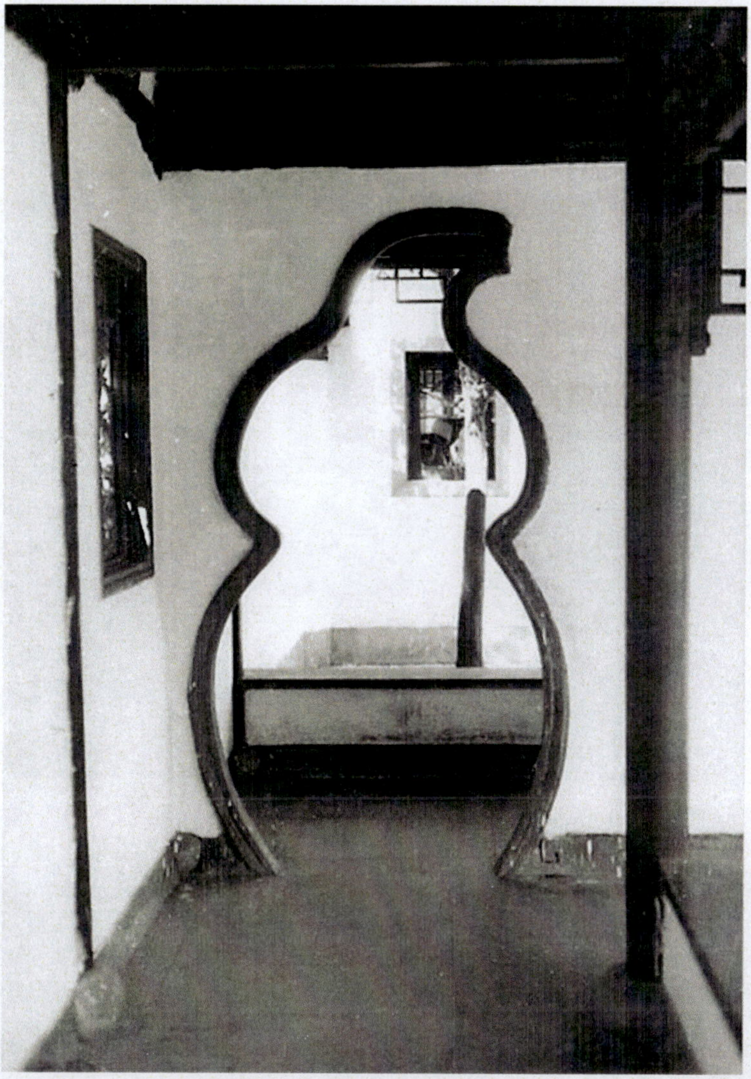

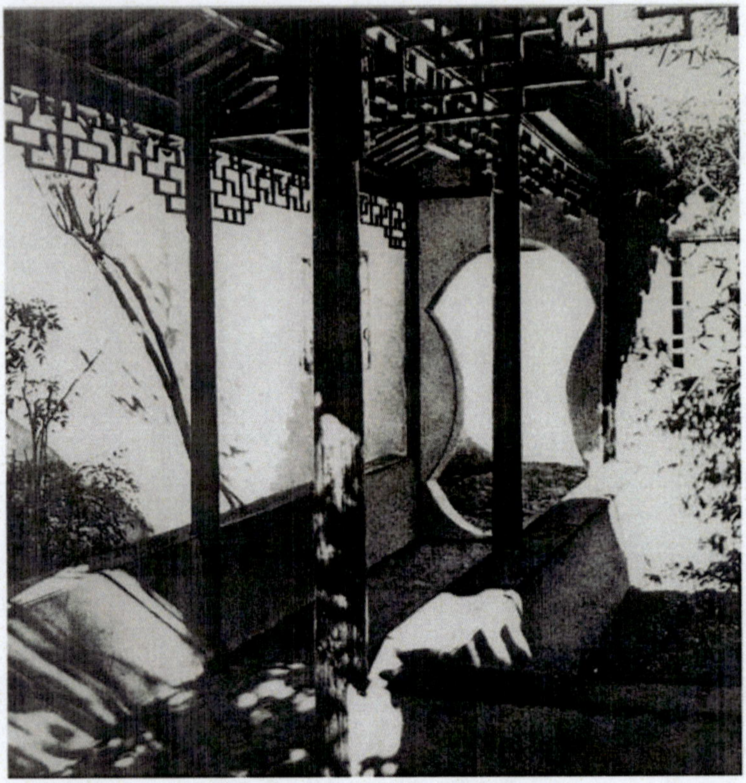

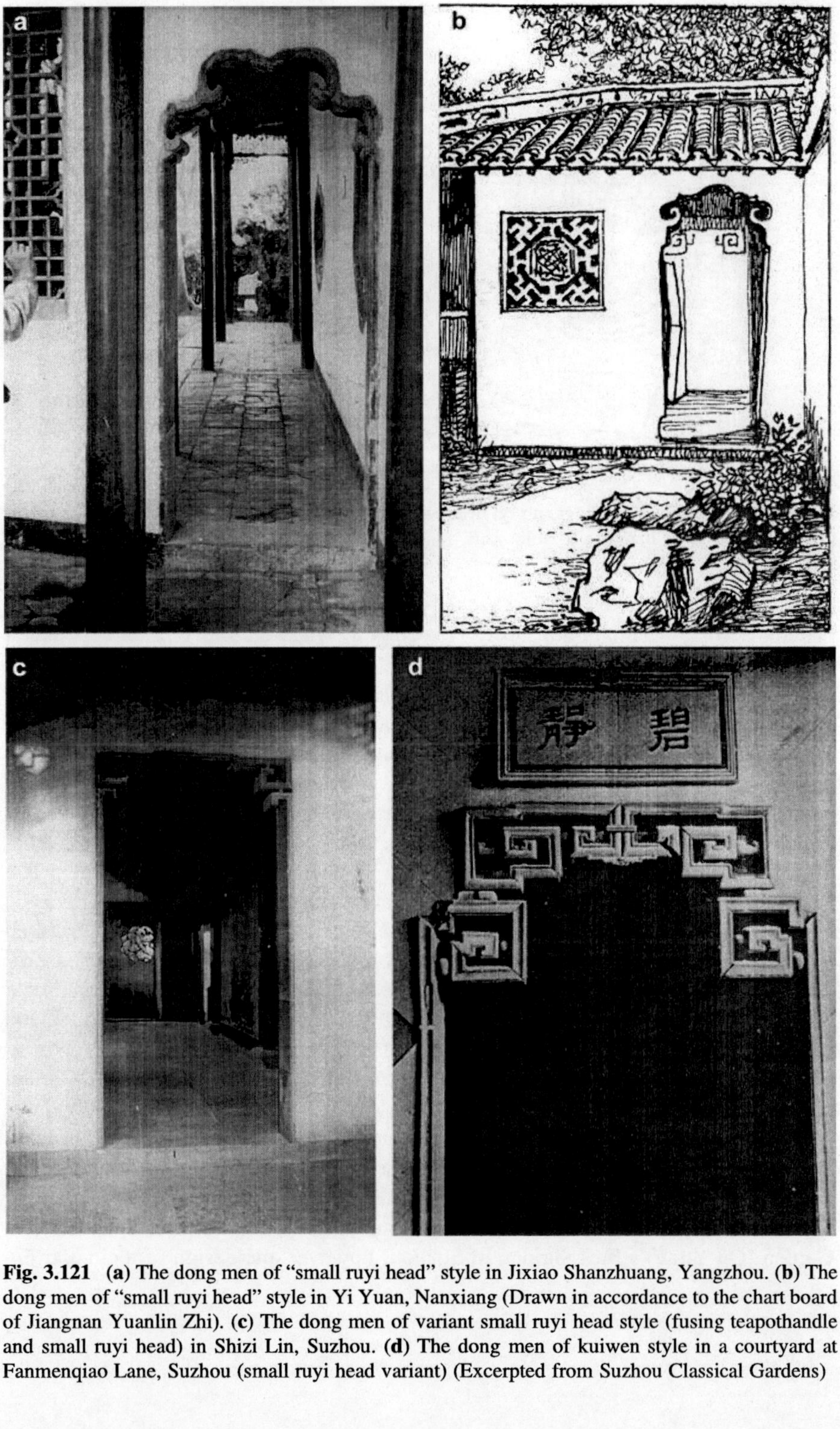

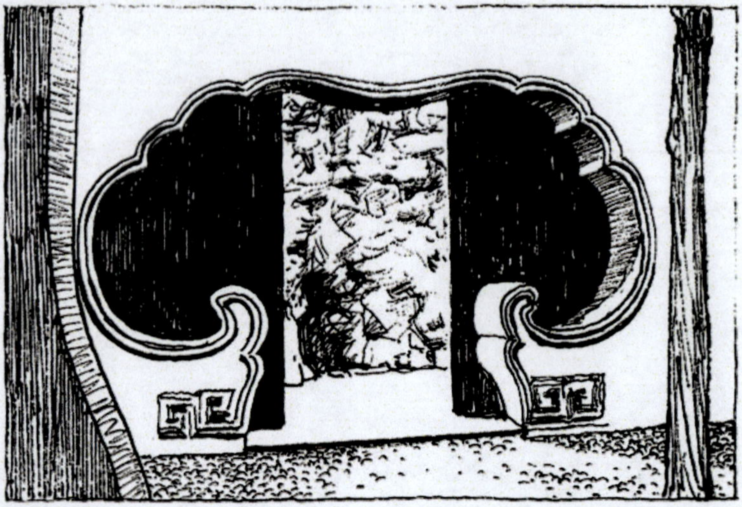

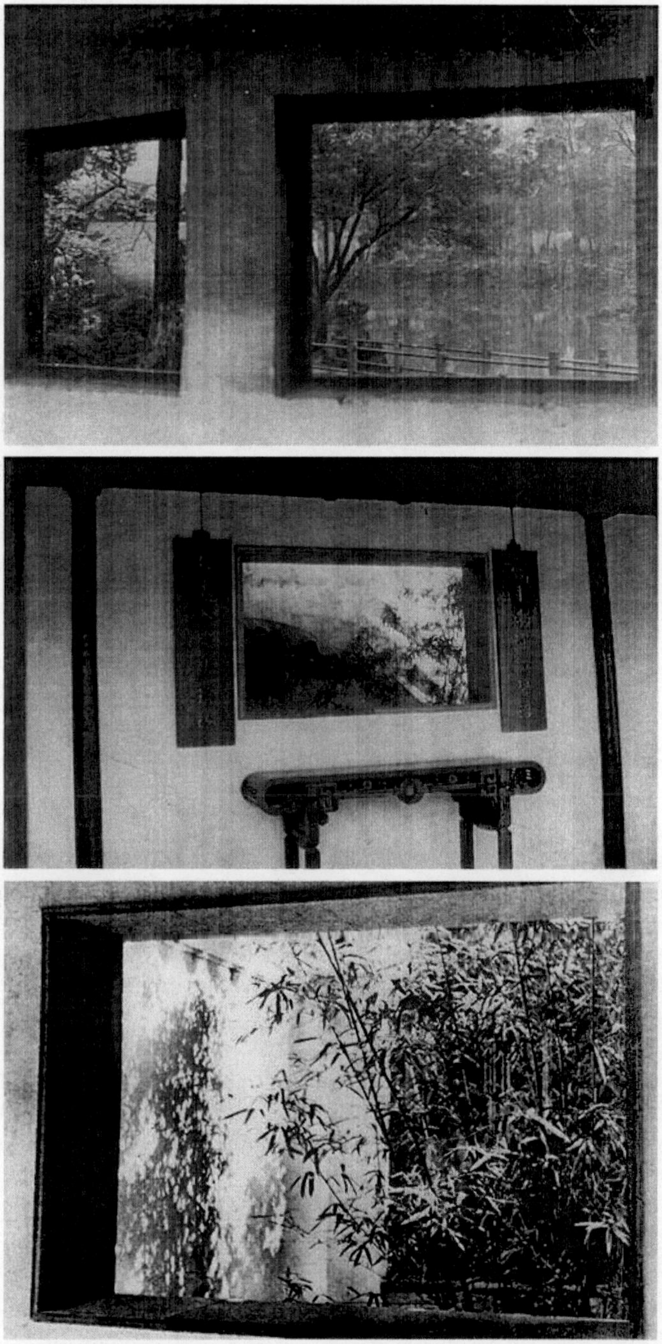

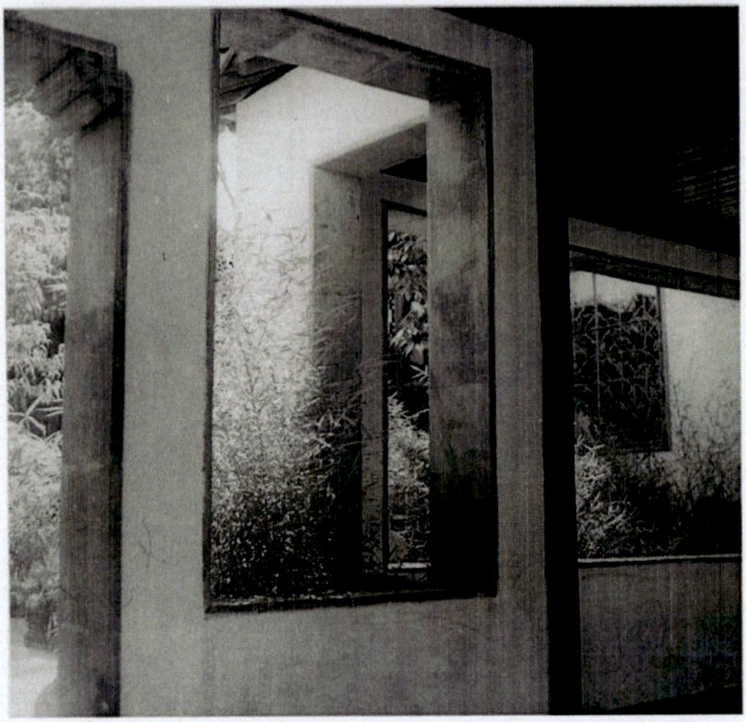

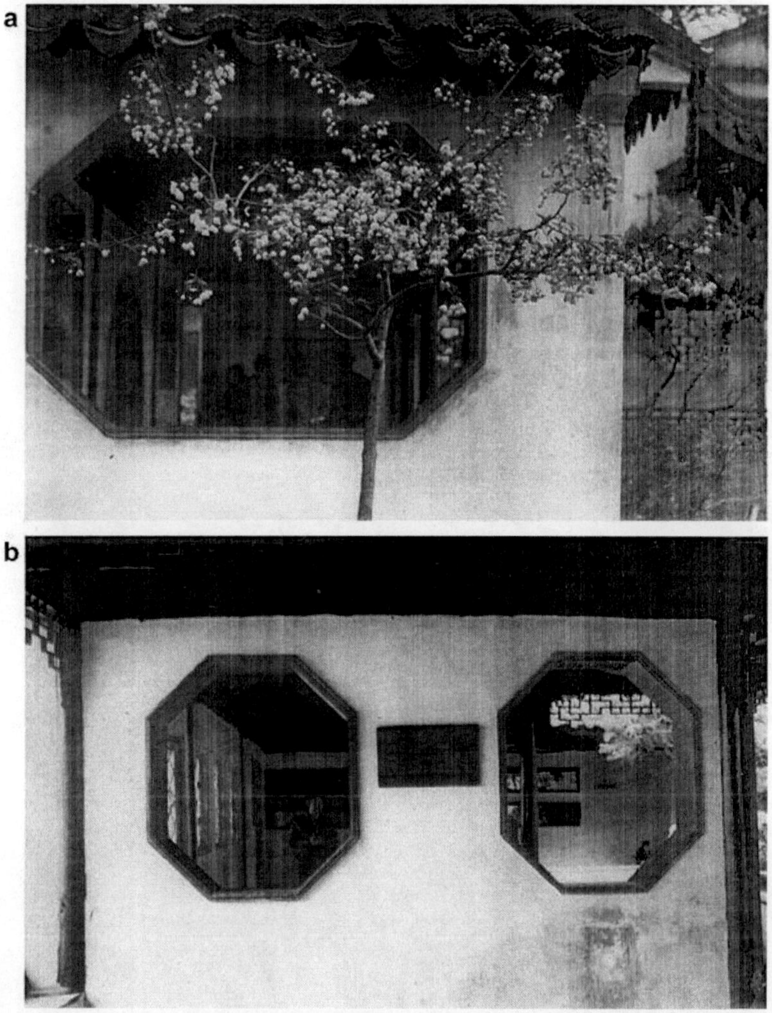

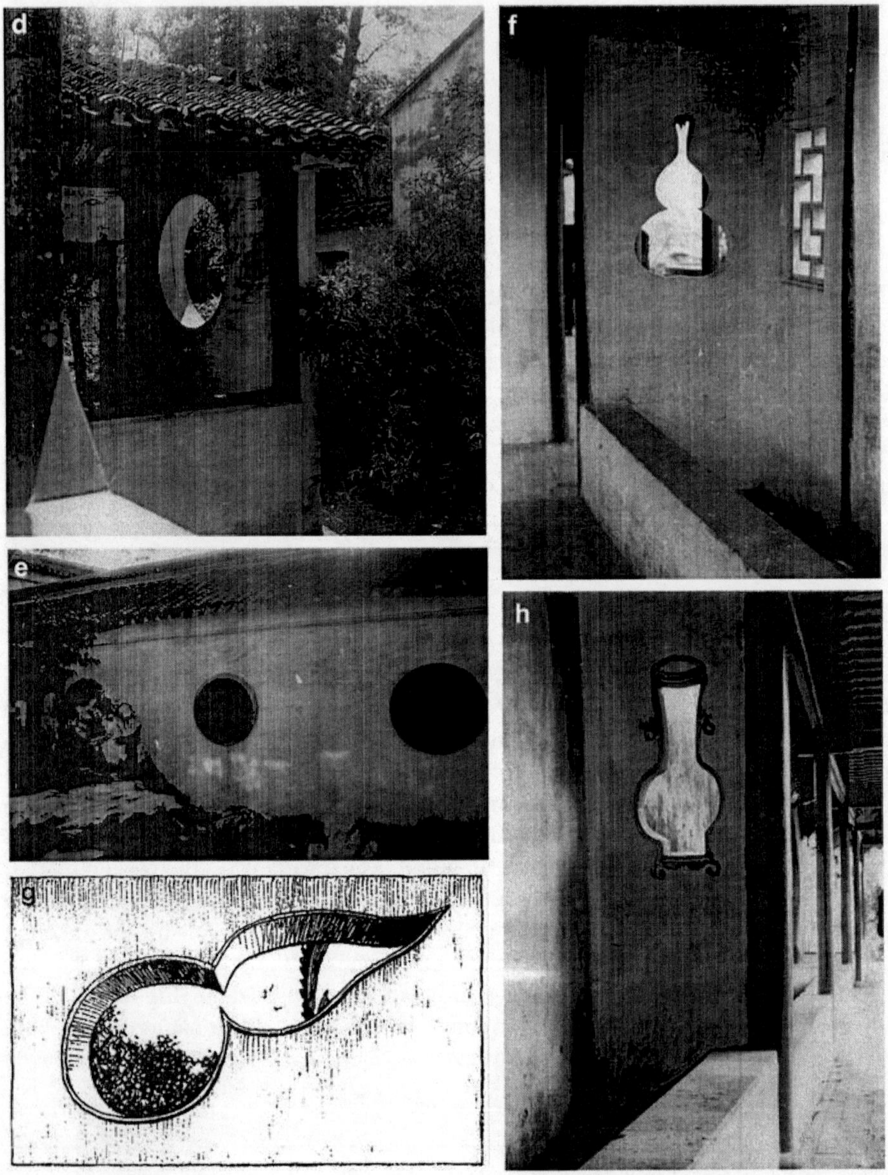

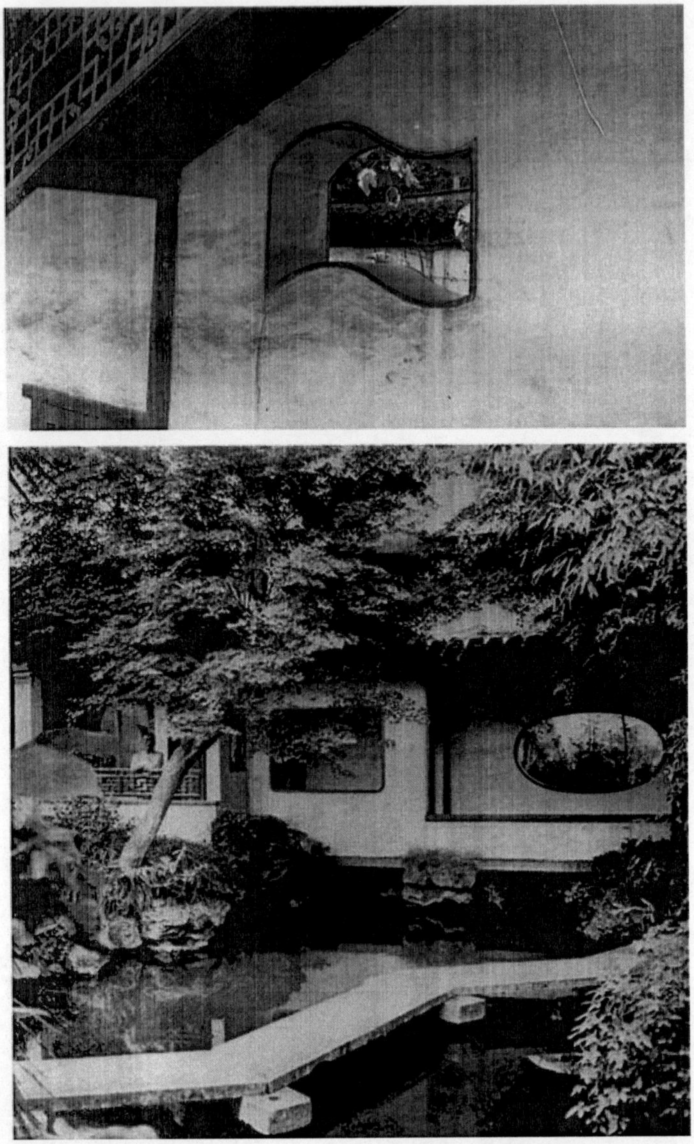

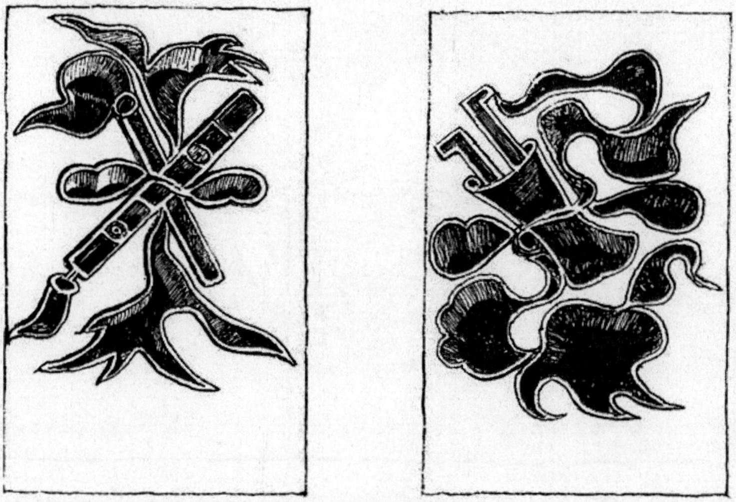

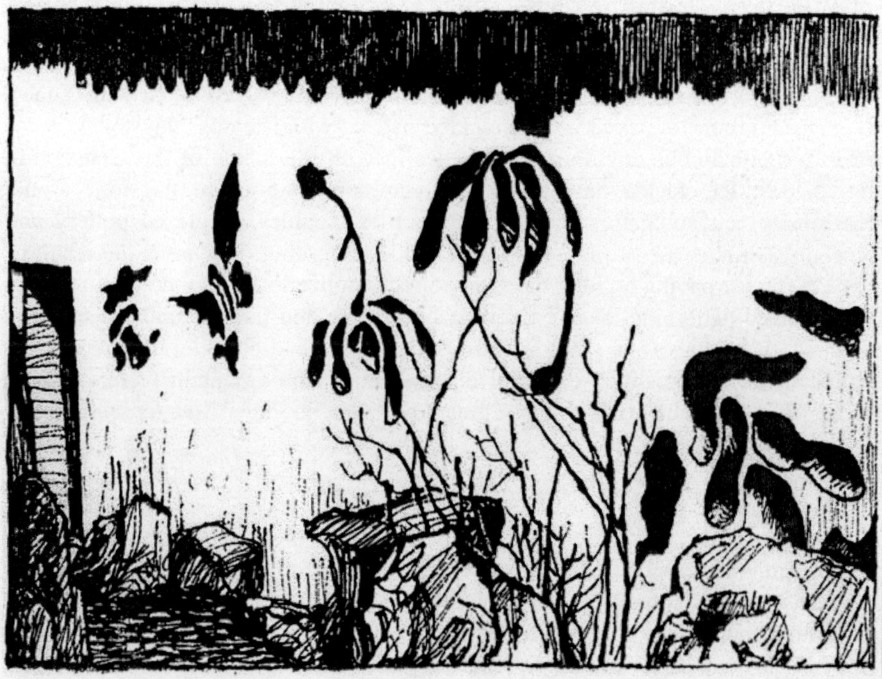

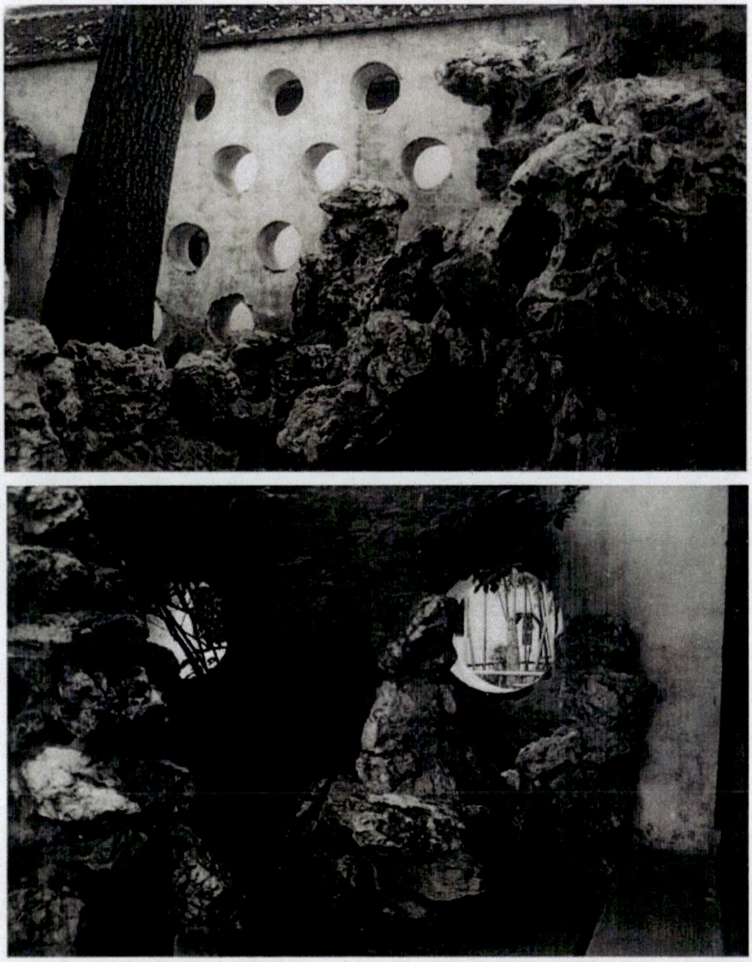

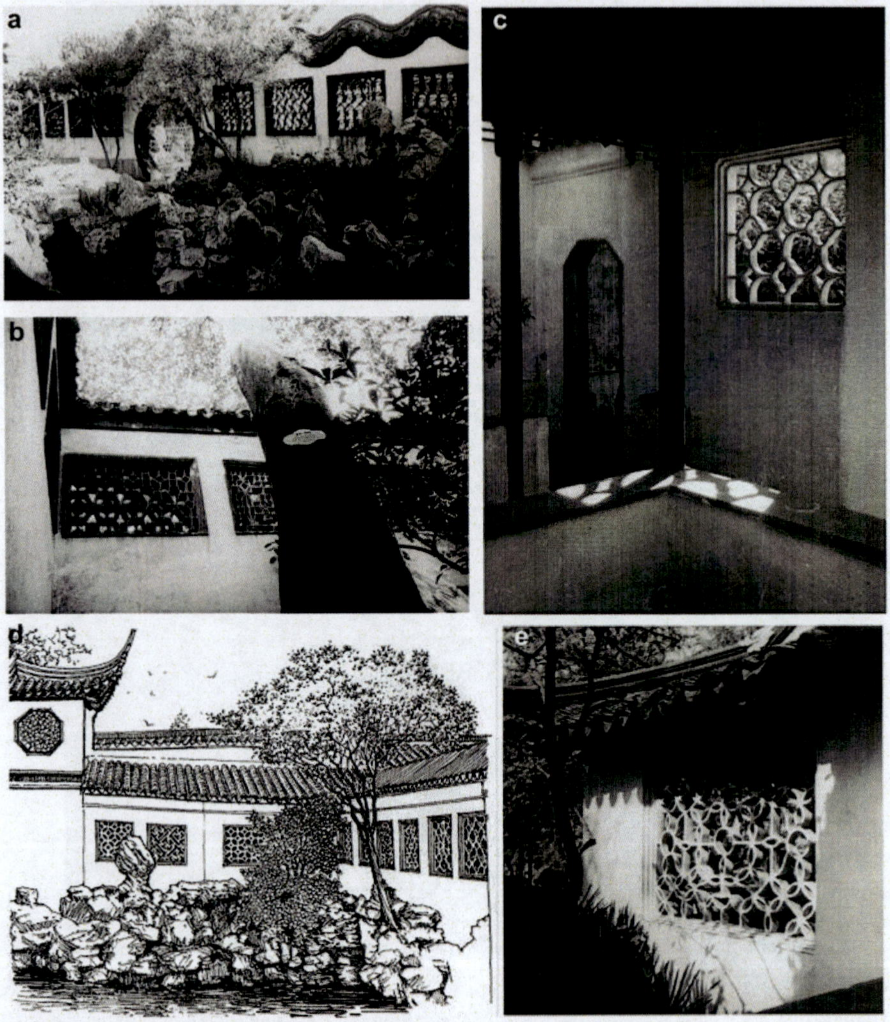

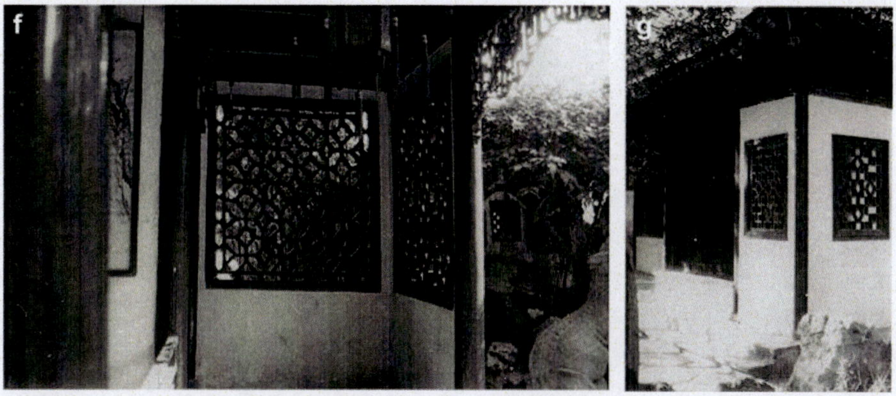

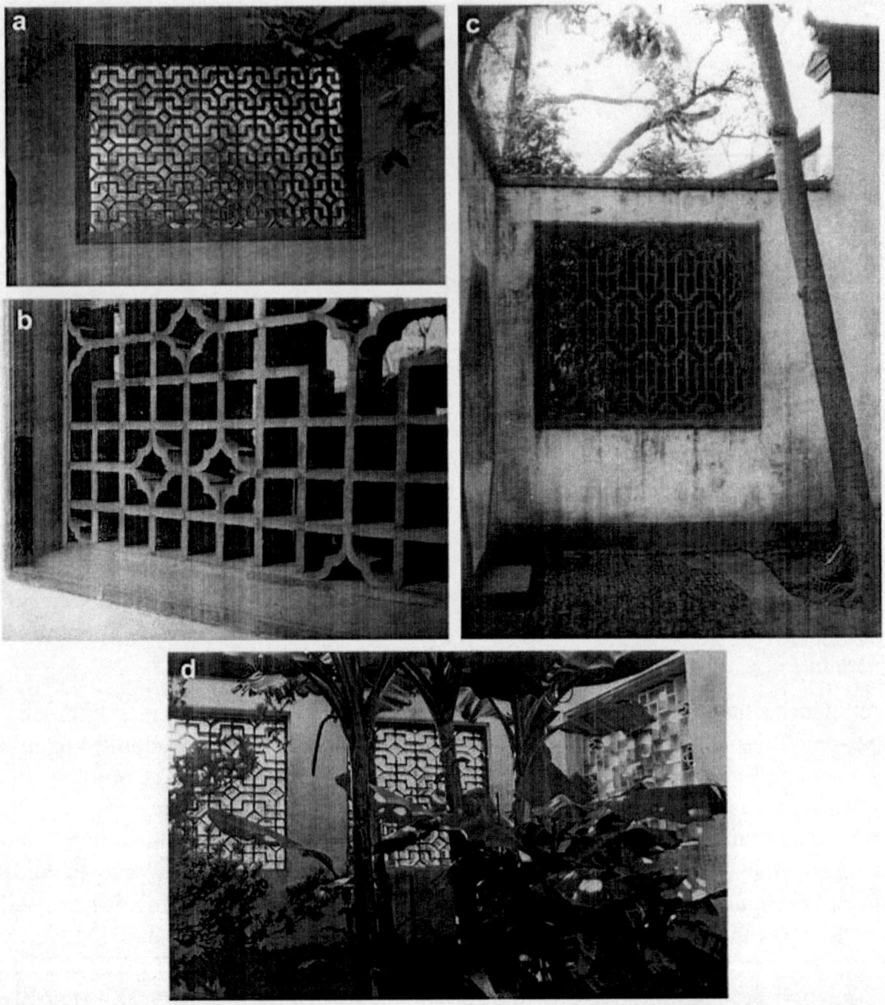

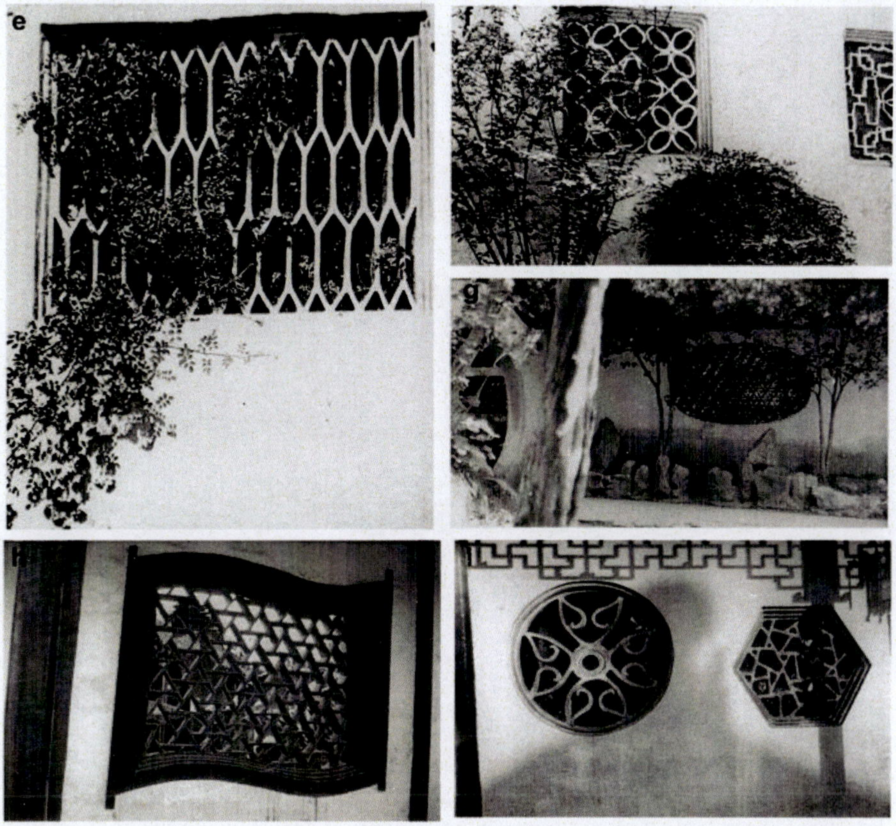

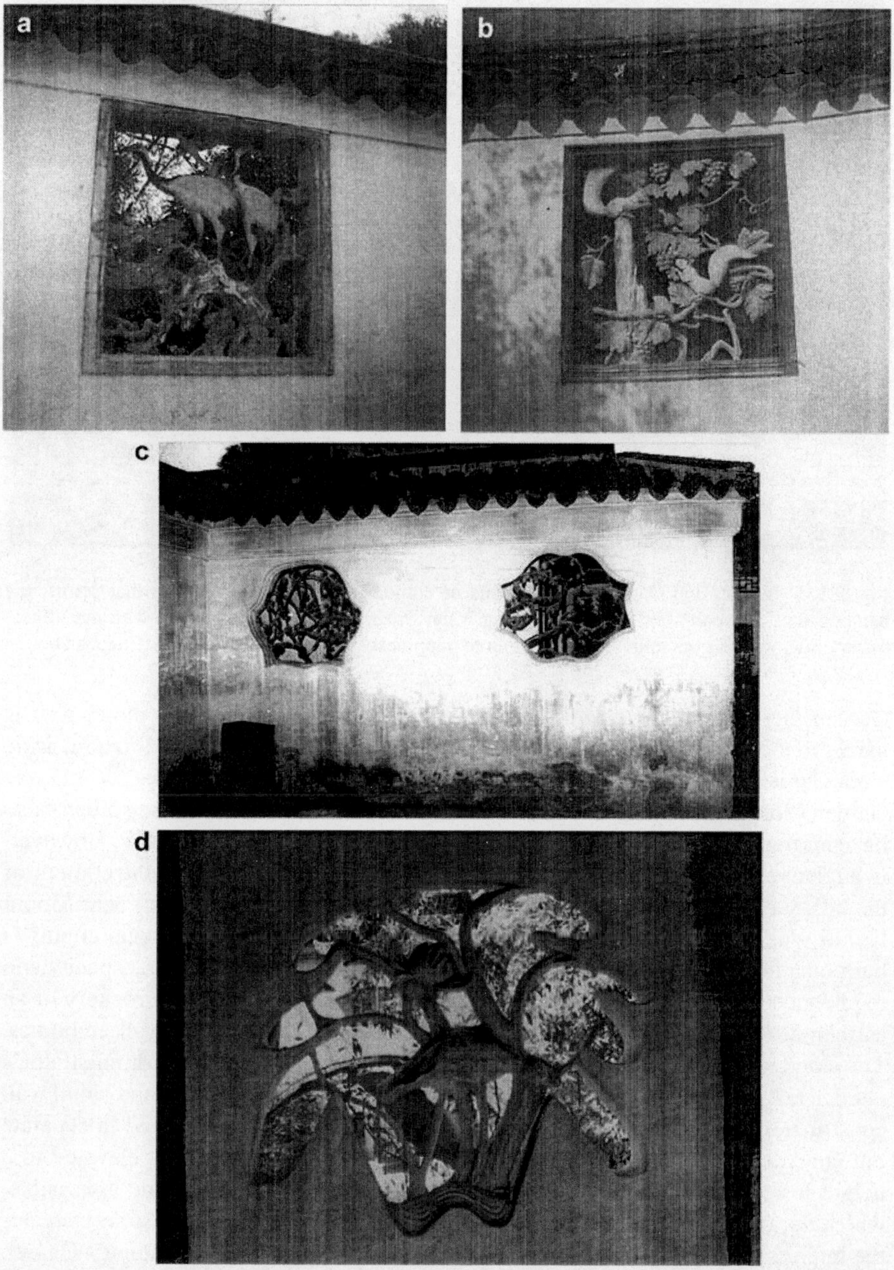

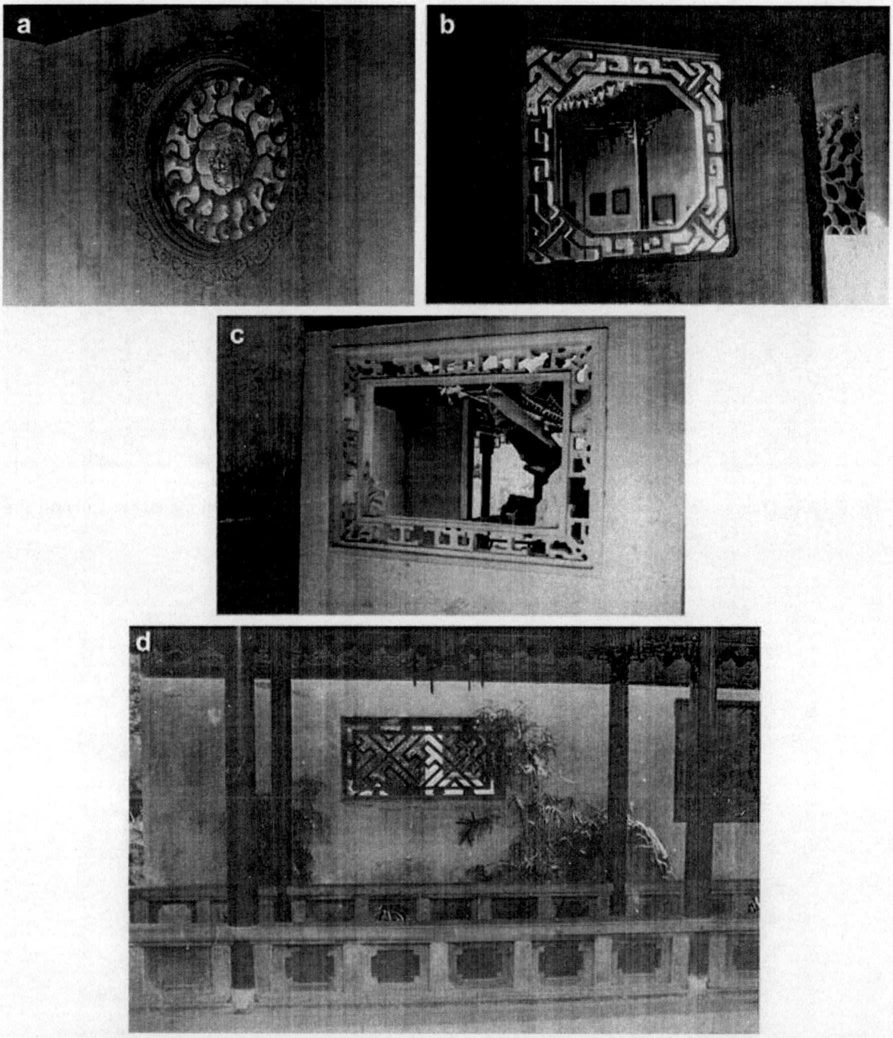

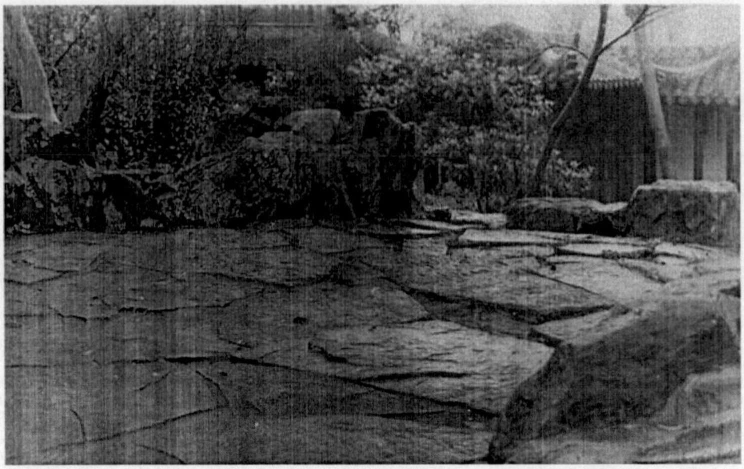

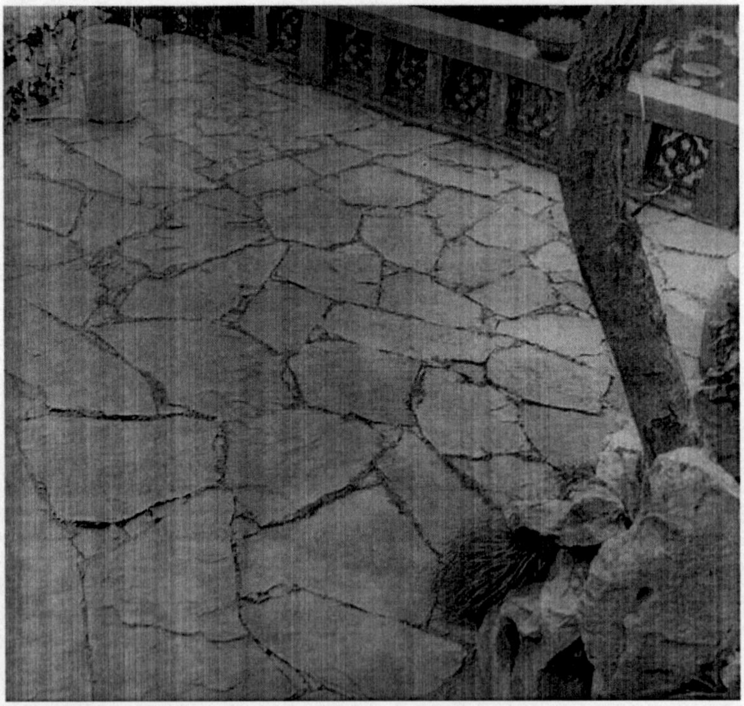

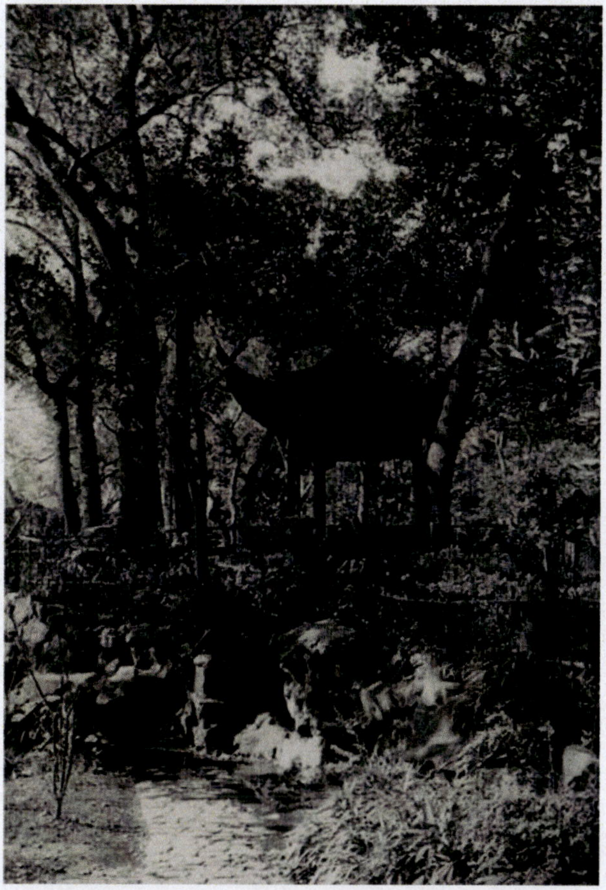

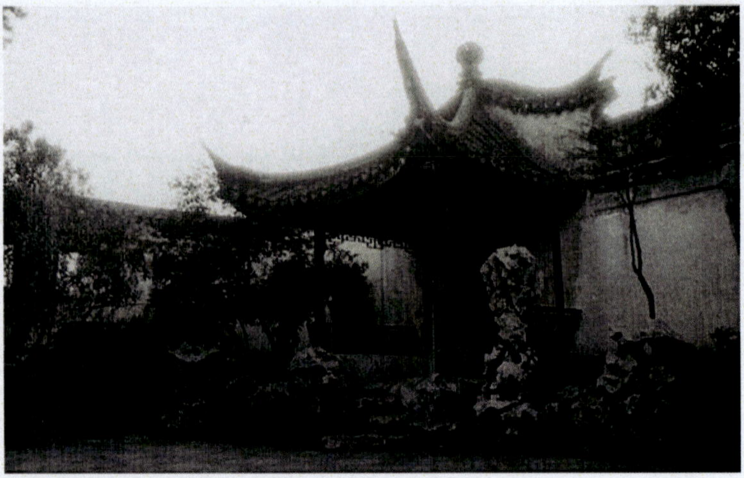

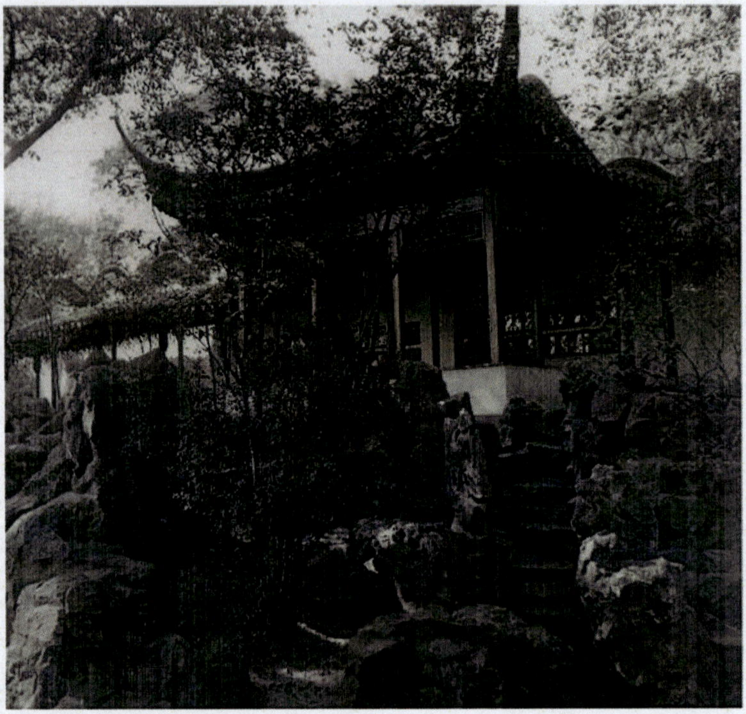

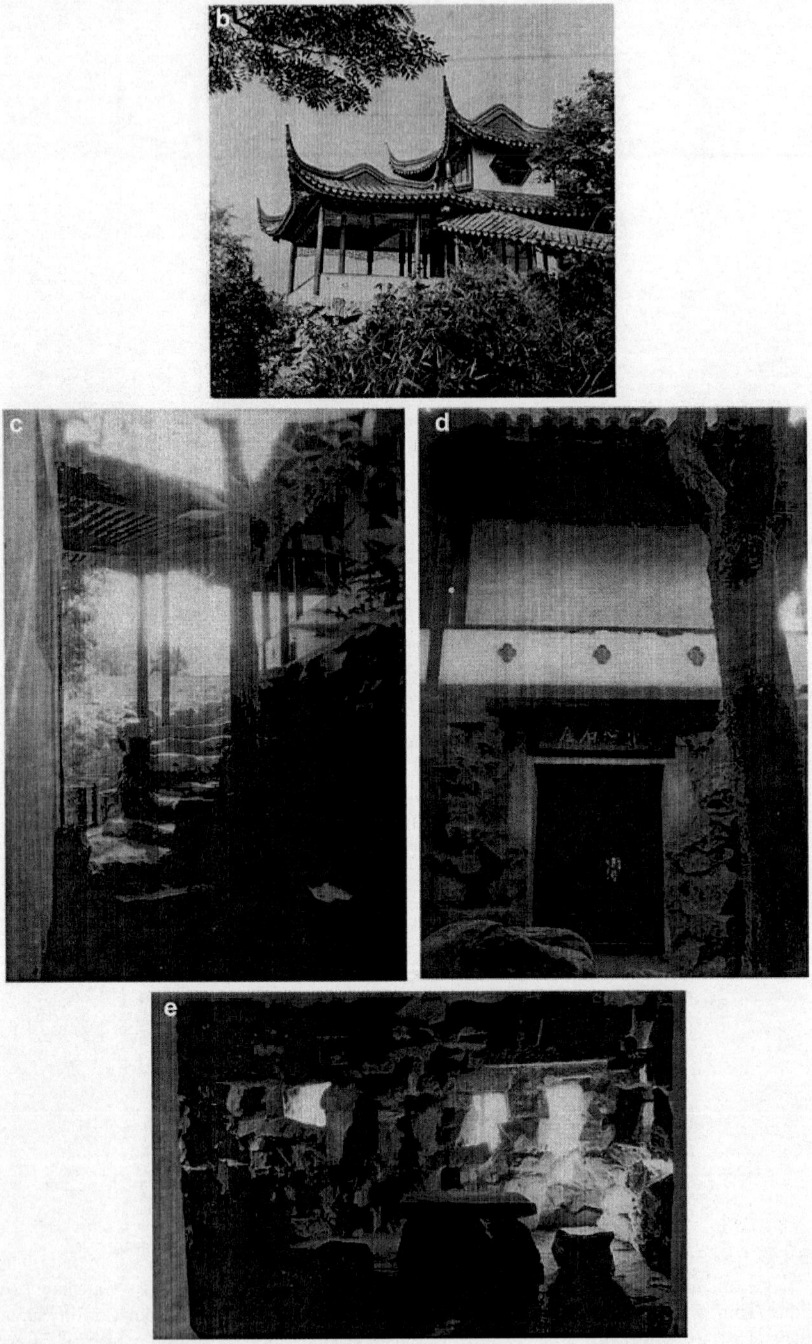

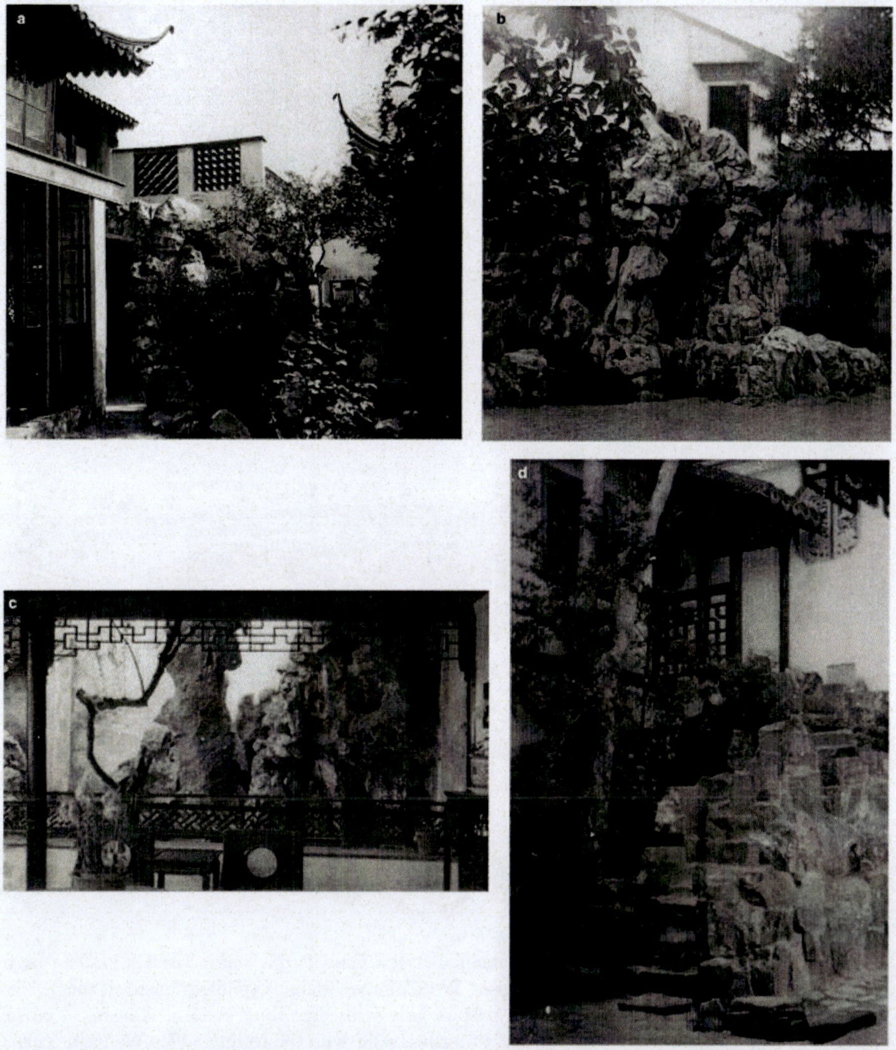

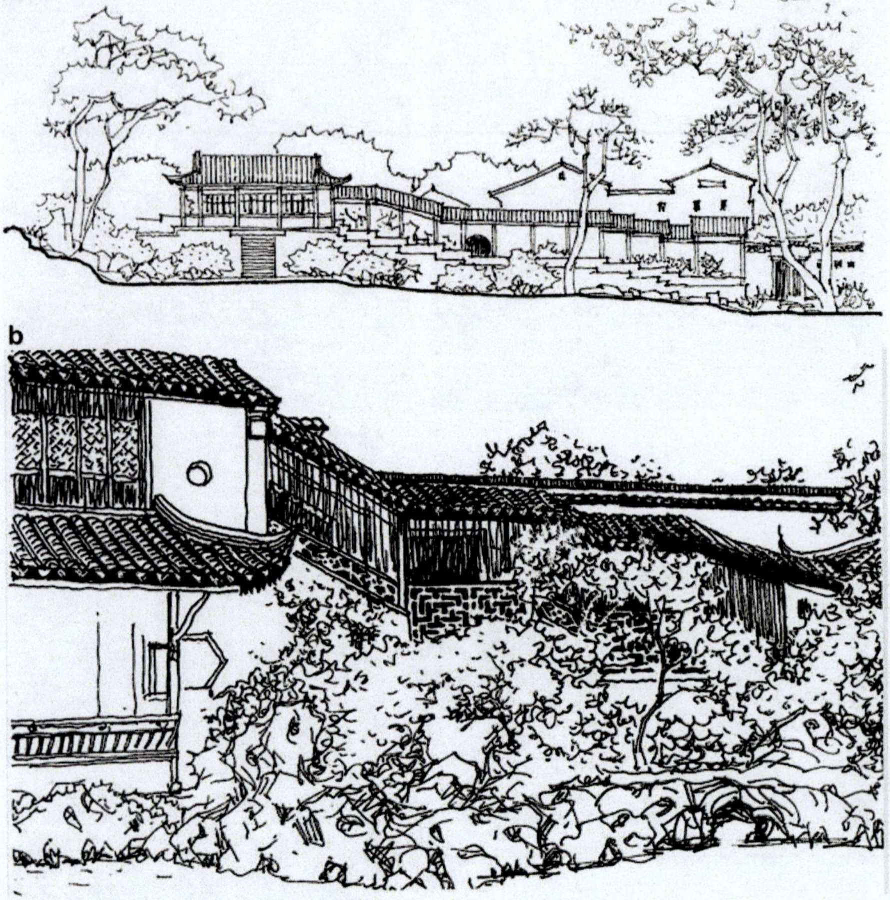

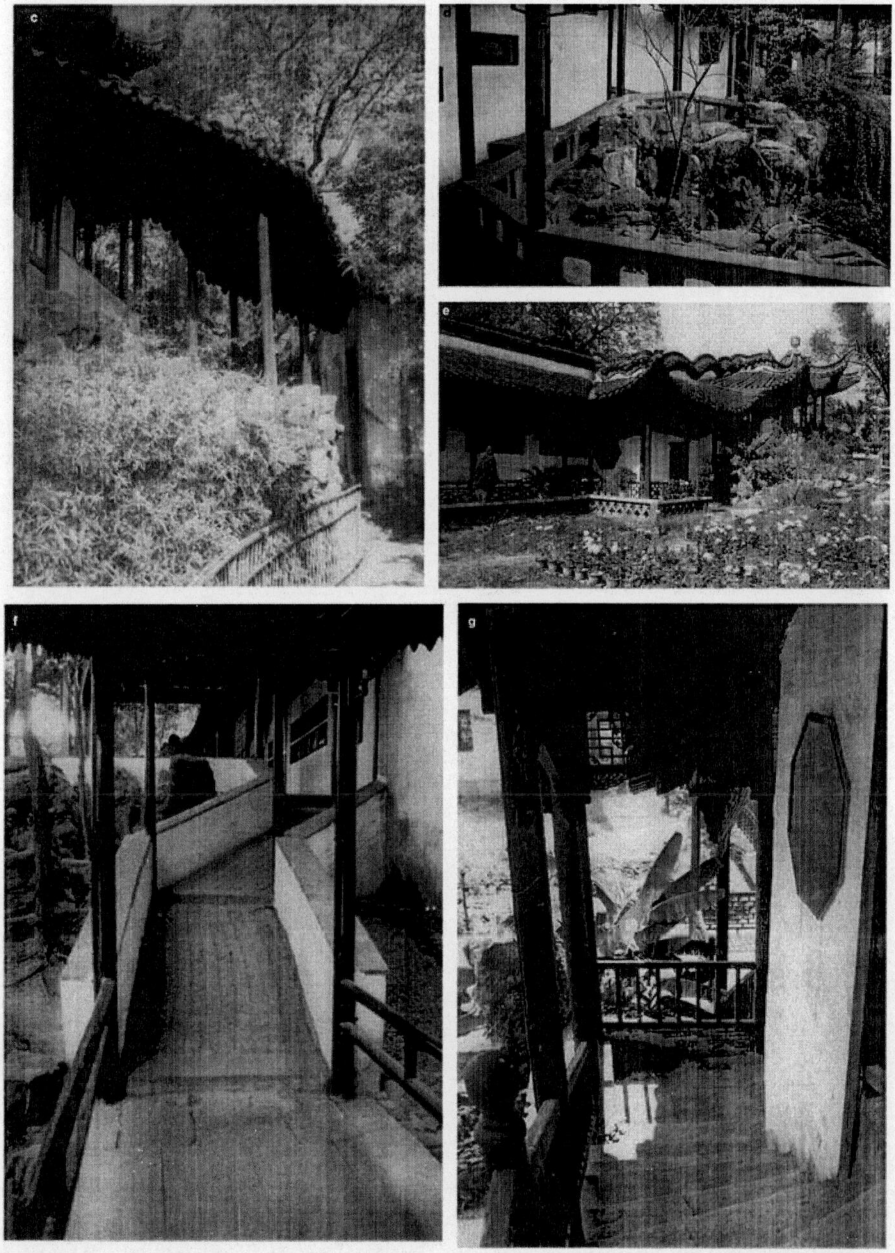

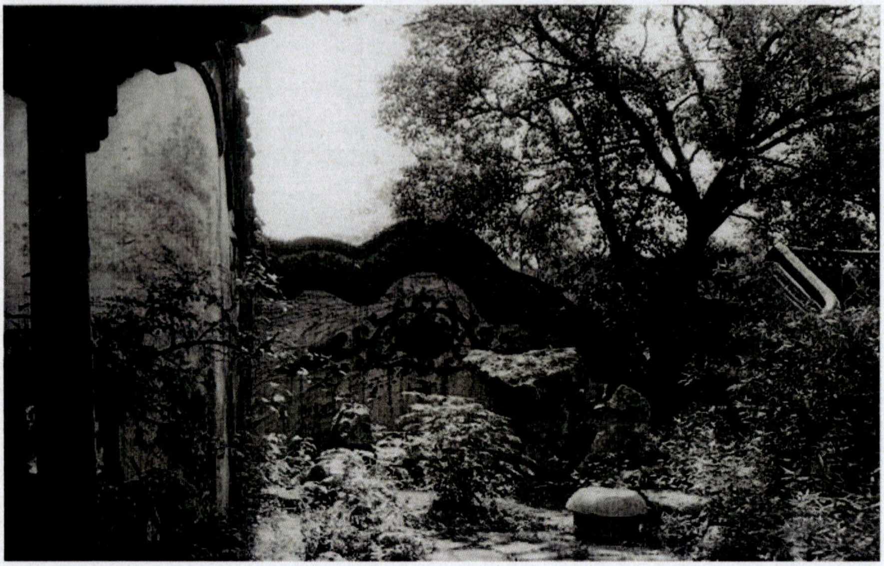

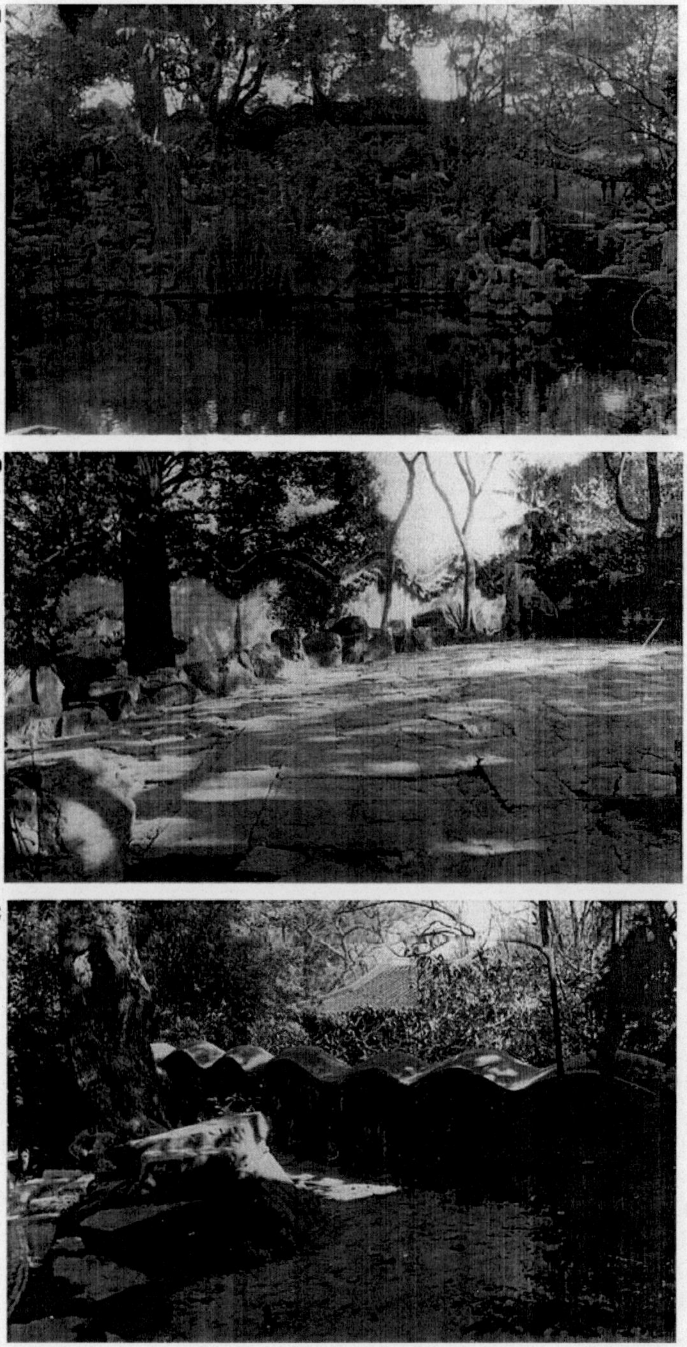

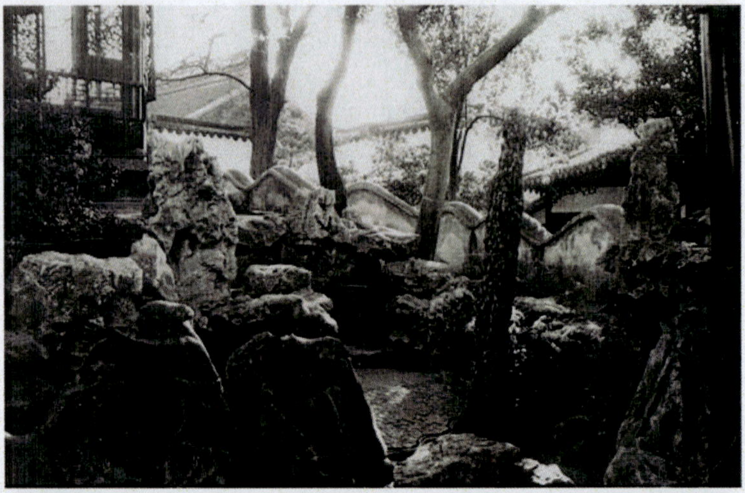

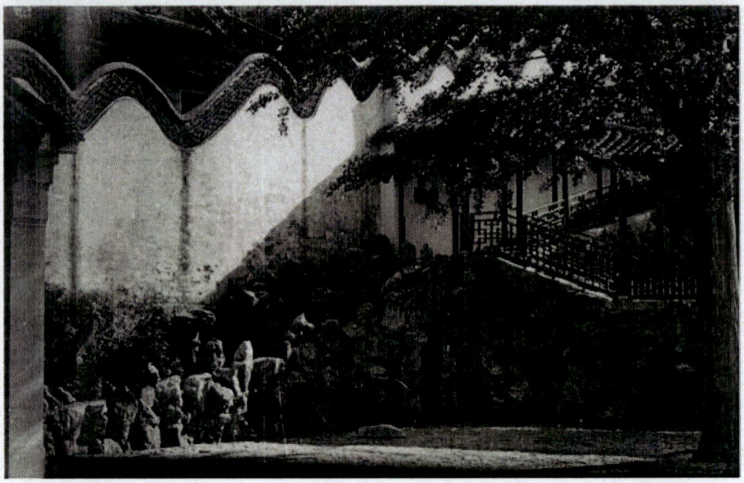

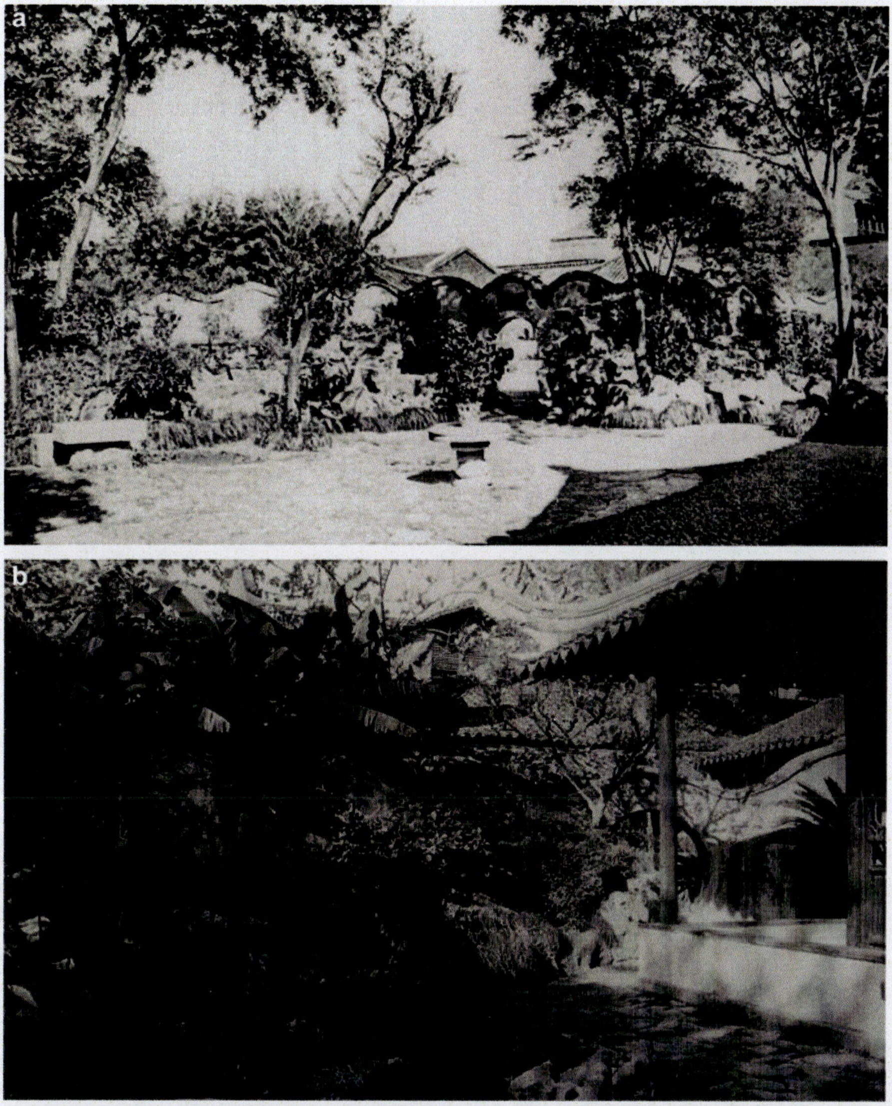

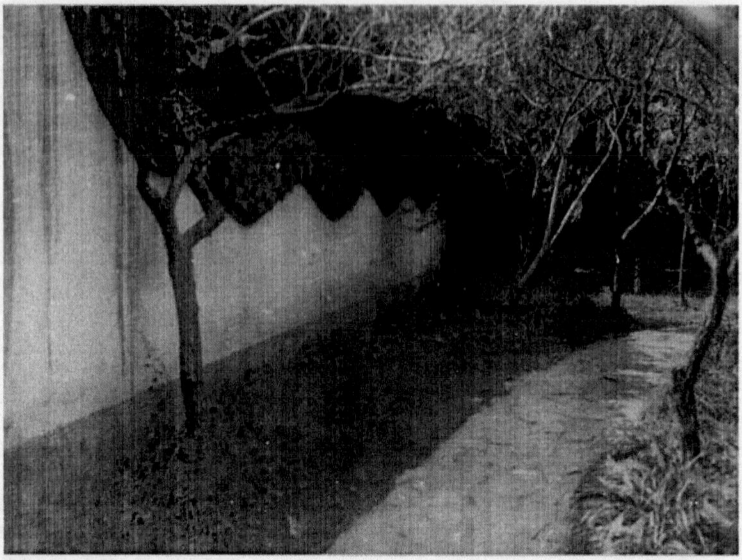

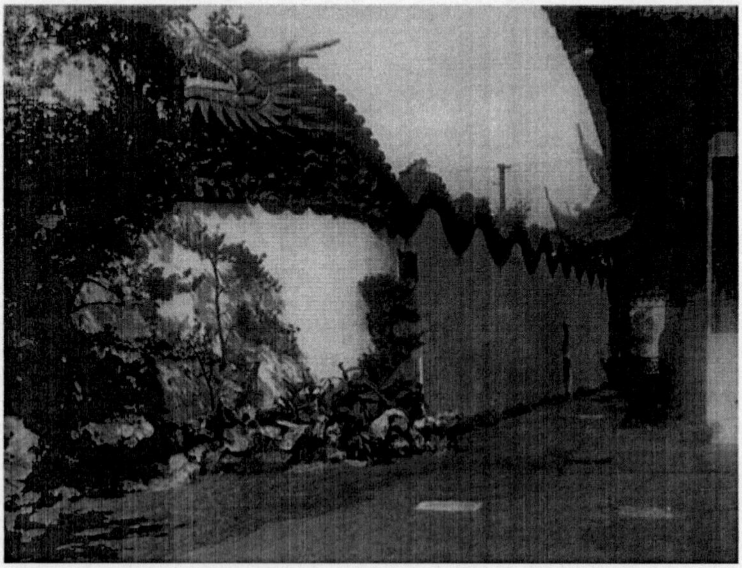

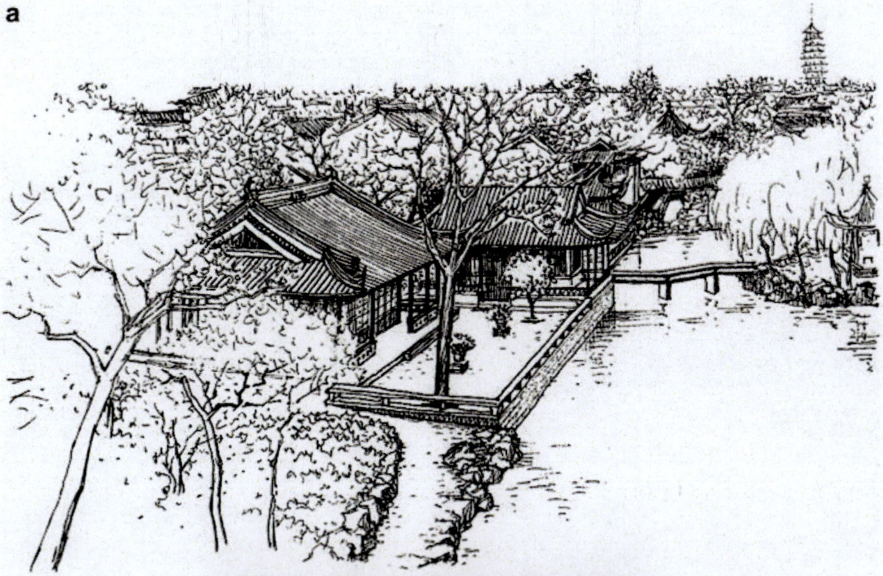

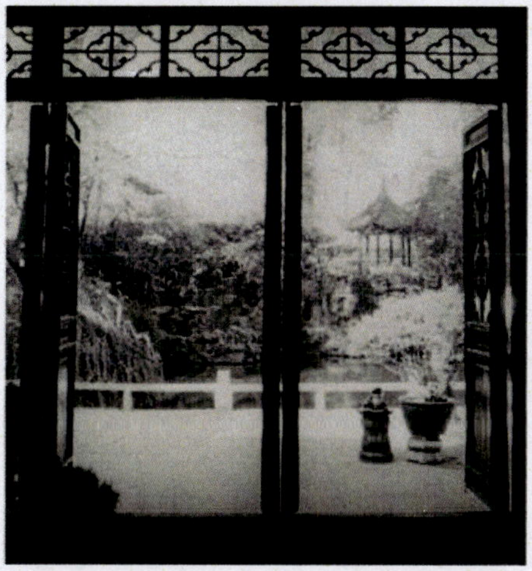

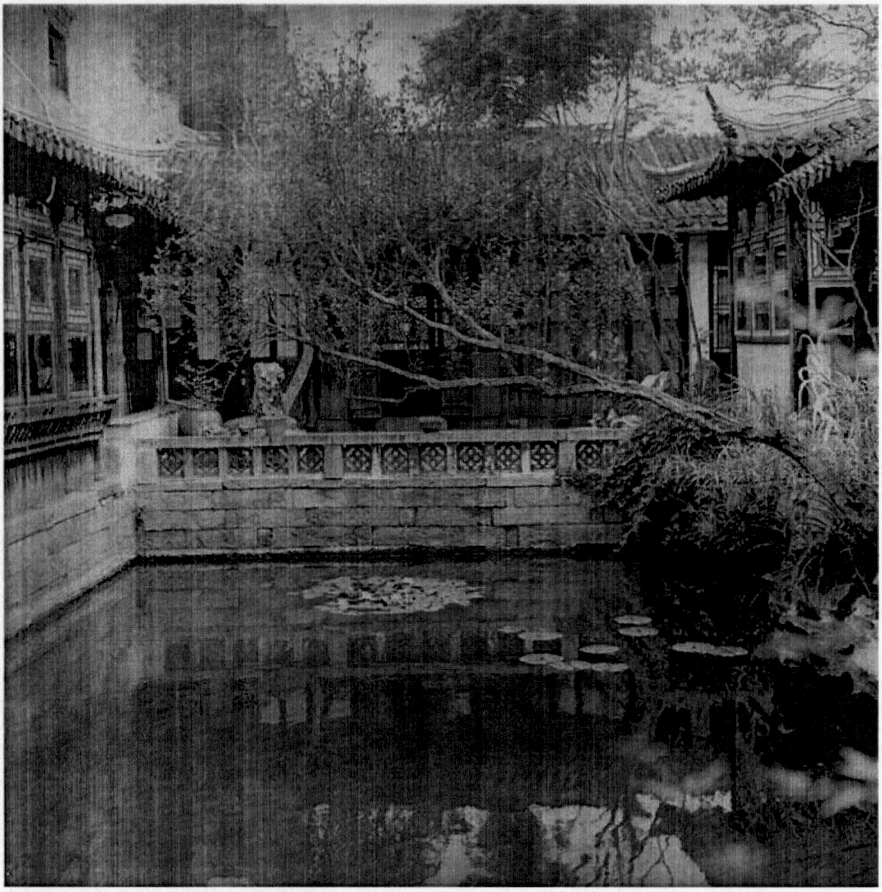

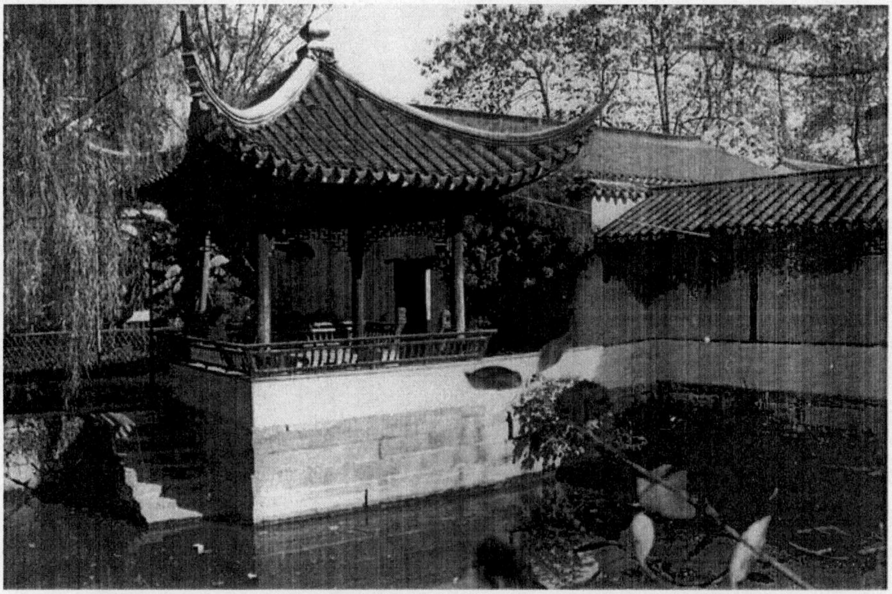

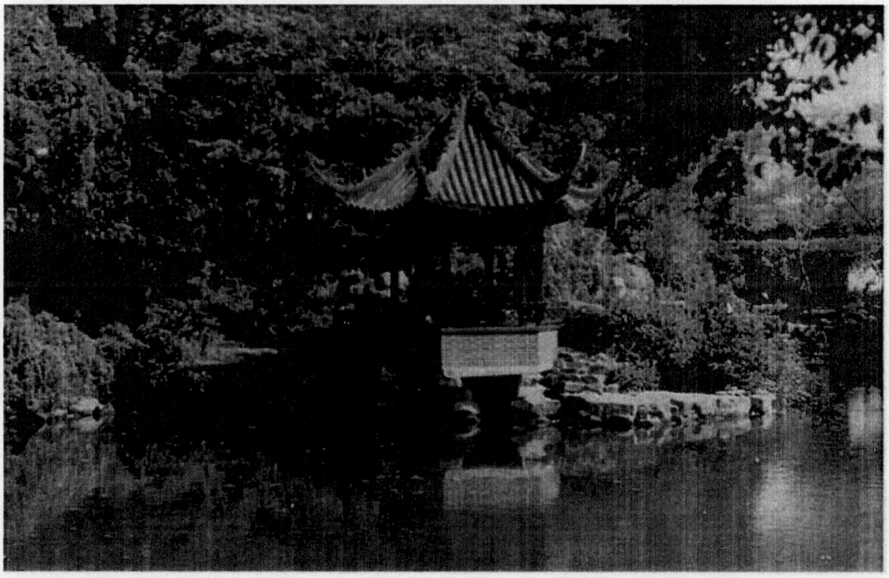

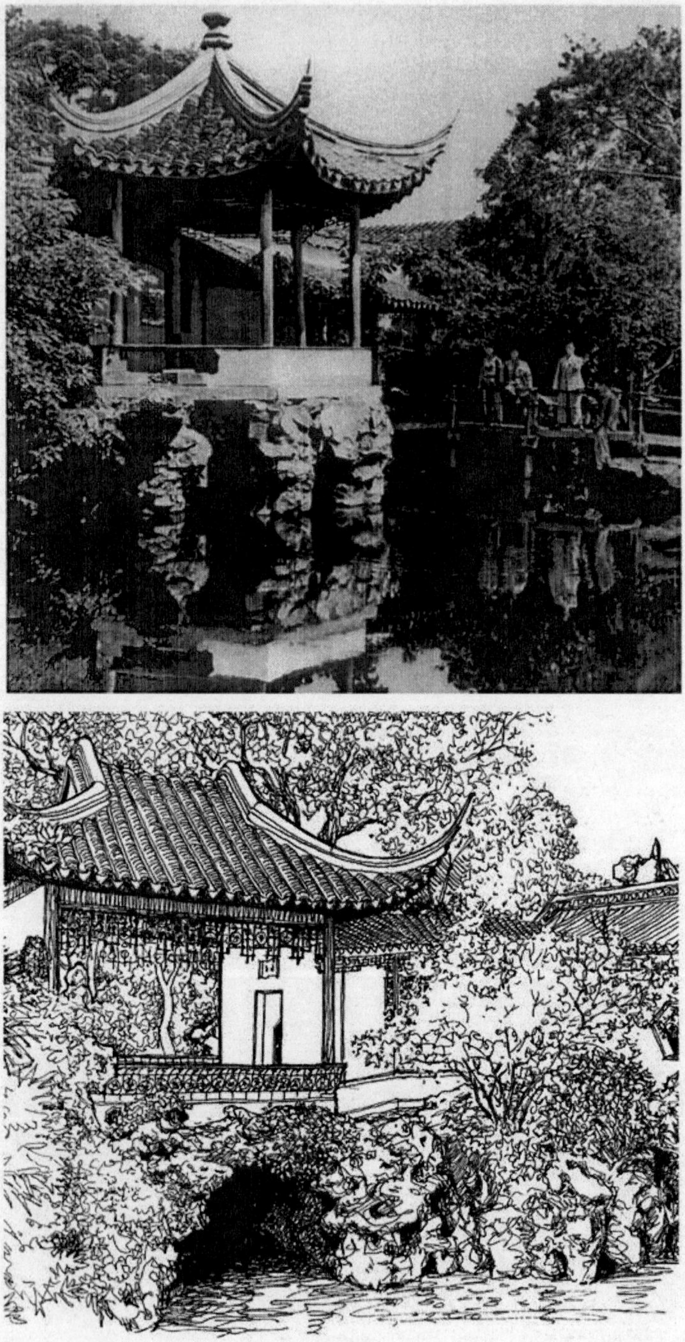

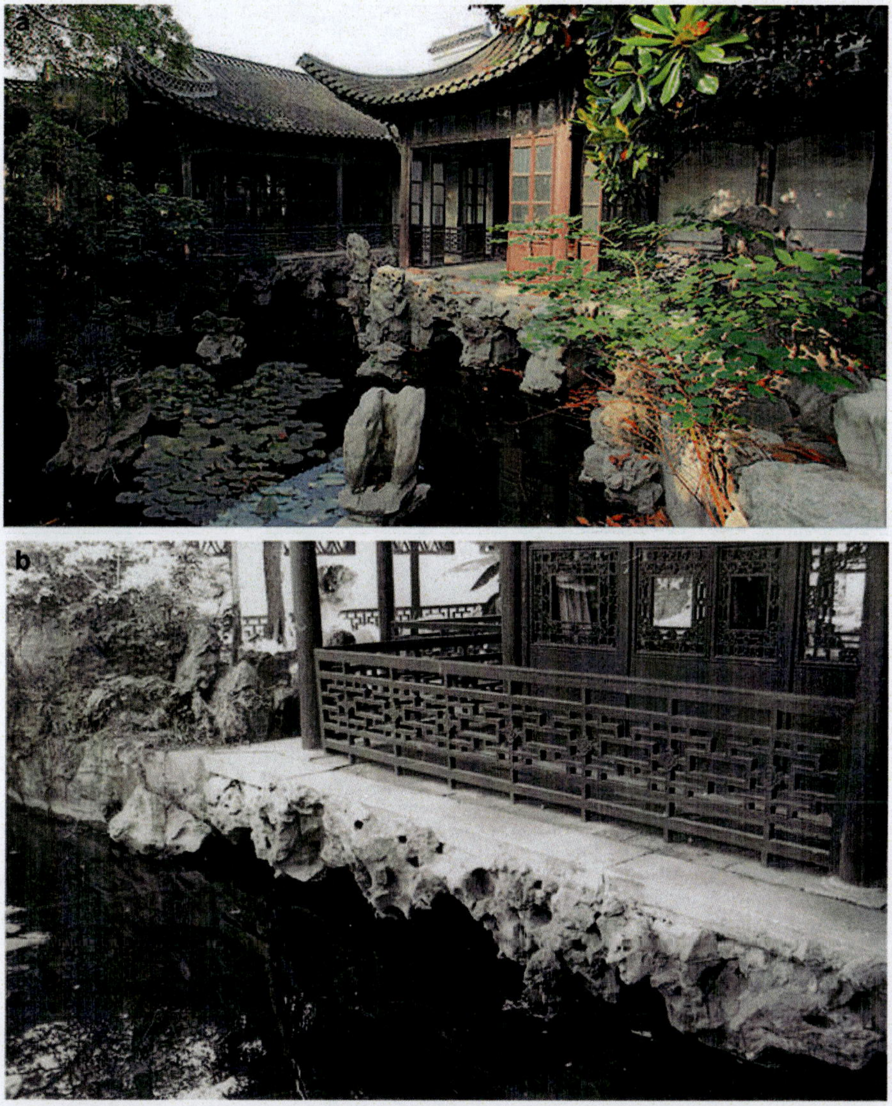

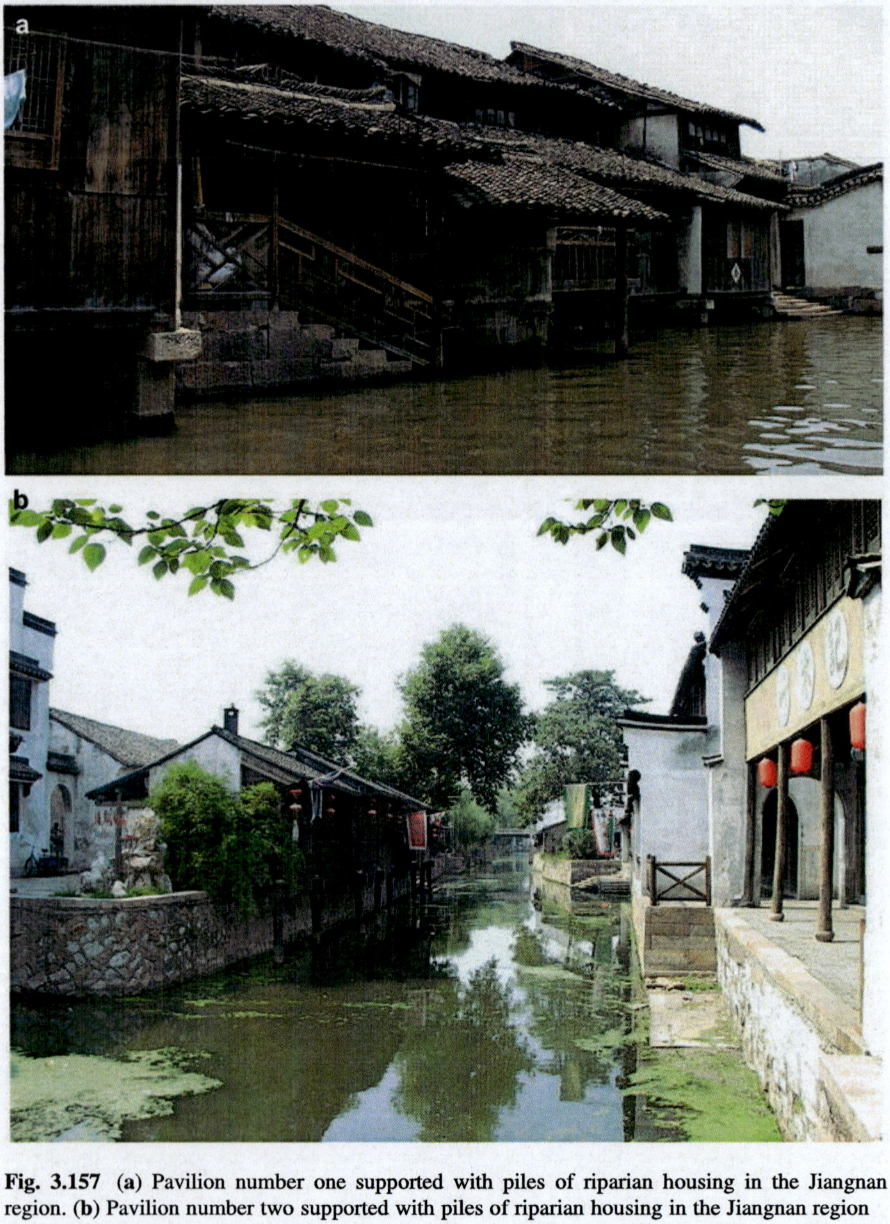

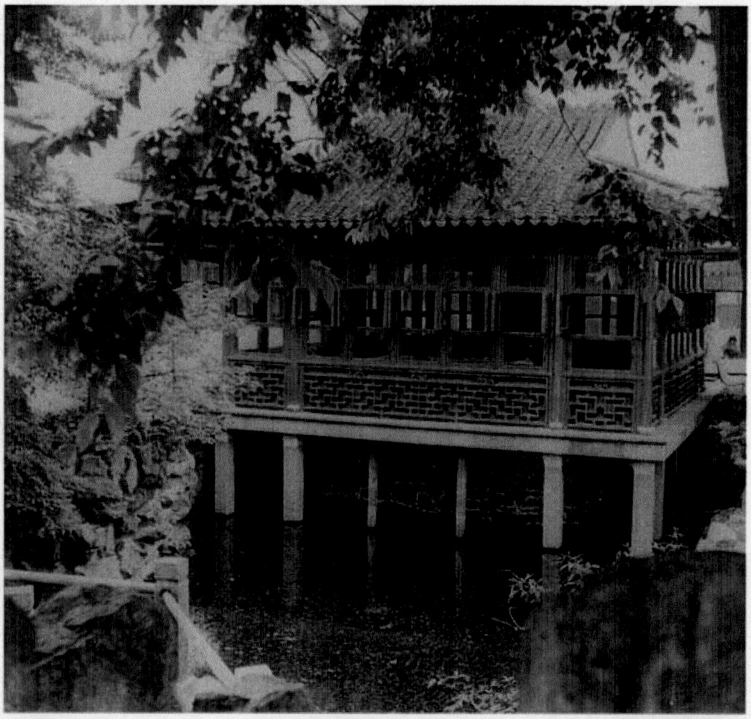

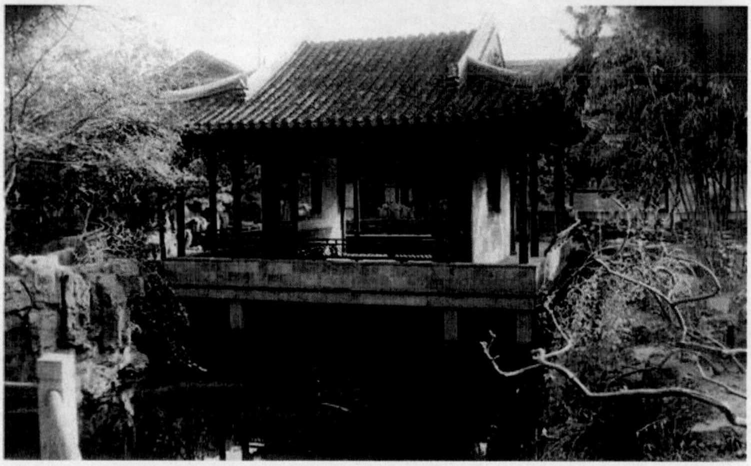

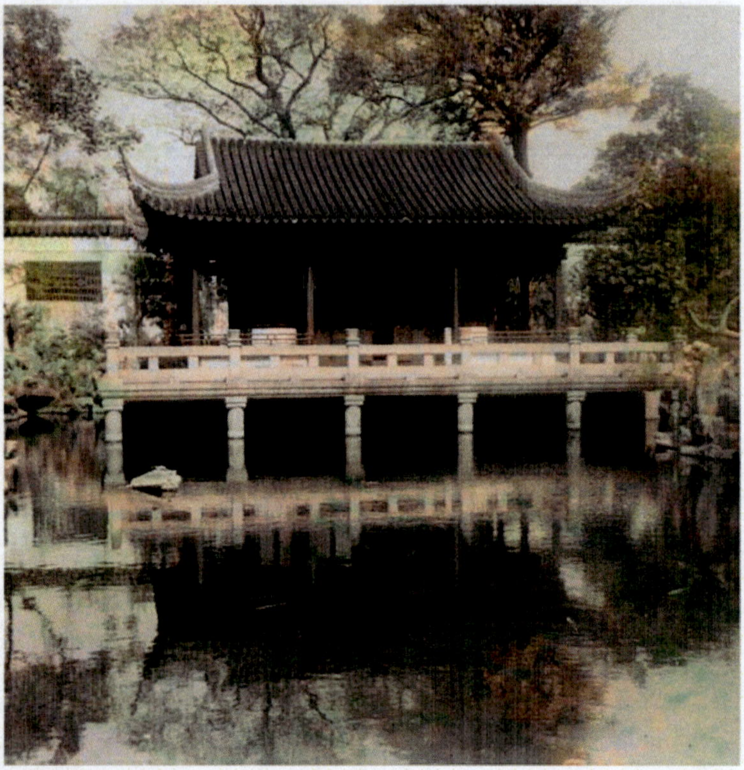

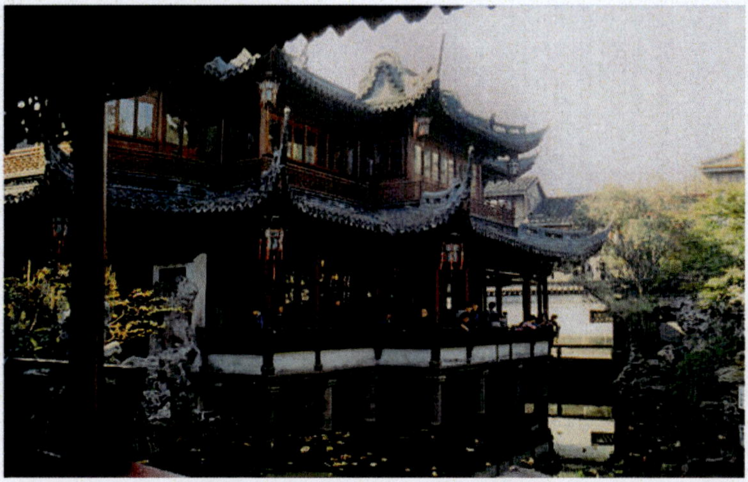

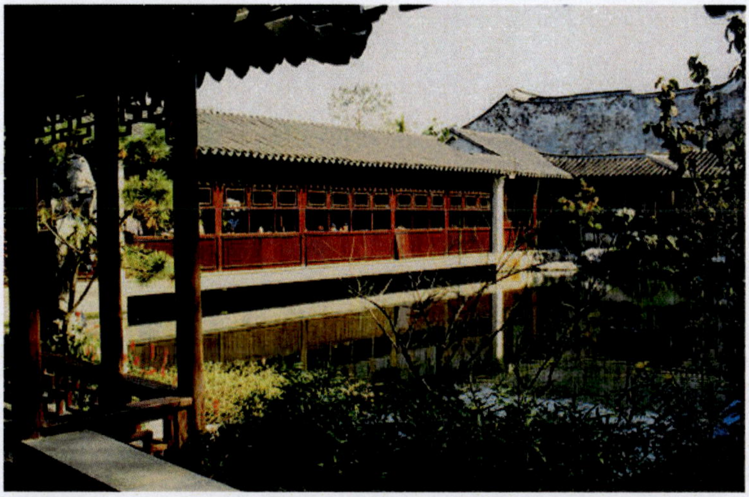

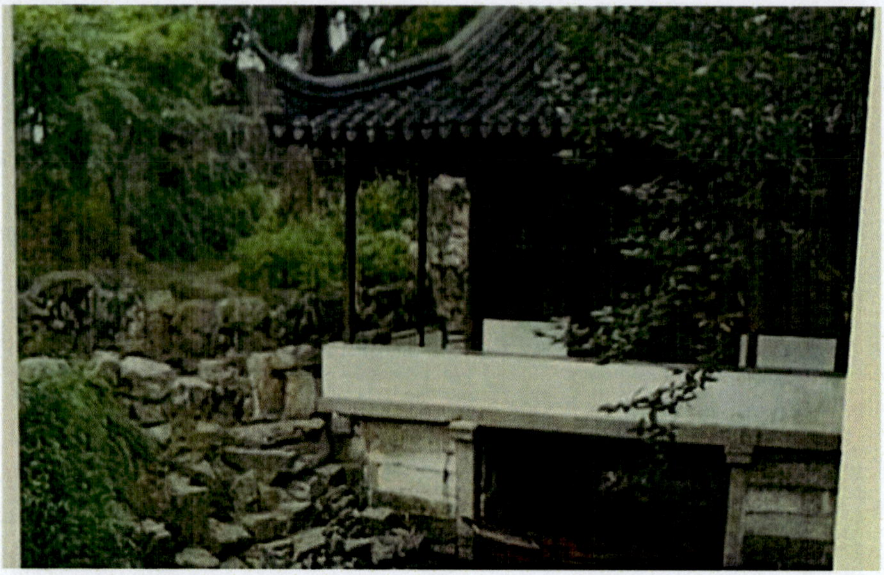

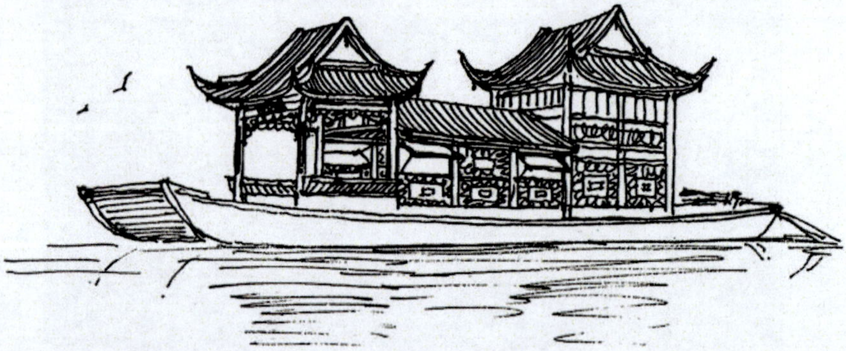

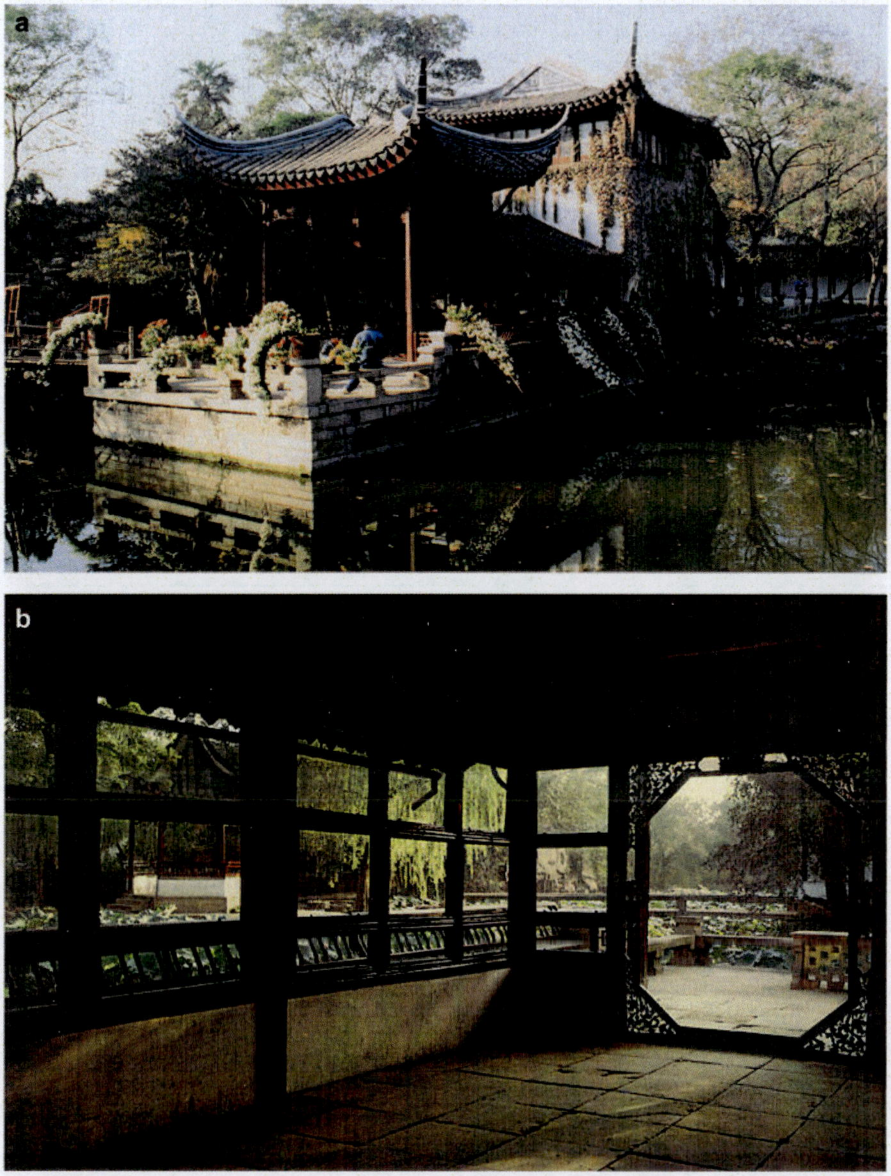

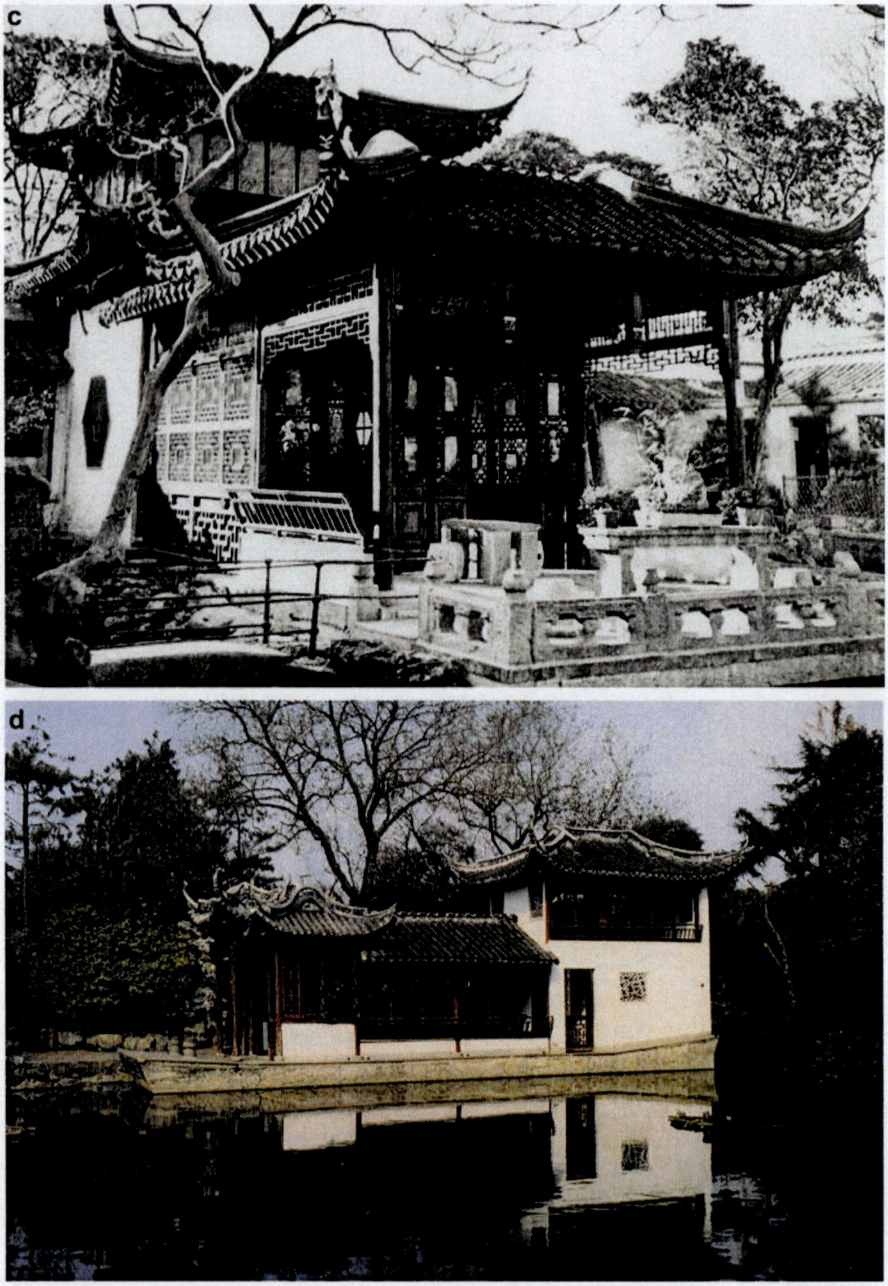

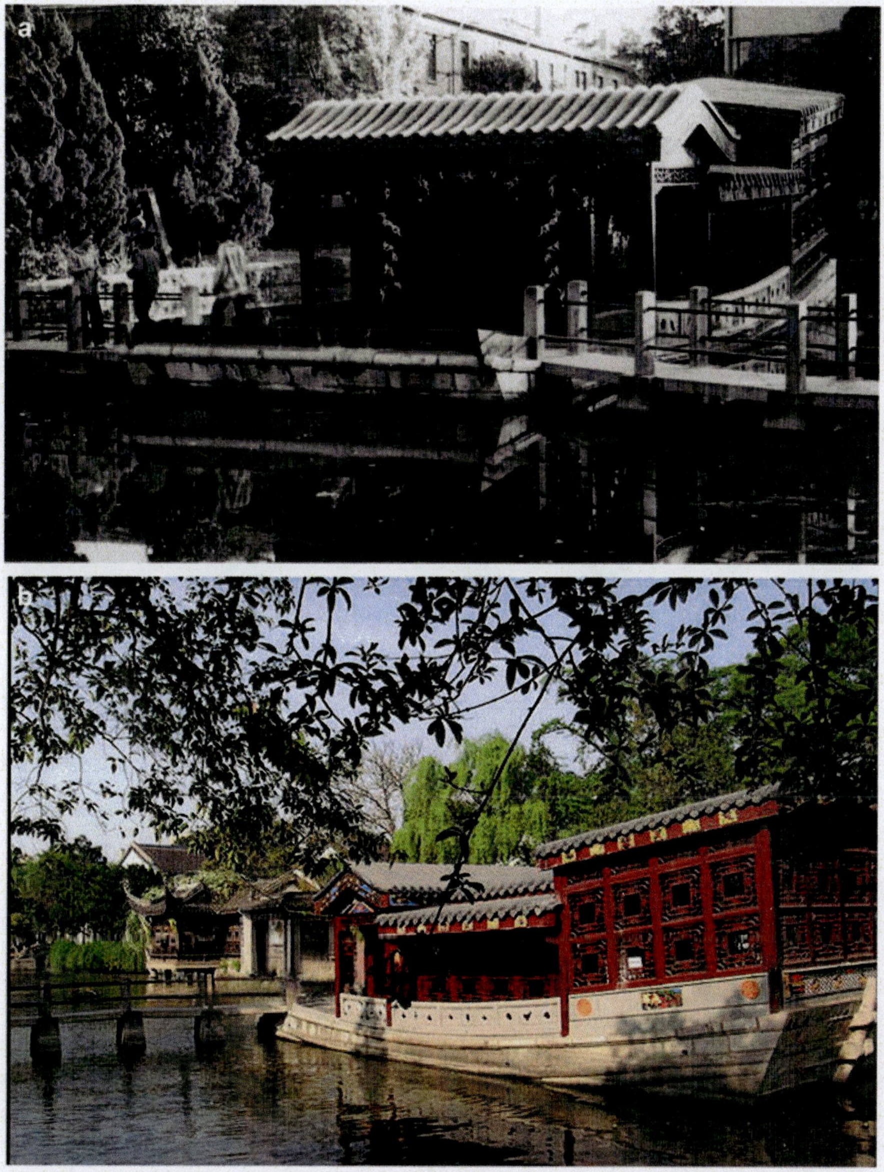

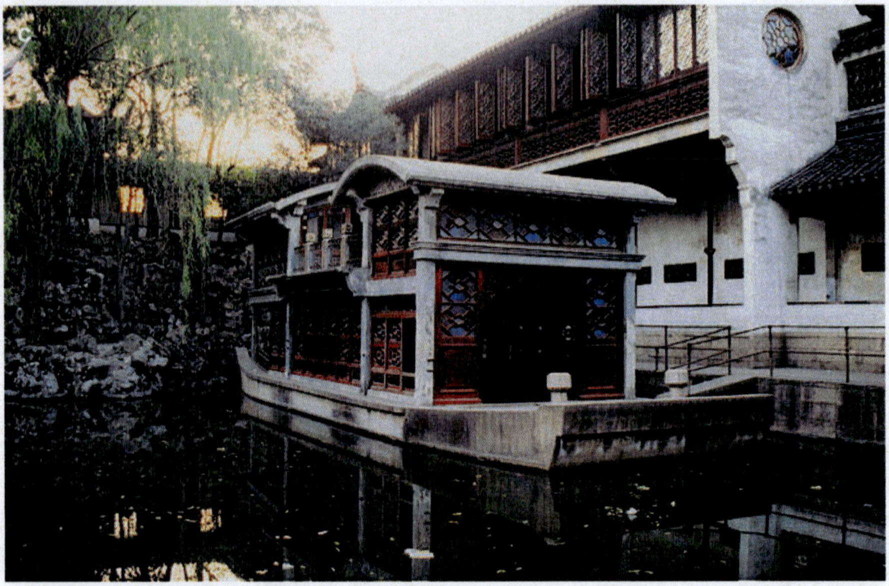

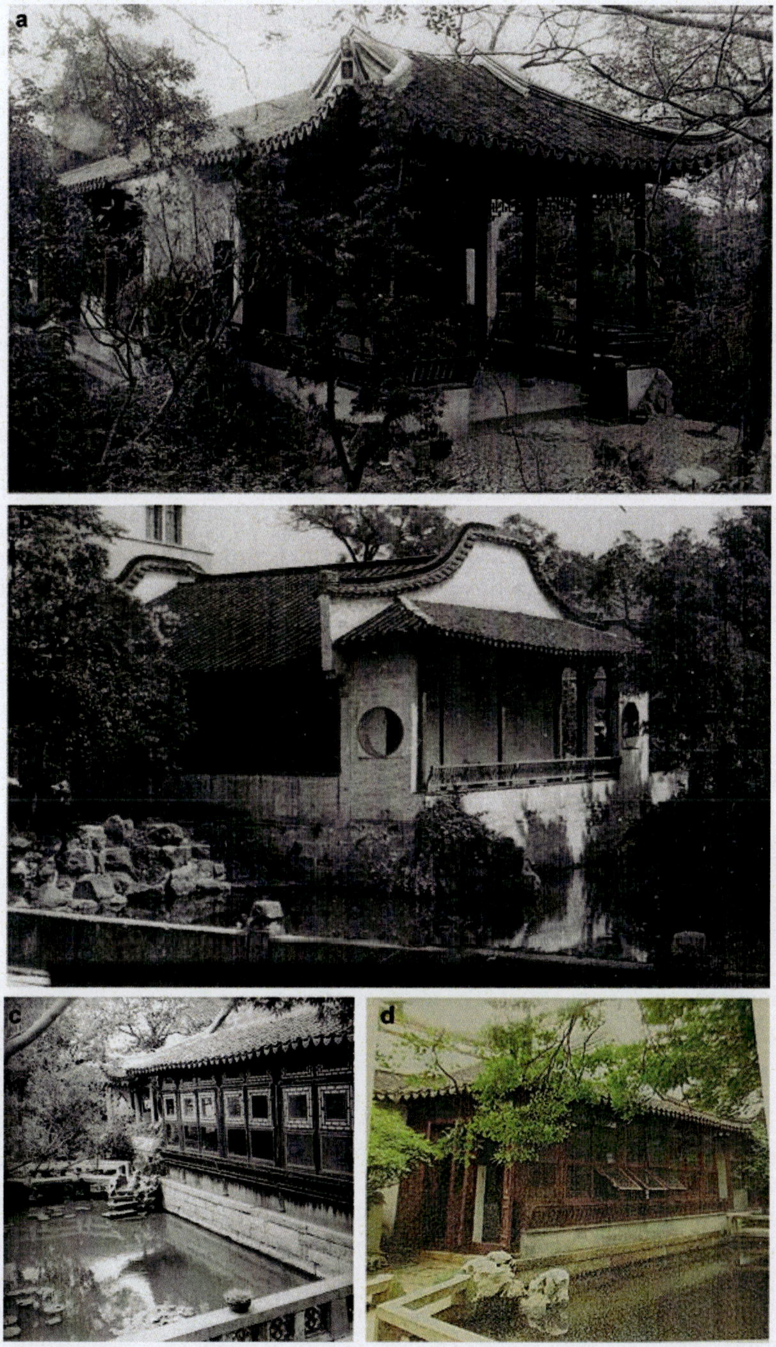

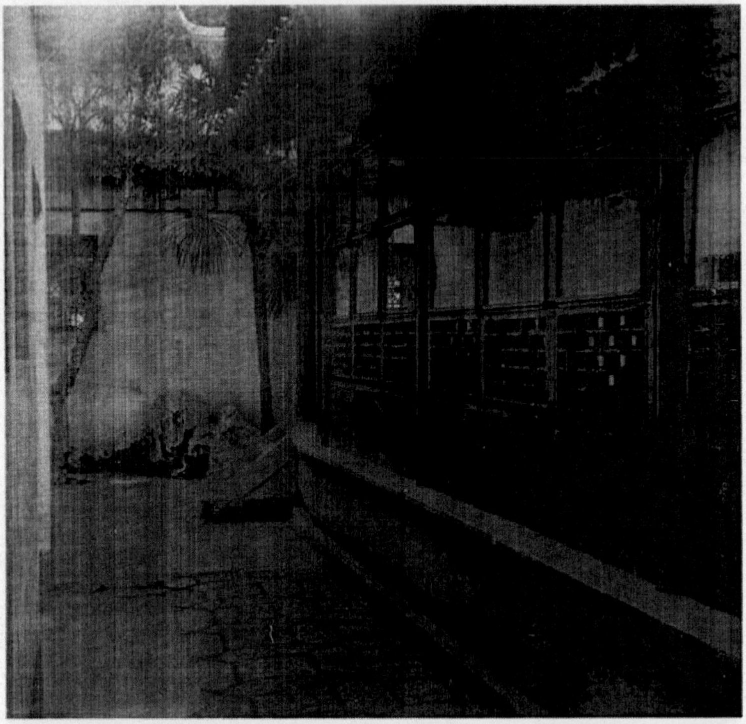

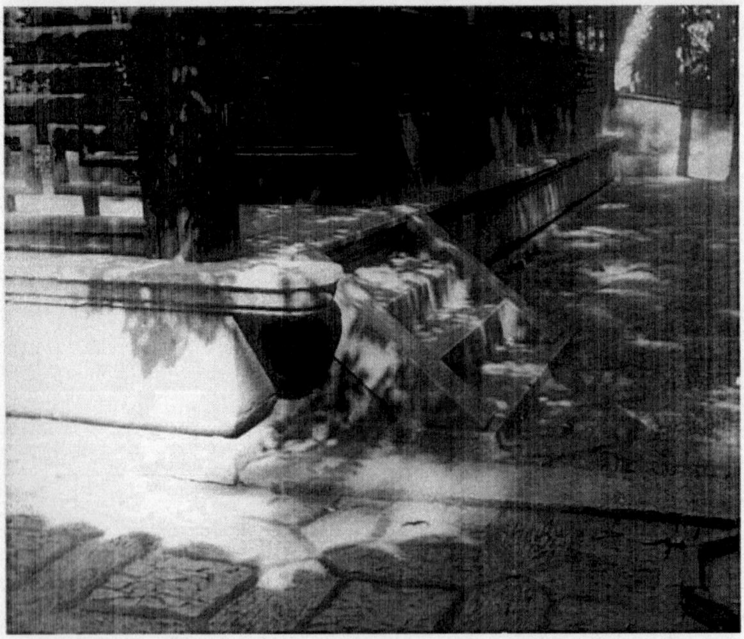

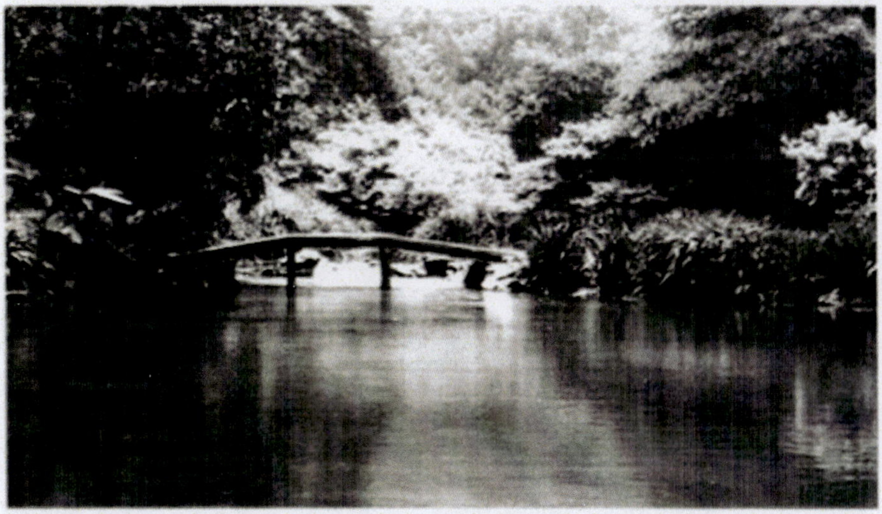

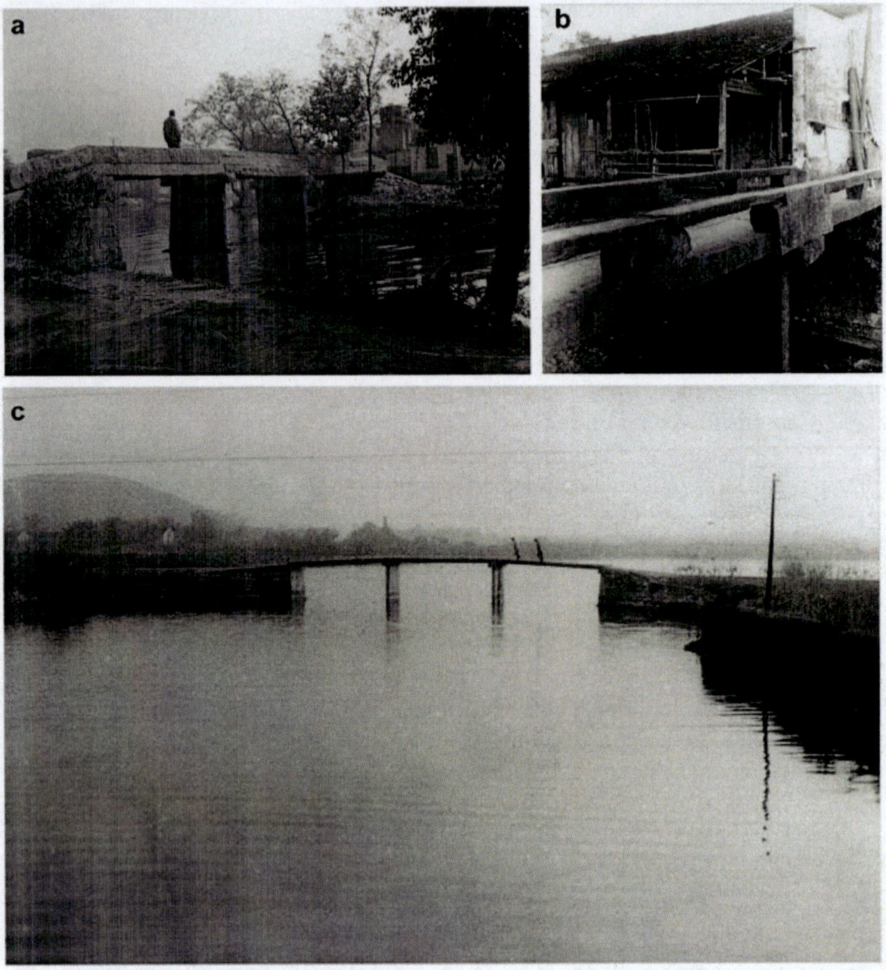

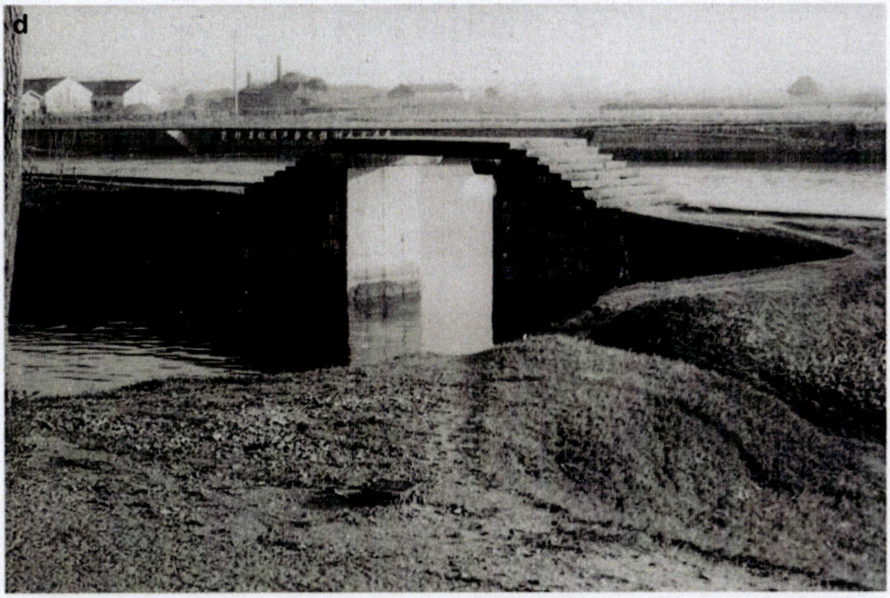

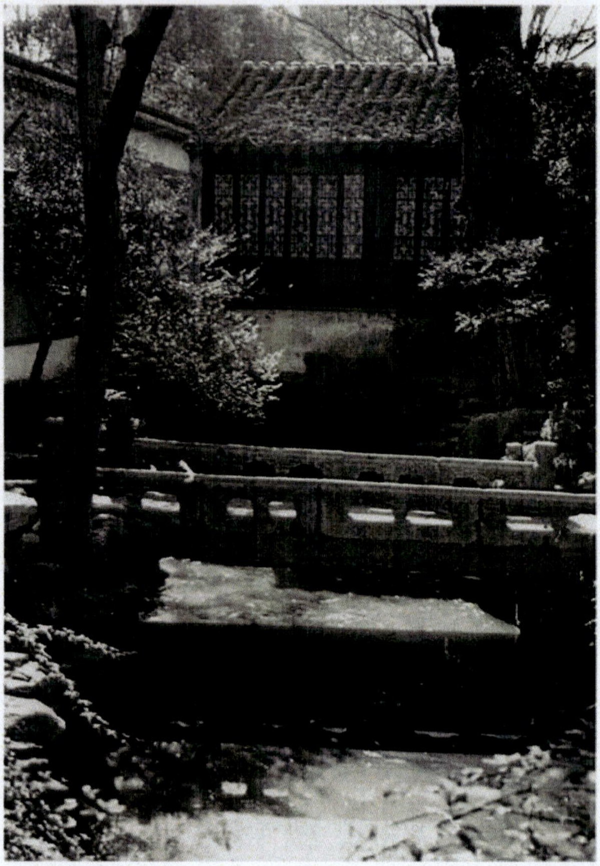

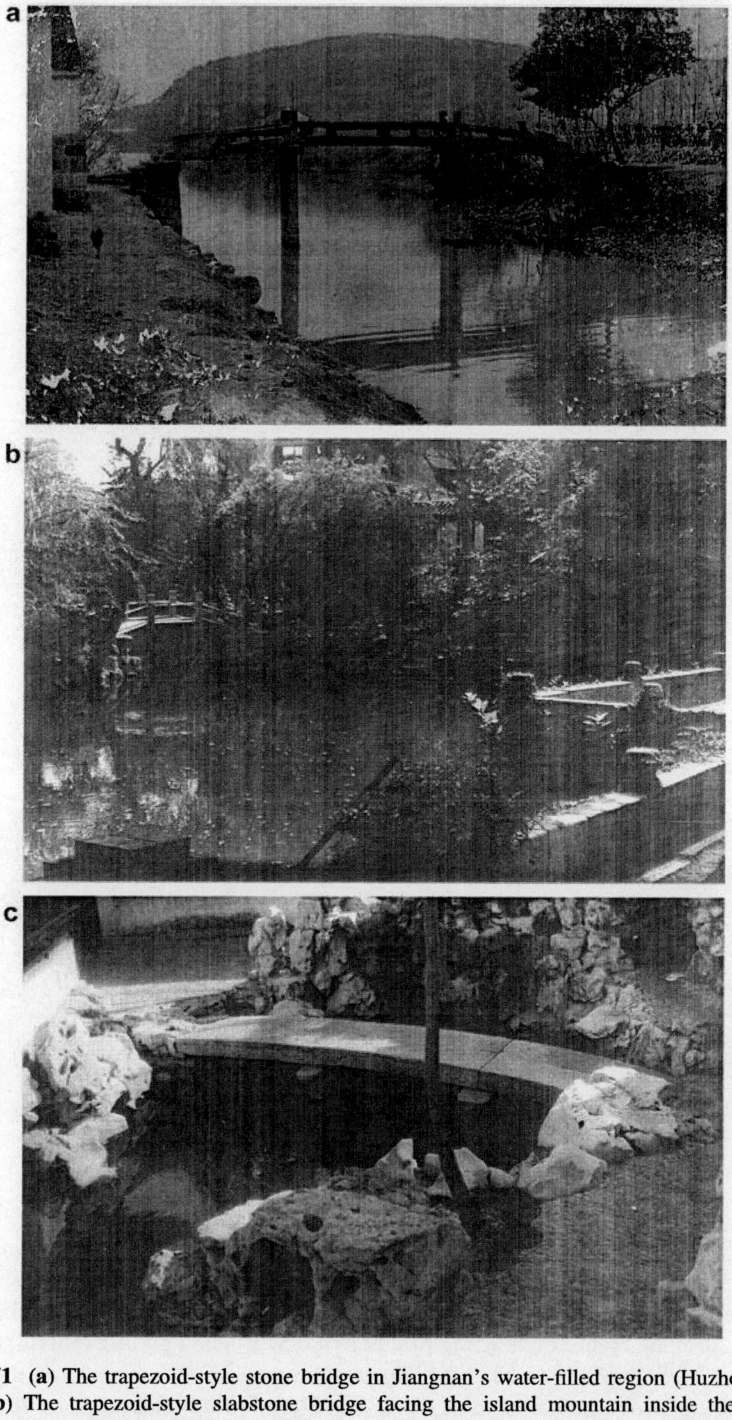

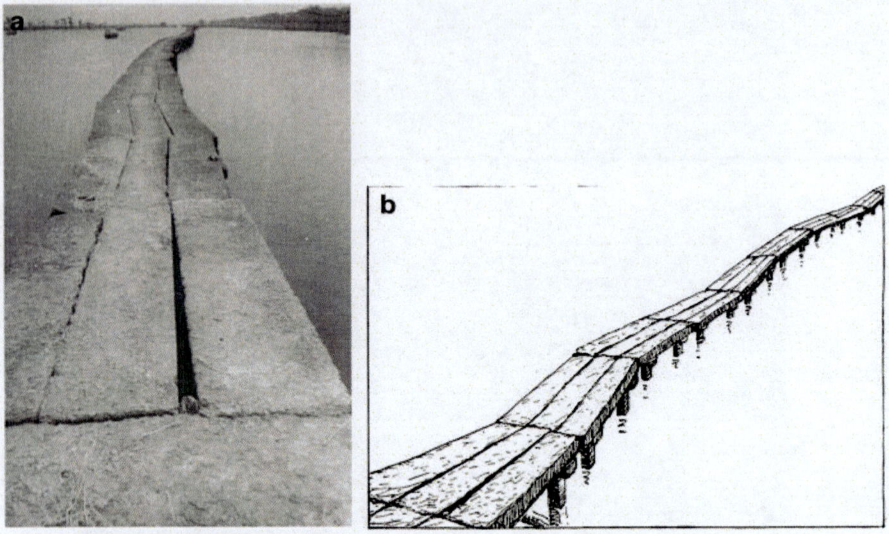

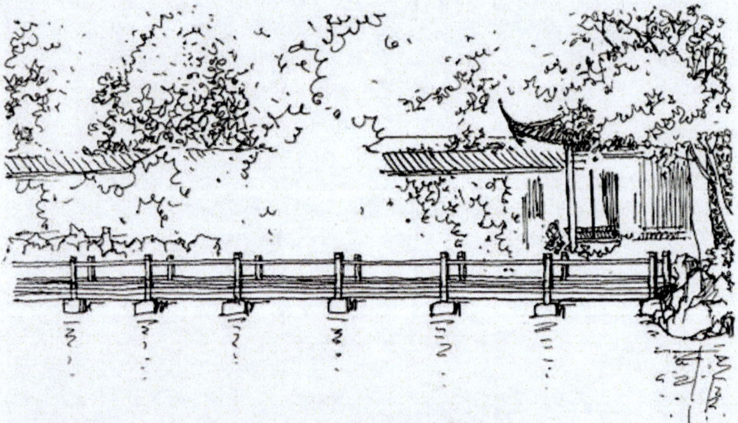

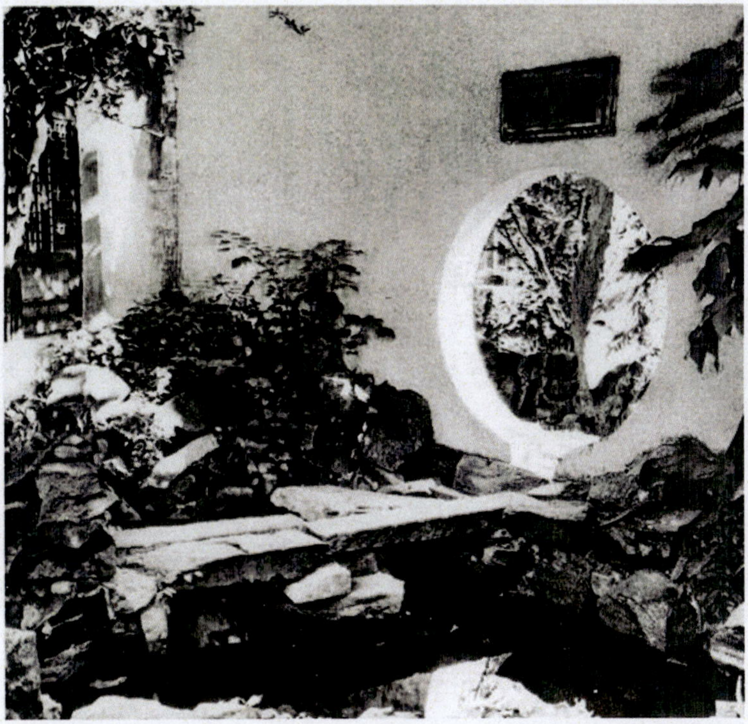

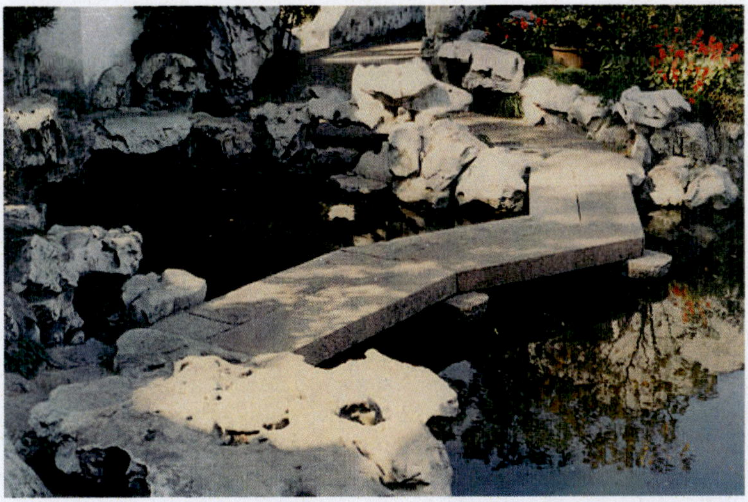

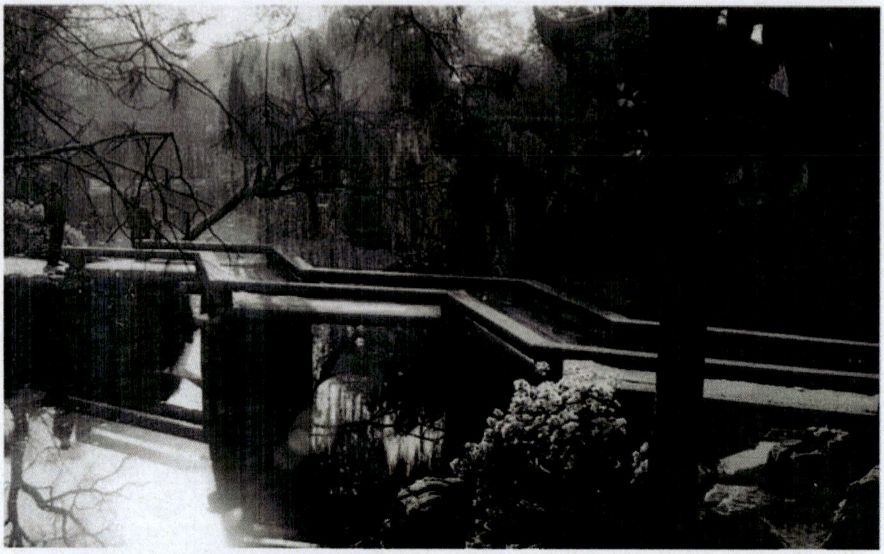

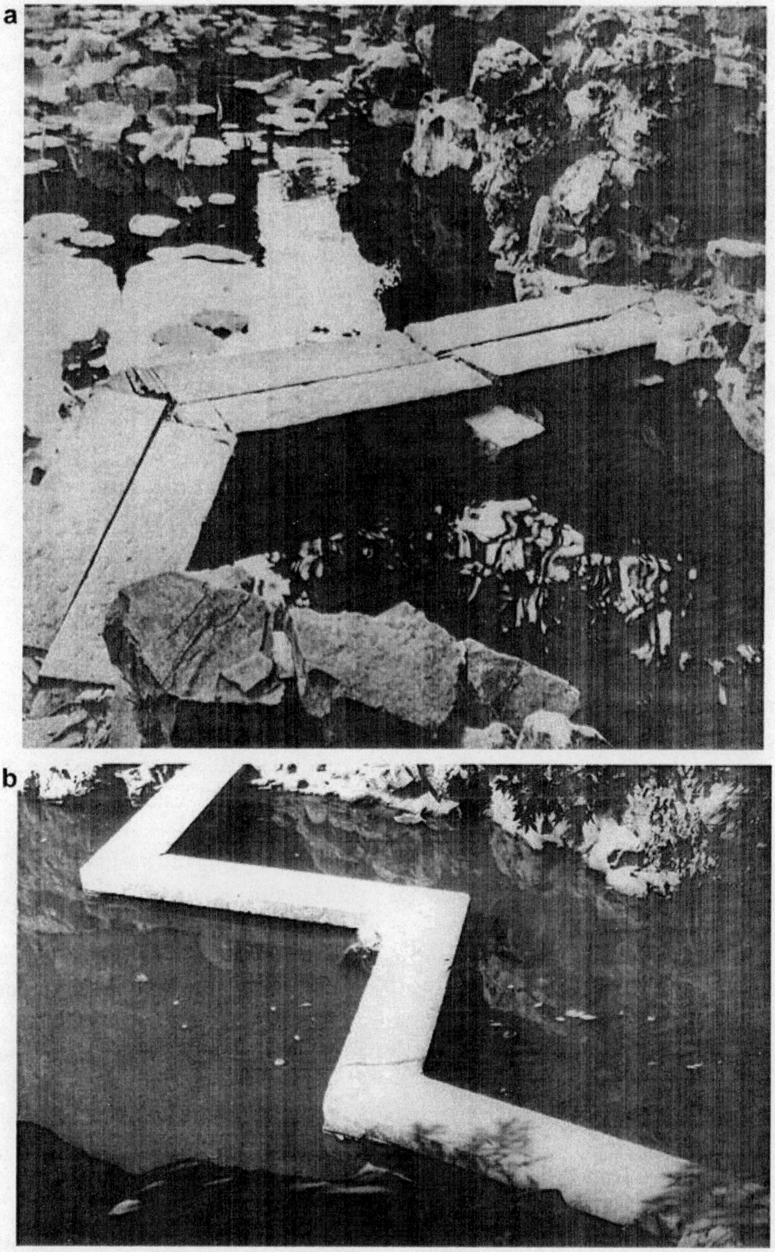

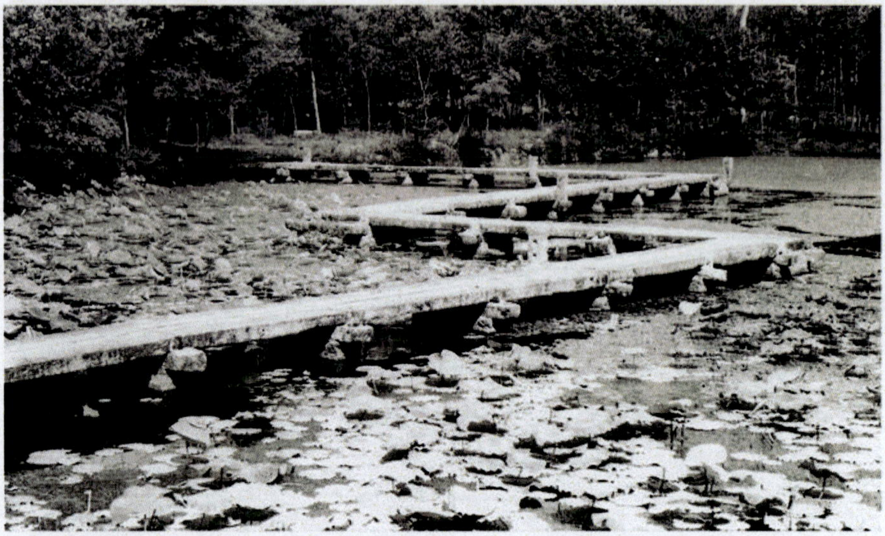

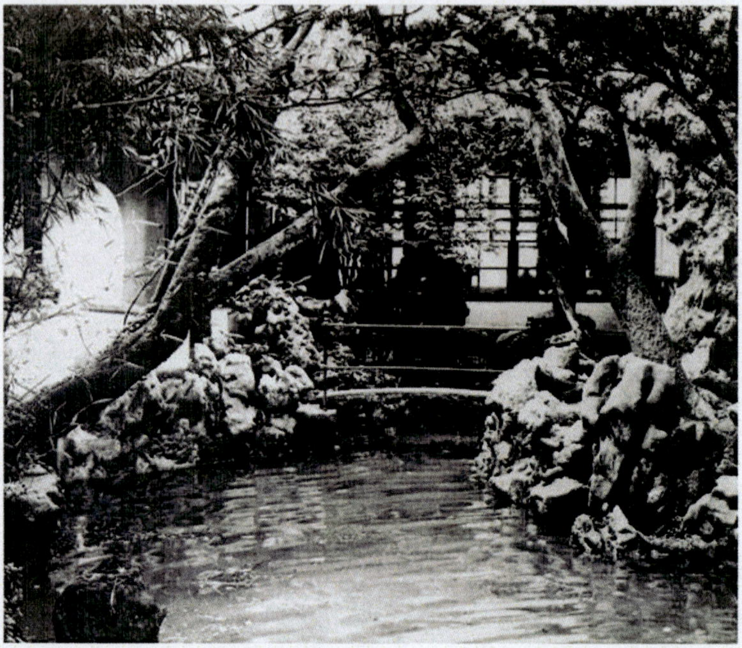

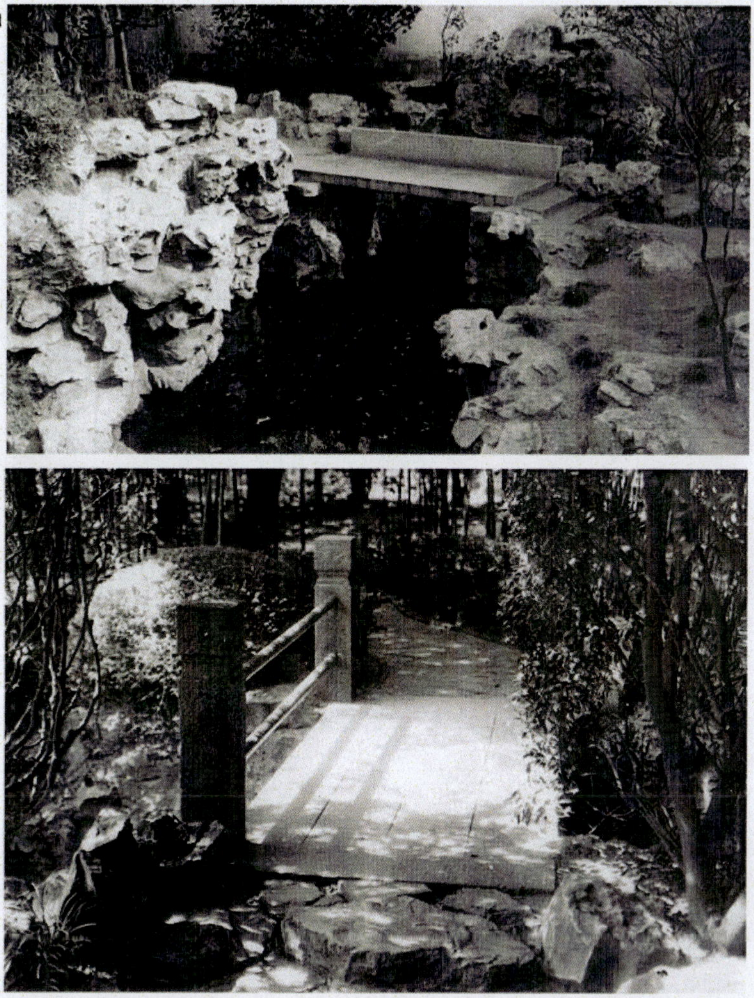

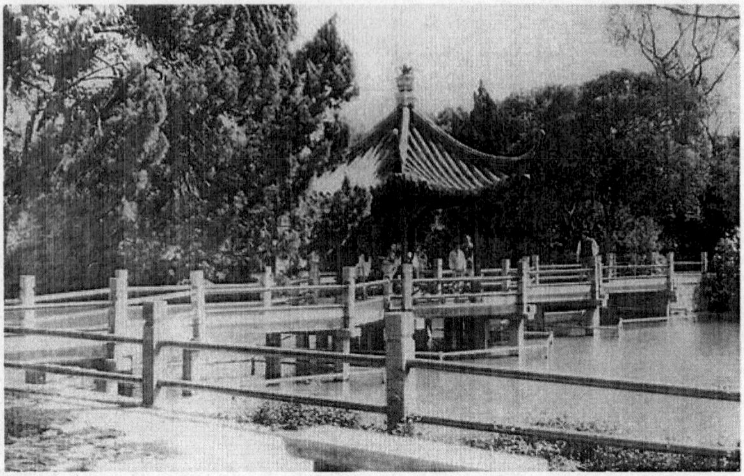

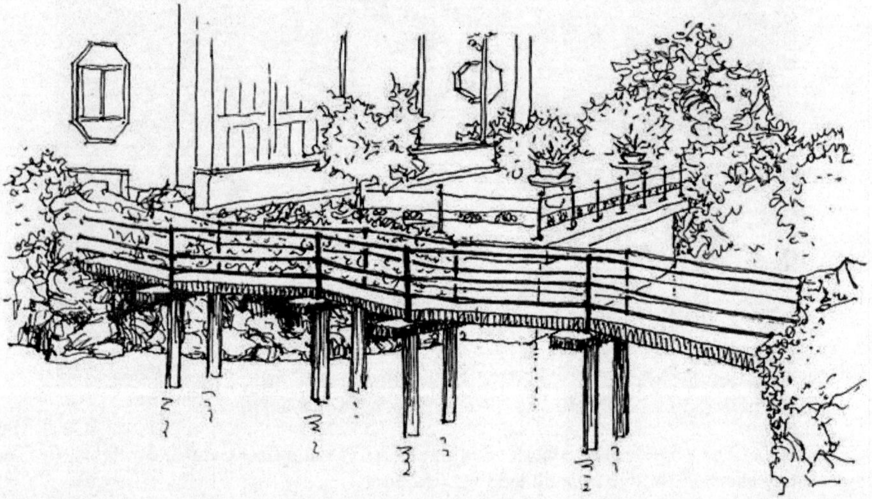

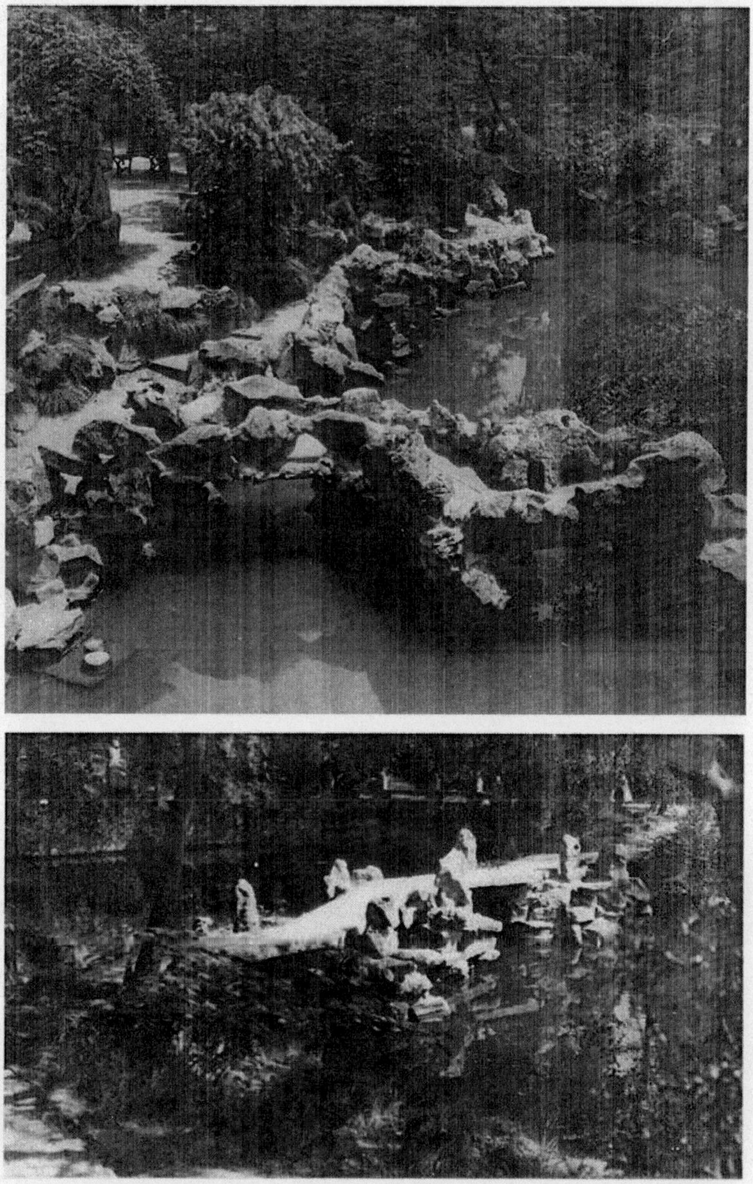

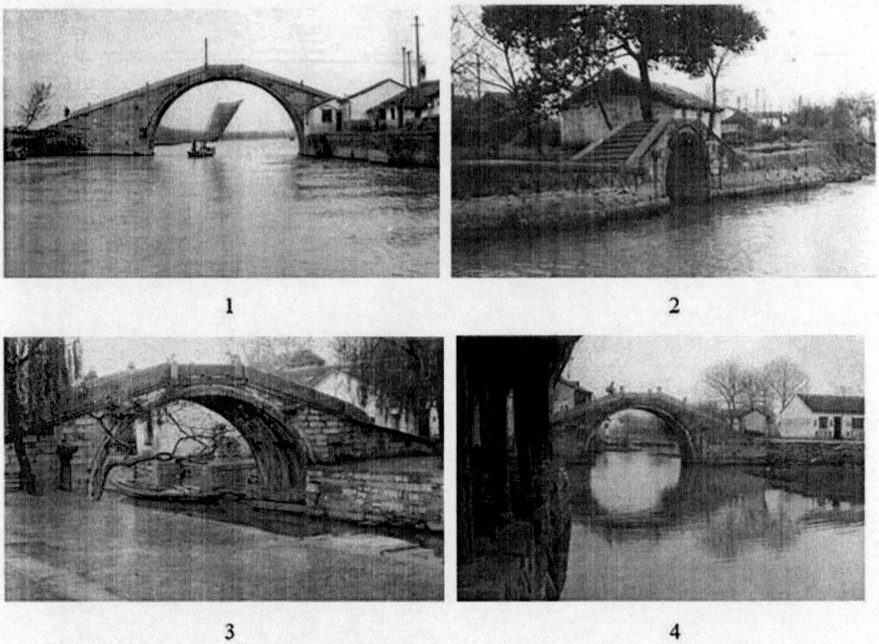

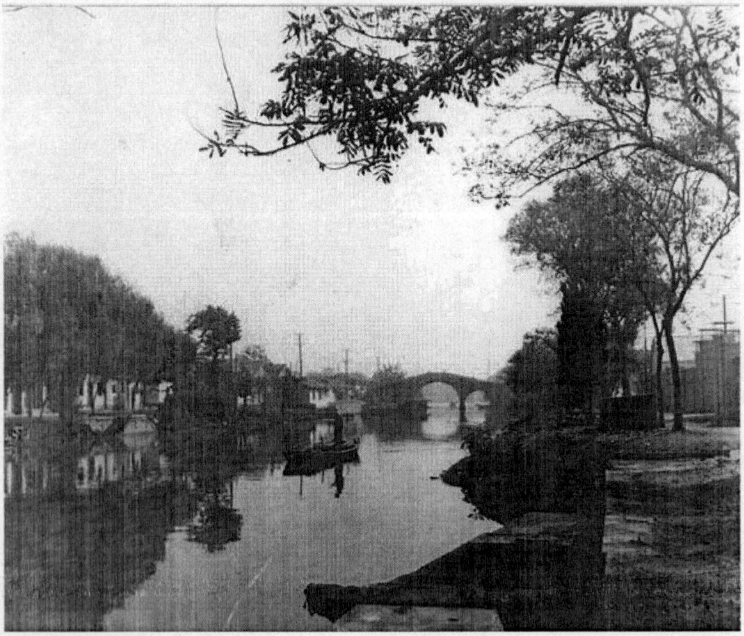

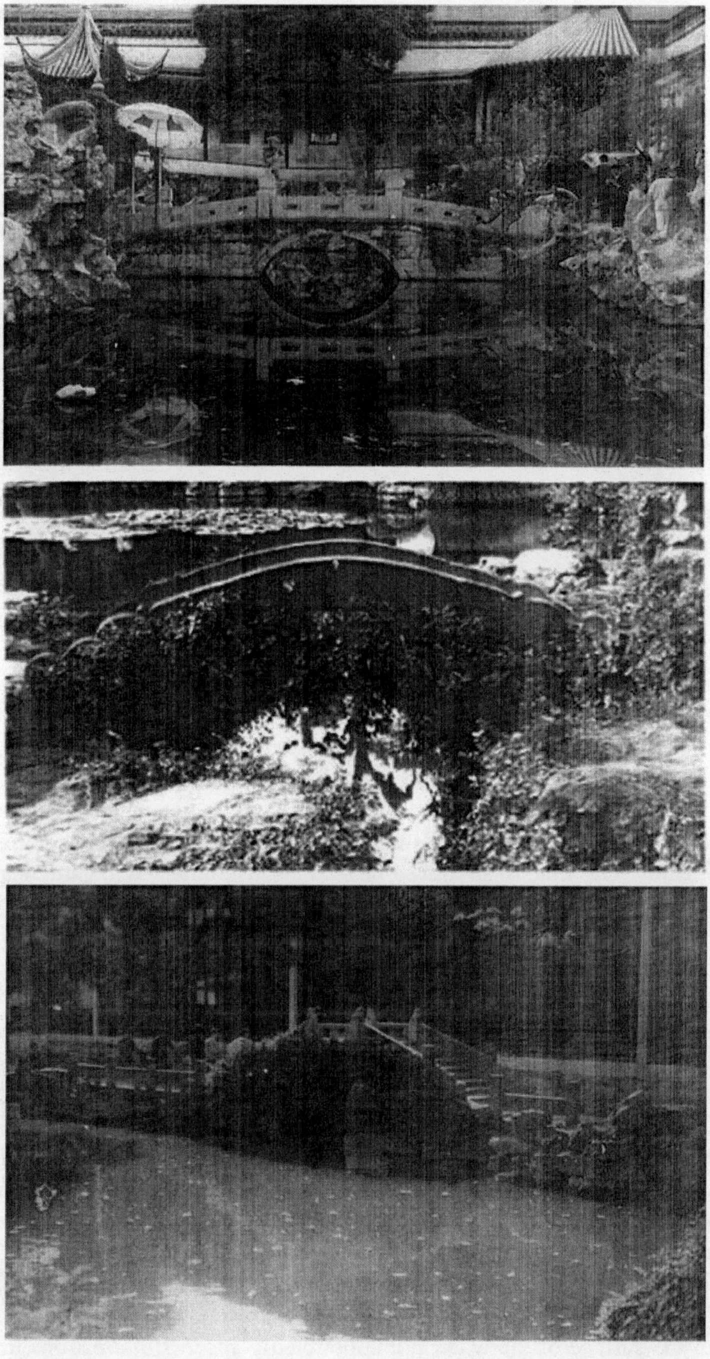

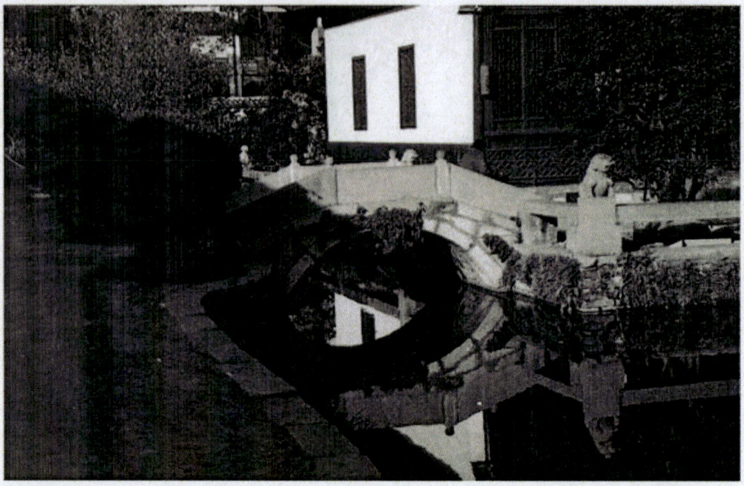

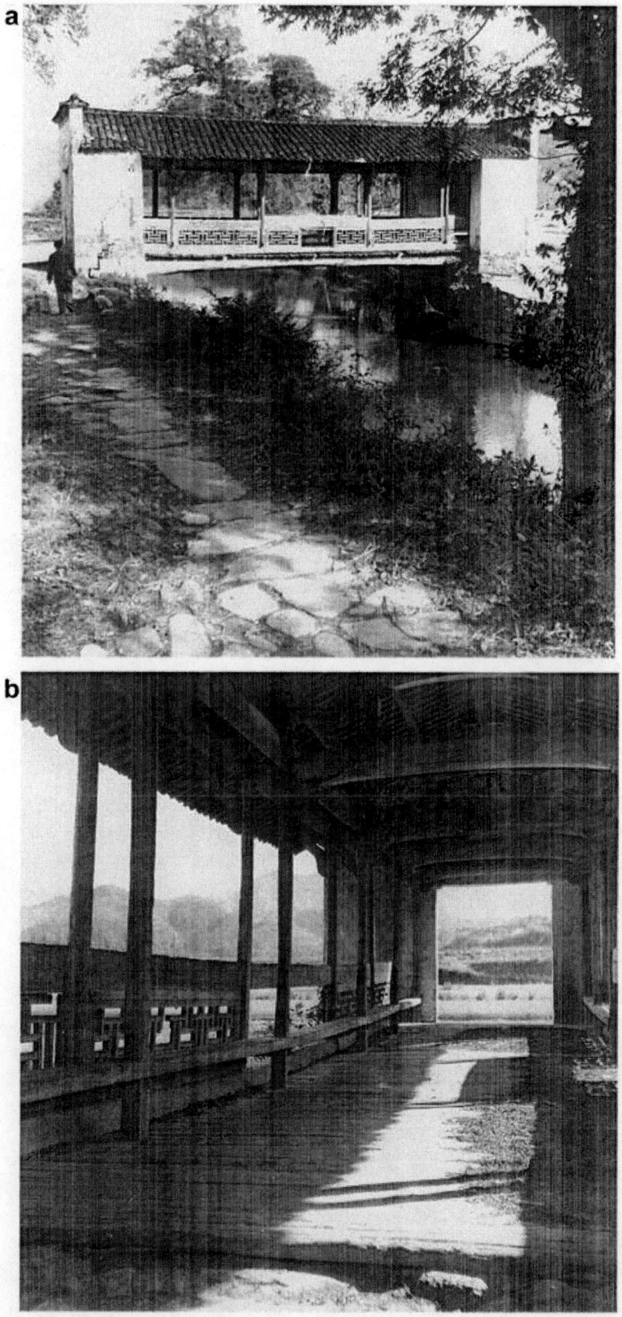

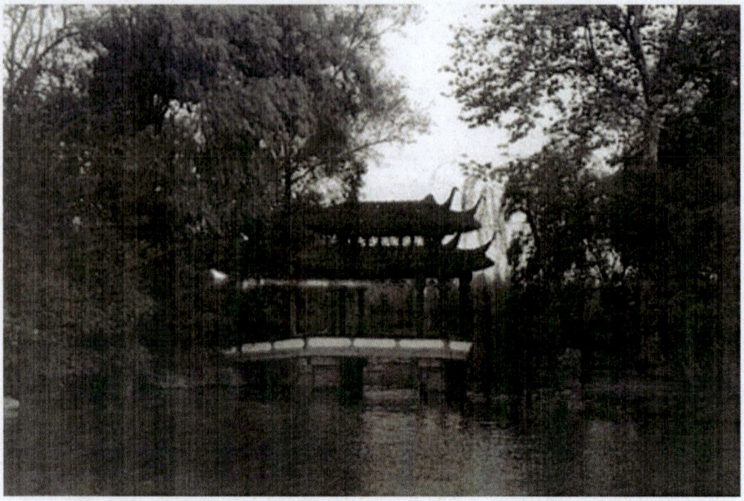

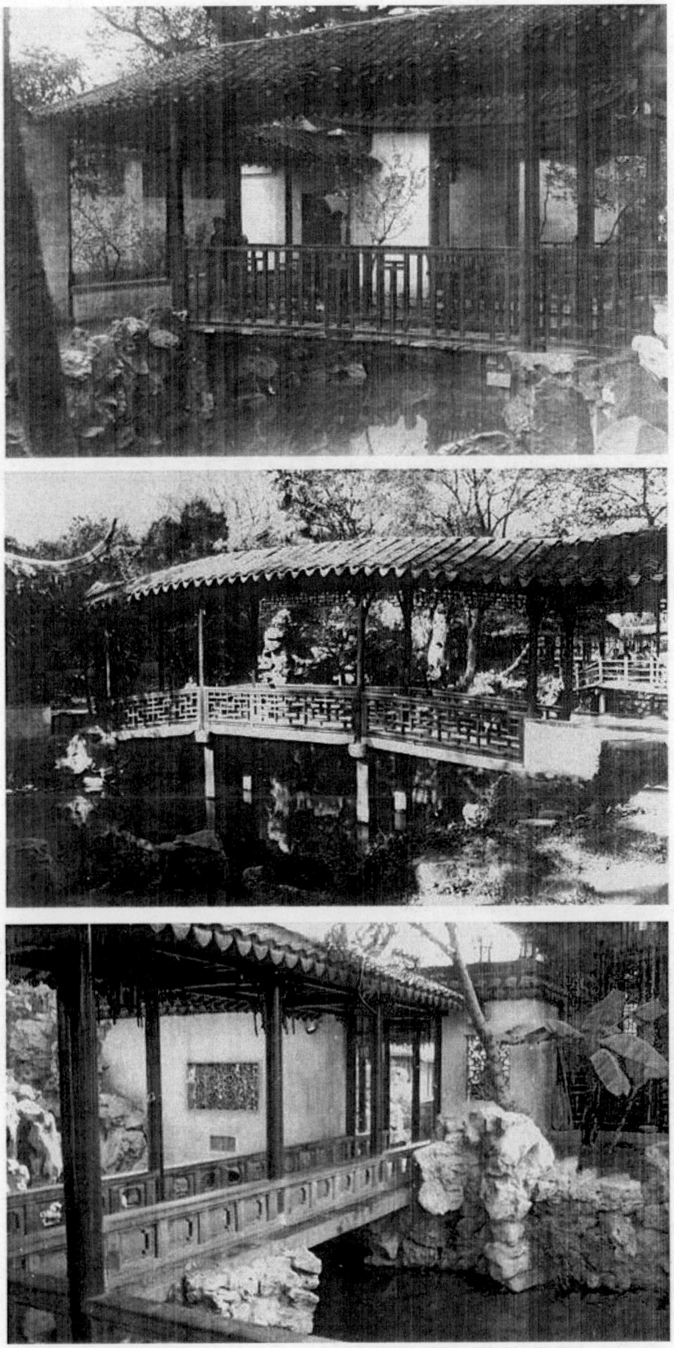

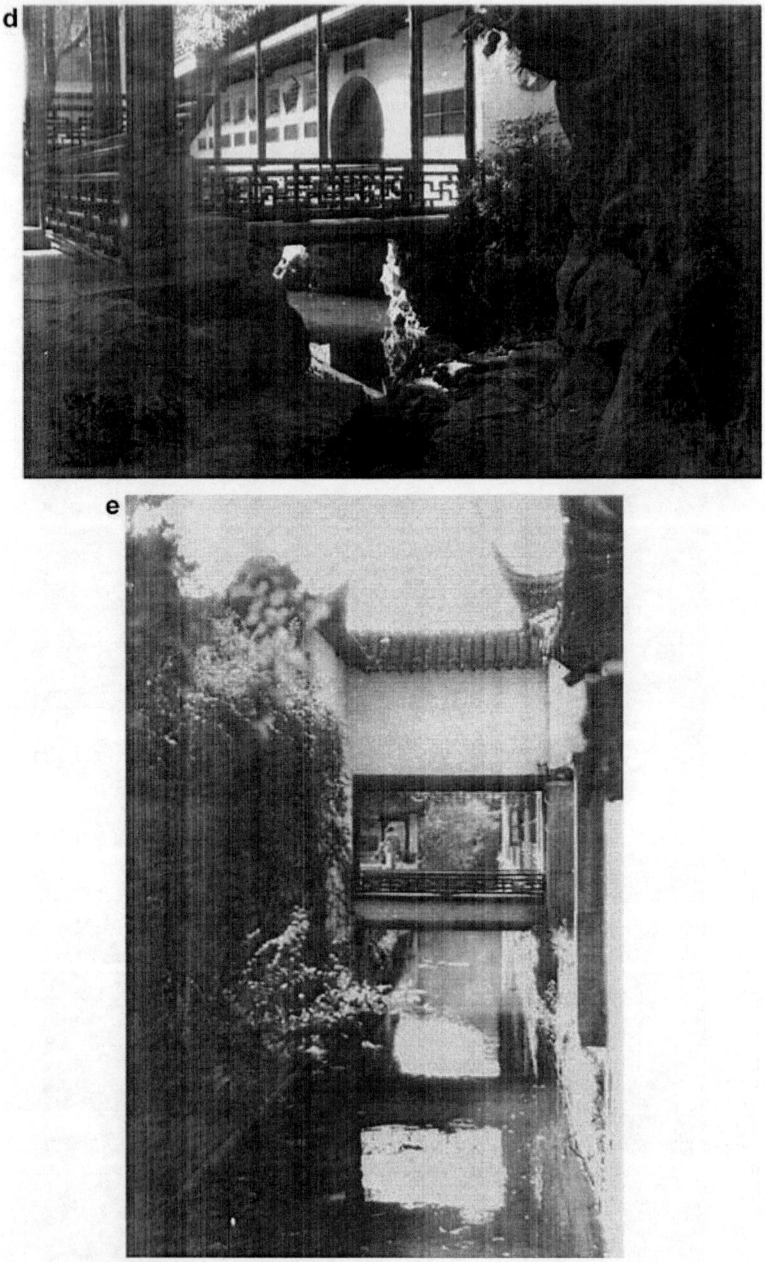

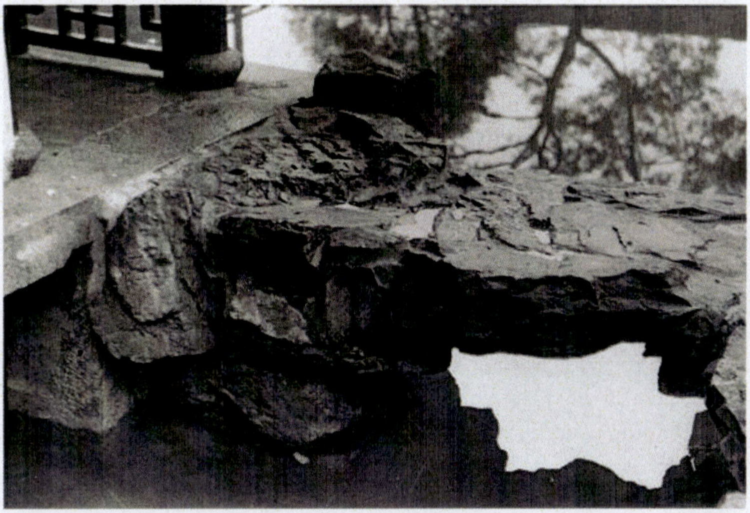

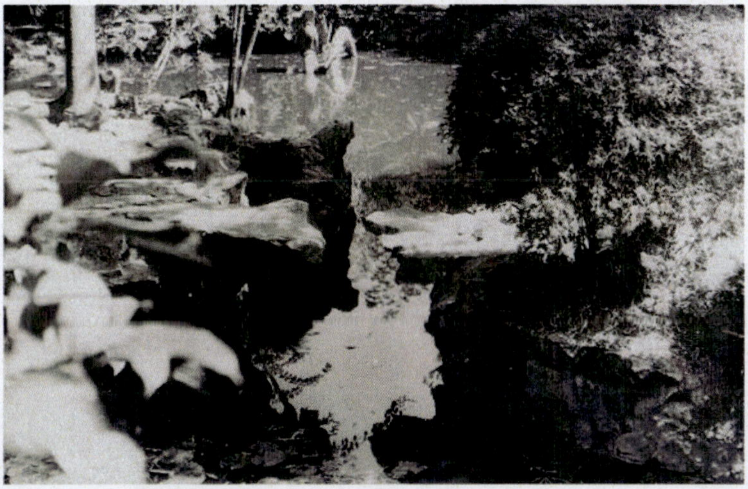

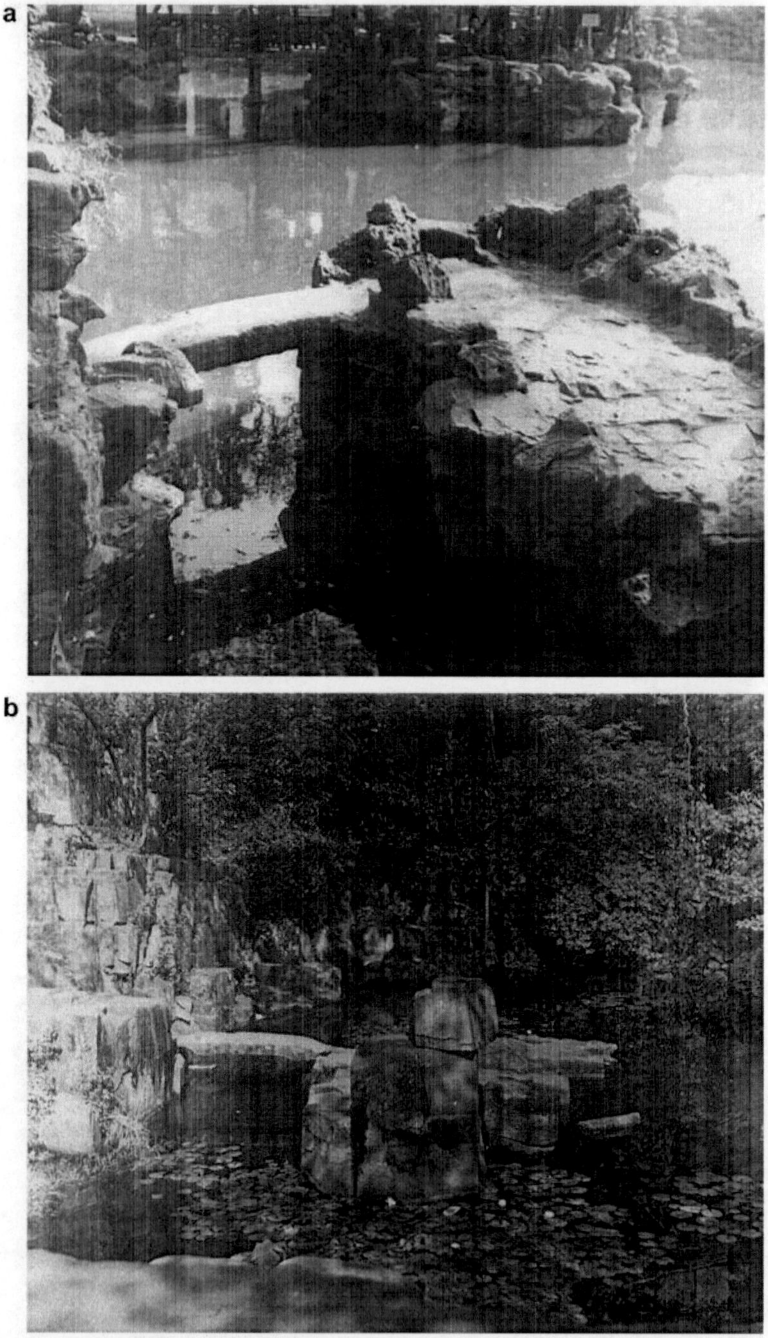

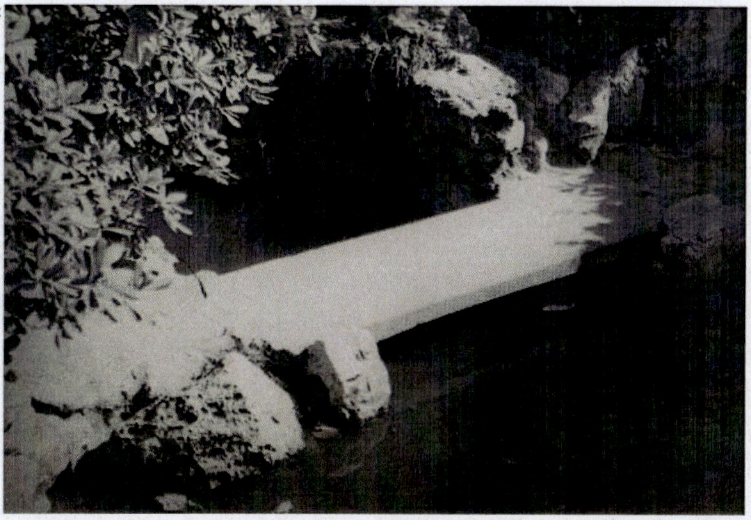

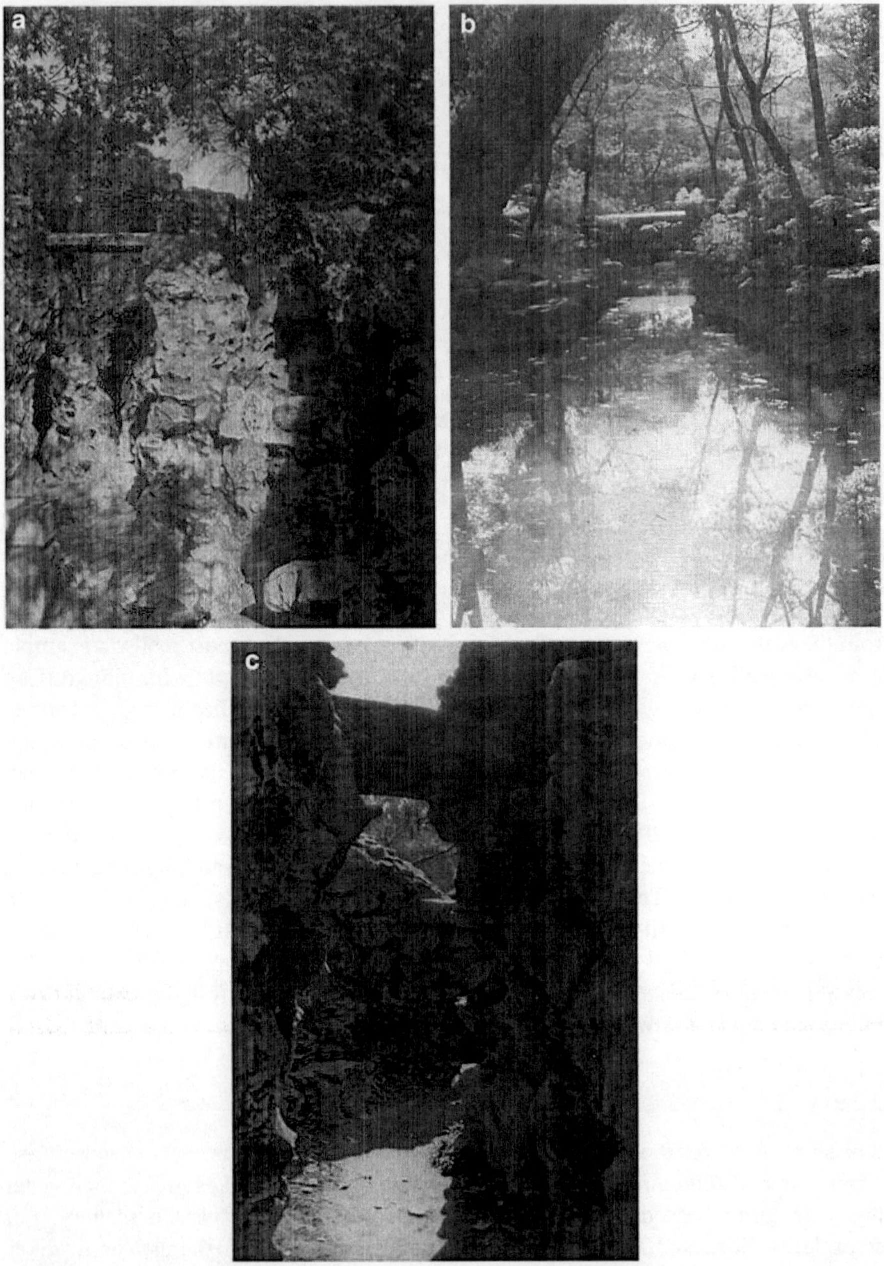

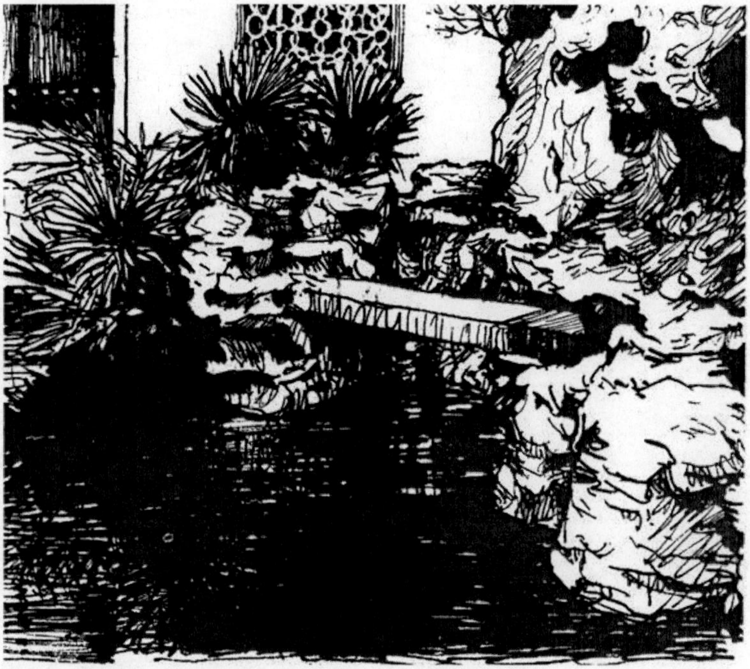

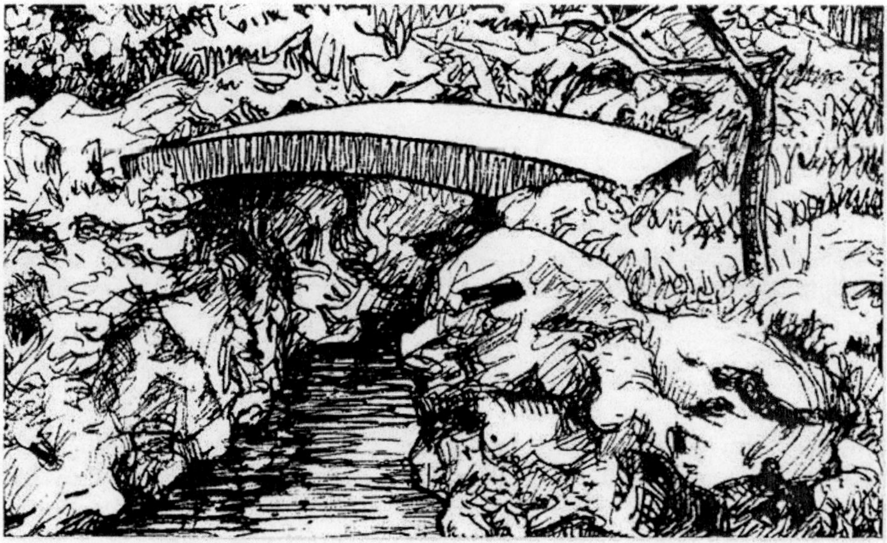

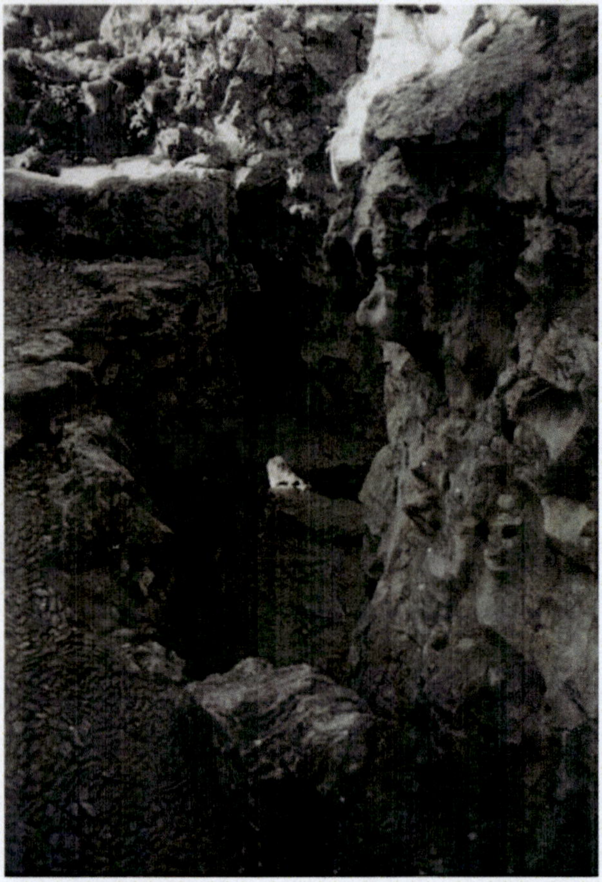

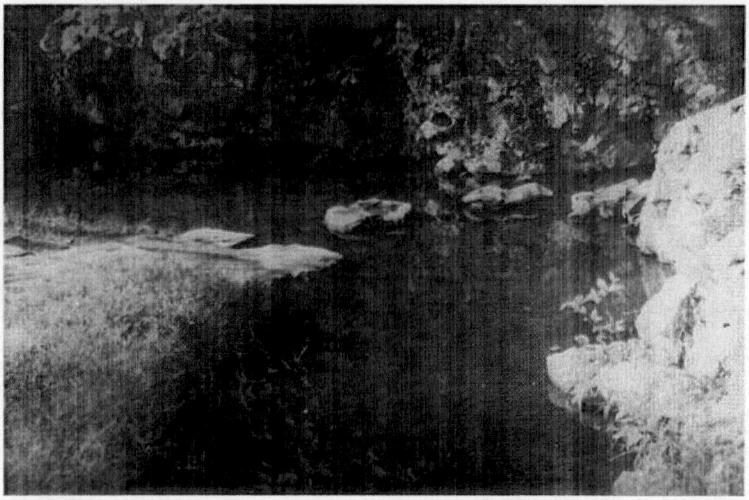

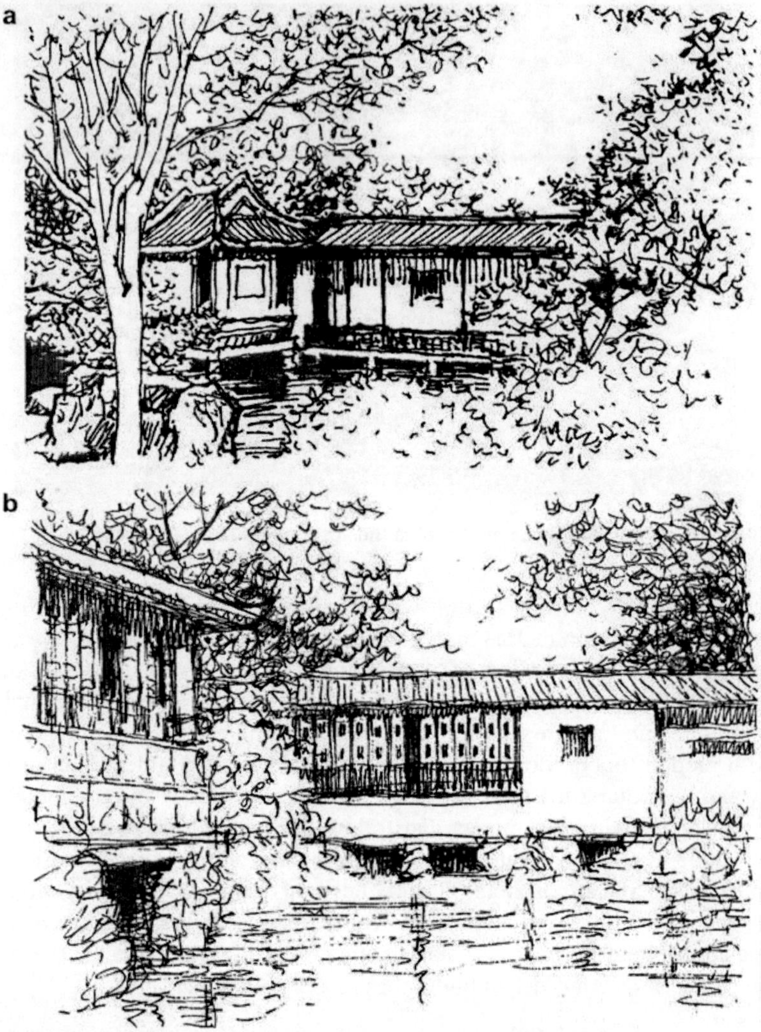

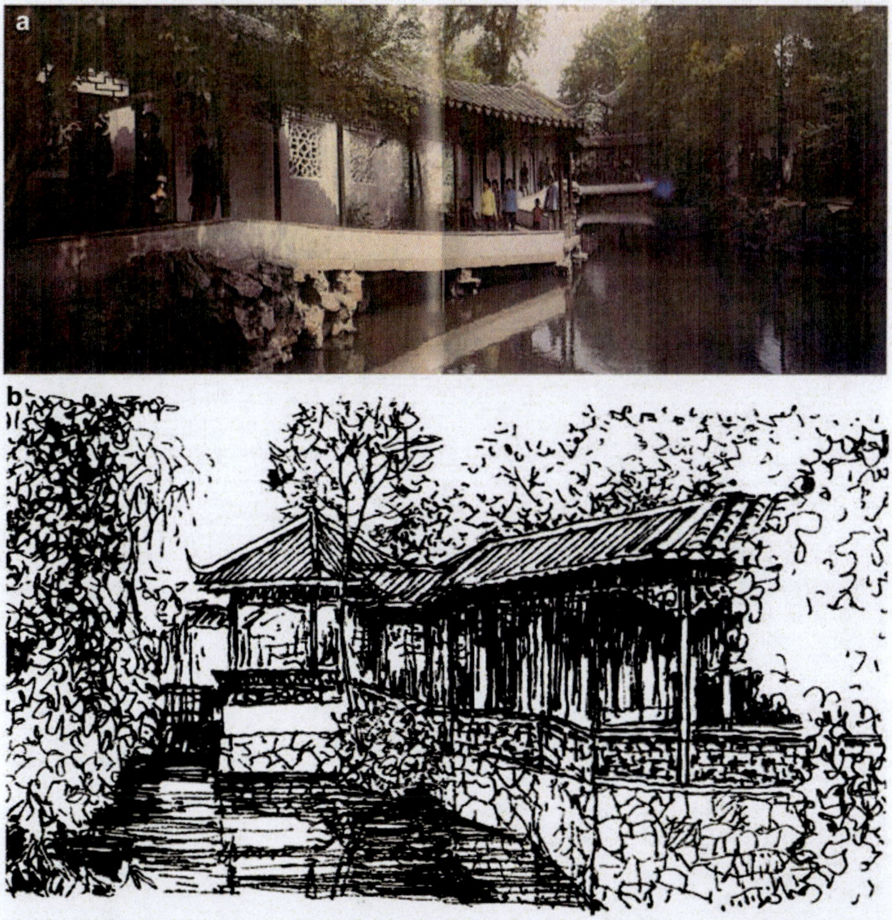

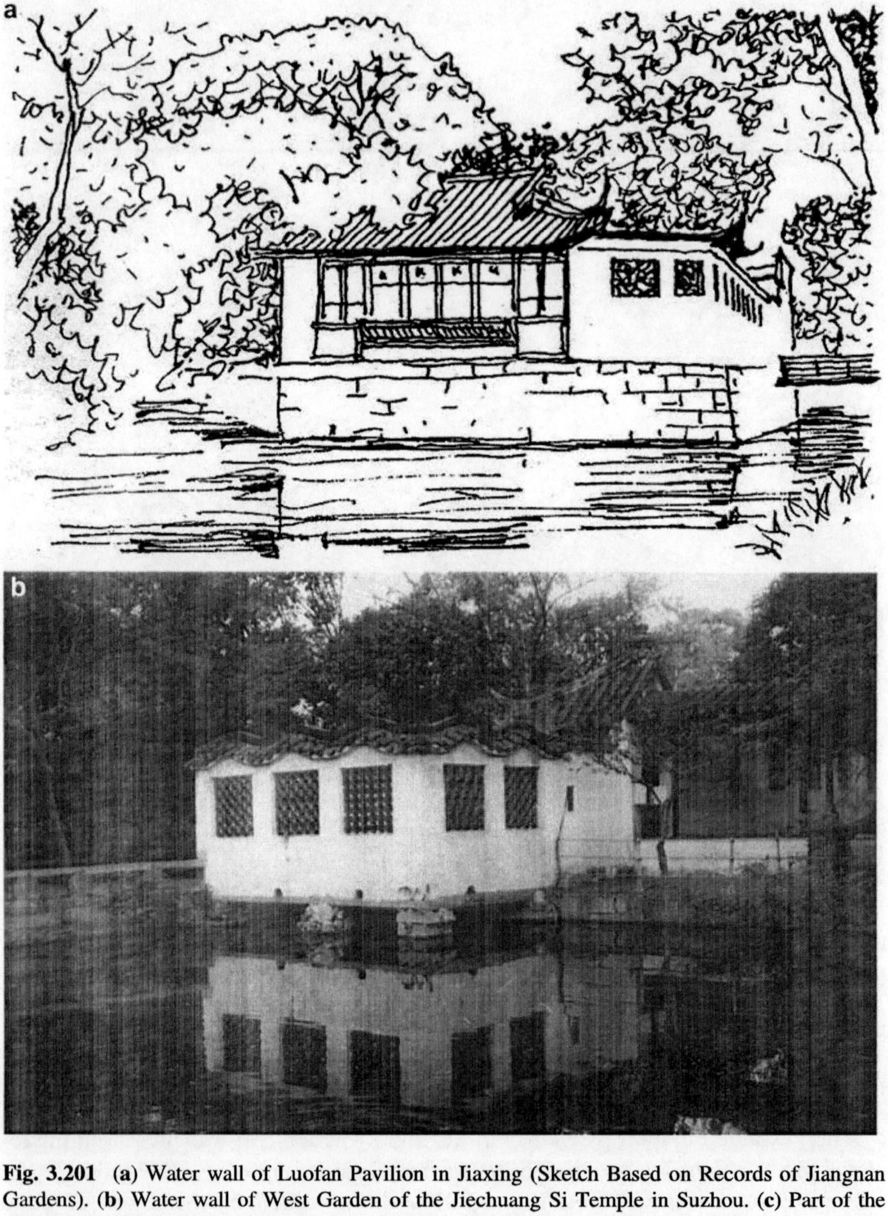

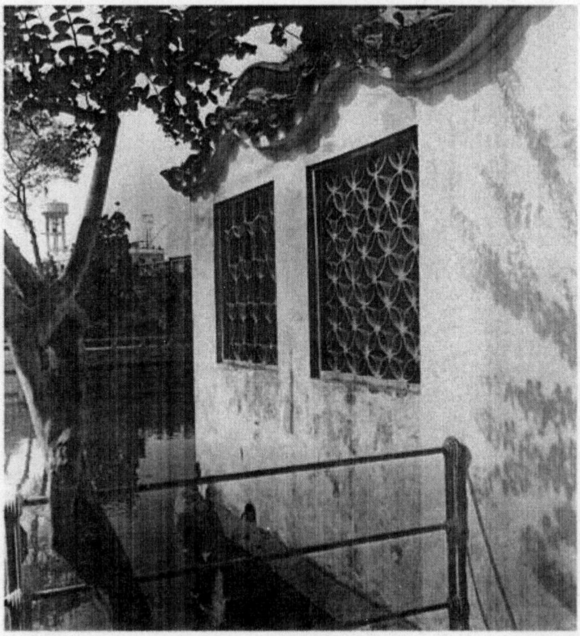

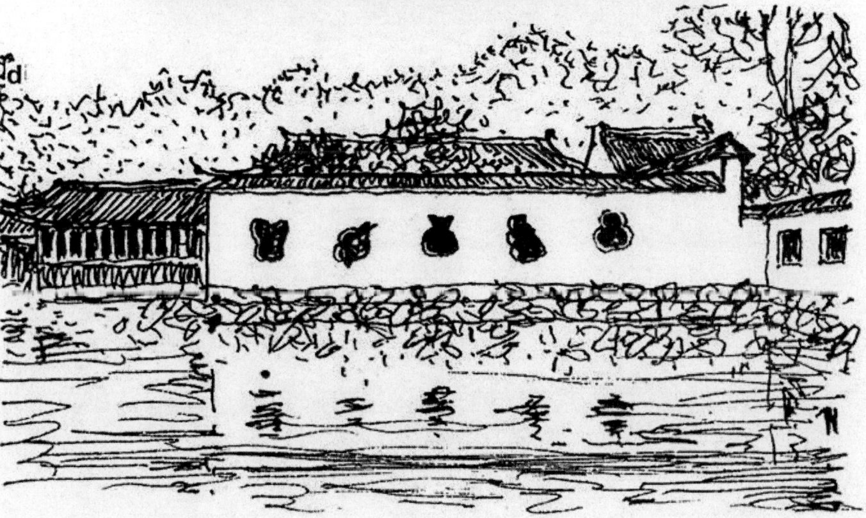

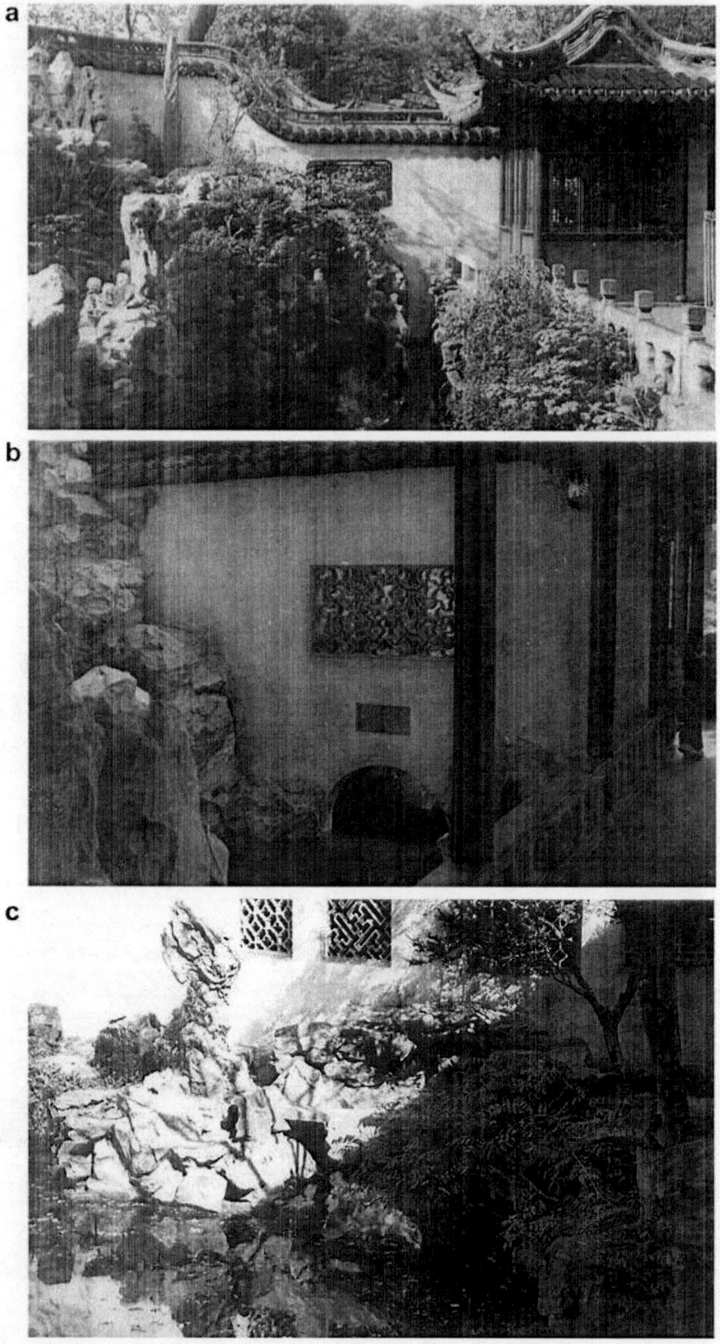

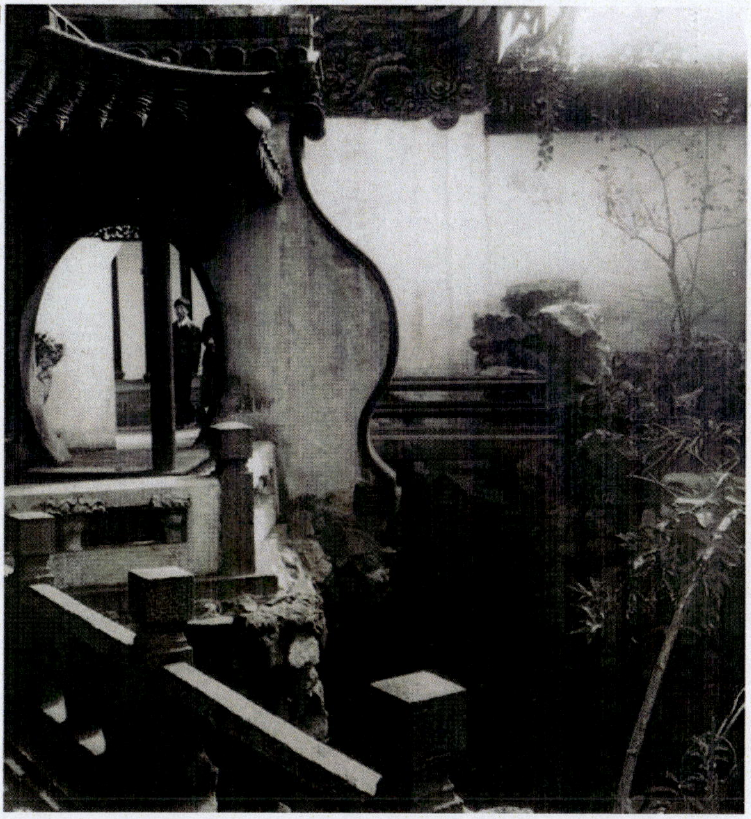

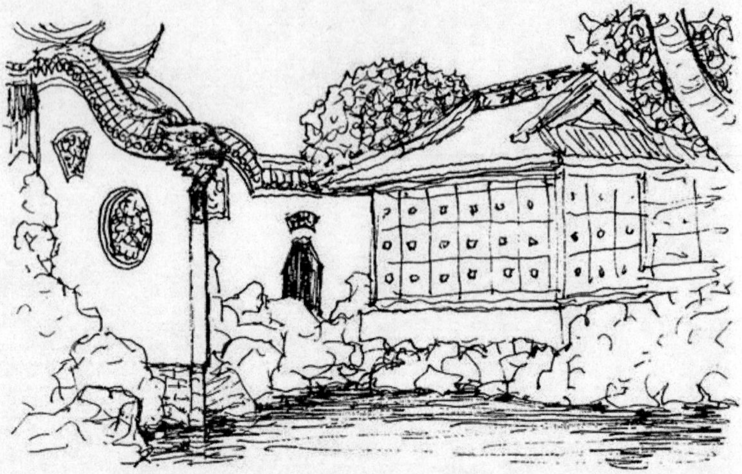

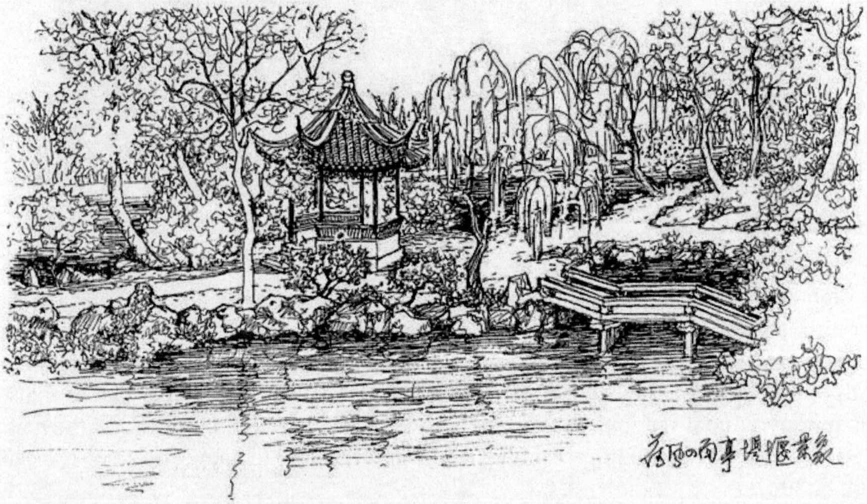

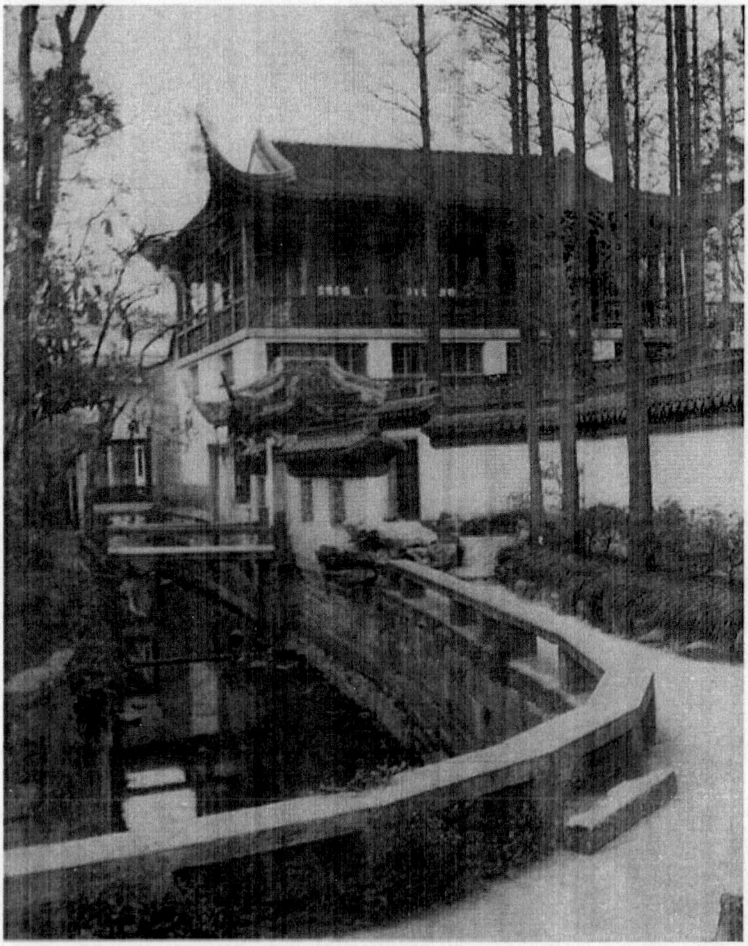

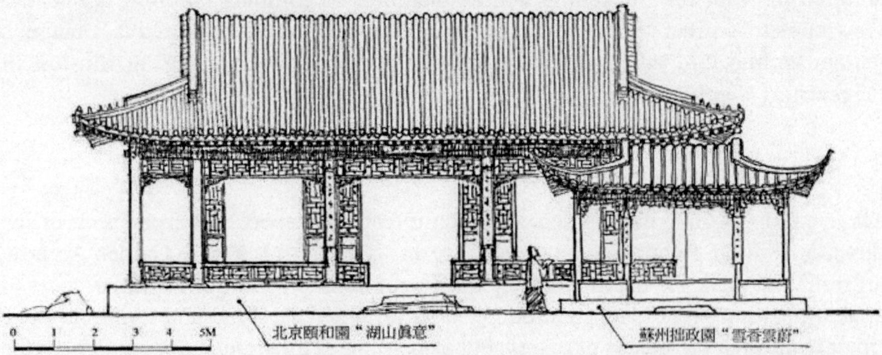

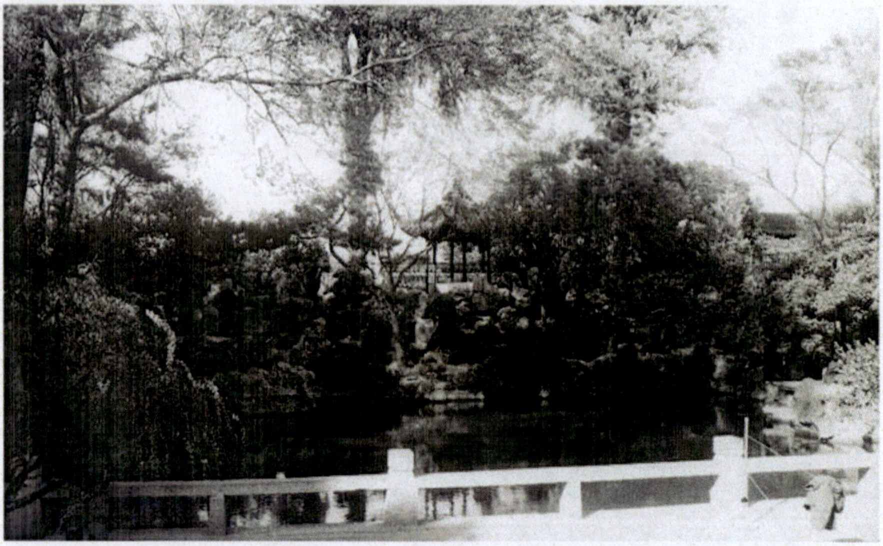

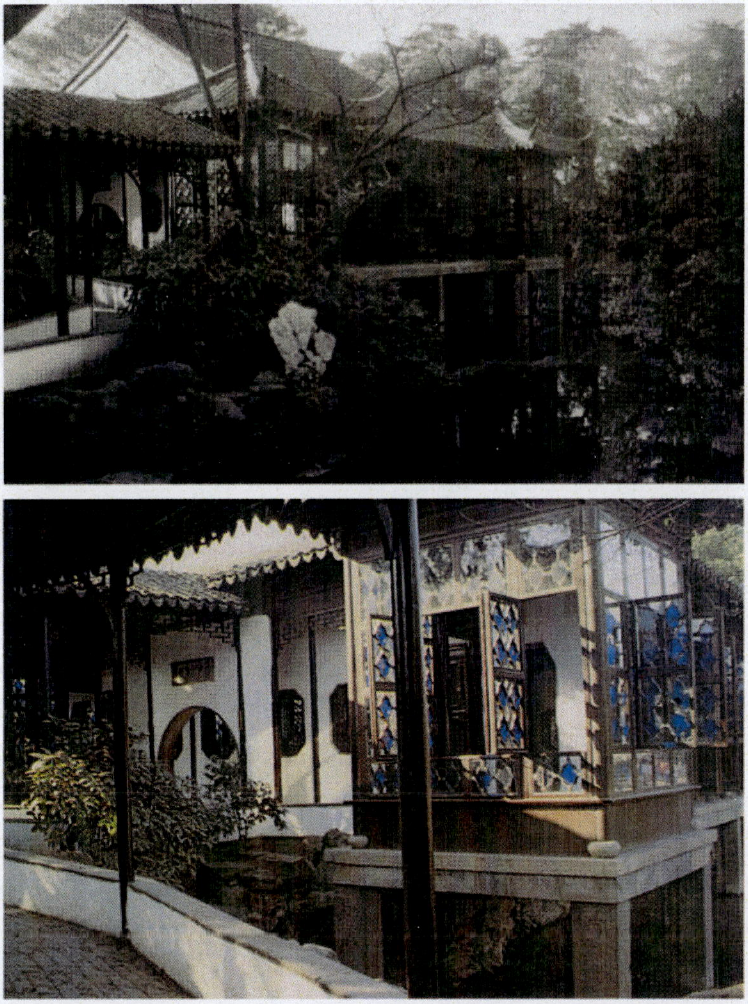

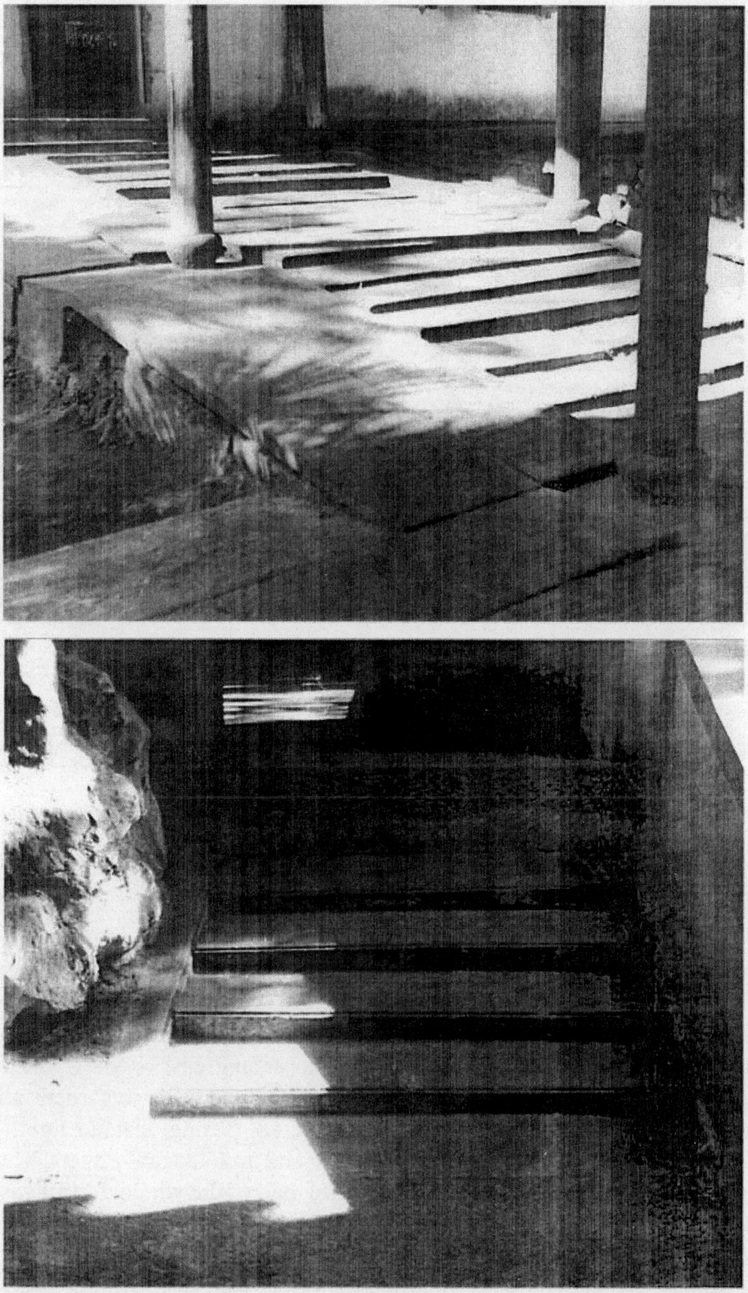

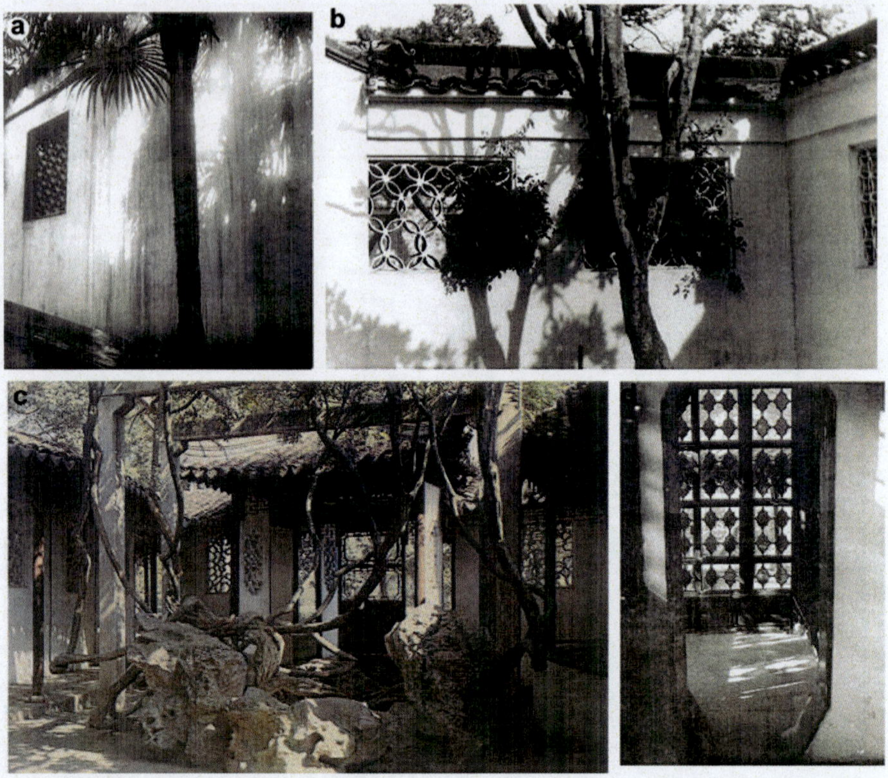

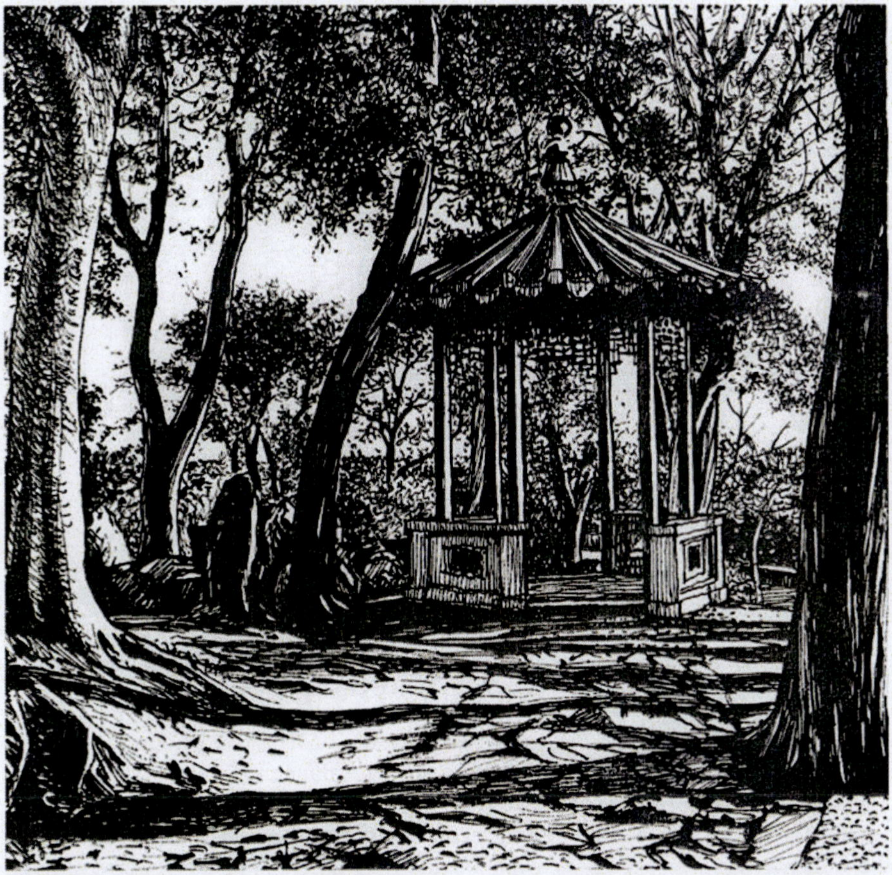

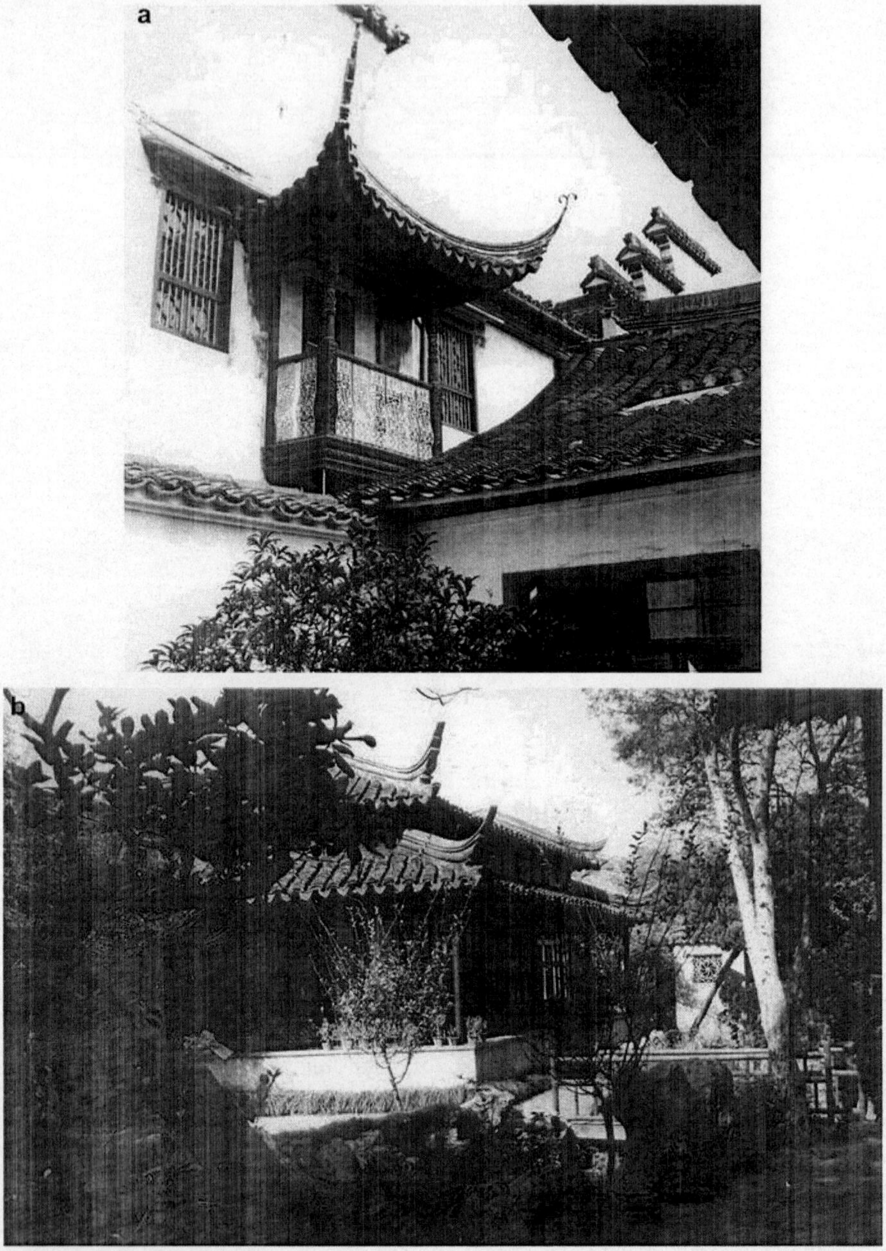

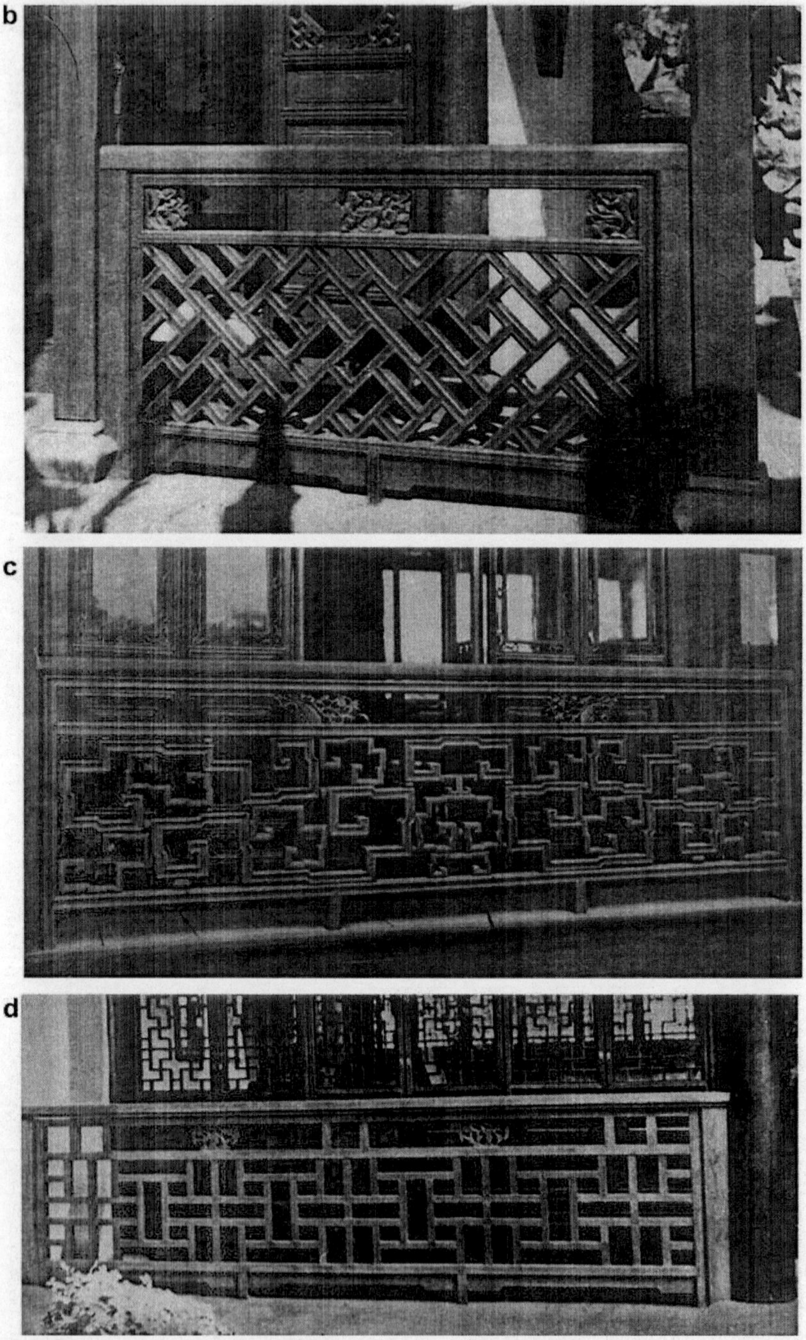

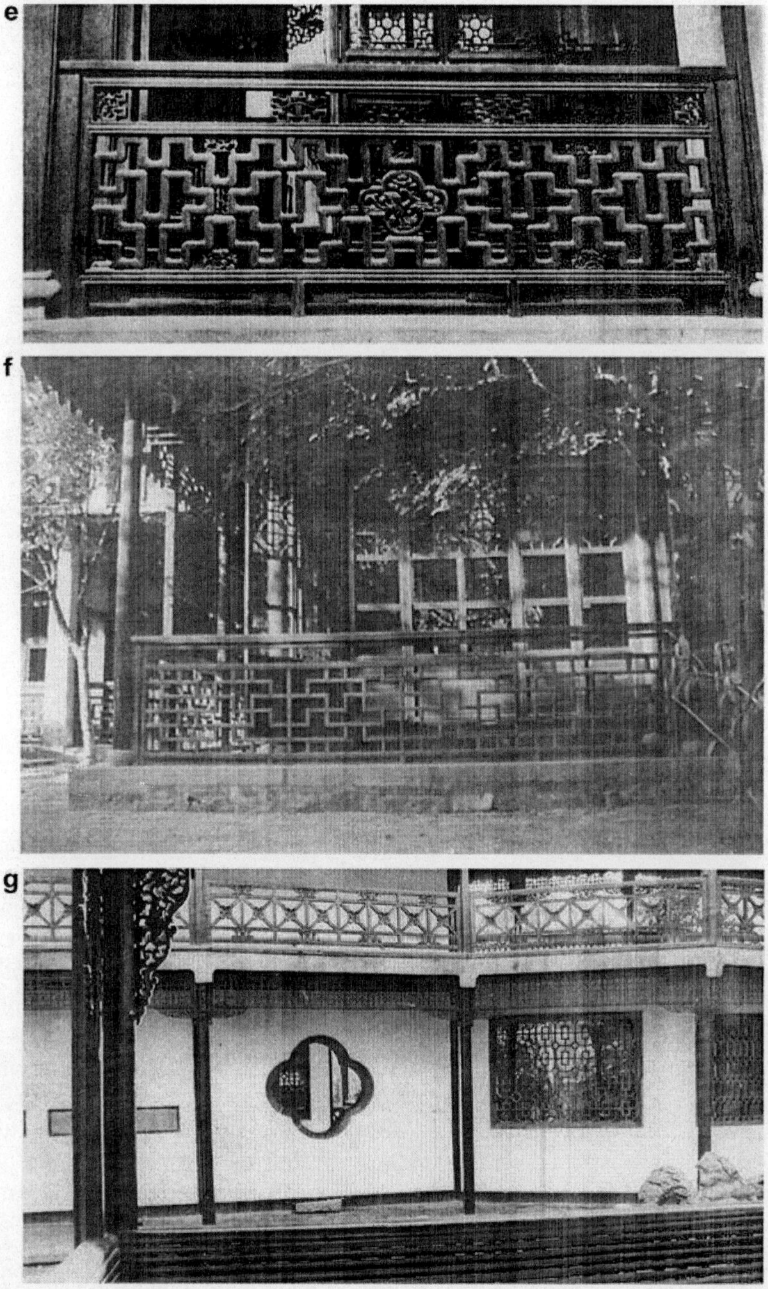

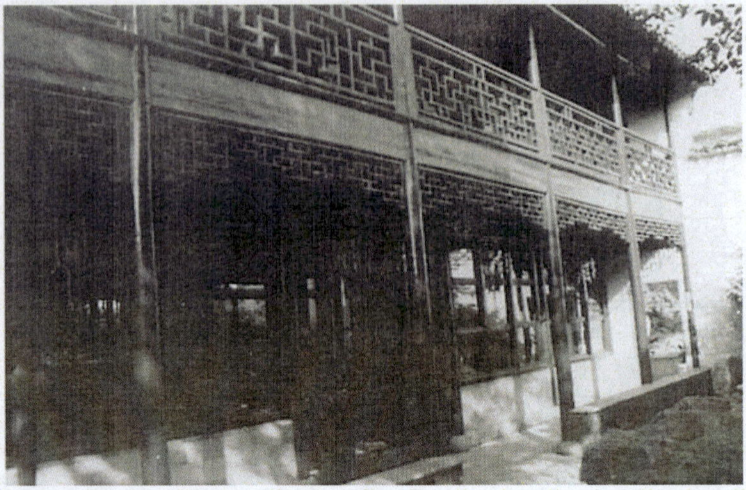

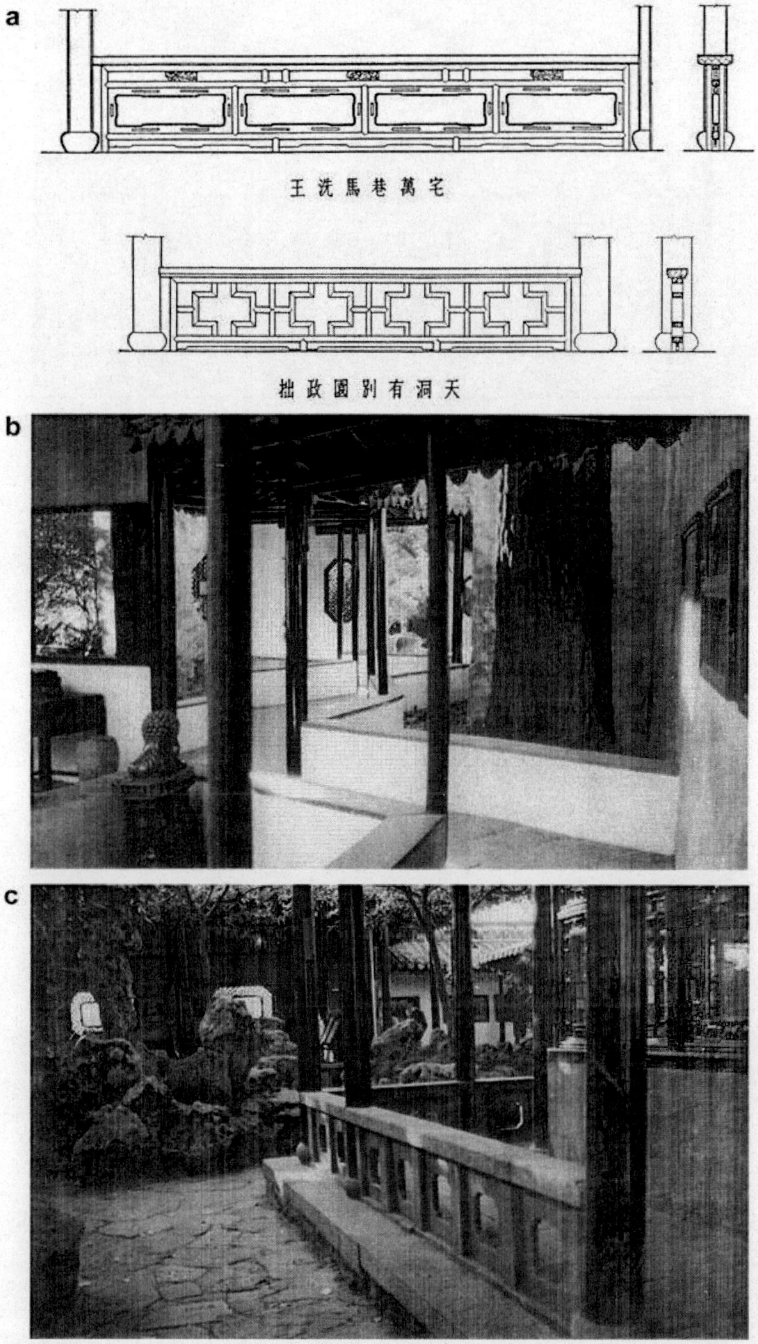

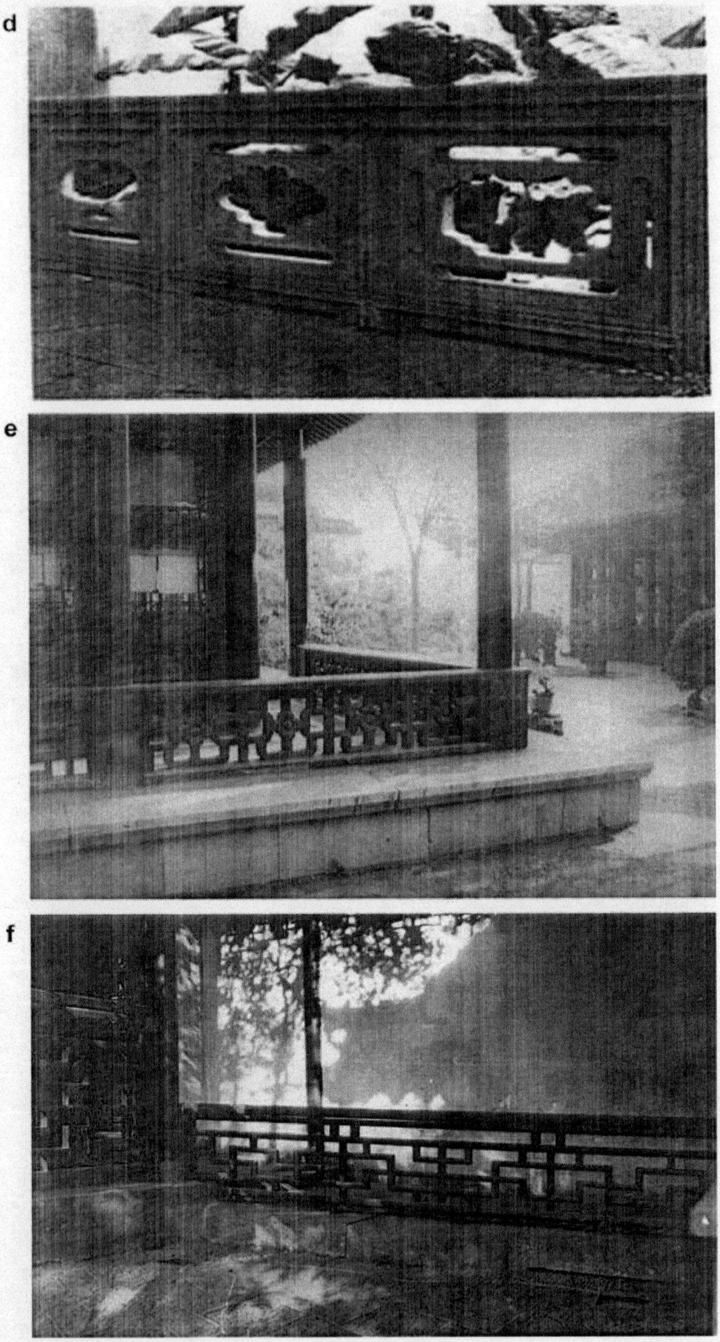

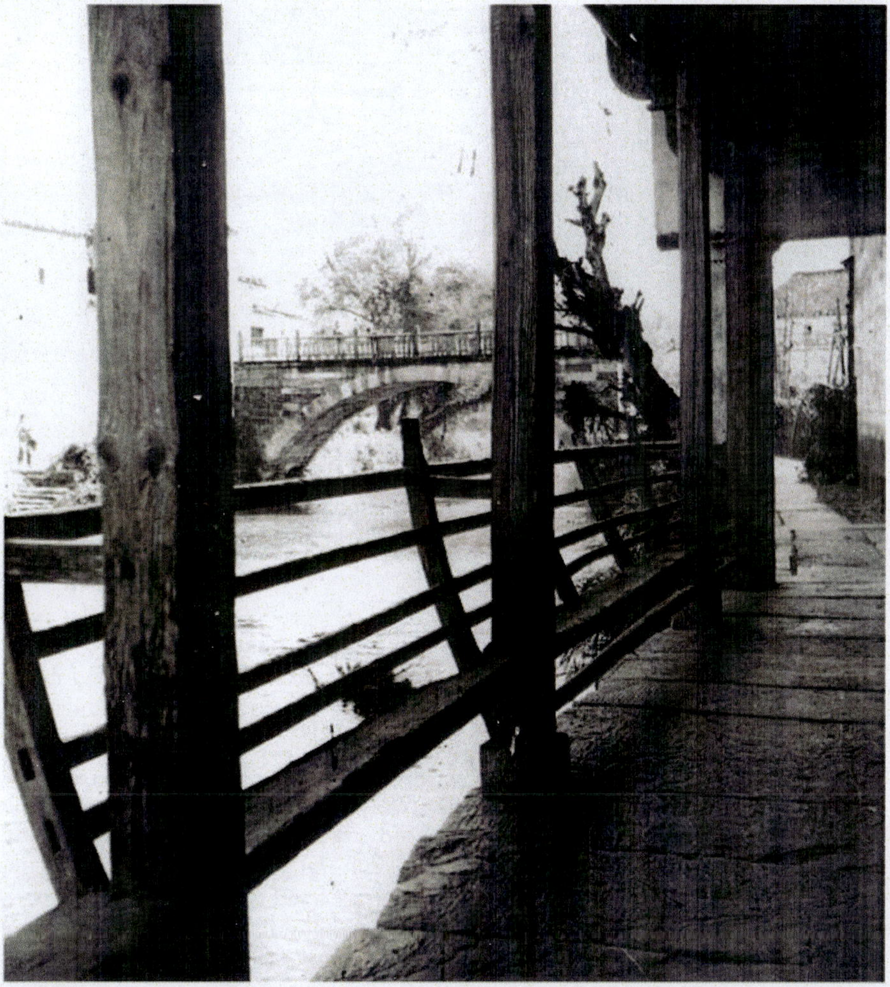

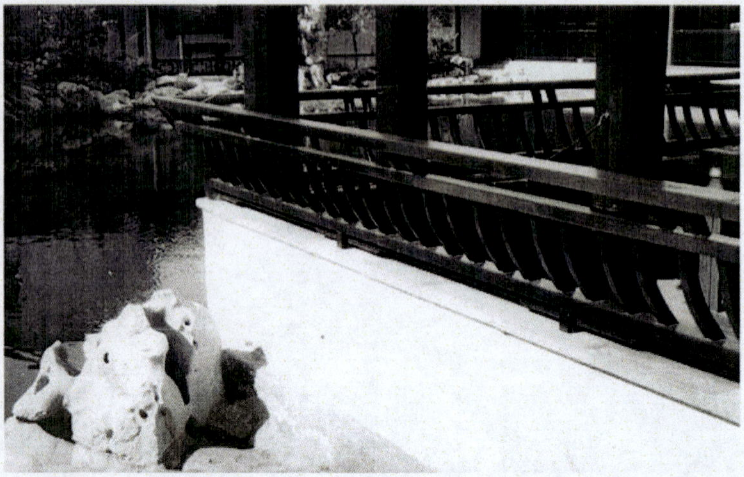

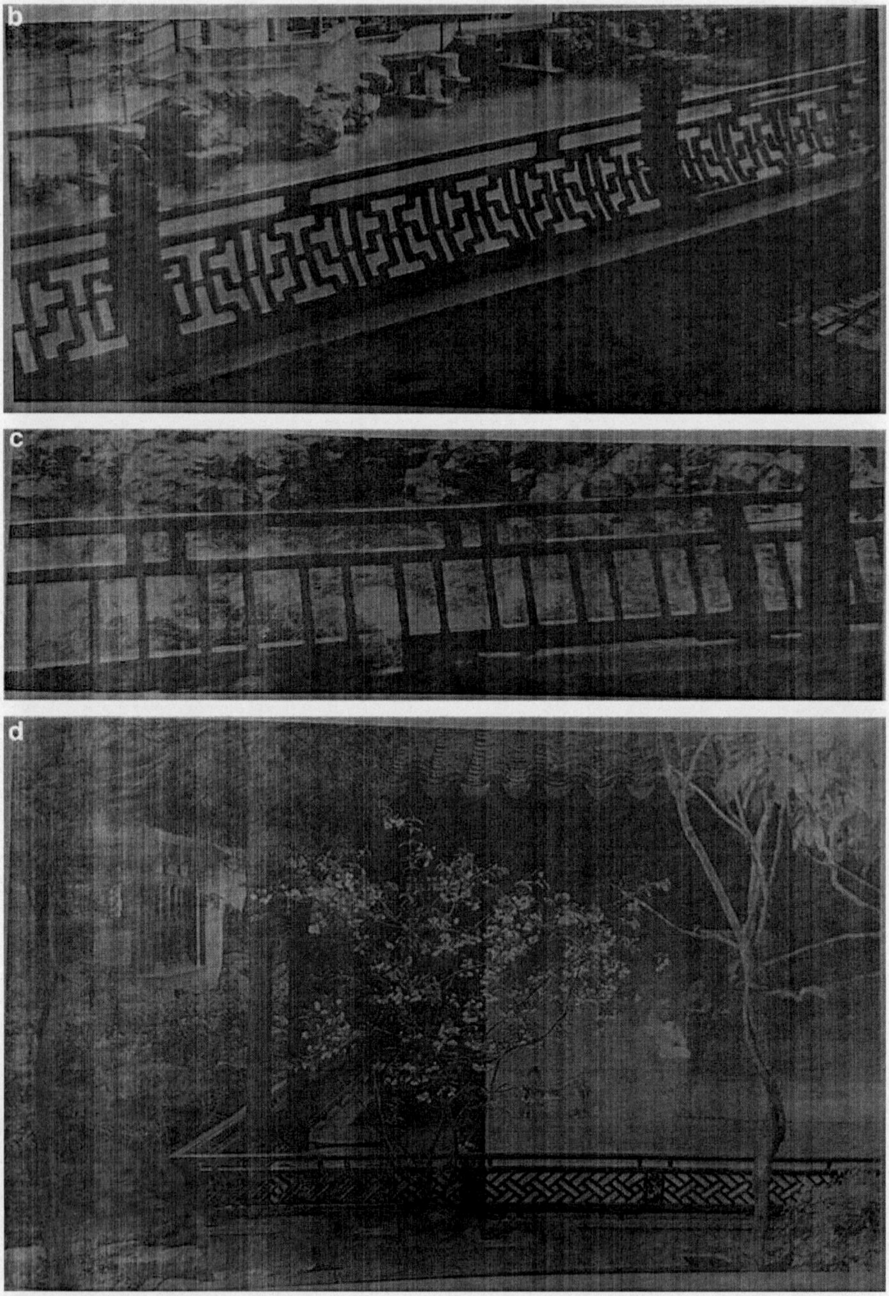

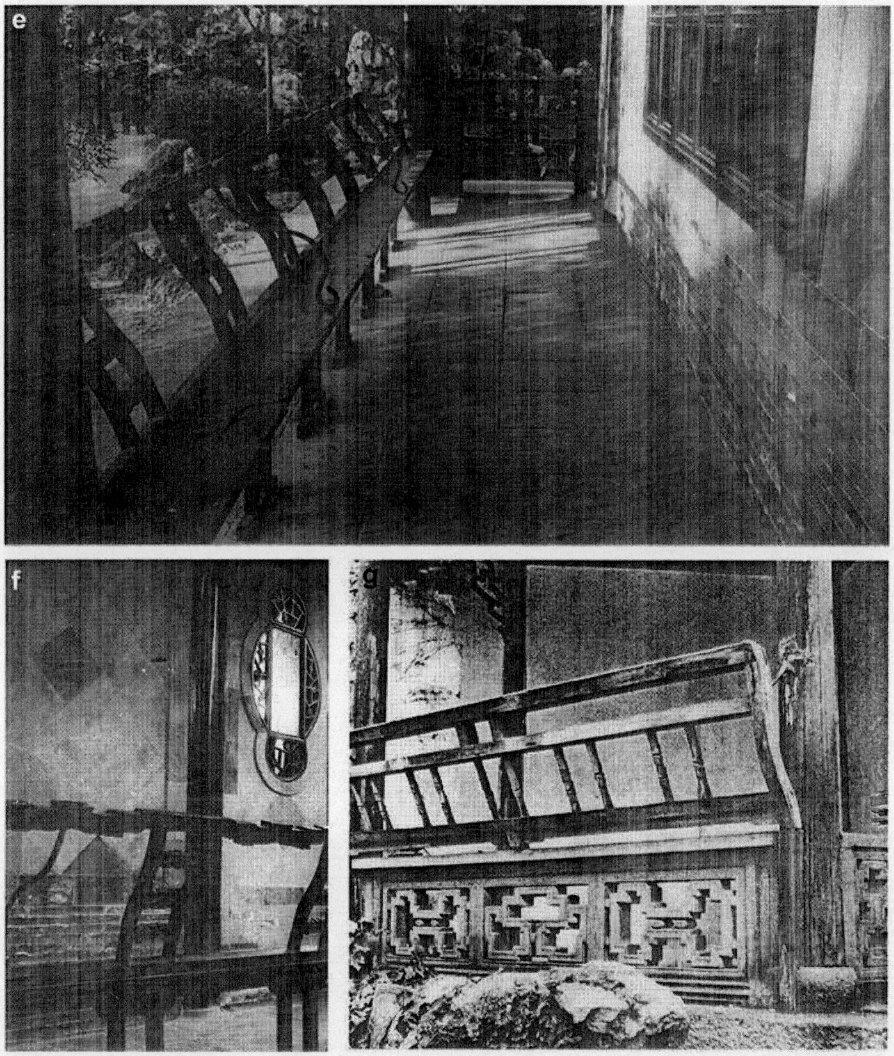

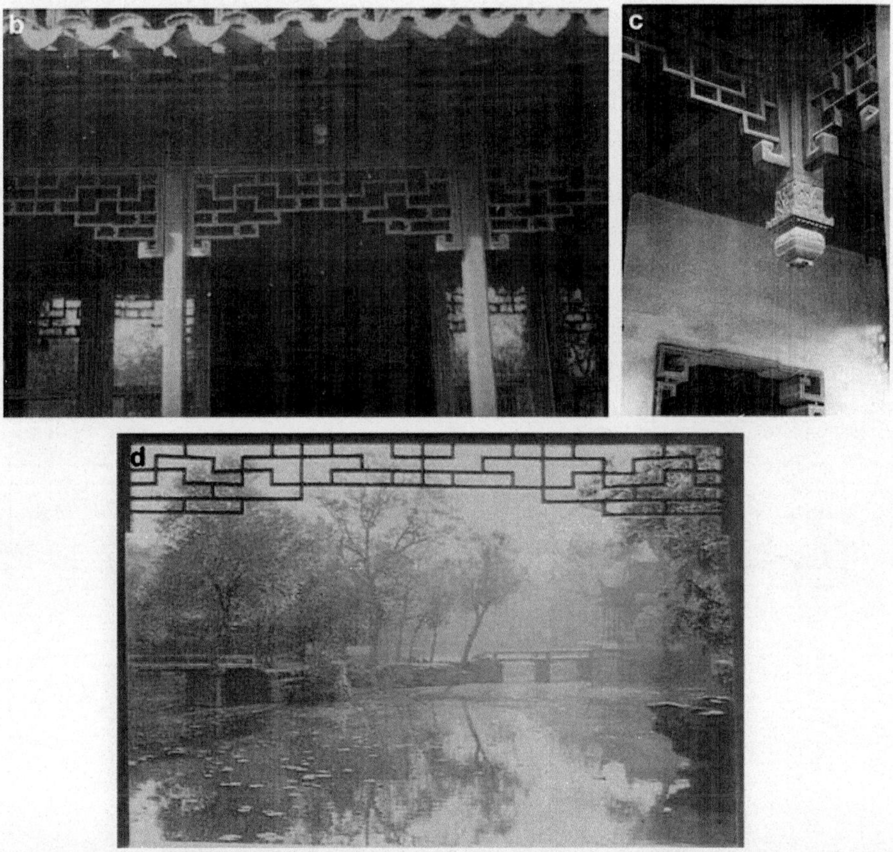

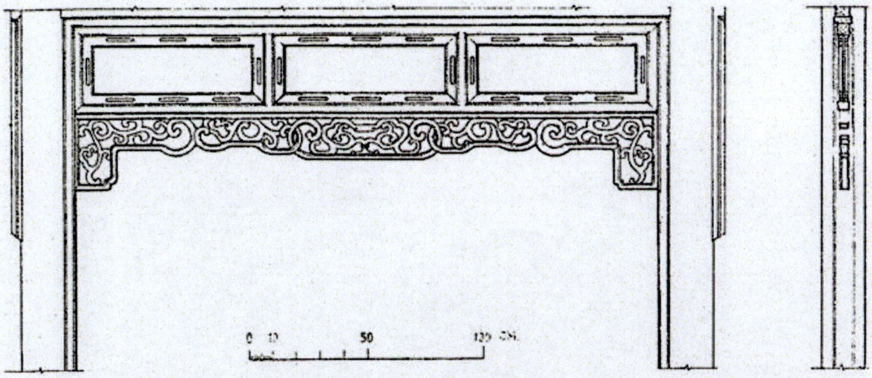

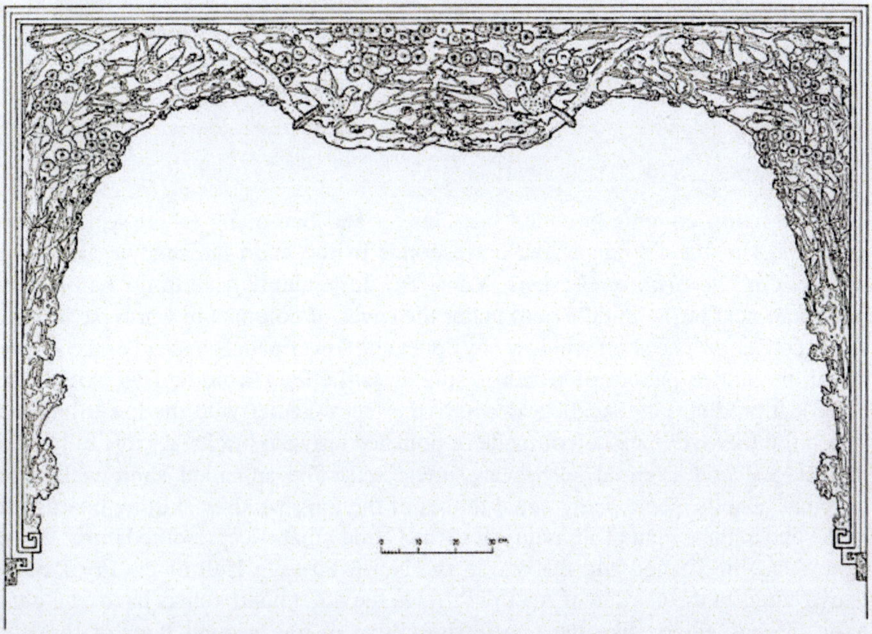

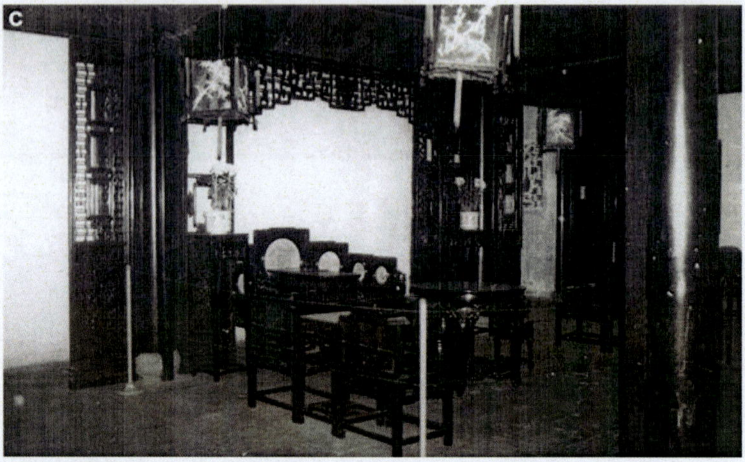

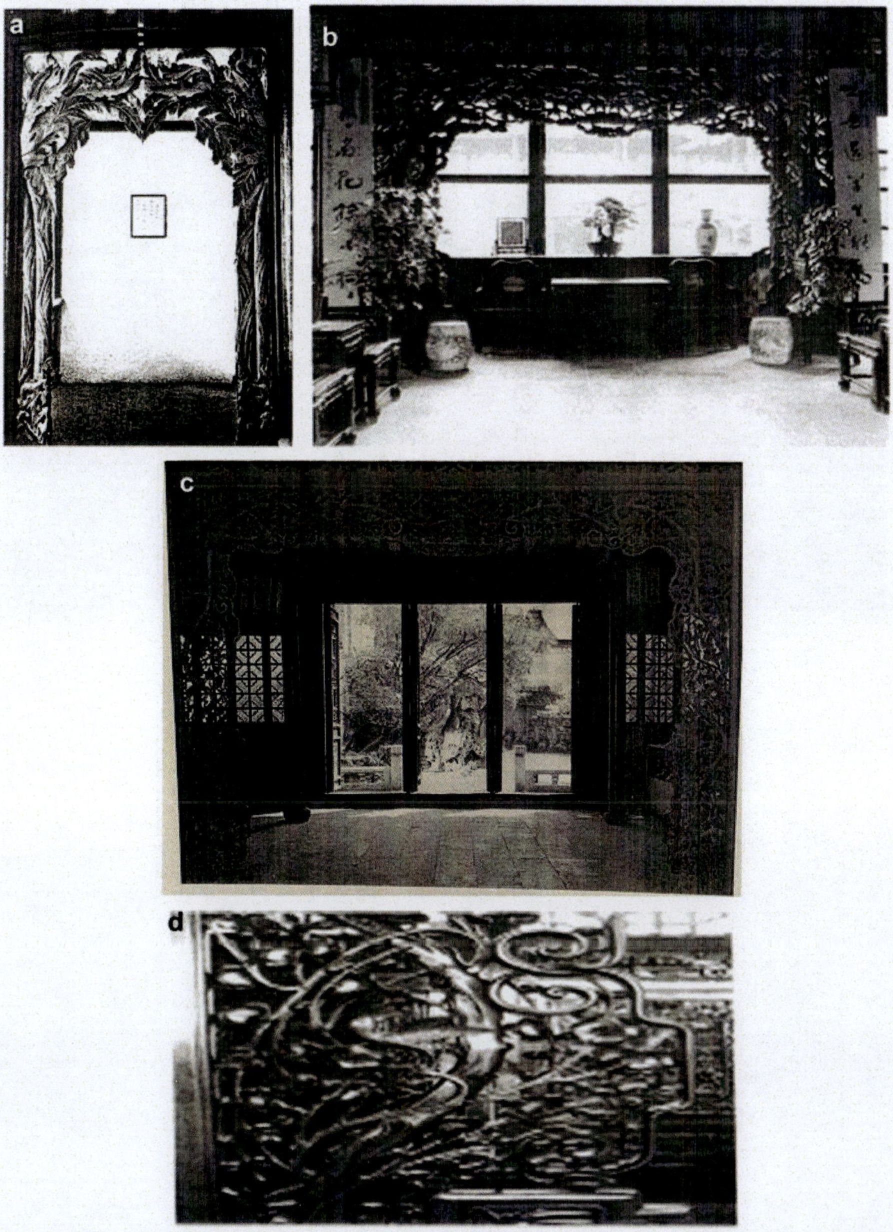

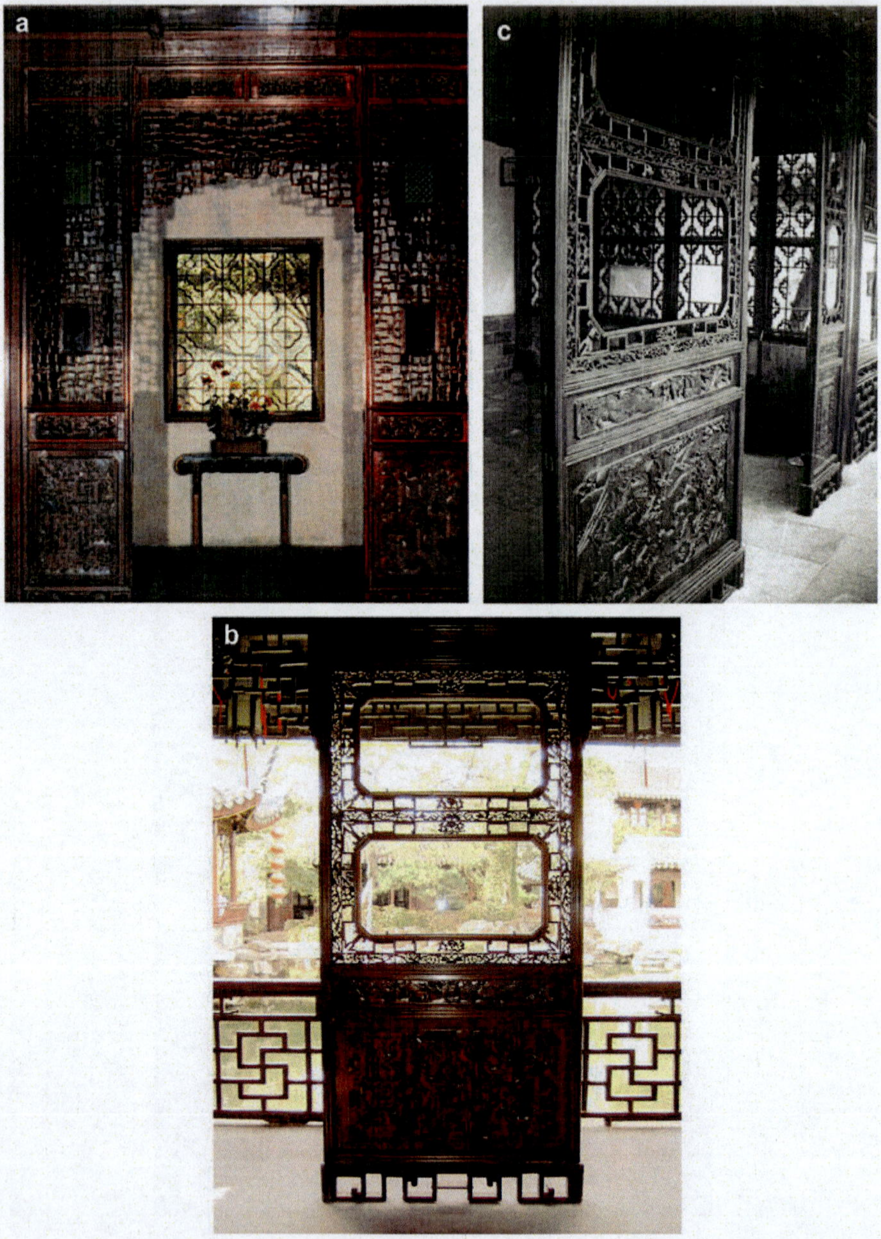

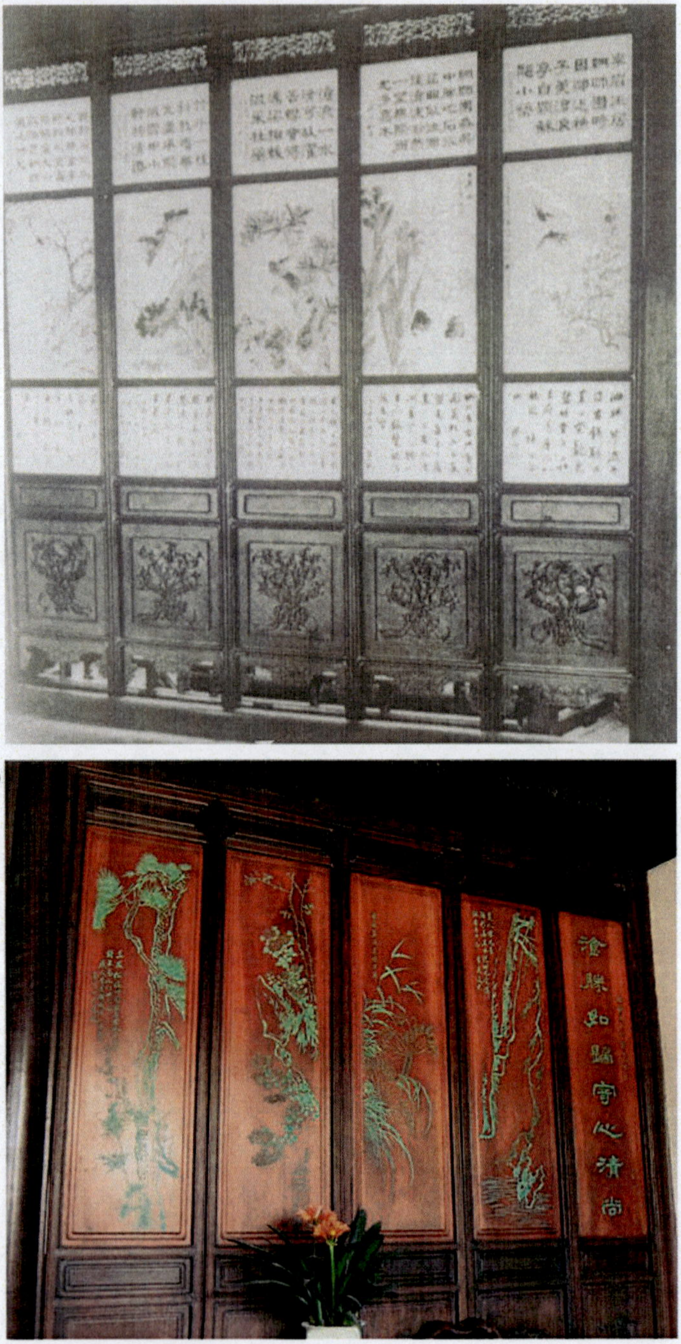

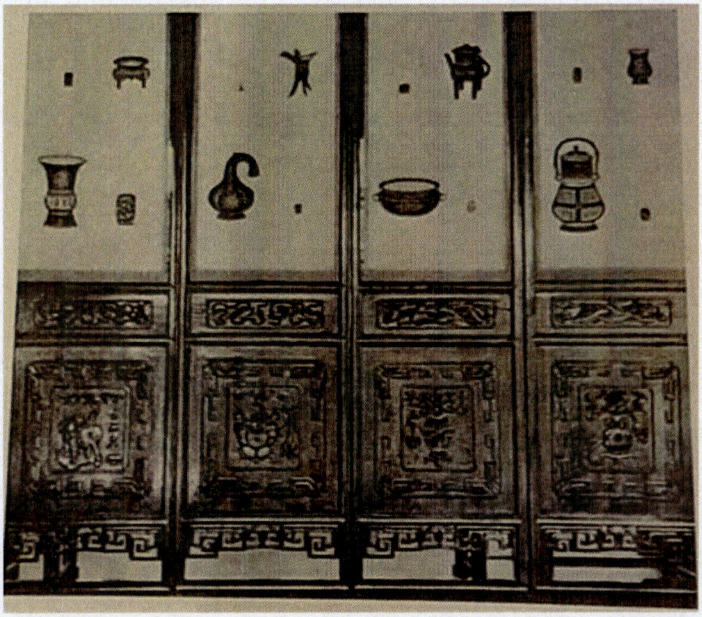

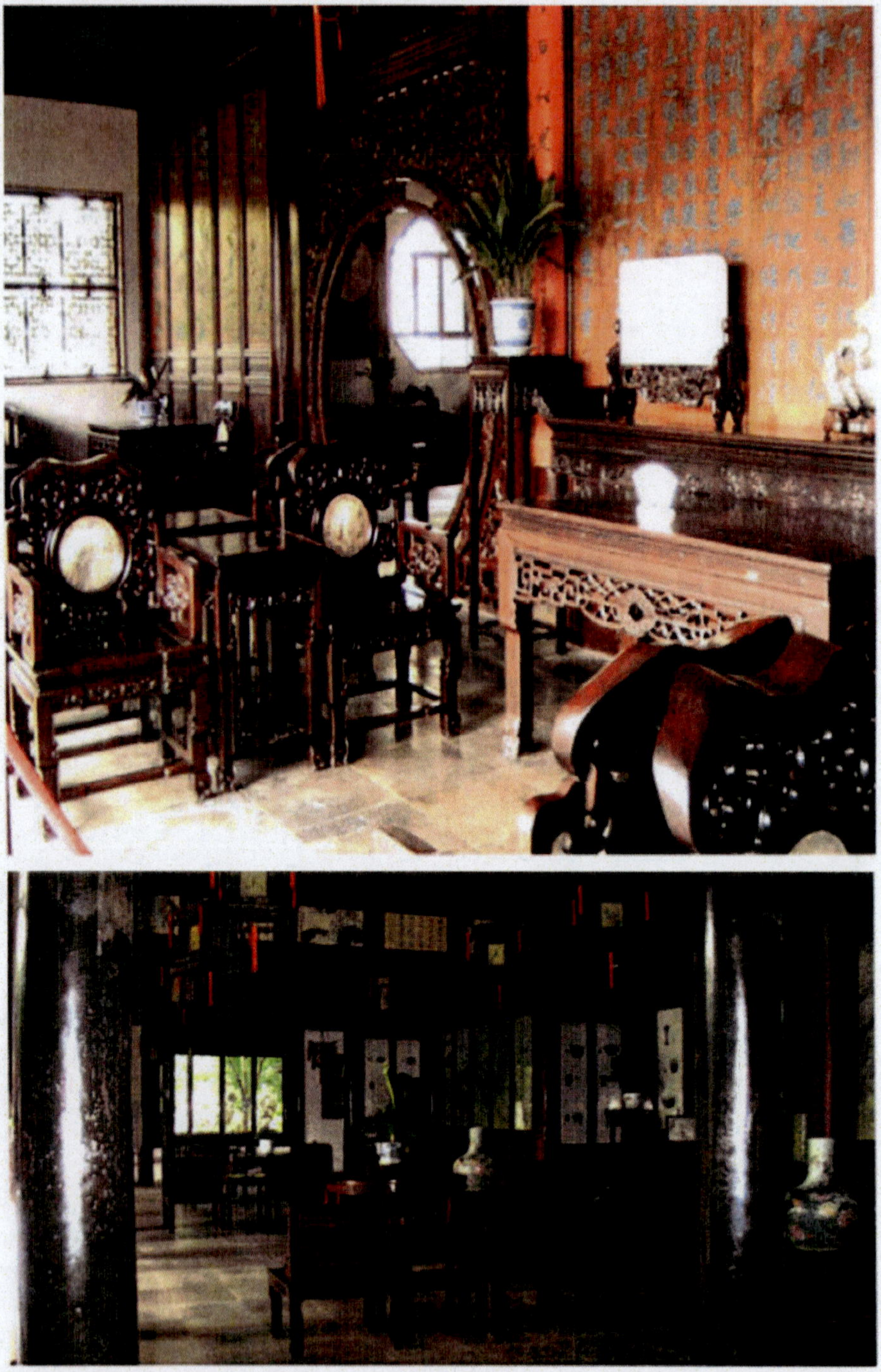

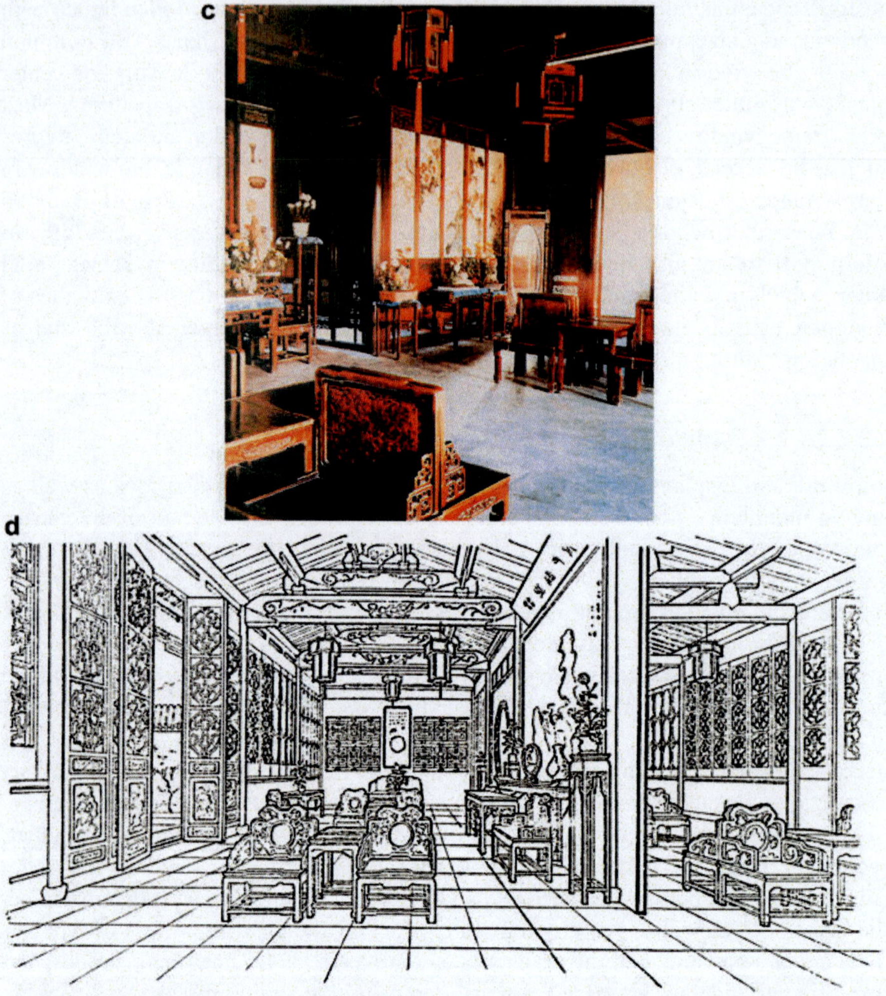

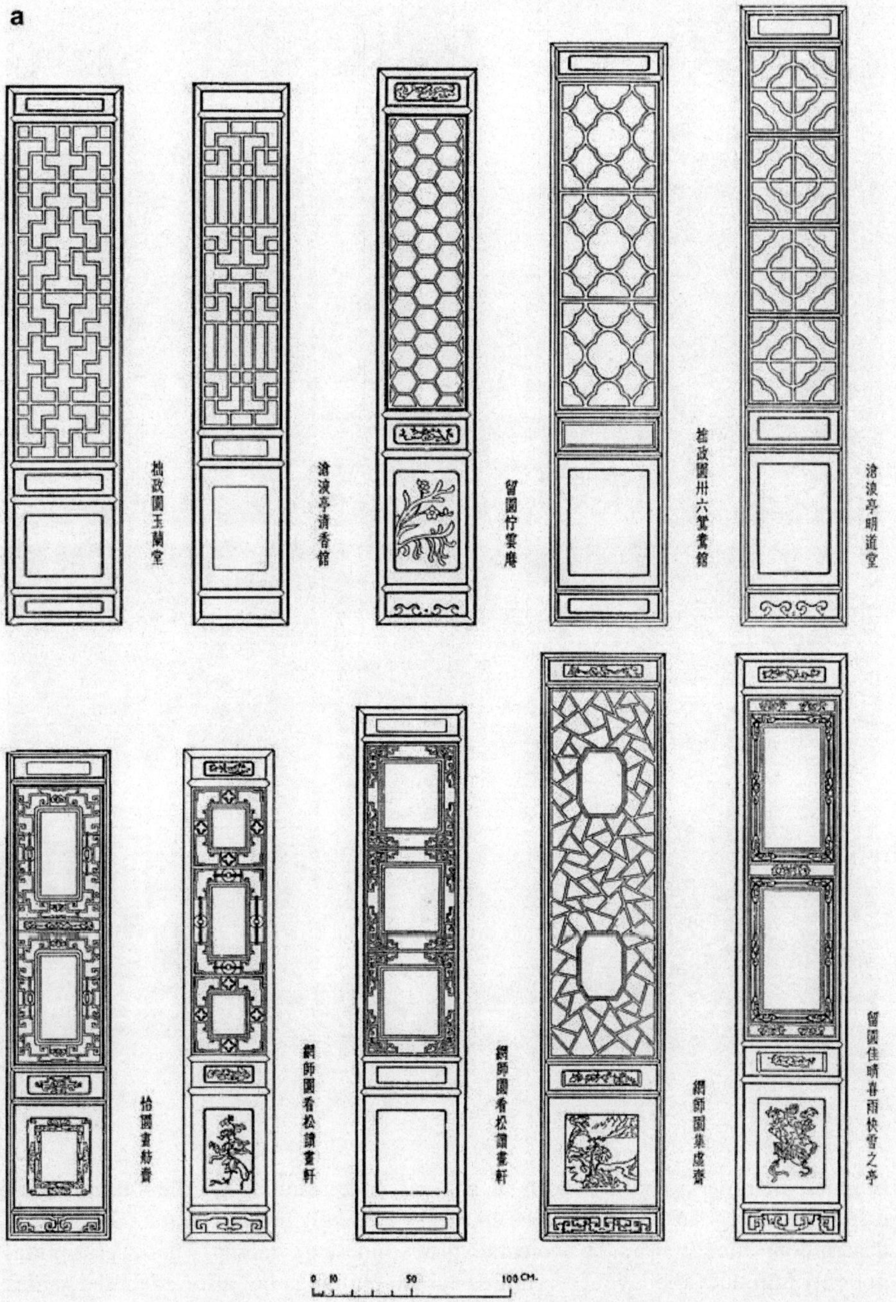

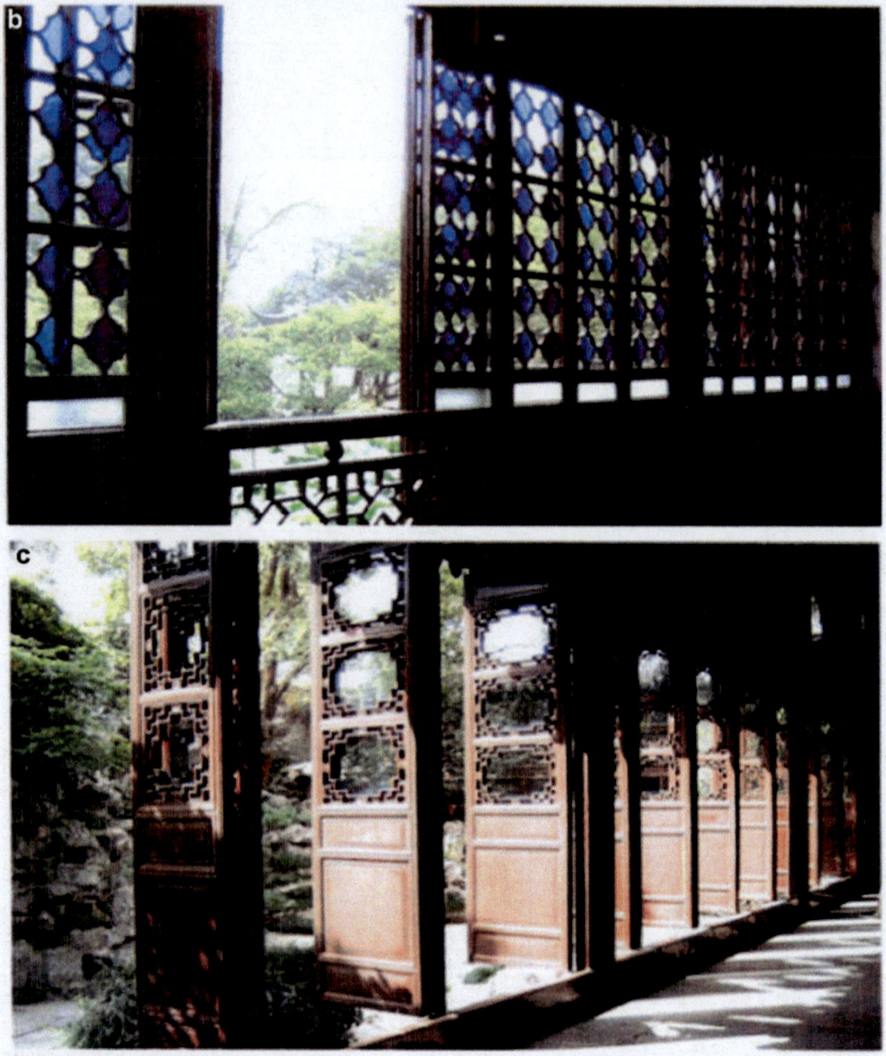

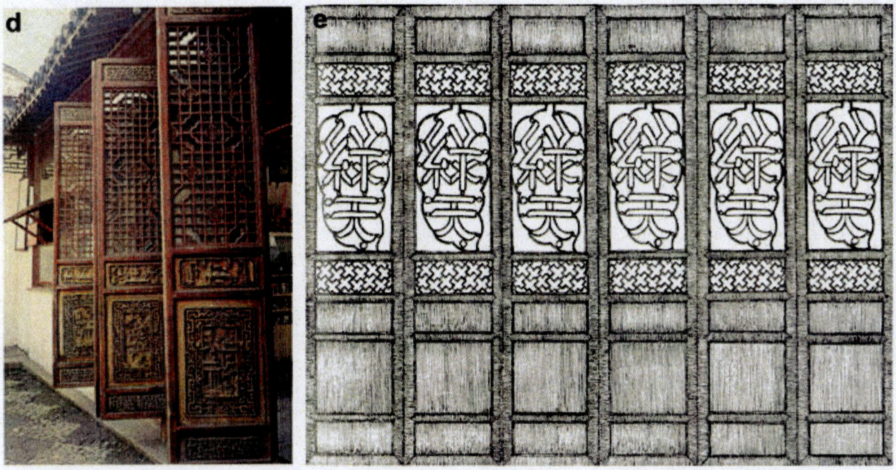

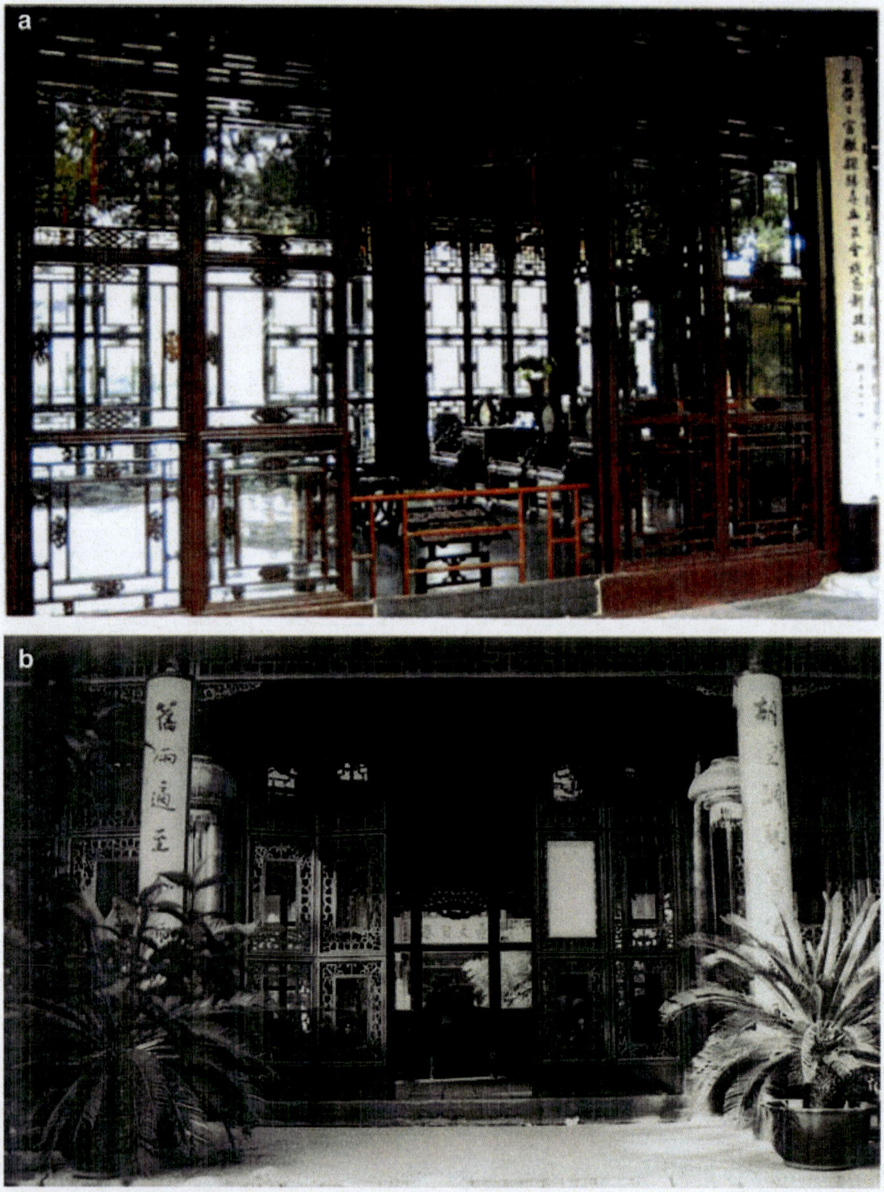

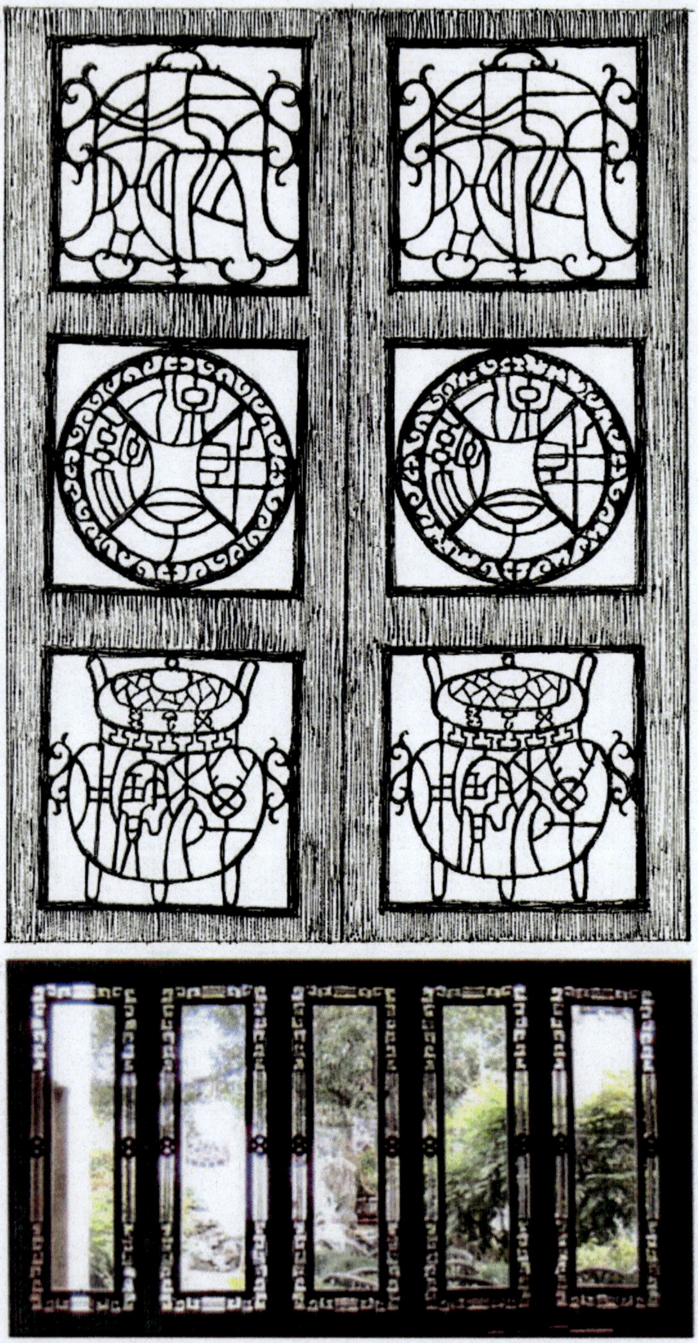

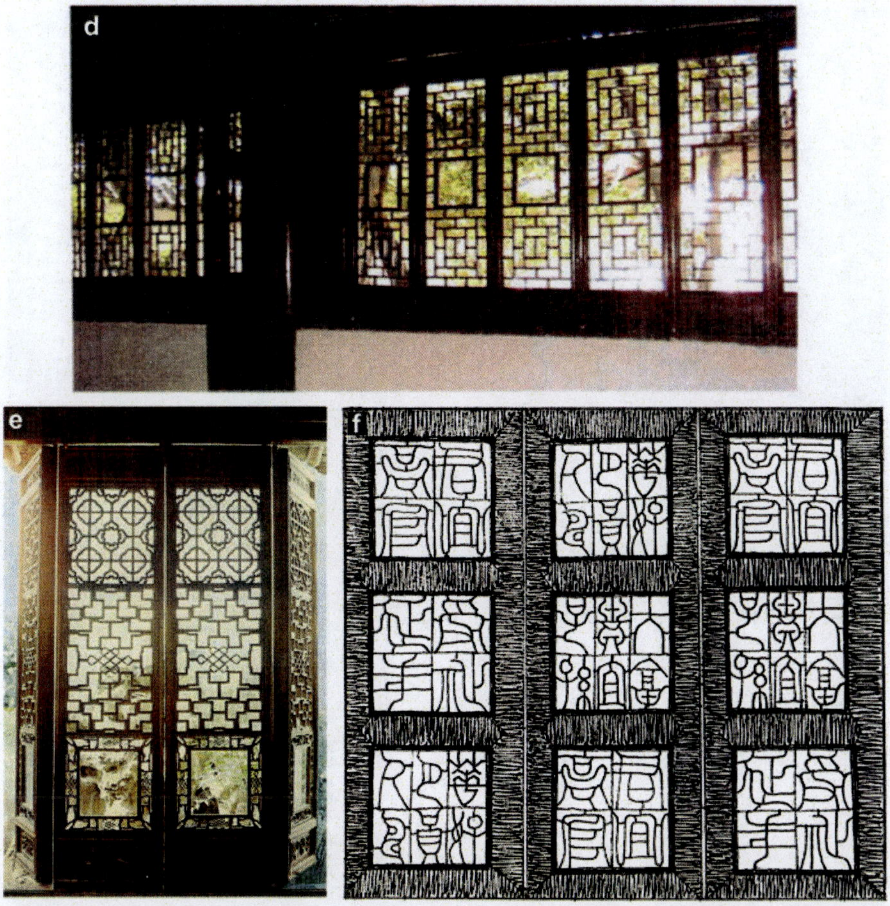

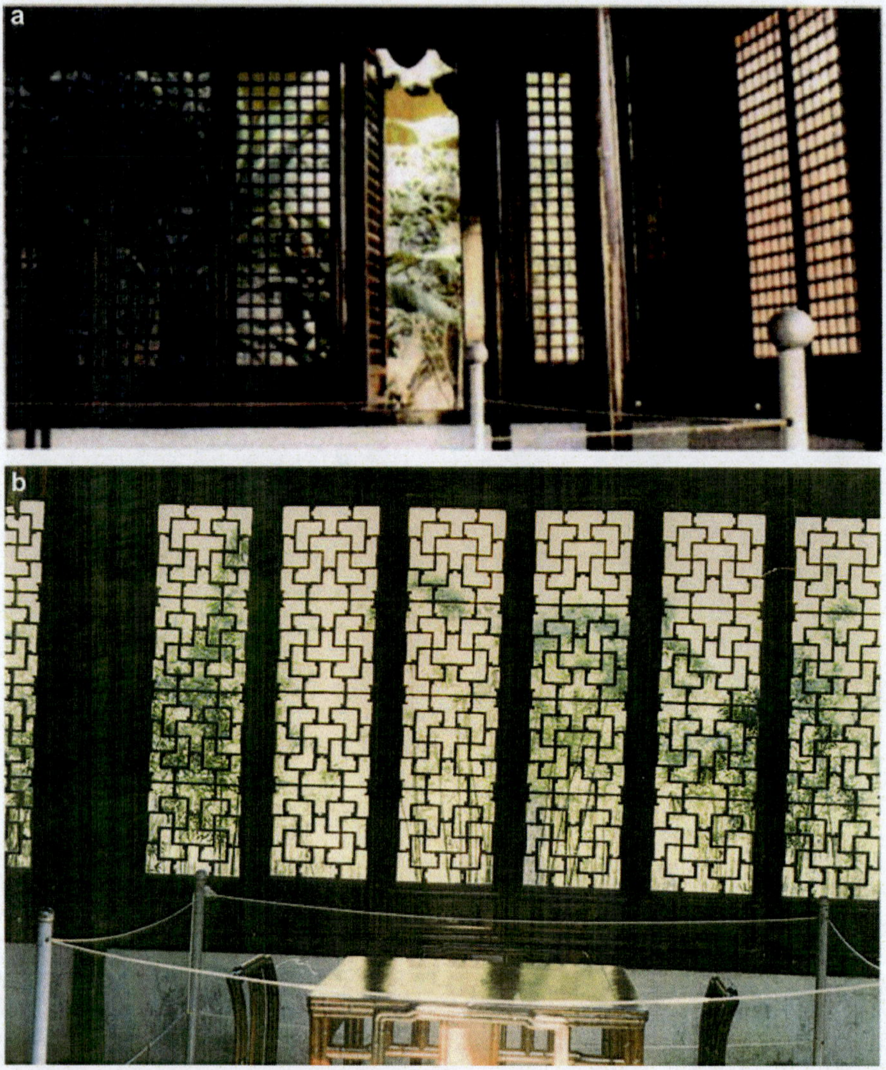

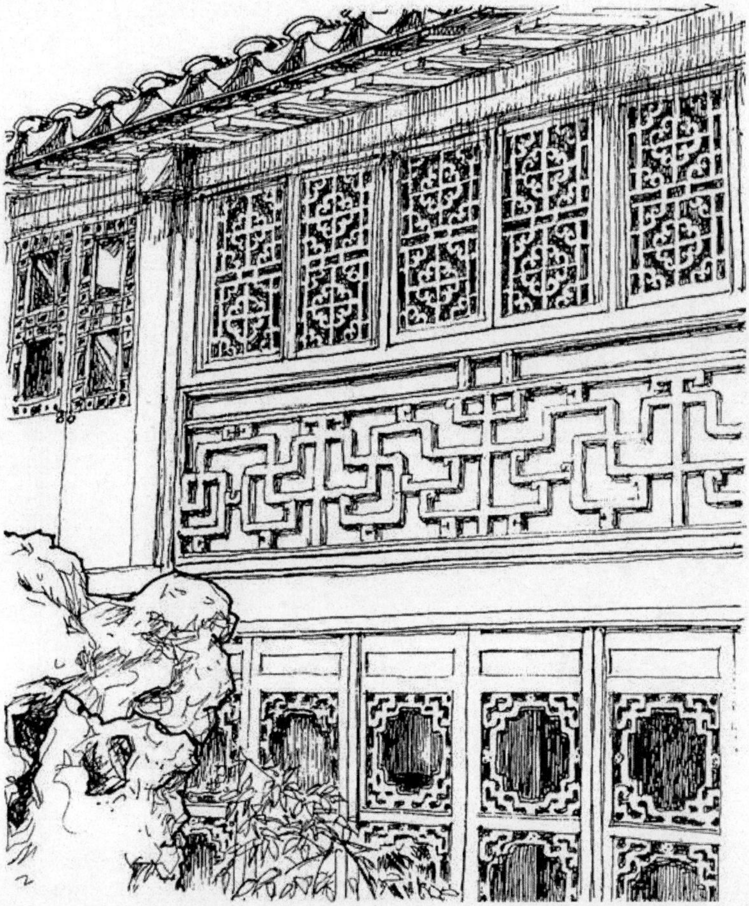

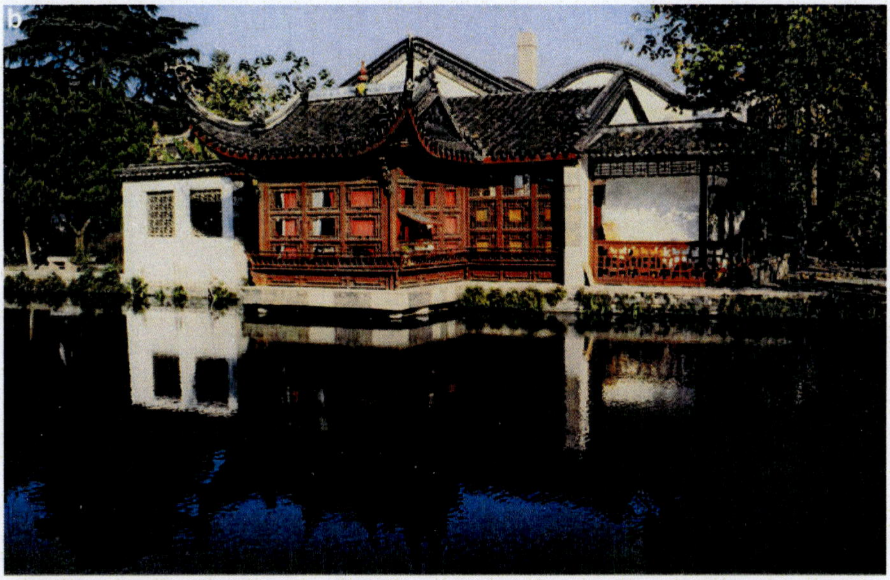

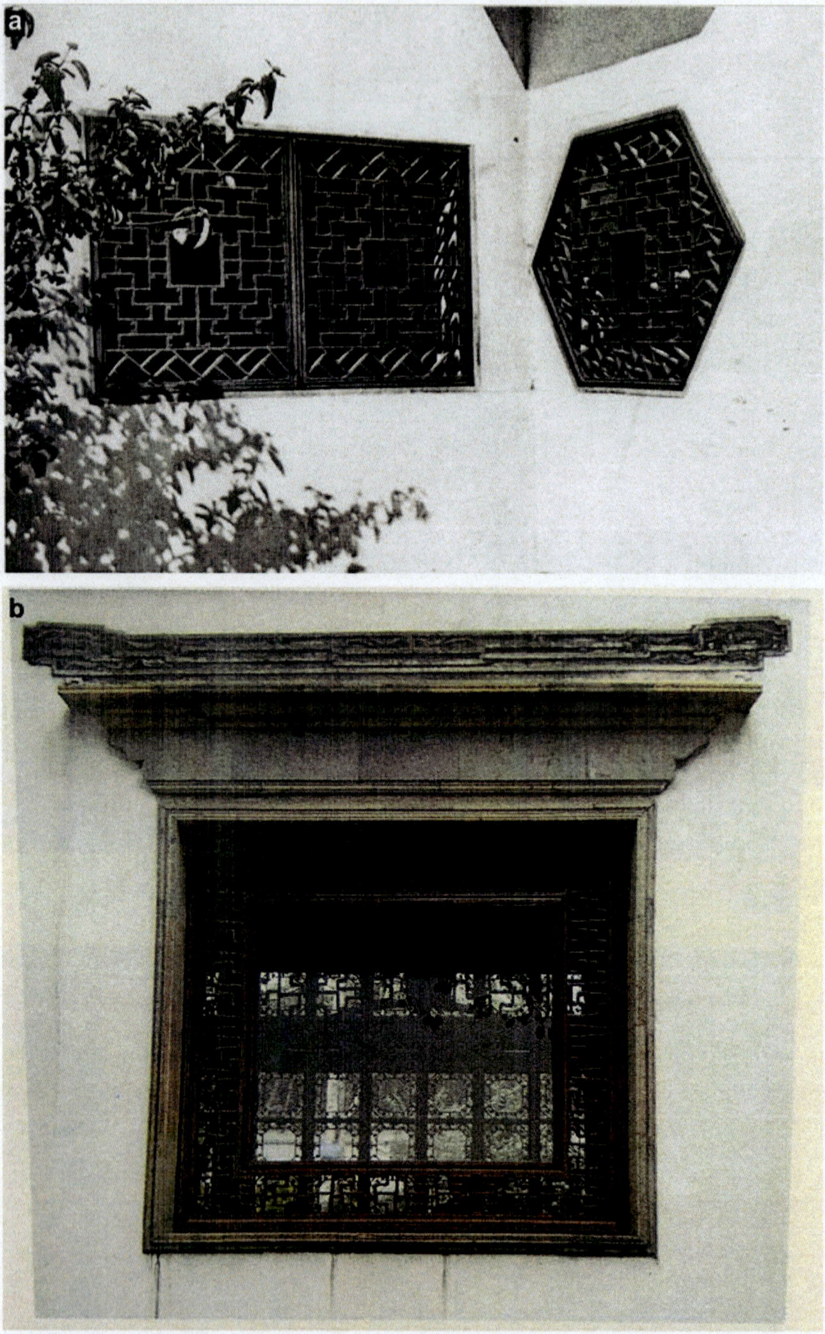

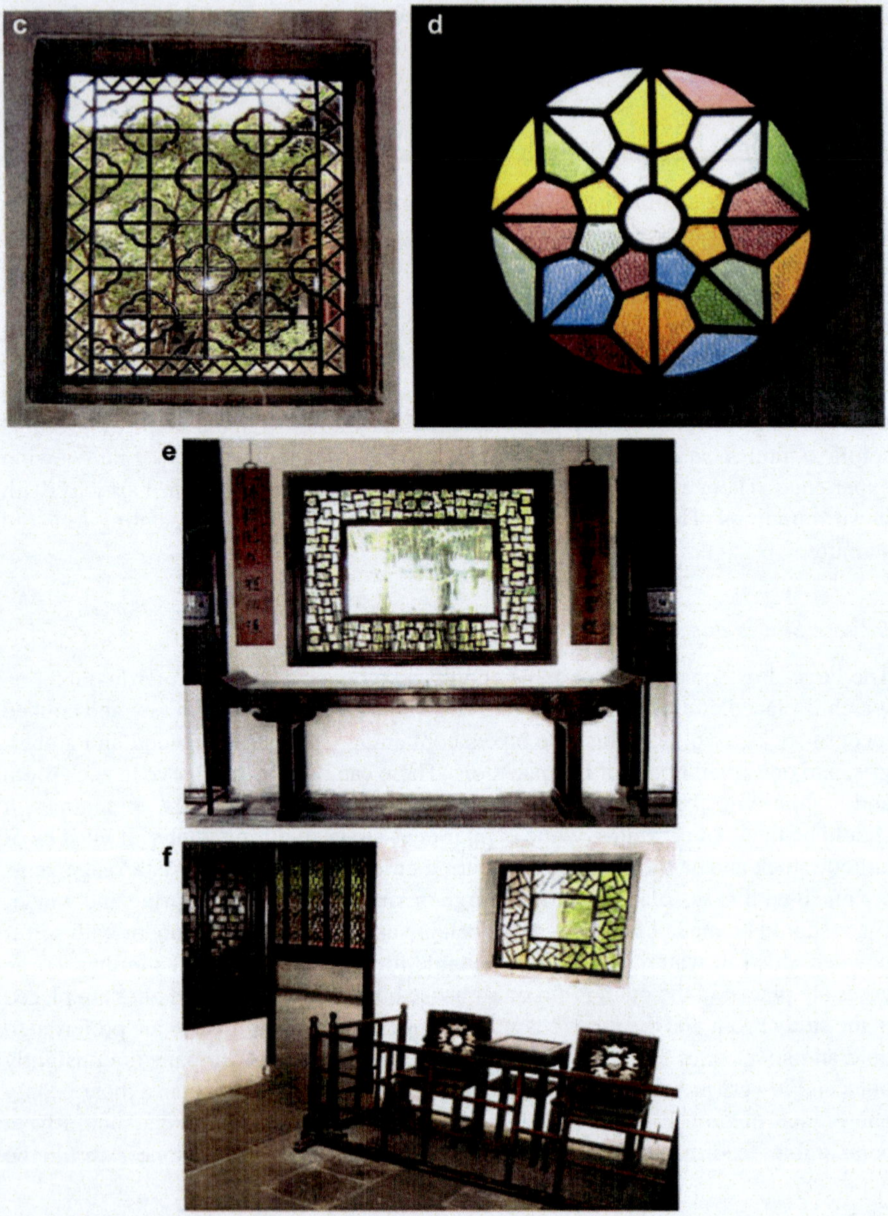

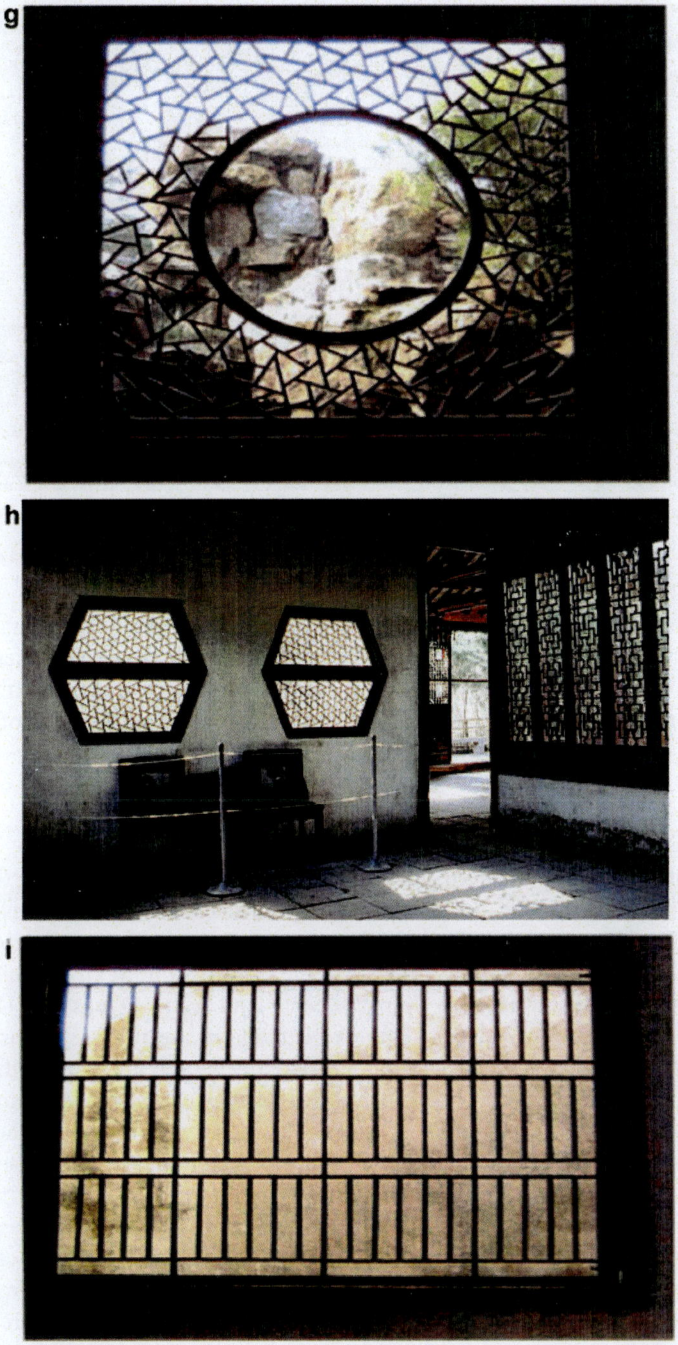

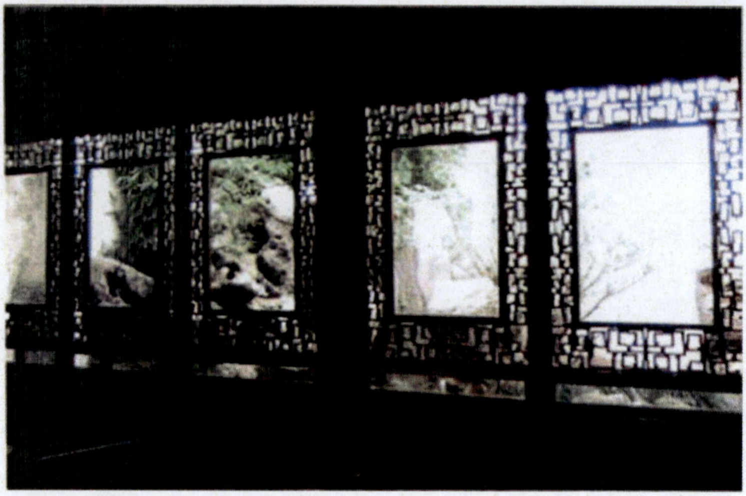

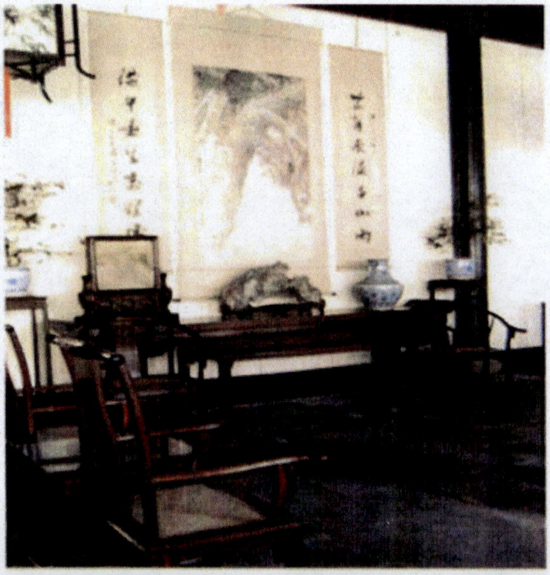

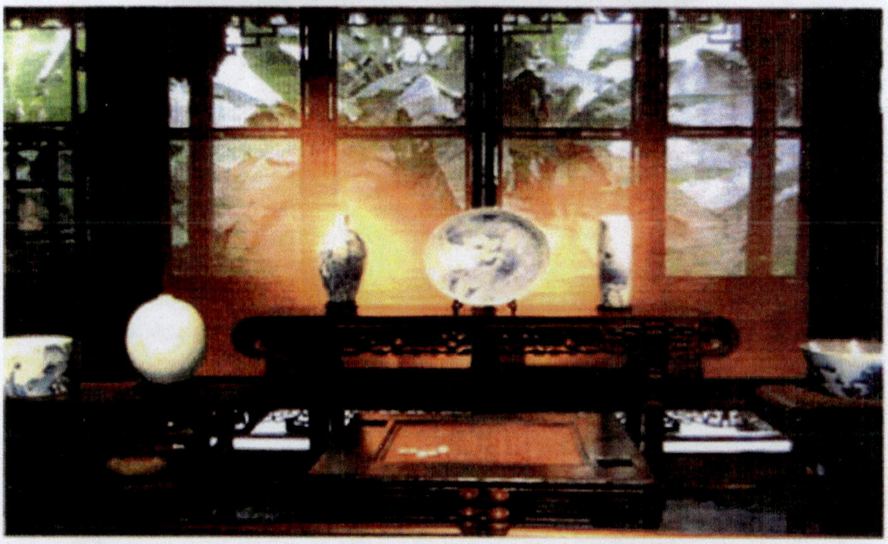

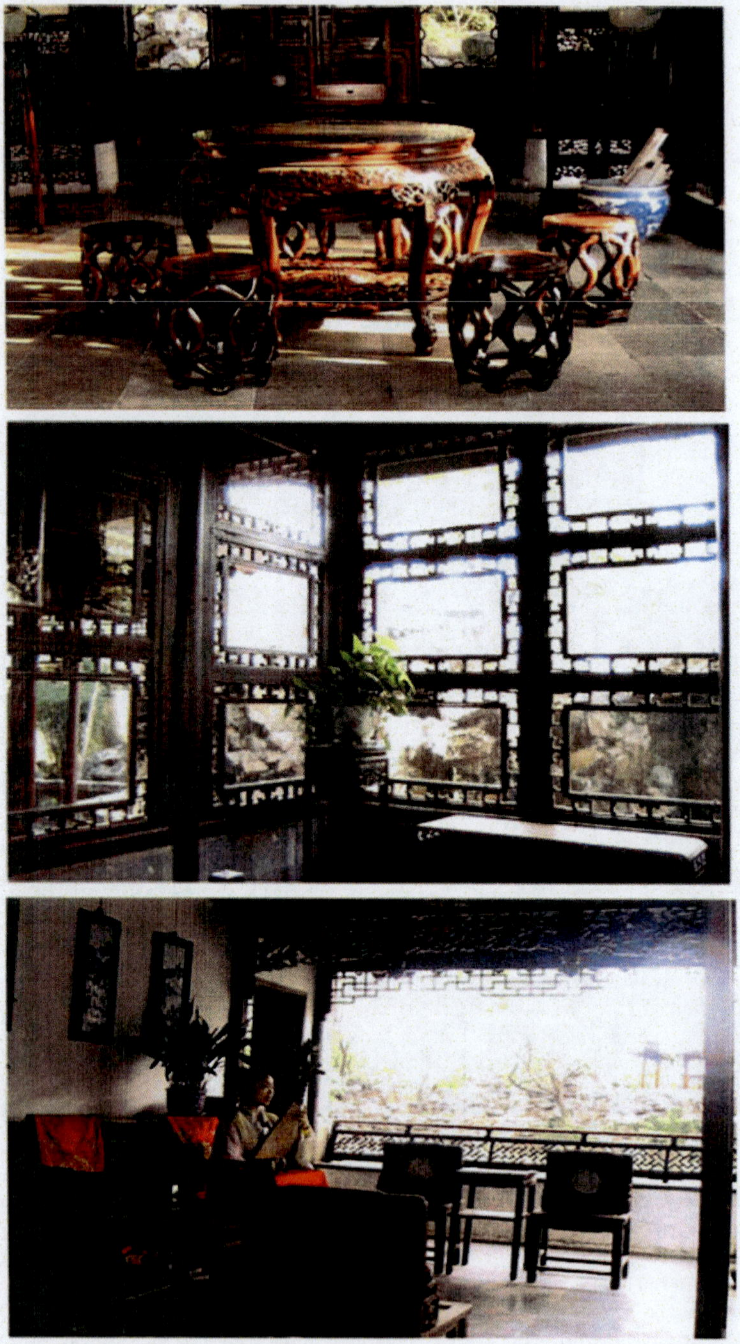

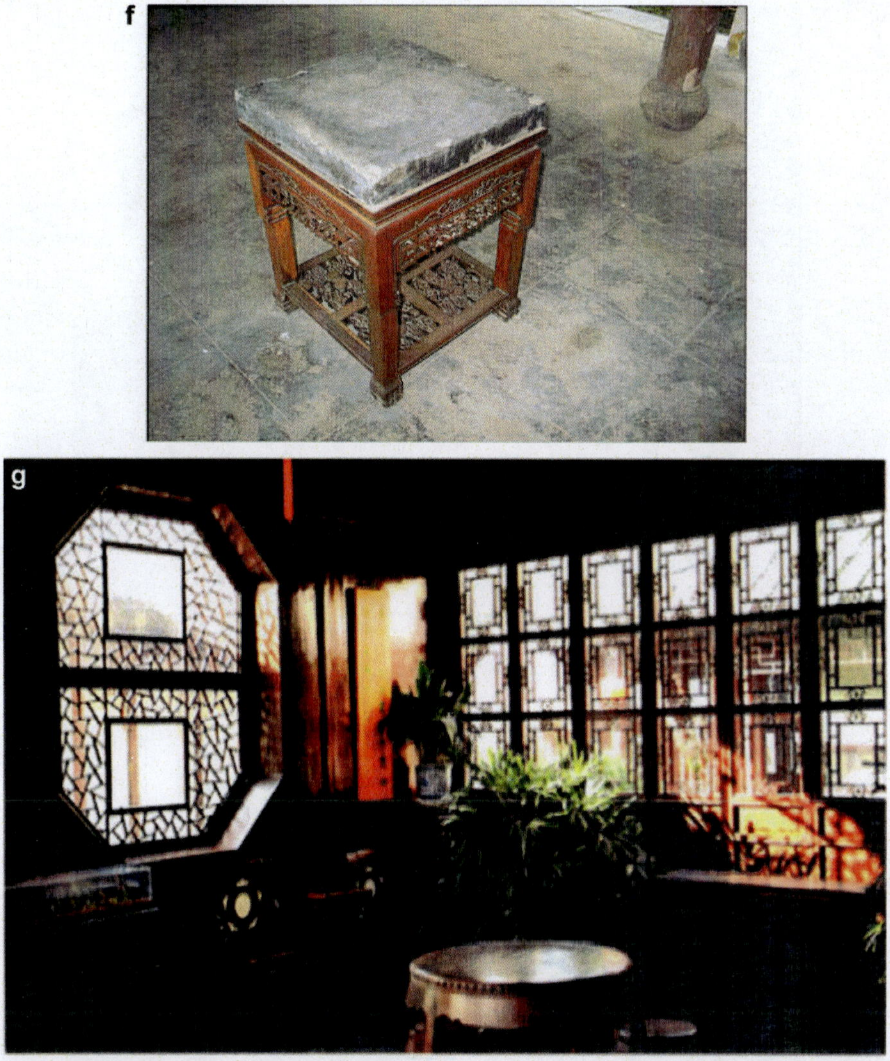

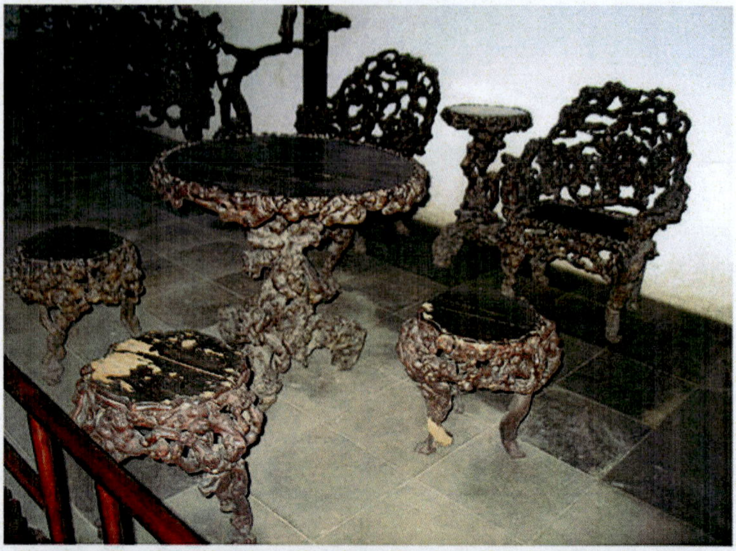

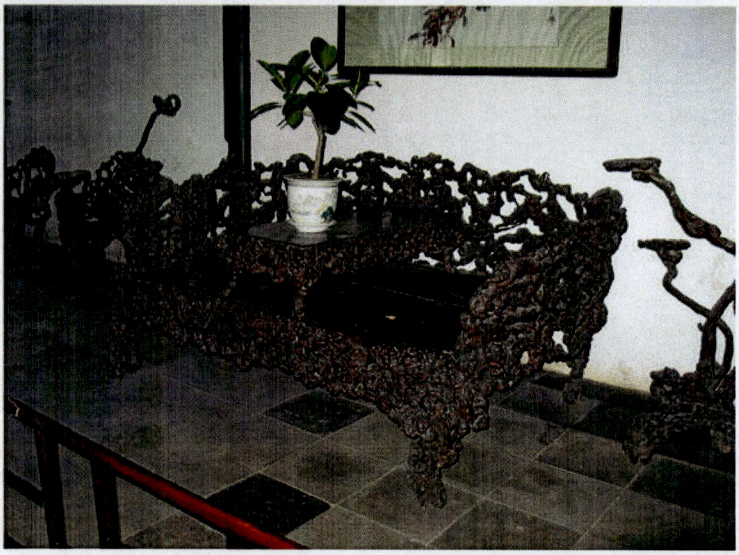

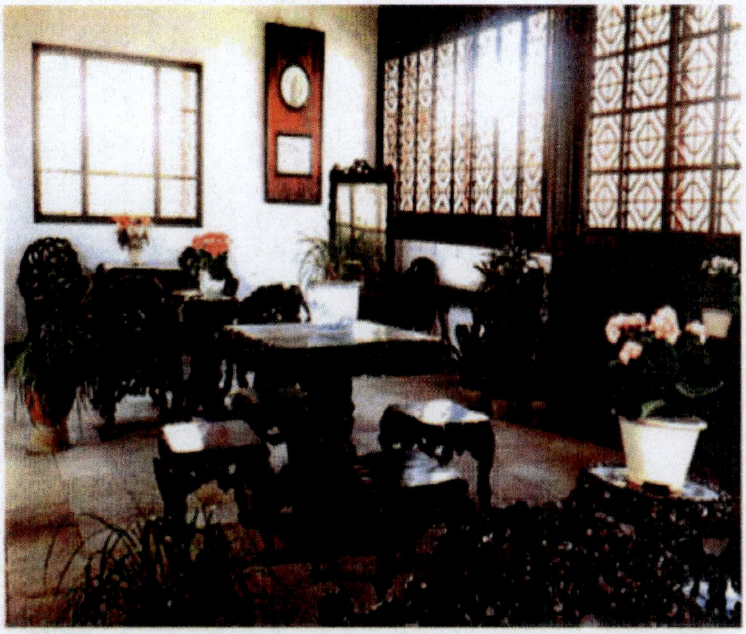

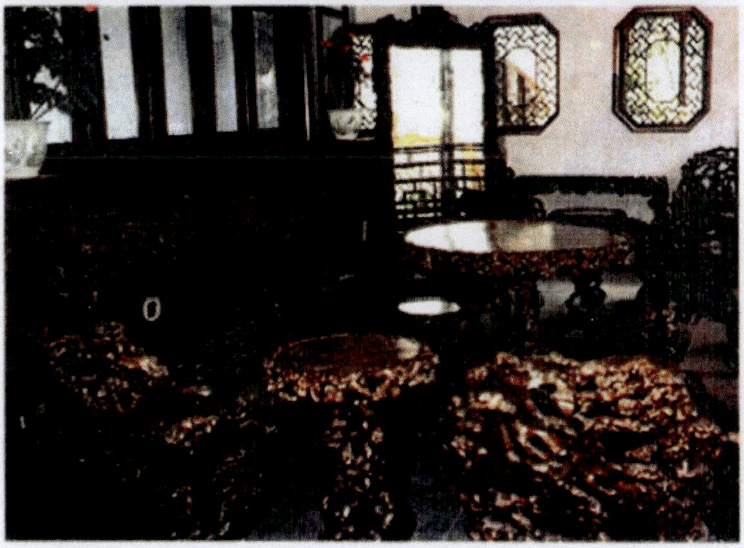

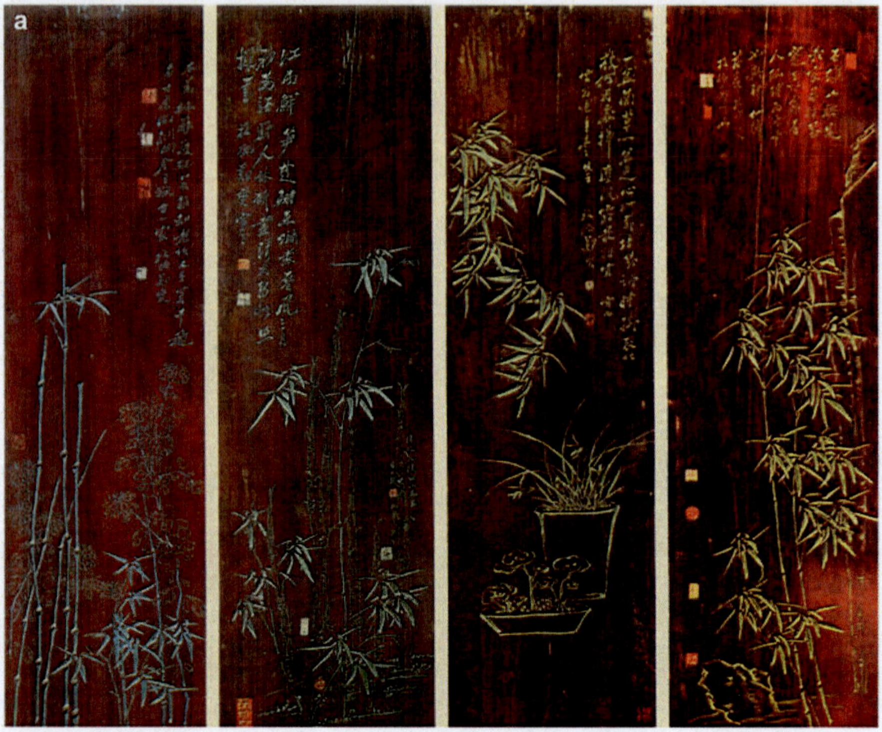

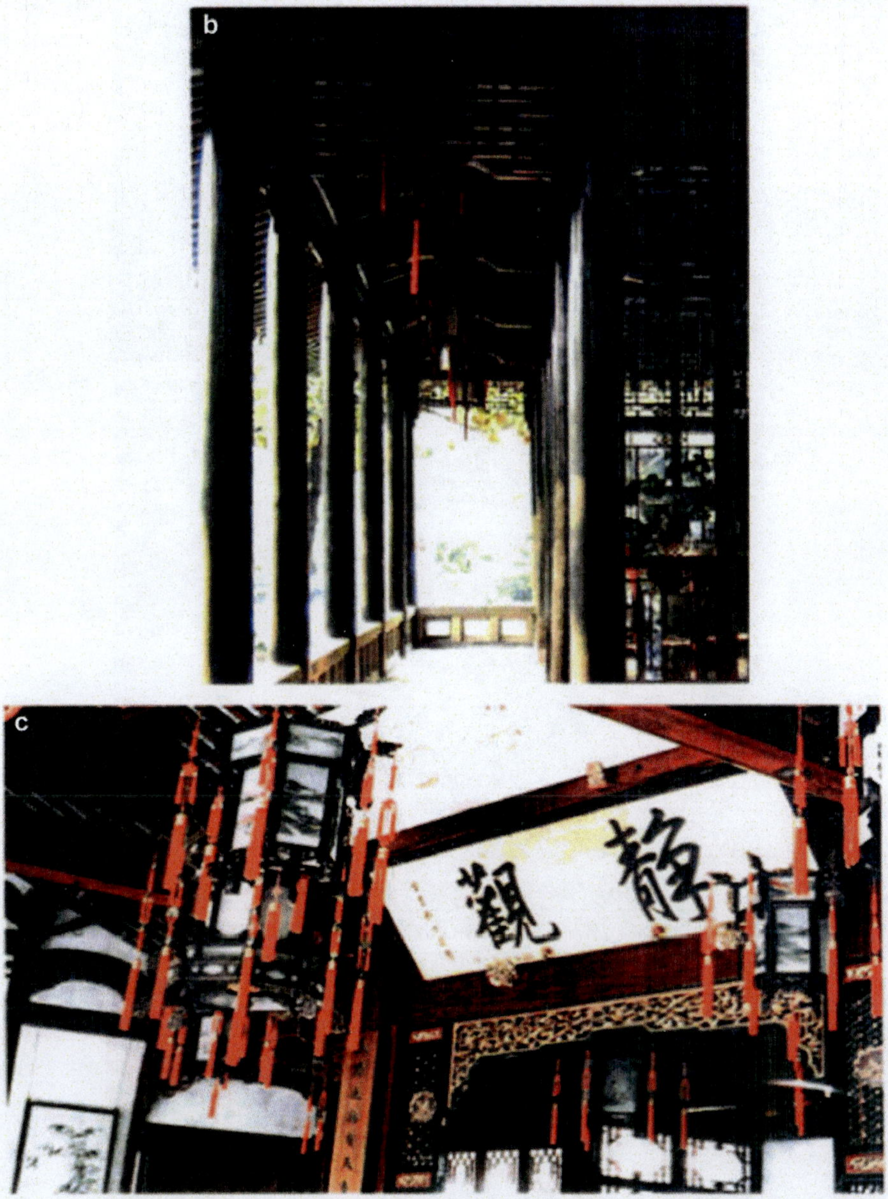

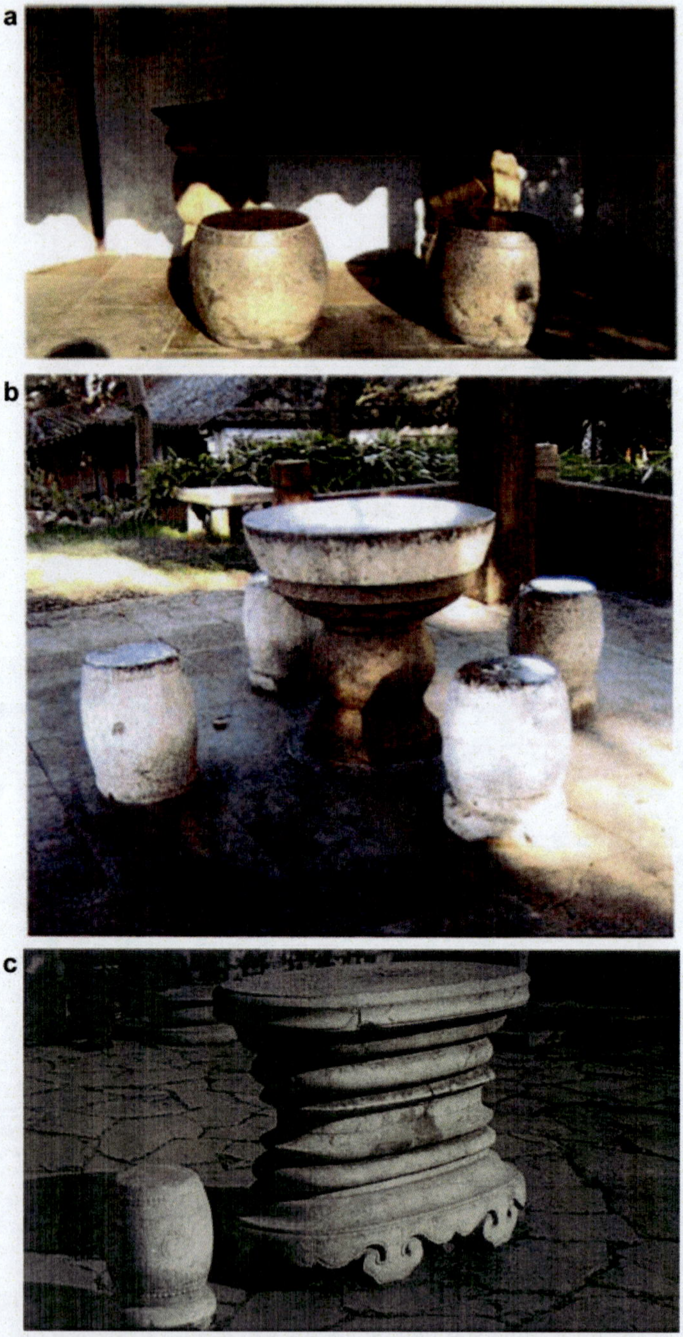

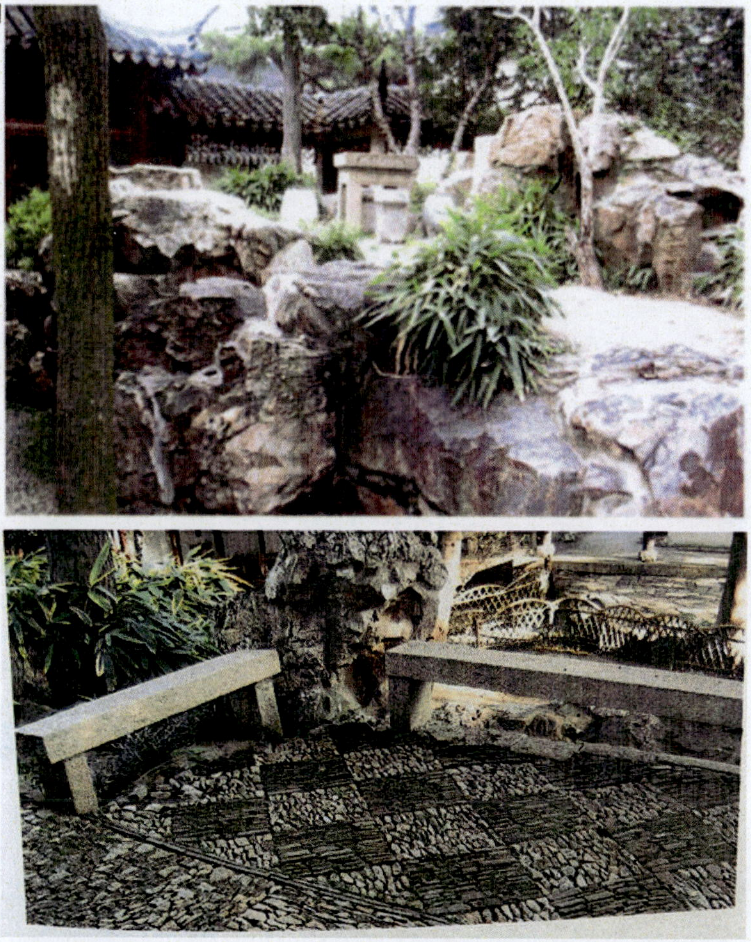

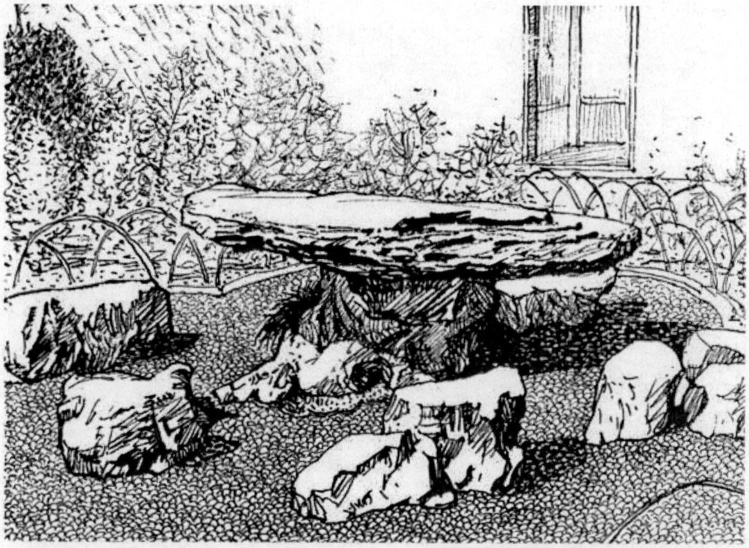

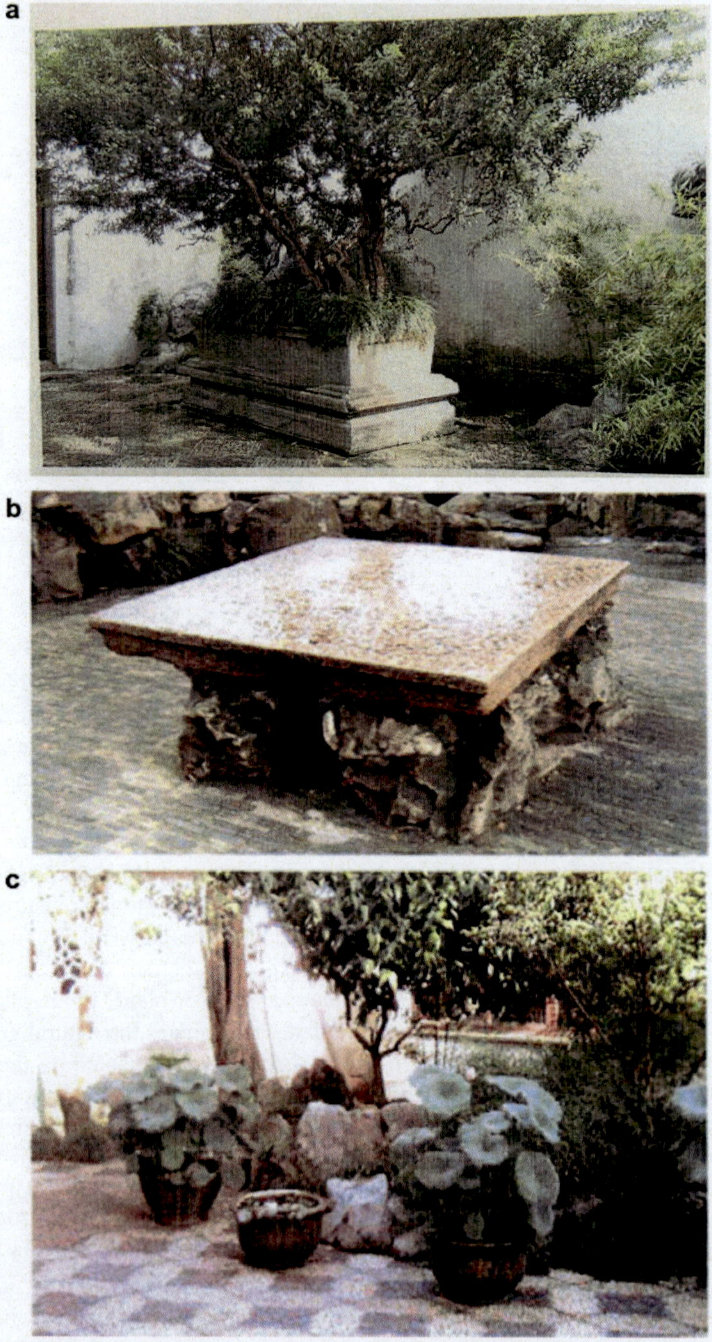

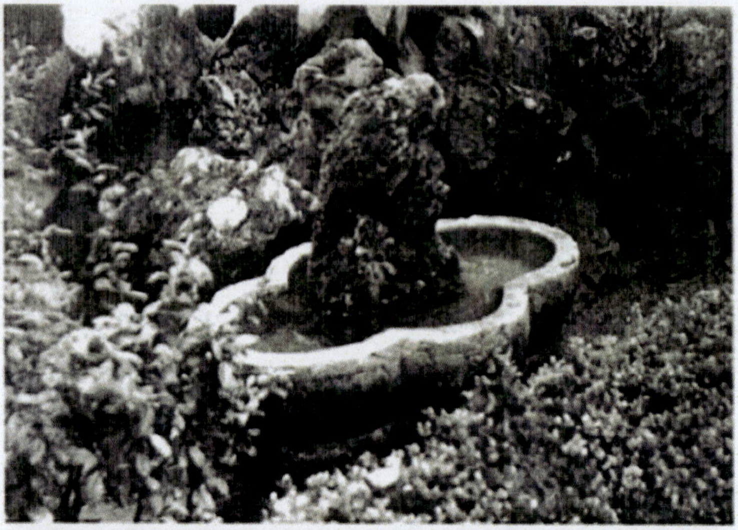
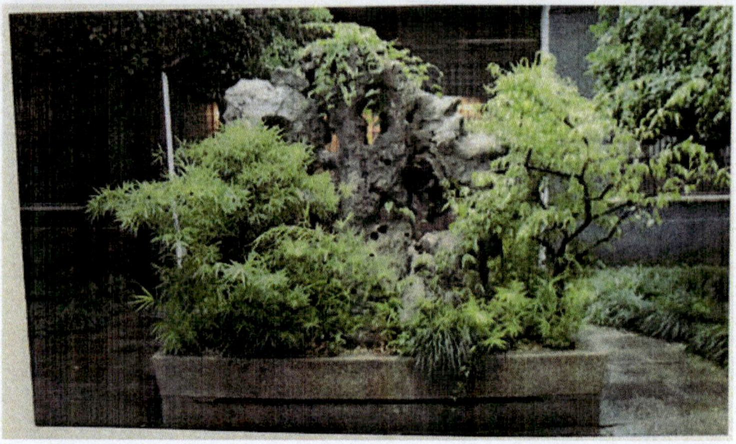

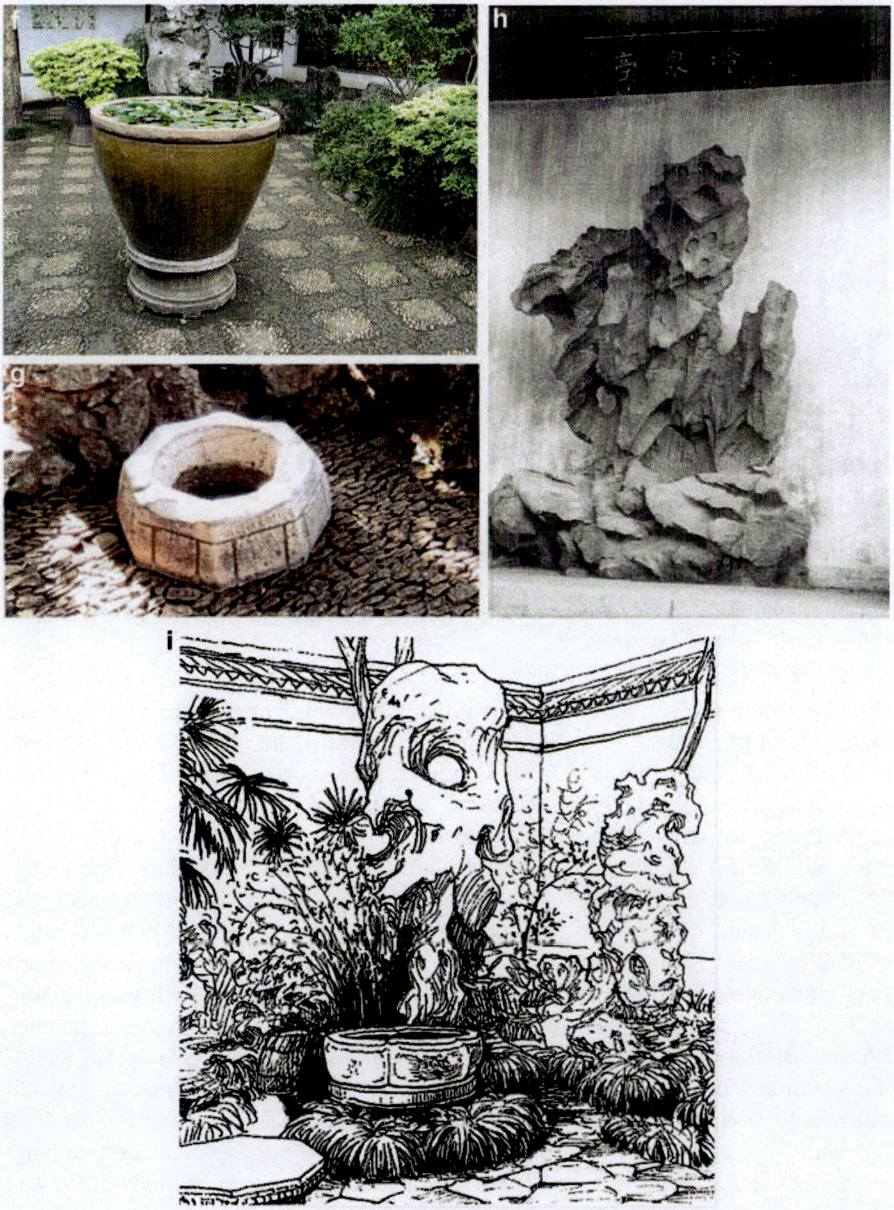

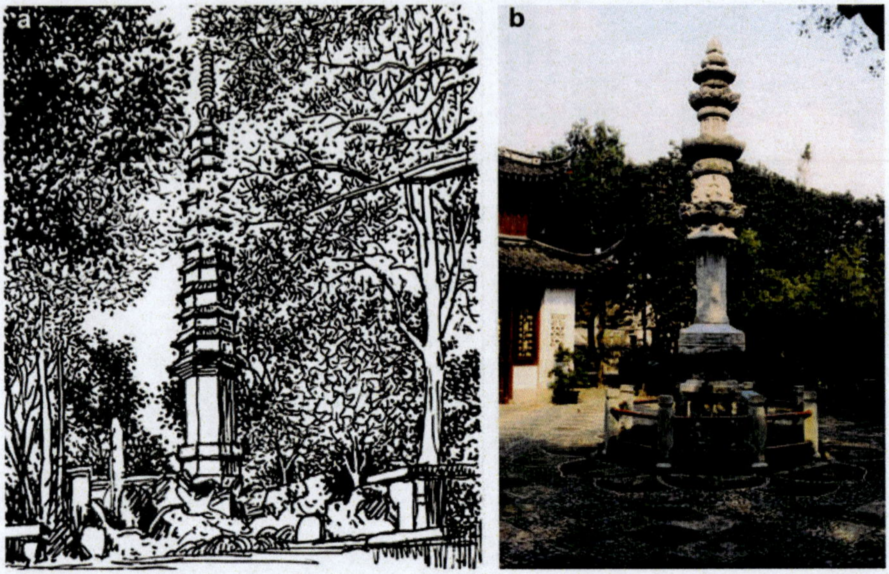

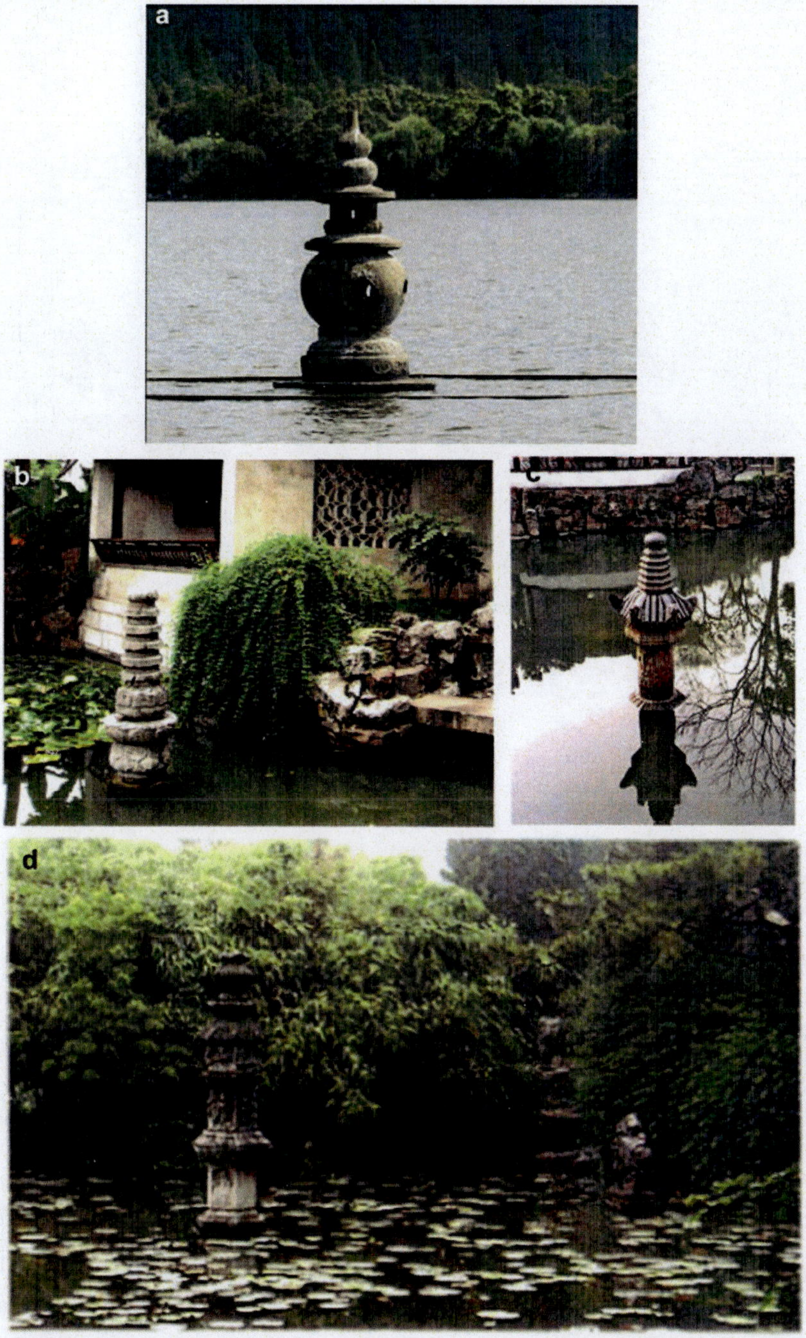

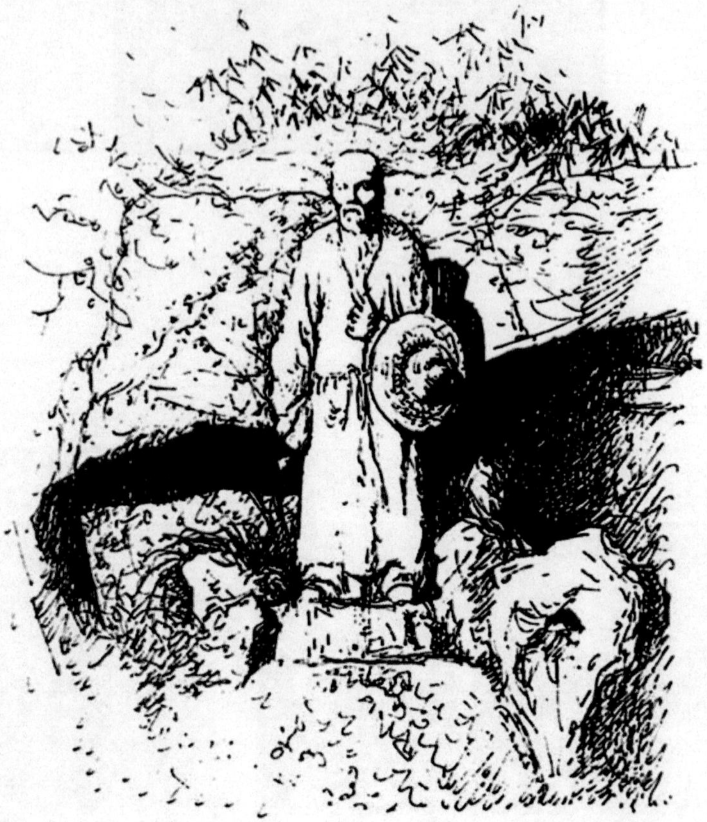

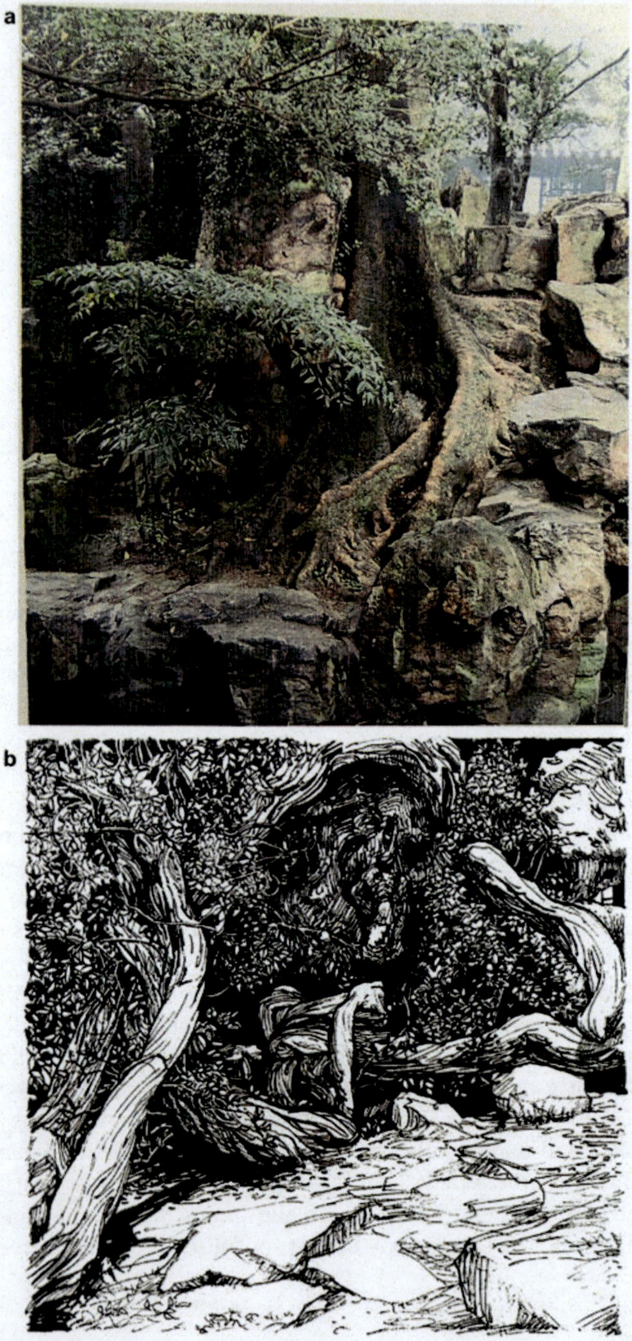

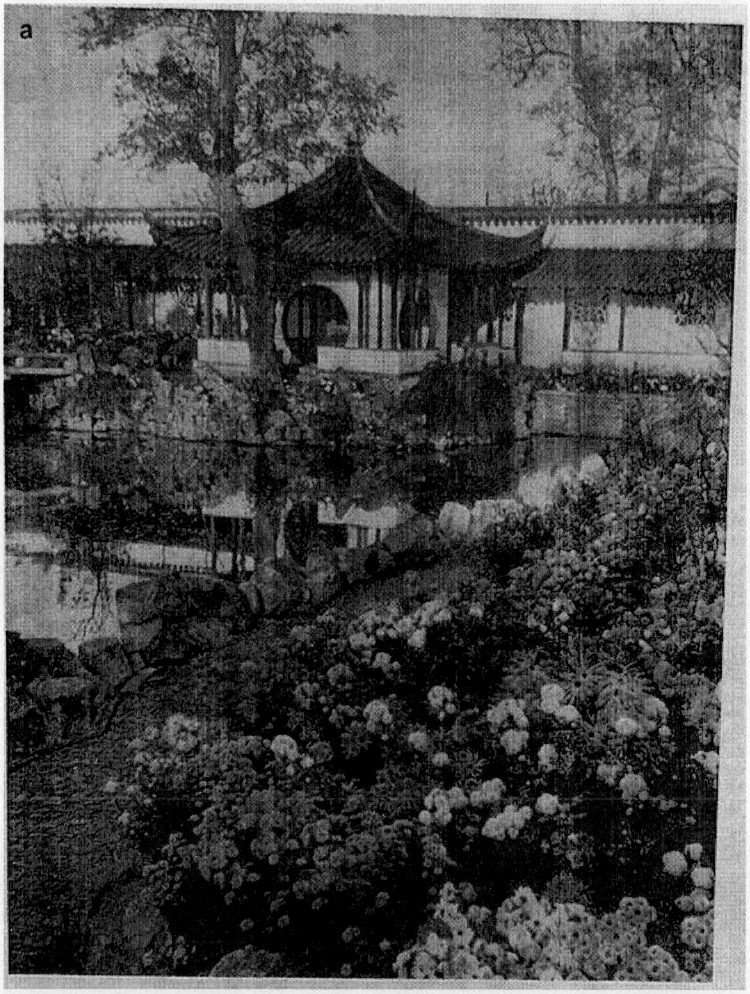

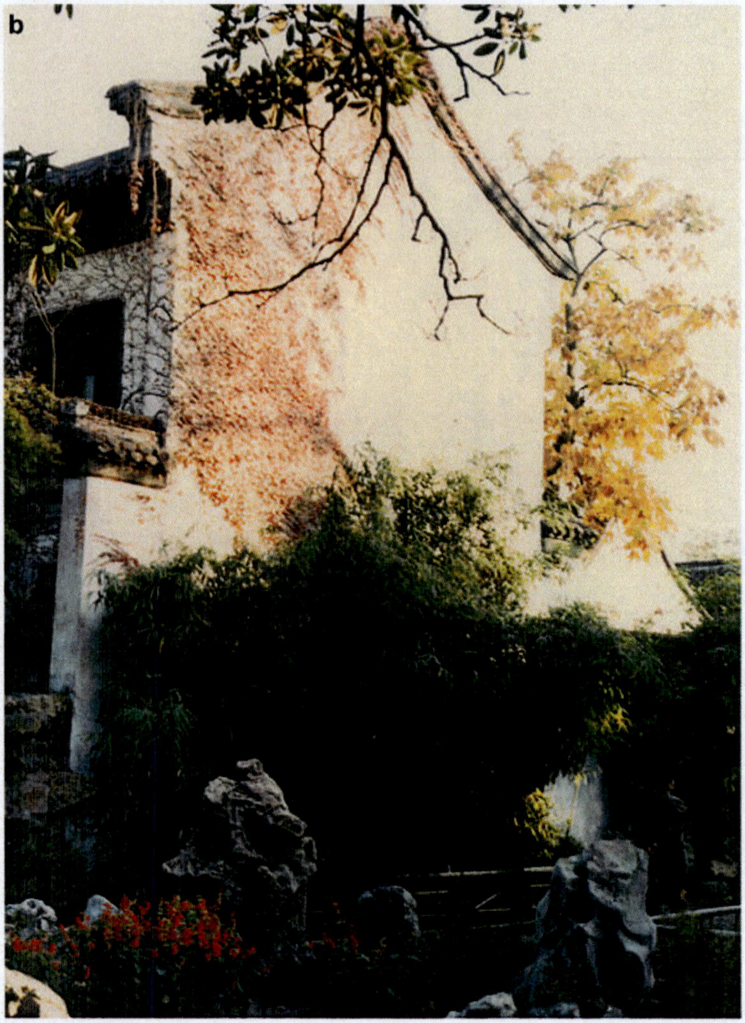

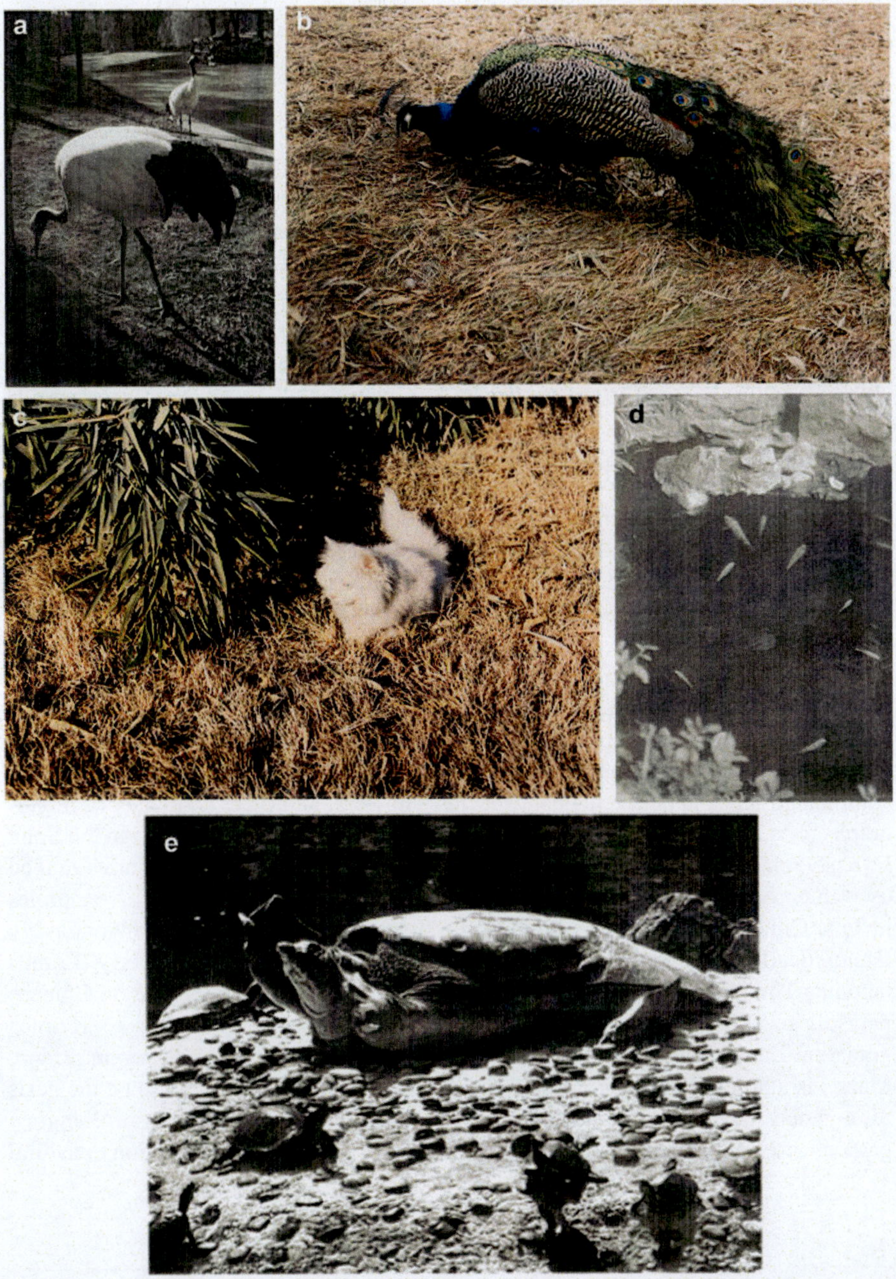

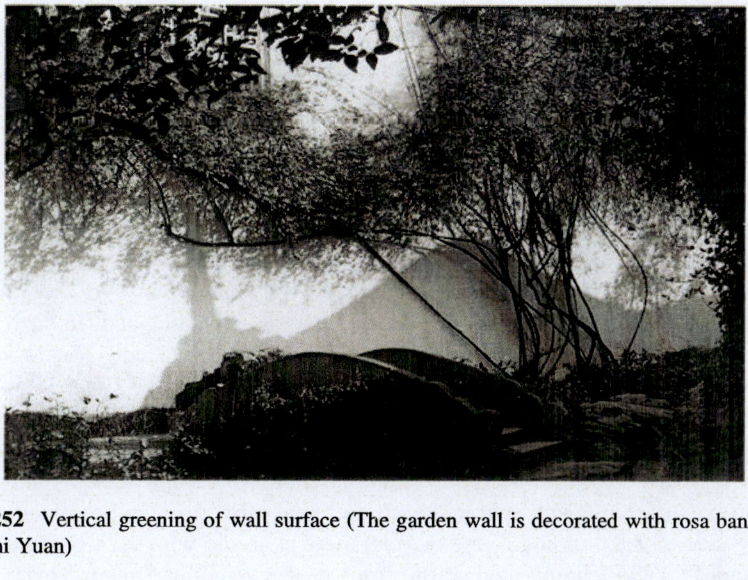

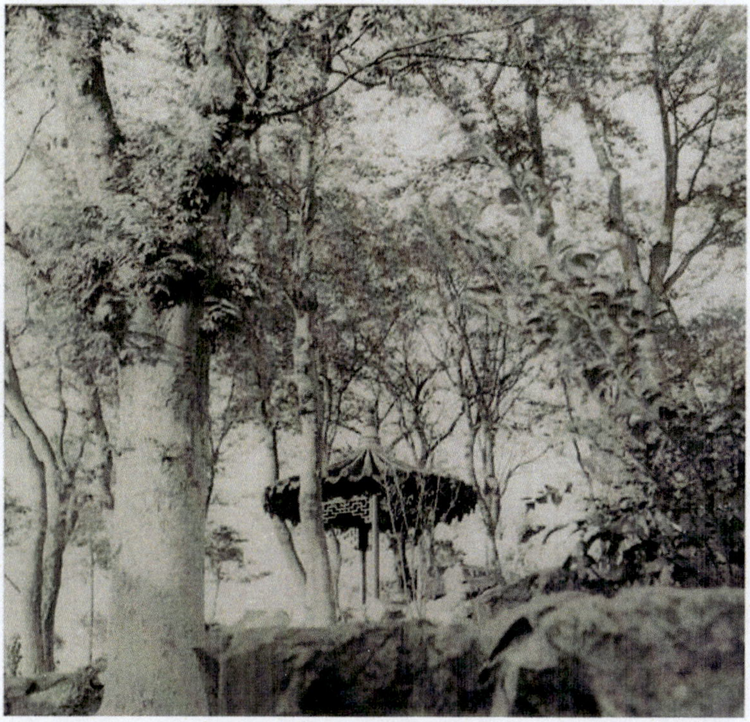

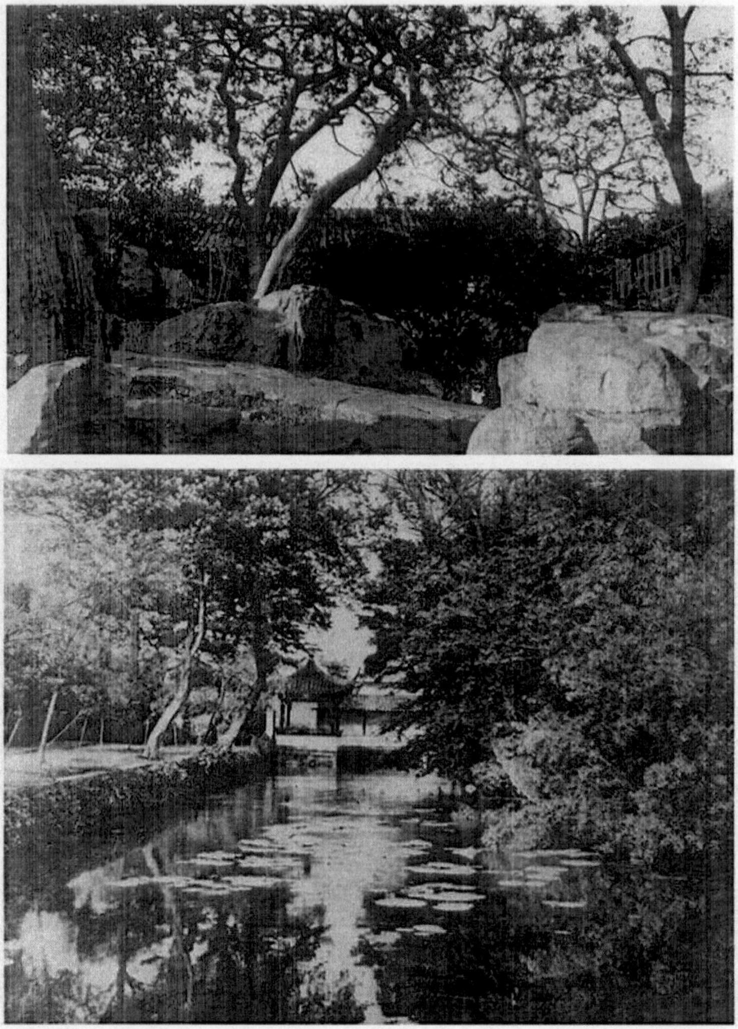

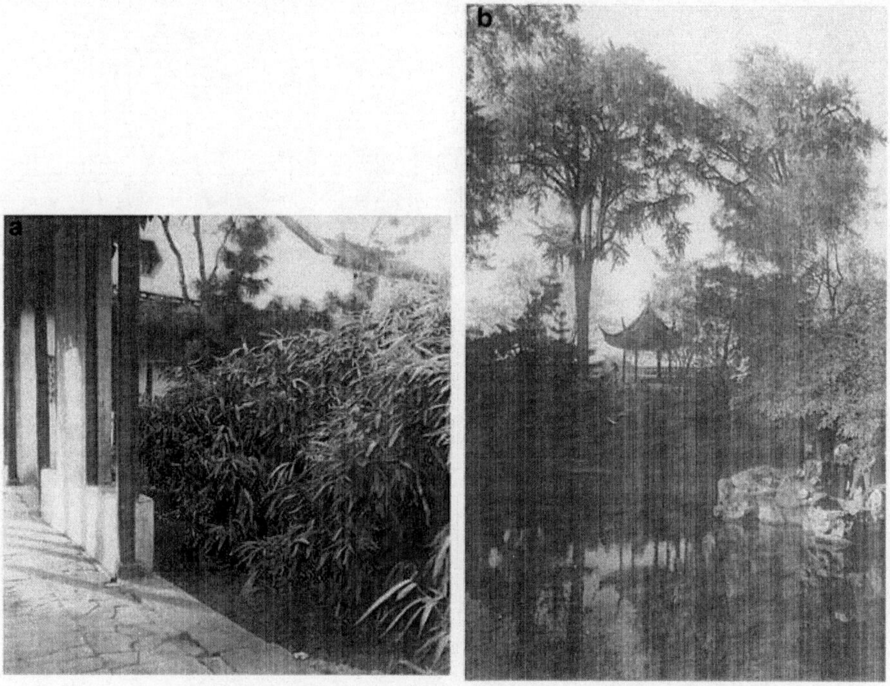

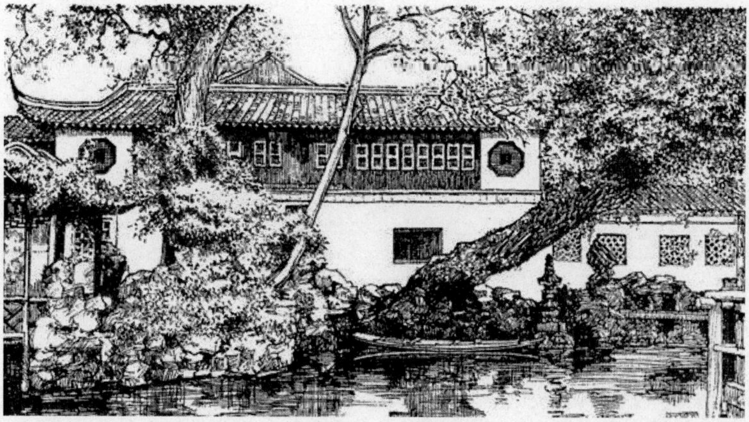

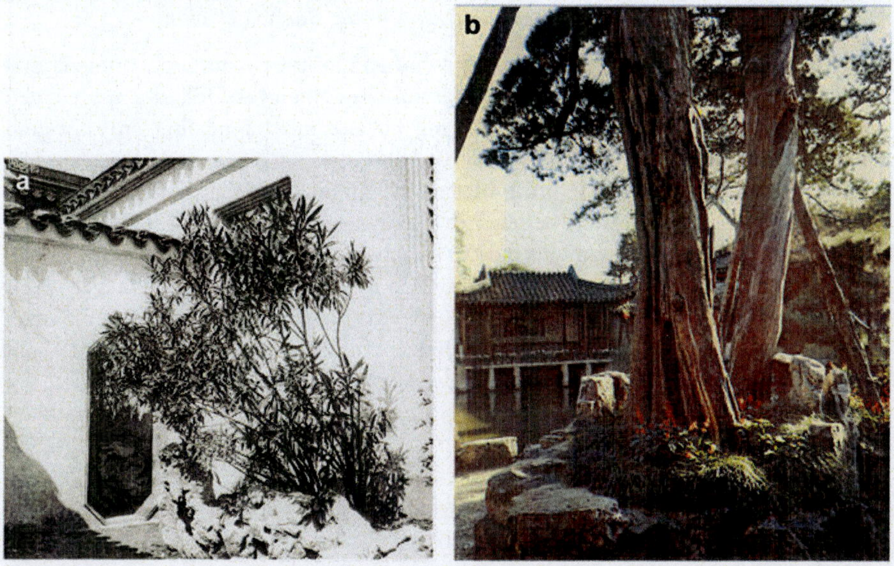

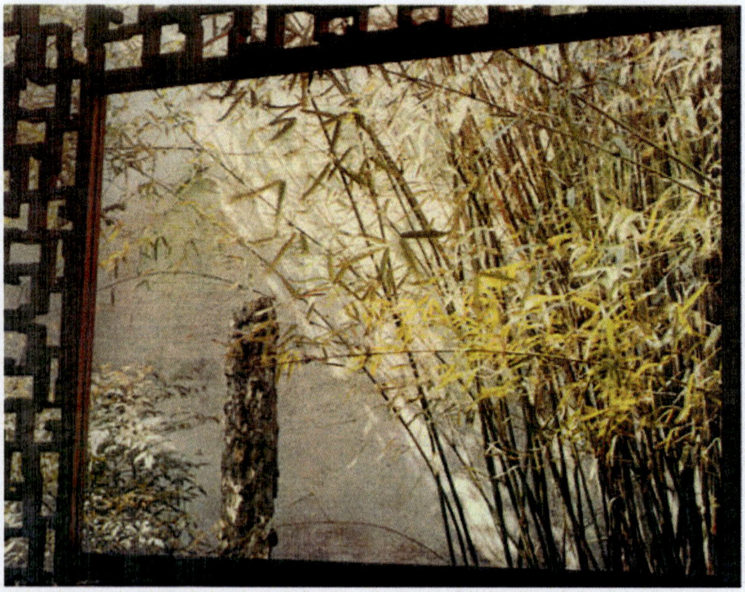

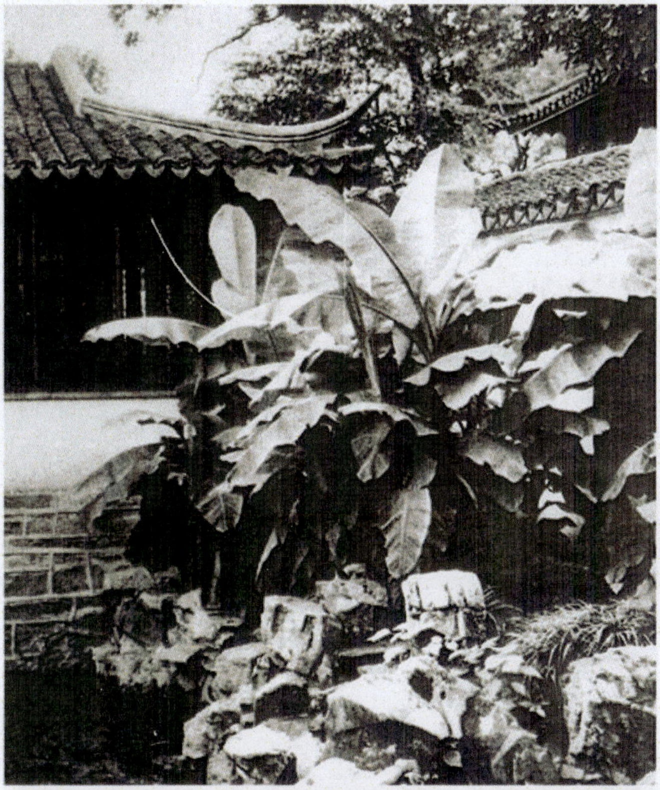

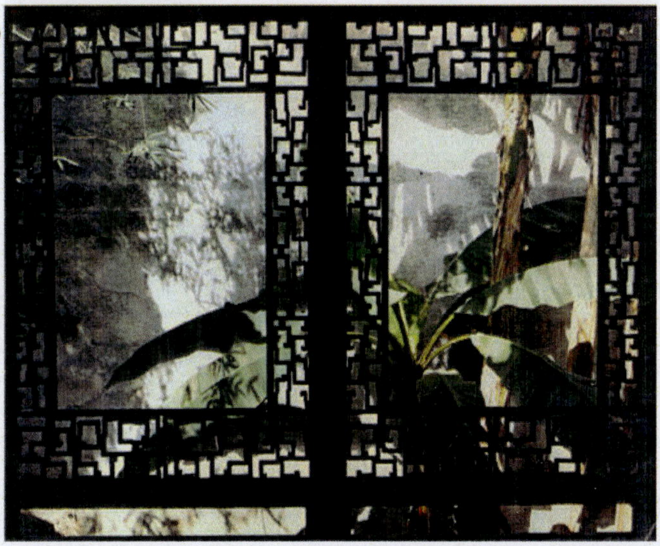

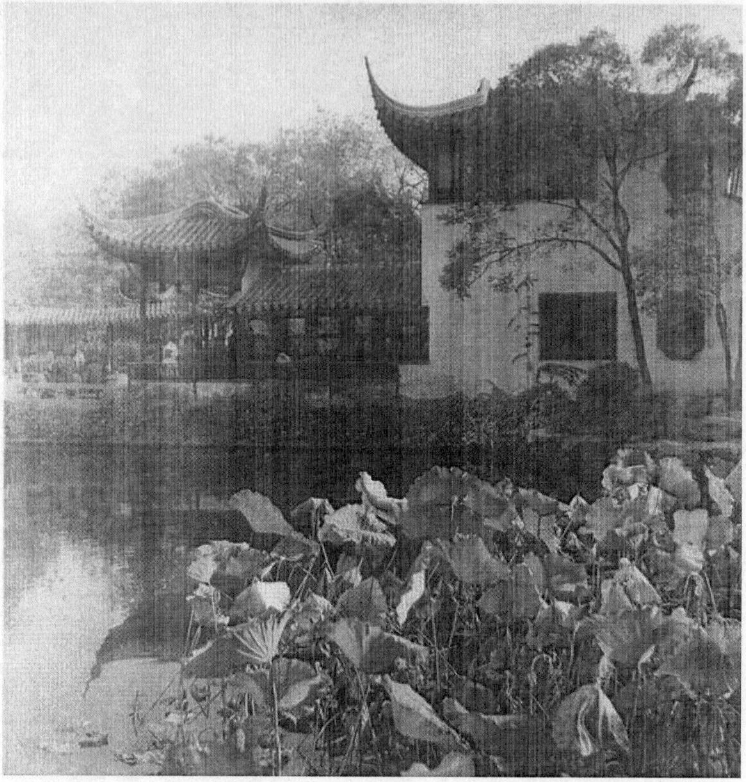

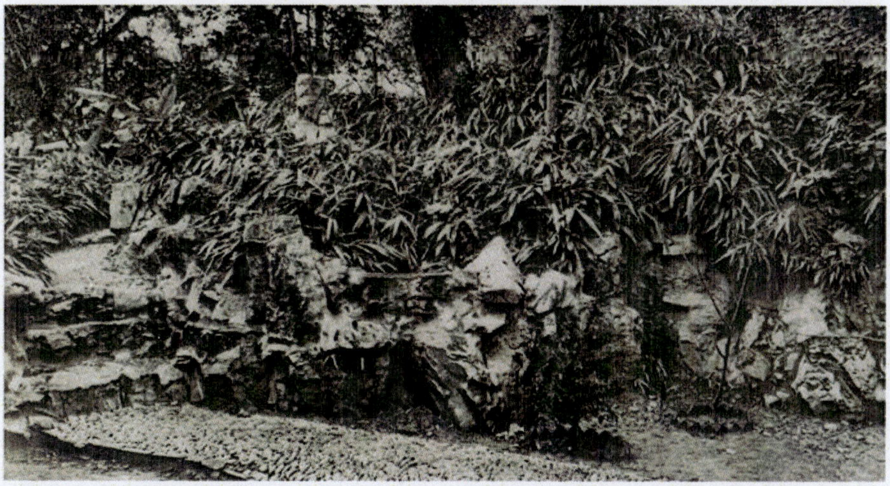

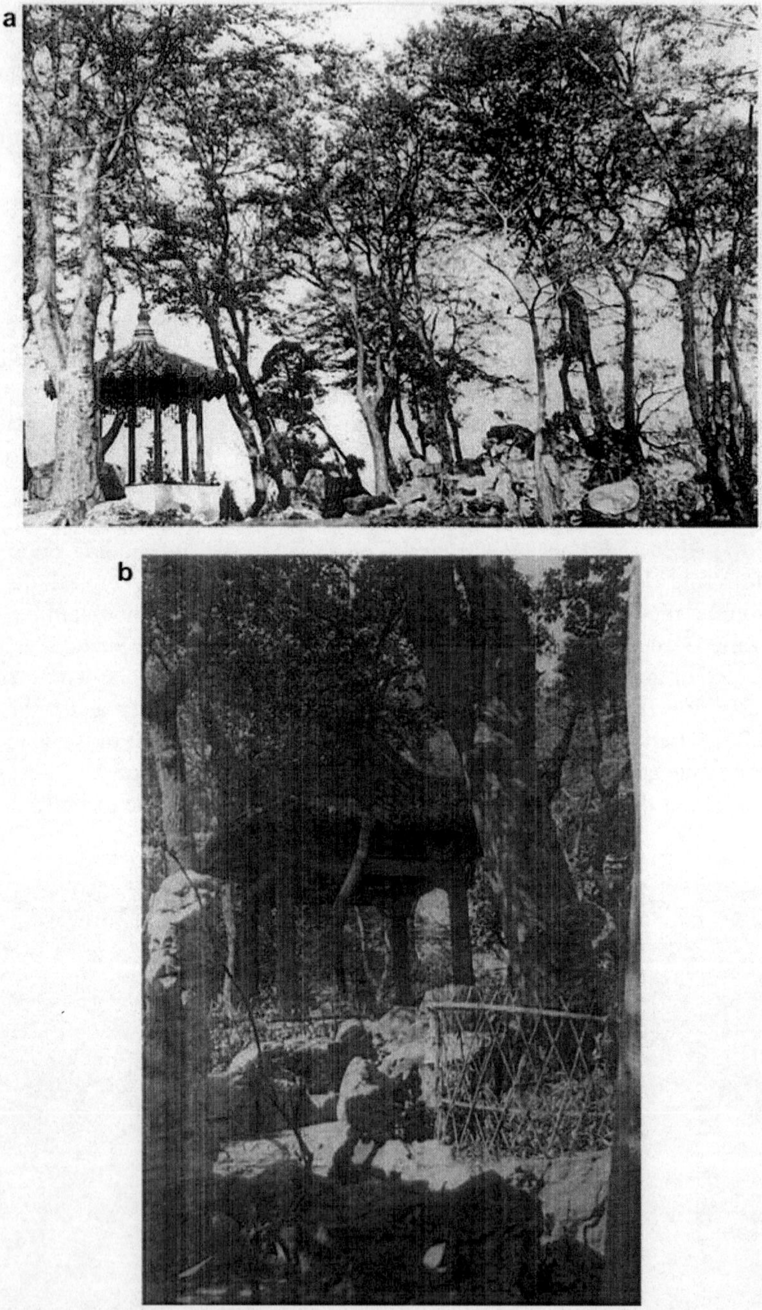

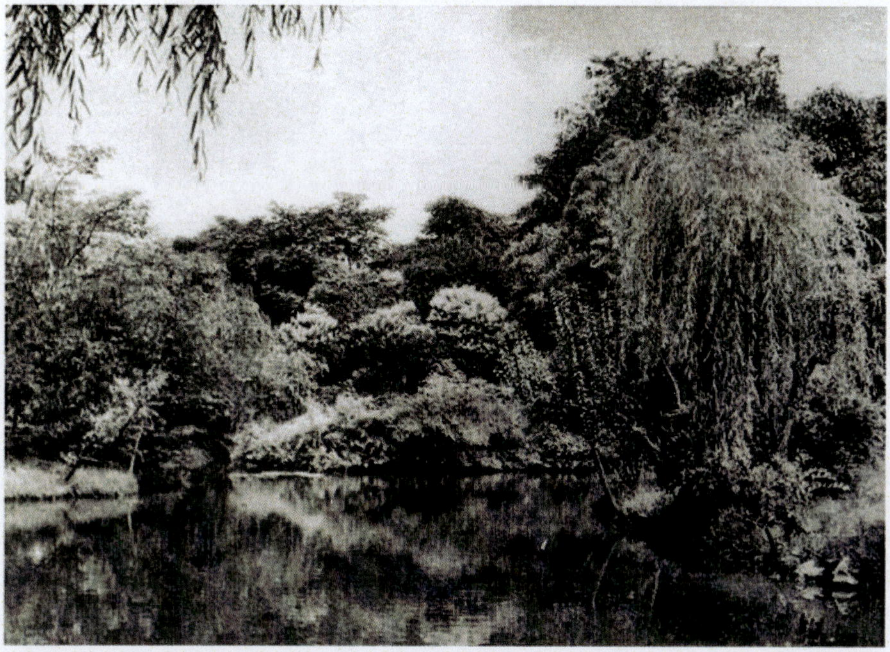

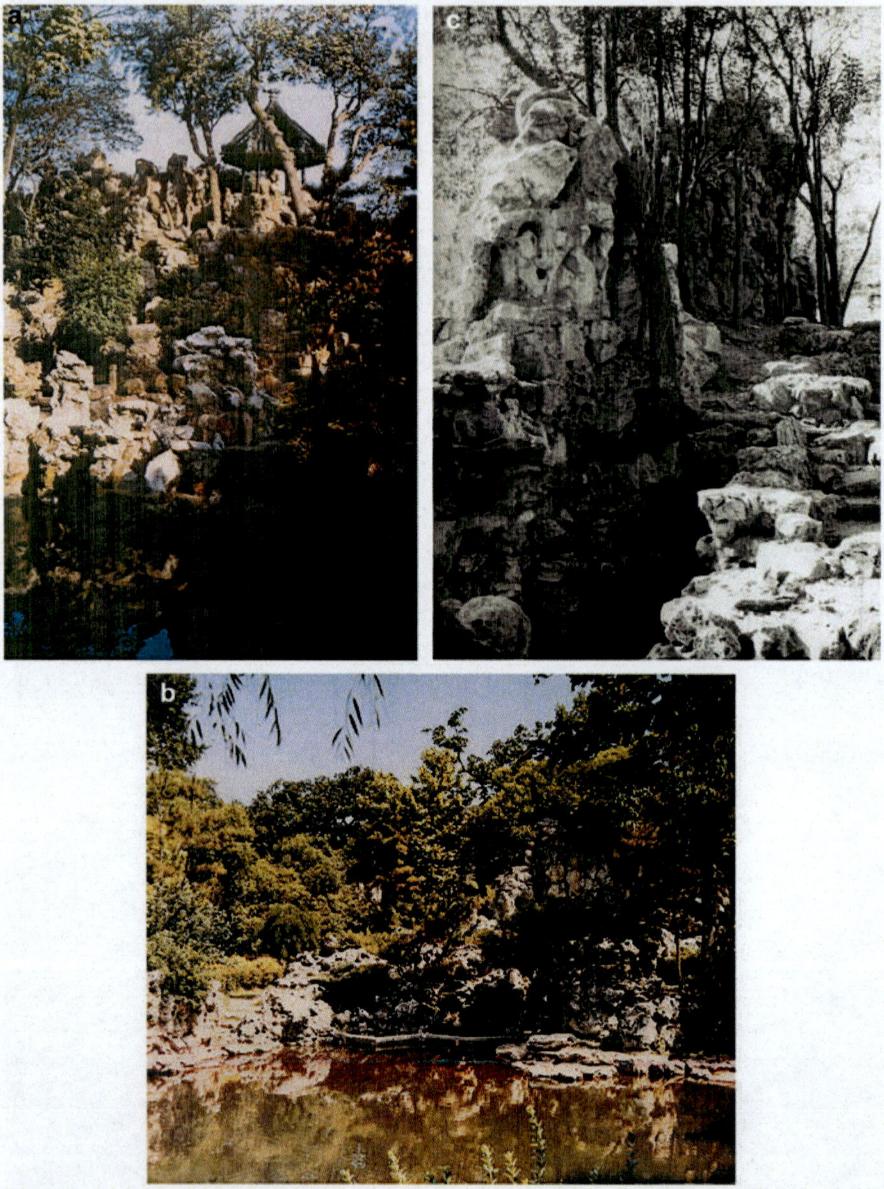

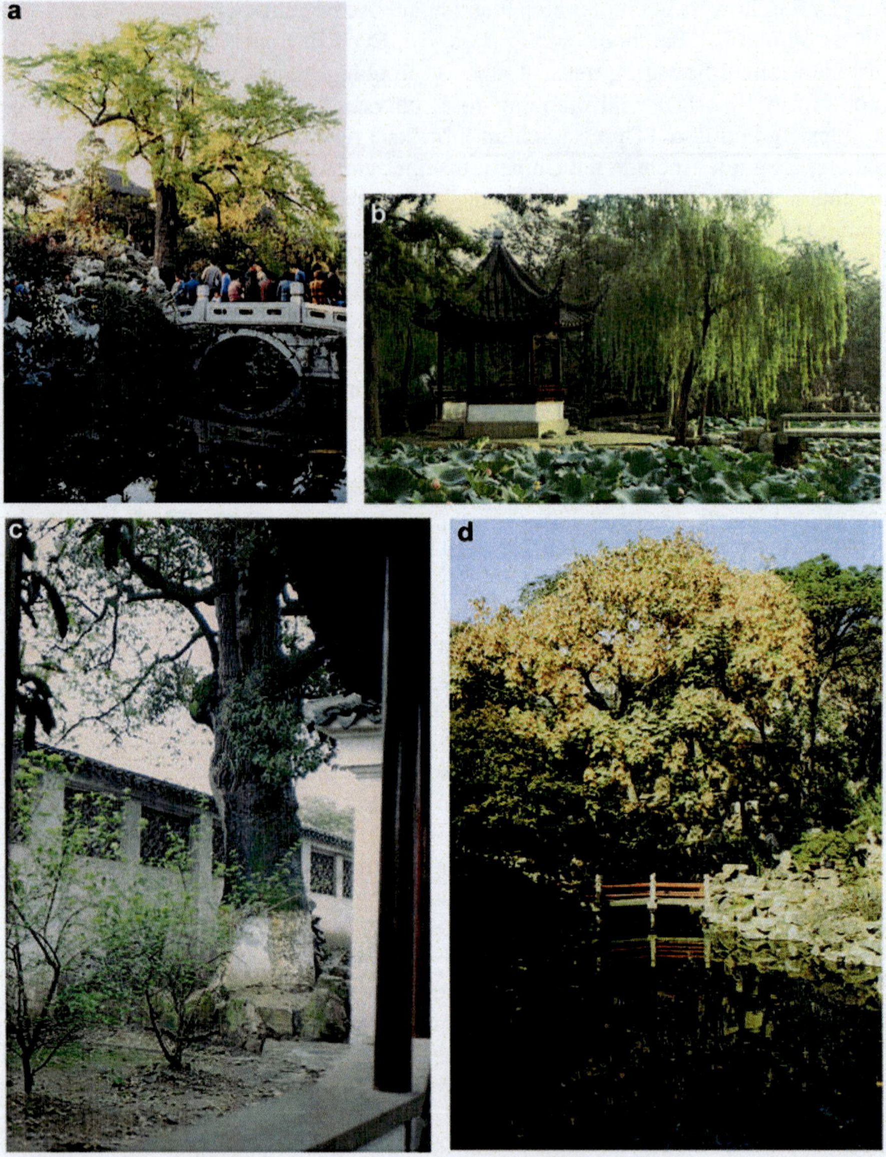

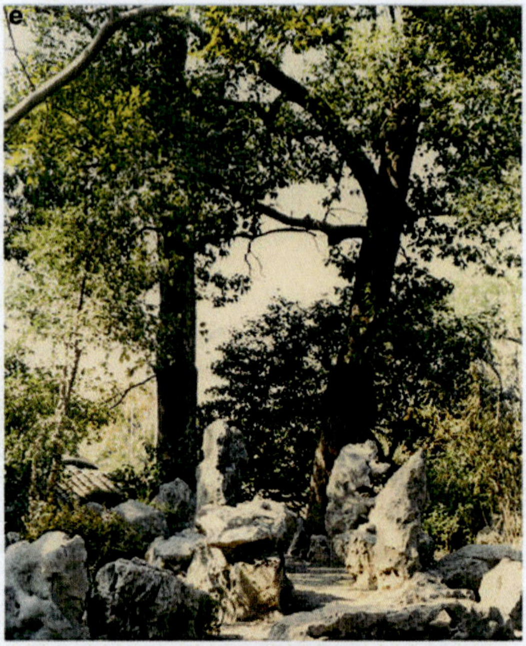

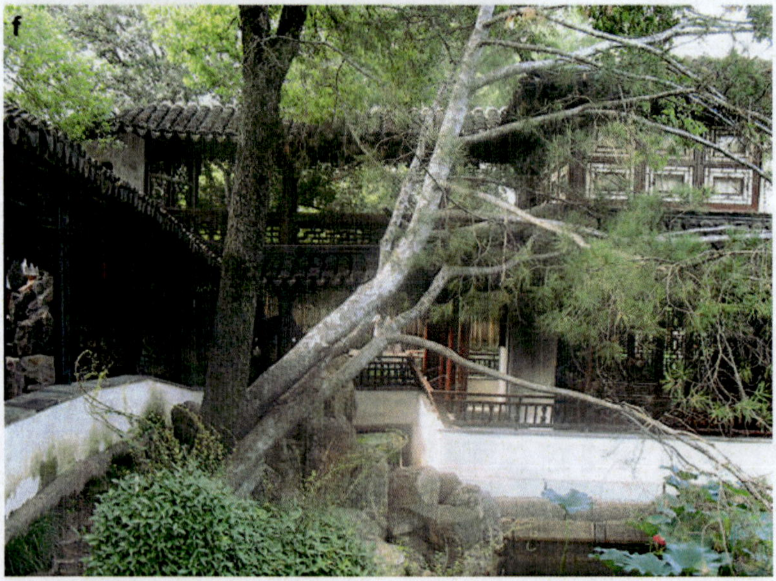

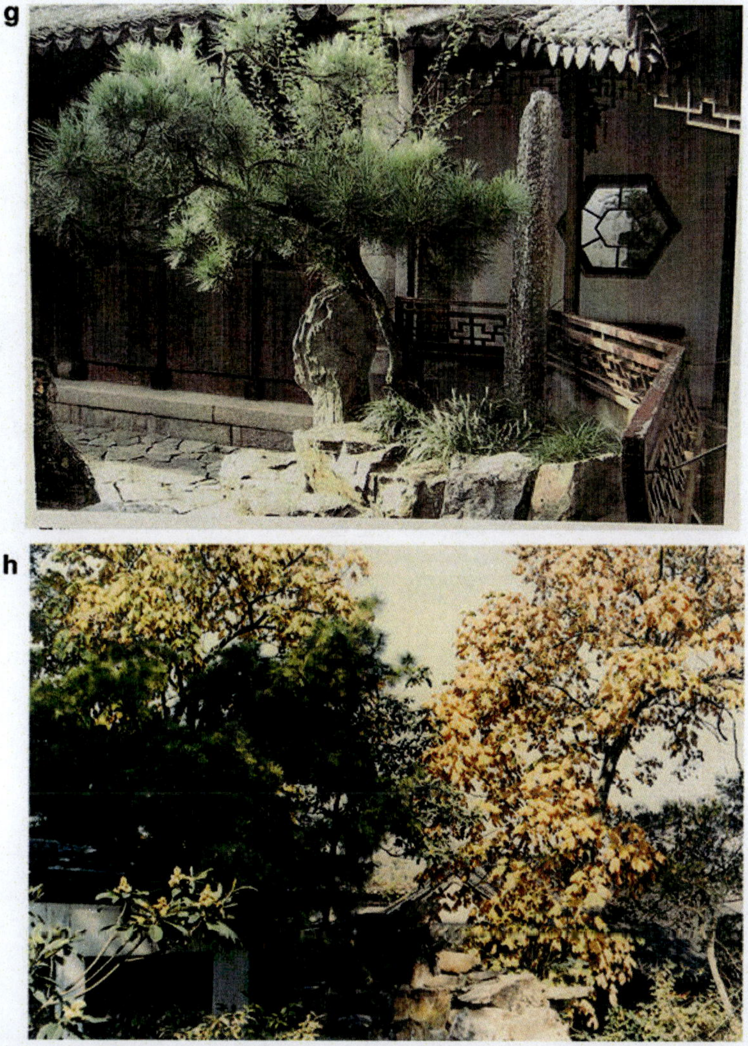

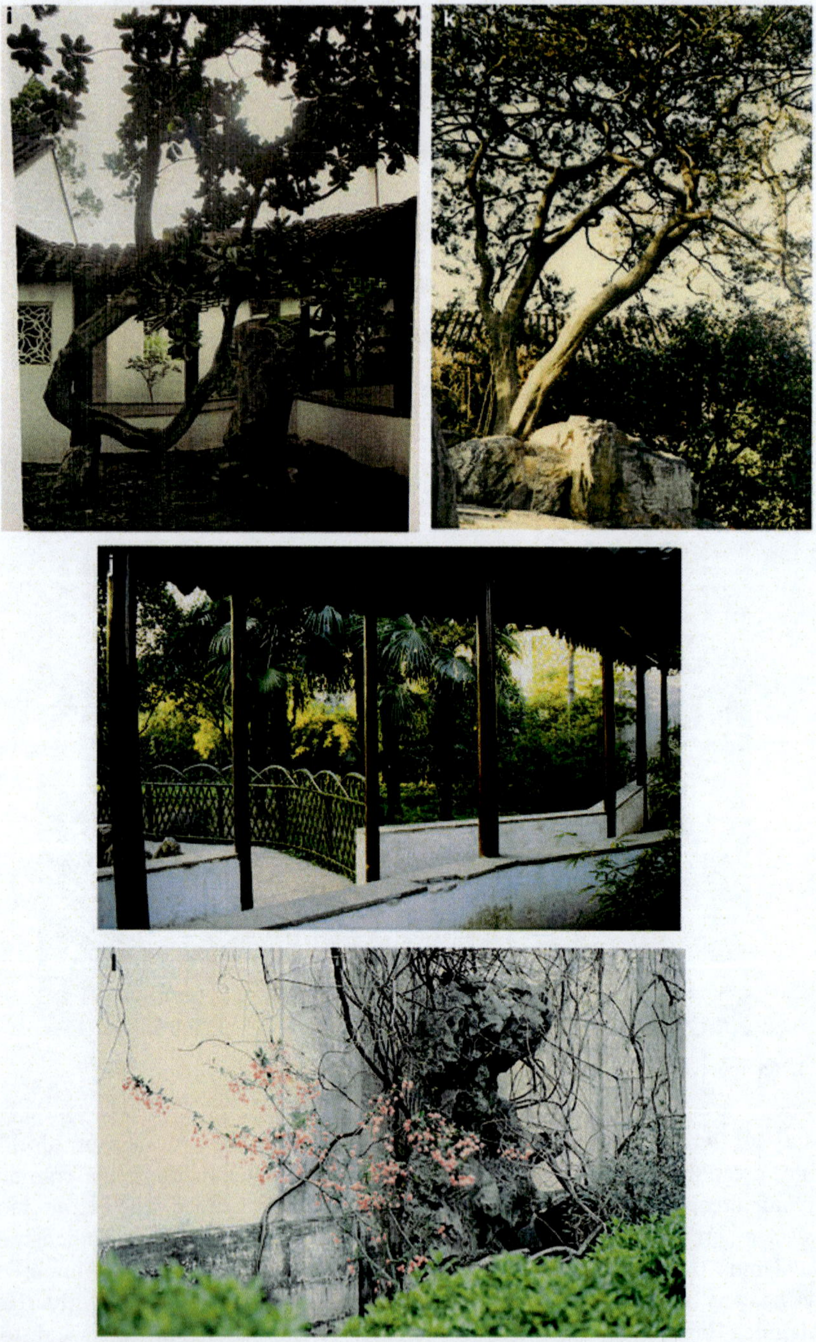

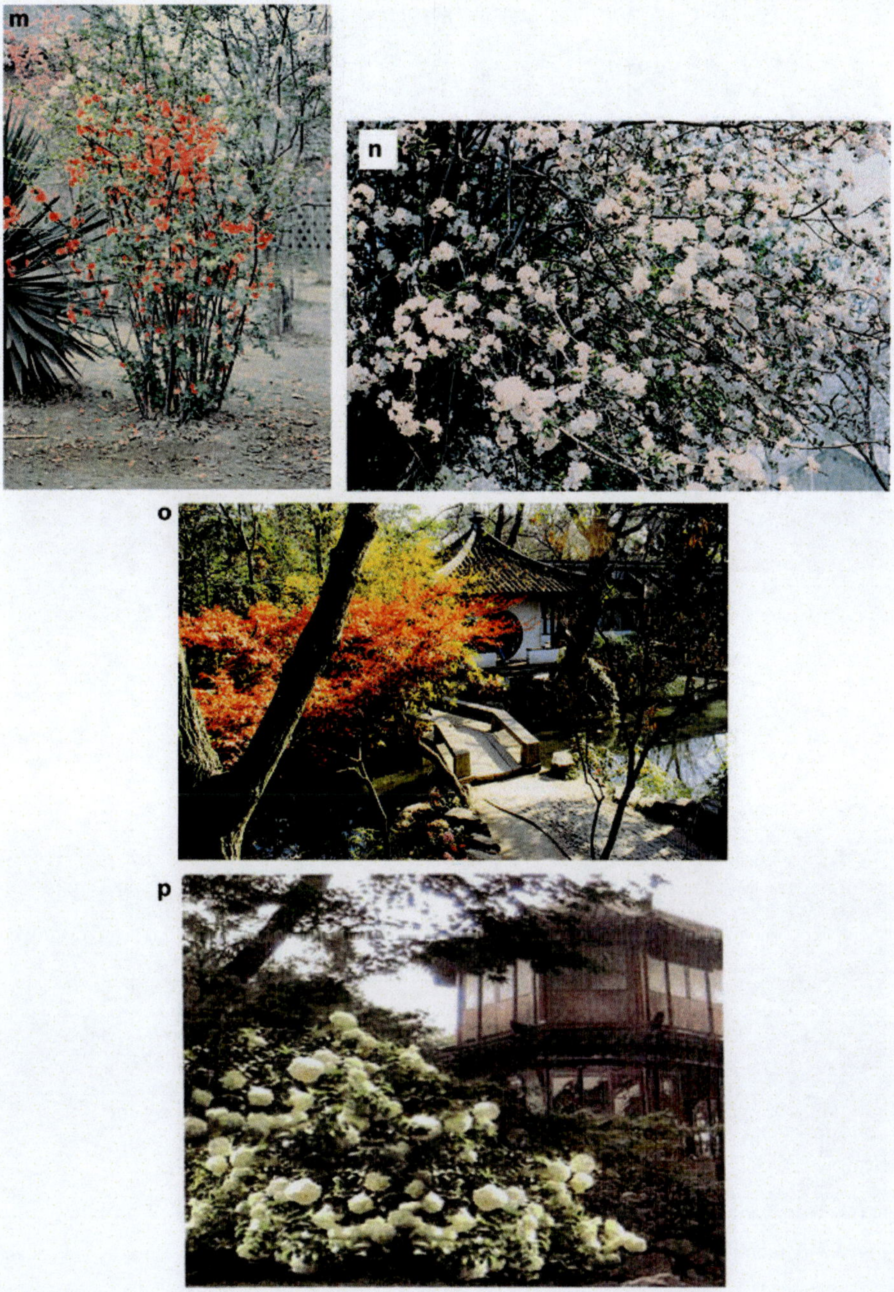

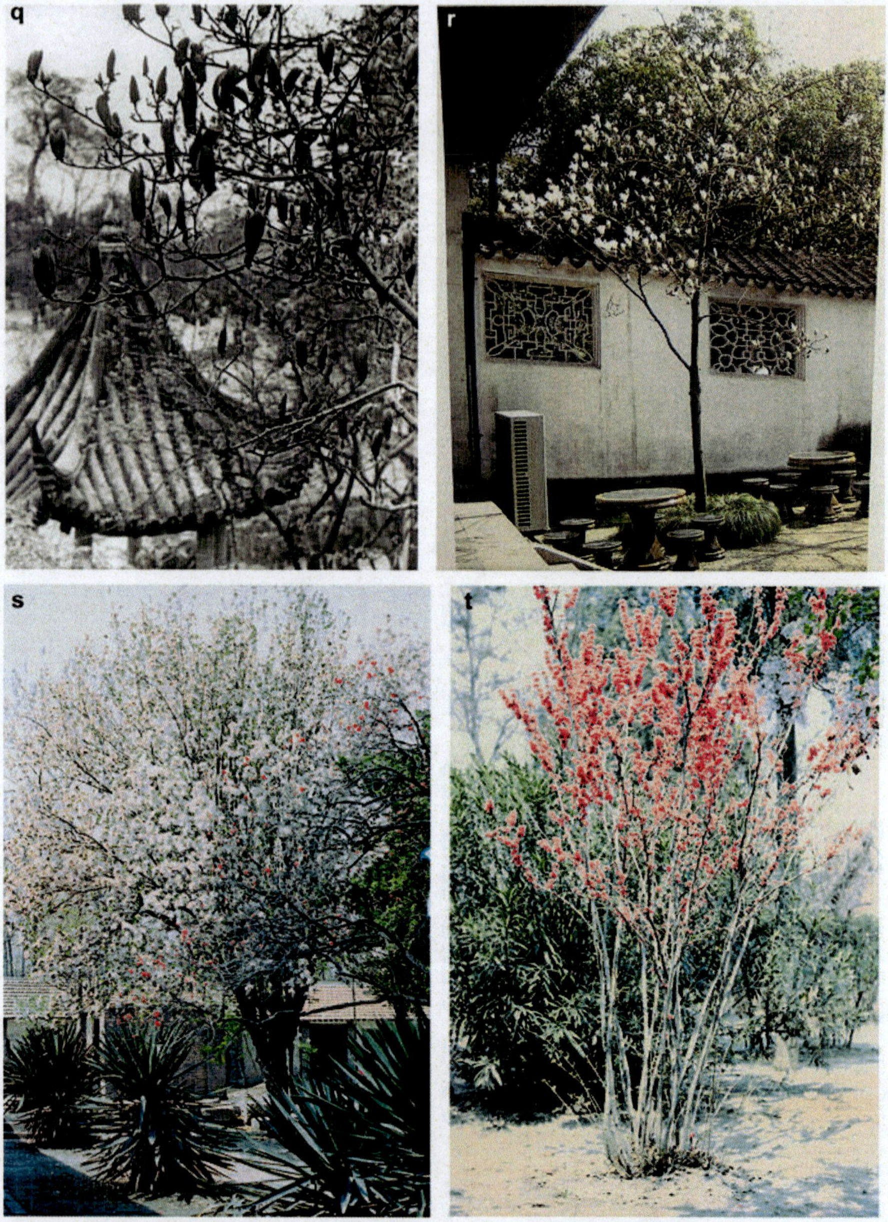

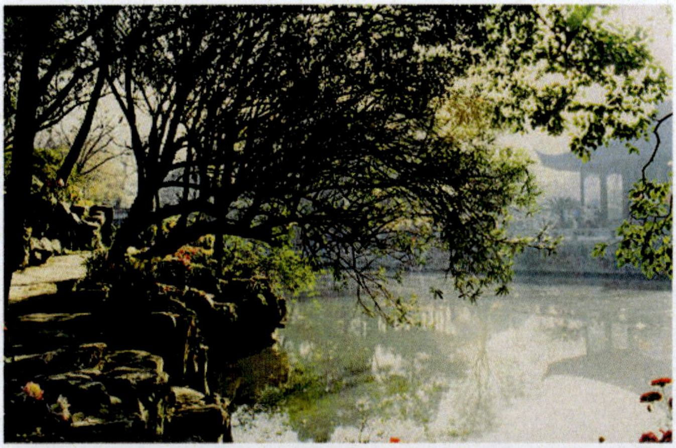

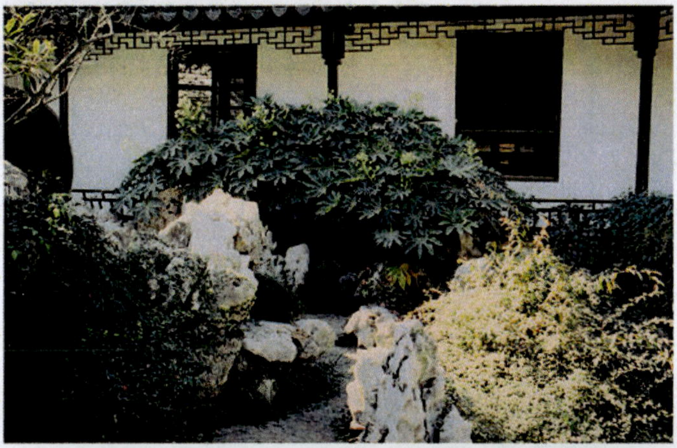

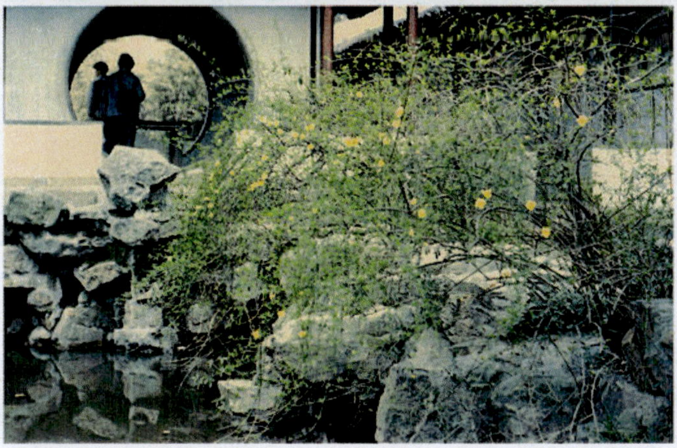

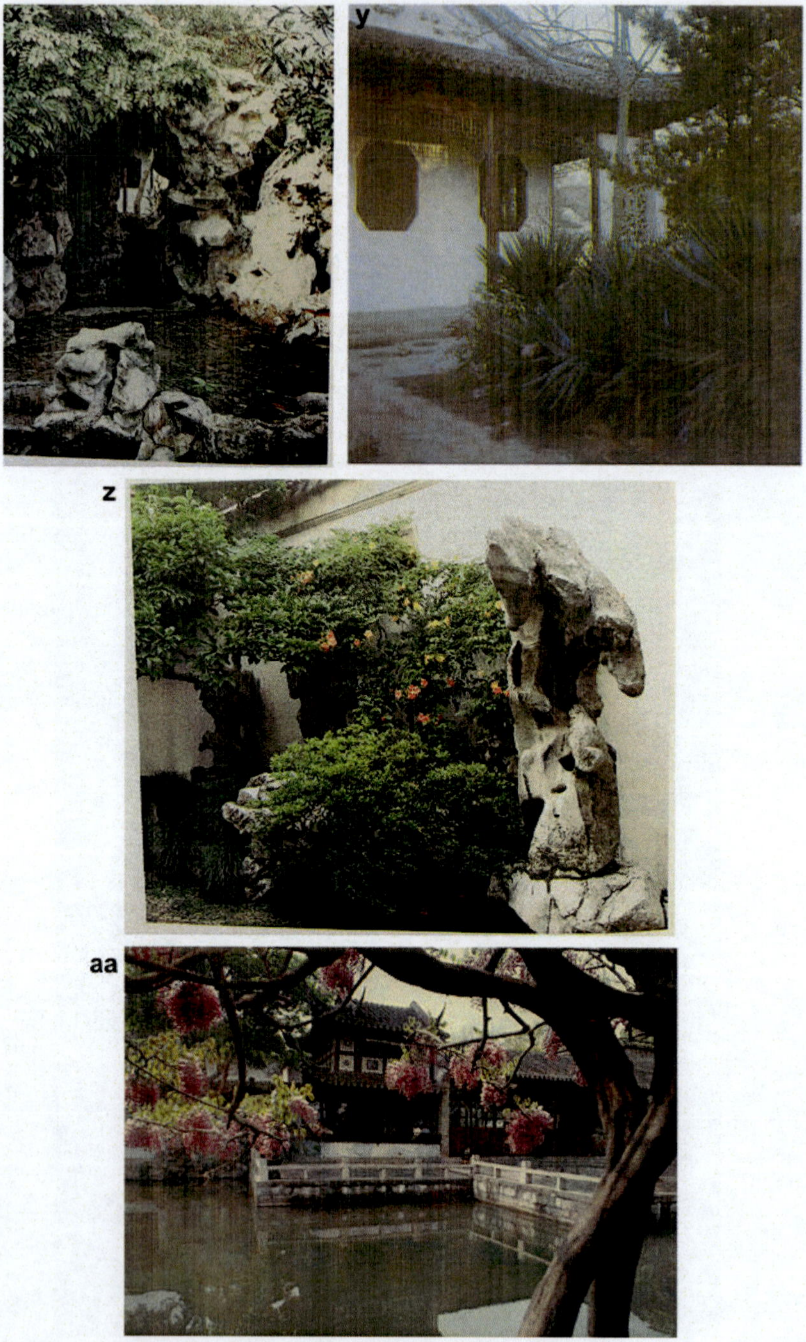

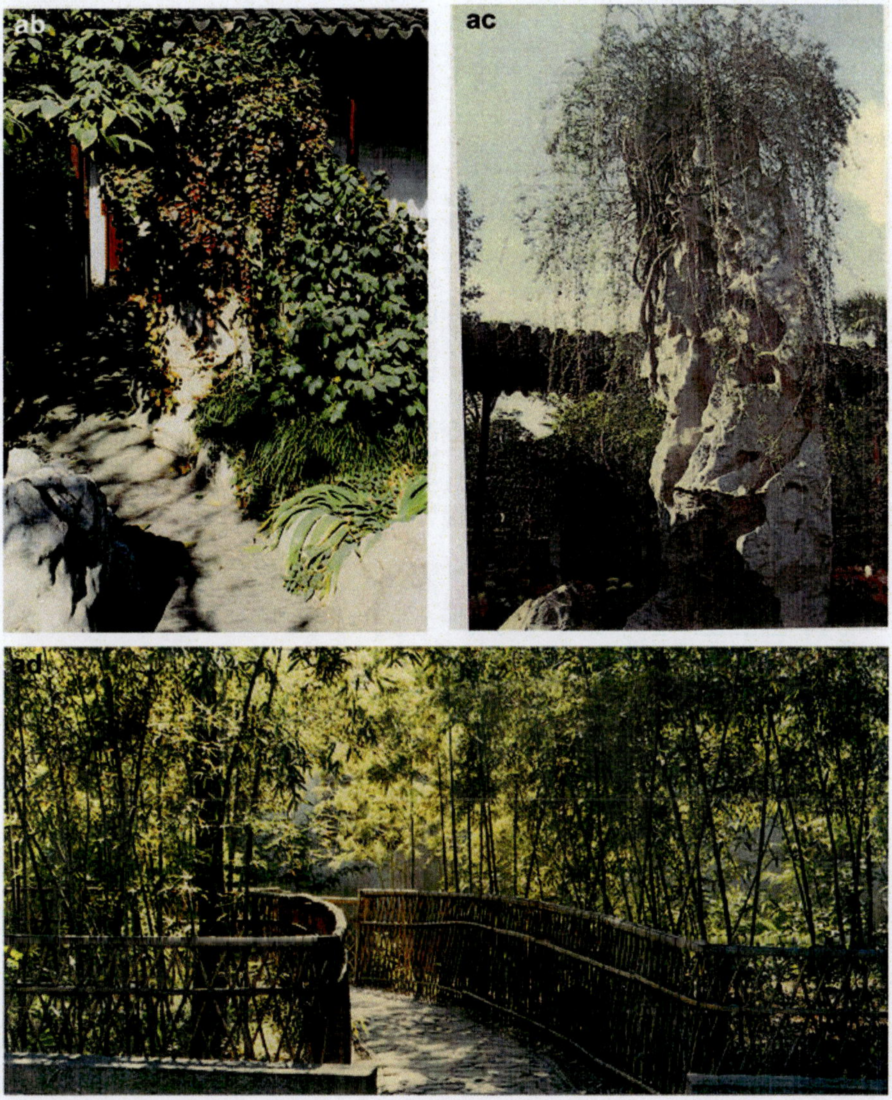

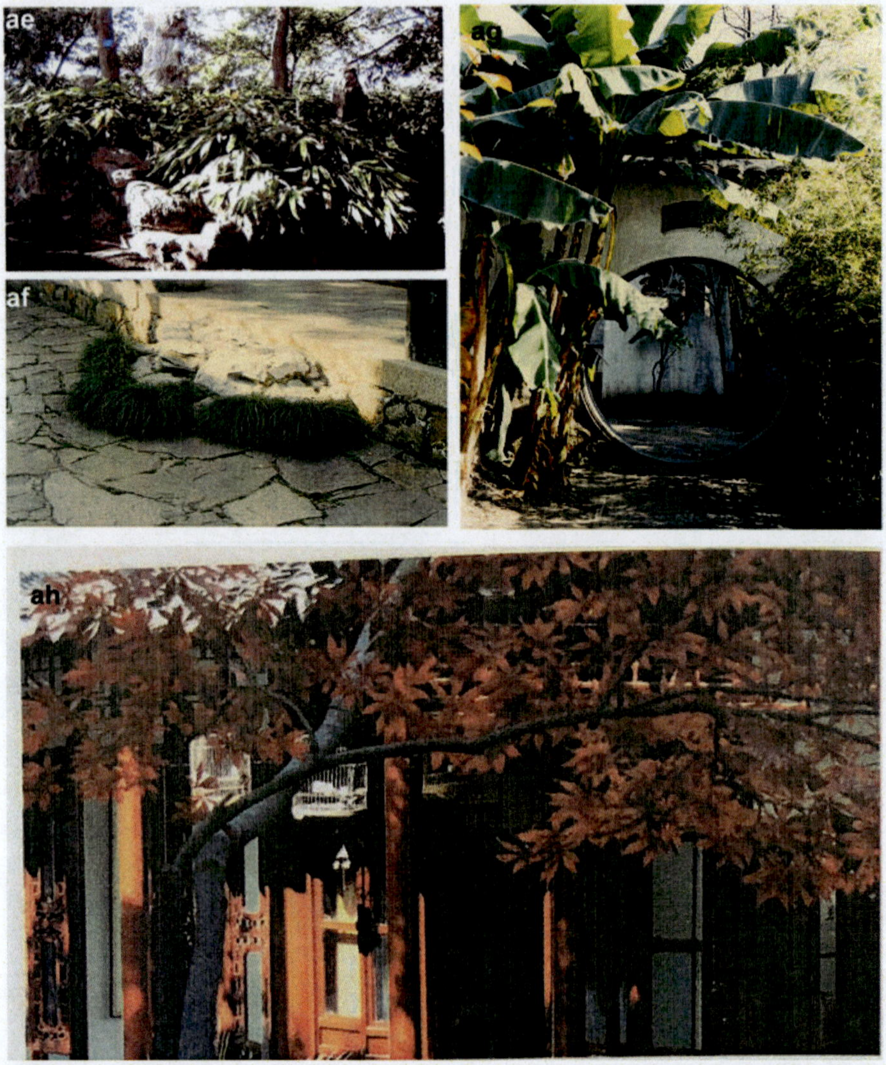

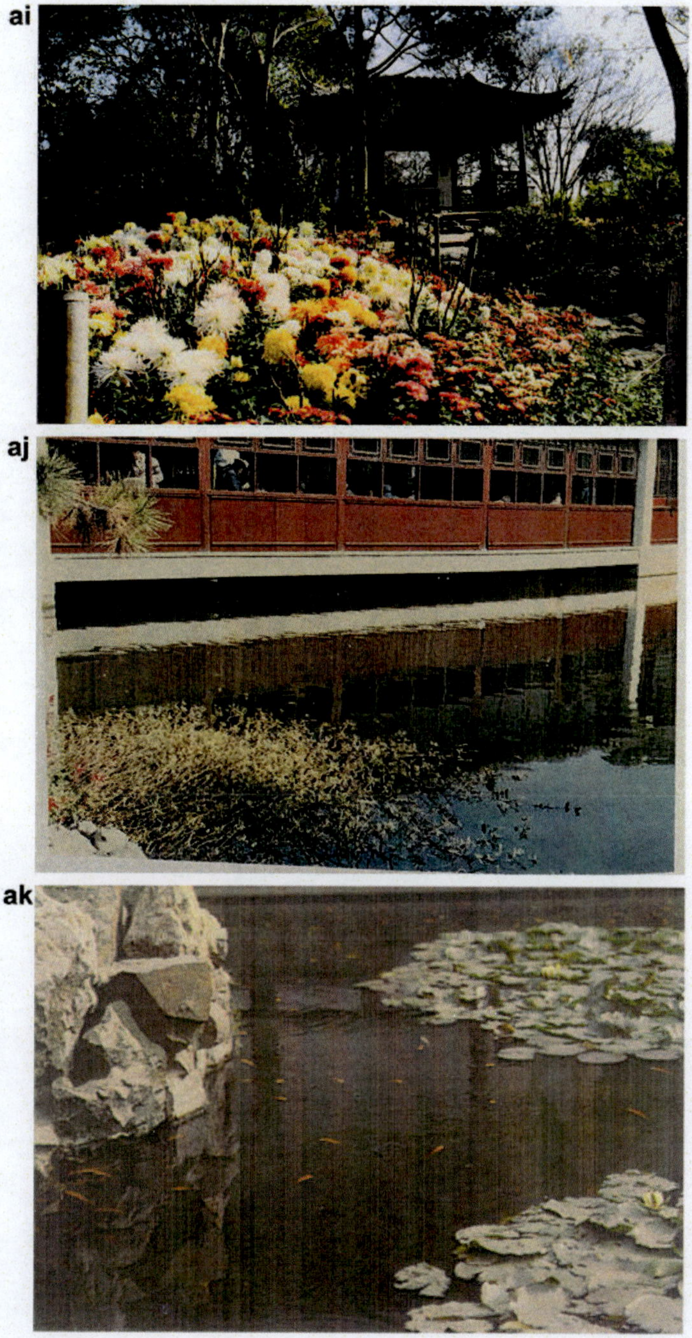

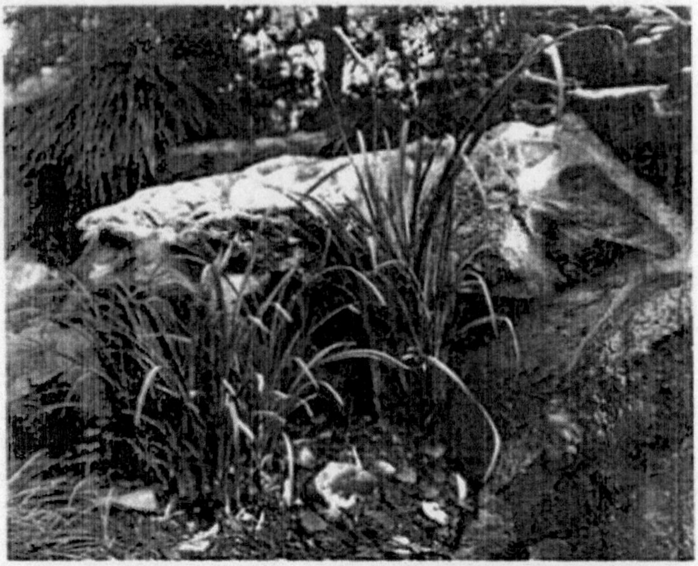

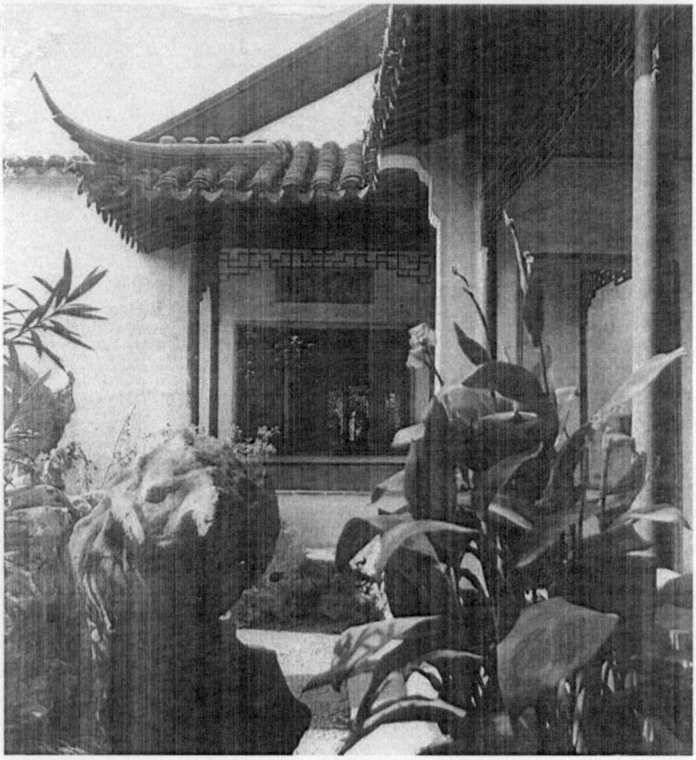

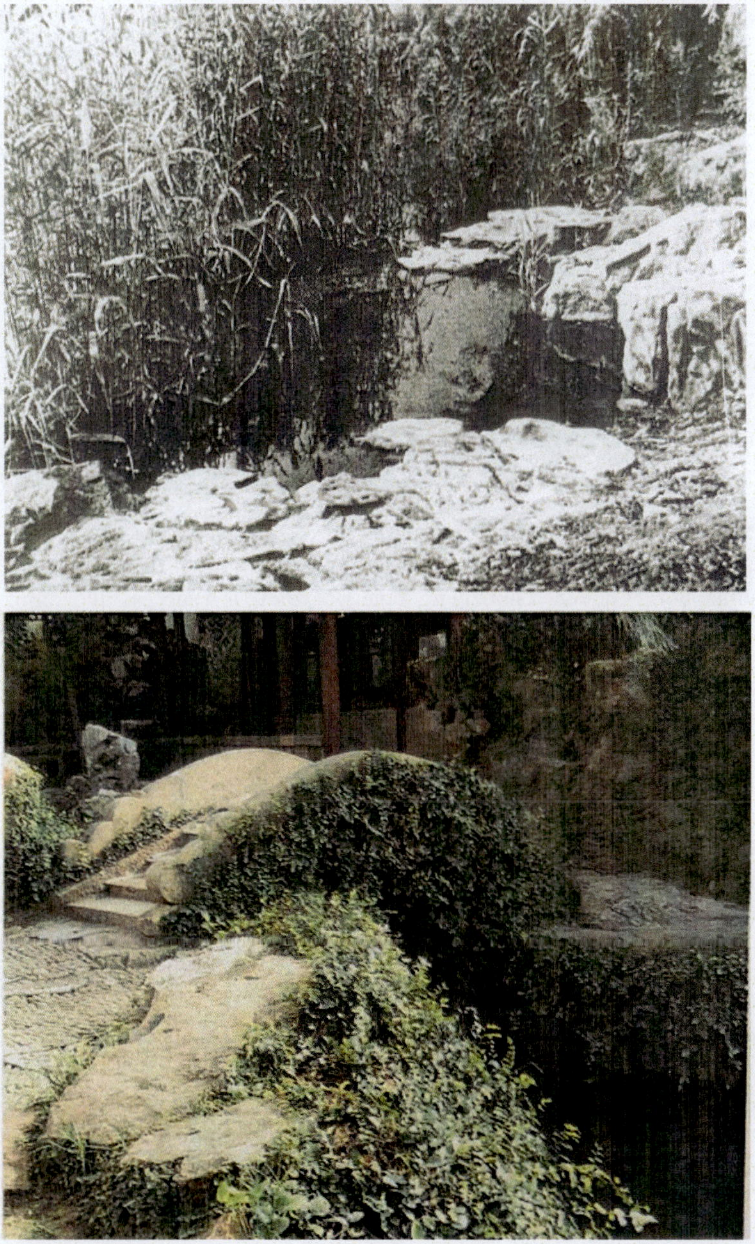

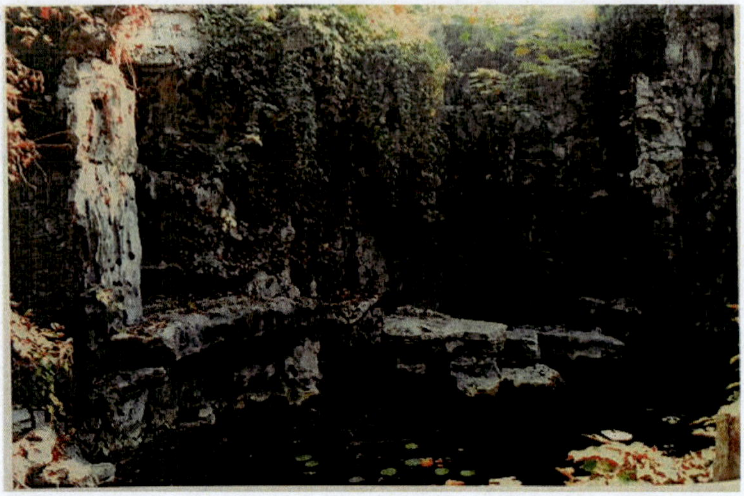

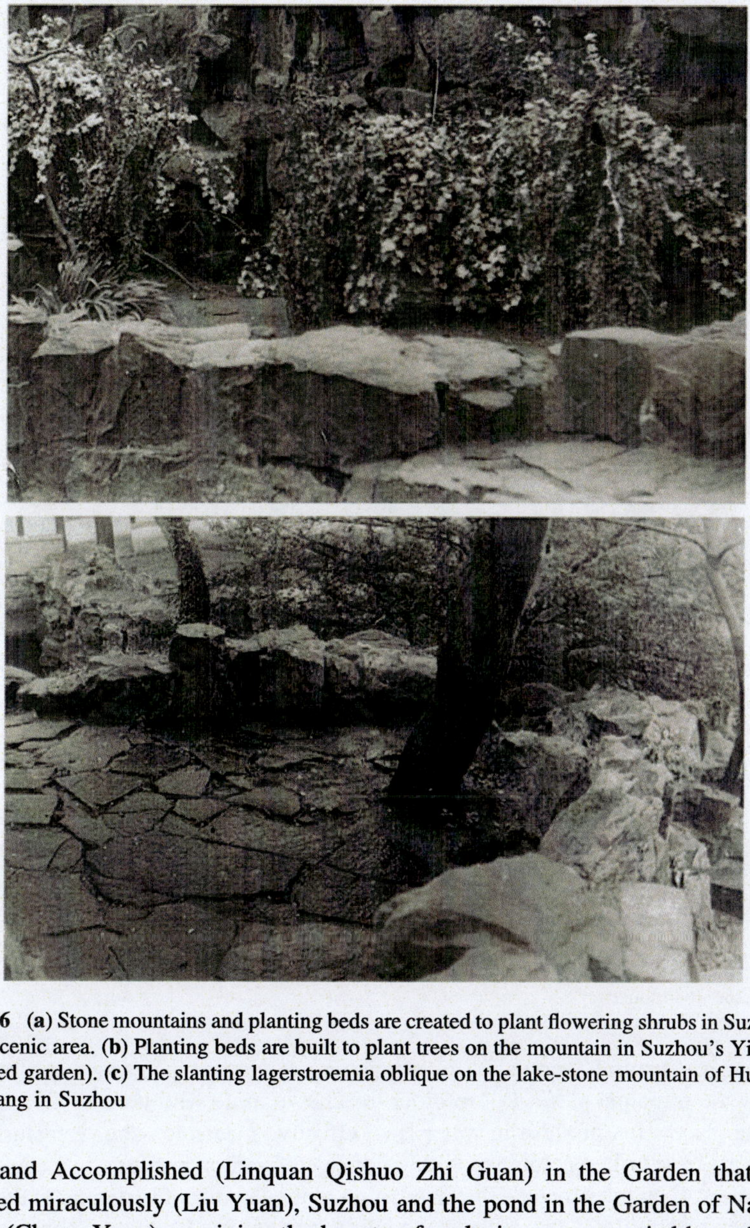

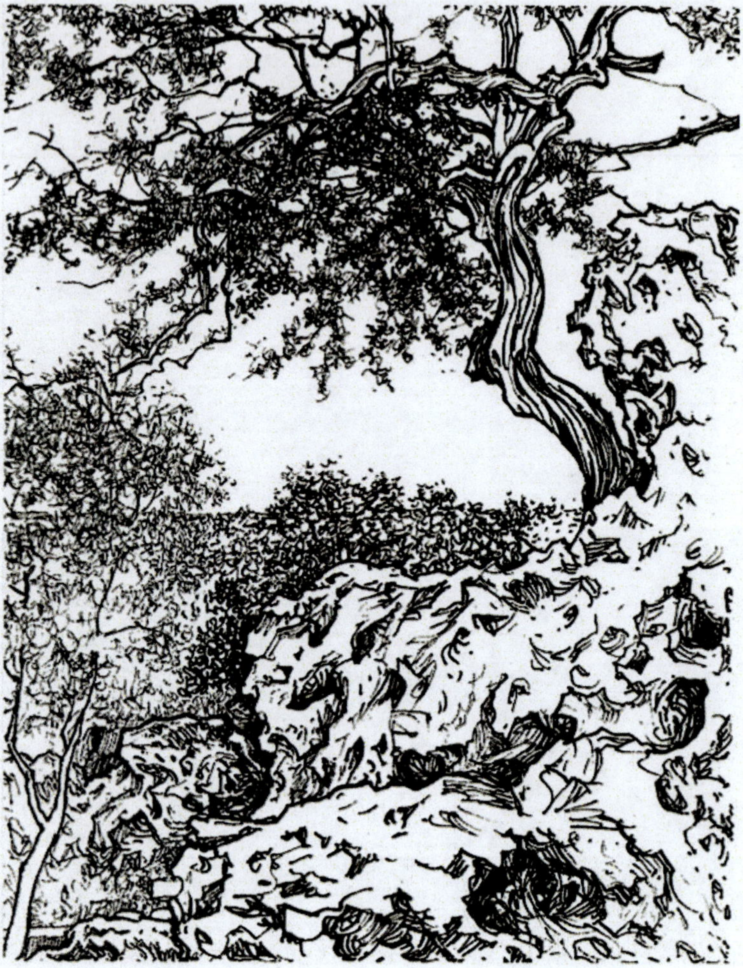

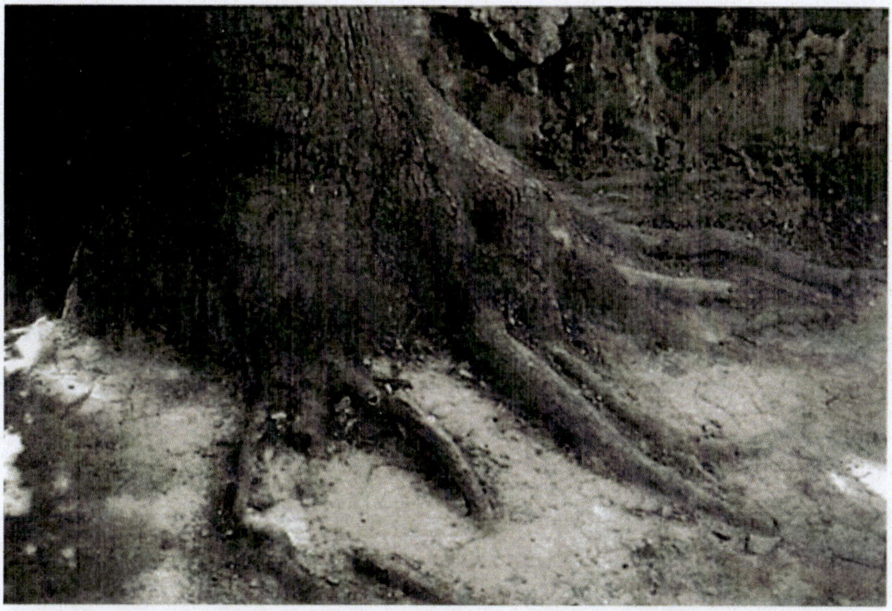

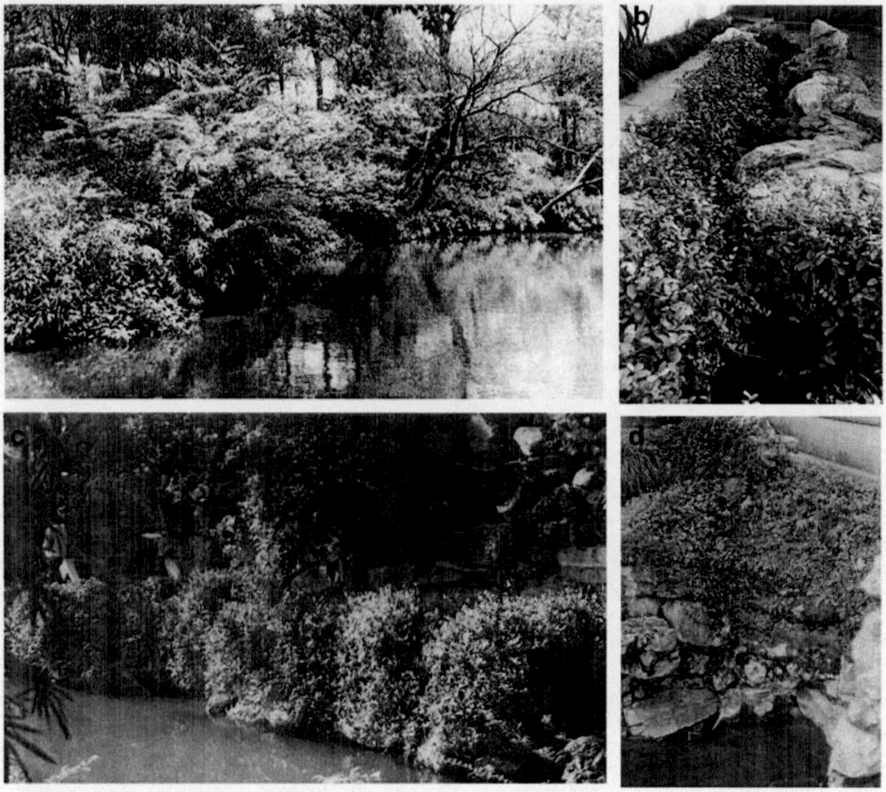

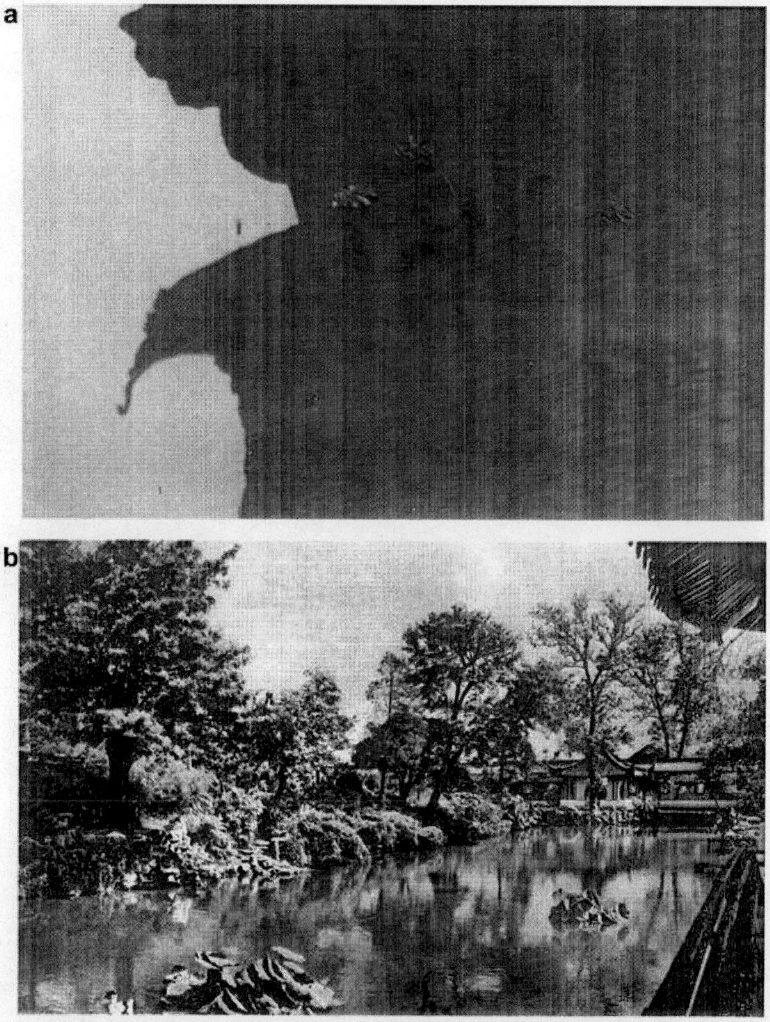

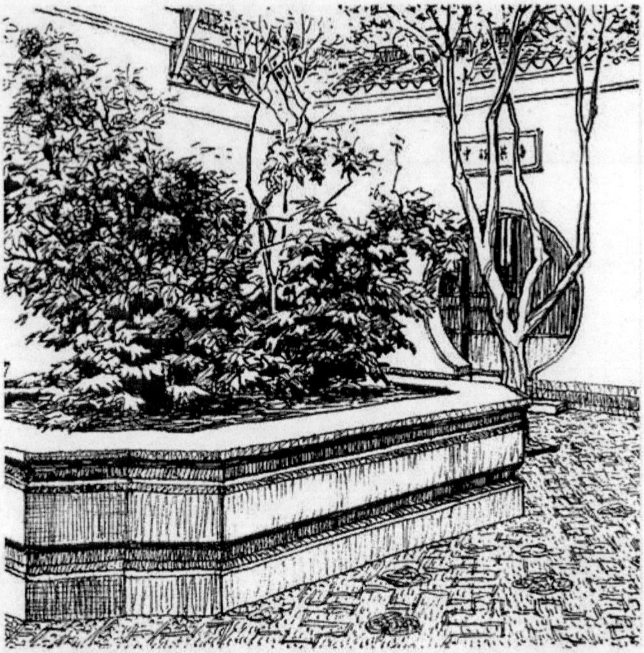

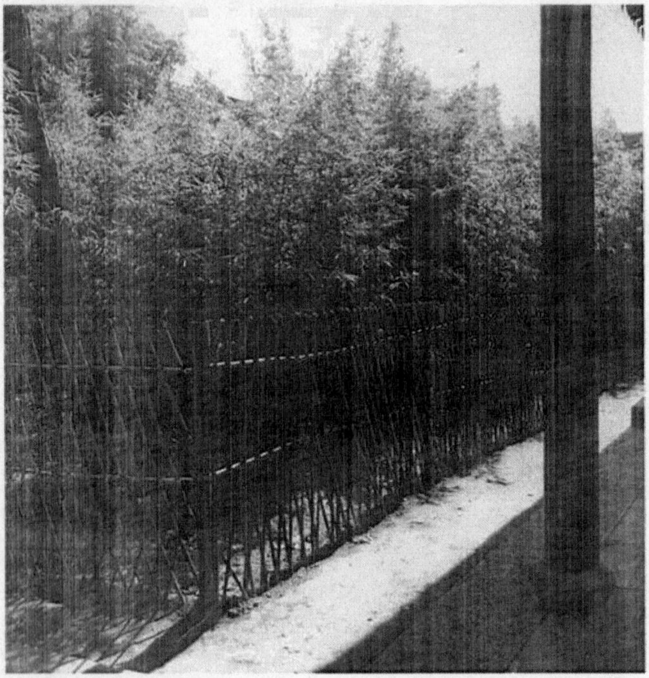

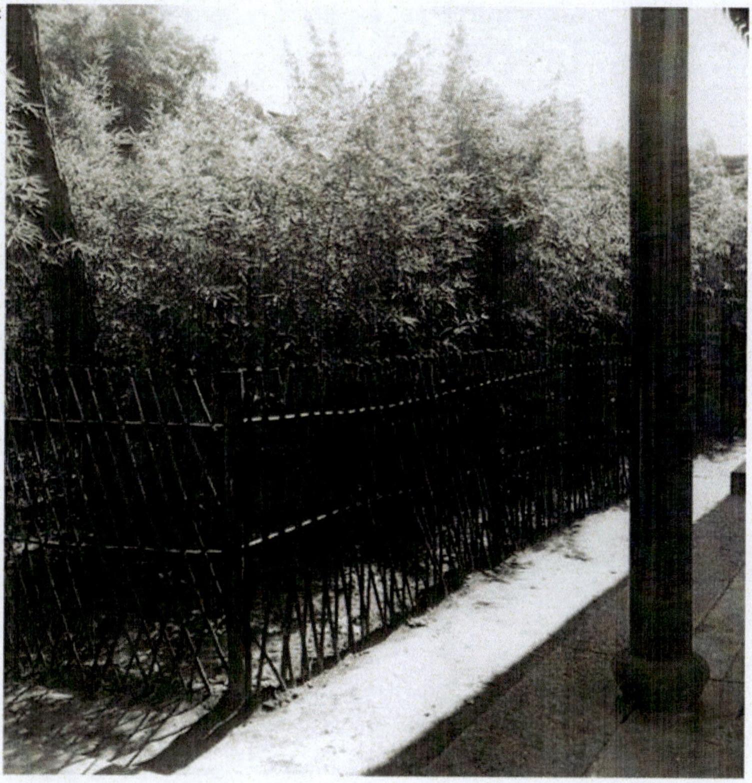

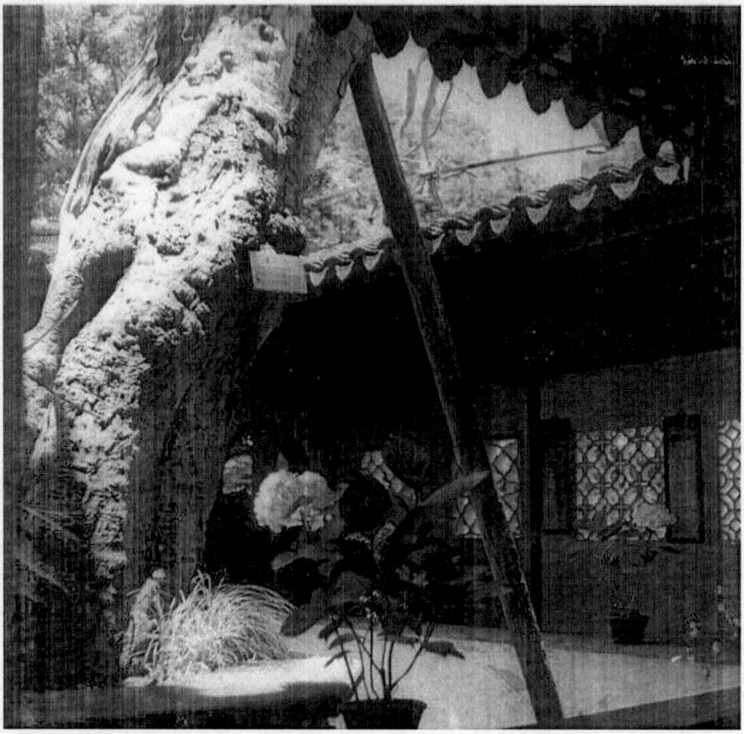

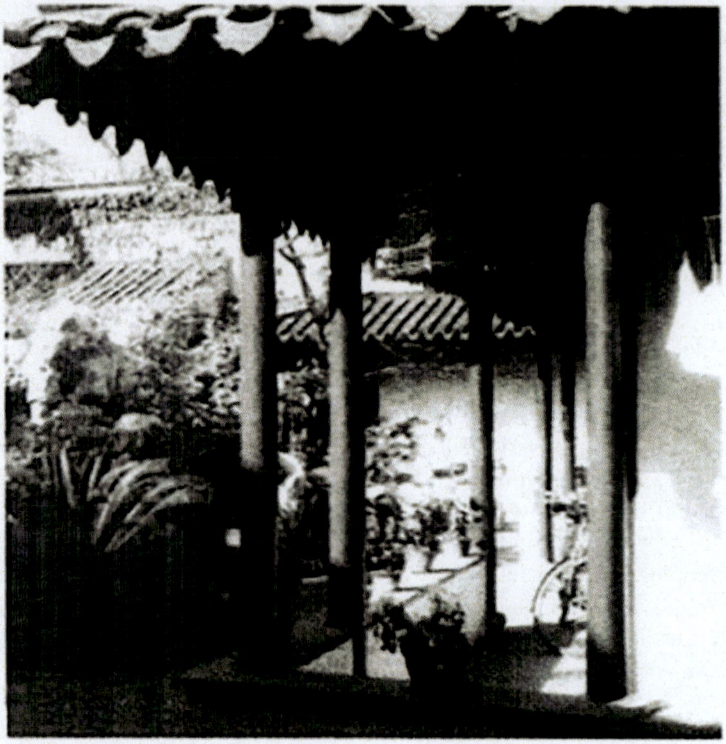

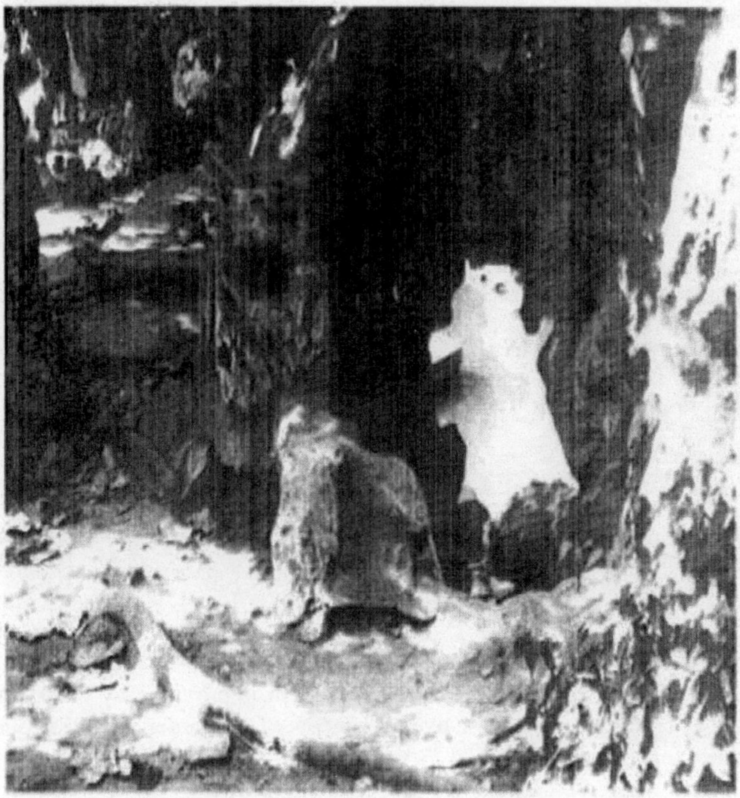

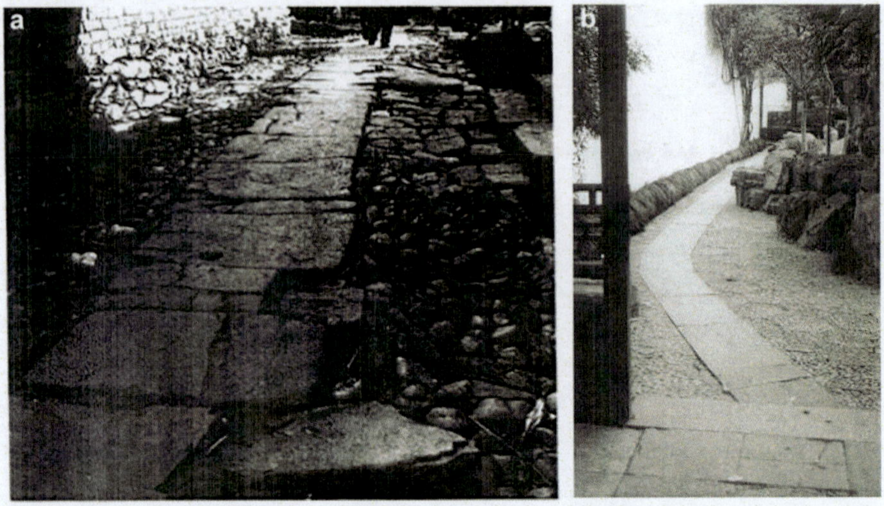

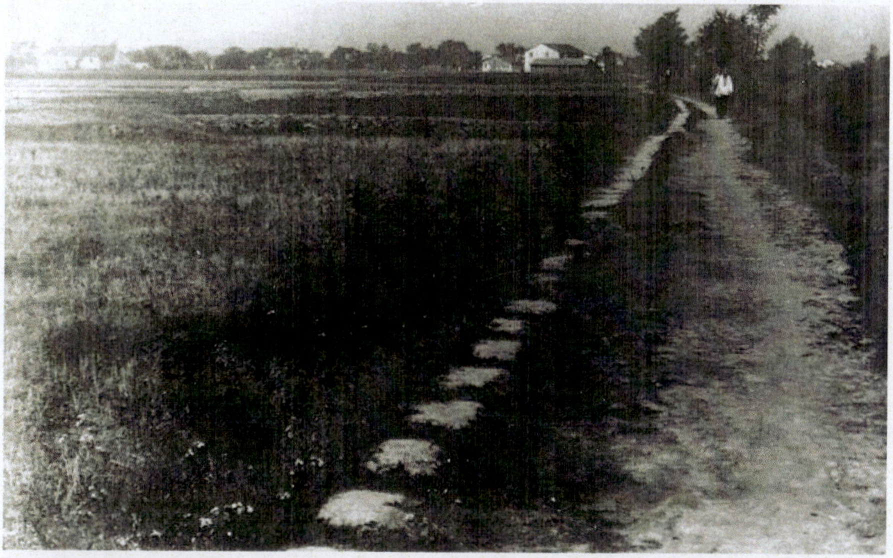

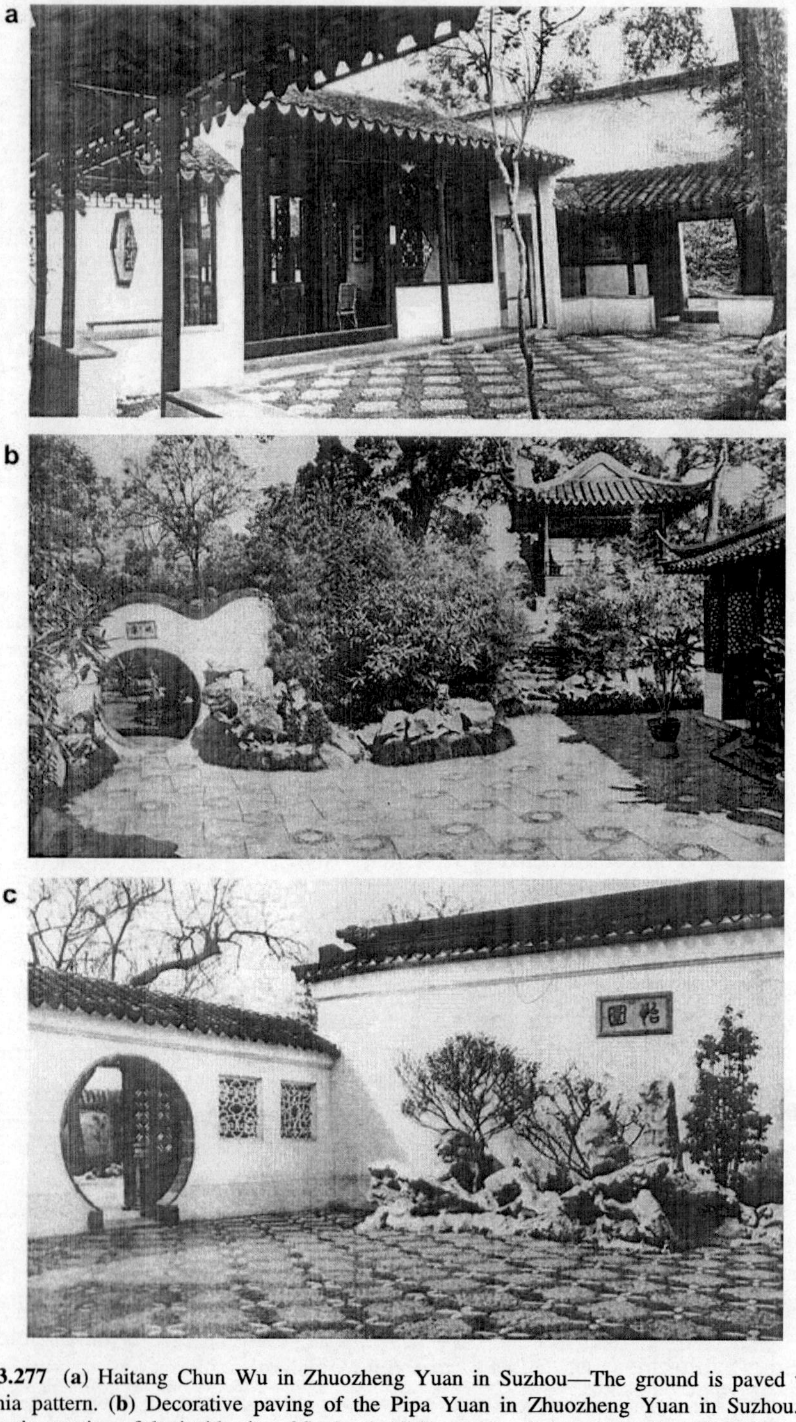

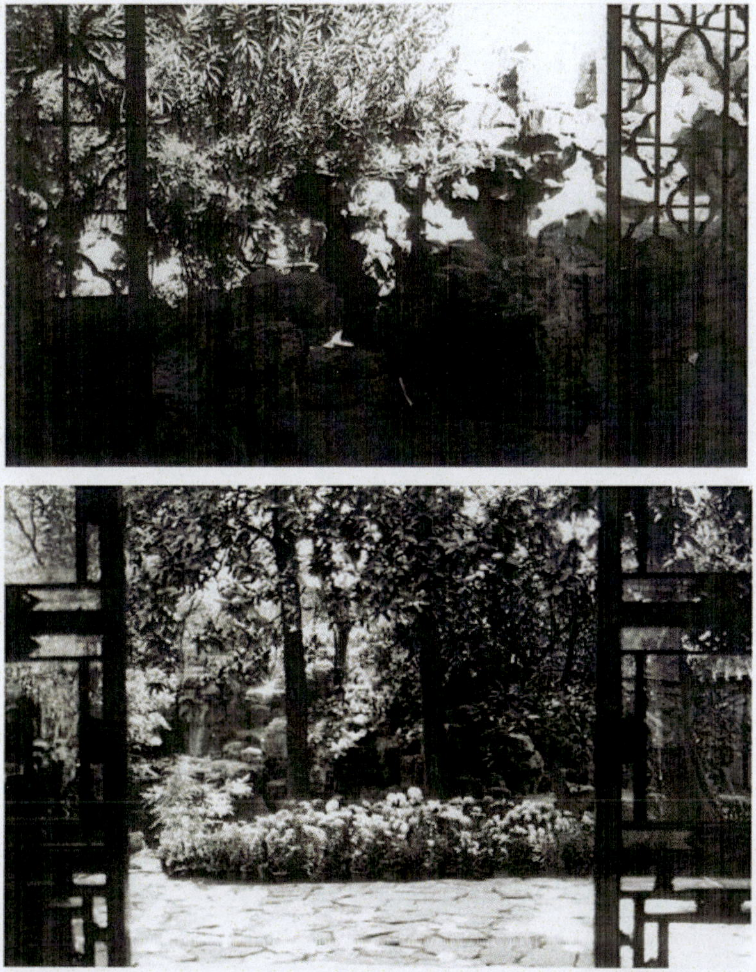

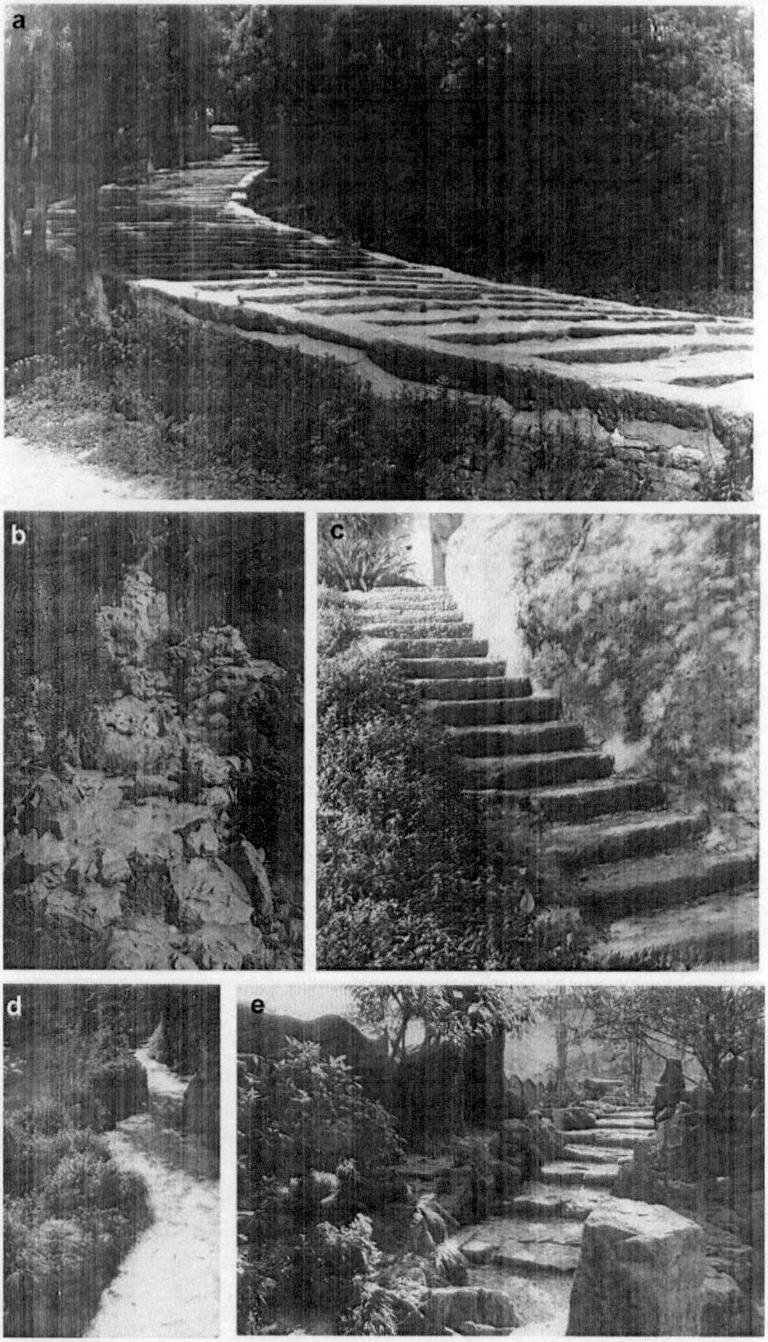

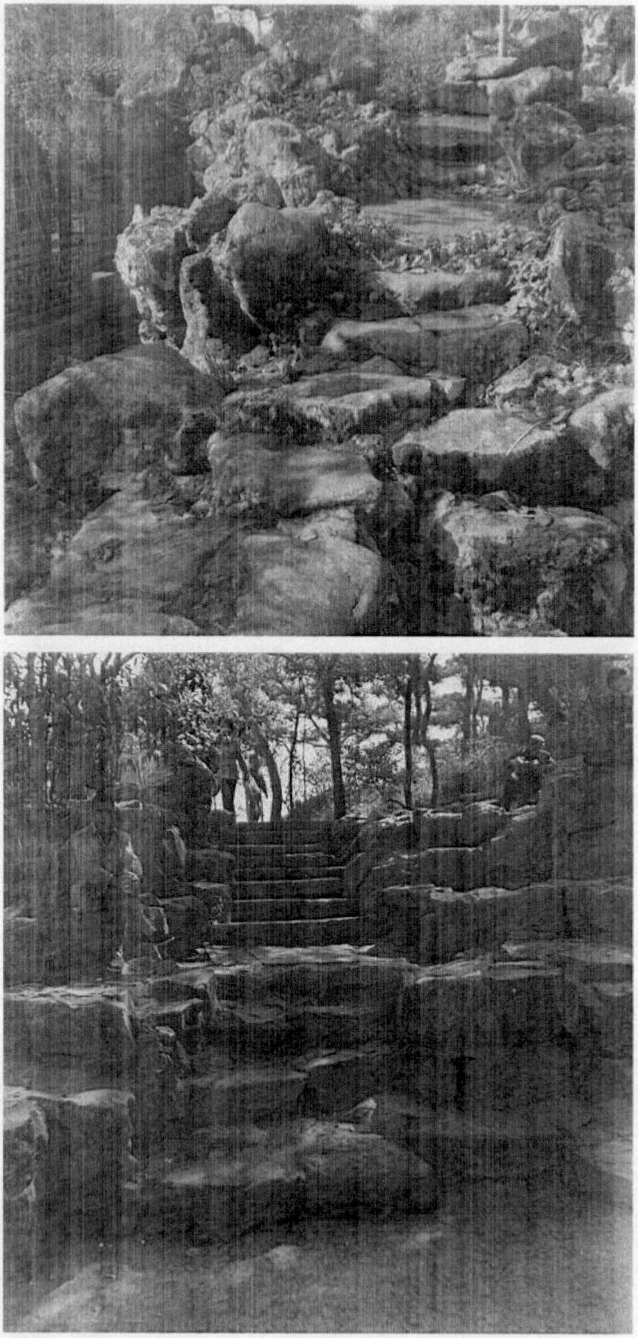

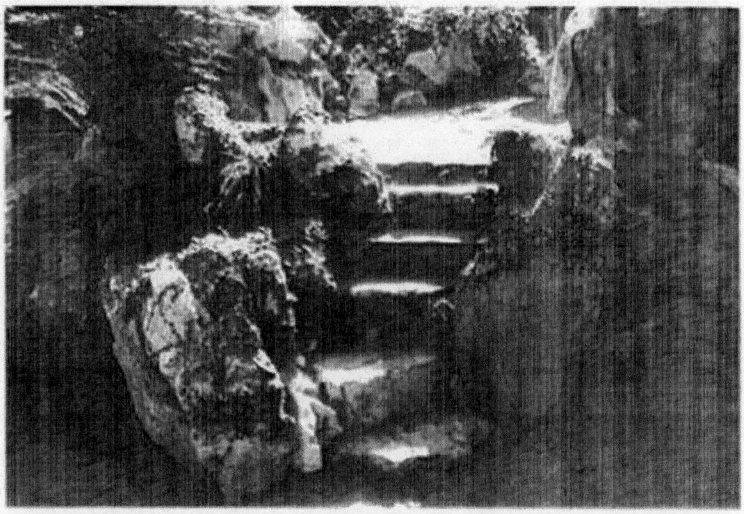

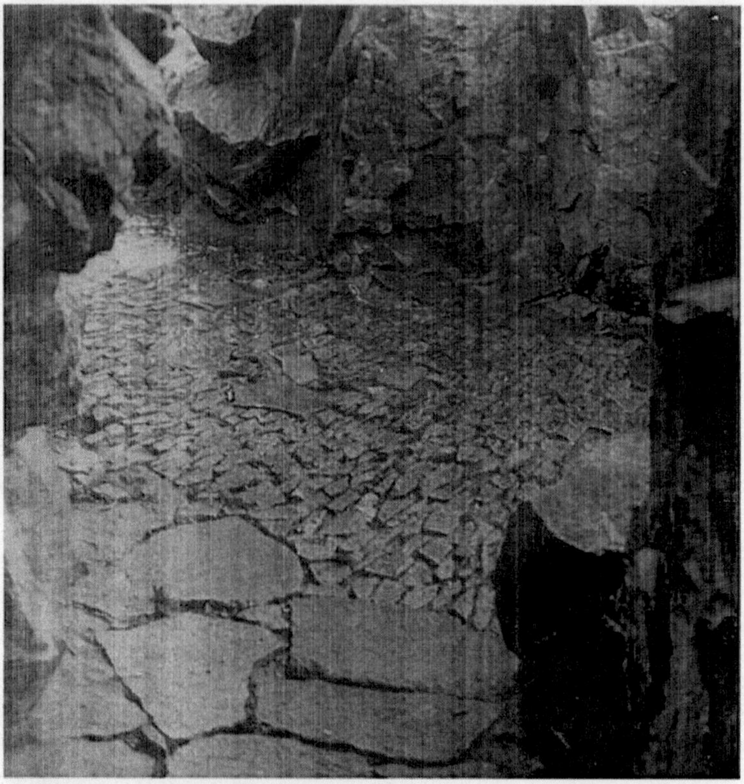

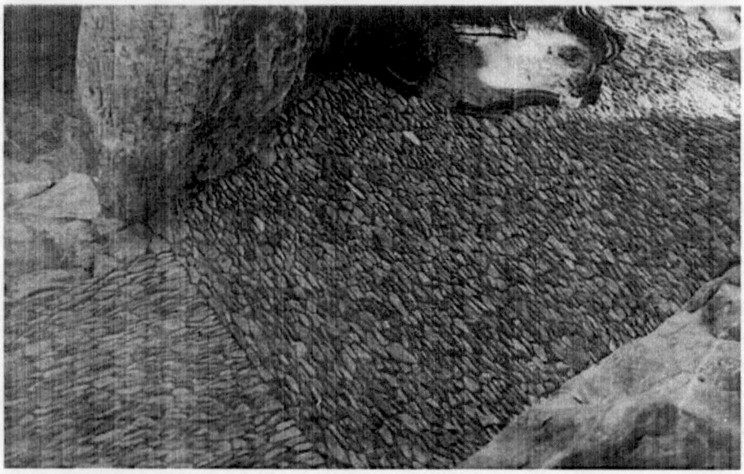

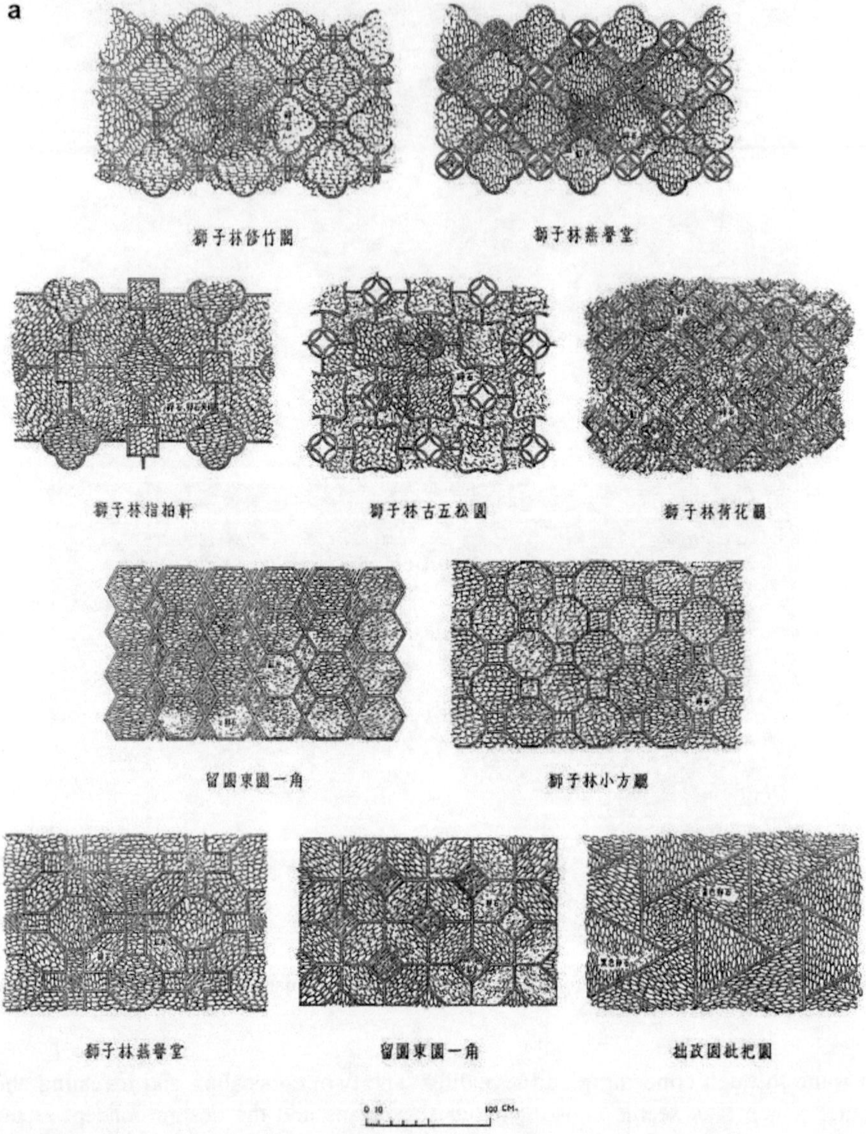

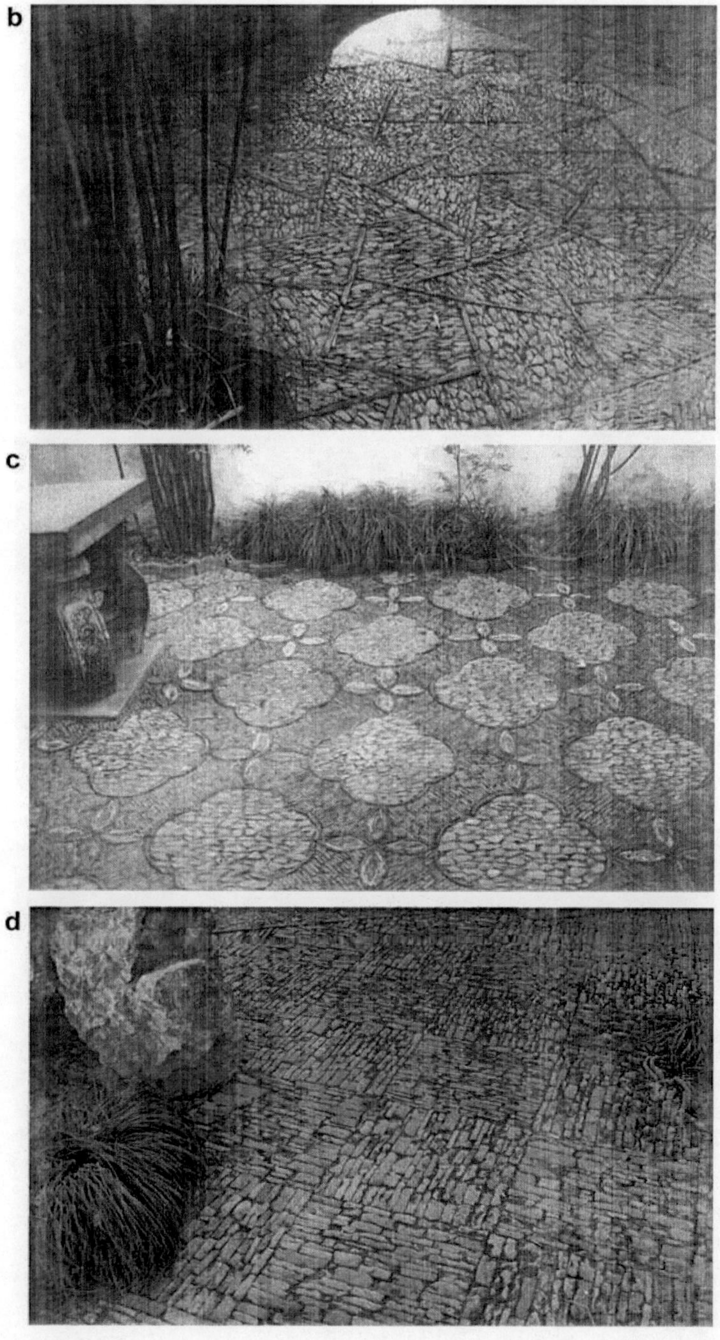

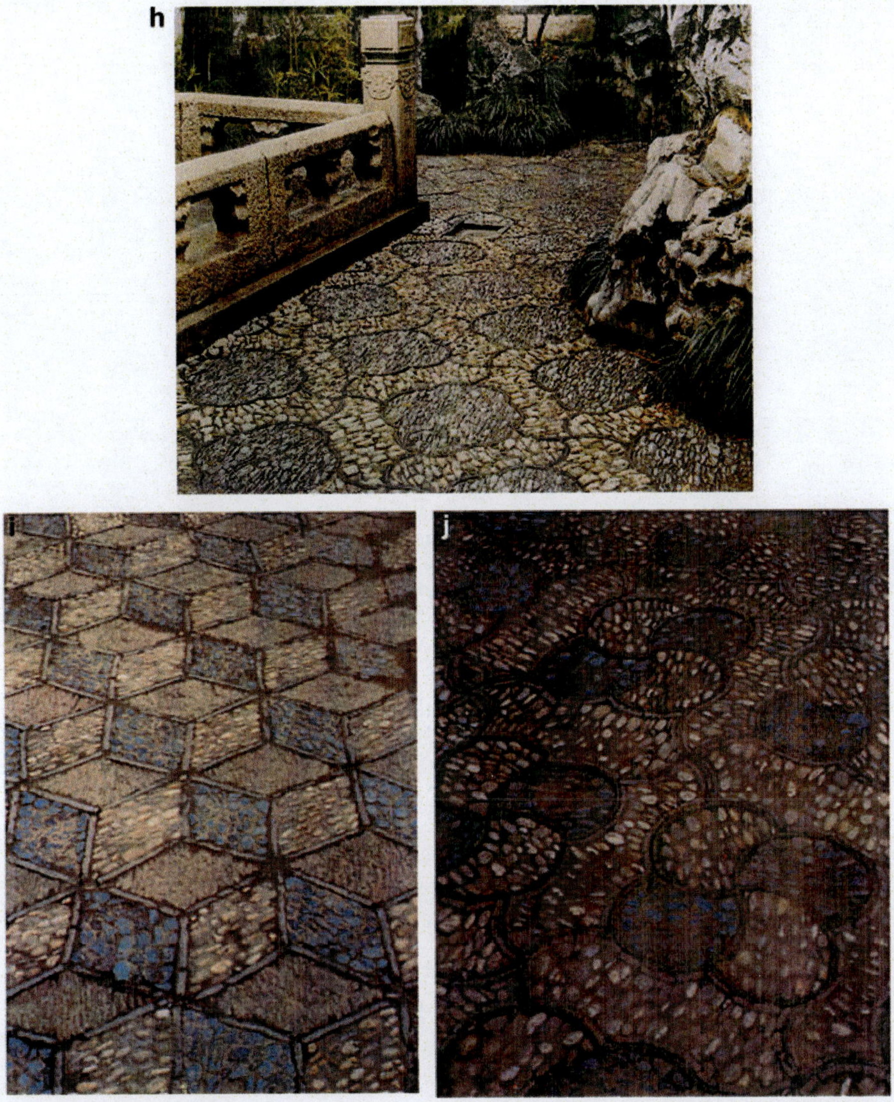

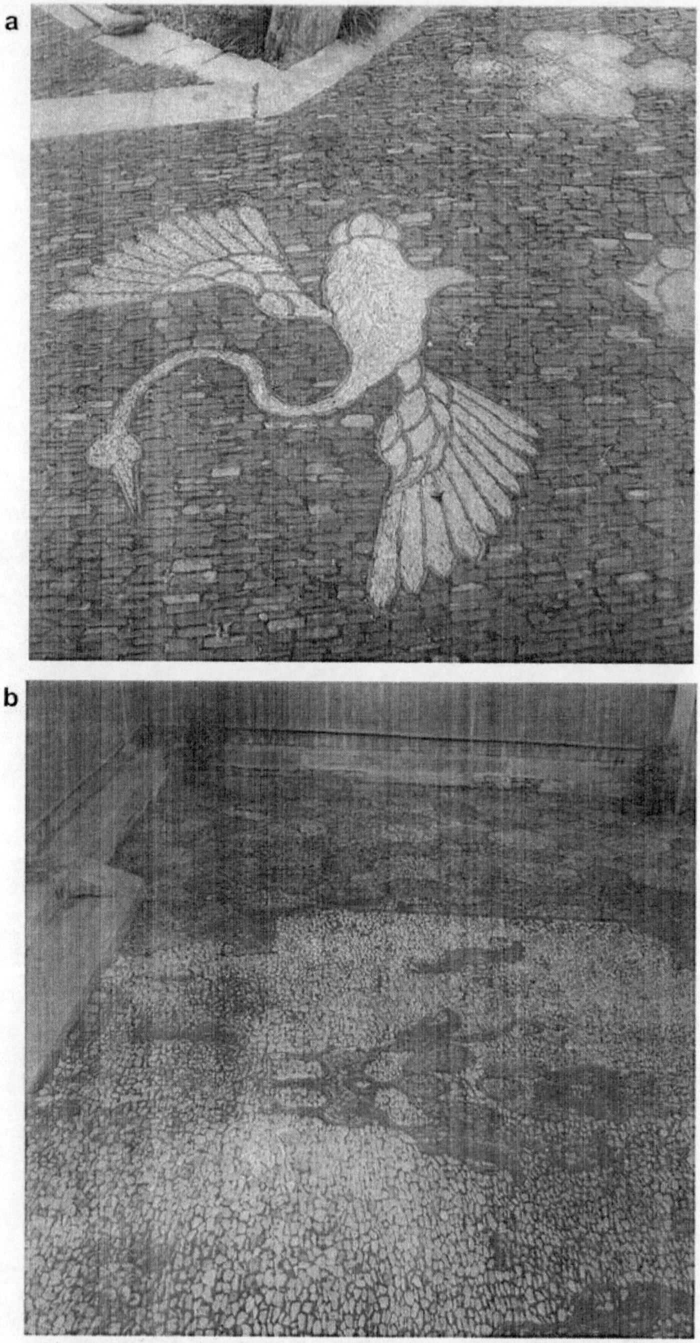

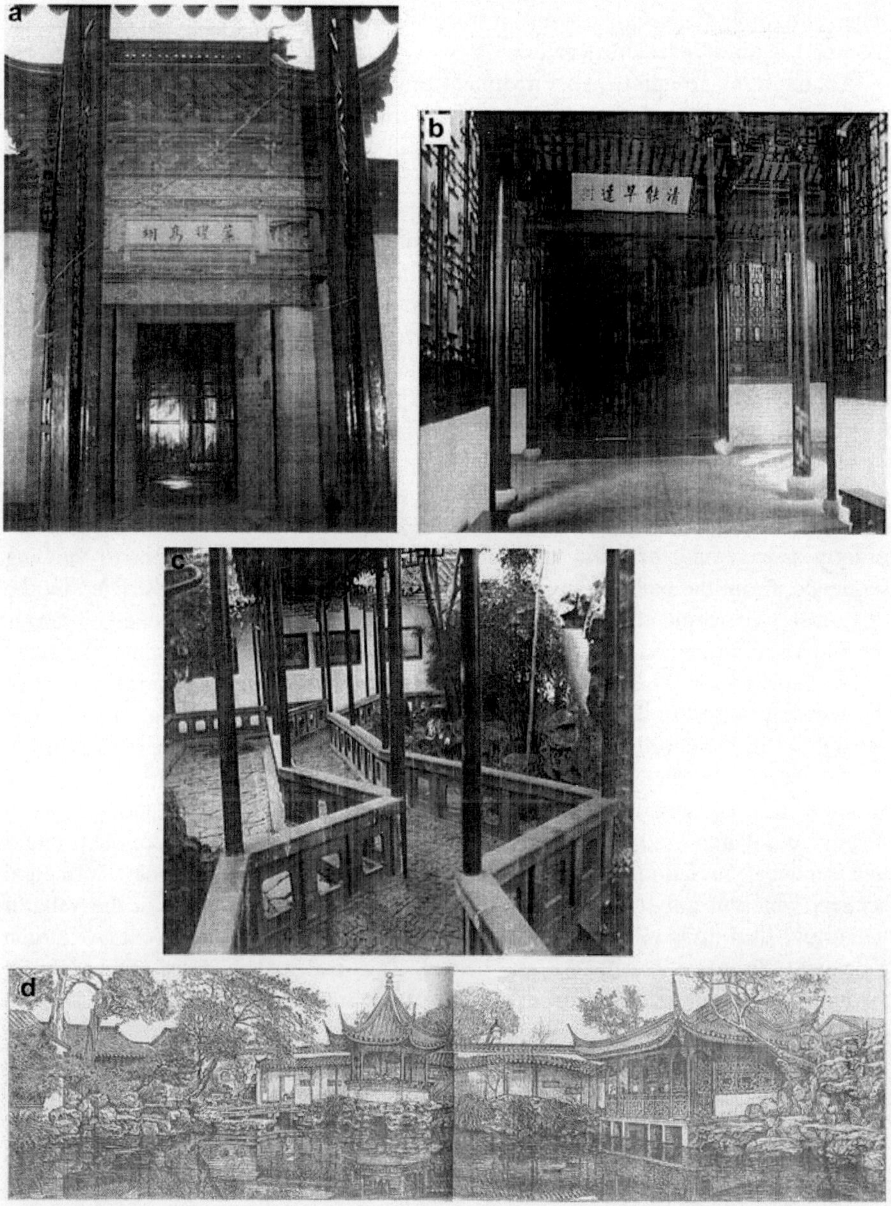

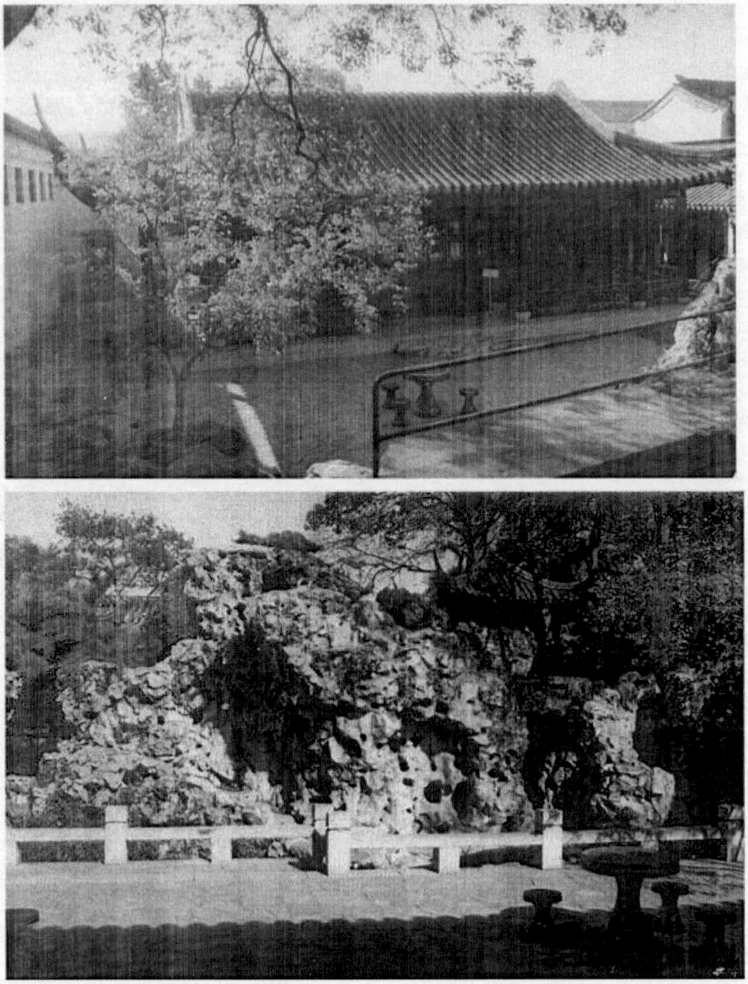

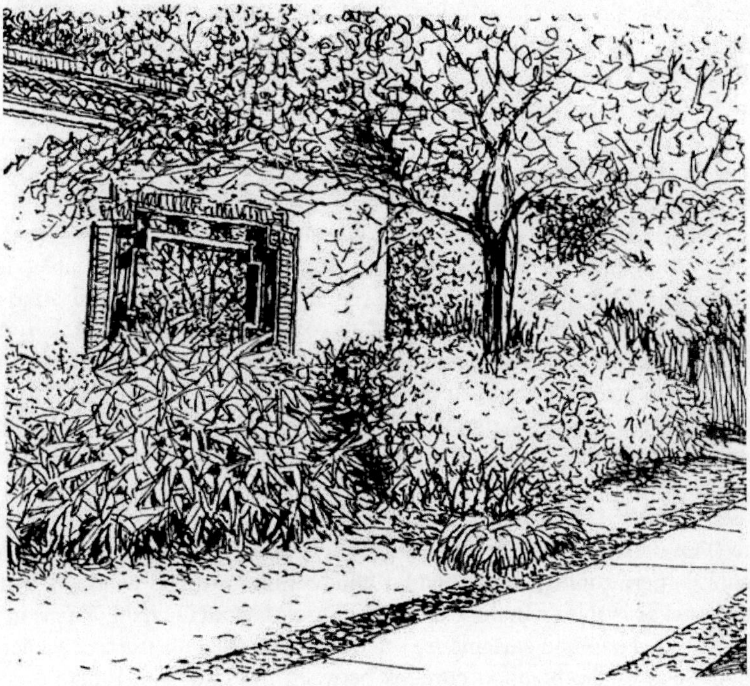

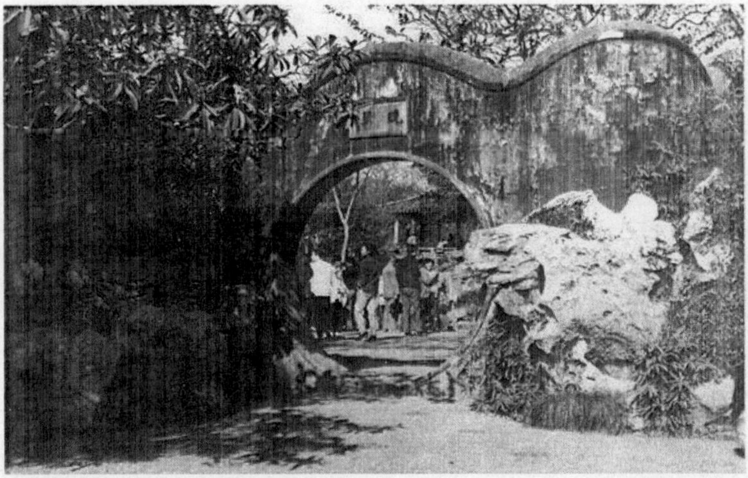

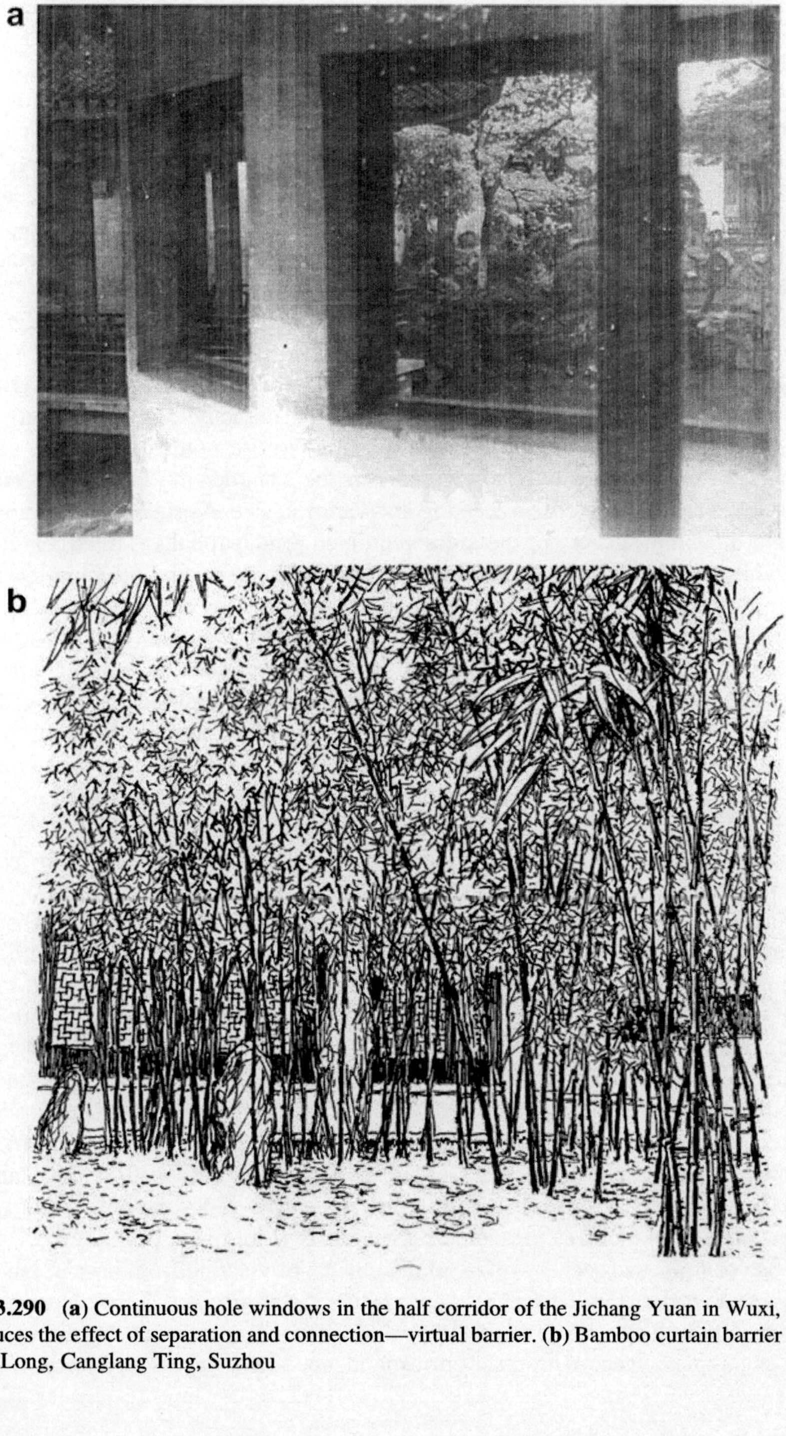

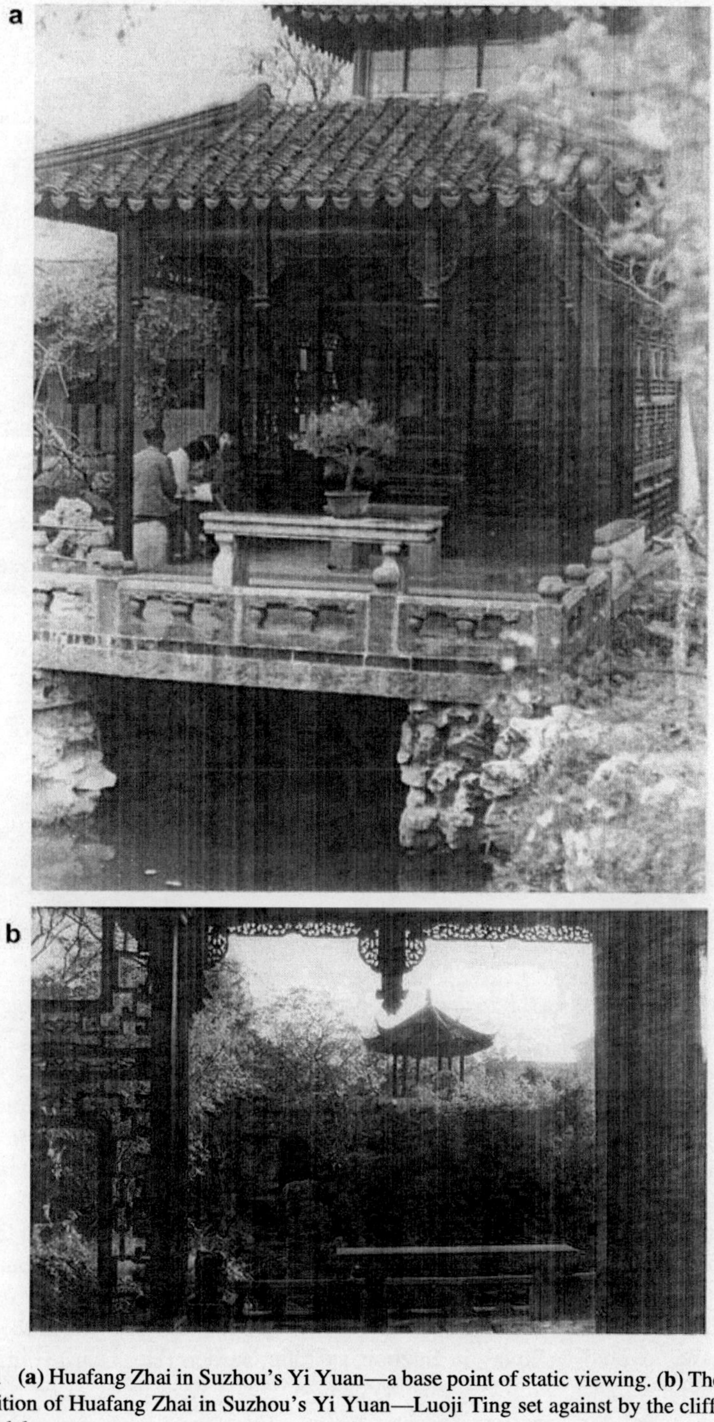

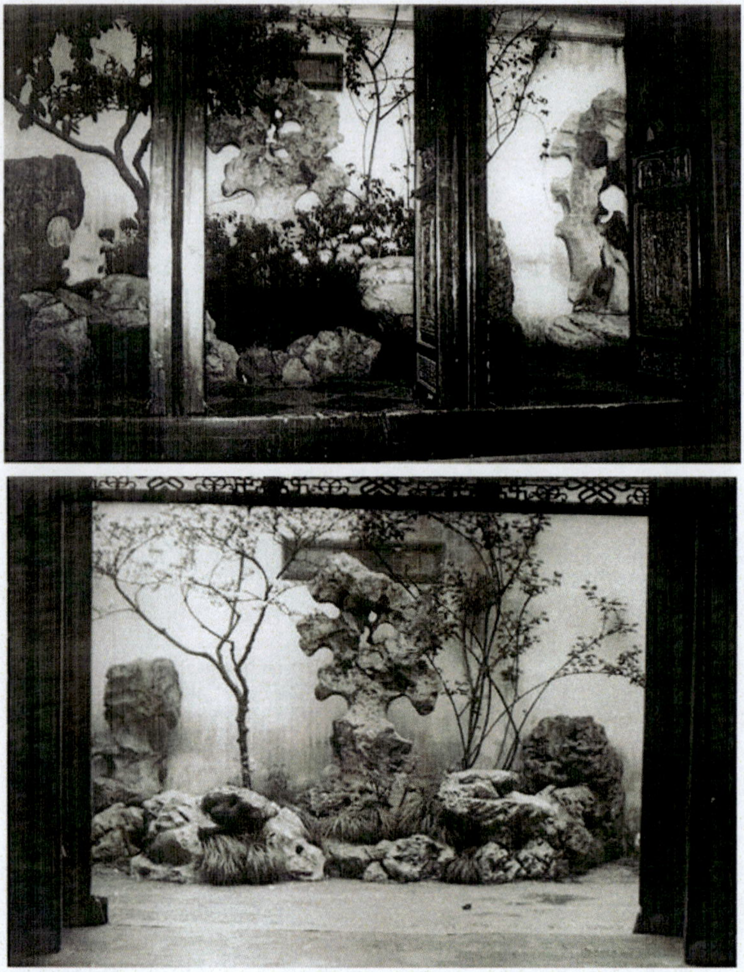

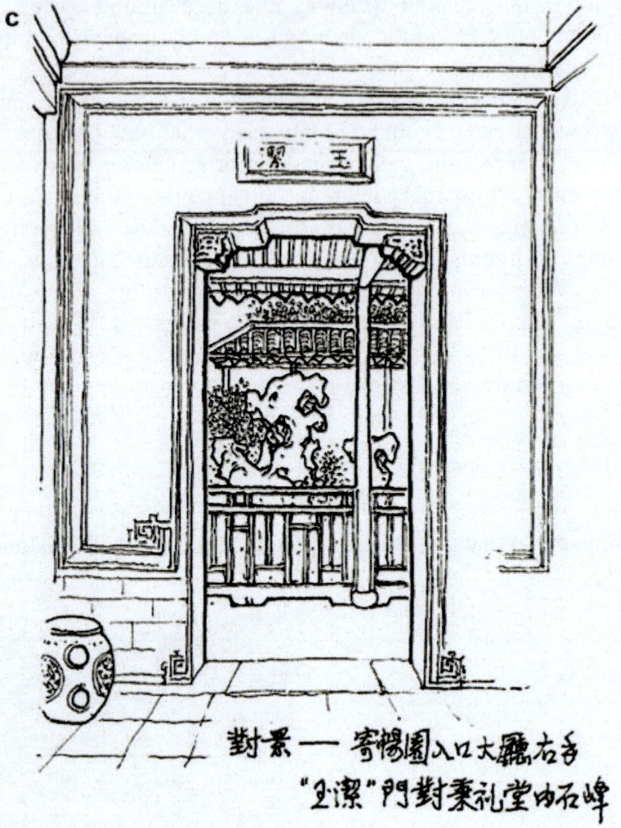

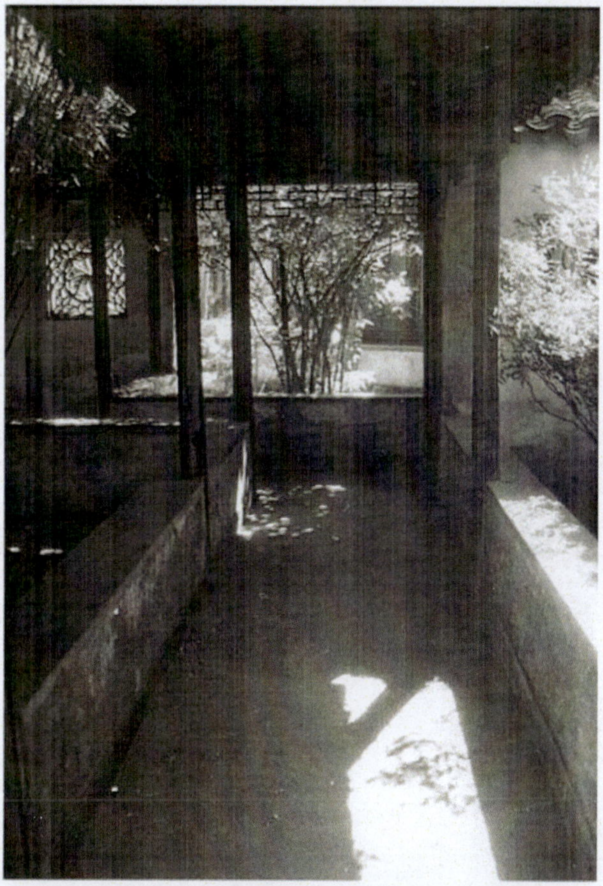

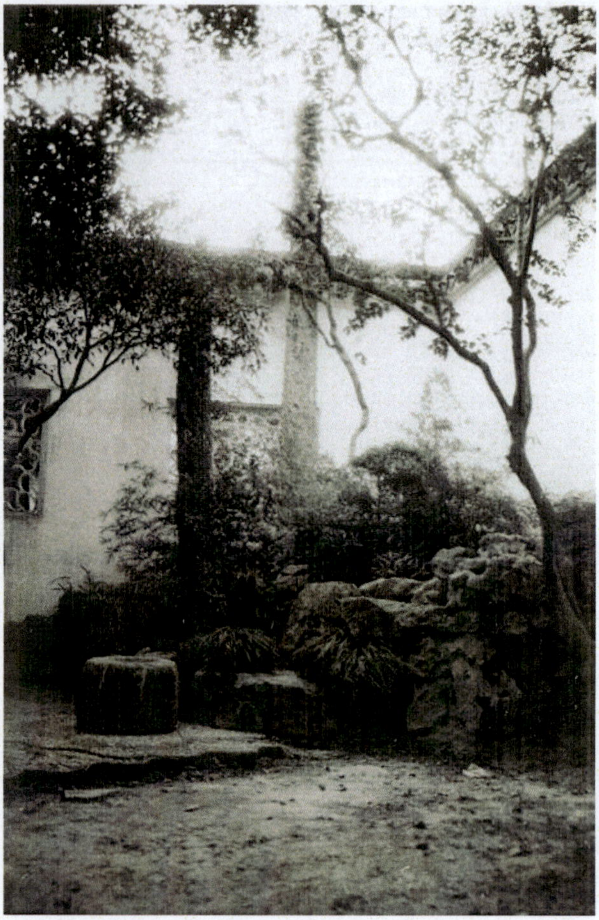

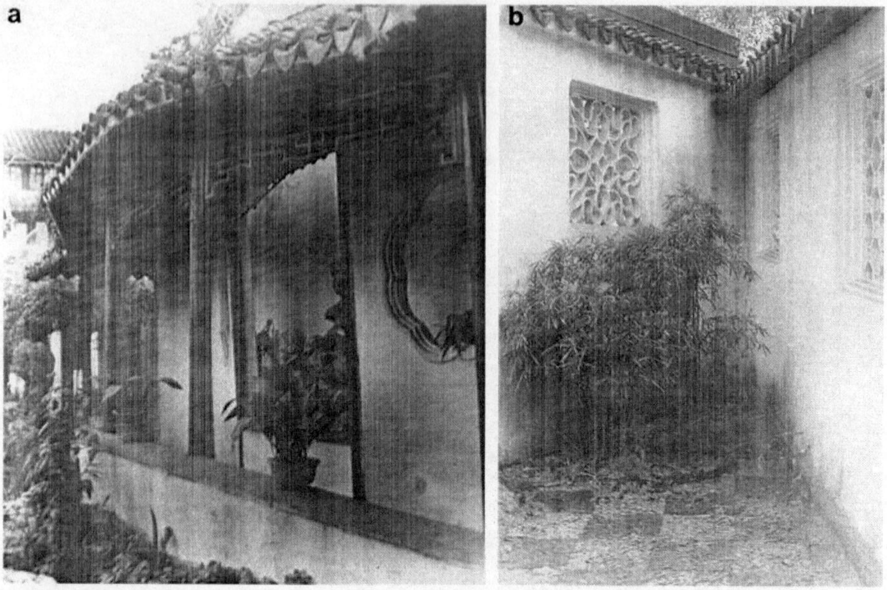

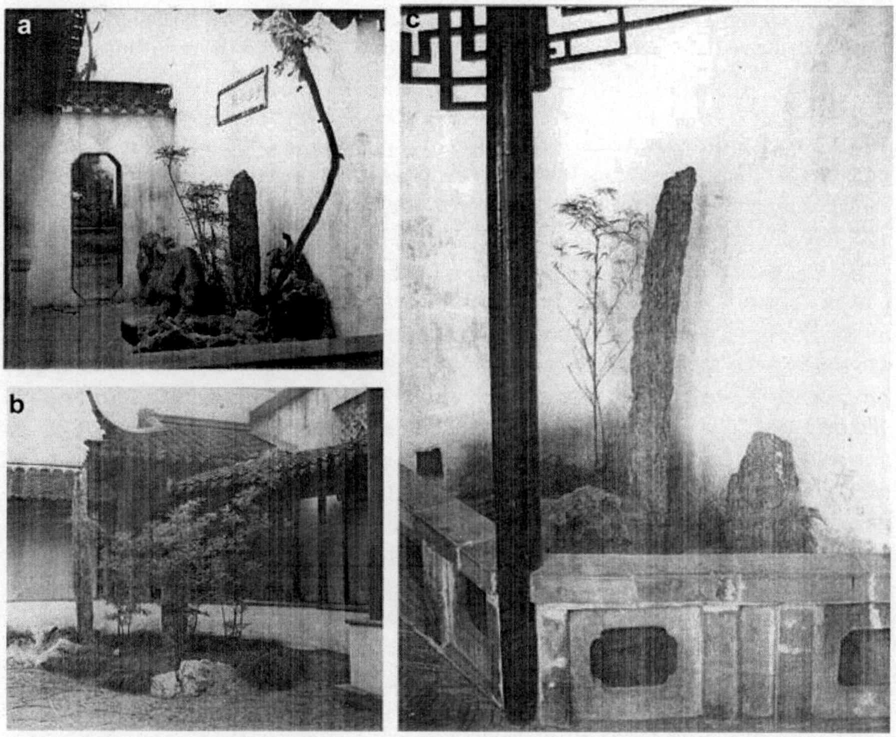

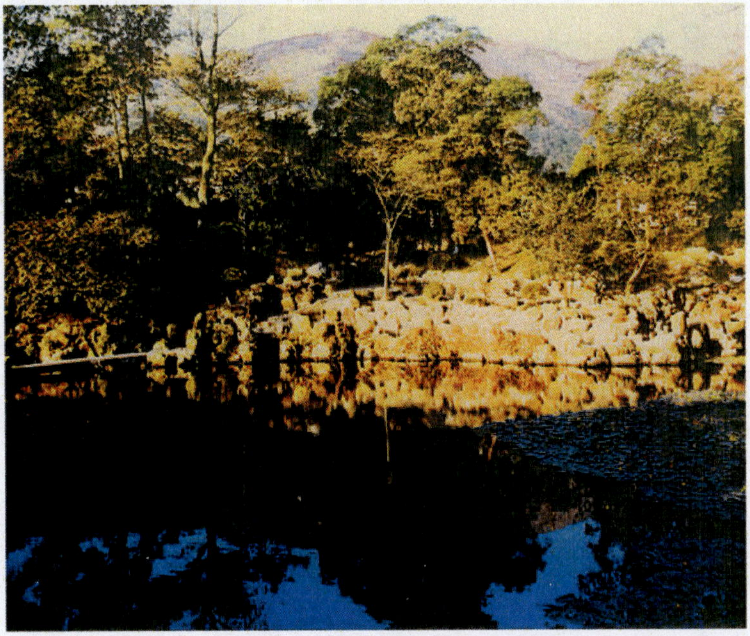

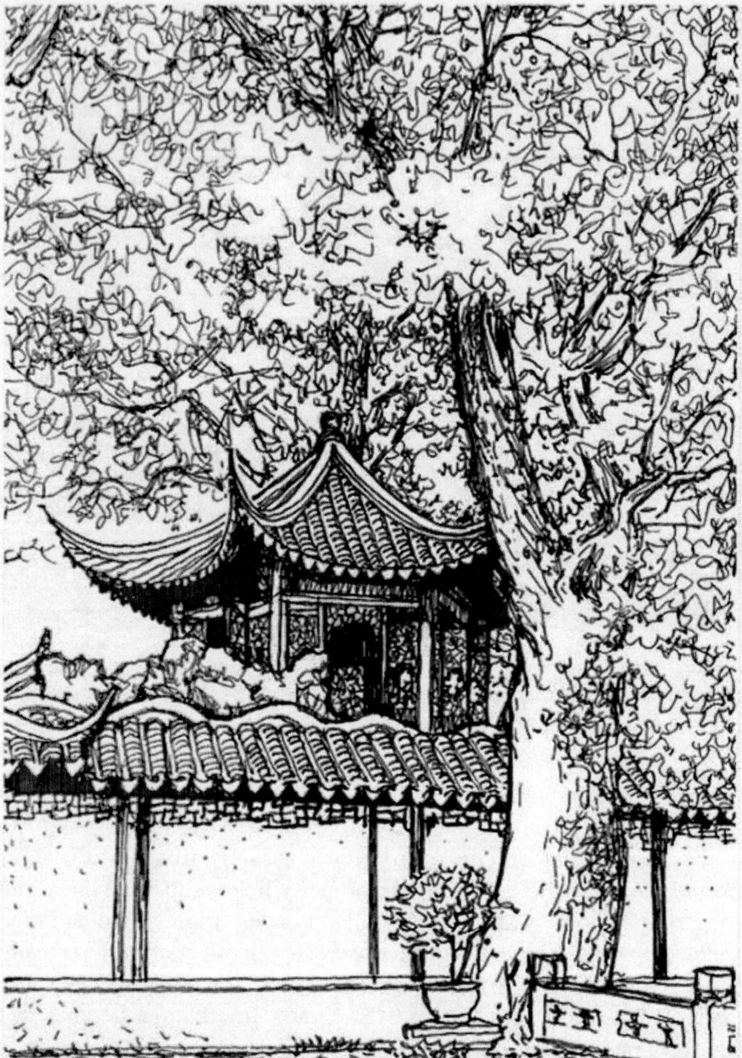

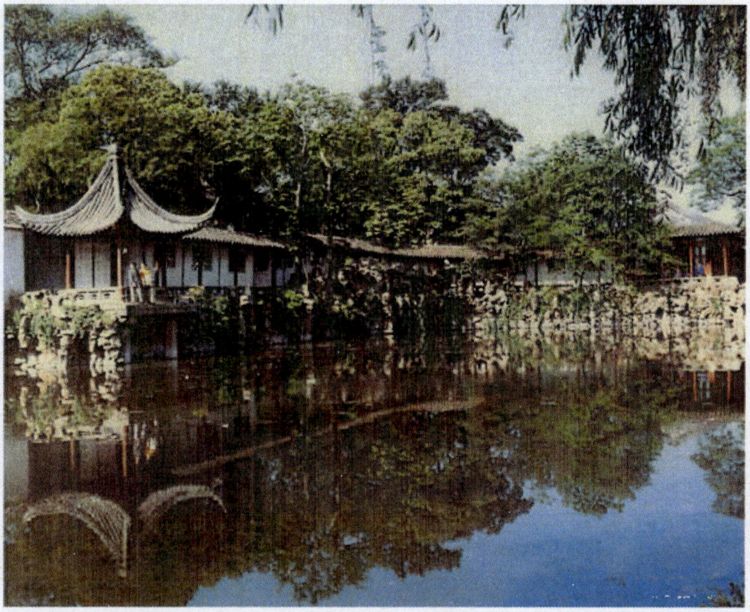

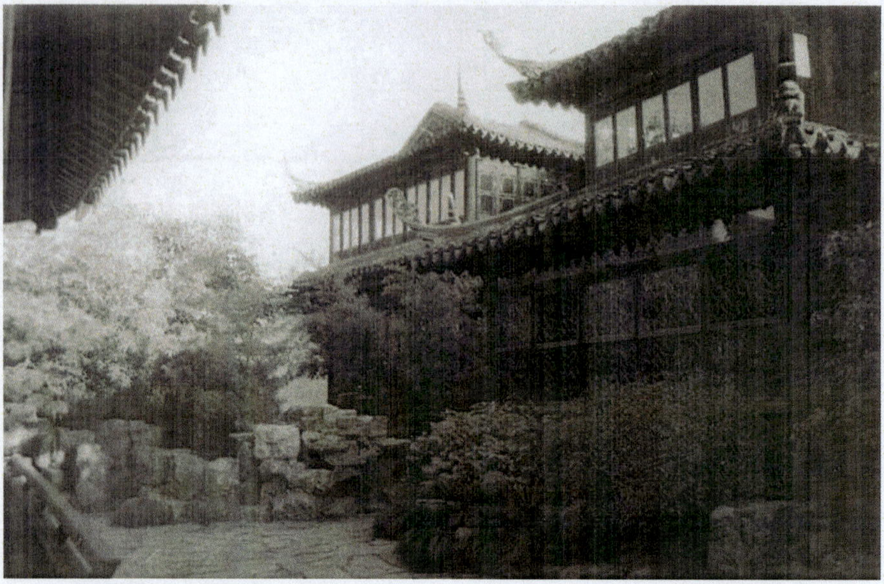

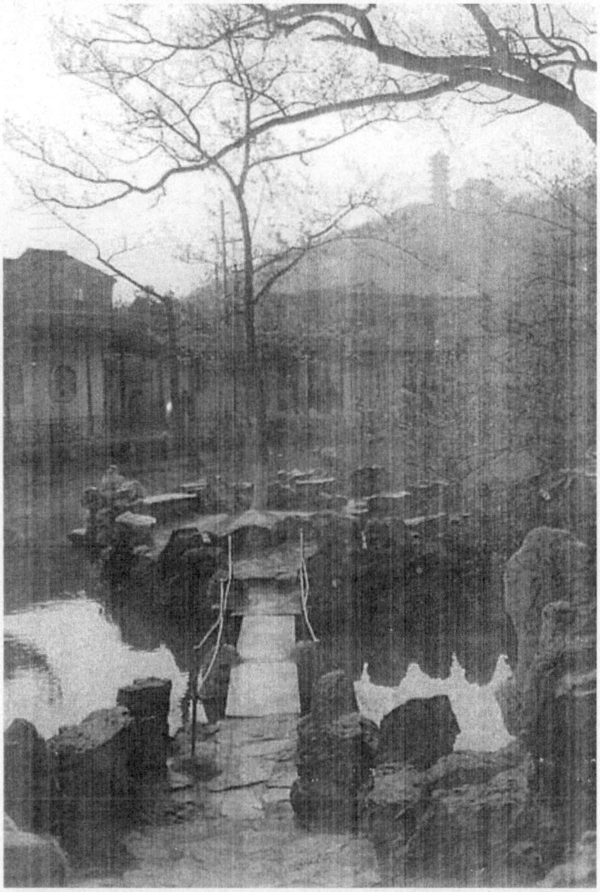

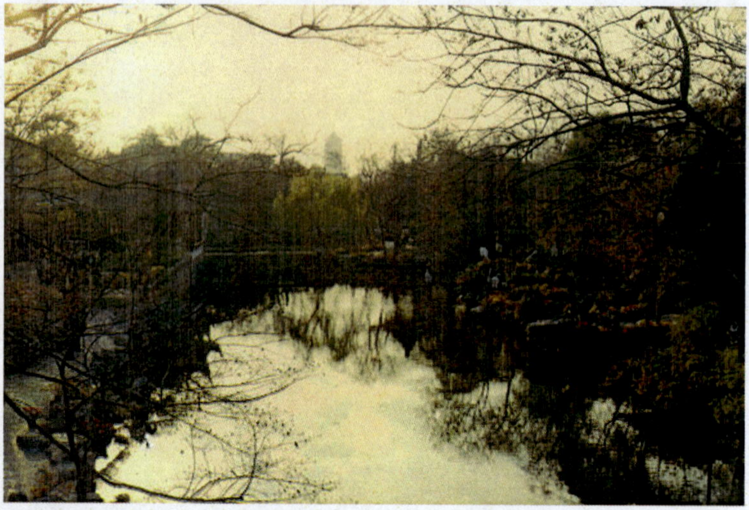

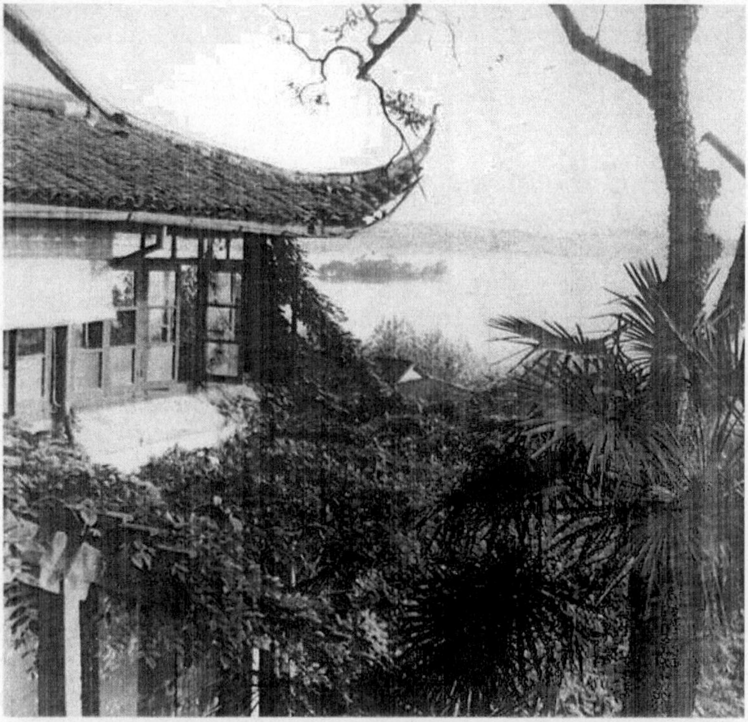

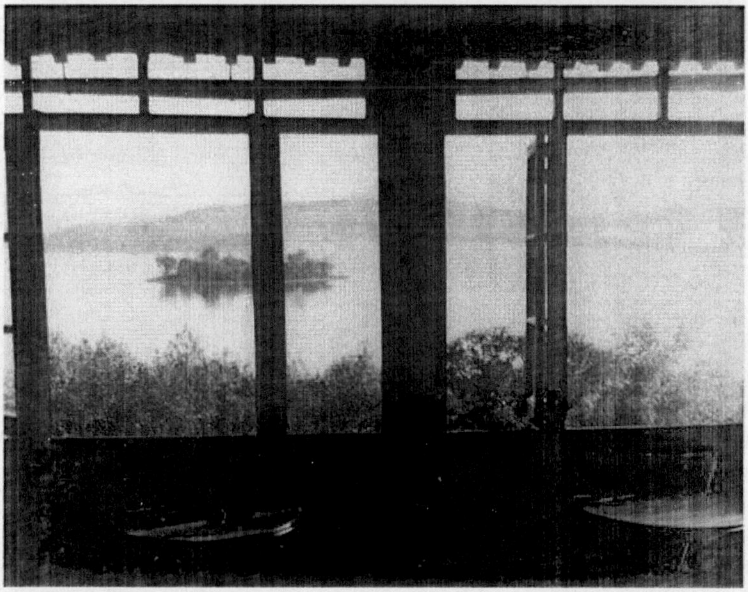

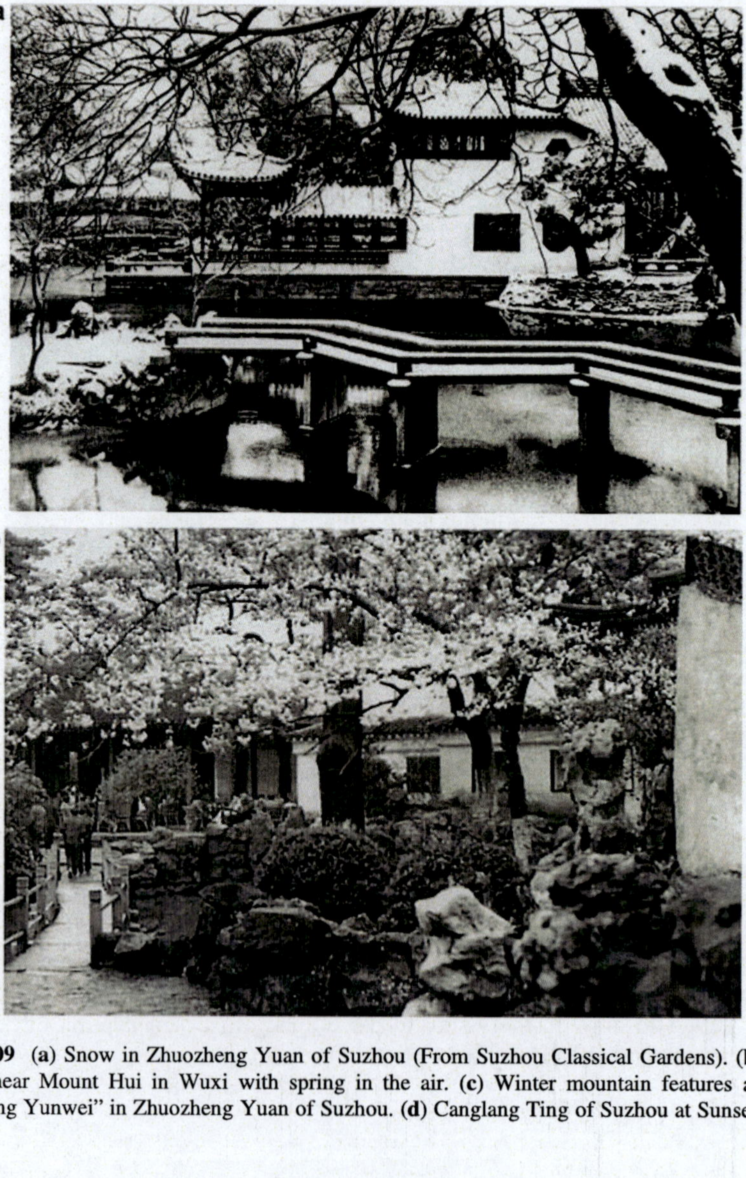

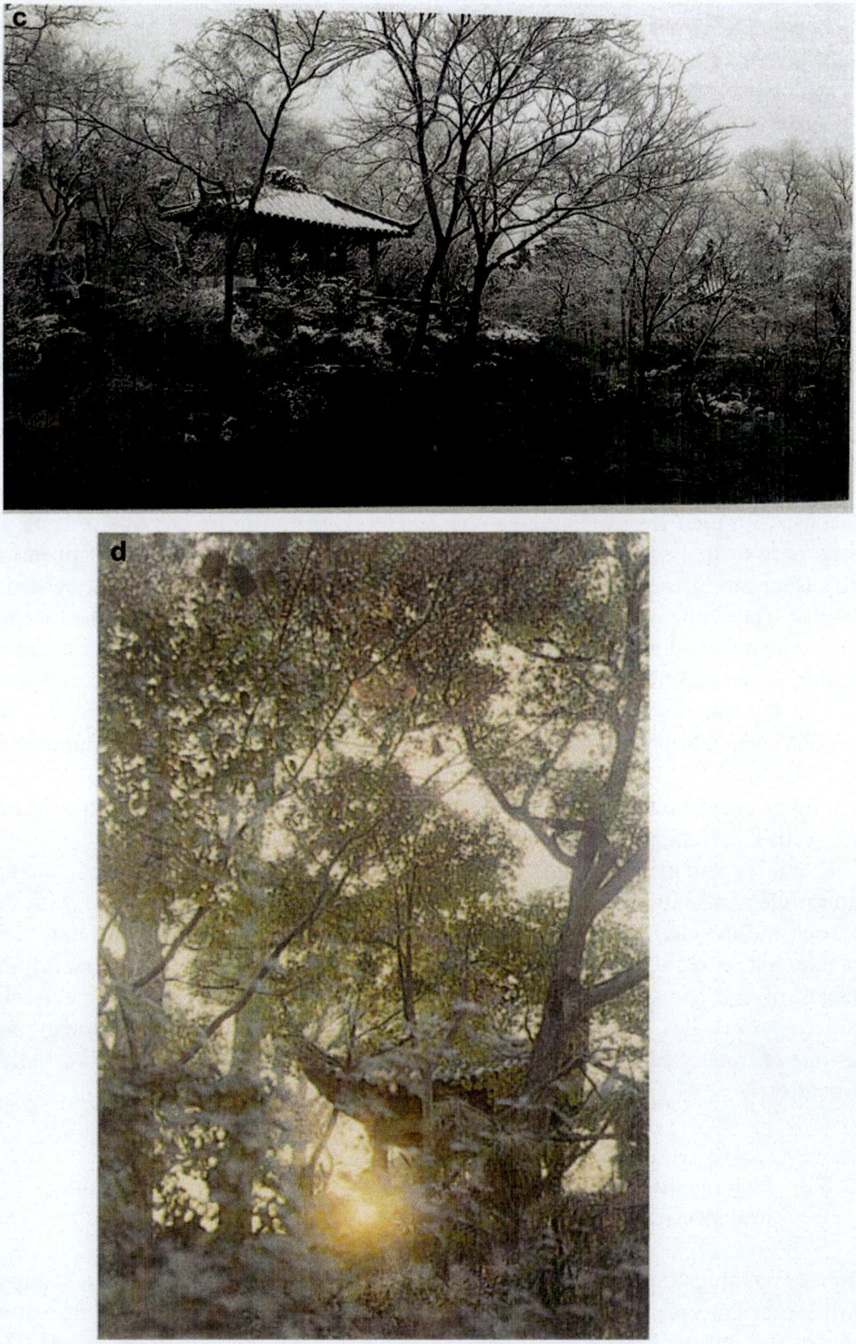

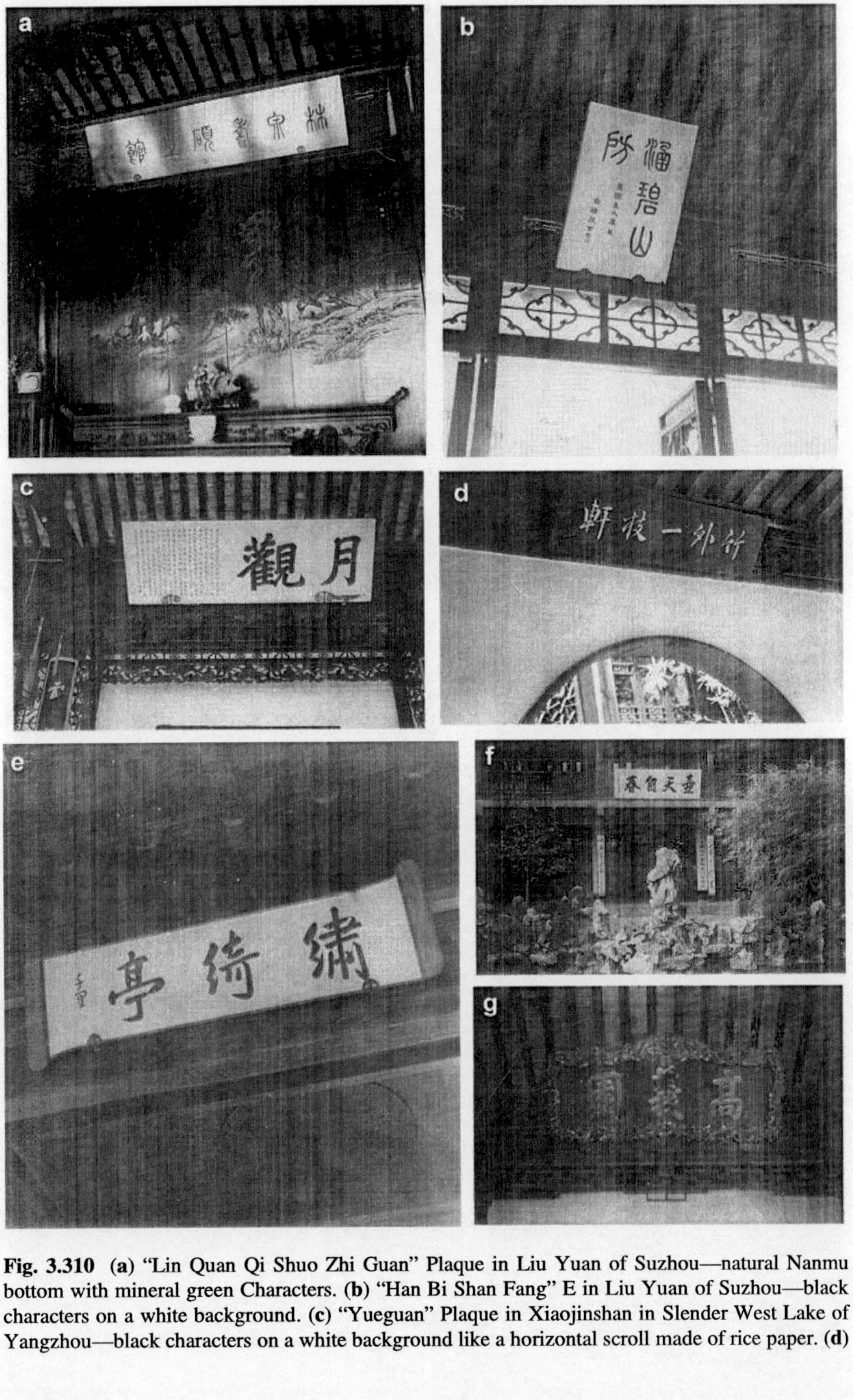

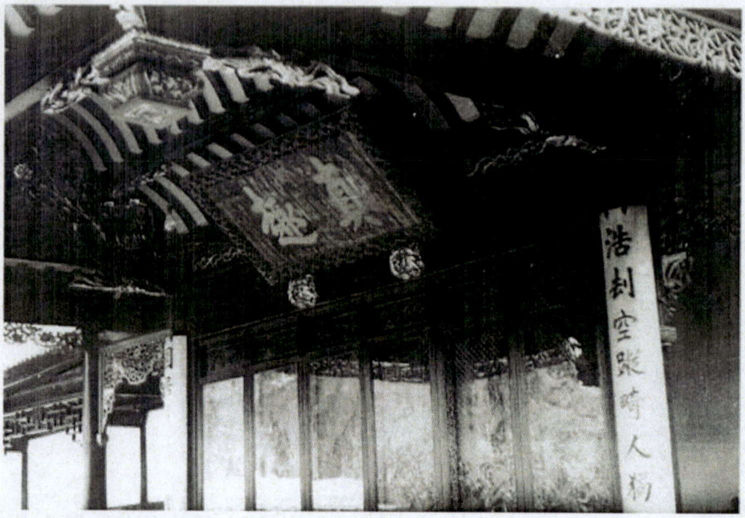

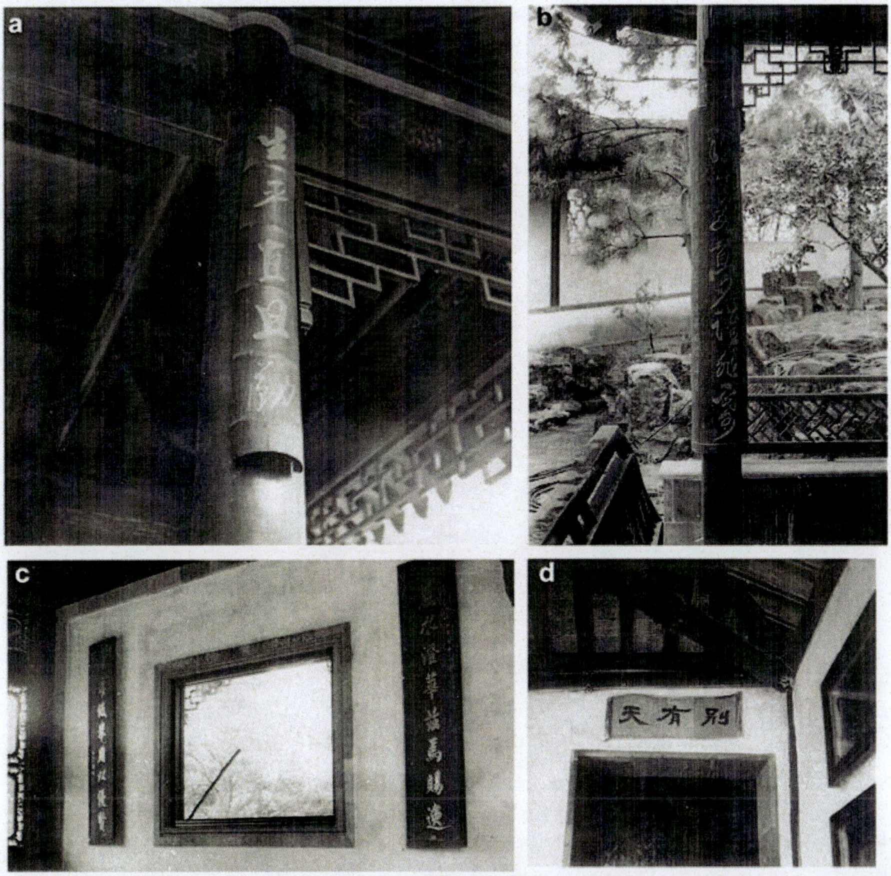

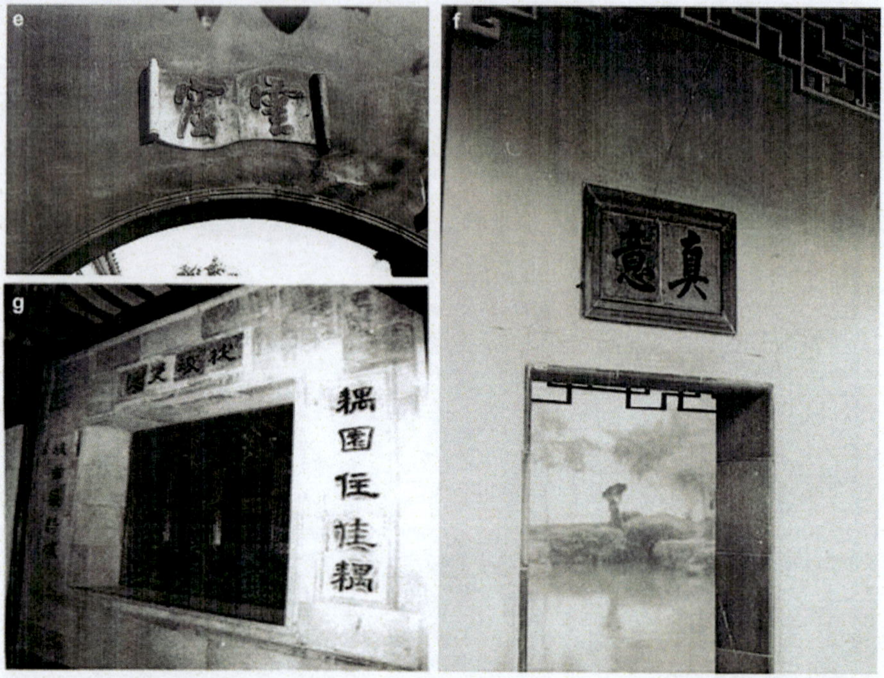

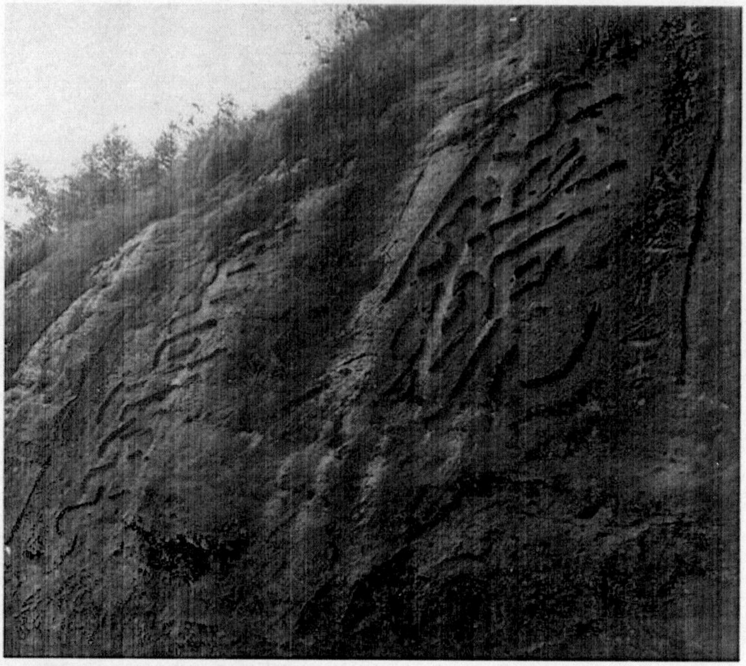

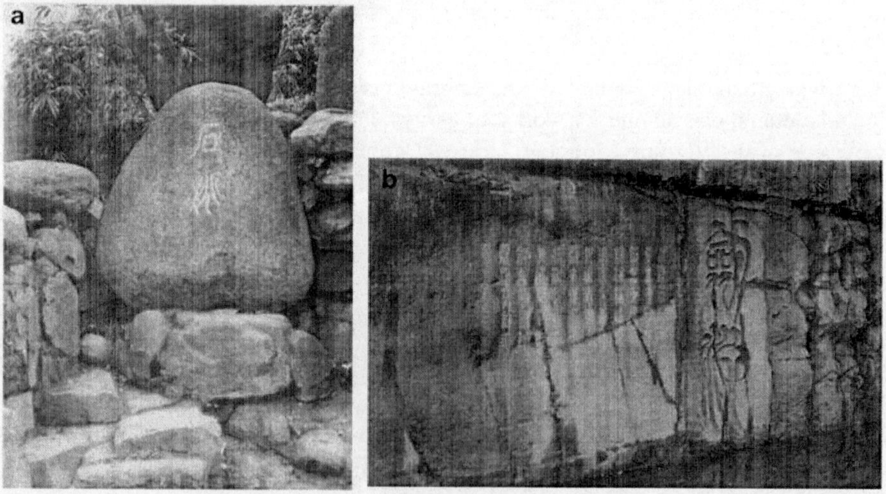

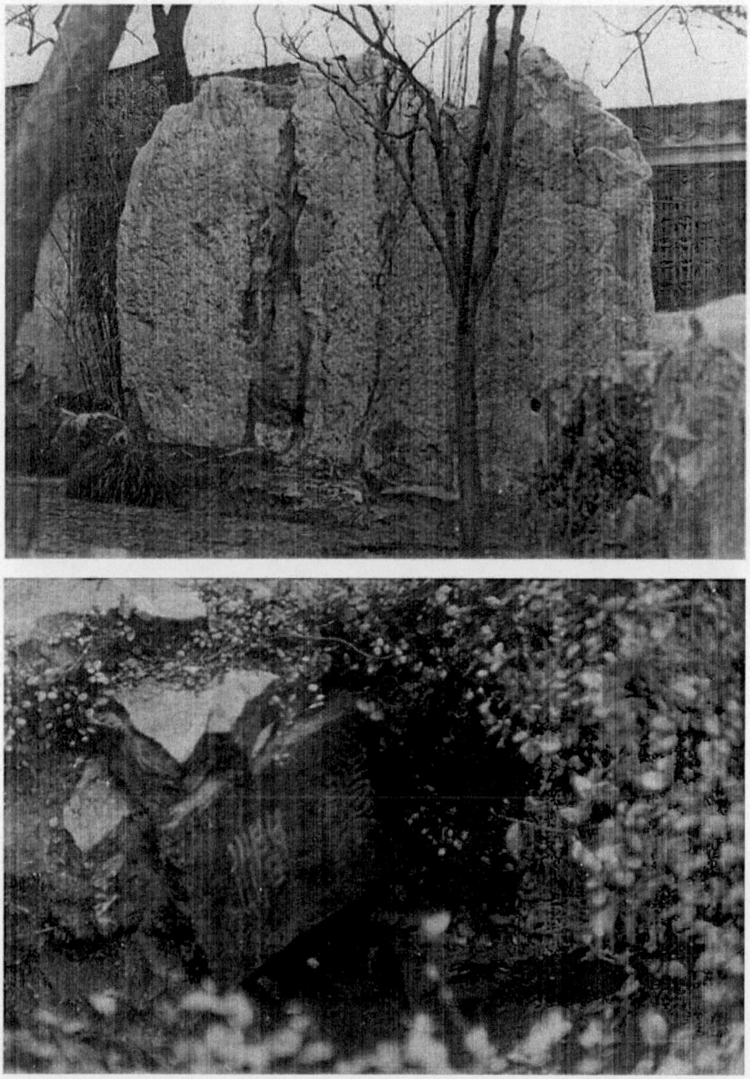

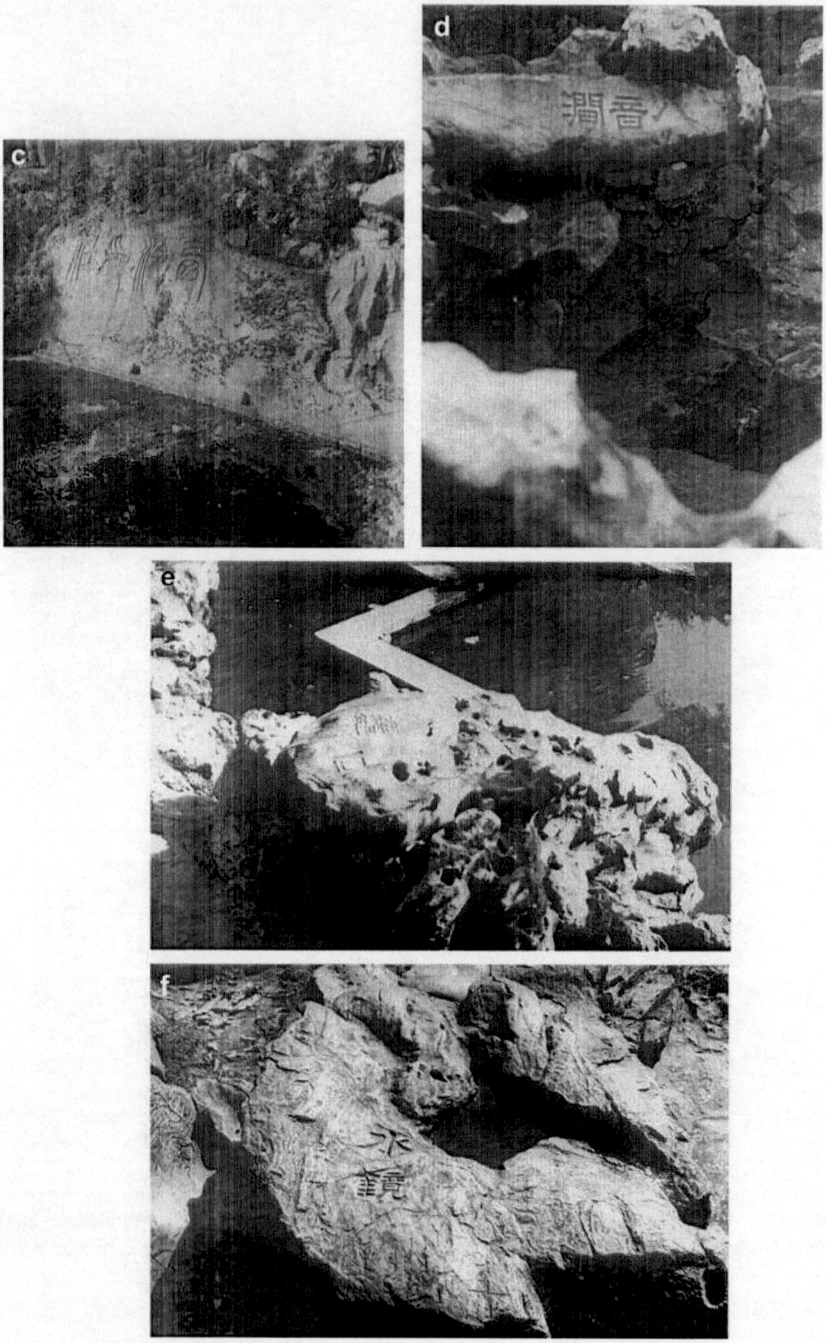

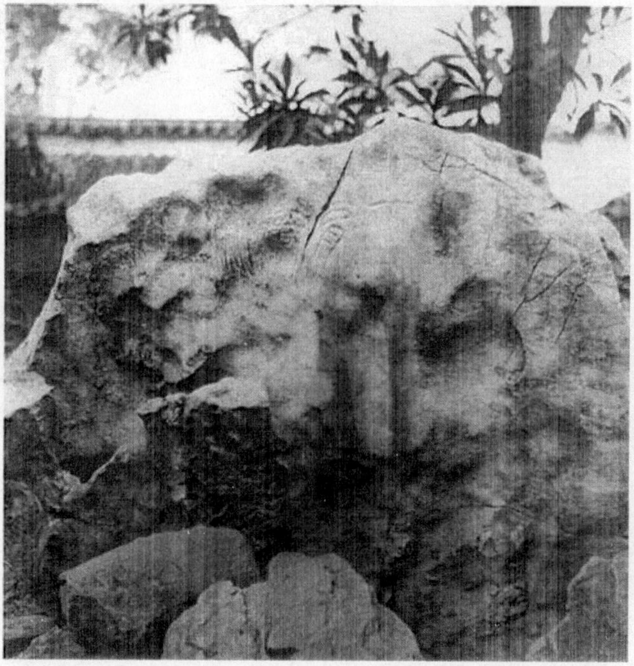

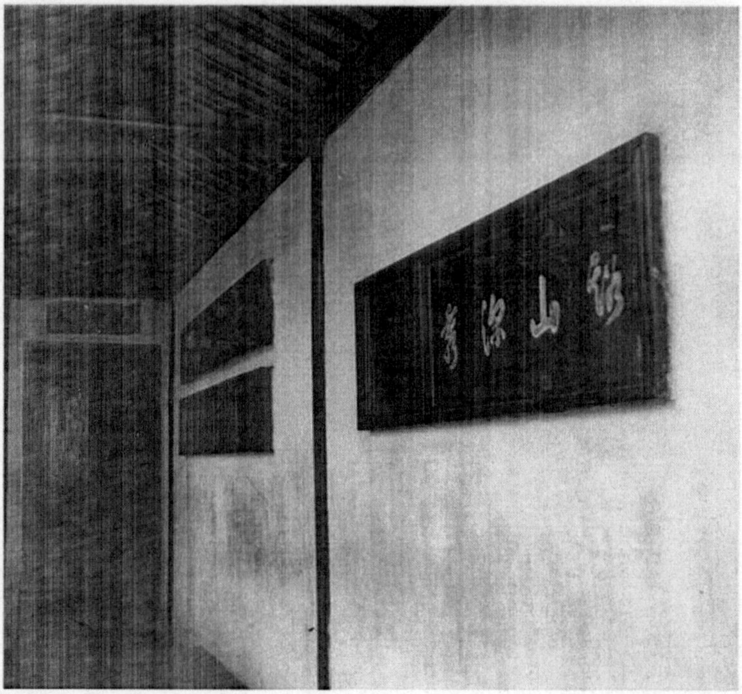

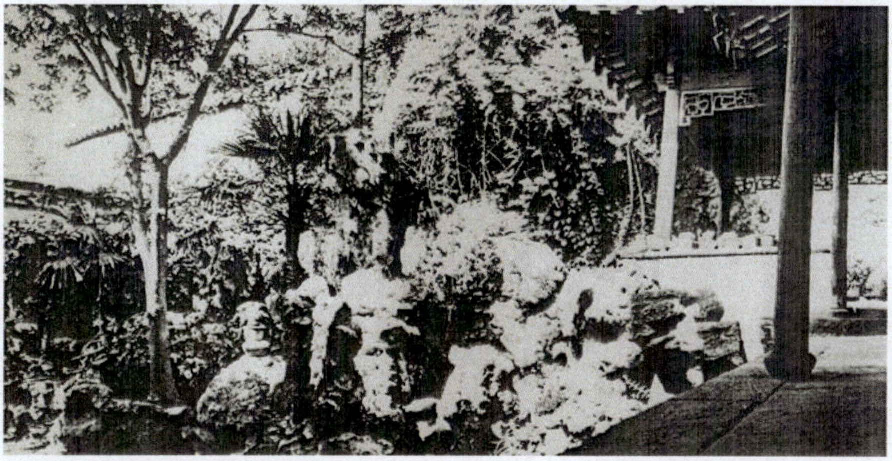

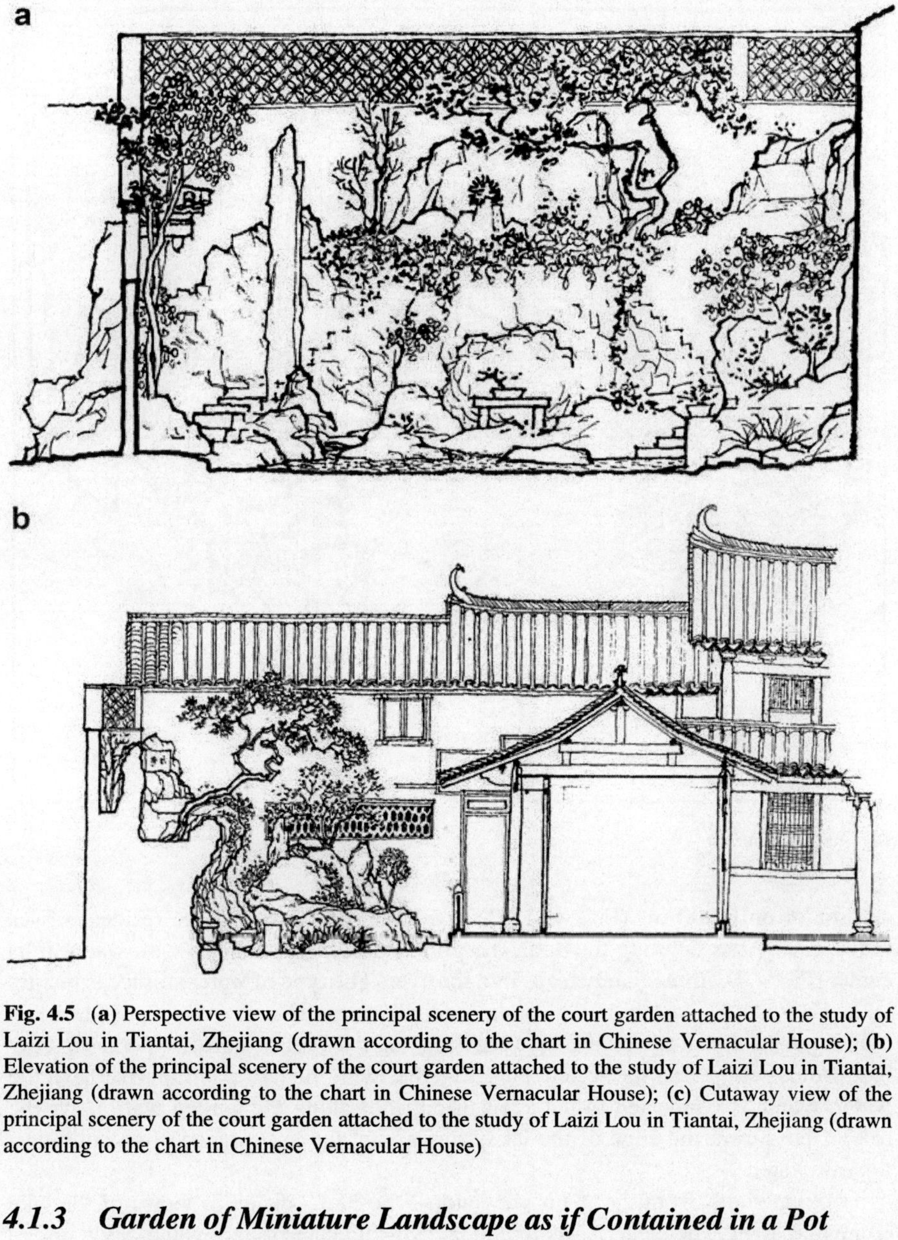

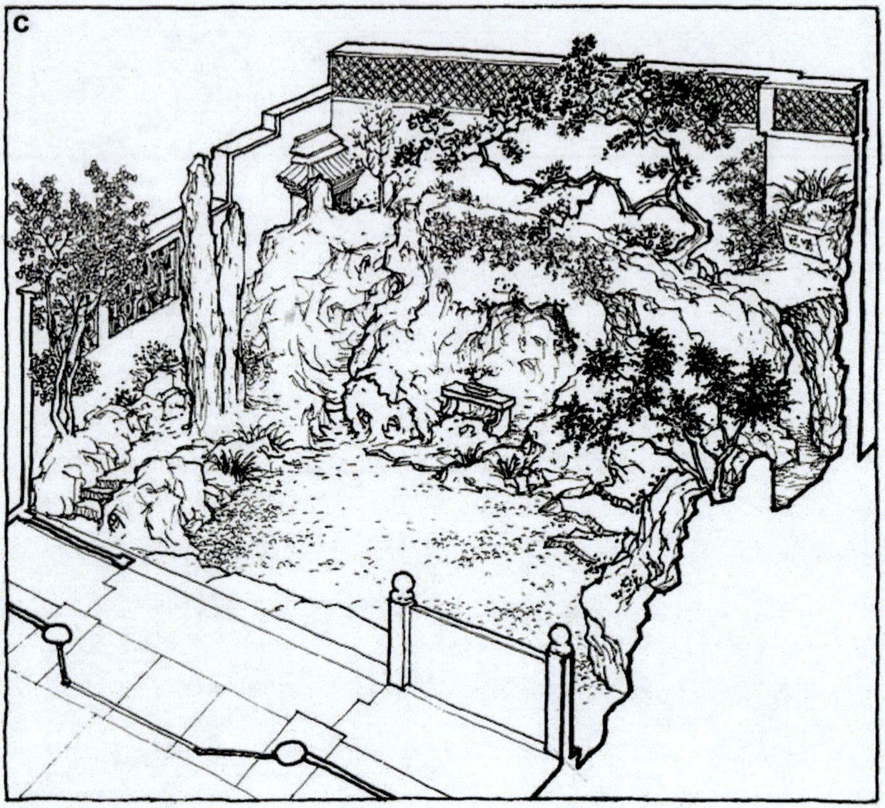

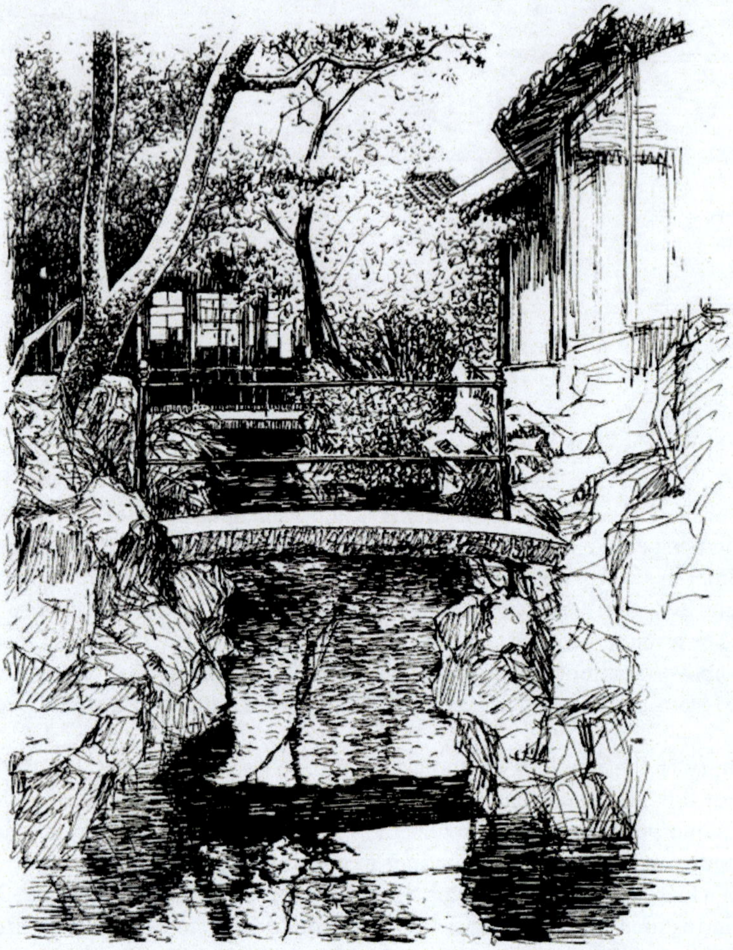

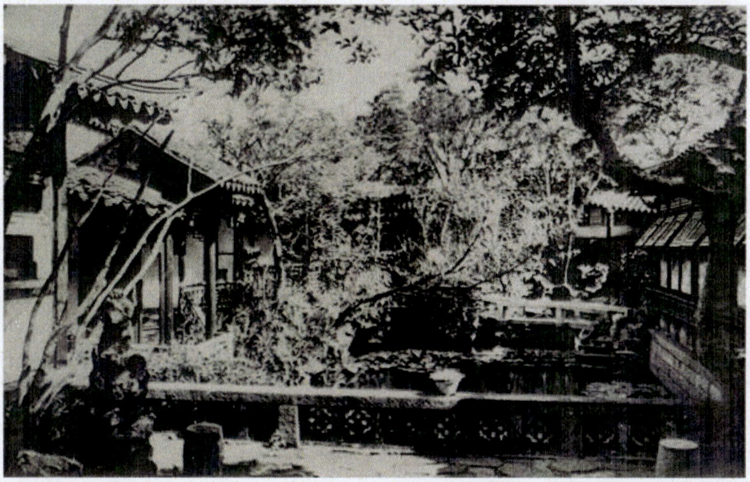

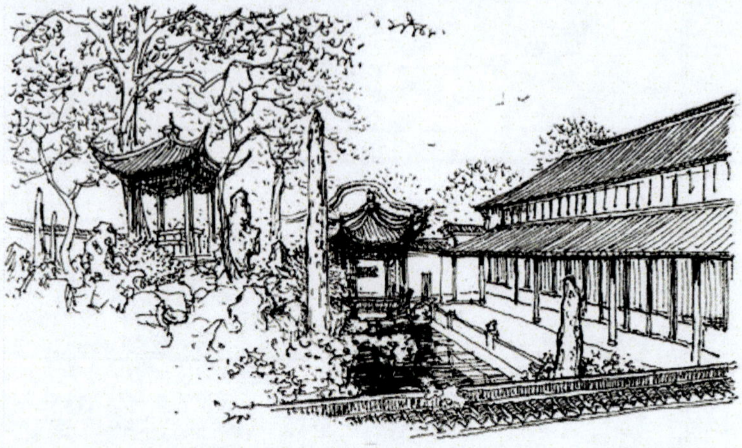

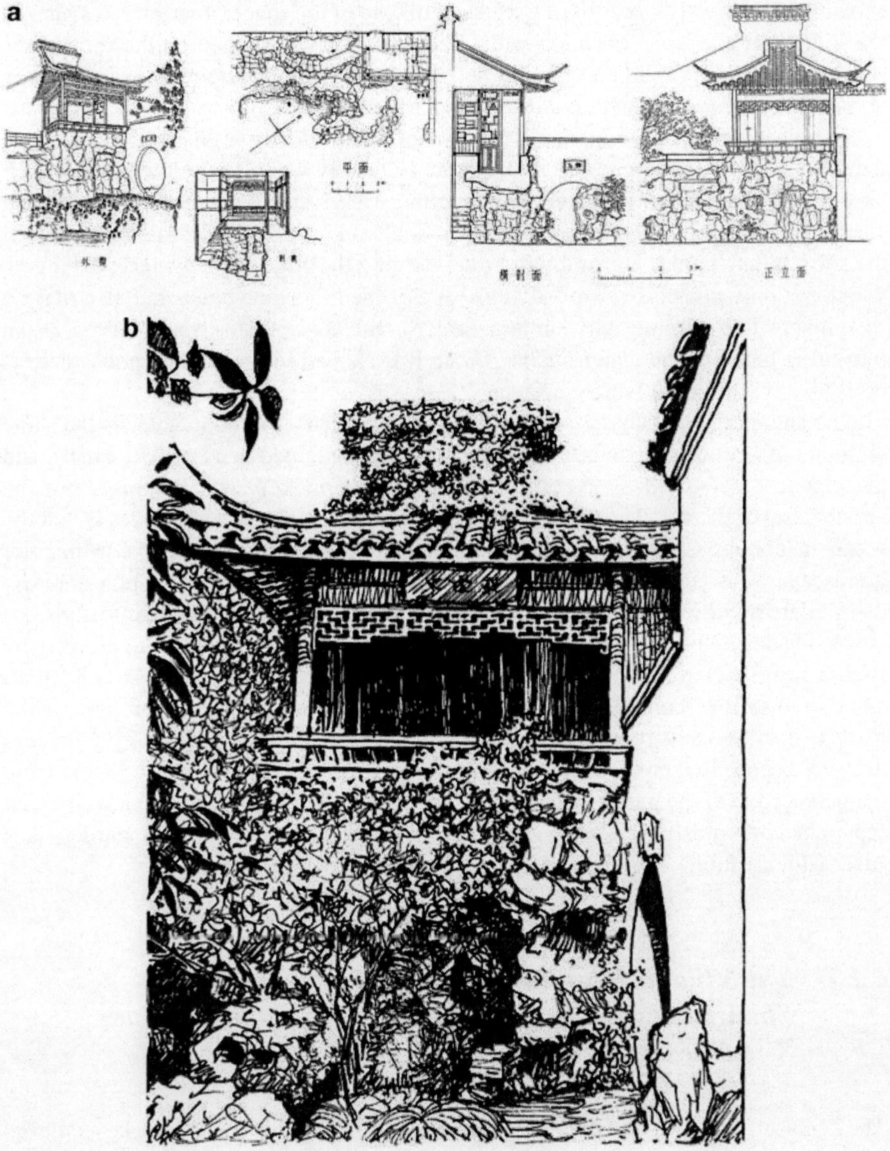

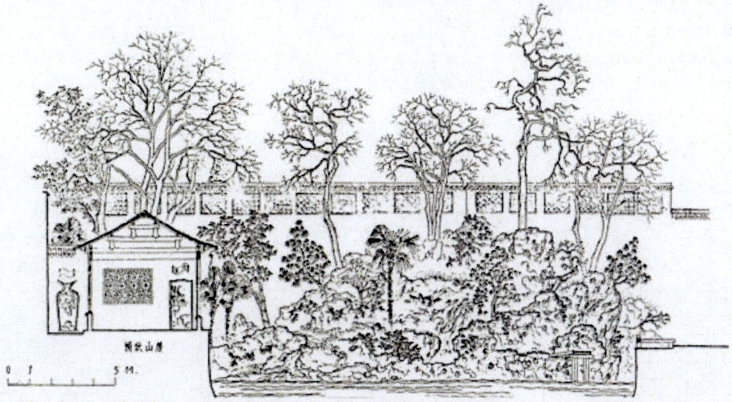

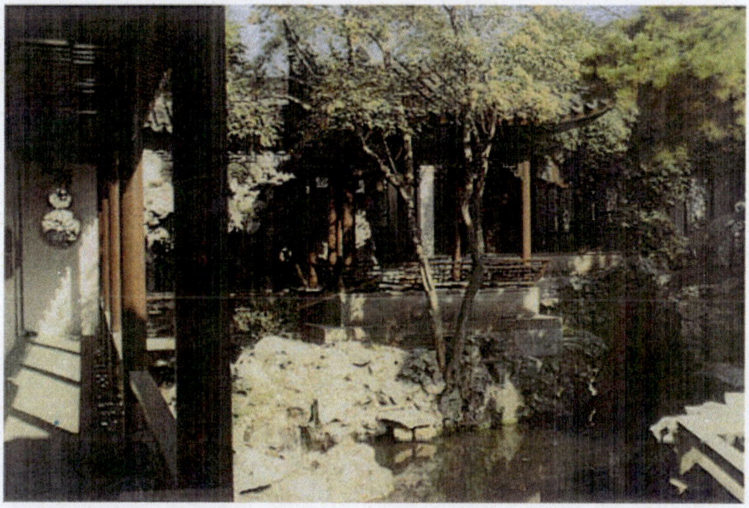

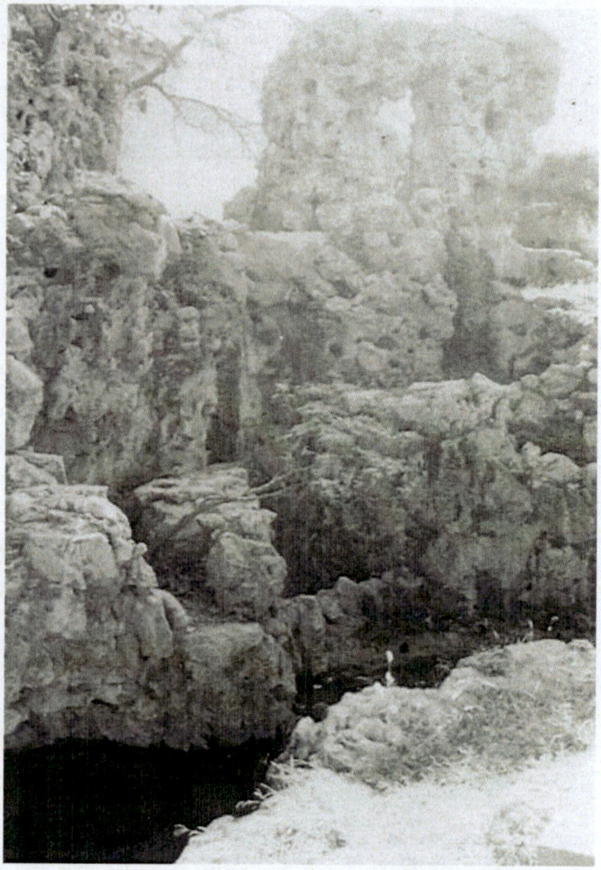

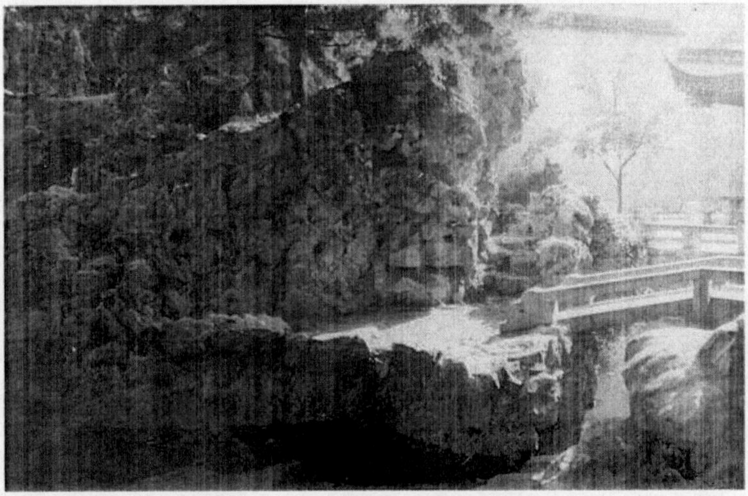

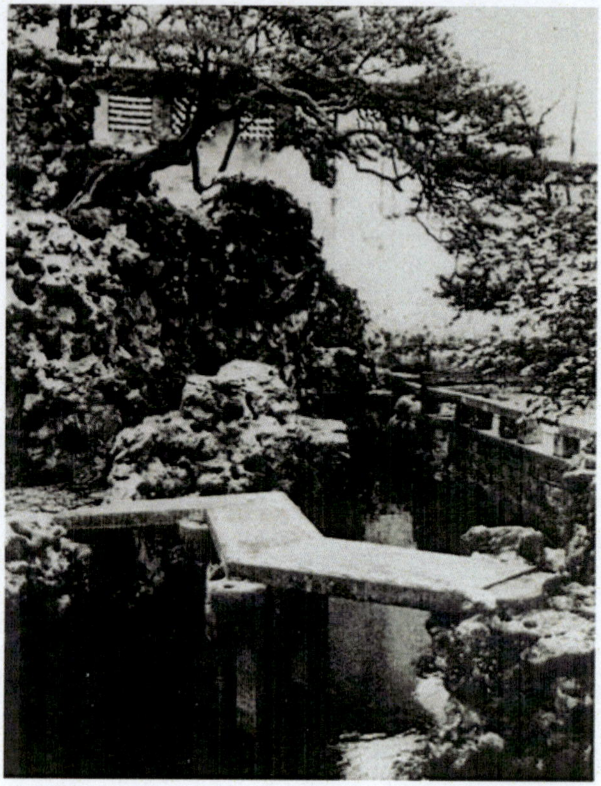

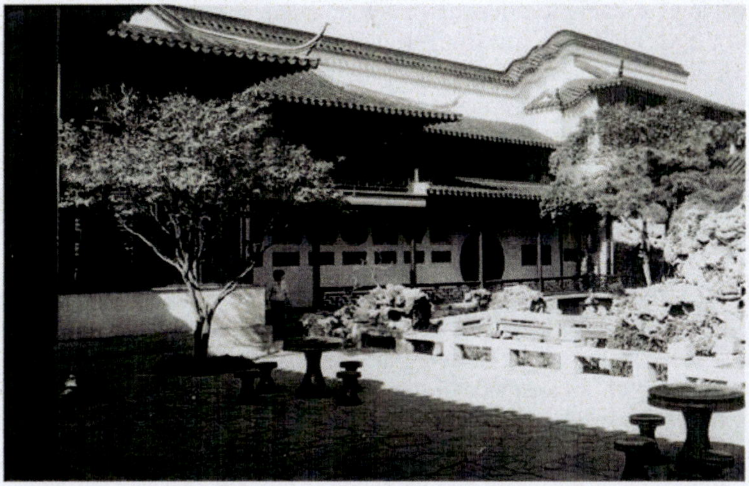

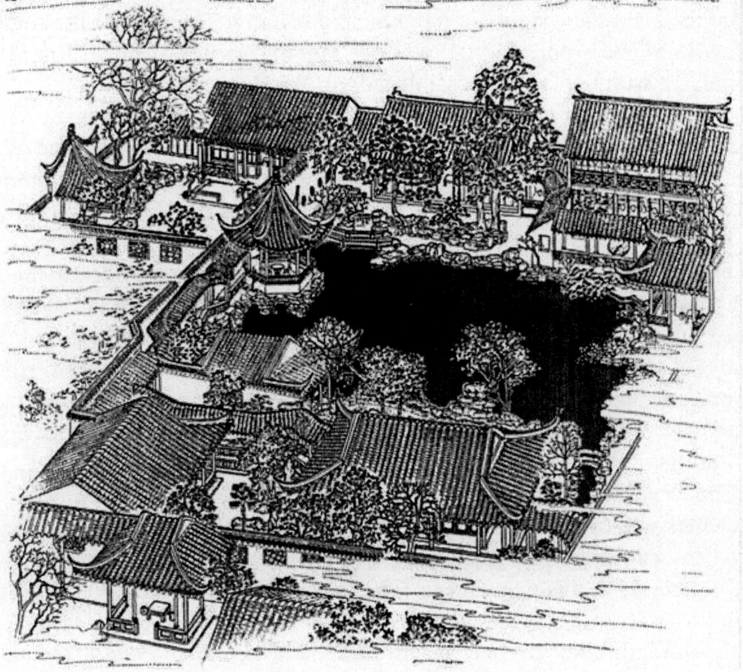

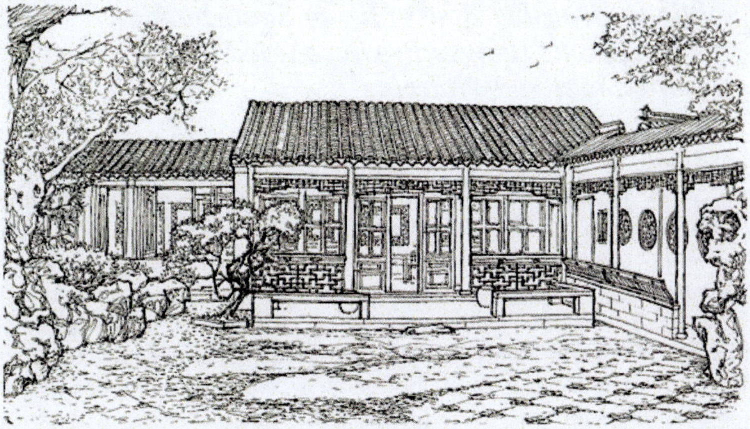

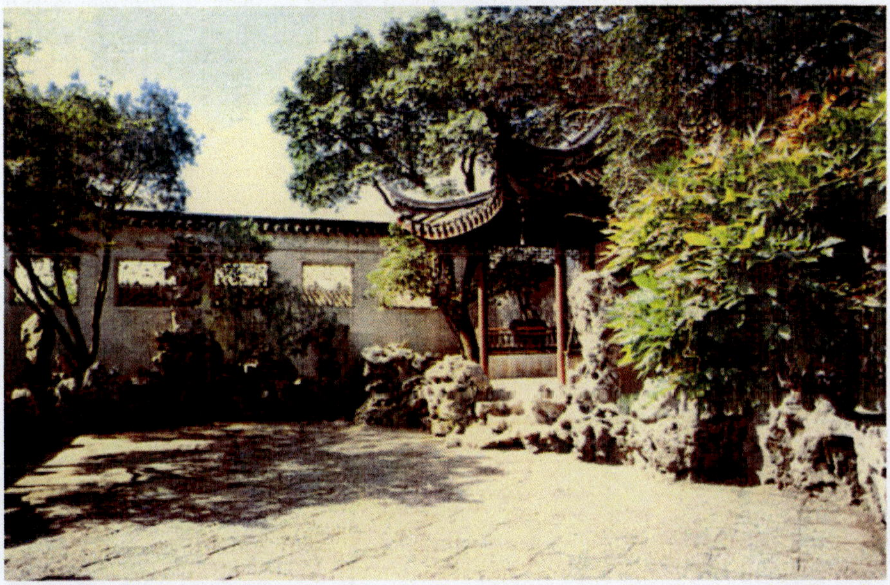

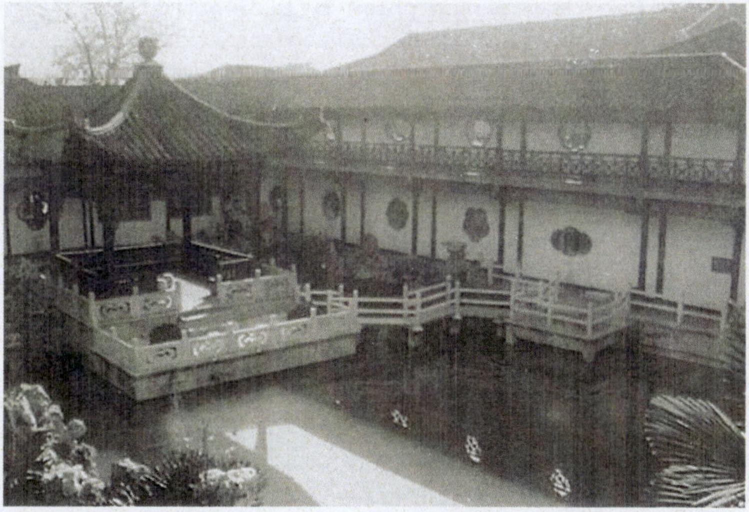

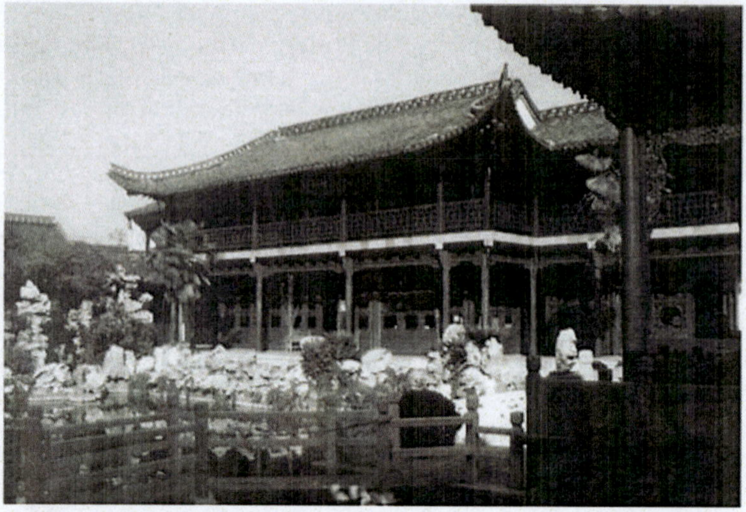

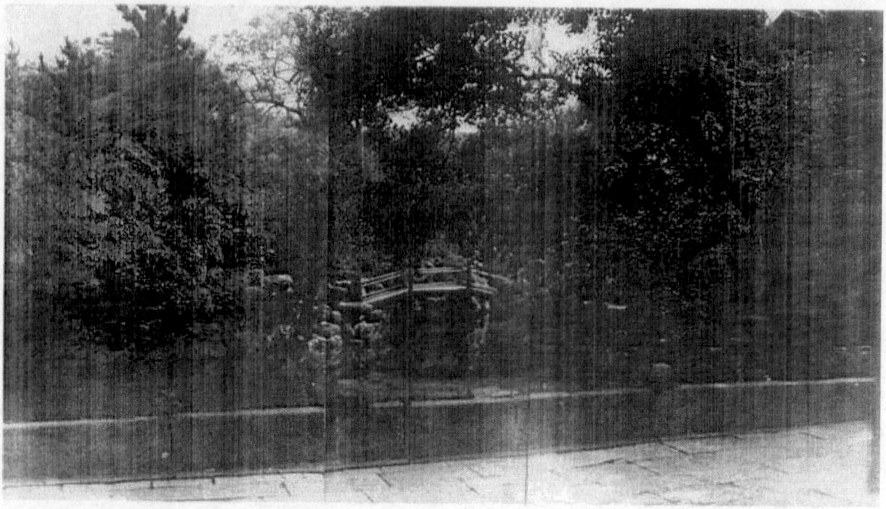

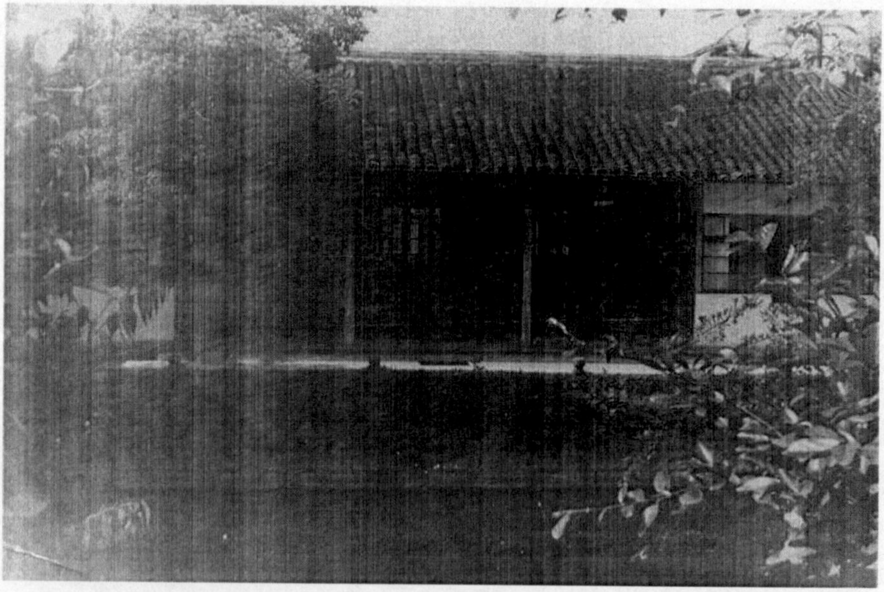

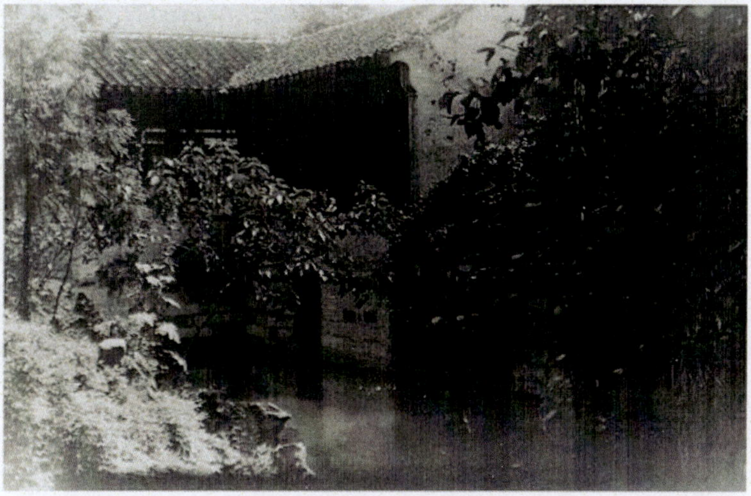

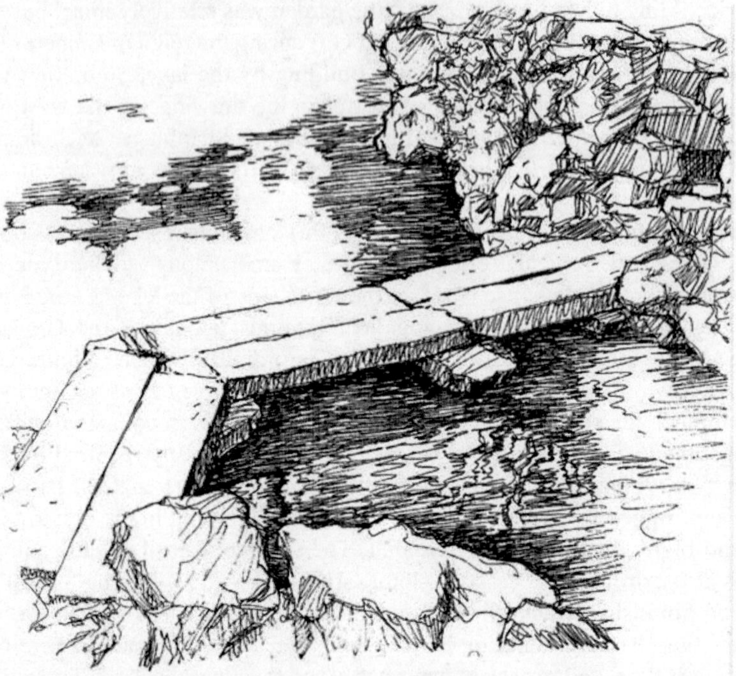

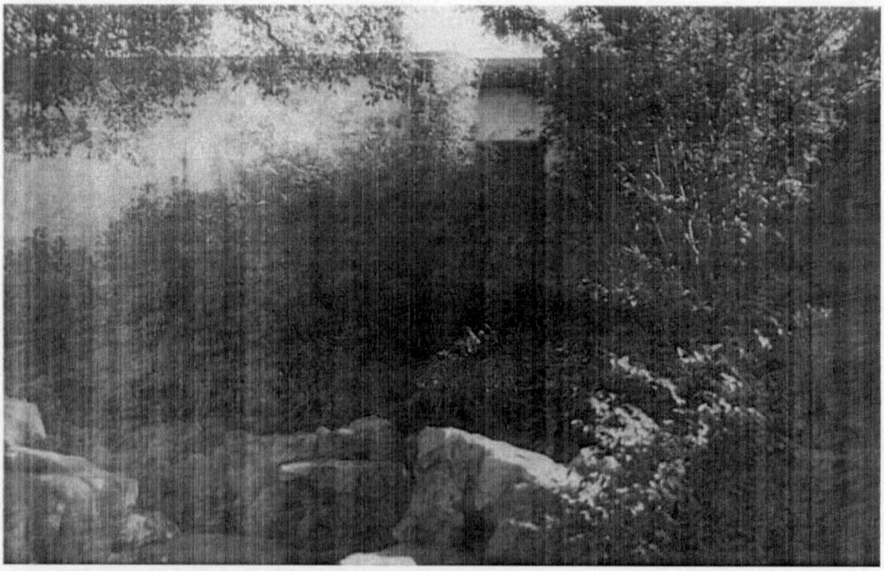

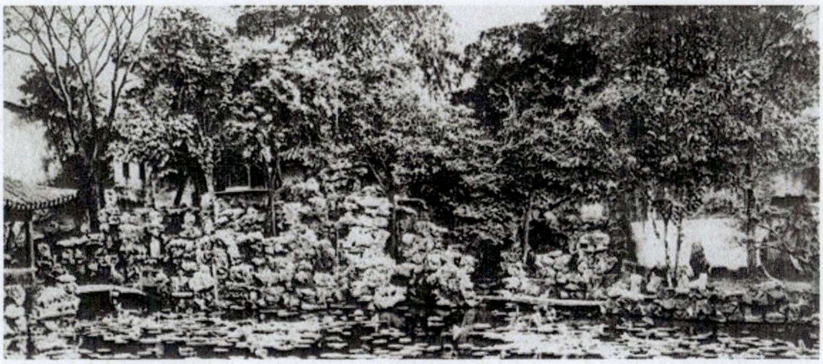

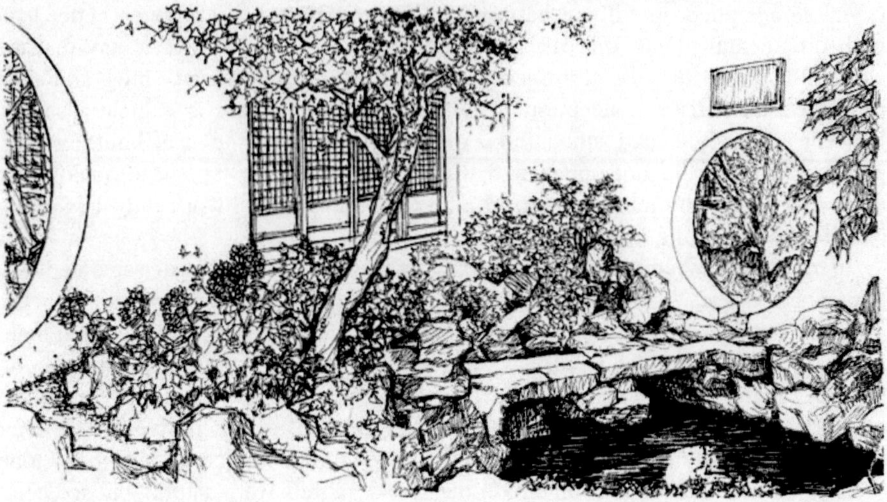

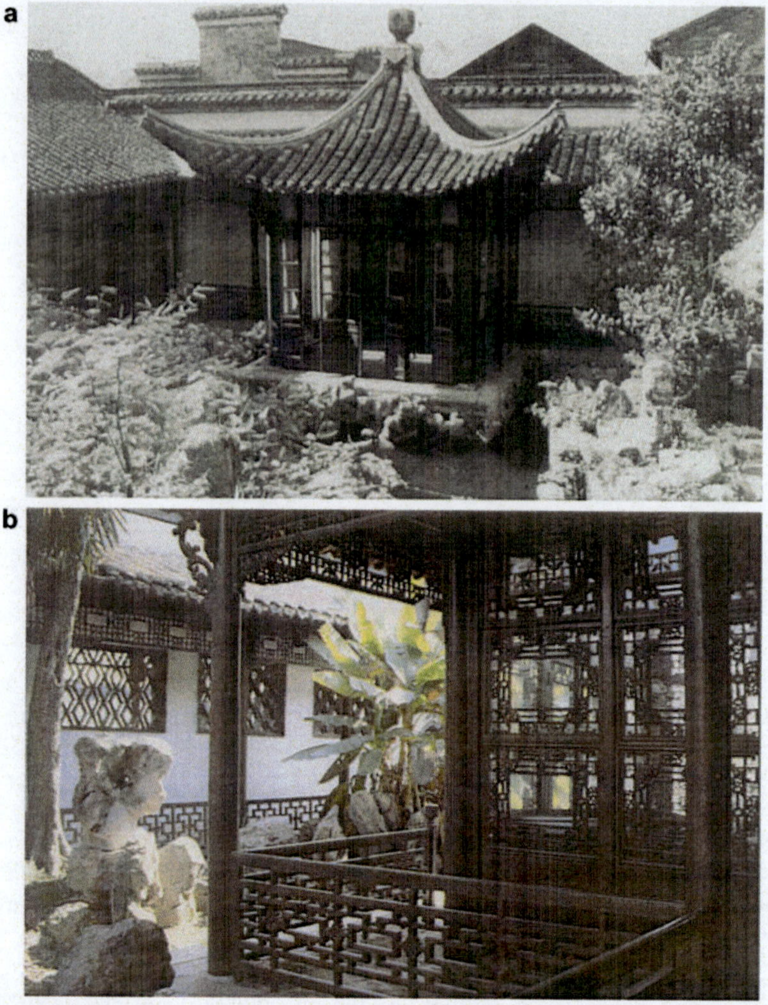

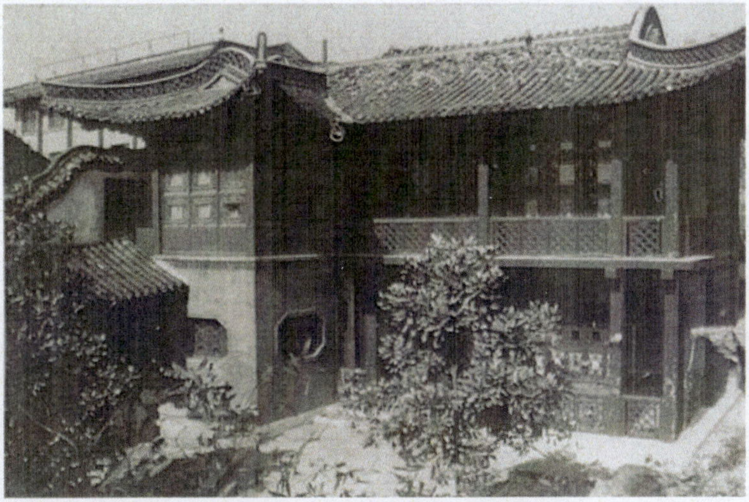

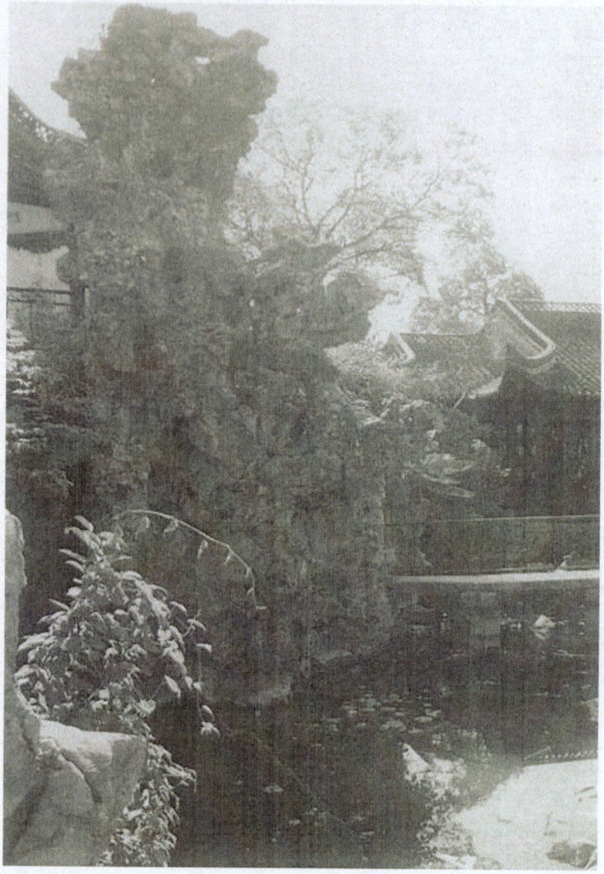

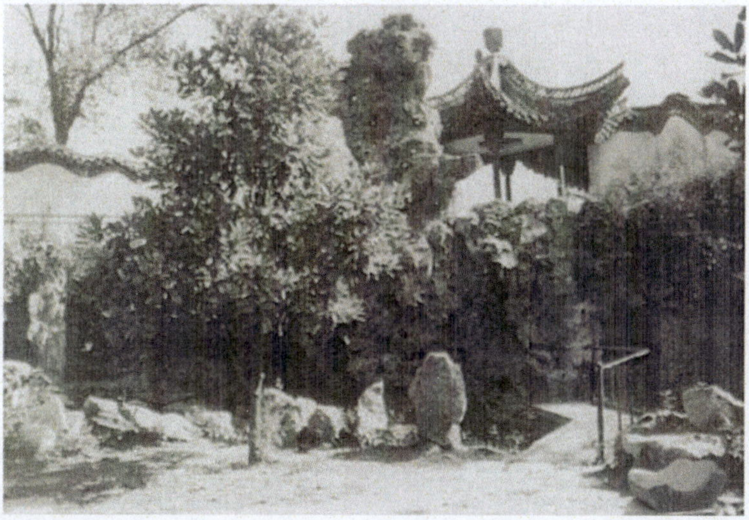

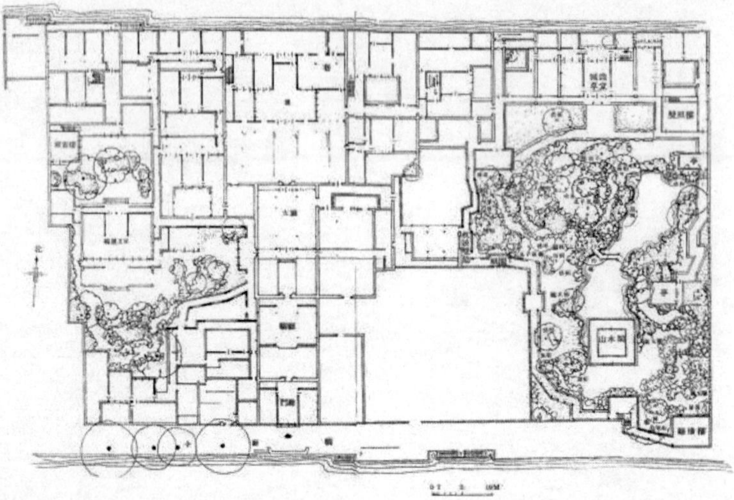

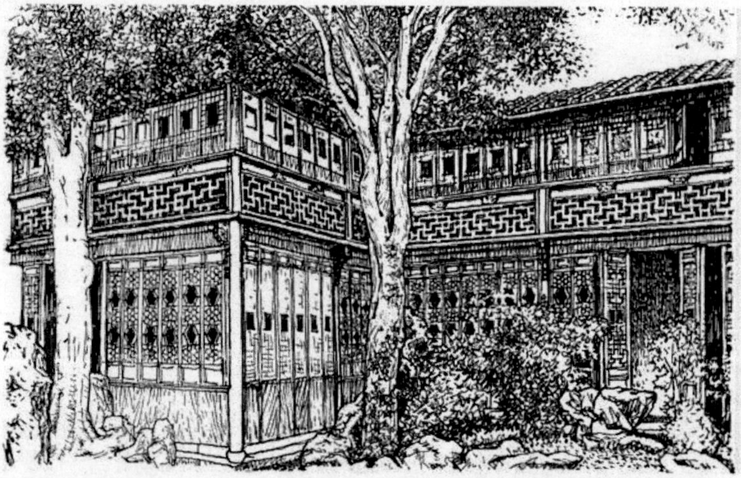

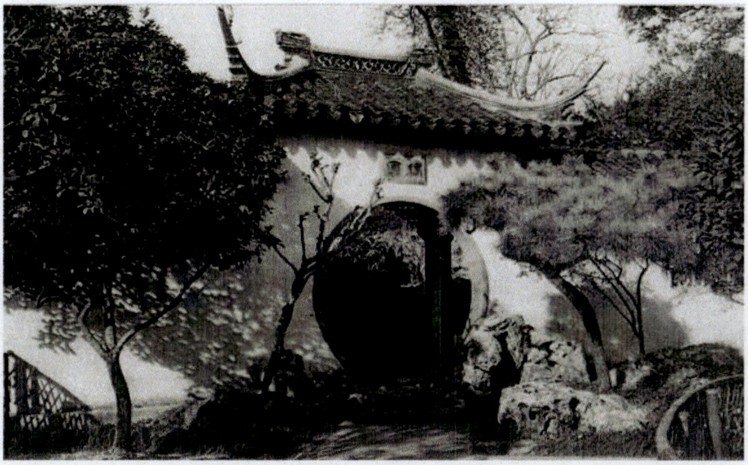

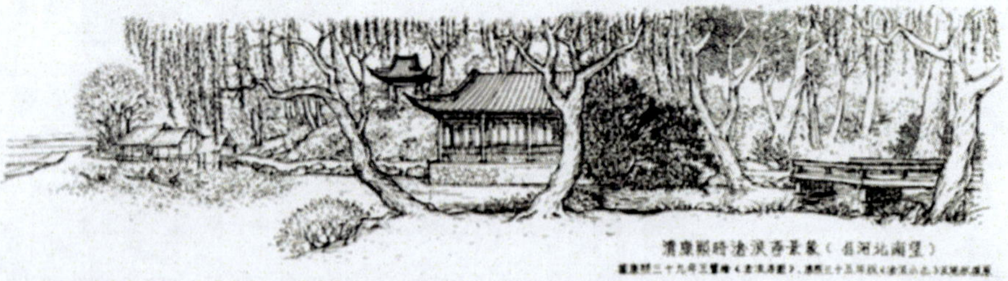

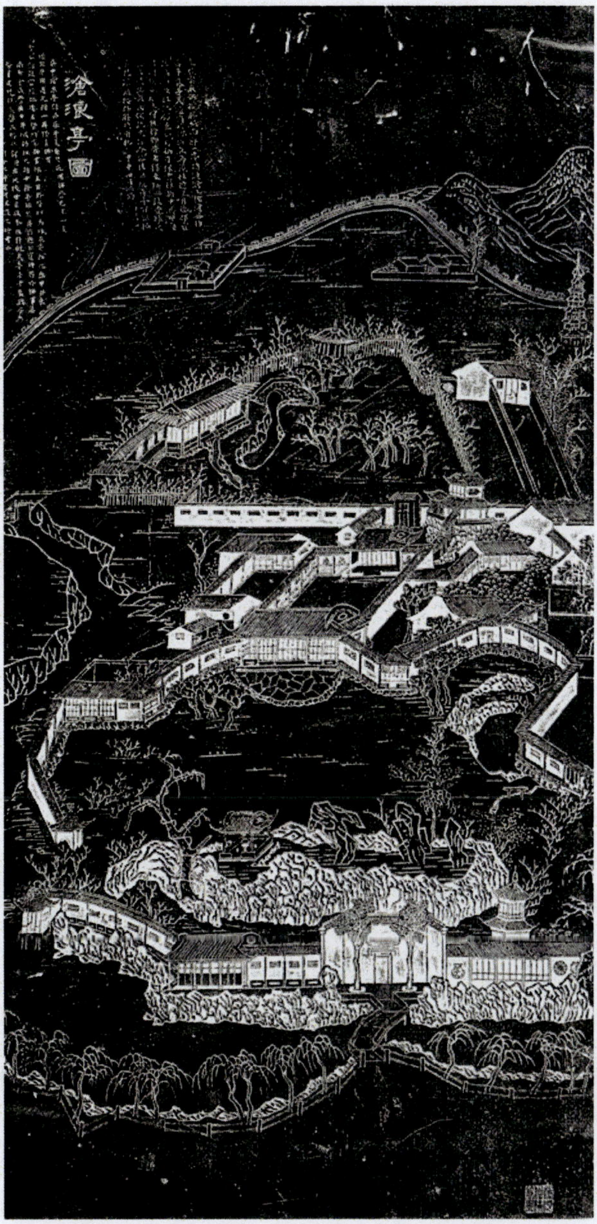

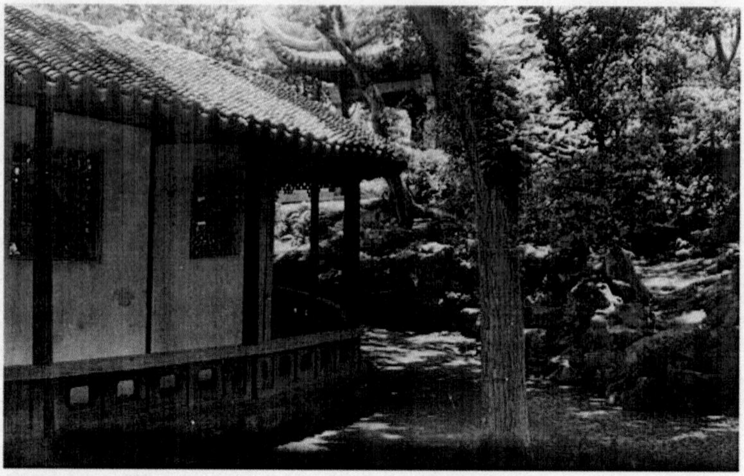

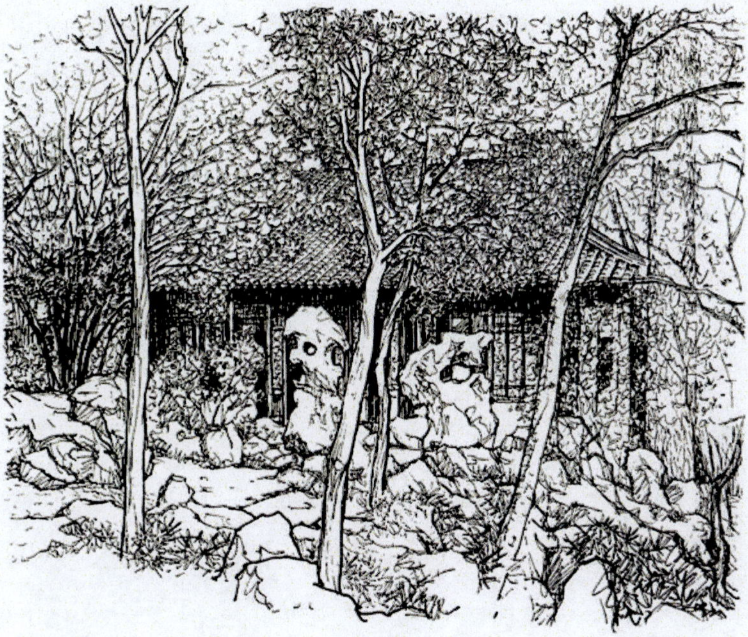

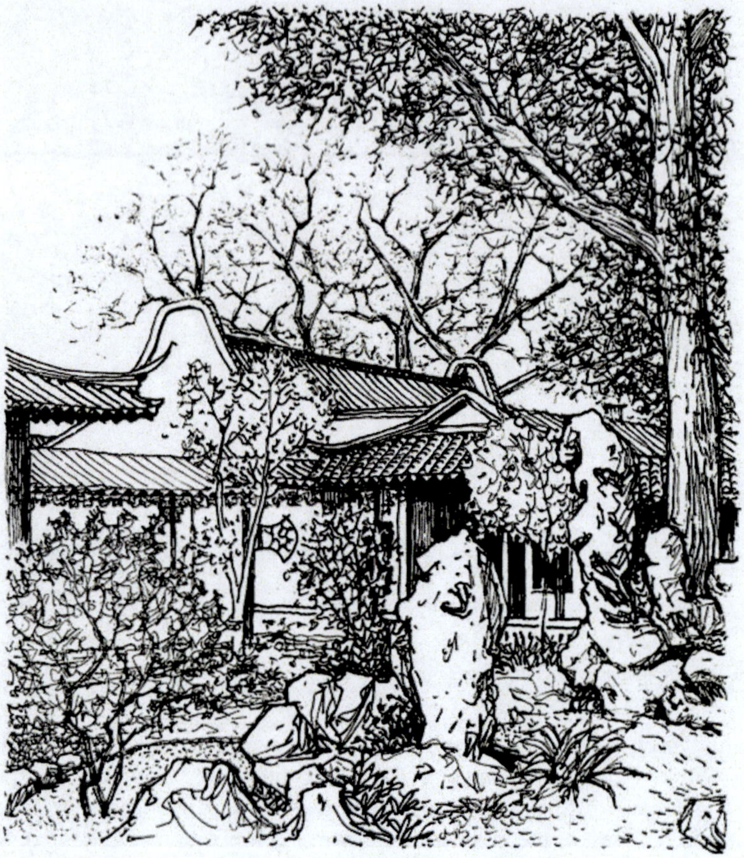

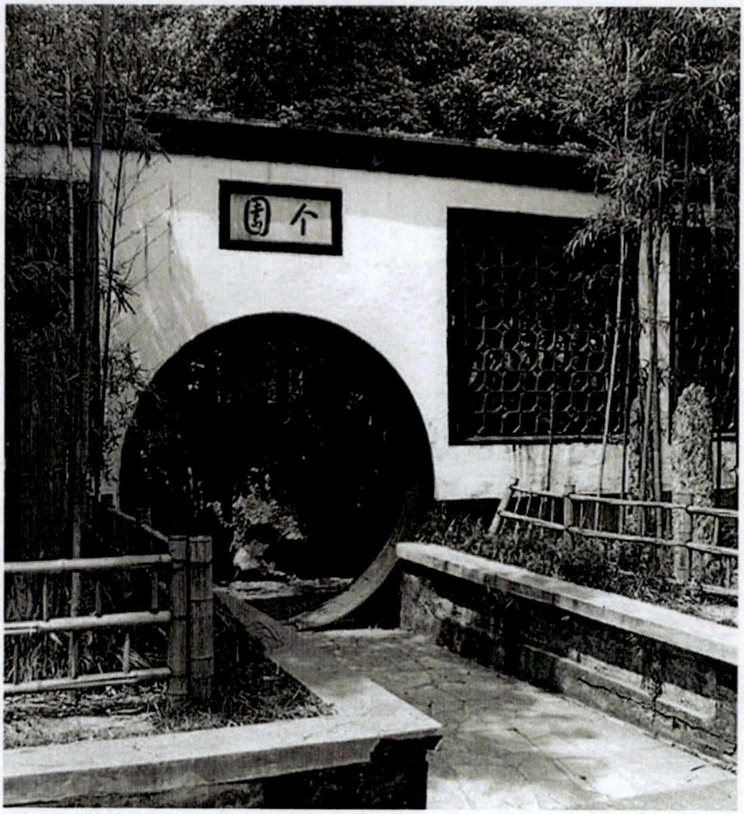

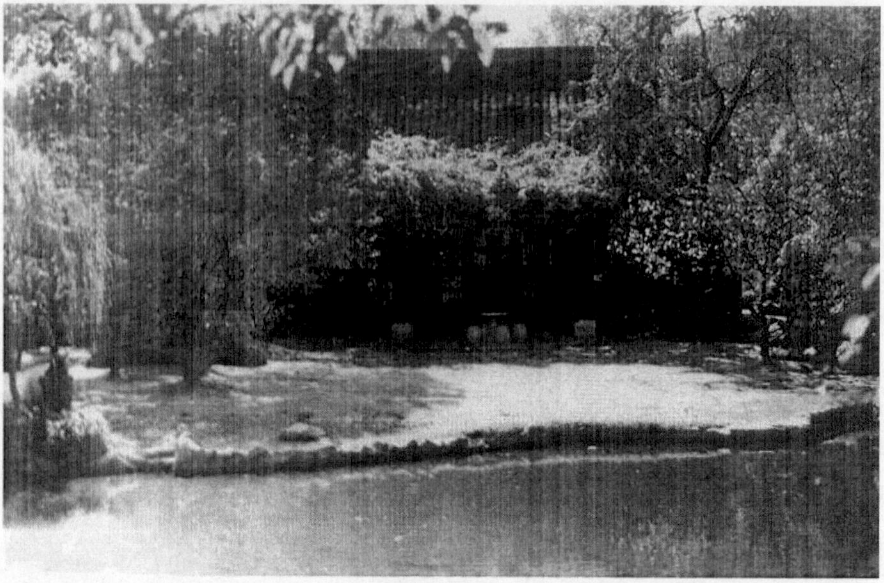

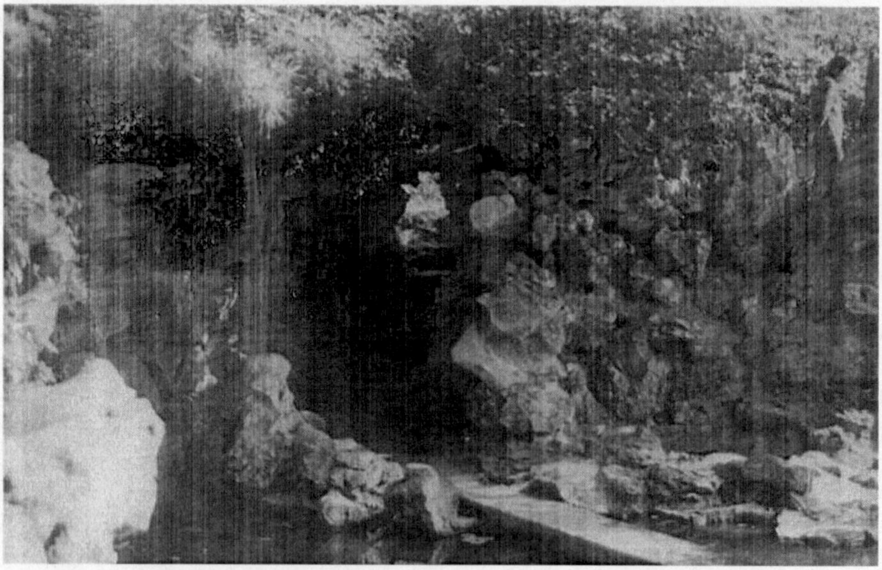

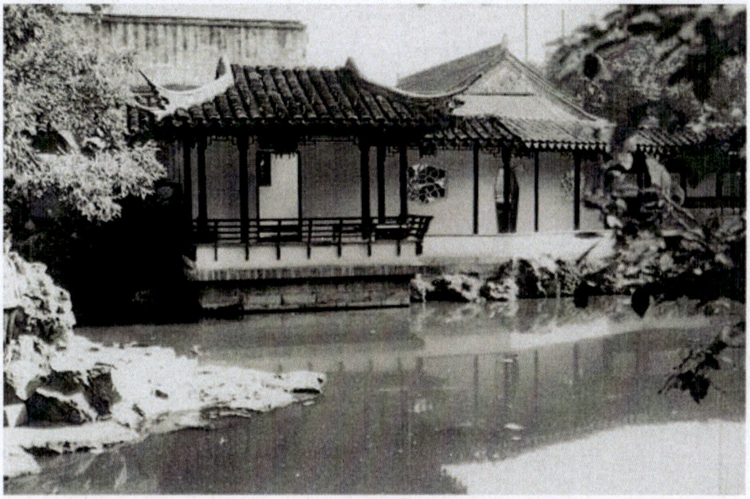

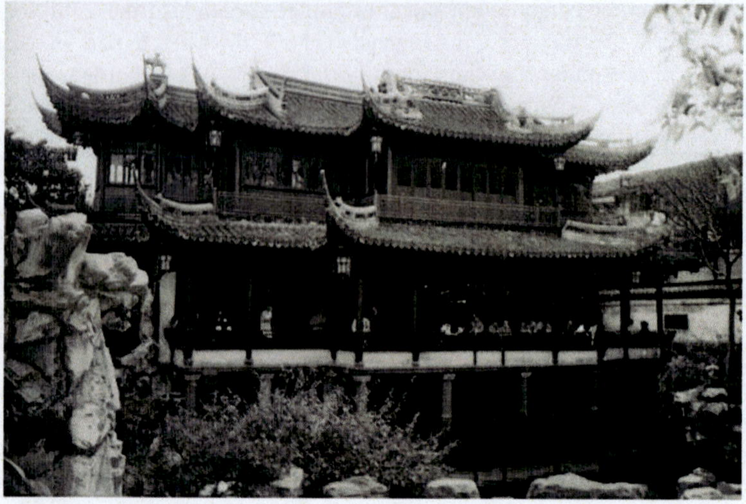

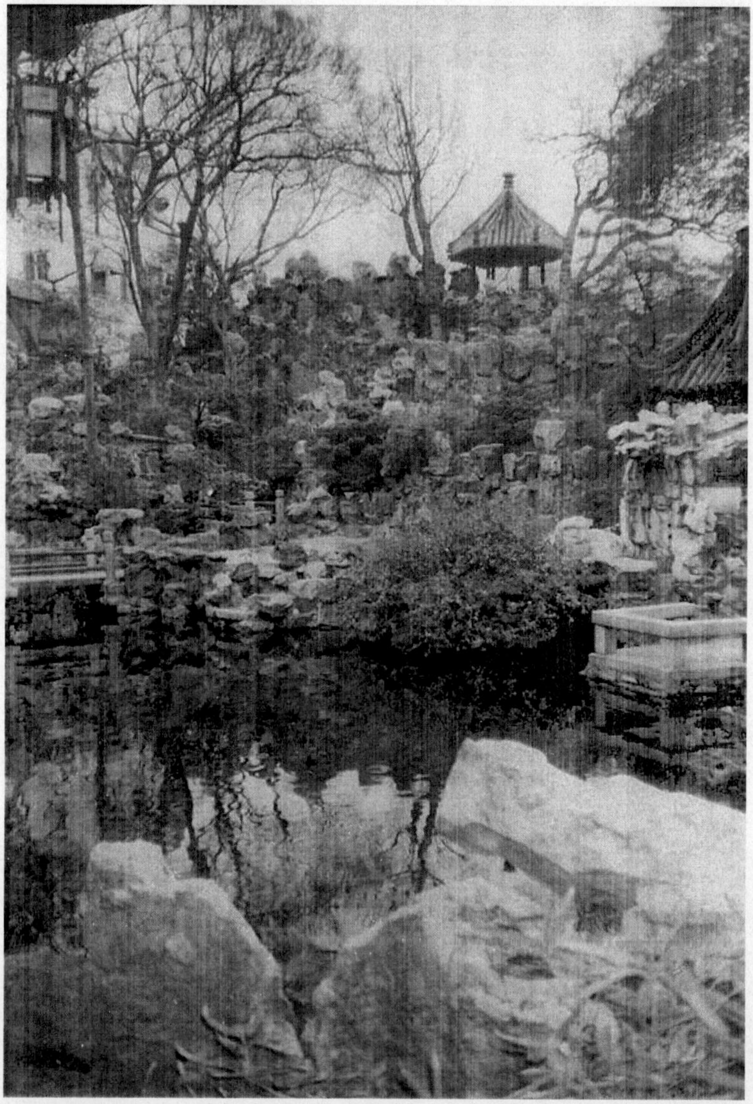

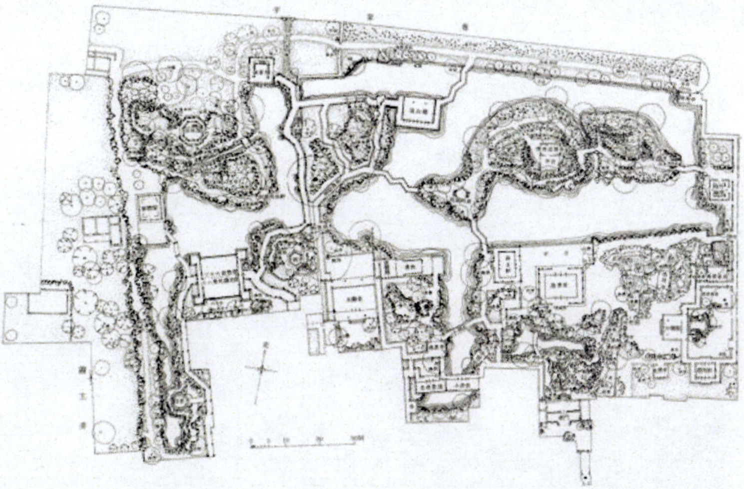

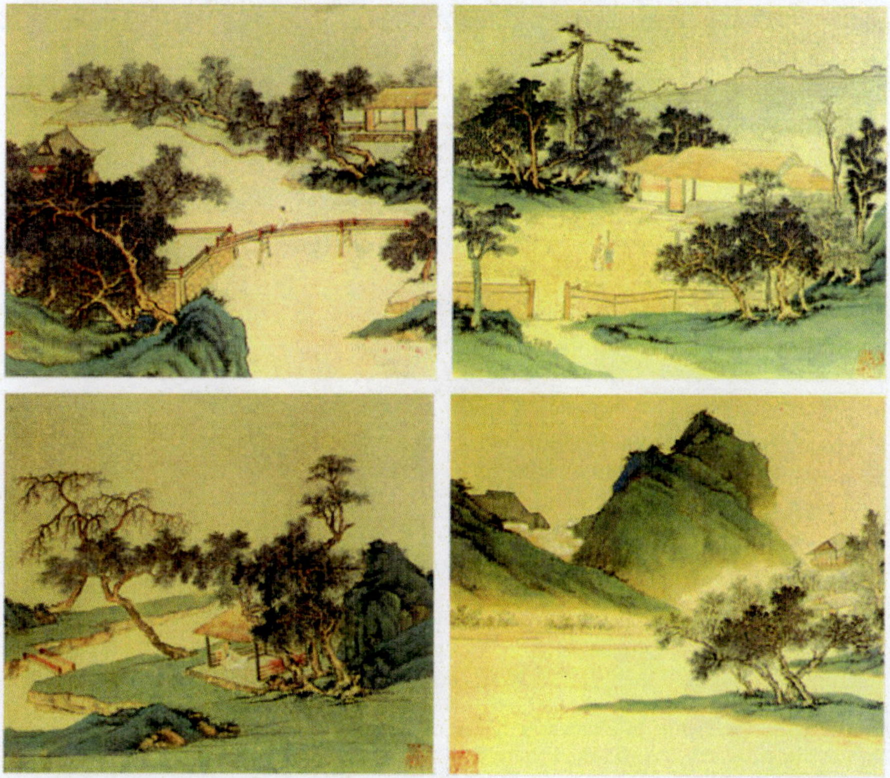

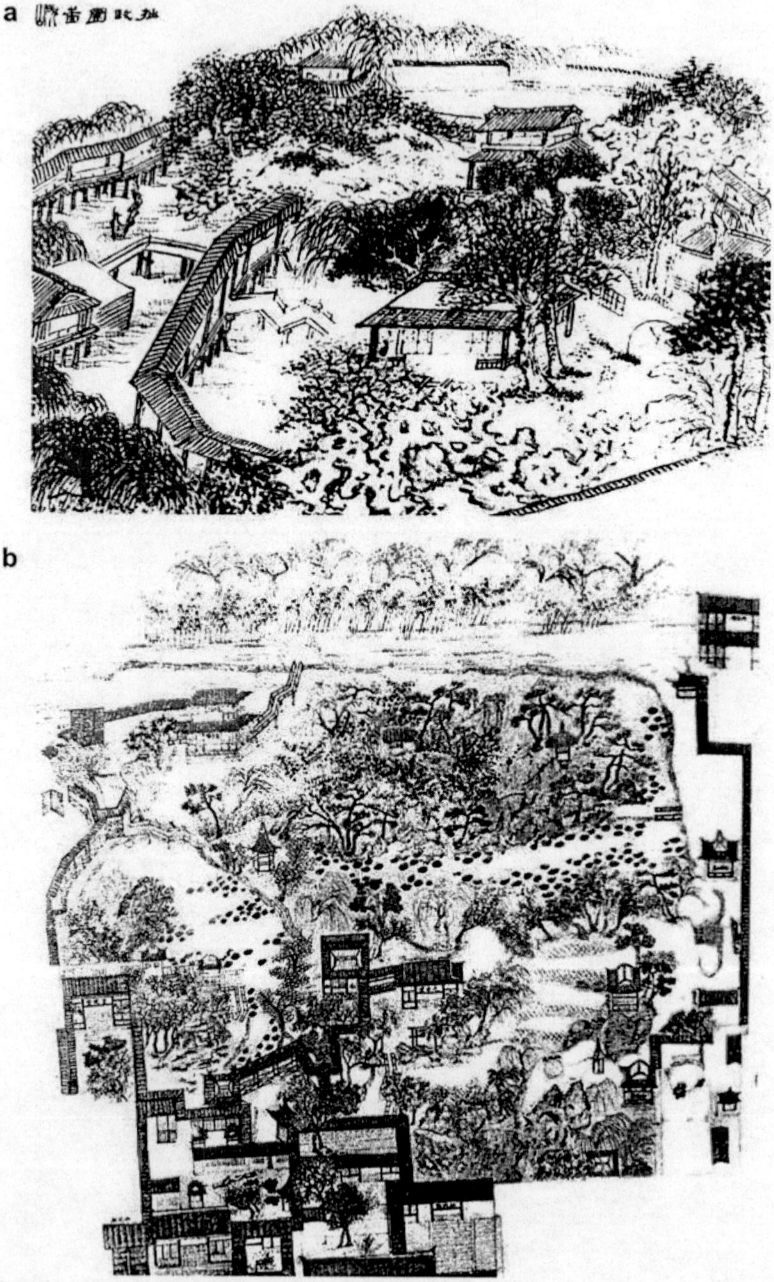

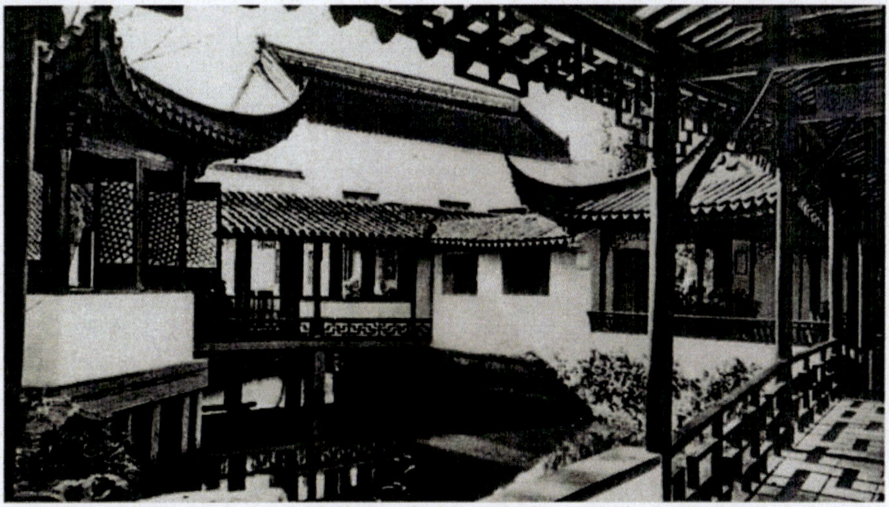

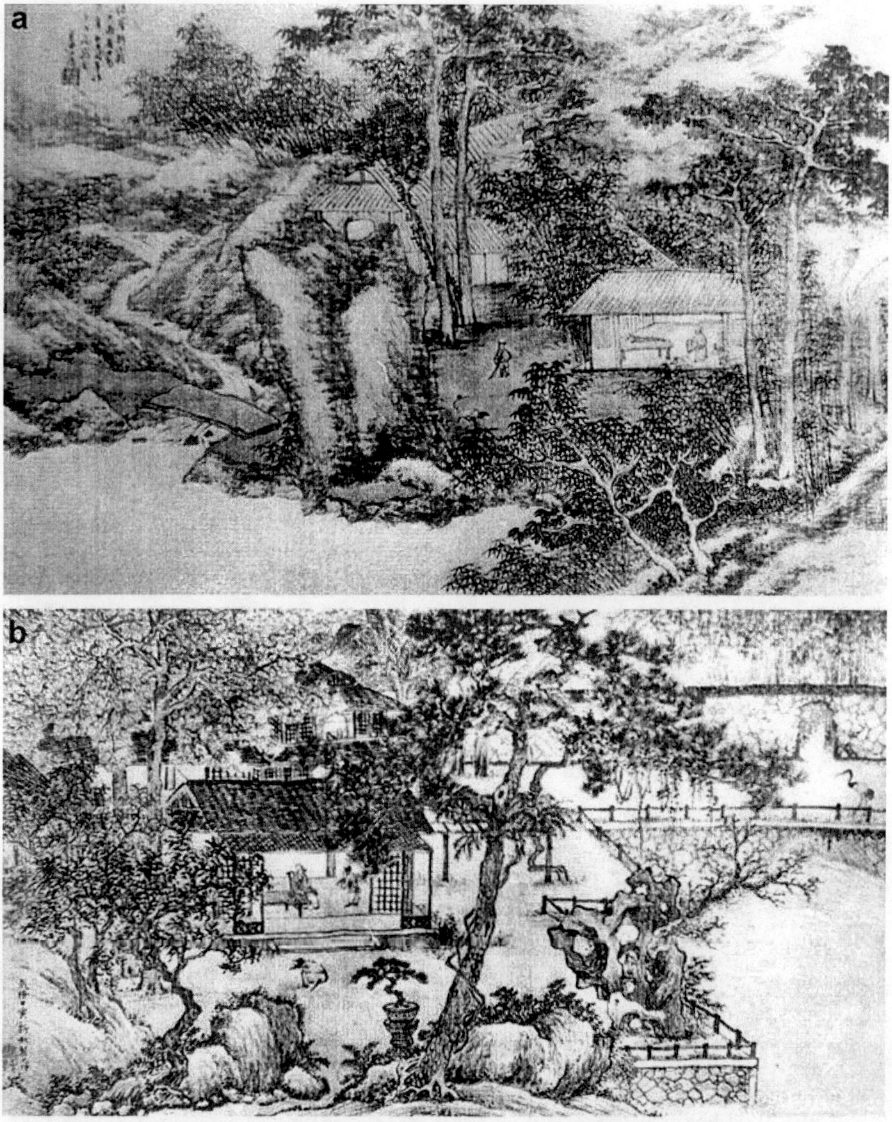

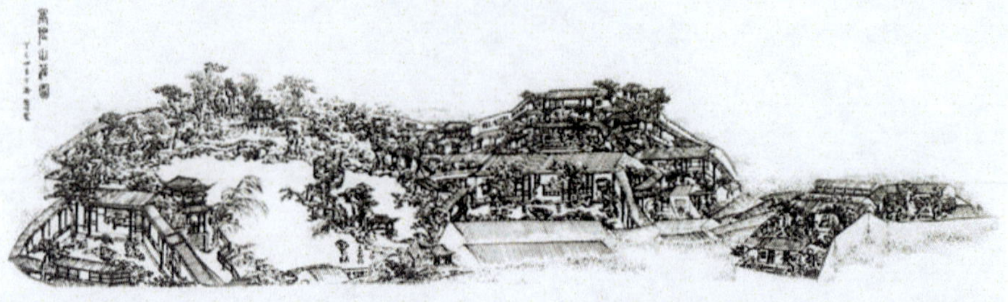

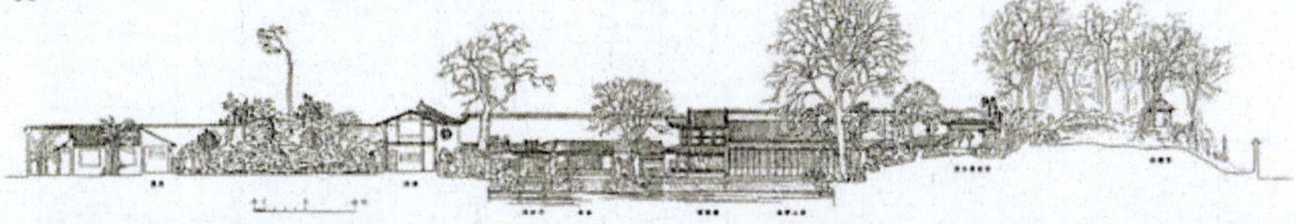

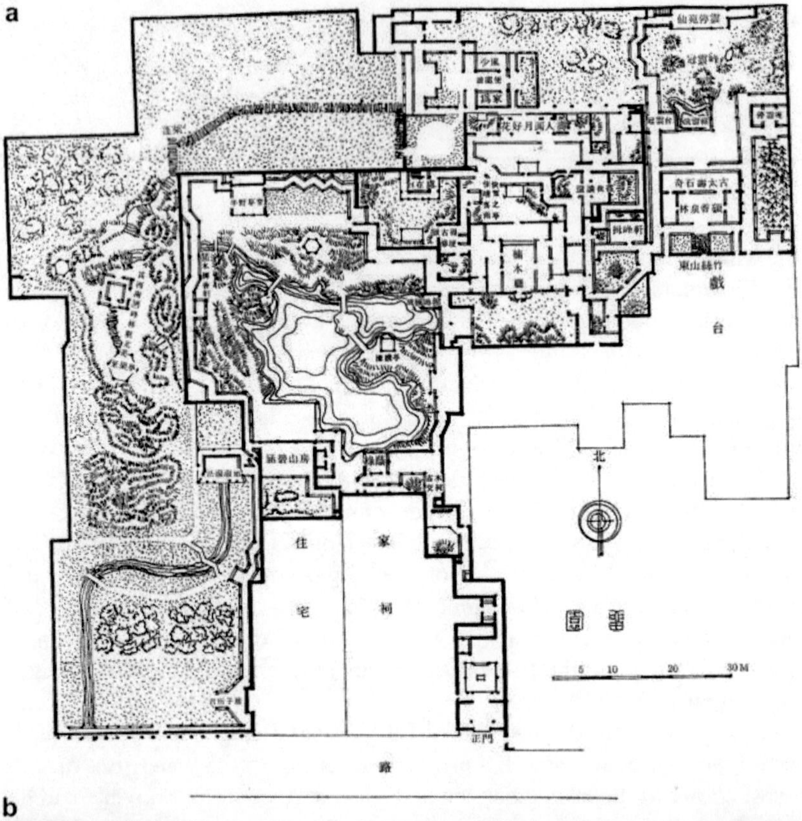

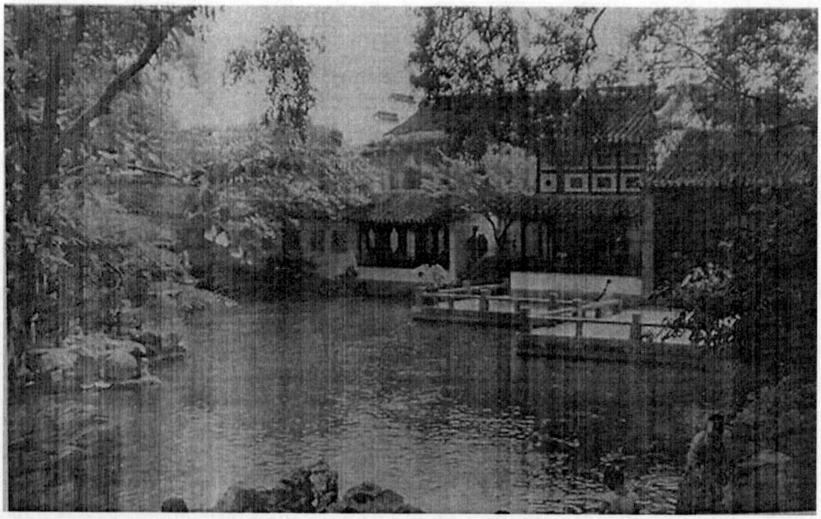

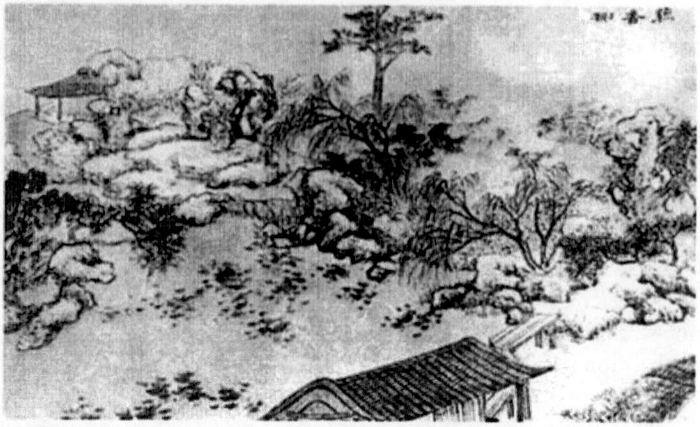

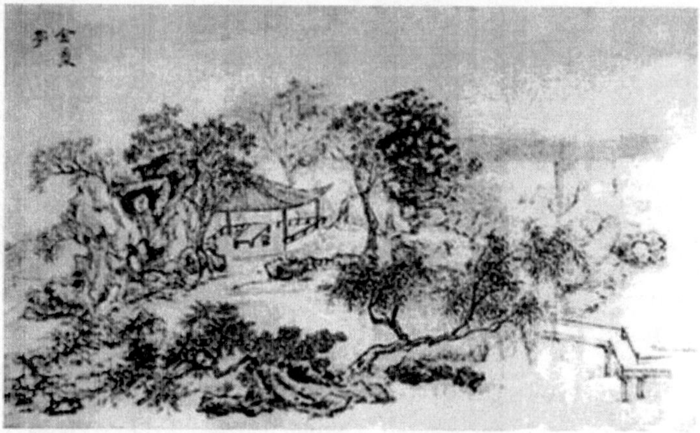

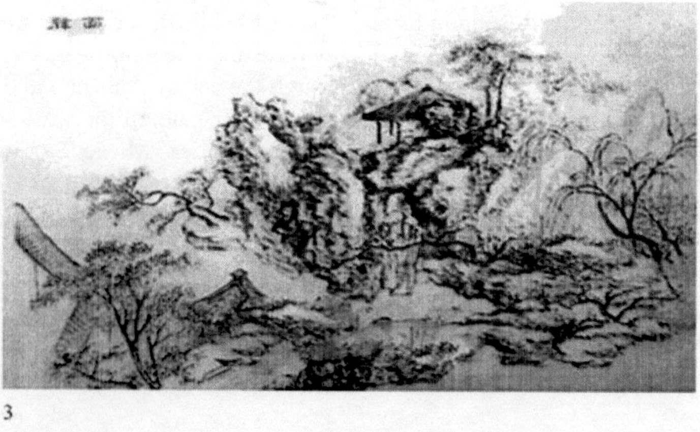

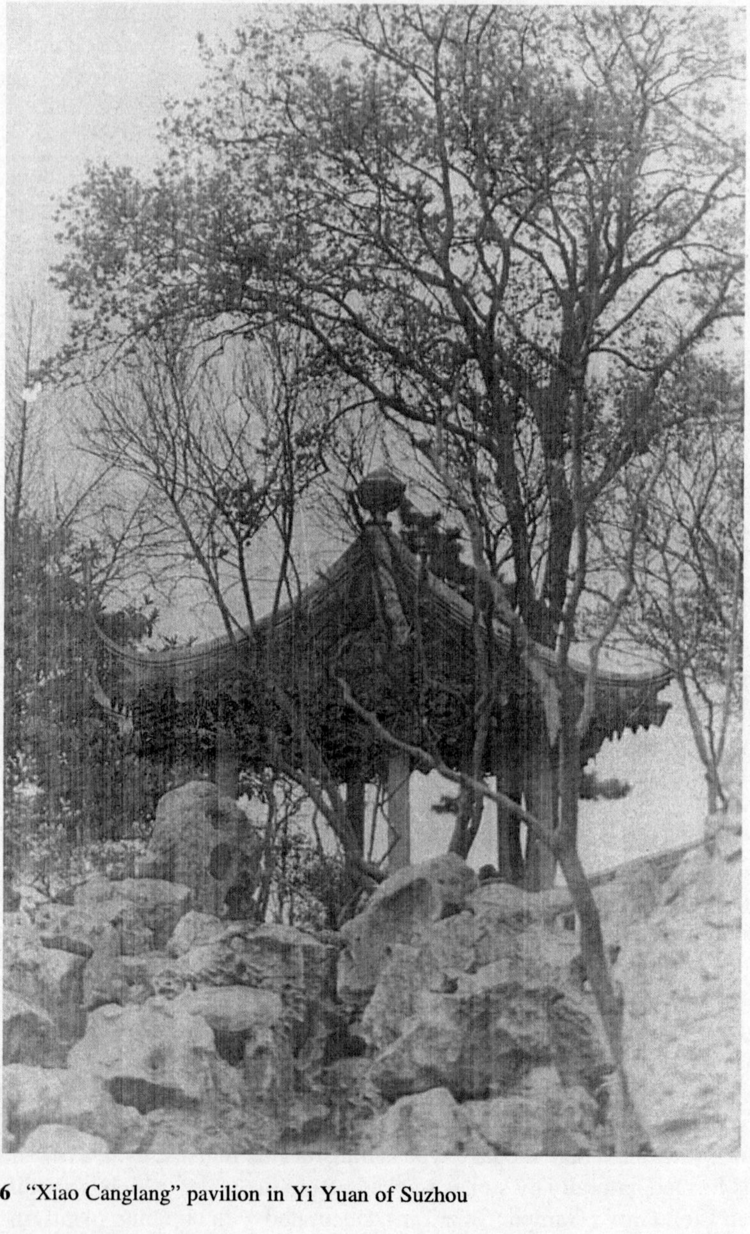

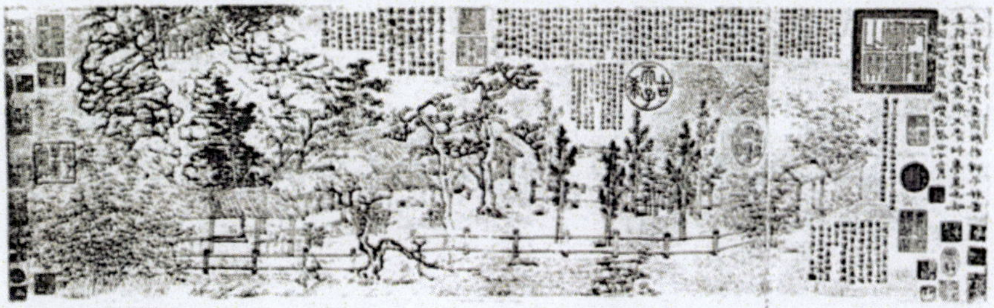

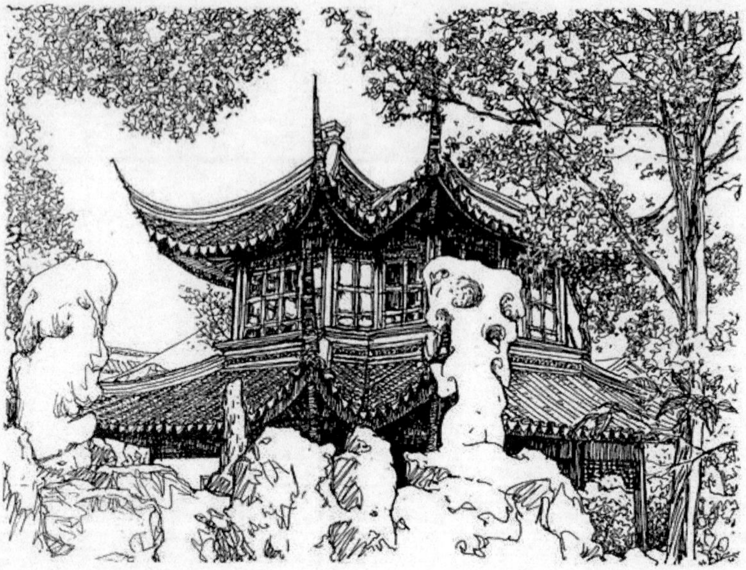

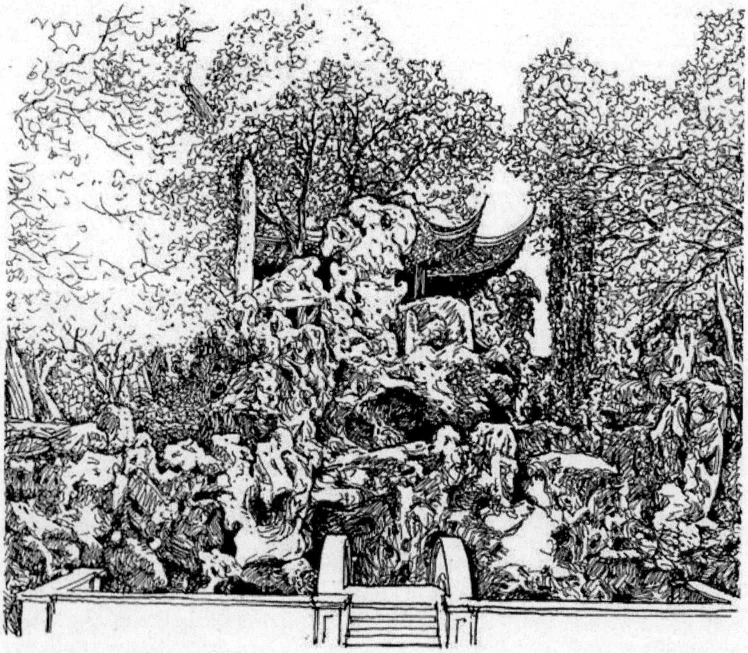

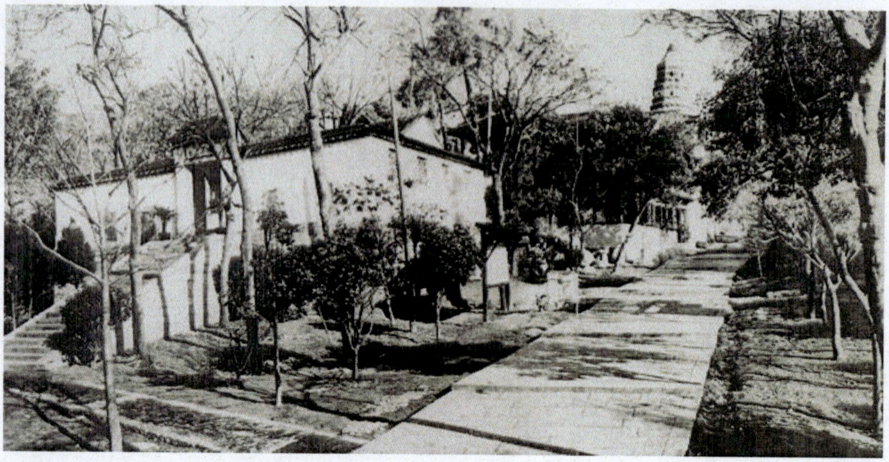

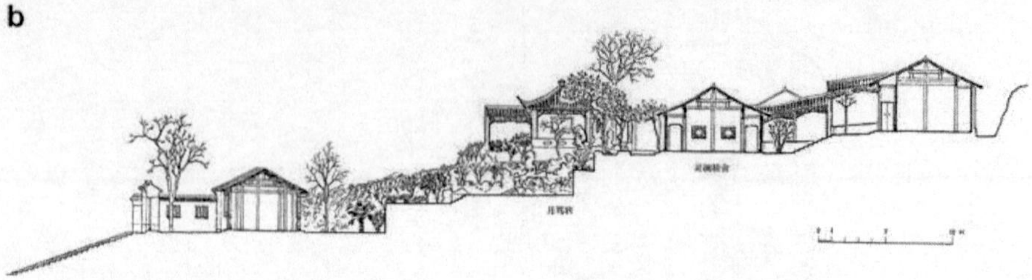

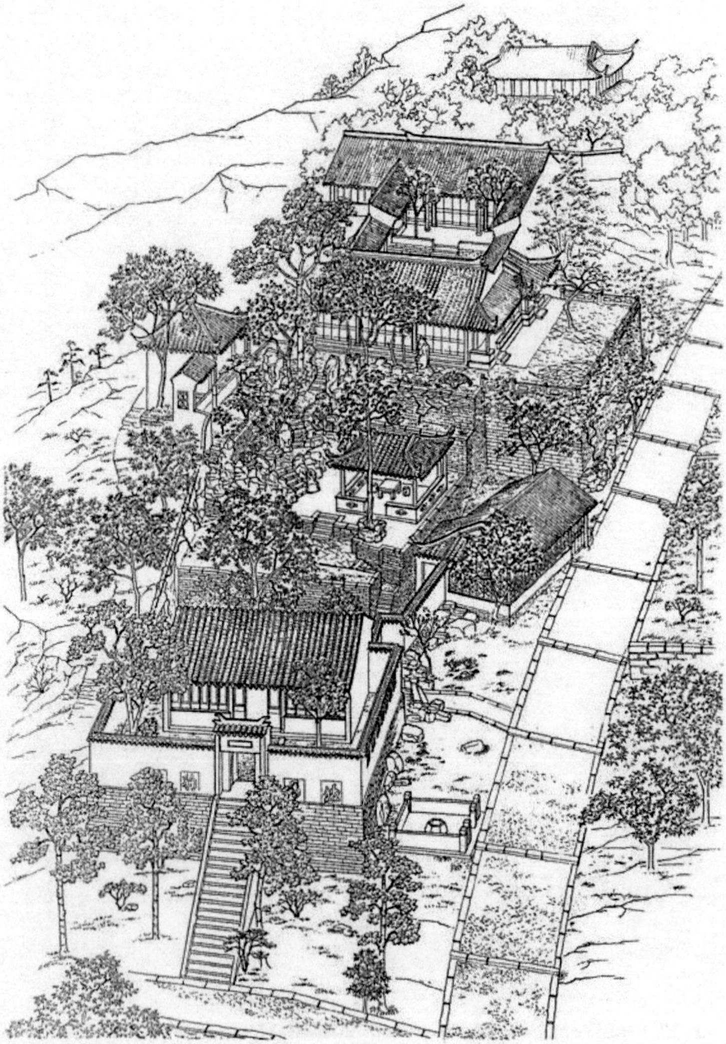

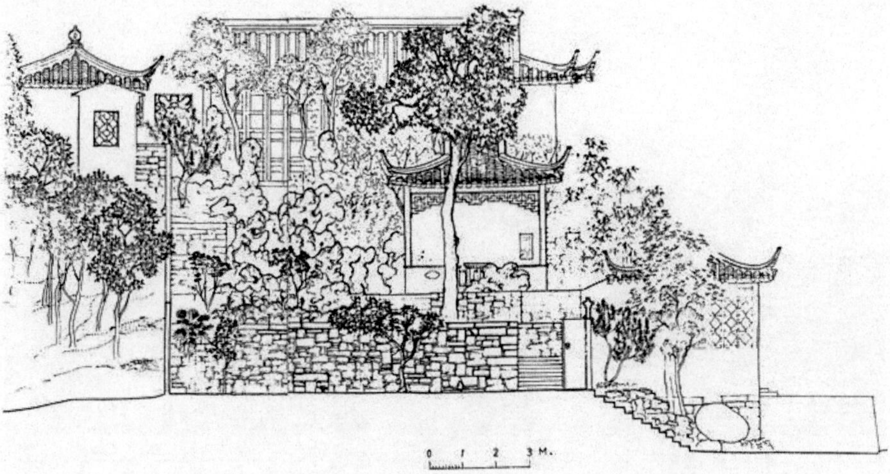

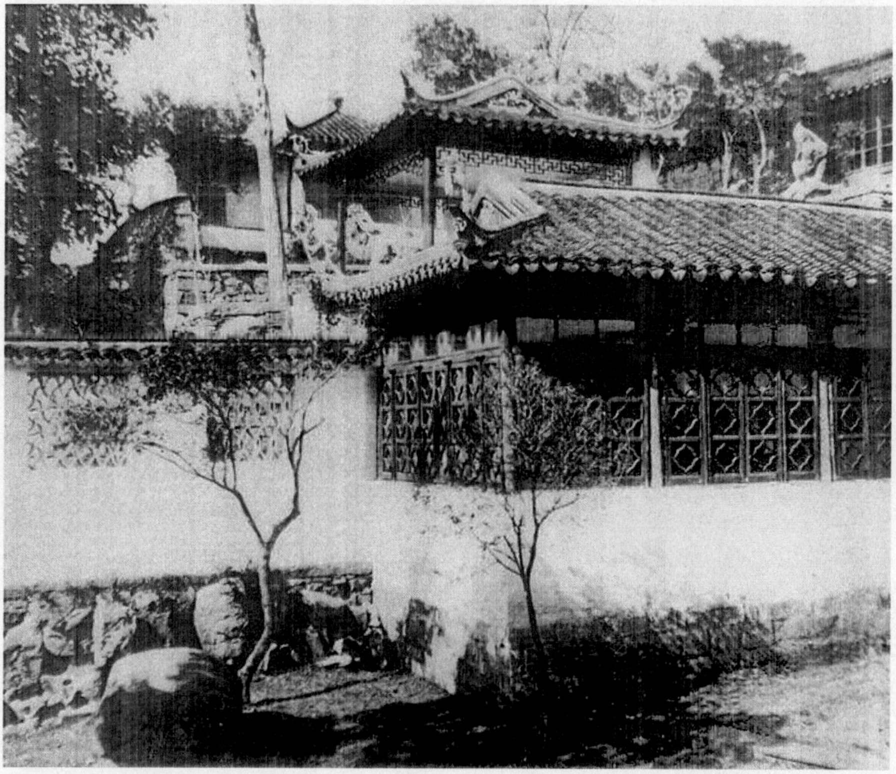

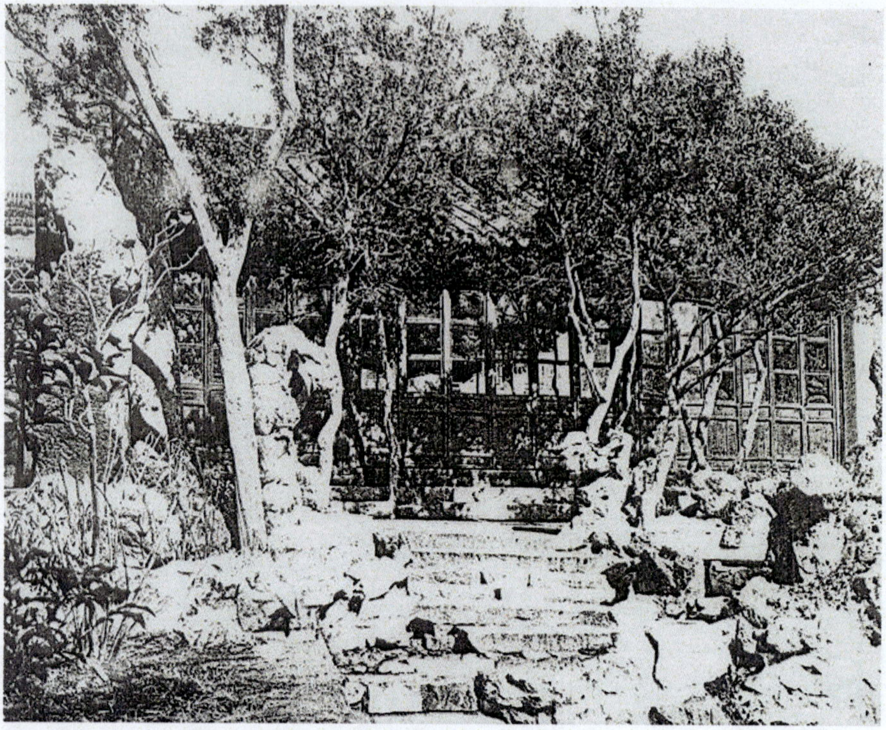

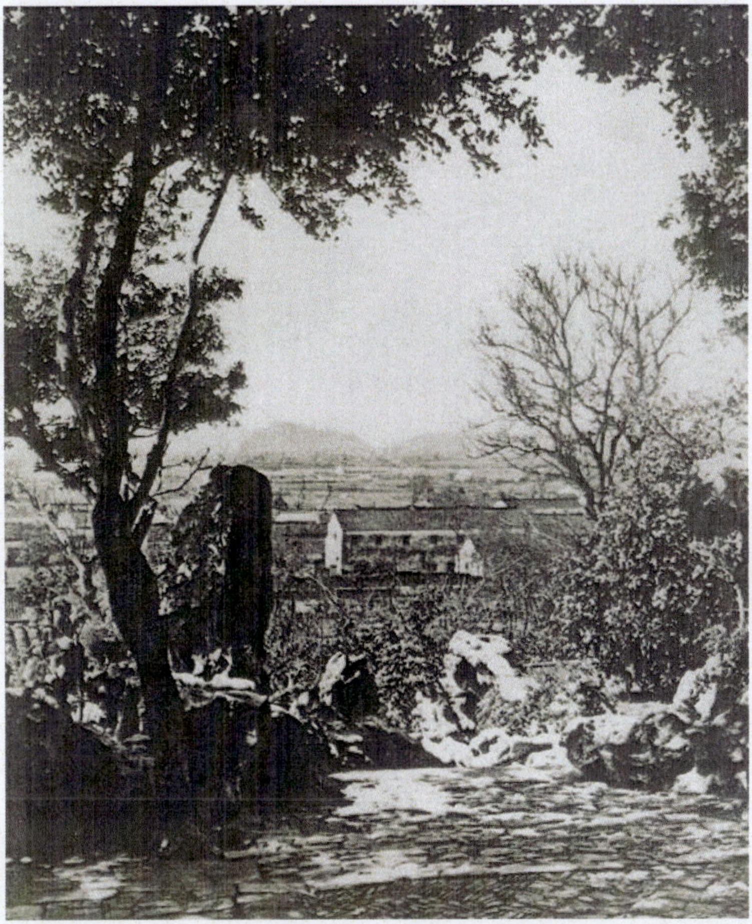

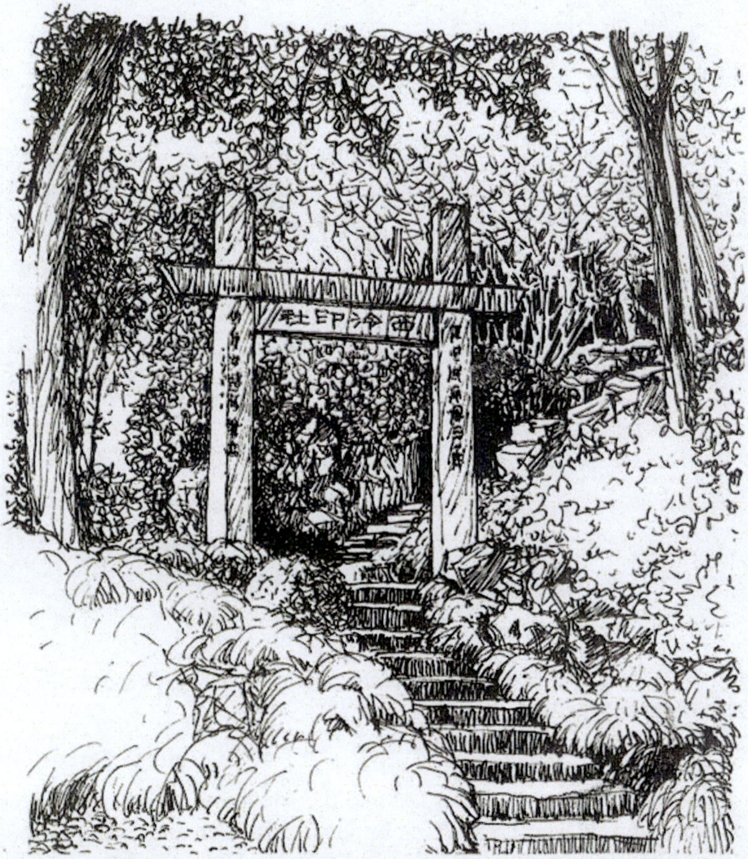

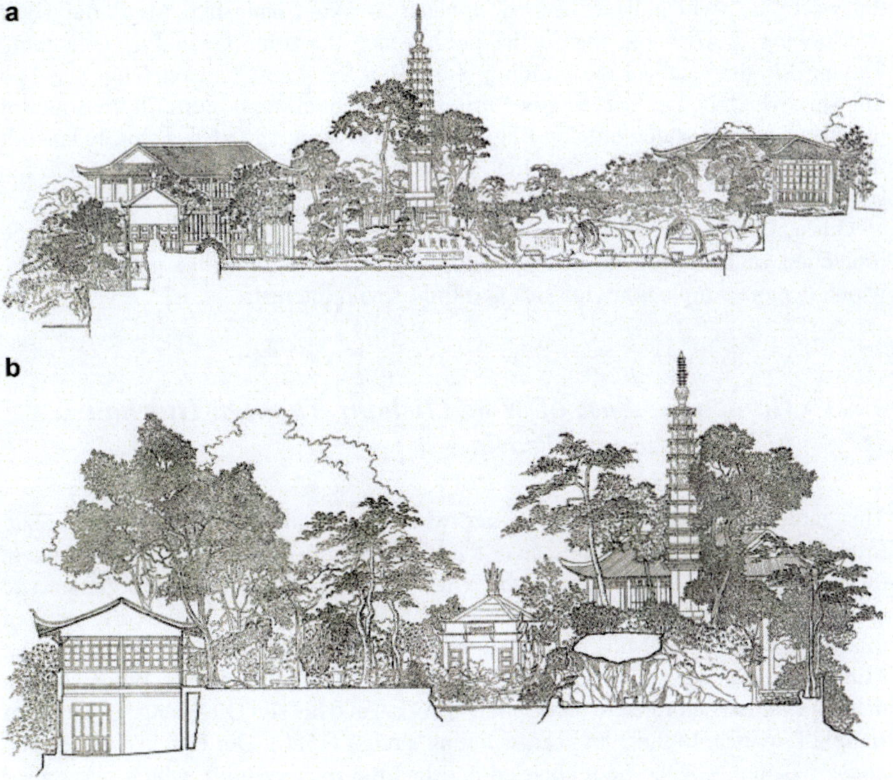

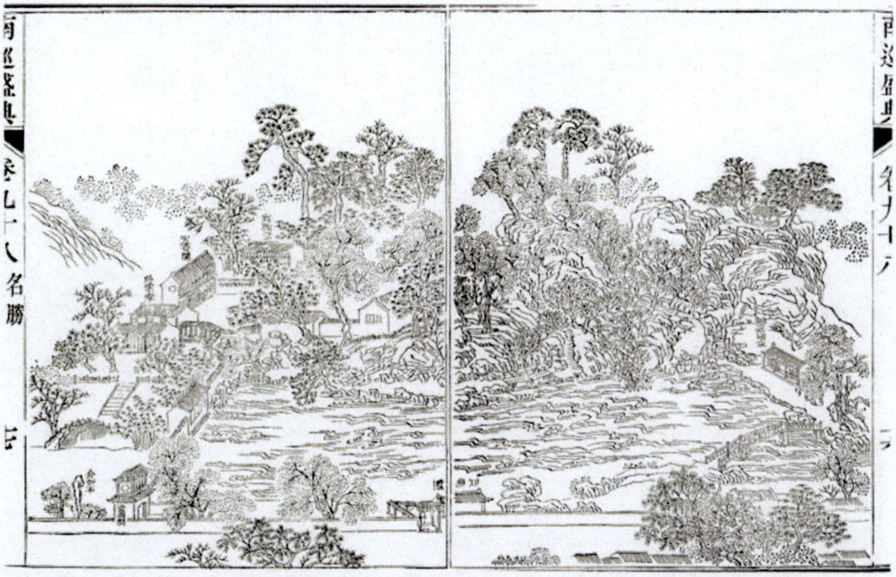

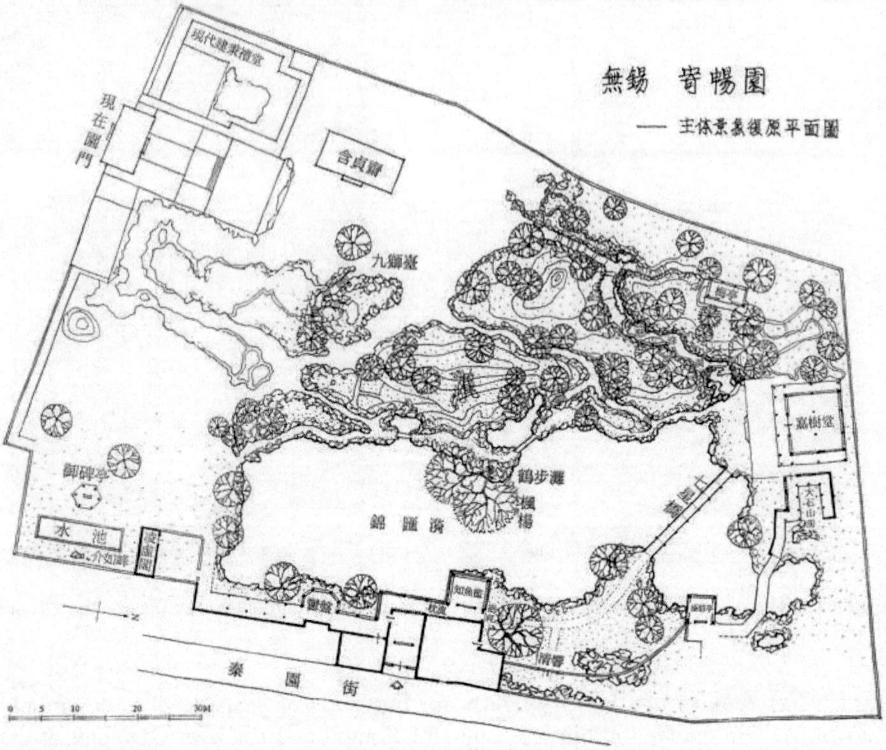

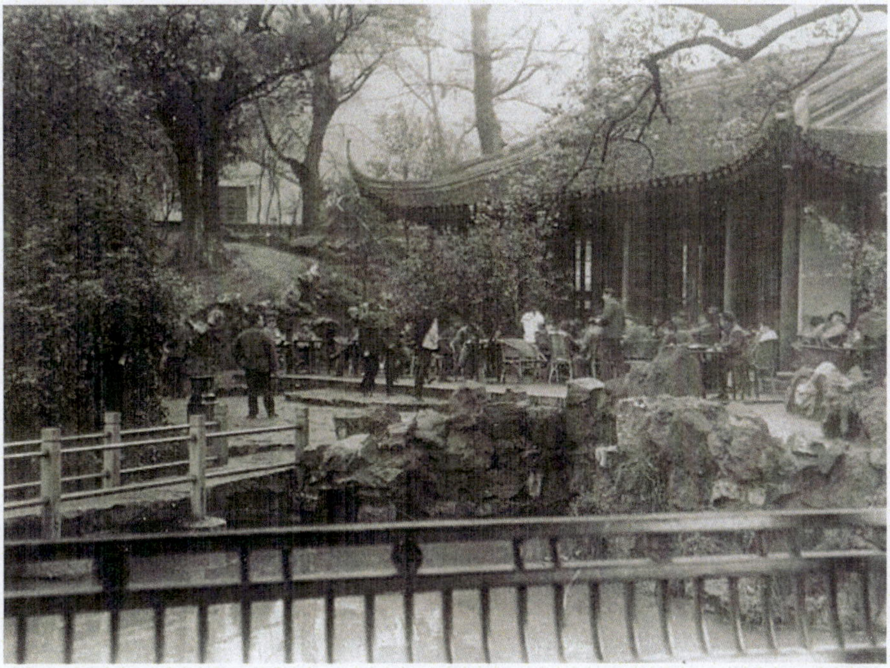

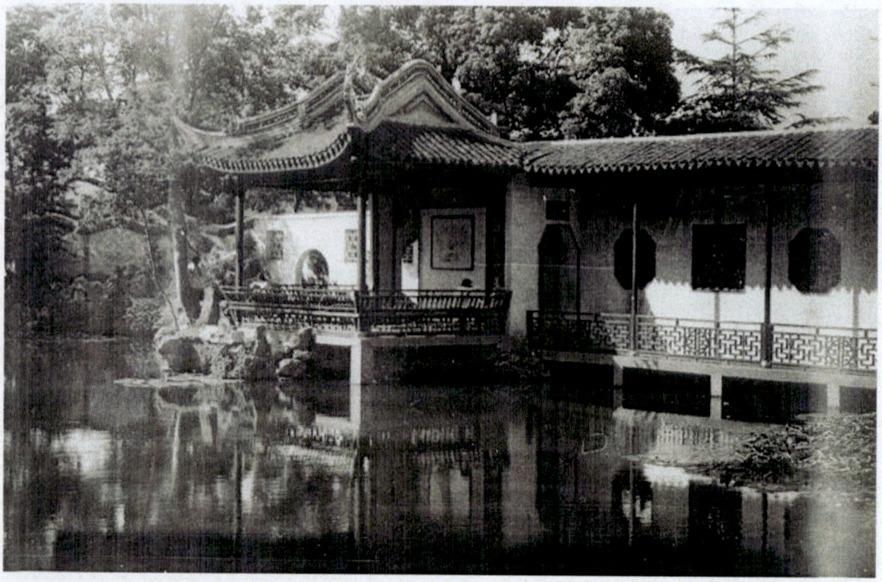

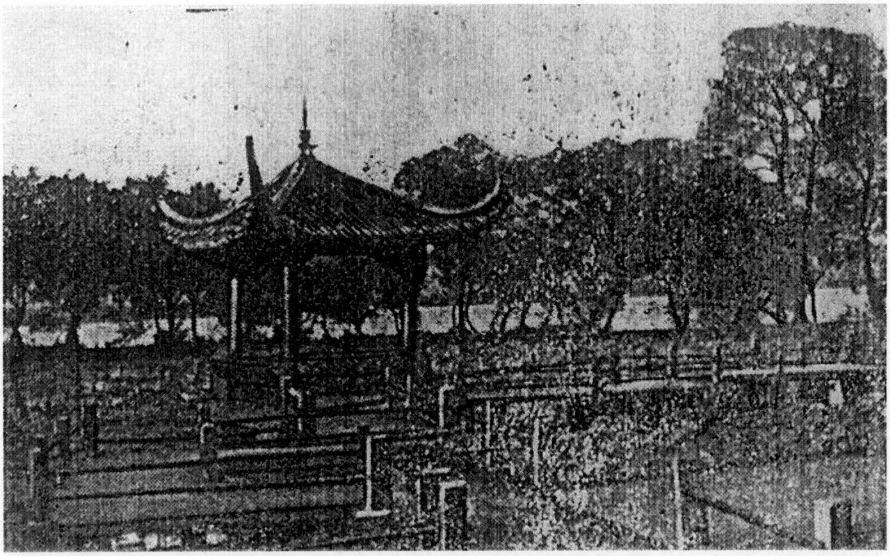

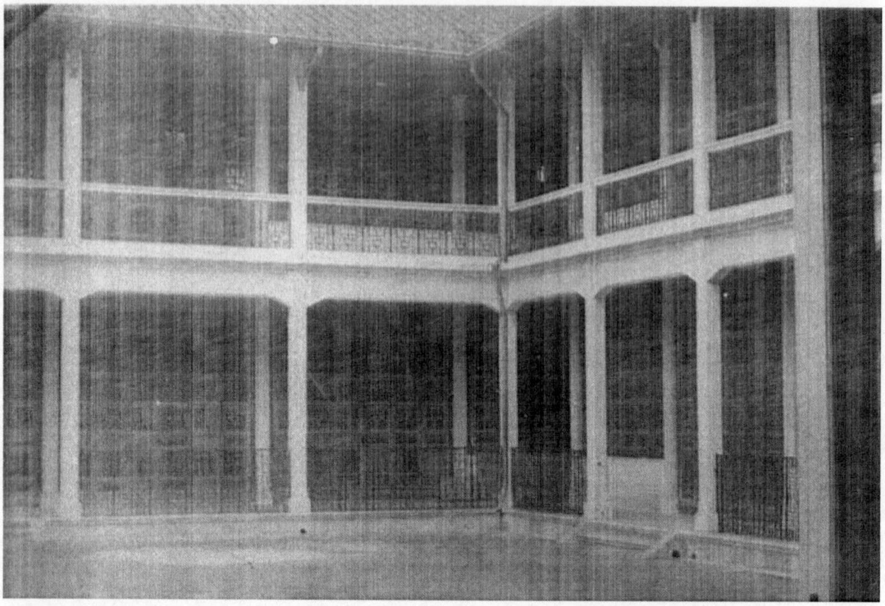

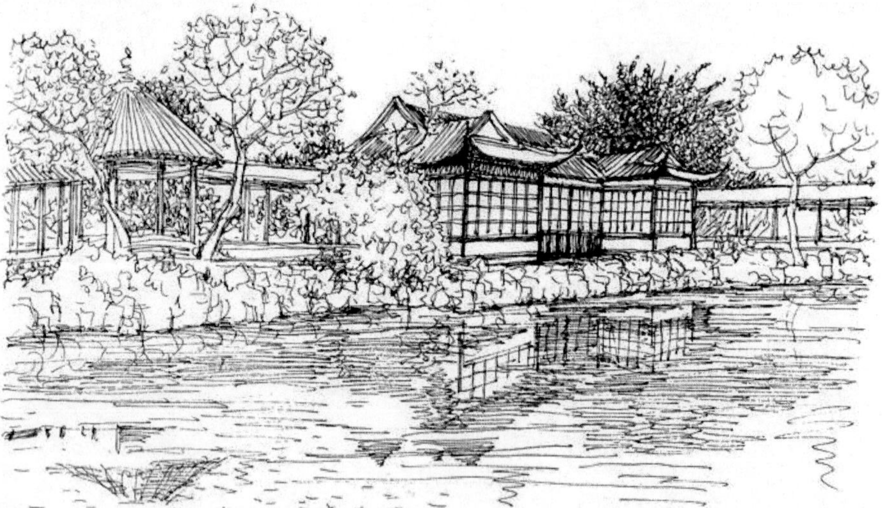

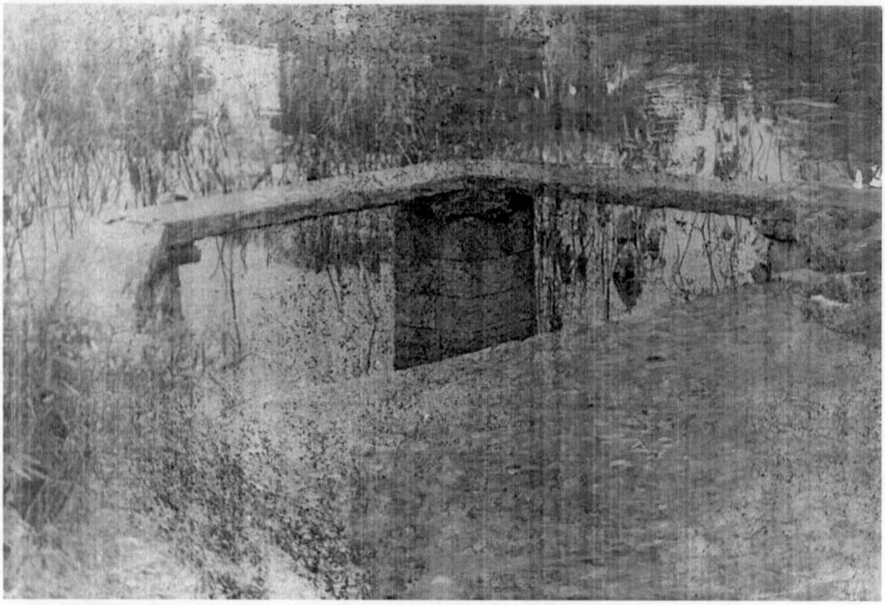

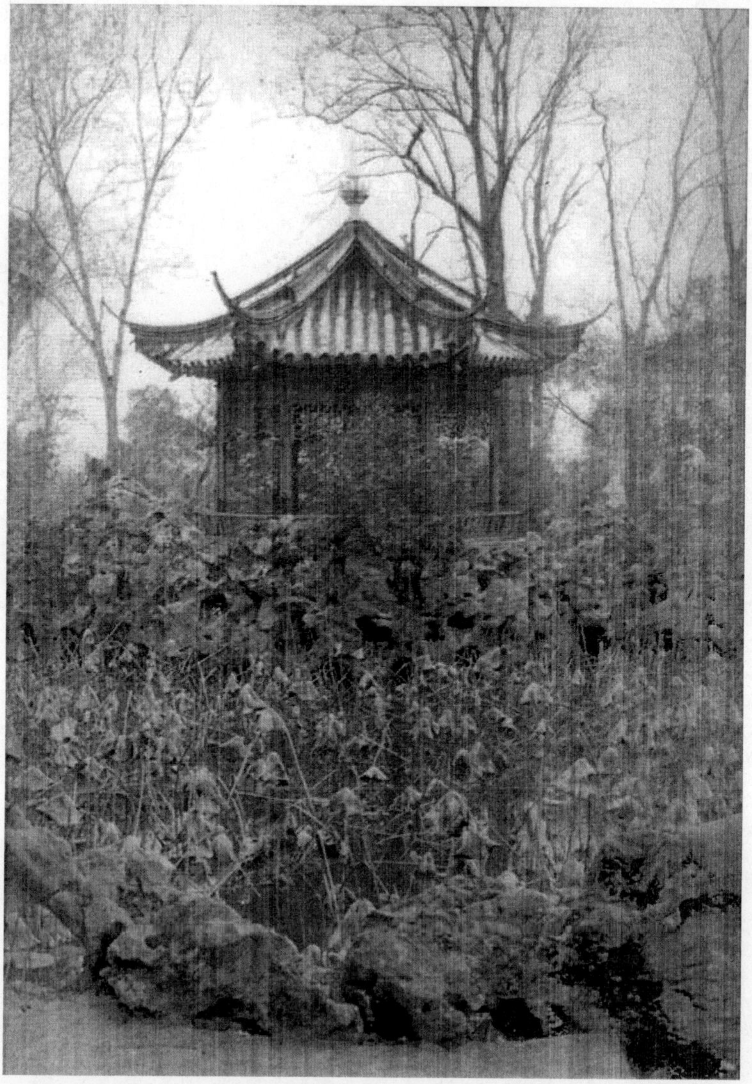

scene takes the role of guiding component whereas the hall becomes the element of the scenic imagery. Take another examples bridge or a pavilion in the garden can be an object of appreciation(with the characteristic of the element of scenic imagery highlighted), as well as the path of touring-a road or a scenic viewpoint(with the characteristic of the guiding component of the scenic imagery highlighted). That is to say, a scenic imagery can be a beautiful object that manifested infinitely many facets and scenes for appreciation, as well as a venue that allows the visitors to step into and make a tour "as if one is in the painting." In another word, scenic imagery, as the object of appreciation, stands in antithesis to its visitors, but at the same time,scenic imagery allows the visitors to stroll within itself, thus it is also in unity with the visitors.(Actually, in this case, the visitors in this scene will be viewed by other visitors as a constituent part of the scenic imagery).

This is precisely why the art of landscape design is far more superior than the art of landscape painting or the miniature landscape basin. It should be pointed out in particular that on the one hand, the elements of scenic imagery carry appreciation value and the guiding component carries practical value, but on the other hard, the elements are the practical basis for the act of appreciating the garden and the guiding component is the organizer of the act of appreciation. Both the elements and the guiding component are organic bodies that are interinclusive, interpenetrating and non-separable. Elements are finite by nature while the guiding component is infinite. In the antithesis and synthesis of the finite and infinite, the resulting scenic imagery is enriched fully. Thus, landscaping art becomes extremely marvelous and magical, and it provokes the sentiment of its visitors greatly. From those extant works of landscaping art Jiangnan of China that are of better quality, one can still observe how in them the elements and the guiding component of the scenic imagery are always interfused into one, thus they provide useful examples for further research.

Conclusion

In the research of the garden of Jiangnan, we make the following observations:

1. Running in the sinews of the objects of appreciation in Jiangnan garden, such as the mountain, water, buildings, plants and animals, is a profound artistic conception. As far as we can discern, this conception centers itself upon the fact that the basic constituent of landscaping art is the scenic imagery of the garden that attempts to express the ideal of "the sentiment of poetry and painting". This scenic imagery is composed of two aspects, the structural basis and the guiding component, which are always inter-dependent and inter-constrained of each other. In particular, the guiding component of the scenic imagery plays the active role of leadership, which can be said to be the soul of the garden. To the visitors of the garden, the guiding component serves as the organizer for the operation of the appreciation viewpoints, but to the internal structure of the scenic imagery, the guiding component not only embodies a spatial relationship, it also embodies a

time relationship;the mutation of space and time of the classical garden of Jiangnan is determined by the guiding component in the main, so is the interfusing of the natural and man made.

2. Under the particular historical condition, the classical garden of Jiangnan was not only an object for appreciation and enjoyment of natural beauty, as an environment for living and play, it is also extremely affectionate and attractive. Until this day, they still retain their artistic charms. The reasons for this are:

   a. The inspiration for the creation oi scenic imagery is taken from nature and from life, out of which these gardens chose those subjects which correspond to the esthetic interest of the time as well as those which are so called the theme of poetry and painting. The mountain, water, trees and flowering plants are elements of natural scenery that are familiar to the people, many of these are further refined by landscape painting and poetry, and fragments of them were selected to be recreated by the landscaping arr. The buildings among the mountain and water serve to further enhance the beauty of life directly. They are related closely with realistic life. For example, the hall, studio, study, tower, pavilion, verandah and room in the garden take the buildings of houses, temples and monasteries as their model for design; the open pavilion, water pavilion and pavilion-bridge or corridor-bridge take after the roadside pavilion in the villages, jetty pavilion and the bridges in the wild that provide shelter from rain and sun: crooked stone bridges take after rope-bridges, the cloud-wall takes after walls that follow mountainous terrain; even the bamboo fence, awning of shed, lattice-window, stone beams, stepping stones on the water surface and patterned floor covering all take after examples of ordinary life with the necessary transformation made. Scenic imagery that are formed by these elements may serve to give the visitors a more concentrated and highlighted aesthetic enjoyment that is full of the intrigue of life.
   b. The practical facilities provided adequately in the classical garden of Jiangnan suit a particular manner of garden living and appreciation very well. The pre-requisite for enjoyment of a garden is comfort, this applies not only to house-garden, but also extends to all gardens alike. If a garden lacks facility for retreat in hot summer, or facility for heating in cold winter, or supply of food and beverages when one most needed them, or facility for rest when one is tired, then, even if the objects for appreciation are extremely pleasant, one would not obtain full esthetic enjoyment out of the garden. The activities of play and appreciation of the classical garden of Jiangnan are activities of rest and past time that are both cultured and entertaining, as such, the gardens are well equipped with relating functional content.

      In accordance with the lifestyle of a gentry, a garden should be furnished with tea-house that allow for brewing of tea, poetry club for recitation and composing of poems, studio for painting, lute-room for playing lute, study in which one can read widely, secret chamber for drinking while embracing greenery, pure abode for *Chan Meditation*, grotto for chess playing and chatting, high tower for day dreaming, hall for entertaining guests, water pavilion for observing fishes, lake, side rock or small boat for fishing, and etc. With all these practical facilities, one can then satisfy the requirements of the gentry-owner of a garden to have an ideal place for enjoyment that is full of the scent of books', a high level and elegant form of enjoyment.

The end result of these provisions in the garden is that the spaces created by the scenic imagery accord well with the sentiment of the owner. Even for the modern visitors, a tour of the garden becomes extremely comfortable with the provision of long corridor that gives shelter for rain and scorching sunshine as well as serving as route for appreciating the garden; and other facilities such as seats along the corridor, rocks in the mountain, near the pool side, in the grotto or beside a path that provides for rest; or even the goose, neck seat under the eave of a building and other indoor furniture in the pavilion, verandah, hall and water pavilion.

c. The scenic imagery of the classical Jiangnan garden is a body uniting both functions of appreciation and utility. In fact, most of the objects for appreciation are also utilitarian facilities. All kinds of bridges in the garden are both decoration of the water-scape as well as path for touring; the buildings among the mountain or by the waterside serve as a constituent part of the artistic picture representing the image of a hill-village or waterside town, as well as places for shelter, rest and holding of play activities. Such organic unity between utility and appreciation represents an effort to fuse the activities of the visitors into the artistic melody of the scenic imagery. Thus the end result is both comfortable, friendly and profoundly sentimental.

3. The classical garden of Jiangnan is a form of space-time art that is also functional. In the spatial aspect, it is an art form that described space with space-recreation of mother nature and living spaces required by the society with the artistic manmade spaces. In the Jiangnan gardens, the natural space blends in with the living space organically, or one can say that natural environment is united artistically with man-made environment. This can be described as the basic achievement of the classical garden of Jiangnan, the essence of the Chinese classical landscaping art.

Man, in the final analysis, is a constituent part of nature. Although man may at times be in antithesis with nature, or at other times be one together with nature, man and nature ultimately share the same destiny. In ancient time, as scientific knowledge is rather primitive, man existed only as a small ring in the whole ecological chain. Our forebear had no power to alter nature, their life were subjected fully to the regulation of nature. Thus they treated nature with awe and worship. From the time of recorded history onwards, though the Chinese Emperors hold absolute power in the world of man, they had also to humble themselves in the face of nature and served it with highest obedience. Therefore, ritual for the worship of heaven, sky, sun, moon, mountain and river had become elaborate national and imperial ceremonies for thousands of years without fail. For the masses, they worshiped gods of mountain and earth deity reverently. Thus in creating man-made environment, man in ancient time would not dare to offence the nature. (Having stated this, one reflects that man always contradicts